Springer Monographs in Mathematics

More information about this series at http://www.springer.com/series/3733

Luis J. Alías • Paolo Mastrolia • Marco Rigoli

Maximum Principles and Geometric Applications

Luis J. Alías
Departamento de Matemáticas
Universidad de Murcia
Murcia, Spain

Paolo Mastrolia
Dipartimento di Matematica
Università degli Studi di Milano
Milan, Italy

Marco Rigoli
Dipartimento di Matematica
Università degli Studi di Milano
Milan, Italy

This work has been partially supported by MINECO/FEDER project MTM2012-34037 and Fundación Séneca project 04540/GERM/06, Spain. This research is a result of the activity developed within the framework of the Programme in Support of Excellence Groups of the Región de Murcia, Spain, by Fundación Séneca, Regional Agency for Science and Technology.

M. Rigoli has been partially supported by MEC Grant SAB2010-0073 and Fundación Séneca Grant 18883/IV/13, Programa Jiménez de la Espada.

ISSN 1439-7382 ISSN 2196-9922 (electronic)
Springer Monographs in Mathematics
ISBN 978-3-319-79605-5 ISBN 978-3-319-24337-5 (eBook)
DOI 10.1007/978-3-319-24337-5

Mathematics Subject Classification: 58-02, 35B50, 53C20, 53C42, 35R01

Springer Cham Heidelberg New York Dordrecht London

Softcover re-print of the Hardcover 1st edition 2016

Printed on acid-free paper

Springer International Publishing AG Switzerland is part of Springer Science+Business Media (www.springer.com)

Contents

List of Symbols

(Σ, g, A)	initial data set for the Einstein equation, page 513
$(M, \langle\, ,\, \rangle, e^{-f}dx)$	weighted Riemannian manifold, page 142
$(M, \langle\, ,\, \rangle, X)$	Ricci soliton structure, page 443
(U, φ)	local chart, page 2
$[\, ,\,]$	Lie bracket, page 10
$\chi_{s,t}$	characteristic function of the annulus $B_t(o)\backslash B_s(o)$, page 69
Δu	Laplacian of the function u, page 32
δ_i^j	suggestive way of writing the Kronecker symbol, page 2
Δ_f	f-Laplacian, page 142
Δ_X	X-Laplacian, page 142
δ_y	Dirac delta centered at y, page 100
δ_{ij}	Kronecker symbol, page 2
$\dot{\gamma}$	tangent vector of the curve γ, page 59
$\ell_Y\langle\, ,\, \rangle$	$(1, 1)$-version of $\mathscr{L}_Y\langle\, ,\, \rangle$, page 448
η	cutoff function, page 117
γ	the matrix of 1-forms (θ_j^i) on U, page 4
λ	soliton constant, page 443
$\Lambda^2(U)$	space of skew-symmetric 2-forms on the open set U, page 13
$\log^{(j)}$	j-th iterated logarithm, page 88
$\mathscr{A}$	m-dimensional area function in the Lorentzian setting, page 511
$\mathscr{C}_{o,\zeta,\theta}$	nondegenerate cone of $\mathbb{R}^n$ with vertex o, direction ζ and width θ, page 273
$\mathscr{G}_k^L$	Green kernel of the operator L on the domain Ω_k, page 127
$\mathscr{L}_X\omega$	Lie derivative of the 1-form ω in the direction of X, page 11
$\mathscr{L}_Xf$	Lie derivative of the function f in the direction of X, page 11
$\mathscr{L}_XY$	Lie derivative of the vector field Y in the direction of X, page 11
$\mathscr{L}_X\langle\, ,\, \rangle$	Lie derivative of the metric $\langle\, ,\, \rangle$ in the direction of X, page 11
$\mathscr{T}$	unit normal $\frac{\partial}{\partial t}$ to the slice $\mathbb{P}_t$, page 52

$\mathscr{T}_p$	the set of all timelike vectors of T_pN, page 501
$\mathbb{N}$	the set of natural numbers $1, 2, \ldots.$, page xv
$\mathbb{P}_t$	slice of the foliation $t \in \mathbb{R} \mapsto \mathbb{P}_t$ of N, page 52
$\mathbb{R}^+$	the set of positive real numbers $(0, +\infty)$, page xvi
$\mathbb{R}_0^+$	the set of nonnegative real numbers $[0, +\infty)$, page xvi
$\mathfrak{o}(m)$	Lie algebra of skew-symmetric matrices, page 4
$\mathfrak{R}$	symmetric endomorphism determined by Riem, page 121
$\mathfrak{X}(M)$	set of all smooth vector fields on M, page 10
$\mathfrak{X}(U)$	set of smooth vector fields on the open set U, page 6
$\mathrm{cut}(o)$	cut locus of the point o, page 58
II	second fundamental tensor of an immersion, page 37
ric	$(1, 1)$-version of Ric, page 448
$-I \times_\rho \mathbb{P}$	generalized Robertson-Walker spacetime, page 504
$\overline{\mathrm{Ric}}$	Ricci curvature of a Lorentzian manifold, page 507
$\mathbb{H}_1^n$	n-dimensional anti-de Sitter space, page 503
$\mathbb{L}^n$	n-dimensional Lorentz-Minkowski space, page 502
$\mathbb{S}_1^n$	n-dimensional de Sitter space, page 502
$\nabla\omega$	covariant derivative of the 1-form ω, page 7
$\nabla d\varphi$	generalized second fundamental tensor, page 45
∇T	covariant derivative of a tensor field T, page 8
∇u	gradient of the function u, page 9
∇X	covariant derivative of the vector field X, page 7
∇	connection induced by the Levi-Civita connection forms, page 6
∇II	covariant derivative of II, page 40
$\nabla_Y\omega$	covariant derivative of ω in the direction of Y, page 8
$\nabla_Y X$	covariant derivative of X in the direction of Y, page 7
ν	unit normal vector field, page 38
$\odot$	symmetric tensor product, page 162
ω_{ik}	covariant derivative of the coefficient ω_i, page 8
ω_m	volume of the unit sphere in $\mathbb{R}^m$, page 57
Deck	covering transformation group, page 489
$\dim M$	dimension of the manifold M, page 2
$\mathrm{dist}_M(x, o)$	Riemannian distance between the points o and x, page 58
$\mathrm{div}\, Z$	divergence of a symmetric $(0, 2)$-tensor field Z, page 9
div Riem	divergence of the Riemann tensor, page 122
$\mathrm{div}\, X$	divergence of a vector field X, page 7
hess	$(1, 1)$-version of the Hessian, page 59
$\mathrm{Hess}(u)$	Hessian of the function u, page 31
$\mathrm{inj}_M(p)$	injectivity radius of the point p in M, page 58
$\mathrm{Lip}_c(M)$	set of Lipschitz functions on M with compact support, page 69
Ric	Ricci tensor, page 17
$\mathrm{Ric}(\nabla r, \nabla r)$	radial Ricci curvature, page 60

Ric_f	Bakry-Emery Ricci tensor, page 444
Ric_X	the tensor $\mathrm{Ric} + \frac{1}{2}\mathscr{L}_X\langle\ ,\ \rangle$, page 456
Riem	$(0,4)$-version of the Riemann curvature tensor, page 14
sgn	signum function, page 176
supp	support of a function, page 117
Tor	torsion tensor, page 10
Tr	trace, page 7
$\mathrm{vol}\,\partial B_R(o)$	volume of the boundary of the geodesic ball $B_R(o)$, page 67
$\mathrm{vol}\,B_R(o)$	volume of the geodesic ball $B_R(o)$, page 67
$\otimes$	tensor product, page 7
$\overline{B_\varepsilon(y)}$	closed ball of center y and radius ε, page 81
$\overline{K}(\Pi)$	sectional curvature of a nondegenerate tangent plane Π in a Lorentzian manifold, page 506
$\overline{R}$	$(1,3)$-curvature tensor for a Lorentzian manifold, page 506
$\overline{S}$	scalar curvature of a Lorentzian manifold, page 507
$\owedge$	Kulkarni-Nomizu product, page 26
$\partial B_R(o)$	boundary of the geodesic ball centered at o with radius R, page 57
$\partial\Omega_t^u$	t-level set of a function u, page 68
$\mathfrak{R}$	curvature operator acting on symmetric $(0,2)$-tensors, page 159
$\sqrt{g}$	square root of the determinant of the metric in polar geodesic coordinates, page 61
τ	time-orientation of a Lorentzian manifold, page 502
$\tau(\varphi)$	tension field of the map φ, page 45
$\mathrm{vol}_f(B_r)$	weighted volume of the geodesic ball B_r, page 475
θ	(column) vector valued 1-form, page 4
θ_j^i	Levi-Civita connections forms, page 3
Θ_j^i	curvature forms, page 12
φ_t	local flow of a vector field, page 12
$\lfloor u\wedge v\rvert^2$	the expression $\langle u,u\rangle\langle v,v\rangle - \langle u,v\rangle^2$, page 19
$\widetilde{N}$	set of all timecones in tangent spaces of N, page 504
$\{\theta^i\}$	(local) orthonormal coframe, page 2
$\{e_i\}$	(local) orthonormal frame, page 2
$\flat : TM \to TM^*$	inverse of the musical isomorphism $\sharp$, page 145
${}^T\gamma$	transpose of the matrix γ, page 4
A	Schouten tensor, page 28
A	change of frames, page 5
A, A_ν	Weingarten endomorphism (or shape operator) in the direction of ν, page 38
A_t	second fundamental tensor in the direction of $-\mathscr{T}$, page 52
$B_R(o)$	geodesic ball centered at o with radius R, page 57
C	Cotton tensor, page 27
$C^\infty(U)$	set of smooth functions on the open set U, page 3

$C_c^\infty(M)$	set of smooth function with compact support on M, page 65
$d\theta$	exterior differential of the 1-form θ, page 10
D_o	the set $M \setminus \mathrm{cut}(o)$, page 58
du	differential of a function u, page 9
$e(\varphi)$	energy density of the map φ, page 45
$E_\Omega(\varphi)$	energy functional of the map φ on the domain Ω, page 45
E_o	maximal starshaped domain with respect to o on which $\exp_o$ is a diffeomorphism, page 58
$\exp_o$	exponential map of M at o, page 58
f	potential of a gradient Ricci soliton, page 443
$f \sim g$ as $x \to \dots$	f asymptotic to g as $x \to \dots$, page 70
f^*	pullback via the map f, page 35
f^*TN	pullback of the bundle TN, page 37
f_*e_i	pushforward of e_i by the map f, page 35
F_X	tidal force operator, page 507
G	Einstein gravitational tensor of a Lorentzian manifold, page 507
G^L	Green kernel of the operator L, page 127
g_{ij}	(local) components of the metric, page 2
h	height function, page 54
h^α_{ij}	coefficients of the second fundamental tensor, page 37
h^ν	mean curvature in the direction of a unit normal vector field ν, page 38
H_k	k-th mean curvatures of a hypersurface, page 38
$H_{T,X}$	differential operator acting on $u \in C^2(M)$ by $H_{T,X}u = T(\mathrm{hess}(u)(\,,\,) + (\mathrm{div}T - X^\flat) \otimes du$, page 144
$I \times_\rho \mathbb{P}$	ρ-warped product of the real interval $I \subseteq \mathbb{R}$, with $0 \in I$, and the Riemannian manifold $(\mathbb{P}, \langle\,,\,\rangle_{\mathbb{P}})$, page 49
$I^+(p)$	chronological future of a point p, page 528
$I^+(S)$	chronological future of a set S, page 527
$I^-(S)$	chronological past of a set S, page 528
I_m	$m \times m$ identity matrix, page 5
J	Jacobi field, page 89
$J^+(p)$	causal future of a point p, page 528
$J^+(S)$	causal future of a set S, page 527
$J^-(S)$	causal past of a set S, page 528
$K(u \wedge v)$, ${}^MK(u \wedge v)$	sectional curvature of the plane $\Pi \subset T_pM$ spanned by u and v, page 18
$K_p(\Pi)$, ${}^MK_p(\Pi)$	sectional curvature of the 2-plane $\Pi \subset T_pM$, page 18
K_{rad}	radial sectional curvature, page 71
L	elliptic operator, page 111
$L_{T,X}$	differential operator acting on $u \in C^2(M)$ by $Lu = \mathrm{div}(T(\nabla u,\,)^\sharp) - \langle X, \nabla u\rangle = \mathrm{Tr}(t \circ \mathrm{hess}(u)) + \mathrm{div}\,T(\nabla u) - \langle X, \nabla u\rangle$, page 142
M_g	model manifold, page 106

$O(m)$	set of $m \times m$ orthogonal matrices, page 5
P	projective curvature tensor, page 31
$p(x, y, t)$	(minimal) positive heat kernel of the Laplace-Beltrami operator, page 100
R	$(1, 3)$-version of the Riemann curvature tensor, page 13
$r(x)$	Riemannian distance function, page 58
$S, {}^{M}S$	scalar curvature (of the manifold M), page 18
$S^2(M)$	space of symmetric $(0, 2)$-tensors, page 160
S_k	k-th elementary symmetric function of the eigenvalues of A, page 38
s_p	Lorentzian cut locus function, page 529
T	stress-energy tensor field, page 512
T	symmetric positive semi-definite $(0, 2)$-tensor field, page 142
T	traceless Ricci tensor, page 22
t	$(1, 1)$-version of the traceless Ricci tensor, page 453
T_p^*M	cotangent space at a point $p \in M$, page 8
T_pM	tangent space at a point $p \in M$, page 8
$TM^{\perp}$	normal bundle, page 35
u^*	the supremum of the function u on the manifold M, page 78
u_*	the infimum of the function u on the manifold M, page 86
u_-	negative part of the function u, page 239
u_i	local components of the differential du, page 9
$V^{\perp}$	the set $\{w \in T_pN : \langle v, w \rangle = 0 \quad \text{for all } v \in V\}$, where V is a linear subspace or a subset, page 501
W	Weyl tensor, page 25
$W^{1,1}(M)$	Sobolev space of functions in $L^1(M)$ with (weak) gradient in $L^1(M)$, page 68
$W_0^{1,2}(M)$	Sobolev space of functions in $L^2(M)$ with (weak) gradient in $L^2(M)$ and zero trace, page 109
X	vector field, page 142
$x^1, \ldots, x^m$	coordinate functions on an open set, page 2
X_k^i	covariant derivative of the coefficient X^i, page 7
X_t	vector field along a smooth curve, page 163
$Z^{\sharp}$	$(1, 1)$-tensor obtained from the symmetric $(0, 2)$-tensor field Z by raising an index, page 9
$(\varphi_t)_*$	push-forward of the flow φ_t, page 12
$(M, \langle\, , \,\rangle)$	Riemannian manifold with metric $\langle\, , \,\rangle$, page 2
$\mathbf{H}$	mean curvature vector field, page 37
${}^{\perp}R^{\alpha}_{\beta ij}$	components of the normal curvature tensor, page 41
$\sharp : T^*M \to TM$	musical isomorphism (sharp map), page 9
$C(u)$	timecone of T_pN determined by u, page 501
OYMP	Omori-Yau maximum principle, page xvi
WMP	Weak maximum principle, page xviii

Introduction

In 1967, studying the possibility of minimally immersing complete submanifolds into Euclidean cones, Omori [210] introduced an important analytical tool nowadays called the *Omori-Yau maximum principle*. The underlying motivation is quite simple and can be illustrated by the following elementary remark. Suppose that a C^2-function on a Riemannian manifold $(M, \langle\, , \rangle)$ attains a maximum at a point $x_0 \in M$; then, at that point,

$$\nabla u(x_0) = 0 \quad \text{and} \quad \mathrm{Hess}(u)(x_0) \leq 0, \tag{1}$$

where with the above notation we mean that the symmetric bilinear form $\mathrm{Hess}(u)(x_0)$ is negative semi-definite. In particular, if the attained maximum is an absolute maximum for u, then

$$u^* = \sup_M u = u(x_0).$$

However, what happens if u is bounded above, that is, $u^* < +\infty$, but u^* is never attained on M? It is not difficult to show (see, for instance, Sect. 2.1) that, if $M = \mathbb{R}^m$ with its canonical flat metric $\langle\, , \rangle$, for any $u \in C^2(\mathbb{R}^m)$ with $u^* < +\infty$, one can always find a sequence of points, call it $\{x_k\}$, with the properties

$$u(x_k) > u^* - \frac{1}{k}; \quad |\nabla u|(x_k) < \frac{1}{k}; \quad \mathrm{Hess}(u)(x_k) < \frac{1}{k}\langle\, , \rangle \tag{2}$$

for each $k \in \mathbb{N}$ (where $\mathbb{N}$ is the set of natural numbers).

On the other hand, it is a simple matter to give examples of manifolds where this property fails. Restricting ourselves to the two-dimensional case for the ease of computations, let us consider $\mathbb{R}^2$ with a fixed origin o and with a metric expressed in polar coordinates $(r, \theta) \in \mathbb{R}^2 \setminus \{o\} = \mathbb{R}^+ \times \mathbb{S}^1$ (where $\mathbb{S}^1$ is the circle of radius 1) in the form

$$\langle\, , \,\rangle = dr^2 + g(r)^2 d\theta^2,$$

where the function $g \in C^\infty(\mathbb{R}_0^+)$ satisfies $g(r) > 0$ on $\mathbb{R}^+$ and

$$g(r) = \begin{cases} r & \text{on } [0,1] \\ r(\log r)^{1+\mu} e^{r^2(\log r)^{1+\mu}} & \text{on } [10, +\infty) \end{cases}$$

for some constant $\mu > 0$. Here, and in what follows, we use the notation $\mathbb{R}^+ = (0, +\infty)$ and $\mathbb{R}_0^+ = [0, +\infty)$. The above metric extends smoothly to o, and its sectional curvature, that is, its Gaussian curvature, is given in $\mathbb{R}^2 \setminus \{o\}$ by

$$K(x) = -\frac{g''(r(x))}{g(r(x))} \sim -c^2 r(x)^2 [\log r(x)]^{2(1+\mu)}$$

as $r(x) \to +\infty$ for some constant $c > 0$. We define a function $u(x)$ by setting

$$u(x) = \int_0^{r(x)} \left(\frac{1}{g(t)} \int_0^t g(s)\, ds \right) dt.$$

Then u is well defined, C^2 on $\mathbb{R}^2$, and it is bounded above since $\mu > 0$. However, computing its Laplacian we find $\Delta u \equiv 1$, showing that the third of the requirements in (2) cannot be fulfilled.

Our considerations point out the need to look for sufficient conditions to guarantee the validity of (2). Omori, as the above examples suggest, focused his attention on curvature conditions; he was able to answer positively to the problem by imposing, besides completeness of the manifold, a constant lower bound on the sectional curvature of $(M, \langle\, , \,\rangle)$.

A few years later, in 1975, the subject was taken up by S.-T. Yau, who modified statement (2) to

$$u(x_k) > u^* - \frac{1}{k}; \quad |\nabla u|(x_k) < \frac{1}{k}; \quad \Delta u(x_k) < \frac{1}{k} \tag{3}$$

for each $k \in \mathbb{N}$. In this relaxed conclusion, to which, from now on, we will refer to as the *Omori-Yau maximum principle* (OYMP for short), he substituted the requirement on Hess(u) with the corresponding requirement on its trace Δu. Considering this new point of view, he provided a sufficient condition for the validity of (3) on complete manifolds in terms of a lower bound on the Ricci curvature that, as it is well known, is obtained by "tracing the curvature." The motivation is loosely tied to the fact that the sectional curvatures are responsible for bounding the Hessian of the (Riemannian) distance function from a fixed point, while the Ricci curvature, for bounding its Laplacian.

Of course a lower bound on Ric is less restrictive than a lower bound on the sectional curvature, and conclusion (3) is very often sufficient to solve interesting geometrical problems, as it was immediately shown by Yau himself in [279] and by Cheng and Yau in [81]. In particular we should mention Yau's version of Schwarz Lemma for holomorphic maps between Kähler manifolds that solved a long-standing problem [281], as well as the solution of the Bernstein problem for maximal spacelike hypersurfaces in the Lorentz-Minkowski space given by Cheng and Yau in [82]. The beautiful initial results of Omori, Yau, and collaborators opened the way to the use of the OYMP in Riemannian geometry.

It is also worth to consider Yau's perspective under a more philosophical respect. Indeed, from the point of view of the analyst, the classical maximum principle is expressed in a different form; precisely, and for the sake of simplicity referring to the Laplace-Beltrami operator (see, for instance, [233, p. 53]), let $u \in C^2(\Omega)$ for some domain $\Omega \subset M$. If u attains its maximum μ at any point $x_0 \in \Omega$ and $\Delta u \geq 0$ on Ω, then $u \equiv \mu$ in Ω. In particular, if Ω is relatively compact, $\partial\Omega \neq \emptyset$ and $u \in C^0\left(\overline{\Omega}\right)$, then $\sup_\Omega u = \sup_{\partial\Omega} u$. It is well known that the proof of this fact is heavily based on property (1) and a famous trick of Hopf (see [233, 235]) consisting in passing from the weak inequality $\Delta u \geq 0$ to the strong inequality $\Delta v > 0$ for an appropriate auxiliary function v. Since the essential steps in the proof of the result are the properties

$$\sup_\Omega u = u(x_0), \quad |\nabla u|(x_0) = 0, \quad \Delta u(x_0) \leq 0 \tag{4}$$

at x_0, Yau calls (4) the *(finite) maximum principle*. This different "pointwise" perspective also justifies the search for sufficient conditions guaranteeing the validity of (3); note that, in fact, in this new version, we do not need u to satisfy, a priori, a certain differential inequality like $\Delta u \geq 0$ as before, and this, conceptually, is a cornerstone. On the other hand, one could think to have lost the "localization" point of view of the analyst. Basically this is not the case, as we shall explain and prove in Chap. 3.

Having realized the above conceptual point, the OYMP became an important tool in the study of the geometry, for instance, of submanifolds, harmonic maps, conformal geometry, and elliptic equations. In the form given in (2) and (3), it rested, respectively, on the assumptions

$$(i)\ {}^MK \geq -B^2, \qquad (ii)\ \mathrm{Ric} \geq -(m-1)B^2 \tag{5}$$

and completeness of $(M, \langle\, ,\, \rangle)$. Here with ${}^MK$ and Ric, we denote the sectional curvature of M and its Ricci tensor. Thus, a reasonable attempt to generalize the principle was in trying to relax conditions (5) *(i)* or (5) *(ii)*. This was achieved, to the best of our knowledge, in two independent papers, [78, 239]. In the latter (5) *(ii)*, for example, was replaced by

$$\mathrm{Ric} \geq -(m-1)G^2(r), \tag{6}$$

where G is in a class of functions displaying a certain behavior at infinity, for instance,

$$G(t) = t\prod_{j=1}^{N} \log^{(j)}(t), \quad t \gg 1,$$

where $\log^{(j)}$ stands for the jth iterated logarithm. More importantly, the proof given in [239] opened the way that led to a new observation: the key auxiliary function γ constructed in the proof to make the argument work did not need, in fact, to come from a distance function on M. Since the behavior of the Laplacian (or of the Hessian) of the latter is governed by curvature conditions, this observation frees us from assumptions of the type (5) or (6) once we can provide γ in some other way, for instance, when $f : M \to N$ is an immersion, via the extrinsic geometry of M (see Chap. 2 for a first example).

In the meanwhile, from a number of geometric applications, it became apparent that the second condition in (2) or (3), that is,

$$|\nabla u|(x_k) < \frac{1}{k} \tag{7}$$

along the sequence, was not always needed to reach the desired geometric conclusions. This suggested the following simpleminded definition introduced in [225]: we say that the *weak maximum principle* (WMP for short) *holds on M for the operator* Δ if for each $u \in C^2(M)$ with the property

$$u^* = \sup_M u < +\infty \tag{8}$$

there exists a sequence $\{x_k\} \subset M$ such that

$$u(x_k) > u^* - \frac{1}{k} \quad \text{and} \quad \Delta u(x_k) < \frac{1}{k}. \tag{9}$$

The unexpected fact is that, as proved in [225], this property is equivalent to stochastic completeness of the manifold or to uniqueness of the solutions of the Cauchy problem for the heat equation (see [227] or [131] for more details in this direction). This has a twofold feedback. On the one hand sufficient conditions to provide stochastic completeness, such as the Khas'minskii test, can be used to guarantee the validity of the WMP; on the other hand, we can use the WMP to investigate probabilistic properties. From this new point of view, can we give simple sufficient conditions coming from geometry to insure the validity of stochastic completeness? Of course the validity of the Omori-Yau maximum principle implies that of the weak, but can we indeed provide genuine sufficient conditions weaker than those guaranteeing the maximum principle in its full strength (2) or (3)? A positive answer is given by the results presented at the end of Chap. 4: the validity of the WMP is obtained under growth conditions on the volume of geodesic balls of the manifold. This is certainly weaker than requiring curvature assumptions that do imply volume growth but which are not implied by the latter. However, one point is still open; that is, also in this case we need completeness of the manifold $(M, \langle\ ,\ \rangle)$. This condition is somehow natural for the OYMP; indeed, if we prove its validity via the function γ that we were talking about before, then two of the conditions on γ imply completeness of the metric (see Remark 2.5 and the proof of Theorem 3.2). However, for the WMP its equivalence with stochastic completeness reveals immediately its independence from the geodesic completeness of the metric. For instance, $\mathbb{R}^2 \setminus \{0\}$ with its canonical Euclidean metric is certainly geodesically incomplete, but it is stochastically complete.

As a final remark we note that, as in the case of the OYMP, the construction of a function γ satisfying Khas'minskii test can often be obtained, for instance, by exploiting solutions of suitable differential inequalities or the extrinsic geometry in case of an immersed manifold. This new point of view enabled Alías et al. [18] to give a positive answer in case of proper immersions to two well-known conjectures of Calabi on minimal submanifolds of Euclidean space; note that an earlier result in this direction appears in [225].

In Riemannian geometry there are many other interesting and natural differential operators besides that of Laplace and Beltrami. Just to mention a few, let us recall the mean curvature operator, the operator associated to the Newton tensors in the geometry of hypersurfaces, the X-Laplacian of generic (i.e., not necessarily gradient) Ricci solitons, and so on. It is therefore quite legitimate to address the problem of generalizing the maximum principles presented so far to a larger class of operators. As expected, the nonlinear case is the most delicate. Yau's original proof or even the more recent approach in [43, 52] (the latter is in fact based on a Euclidean argument presented in [227]) do not go through in this new setting to prove the corresponding form of the OYMP, while the extension of the WMP turns out to be simpler. However, both problems have been recently solved in [5]. The family of operators L considered is as follows (in fact in Chap. 4 we present an enlarged class): L acts on, say, $u \in C^2(M)$ by

$$Lu = \operatorname{div}\left(|\nabla u|^{-1}\varphi(|\nabla u|)T(\nabla u, \cdot)^\sharp\right) - \langle X, \nabla u\rangle. \tag{10}$$

Here T is a generic symmetric, positive definite (or semi-definite) $(0, 2)$-tensor, $^\sharp : TM^* \to TM$ is the standard musical isomorphism, X a vector field on M, and

$\varphi : \mathbb{R}_0^+ \to \mathbb{R}_0^+$ a function satisfying $\varphi \in C^0(\mathbb{R}_0^+) \cap C^1(\mathbb{R}^+)$ and the structural conditions

$$(i)\ \varphi(0) = 0; \quad (ii)\ \varphi(s) > 0 \text{ for } s \in \mathbb{R}^+; \quad (iii)\ \varphi(s) \le As^\delta \text{ on } \mathbb{R}^+ \tag{11}$$

for some constants $A, \delta > 0$.
These operators with $X \equiv 0$ have been considered for the WMP in a series of papers by Pigola, Rigoli, and Setti, see, e.g., [226, 229] and also [245]. Note, for instance, that for $X \equiv 0$ and $T = \langle\, , \,\rangle$, the choices

$$\varphi(s) = \frac{s}{\sqrt{1+s^2}} \quad \text{or } \varphi(s) = s^{p-1},$$

respectively, give the mean curvature operator and the p-Laplacian; for $\varphi(s) = s$ and $X = (\operatorname{div} T)^\sharp$, we obtain the trace operator $\operatorname{Tr}(t \circ \operatorname{hess}(u))$, where $\operatorname{hess}(u), t : TM \to TM$ are defined by

$$\operatorname{hess}(u)(Y) = \operatorname{Hess}(u)(Y, \cdot)^\sharp,$$
$$t(Y) = T(Y, \cdot)^\sharp$$

for every vector field Y on M.

So far we have always tacitly understood to consider the case $u^* = \sup_M u < +\infty$; can we say something in case where $u^* = +\infty$ but u has a controlled behavior from above at infinity? In this case, for the WMP we have a positive answer under an assumption relating the growth of u at infinity with that of the volume of geodesic balls (and with δ in *(iii)* above in case of a general operator). The importance of this type of result is manifest: for instance, its use allows us to obtain a comparison result for nonnegative solutions u and v of a Yamabe-type equation of the form

$$\Delta u + a(x)u - b(x)u^\sigma = 0, \quad \sigma > 1,$$

with $b(x) > 0$, under some mild conditions on the manifold, the coefficients $a(x)$ and $b(x)$ and in the sole requirement

$$v(x) \ge C_1 r(x)^\tau, \quad u(x) \le C_2 r(x)^\tau,$$

with $\tau \ge 0$, for the behavior of u and v at infinity. For details, see Theorem 5.5 of [189] or [244] and also Sect. 4.3.1 and [4] for a similar equation. On the other hand, the version of the OYMP when $u^* = +\infty$ remains in some sense an open question, but see, for instance, Theorem 3.5.

Going back to the WMP for the Laplace-Beltrami operator Δ, we realize (see Chap. 2 for details) that a slightly stronger form of the principle is equivalent to the usual notion of *parabolicity*. Somehow this explains why results related, respectively, to stochastic completeness and *recurrence*, the probabilistic version of parabolicity that resembles each other in many circumstances. This observation will enable us to extend the notion of parabolicity, in the form of a Liouville-type result, to a new notion that we call *strong parabolicity* and which turns out to be equivalent to the former for a large class of operators. After having introduced the above analytical tools, to show their effectiveness, we dedicate the second part of the book to some applications. We have chosen to concentrate on the geometry of submanifolds, and in particular on that of hypersurfaces, but we also illustrate their usefulness in dealing with some elliptic PDE problem and the geometry of Ricci solitons. Further applications are given in the setting of Lorentzian geometry. This material will be discussed more appropriately in a short while when we will describe the content of the various chapters. We only like to underline that the results presented are quite recent and belong to an active field of actual research; thus, sometimes we comment with open questions and problems that to the best of our present knowledge have not yet been completely answered or solved. In this sense our work introduces the reader to active research topics, and we hope to provide her/him with quite efficient technical tools to move toward their solutions and that of other related geometric and analytic problems.

We now outline in some detail the content of the various chapters of the book, pointing out that at the beginning of each one of them, the reader is guided by a short initial introduction focusing on the various themes.

In Chap. 1 we present a "crash course" in Riemannian geometry, with two purposes: the first is to fix definitions and notation and to get the reader acquainted with the formalism of the moving frame to perform computation by her-himself. The material we present will be used in the subsequent chapters, especially in the geometric applications. We like to mention that we also prove some properties of "curvature" tensors (Weyl, Cotton, projective, etc.) and some commutation relations of their covariant derivatives that are not easily available in the literature. We end the chapter with some brief considerations on the Laplacian and Hessian comparison theorems that constitute an essential tool in geometric analysis. A very detailed exposition of these topics without the use of Jacobi fields, together with a preliminary discussion on the cut locus of a point (and more generally on the focal locus), can be found in the recent paper [44].

The second purpose concerns the aim of the book to be as self-contained as possible, with the intent to quickly introduce the young reader to current research topics. We feel that the material presented is certainly sufficient for the understanding of the rest of the chapters, but we also hope that it will act as a stimulus to deepen the knowledge of the subject on standard treatises such as [73, 102, 156, 232] and so on. However, although we present the method of the moving frame in some detail, and this certainly will be essential for some computations (for instance, in Chap. 8), we shall also use freely Koszul formalism. This is quite standard in the differential geometry community, and it will help readers unwilling to spend some time to master a different formalism that is still not so loved by many people but that is undeniably effective in many situations, as we shall see.

Chapter 2 introduces and motivates the OYMP and the WMP for the Laplacian and the Hessian operators. We point out the mutual relations between the conditions in (3), proving in particular Ekeland quasi-minimum principle in the form of Proposition 2.2 that shall be used in this chapter and in Chap. 3 (note that some similar form of Proposition 2.2 is well-known *folklore* in geometry, see also [227]). Next we introduce the original statement of Omori and Yau for the maximum principle and the generalized OYMP with an auxiliary function γ satisfying the conditions listed in Theorem 2.4; in this situation we assume $\gamma \in C^2(M)$, but we show in Theorem 2.5 that γ can be built starting from the distance function from a fixed origin when $(M, \langle\ ,\ \rangle)$ is complete. Here the point is delicate, since γ is only $\mathrm{Lip}_{loc}(M)$: to solve the problem we elaborate on a trick of Calabi [55]. However, we also point out that since γ is a solution of a certain differential inequality in the Lip_{loc} sense, we can in fact use, in the case of the Laplacian, an alternative way based on a comparison result proved in [227] (see Remark 2.8, following the proof of Theorem 2.5, and Proposition 3.1 with Remark 3.9). This is an important alternative that shall be used in the nonlinear case but in the $C^1(M)$ class, since we need a strengthened version of Proposition 3.1 (see Theorem 3.9 in Chap. 3). We also construct γ from the extrinsic geometry of an immersed submanifold $f : M \to N$ under some mild geometrical restrictions, mainly on the mean curvature of the immersion (another construction of a similar kind is given, for instance, in Sect. 5.4 of Chap. 5).

The further step is to introduce the WMP and its equivalence with stochastic completeness and with other analytic properties; see Theorems 2.7 and 2.9. We then discuss some properties and results related to stochastic completeness to show how this equivalence can be used to prove them in a different and particularly simple way. In Sect. 2.4 we present two applications of stochastic completeness to a curvature problem and a Liouville-type theorem (see, respectively, Theorems 2.17 and 2.18).

The chapter ends showing the relation between parabolicity of $(M, \langle\ ,\ \rangle)$ and a stronger form of the WMP. The observations presented here justify our extension of the notion of parabolicity that we call strong parabolicity that will appear at the end of Chap. 4.

Having provided some initial motivations in the simplest cases of the Laplace-Beltrami operator and of the Hessian, in Chap. 3 we deal with some new forms of the maximum principle both in the linear and nonlinear case. Motivated by the results of Chap. 2, we present the weak and the Omori-Yau maximum principles with the aid of an auxiliary function γ (see, for instance, the statement of Theorem 3.1). In the linear case, once we have chosen the class of operators as in (3.1), that is,

$$Lu = \mathrm{div}(T(\nabla u,\ )^\sharp) - \langle X, \nabla u\rangle,$$

we face fewer technical difficulties than in the nonlinear case, and guided by the insights of Chap. 2 we can extend, for example, Theorems 3.1 and 3.2 to the case where γ is the composition of the distance function from a fixed origin o in the complete manifold $(M, \langle\ ,\ \rangle)$ with an appropriate function, and, therefore, γ is only $\mathrm{Lip}_{loc}(M)$. It is worth to observe that, in proving Theorem 3.1 that corresponds to the WMP, we do not need any use of the Ekeland quasi-minimum principle: indeed, in this case $(M, \langle\, , \,\rangle)$ need not even be geodesically complete. On the other hand, Proposition 2.2 is needed in the proof of Theorem 3.2 for which first we show that the metric is complete. This latter fact is implied by two of the conditions on γ, but, as a matter of fact, the validity of the Ekeland quasi-minimum principle for a certain class of functions on a metric space implies completeness of the latter and therefore, in our case, of the manifold (see Remark 2.1). As already observed this fact is related to the first two properties in (3) for the validity of the OYMP. However, see also the considerations after Example 2.2 in Chap. 2. We complete our discussion of the linear case by extending an L^∞ a priori estimate that Cheng and Yau [81] and Motomiya [199] (but the proof in the latter is incorrect) proved for the Laplacian. In fact we deal with solutions u of differential inequalities of the type

$$g(x)\Big[\operatorname{div}\Big(T(\nabla u, \cdot)^\sharp\Big) - \langle X, \nabla u\rangle.\Big] \geq \varphi(u, |\nabla u|). \tag{12}$$

Without entering into further details, the main assumption on the right-hand side of (12) is that $f(t) = \varphi(t, 0) > 0$ for $t \gg 1$ and $F(t) = \int_a^t f(s)\,ds$, $a \gg 1$, satisfying

$$\frac{1}{\sqrt{F(t)}} \in L^1(+\infty). \tag{13}$$

The alert reader will immediately recognize (13) as the classical Keller-Osserman condition; in particular, it is a sharp condition for the validity of an L^∞ upper bound on u. We shall come back on this when describing a corresponding L^∞-estimate in the nonlinear case. We note that the value of this type of estimate can be hardly overestimated both in the geometric and analytic setting.

We then come to discuss the nonlinear case. Here the situation is much subtler, and due to the type of operators we consider, that is,

$$Lu = \operatorname{div}\Big(|\nabla u|^{-1}\varphi(|\nabla u|)T(\nabla u, \cdot)^\sharp\Big) - \langle X, \nabla u\rangle,$$

with φ as in (11), we restrict our attention to C^1 functions u; of course the above operator has to be interpreted in the weak sense. The validity of the corresponding maximum principles, weak and Omori-Yau, is obtained, respectively, in Theorems 3.11 and 3.13 whose statements and proofs are similar to those of Theorems 3.1 and 3.2. However, the analytical difficulties that we are now facing with this large class of operators are definitely deeper. Thus, we devote an entire section to a careful proof of the auxiliary analytical results in the form that we shall need for our purposes (this part, Sect. 3.3.1, is based on Pucci and Serrin [234] and Pucci et al. [236]). In fact the results we obtain are more general than what is strictly needed. We do this for two reasons: the first is that these results are interesting in their own and have a wide range of applicability; the second is that in the maximum principle, we prove (see Theorem 3.10) there appears a somewhat dual form of (13). In fact there are four conditions, which are intertwined and related to (13), that are responsible for the maximum principle, the compact support principle, and L^∞ a priori estimates. This is partly clarified in [183, 236] and the very recent [45].

In Sect. 4.1 of Chap. 4, we prove an L^∞ a priori estimate and a WMP under a volume growth condition for geodesic balls. In this more general nonlinear case, to obtain an upper bound for u, solution of a certain differential inequality of the form

$$Lu = \operatorname{div}\left(|\nabla u|^{-1}\varphi(|\nabla u|)T(\nabla u, \cdot)^\sharp\right) \geq b(x)f(u),$$

we need to impose the condition

$$\liminf_{t\to+\infty} \frac{f(t)}{t^\sigma} > 0$$

for some $\sigma > \delta$, where δ is the structural constant in (11) *(iii)*. This condition is slightly stronger than the corresponding nonlinear Keller-Osserman requirement; we feel that the latter should suffice, but this remains an open problem.

The next item is a proof of a *controlled growth weak maximum principle*, meaning with this that a form of the WMP can be obtained even for an unbounded u provided we have a control of the type

$$\limsup_{r(x)\to+\infty} \frac{u(x)}{r(x)^\sigma} < +\infty.$$

For the precise statement of the result, we refer to Theorem 4.4: to show its usefulness we also give an application to Killing graphs.

As mentioned before we then localize the WMP to the family of open sets Ω of M with $\partial\Omega \neq \emptyset$. For an operator as in (10), the principle becomes equivalent to: *for each Ω as above, $f \in C^0(\mathbb{R})$ and $v \in C^0(\overline{\Omega}) \cap C^1(\Omega)$ satisfying*

$$\begin{cases} Lv & \geq f(v) \quad \text{on } \Omega, \\ \sup_\Omega v & < +\infty \end{cases}$$

we have either $f(v) \leq 0$ or $\sup_\Omega v = \sup_{\partial\Omega} v$. In particular: *for each Ω as above, $\beta \in \mathbb{R}^+$ and $v \in C^0(\overline{\Omega}) \cap C^1(\Omega)$ satisfying*

$$\begin{cases} Lv & \geq \beta \quad \text{on } \Omega, \\ \sup_\Omega v & < +\infty \end{cases} \tag{14}$$

we have

$$\sup_\Omega v = \sup_{\partial\Omega} v. \tag{15}$$

This last formulation is in the vein of classical analysis, and it reveals extremely powerful in applications (for its use in some geometric setting, see Chaps. 5 and 7). This new form also justifies the introduction of a new concept, i.e., strong parabolicity. We compare this with the usual one given as the validity of a Liouville-type theorem; we also provide sufficient conditions for the validity of both. We end the chapter with an application of Theorems 4.1 and 4.2 and of these last concepts to generalize to our family of operators a result of Dancer and Du [97], which in the parabolic case on $\mathbb{R}^m$ is a consequence of the "hair trigger" effect of Aronson and Weinberger [32]. In fact we improve on it even in the case of the Laplace-Beltrami operator; this part is based on [226].

Chapter 5 is devoted to the applications of the material (mainly) of Chap. 4, to the study of the geometry of submanifolds. In the first result, Theorem 5.1, we improve on Omori [210] on the impossibility of immersing minimally a submanifold into a nondegenerate cone of Euclidean space. Following Mari and Rigoli [181], we then generalize this result to smooth maps $\varphi : M \to \mathbb{R}^n$ by providing a sharp upper estimate for the width of the cone containing the image $\varphi(M)$ in terms of the distance of $\varphi(M)$ to a certain hyperplane, the norm of the tension field of φ and its energy density.

We then continue in this spirit and prove various (generalizations of) classical theorems on the impossibility of isometrically immerse given manifolds in Euclidean space or in cones. As it is well known, this subject of investigation goes back to Tompkins [264], Chern and Kuiper [86], Otsuki [215], and so on, up to the work of Jorge and Koutrofiotis [154]. In fact we extend this latter result to immersions into cones by providing again a sharp upper estimate of the width of the cone.

The next step, suggested by one of Calabi's conjecture on minimal hypersurfaces in $\mathbb{R}^n$, is to consider cylindrically bounded submanifolds (see Sect. 5.4). Calabi's conjecture asserts that any complete nonflat minimal hypersurface in $\mathbb{R}^n$ has unbounded projection in every $(n-2)$-dimensional subspace of $\mathbb{R}^n$; in this case we can indeed fully appreciate the function theoretic form of the WMP. The function γ of Theorem 3.1, for instance, is constructed via the projection of the immersion on the unbounded component of $f(M) \subset \mathbb{R}^l \times B_R$, where B_R stands for a geodesic ball of radius R in N and $f : M \to \mathbb{R}^l \times N$ (see Theorem 5.9, Corollary 5.8, and related results in Sects. 5.4.1 and 5.4.2).

On the other hand, it is a well-known result of Ruh and Vilms [249], that the Gauss map $\gamma_f : M \to G_m(\mathbb{R}^n)$ of an isometric immersion $f : M \to \mathbb{R}^n$ is harmonic if and only if f has parallel mean curvature. We study some consequences of this fact with the aid of our analytic machinery (see, for instance, Theorems 5.11 and 5.12)

In Chap. 6 we focus our attention to the applications of these techniques to the study of the geometry of hypersurfaces. In particular, we begin by considering complete hypersurfaces immersed with constant mean curvature into Riemannian space forms and deriving sharp estimates for the infimum and the supremum of their scalar curvature, classifying the cases of equality. Similarly, we consider the case of hypersurfaces with constant scalar curvature into Riemannian space forms; this forces us to use the well-known differential operator introduced by Cheng and Yau [83] in order to give estimates on the second fundamental form. We are able to characterize the cases of equality also in this situation. After this, to extend our investigation to a larger class of Riemannian ambient spaces, it appears convenient to consider manifolds with a sufficiently large family of complete embedded constant mean curvature hypersurfaces. Such a family plays the role of the umbilical hypersurfaces in space forms. A natural class of ambient manifolds where this happens is that of warped products that are foliated by totally umbilical leaves. In this setting, in Chap. 7 we derive higher-order mean curvature estimates for complete immersed hypersurfaces Σ and determine sufficient conditions in case of constant higher-order mean curvatures to guarantee that if the image is contained in a slab, then it is a leaf of the foliation. In doing so we need to consider quite general operators that come from the Newton tensors and some appropriate linear combinations of them. Finally, as an application of our localized form of the WMP, we give height estimates for hypersurfaces in a product space, where the results appear in a quite neat form.

In Chap. 8 we study Ricci solitons. Our emphasis is on *generic* Ricci solitons, that is, Riemannian manifolds $(M, \langle\, ,\, \rangle)$ for which there exist a constant $\lambda \in \mathbb{R}$ and a vector field X satisfying

$$\mathrm{Ric} + \frac{1}{2}\mathscr{L}_X \langle\, ,\, \rangle = \lambda \langle\, ,\, \rangle,$$

where $\mathscr{L}_X \langle\, ,\, \rangle$ is the Lie derivative of the metric $\langle\, ,\, \rangle$ in the direction of X. When $X = \nabla f$ for some potential f, the above equation assumes the form

$$\mathrm{Ric} + \mathrm{Hess}(f) = \lambda \langle\, ,\, \rangle$$

and the soliton is called a *gradient* Ricci soliton. The case of generic Ricci solitons is quite heavy from the computational point of view, so that the first part of the chapter is dedicated to various useful calculations. In the second part first we analyze the validity of the weak and strong maximum principle for the operator $\Delta_X = \Delta - \langle X, \nabla \rangle$ and for the symmetric diffusion operator $\Delta_f = e^f \operatorname{div}\left(e^{-f}\nabla\right) = \Delta - \langle \nabla f, \nabla \rangle$. In both cases the validity of the OYMP is granted by the structure with no further assumptions besides that of completeness of $(M, \langle\, ,\, \rangle)$. We combine this fact with the a priori estimate of Theorem 3.6 to obtain, in Theorem 8.2, lower and upper bounds for the infimum of the scalar curvature of the soliton; similarly, in Theorem 8.3 we provide a lower bound for the supremum of the norm of the traceless Ricci tensor. For further results we refer to [188].

The above results are then refined and further analyzed in the gradient case, classifying in particular some classes of solitons (see [224]). We end the chapter with a very recent result on generic Ricci solitons based on [68] for which we use a sufficient condition for strong parabolicity given in Sect. 4.4.

The final chapter, Chap. 9, is devoted to some applications to spacelike hypersurfaces in Lorentzian spacetimes. After some basic preliminaries on their geometry, we give a proof of the celebrated Bernstein-type theorem of Cheng and

Yau [82], which states that the only complete maximal hypersurfaces in the Lorentz-Minkowski space are the spacelike hyperplanes; this is followed by some other related results. Next, using the corresponding comparison theorems for the Lorentzian distance function, we extend our study to the case of spacelike hypersurfaces in spacetimes obtaining sharp estimates for the mean curvature and more generally for higher-order mean curvatures, of such hypersurfaces. We also consider the case of spacelike hypersurfaces immersed into Lorentzian warped product manifolds, called here *generalized Robertson-Walker spacetimes*, obtaining in this context height estimates and rigidity results.

We conclude this introduction by underlining the important role of the interplay between the analytic and geometric points of view. The OYMP and the WMP are good examples of how this interplay can help in solving geometric problems with the aid of analytic tools and how geometric problems force us to consider new analytic open questions that are naturally posed by them. One of the aims of this book is to clarify this relationship, trying to get to the core of the problems and, as a consequence, to provide what we believe are among the most efficient tools to deal with them.

Chapter 1
A Crash Course in Riemannian Geometry

This chapter is devoted to a quick review of some results in Riemannian geometry using the moving frame formalism. While we assume basic knowledge of the general subject as presented, for instance, in the standard references [51, 121, 156, 170, 171, 219, 272], several computations will be carried out in full detail in order to acquaint the reader with notation.

After having introduced the notion of coframes and frames we describe the Levi-Civita connection and curvature in terms of connection and curvature forms via E. Cartan first and second structure equations.

Symmetries and various properties of the curvature tensors (Riem, Ric, Weyl, Projective,...) are described at length together with a number of identities repeatedly used in the sequel. In particular we obtain some commutation rules for covariant derivatives of functions and tensors up to a certain order, also with the aim of pointing out the general procedure to determine them when needed in other situations.

Next we give a description of the geometry of submanifolds and of hypersurfaces with some attention to the case where the ambient space is a warped product; along the way we obtain relevant formulas that will appear in Chaps. 6 and 8. We also provide a brief introduction to the geometry of smooth maps between Riemannian manifolds; in particular, we introduce the generalized second fundamental tensor. The vanishing of its trace, the well-known "tension field", characterizes harmonic maps.

At the end of the chapter we describe some basic results on the Riemannian distance function from a fixed reference point $o \in M$; we briefly discuss the cut locus of o and some of its properties. We then describe comparison results for the Laplacian and the Hessian of the distance function, and for the volume of geodesic balls in terms of bounds on the appropriate curvature.

Here and in the rest of the book all manifolds are assumed to be connected, unless otherwise stated.

L.J. Alías et al., *Maximum Principles and Geometric Applications*,
Springer Monographs in Mathematics, DOI 10.1007/978-3-319-24337-5_1

1.1 Moving Frames, Levi-Civita Connection Forms and the First Structure Equation

Let $(M, \langle\,,\,\rangle)$ be a Riemannian manifold of dimension $m = \dim M$ with metric $\langle\,,\,\rangle$. Let $p \in M$ and let (U, φ) be a *local chart* such that $p \in U$. Denote by $x^1, \ldots, x^m$ the coordinate functions on U. Then, at any $q \in U$ we have

$$\langle\,,\,\rangle = g_{ij}\, dx^i \otimes dx^j, \tag{1.1}$$

where dx^i denotes the differential of the function x^i and g_{ij} are the (local) components of the metric defined by $g_{ij} = \langle \frac{\partial}{\partial x^i}, \frac{\partial}{\partial x^j} \rangle$. In Eq. (1.1), and throughout this book, we adopt the Einstein summation convention over repeated indices. Applying in q the Gram-Schmidt orthonormalization process we can find linear combinations of the 1-forms dx^i, that we will call θ^i, $i = 1, \ldots, m$, such that (1.1) takes the form

$$\langle\,,\,\rangle = \delta_{ij}\theta^i \otimes \theta^j, \tag{1.2}$$

where δ_{ij} is the Kronecker symbol. Since, as q varies in U, the previous process gives rise to coefficients that are C^∞ functions of q, the set of 1-forms $\{\theta^i\}$ defines an orthonormal system on U for the metric $\langle\,,\,\rangle$, that is, a (local) *orthonormal coframe*. It is usual to write

$$\langle\,,\,\rangle = \sum_{i=1}^{m} (\theta^i)^2,$$

instead of (1.2). We also define the (local) *dual orthonormal frame* $\{e_i\}$, $i = 1, \ldots, m$, as the set of vector fields on U satisfying

$$\theta^j(e_i) = \delta_i^j \tag{1.3}$$

(where δ_i^j is just a suggestive way of writing the Kronecker symbol, reflecting the position of the indices in the pairing of θ^j and e_i).

Proposition 1.1 *Let $\{\theta^i\}$ be a local orthonormal coframe defined on the open set $U \subset M$; then on U there exist unique 1-forms $\left\{\theta_j^i\right\}$, $i, j = 1 \ldots, m$, such that*

$$d\theta^i = -\theta_j^i \wedge \theta^j \tag{1.4}$$

and

$$\theta_j^i + \theta_i^j = 0. \tag{1.5}$$

The forms θ^i_j are called the *Levi-Civita connections forms* associated to the orthonormal coframe $\{\theta^i\}$.

Proof Assume the existence of the forms θ^i_j satisfying (1.4) and (1.5) and let us determine their expression. Of course

$$\theta^i_j = a^i_{jk}\theta^k$$

for some $a^i_{jk} \in C^\infty(U)$ and (1.5) is equivalent to

$$a^i_{jk} + a^j_{ik} = 0. \tag{1.6}$$

The 2-forms $d\theta^i$ can be written, for some (unique) coefficients $b^i_{jk} \in C^\infty(U)$, as

$$d\theta^i = \frac{1}{2}b^i_{jk}\theta^j \wedge \theta^k, \qquad \text{with } b^i_{jk} + b^i_{kj} = 0.$$

Since (1.4) must hold we have

$$\frac{1}{2}b^i_{jk}\theta^j \wedge \theta^k = -a^i_{jk}\theta^k \wedge \theta^j = a^i_{jk}\theta^j \wedge \theta^k = \frac{1}{2}(a^i_{jk} - a^i_{kj})\theta^j \wedge \theta^k.$$

It follows that

$$b^i_{jk} = a^i_{jk} - a^i_{kj}. \tag{1.7}$$

Cyclic permutations of the indices i, j, k and the use of (1.6) and (1.7) yield

$$b^k_{ij} = a^k_{ij} - a^k_{ji} = -a^i_{kj} + a^j_{ki}, \tag{1.8}$$

and

$$b^j_{ki} = a^j_{ki} - a^j_{ik} = a^j_{ki} + a^i_{jk}. \tag{1.9}$$

Adding (1.7) to (1.9) and subtracting (1.8) we obtain

$$a^i_{jk} = \frac{1}{2}(b^i_{jk} - b^k_{ij} + b^j_{ki}). \tag{1.10}$$

The previous relation determines the expression of the forms θ^i_j and also proves uniqueness. Now define

$$\theta^i_j = \frac{1}{2}(b^i_{jk} - b^k_{ij} + b^j_{ki})\theta^k, \tag{1.11}$$

where the b^i_{jk}'s satisfy

$$b^i_{jk} + b^i_{kj} = 0.$$

It is clear that

$$a^j_{ik} = \frac{1}{2}(b^j_{ik} - b^k_{ji} + b^i_{kj}) = -\frac{1}{2}(b^i_{jk} - b^k_{ij} + b^j_{ki}) = -a^i_{jk},$$

thus (1.6) is met, and then the θ^i_j's defined in (1.11) satisfy (1.5); it is also immediate to verify that they satisfy (1.4). □

Equation (1.4), that is,

$$d\theta^i = -\theta^i_j \wedge \theta^j,$$

is called the *first structure equation*. We shall see in a short while the geometric meaning of (1.5).

1.2 Covariant Derivative of Tensor Fields, Connection and Meaning of the First Structure Equation

A matrix notation is sometimes useful when performing computations with moving frames. On the open set U let θ be the (column) vector valued 1-form whose components are $(\theta^1, \dots, \theta^m)$, and let γ be the matrix of 1-forms (θ^i_j) on U. Then, (1.5) becomes ${}^T\gamma = -\gamma$ (where ${}^T\gamma$ denote the transpose of γ), that is, γ takes values in the Lie algebra $\mathfrak{o}(m)$ of skew-symmetric matrices, and the first structure equation reads

$$d\theta = -\gamma \wedge \theta.$$

We want to focus our attention on the change of $\{\theta^i_j\}$ while changing the orthonormal coframe $\{\theta^i\}$. First, we need a simple

Lemma 1.1 (Cartan's Lemma) *Let $U \subset M$ be an open set of the Riemannian manifold $(M, \langle\, ,\, \rangle)$. Let $\{\theta^i\}$ be a local basis of T^*U, and assume that a set of 1-forms $\{\omega^i_\lambda\}$ on U, with $\lambda \in \Lambda$ and where Λ is any set of indexes, satisfies $\sum_i \omega^i_\lambda \wedge \theta^i = 0$. Then, there exist smooth functions $b^i_{\lambda,k}$ on U such that*

$$\omega^i_\lambda = b^i_{\lambda,k}\theta^k \qquad \text{and} \qquad b^i_{\lambda,k} = b^k_{\lambda,i},$$

that is, the matrix $B = (b^i_{\lambda,k})^i_k$ is an $m \times m$ symmetric matrix.

Proof We can write ω^i_λ as $\omega^i_\lambda = b^i_{\lambda,k}\theta^k$ for some smooth functions $b^i_{\lambda,k}$ on U. Then from $\sum_i \omega^i_\lambda \wedge \theta^i = 0$ we deduce

$$0 = \sum_{i,k} b^i_{\lambda,k}\theta^k \wedge \theta^i = \sum_{i<k}(b^i_{\lambda,k} - b^k_{\lambda,i})\theta^k \wedge \theta^i,$$

which easily implies the thesis. □

In the next proposition we show how the Levi-Civita connection forms change when the frame changes; we denote by $O(m)$ the set of $m \times m$ orthogonal matrices.

Proposition 1.2 *Let $\{e_i\}$ and $\{\widetilde{e_i}\}$ be two orthonormal frames, respectively defined on the open sets U and $\widetilde{U}$ with $U \cap \widetilde{U} \neq \emptyset$, and let $A : U \cap \widetilde{U} \to O(m)$ be a (smooth)* change of frames, *that is,*

$$\widetilde{e_i} = A^j_i e_j \qquad \text{on } U \cap \widetilde{U},\ i, j \in \{1, \ldots, m\} \tag{1.12}$$

(with a slight abuse of notation, we write $\widetilde{e} = eA$, where e (resp. $\widetilde{e}$) is the matrix having e_i (resp. $\widetilde{e_i}$) as i-th column). Then, the matrix γ transforms according to

$$\widetilde{\gamma} = A^{-1}\gamma A + A^{-1}dA, \tag{1.13}$$

or, in components,

$$\widetilde{\theta}^i_j = (A^{-1})^i_k \theta^k_t A^t_j + (A^{-1})^i_k dA^k_j \qquad i, j, k, t \in \{1, \ldots, m\}.$$

Proof We adopt the matrix notation for simplicity. By (1.12) the corresponding coframes $\{\theta^i\}, \{\widetilde{\theta}^i\}$ change according to

$$\widetilde{\theta}^i = (A^{-1})^i_j \theta^j \qquad \text{or, in matrix notation,} \quad \widetilde{\theta} = A^{-1}\theta. \tag{1.14}$$

Differentiating (1.14) and using the first structure equation we get

$$d\widetilde{\theta} = dA^{-1} \wedge \theta + A^{-1}d\theta = -A^{-1}dA\,A^{-1} \wedge \theta - A^{-1}\gamma \wedge \theta, \tag{1.15}$$

where

$$d(A^{-1}) = -A^{-1}dA\,A^{-1} \tag{1.16}$$

follows differentiating the components of $A^{-1}A = I_m$, I_m being the $m \times m$ identity matrix. Again from the first structure equation

$$d\widetilde{\theta} = -\widetilde{\gamma} \wedge \widetilde{\theta} = -\widetilde{\gamma} \wedge A^{-1}\theta = -\widetilde{\gamma}A^{-1} \wedge \theta. \tag{1.17}$$

Putting together (1.15) and (1.17) we get

$$\widetilde{\gamma}A^{-1} \wedge \theta = A^{-1}\gamma \wedge \theta + A^{-1}dA\,A^{-1} \wedge \theta,$$

hence multiplying by A on the right we get

$$\left(\widetilde{\gamma} - A^{-1}\gamma A - A^{-1}dA\right) \wedge \theta = 0.$$

Define now $B = (B^i_j) = \widetilde{\gamma} - A^{-1}\gamma A - A^{-1}dA$. By Cartan's Lemma 1.1, $B^i_j = b^i_{jk}\theta^k$ with $b^i_{jk} = b^i_{kj}$; observe that B is skew-symmetric in the indices i and j: indeed, it is easy to prove that both $A^{-1}\gamma A$ and $A^{-1}dA$ are skew-symmetric by mere computation, using that $A^{-1} = {}^TA$ and ${}^T\gamma = -\gamma$. From the above we have the symmetries

$$b^i_{jk} = -b^j_{ik} = -b^j_{ki} = b^k_{ji} = b^k_{ij} = -b^i_{kj} = -b^i_{jk},$$

so that B vanishes identically, and this proves the proposition. □

Remark 1.1 We observe *en passant* that the last part of the proof of Proposition 1.2 hides the deep reason beyond the fact that both $A^{-1}\gamma A$ and $A^{-1}dA$ are indeed $\mathfrak{o}(n)$-valued matrices of 1-forms. This reason is apparent for those with some familiarity in Lie group theory: indeed, $A^{-1}\gamma A$ is the composition of the $\mathfrak{o}(m)$-valued 1-form γ with the *adjoint action*

$$\mathrm{Ad}(A^{-1}) \in GL(\mathfrak{o}(m)),$$

and $A^{-1}dA$ is the pullback of the *Maurer-Cartan form* of $O(n)$ via $A : U \to O(m)$. We refer the interested reader to the beautiful book [256].

Starting from the Levi-Civita connection forms, we can define a *covariant derivative* ∇ on every tensor bundle. Let $\{e_i\}, \{\theta^i\}$ be an orthonormal frame and its dual coframe on the open set U. The *connection* ∇ *induced by the Levi-Civita connection forms* is defined by

$$\nabla e_i = \theta^j_i \otimes e_j, \tag{1.18}$$

and, for every $X, Y \in \mathfrak{X}(U)$ (where $\mathfrak{X}(U)$ is the set of smooth vector fields on the open set U), $f \in C^\infty(U)$, by the rules

$$\nabla(X + Y) = \nabla X + \nabla Y, \qquad \nabla(fX) = df \otimes X + f\nabla X; \tag{1.19}$$

the *dual connection*, still denoted with ∇, is given by the formula

$$\nabla\theta^i = -\theta^i_j \otimes \theta^j$$

(which follows imposing the condition $\nabla\theta^i(e_j) + \theta^i(\nabla e_j) = \nabla(\theta^i(e_j)) = d(\theta^i(e_j)) = 0$; see below for the relation between the covariant derivative and the differential of a function). The connection ∇ is globally defined, and independent of the chosen frame $\{e_i\}$. Indeed, if $\widetilde{e_i} = A_i^j e_j$ on the intersection of two open sets U, $\widetilde{U}$, then, using (1.13),

$$\begin{aligned}\nabla\widetilde{e_i} &= \nabla(A_i^j e_j) = dA_i^k \otimes e_k + A_i^j\theta_j^k \otimes e_k = dA_i^k \otimes (A^{-1})_k^t\widetilde{e_t} + A_i^j\theta_j^k \otimes (A^{-1})_k^t\widetilde{e_t} \\ &= \left[(A^{-1})_k^t dA_i^k + (A^{-1})_k^t\theta_j^k A_i^j\right] \otimes \widetilde{e_t} = \widetilde{\theta}_i^t \otimes \widetilde{e_t},\end{aligned}$$

and the same for $\nabla\theta^i$.

For a *vector field* $X \in \mathfrak{X}(M)$, which can be locally written as $X = X^i e_i$, the covariant derivative ∇X is the tensor field of type $(1, 1)$

$$\nabla X = (dX^i) \otimes e_i + X^i\nabla e_i = (dX^i + X^j\theta_j^i) \otimes e_i.$$

Setting

$$X_k^i\theta^k = dX^i + X^j\theta_j^i,$$

∇X can be written as

$$\nabla X = X_k^i\theta^k \otimes e_i,$$

and X_k^i is said to be the covariant derivative of the coefficient X^i. If $Y \in \mathfrak{X}(M)$ we define the *covariant derivative of* X *in the direction of* Y as the vector field

$$\nabla_Y X = \nabla X(Y),$$

which in components reads as

$$\nabla_Y X = X_k^i\theta^k(Y)e_i = X_k^i Y^k e_i.$$

We also recall that the *divergence* of the vector field $X \in \mathfrak{X}(M)$ is the trace of ∇X, that is,

$$\operatorname{div} X = \operatorname{Tr}(\nabla X) = \langle\nabla_{e_i} X, e_i\rangle = X_i^i. \tag{1.20}$$

For a 1*-form* ω, which can be written locally as $\omega = \omega_i\theta^i$, the covariant derivative $\nabla\omega$ is the tensor field of type $(0, 2)$

$$\nabla\omega = (d\omega_i) \otimes \theta^i + \omega_i\nabla\theta^i = \left(d\omega_i - \omega_j\theta_i^j\right) \otimes \theta^i.$$

Setting

$$\omega_{ik}\theta^k = d\omega_i - \omega_j\theta^j_i,$$

it follows that $\nabla\omega$ can be written as

$$\nabla\omega = \omega_{ik}\theta^k \otimes \theta^i.$$

If $Y \in \mathfrak{X}(M)$ we define the *covariant derivative of ω in the direction of Y* as the 1-form

$$\nabla_Y\omega = \nabla\omega(Y),$$

which in components reads as

$$\nabla_Y\omega = \omega_{ik}\theta^k(Y)\theta^i = \omega_{ik}Y^k\theta^i.$$

The extension of ∇ to a generic tensor field T is done via the Leibniz rule. We recall that a *tensor field of* of type (r, s) is a law that assigns to each point $p \in M$ a multilinear map

$$T_p : \underbrace{T^*_pM \times \cdots \times T^*_pM}_{r \text{ times}} \times \overbrace{T_pM \times \cdots \times T_pM}^{s \text{ times}} \to \mathbb{R},$$

where T_pM and T^*_pM are, respectively, the tangent and the cotangent space of M at p with the usual differentiability requirement with respect to the variable p (see for instance [171]). Thus for a local orthonormal coframe $\{\theta^i\}$ with dual frame $\{e_i\}$ on the open set U we have

$$T = T^{i_1\ldots i_r}_{j_1\ldots j_s}\theta^{j_1} \otimes \ldots \otimes \theta^{j_s} \otimes e_{i_1} \otimes \ldots \otimes e_{i_r}.$$

The *covariant derivative* of T, ∇T, is then defined on U as the $(r, s+1)$ tensor field

$$\nabla T_U = T^{i_1\ldots i_r}_{j_1\ldots j_s,k}\theta^k \otimes \theta^{j_1} \otimes \cdots \otimes \theta^{j_s} \otimes e_{i_1} \otimes \cdots \otimes e_{i_r}$$

where the coefficients are

$$\begin{aligned} T^{i_1\ldots i_r}_{j_1\ldots j_s,k}\theta^k = {} & dT^{i_1\ldots i_r}_{j_1\ldots j_s} - T^{i_1\ldots i_r}_{hj_2\ldots j_s}\theta^h_{j_1} - \ldots - T^{i_1\ldots i_r}_{j_1\ldots j_{s-1}h}\theta^h_{j_s} \\ & + T^{hi_2\ldots i_r}_{j_1\ldots j_s}\theta^{i_1}_h + \cdots + T^{i_1\ldots i_{r-1}h}_{j_1\ldots j_s}\theta^{i_r}_h. \end{aligned}$$

We want to highlight the fact that, by the discussion above, the tensor field ∇T is *globally* defined. We remark that the operator ∇ so defined satisfies by definition the Leibniz rule and other nice properties like the commutativity with the trace of any pair of indices. Indeed, one can verify that the previous definition matches the "canonical" one usually given in terms of the Koszul formalism (see for example [170, 219] and Remark 1.4 below).

Note also that, for a *function* $u \in C^\infty(M)$, the covariant derivative coincides with the differential, i.e.

$$\nabla u = u_{,i}\theta^i = du.$$

Indeed, by definition, thinking of u as a $(0,0)$-tensor field,

$$u_{,i}\theta^i = du;$$

from now we will simply write

$$du = u_i\theta^i. \tag{1.21}$$

Remark 1.2 The notation for the covariant derivative of a function may give rise to some ambiguity; indeed, in the literature (and also in the rest of this book) ∇u often denotes the *gradient* of u, that is the vector field dual to the 1-form du: more explicitly, $\nabla u = (du)^\sharp$, where $\sharp$ is the *musical isomorphism* $^\sharp : T^*M \to TM$ (also called *sharp map*) defined by

$$\left\langle (du)^\sharp, Y\right\rangle = \langle \nabla u, Y\rangle = du(Y) = Y(u),$$

for all $Y \in \mathfrak{X}(M)$. Note also that, in components, we have $(\nabla u)^i = \delta^{ij}(du)_j = \delta^{ij}u_j = u_i$, that is, in an orthonormal frame, differential and gradient of a function have the same coefficients with respect to the (dual) bases $\{\theta^i\}$ and $\{e_i\}$. It is not difficult to see that this turns to be true also when we "raise an index" or "lower an index" for higher order tensors (see e.g. [170]): in a orthonormal frame, writing an index "up" or "down" doesn't change the numerical value of a component of a tensor (note that this is in contrast with the case of a nonorthonormal frame, see again [170]). In the rest of the book we choose to maintain the "correct" positions of the indexes only to keep in mind the type of the tensors involved in our computations.

Remark 1.3 Since it will be used in the sequel (e.g. in Chap. 8), we recall here the definition of *divergence of a symmetric* $(0,2)$-*tensor field*. To this purpose, let Z be a symmetric $(0,2)$-tensor field, which locally can be written as $Z = Z_{ij}\theta^j \otimes \theta^i = Z_{ji}\theta^j \otimes \theta^i$. The divergence of Z, $\operatorname{div} Z$, is the 1-form

$$\operatorname{div} Z = \operatorname{Tr}(\nabla Z^\sharp), \tag{1.22}$$

where $Z^\sharp$ is the $(1,1)$-tensor obtained from Z by raising an index (since Z is symmetric, the choice of the index is arbitrary) and the trace is with respect to the "new" index induced by the covariant derivative and one of the "old" ones.

In components we have $Z^\sharp = Z^i_j\theta^j \otimes e_i = Z_{ij}\theta^j \otimes e_i$ (by Remark 1.2), $\nabla Z^\sharp = Z^i_{j,k}\theta^k \otimes \theta^j \otimes e_i = Z_{ij,k}\theta^k \otimes \theta^j \otimes e_i$, thus

$$\operatorname{div} Z = Z_{ij,i}\theta^j = Z_{ji,i}\theta^j. \tag{1.23}$$

Consider now the metric tensor $\langle\,,\,\rangle$ (on the open set U)

$$\langle\,,\,\rangle = \delta_{ij}\theta^i \otimes \theta^j.$$

Then

$$\delta_{ij,k}\theta^k = d\delta_{ij} - \delta_{lj}\theta^l_i - \delta_{il}\theta^l_j = -(\theta^j_i + \theta^i_j).$$

Therefore $\nabla\langle\,,\,\rangle \equiv 0$ if and only if (1.5) holds. In other words, (1.5) expresses the *"compatibility" of the covariant derivative with the metric* (equivalently, the *parallelism* of the metric with respect to ∇).

We also observe that the validity of (1.4) is equivalent to the validity of

$$[X, Y] = \nabla_X Y - \nabla_Y X \qquad \forall\, X, Y \in \mathfrak{X}(M) \tag{1.24}$$

(where $[\,,\,]$ is the Lie bracket and $\mathfrak{X}(M)$ is the set of all smooth vector fields on M). One refers to (1.24) as to the fact that the Levi-Civita connection is *torsion-free*. Note that the left-hand side of (1.24) is independent of the choice of a metric on M. Since the *torsion* of a generic (linear) connection ∇ on M is the $(1, 2)$ tensor field

$$\operatorname{Tor}(X, Y) = \nabla_X Y - \nabla_Y X - [X, Y],$$

this justifies the expression "torsion-free" used above. To prove the equivalence, recall that the exterior differential of a 1-form θ is intrinsically defined by

$$d\theta(X, Y) = X(\theta(Y)) - Y(\theta(X)) - \theta([X, Y]) \qquad \forall\, X, Y \in \mathfrak{X}(M); \tag{1.25}$$

moreover, as a consequence of the definition of covariant derivative,

$$(\nabla_X\theta)(Y) = \nabla_X(\theta(Y)) - \theta(\nabla_X Y) = X(\theta(Y)) - \theta(\nabla_X Y), \tag{1.26}$$

so that

$$X\big(\theta^i(Y)\big) - \theta^i(\nabla_X Y) = (\nabla_X\theta^i)(Y) = -\theta^i_j(X)\theta^j(Y),$$

that is,

$$X\big(\theta^i(Y)\big) + \theta^i_j(X)\theta^j(Y) = \theta^i(\nabla_X Y).$$

Then we compute $\left(d\theta^i + \theta^i_j \wedge \theta^j\right)(X,Y)$, that is

$$\begin{aligned} d\theta^i(X,Y) + \theta^i_j \wedge \theta^j(X,Y) &= X(\theta^i(Y)) - Y(\theta^i(X)) - \theta^i([X,Y]) \\ &+ \theta^i_j(X)\theta^j(Y) - \theta^i_j(Y)\theta^j(X) \\ &= X(\theta^i(Y)) + \theta^i_j(X)\theta^j(Y)) - Y(\theta^i(X)) - \theta^i_j(Y)\theta^j(X)) - \theta^i([X,Y]) \\ &= \theta^i(\nabla_X Y - \nabla_Y X - [X,Y]) \\ &= \theta^i(\mathrm{Tor}(X,Y)), \end{aligned}$$

and the claim follows.

Remark 1.4 By the fundamental theorem of Riemannian geometry (see for instance [170] or [219]), we deduce that the connection ∇ coincides, as we said previously, with the *Levi-Civita connection* of the metric $\langle\ ,\ \rangle$.

1.3 Lie Derivatives, the Second Structure Equation and Curvature(s)

We now define the *Lie derivative* of Y in the direction of X to be $\mathscr{L}_X Y = [X,Y]$, so that condition (1.24) can be written in the form

$$\mathscr{L}_X Y = \nabla_X Y - \nabla_Y X. \tag{1.27}$$

Setting also

$$\mathscr{L}_X f = X(f) \tag{1.28}$$

for $f \in C^\infty(M)$, and

$$(\mathscr{L}_X \omega)(Y) = \mathscr{L}_X(\omega(Y)) - \omega(\mathscr{L}_X Y), \tag{1.29}$$

if ω is a 1-form, we can extend $\mathscr{L}_X$ to a generic tensor field requiring $\mathbb{R}$-linearity and the validity of the Leibniz rule (see also [171, 219]). Using (1.26), we compute the Lie derivative of the metric in the direction of X, $\mathscr{L}_X\langle\ ,\ \rangle$ (note that the latter has to be a covariant tensor of order 2, that is, a $(0,2)$-tensor):

$$\begin{aligned} (\mathscr{L}_X\langle\ ,\ \rangle)(Y,Z) &= ((\mathscr{L}_X\theta^i) \otimes \theta^i + \theta^i \otimes (\mathscr{L}_X\theta^i))(Y,Z) = \\ &= \theta^i(Z)(\mathscr{L}_X\theta^i)(Y) + \theta^i(Y)(\mathscr{L}_X\theta^i)(Z) = \\ &= \theta^i(Z)\big[\mathscr{L}_X(\theta^i(Y)) - \theta^i(\mathscr{L}_X Y)\big] + \end{aligned}$$

$$
\begin{aligned}
&+ \theta^i(Y)\big[\mathscr{L}_X(\theta^i(Z)) - \theta^i(\mathscr{L}_X Z)\big] = \\
&= \theta^i(Z)X(\theta^i(Y)) - \theta^i(Z)\theta^i(\nabla_X Y - \nabla_Y X) + \\
&+ \theta^i(Y)X(\theta^i(Z)) - \theta^i(Y)\theta^i(\nabla_X Z - \nabla_Z X) = \\
&= \theta^i(Z)(\nabla_X \theta^i)(Y) + \theta^i(Y)(\nabla_X \theta^i)(Z) + \\
&+ \theta^i(Z)\theta^i(\nabla_Y X) + \theta^i(Y)\theta^i(\nabla_Z X) = \\
&= (\nabla_X \theta^i \otimes \theta^i + \theta^i \otimes \nabla_X \theta^i)(Y,Z) + \langle \nabla_Y X, Z\rangle + \langle Y, \nabla_Z X\rangle = \\
&= (\nabla_X \langle\, ,\, \rangle)(Y,Z) + \langle \nabla_Y X, Z\rangle + \langle Y, \nabla_Z X\rangle = \\
&= \langle \nabla_Y X, Z\rangle + \langle Y, \nabla_Z X\rangle,
\end{aligned}
$$

where in the last equality we have used the fact that the metric is parallel with respect to the Levi-Civita connection. Thus, we have proved the useful identity

$$
(\mathscr{L}_X \langle\, ,\, \rangle)(Y,Z) = \langle \nabla_Y X, Z\rangle + \langle Y, \nabla_Z X\rangle \tag{1.30}
$$

for all $X, Y, Z \in \mathfrak{X}(M)$. Note that Eq. (1.30) in components reads as

$$
(\mathscr{L}_X \langle\, ,\, \rangle)_{ij} = \langle \nabla_{e_i} X, e_j \rangle + \langle e_i, \nabla_{e_j} X\rangle = X^j_i + X^i_j. \tag{1.31}
$$

We also recall that a vector field X is said to be a *Killing field* if $\mathscr{L}_X \langle\, ,\, \rangle = 0$.

It can be proved that the Lie derivative of Y in the direction of X has the following geometric meaning (see e.g. [171]):

$$
(\mathscr{L}_X Y)_p = \left.\frac{d}{dt}\right|_{t=0} (\varphi_{-t})_* Y_{\varphi_t(p)} = \lim_{t \to 0} \frac{(\varphi_{-t})_* Y_{\varphi_t(p)} - Y_p}{t},
$$

where φ_t is the local flow generated by X and $(\varphi_t)_*$ is the push-forward. The analogous applies to $\mathscr{L}_X h$, with h a generic tensor field (see also Chap. 2 for the special case of $\mathscr{L}_X \langle\, ,\, \rangle$).

We now consider the second structure equation. With the above notations we introduce a family of 2-forms, the *curvature forms* $\{\Theta^i_j\}$ associated to the orthonormal coframe $\{\theta^i\}$ via the *second structure equation*

$$
d\theta^i_j = -\theta^i_k \wedge \theta^k_j + \Theta^i_j, \tag{1.32}
$$

which in matrix notation becomes

$$
d\gamma = -\gamma \wedge \gamma + \Theta.
$$

Because of (1.5) it follows immediately that

$$
\Theta^i_j + \Theta^j_i = 0. \tag{1.33}
$$

Using the basis $\{\theta^i \wedge \theta^j\}_{1\leq i<j\leq m}$ of the space of skew-symmetric 2-forms $\Lambda^2(U)$ on the open set U, we may write

$$\Theta^i_j = \frac{1}{2} R^i_{jkt} \theta^k \wedge \theta^t \tag{1.34}$$

for some coefficients $R^i_{jkt} \in C^\infty(U)$ satisfying

$$R^i_{jkt} + R^i_{jtk} = 0. \tag{1.35}$$

Furthermore, note that (1.33) implies

$$R^i_{jkt} + R^j_{ikt} = 0. \tag{1.36}$$

We now show that the coefficients R^i_{jkt} are precisely the coefficients of the ((1, 3)-version of the) *Riemann curvature tensor* R, that in global notation is defined by

$$R(X, Y)Z = \nabla_X(\nabla_Y Z) - \nabla_Y(\nabla_X Z) - \nabla_{[X,Y]}Z \qquad \forall\, X, Y, Z \in \mathfrak{X}(M). \tag{1.37}$$

Remark 1.5 Some authors choose the opposite convention, defining $R(X, Y)Z = \nabla_Y(\nabla_X Z) - \nabla_X(\nabla_Y Z) + \nabla_{[X,Y]}Z$.

Remark 1.6 To be clear, please note that in this special case the position of the indexes in the coefficients R^i_{jkt} does not reflect the effective position of the entries: in other words, we have $R^i_{jkt} e_i = R(e_k, e_t)e_j$ instead of the expected formula $R^i_{jkt} e_i = R(e_j, e_k)e_t$ (see the discussion below). This is due to historical reasons.

We write $\nabla_i e_j$ to abbreviate $\nabla e_j(e_i)$. By definition, using properties (1.18), (1.19) and (1.25) we argue that

$$\begin{aligned}
R(e_k, e_t)e_j &= \nabla_k(\nabla_t e_j) - \nabla_t(\nabla_k e_j) - \nabla_{[k,t]} e_j \\
&= \nabla_k(\theta^r_j(e_t)e_r) - \nabla_t(\theta^r_j(e_k)e_r) - \theta^i_j([e_k, e_t])e_i \\
&= \theta^r_j(e_t)\nabla_k e_r + e_k(\theta^r_j(e_t))e_r - \theta^r_j(e_k)\nabla_t e_r \\
&\quad - e_t(\theta^r_j(e_k))e_r - \theta^i_j([e_k, e_t])e_i \\
&= \theta^r_j(e_t)\theta^i_r(e_k)e_i + e_k(\theta^i_j(e_t))e_i - \theta^r_j(e_k)\theta^i_r(e_t)e_i - e_t(\theta^i_j(e_k))e_i \\
&\quad - \theta^i_j([e_k, e_t])e_i \\
&= \left(\theta^i_r \wedge \theta^r_j\right)(e_k, e_t)e_i + d\theta^i_j(e_k, e_t)e_i.
\end{aligned}$$

Therefore we deduce

$$\langle R(e_k, e_t)e_j, e_i\rangle = \left(d\theta^i_j + \theta^i_r \wedge \theta^r_j\right)(e_k, e_t), \tag{1.38}$$

which proves our claim.

Using (1.38) we then have

$$R^i_{jkt} = \Theta^i_j(e_k, e_t) = \big(d\theta^i_j + \theta^i_k \wedge \theta^k_j\big)(e_k, e_t) = \langle R(e_k, e_t)e_j, e_i\rangle,$$

hence the $(1, 3)$-Riemann curvature tensor (1.37) can be written in components as

$$R = \langle R(e_k, e_t)e_j, e_i\rangle \theta^k \otimes \theta^t \otimes \theta^j \otimes e_i = R^i_{jkt}\theta^k \otimes \theta^t \otimes \theta^j \otimes e_i. \tag{1.39}$$

The $(0, 4)$-*version* of R is defined by $\operatorname{Riem}(X, Y, Z, W) = \langle R(Z, W)Y, X\rangle$, so that its coefficients R_{ijkt} satisfy

$$R_{ijkt} = \operatorname{Riem}(e_i, e_j, e_k, e_t) = \langle R(e_k, e_t)e_j, e_i\rangle = R^i_{jkt}$$

and

$$\operatorname{Riem} = R_{ijkt}\theta^i \otimes \theta^j \otimes \theta^k \otimes \theta^t. \tag{1.40}$$

This shows that R_{ijkt} is simply obtained performing the operation of *lowering the index* i in the first position using the metric tensor:

$$R_{ijkt} = \delta_{ir}R^r_{jkt} = R^i_{jkt},$$

Remark 1.7 We warn the reader that there is a number of different conventions for the $(0, 4)$-Riemann curvature tensor (see the discussion in [170]).

Observe that, although the curvature tensor is everywhere defined, this is not true for the curvature forms.

Proposition 1.3 *The matrix of curvature 2-forms* $\Theta = (\Theta^i_j)$ *takes values in* $\mathfrak{o}(n)$ *and, if* $\tilde{e} = eA$ *is a (local) change of orthonormal frame with* $A : U \to O(n)$*, then* Θ *varies according to*

$$\widetilde{\Theta} = A^{-1}\Theta A \tag{1.41}$$

Proof First of all, (1.33) shows that Θ is an $\mathfrak{o}(n)$-valued 2-form. Using the second structure equation (1.32), (1.13) and (1.16) we get

$$\begin{aligned}
\widetilde{\Theta} &= d\widetilde{\gamma} + \widetilde{\gamma} \wedge \widetilde{\gamma} = d(A^{-1}\gamma A + A^{-1}dA) + (A^{-1}\gamma A + A^{-1}dA) \wedge (A^{-1}\gamma A + A^{-1}dA) \\
&= (-A^{-1}dA\,A^{-1}) \wedge \gamma A + A^{-1}d\gamma A - A^{-1}\gamma \wedge dA + d(A^{-1}) \wedge dA + A^{-1}(\gamma \wedge \gamma)A \\
&\quad + A^{-1}\gamma \wedge dA + (A^{-1}dA\,A^{-1}) \wedge \gamma A + A^{-1}dA \wedge A^{-1}dA \\
&= A^{-1}(d\gamma + \gamma \wedge \gamma)A + d(A^{-1}) \wedge dA + A^{-1}dA \wedge A^{-1}dA \\
&= A^{-1}\Theta A - A^{-1}dA\,A^{-1} \wedge dA + A^{-1}dA \wedge A^{-1}dA = A^{-1}\Theta A,
\end{aligned}$$

and this proves the proposition. □

The *Bianchi identities* and the *symmetries of the curvature tensor* can be easily deduced from the structure equations: indeed, as we already observed before, (1.33) implies (1.36), that is

$$R^i_{jkt} + R^j_{ikt} = 0;$$

therefore we have the symmetries

$$R^i_{jkt} = -R^i_{jtk} = -R^j_{ikt} \tag{1.42}$$

(and the corresponding symmetries for the $(0, 4)$ version). Differentiating the first structure equations (1.4) and using (1.32) we deduce

$$\begin{aligned} 0 &= d(d\theta^i) = -d(\theta^i_j \wedge \theta^j) = -d\theta^i_j \wedge \theta^j + \theta^i_j \wedge d\theta^j \\ &= \theta^i_k \wedge \theta^k_j \wedge \theta^j - \Theta^i_j \wedge \theta^j - \theta^i_j \wedge \theta^j_k \wedge \theta^k \\ &= \theta^i \wedge \Theta^i_j, \end{aligned}$$

that is, renaming indices,

$$\theta^j \wedge \Theta^i_j = 0. \tag{1.43}$$

This identity goes under the name of *first Bianchi identity*. Using (1.34) and skew-symmetrizing we obtain

$$0 = R^i_{jkt}\theta^j \wedge \theta^k \wedge \theta^t = \frac{1}{6}(R^i_{jkt} - R^i_{jtk} + R^i_{tjk} - R^i_{tkj} + R^i_{ktj} - R^i_{kjt})\theta^j \wedge \theta^k \wedge \theta^t.$$

Thus using (1.35) we deduce the first Bianchi identity in the classical form

$$R^i_{jkt} + R^i_{ktj} + R^i_{tjk} = 0 \qquad \text{(equivalently: } R_{ijkt} + R_{iktj} + R_{itjk} = 0\text{)}. \tag{1.44}$$

Note that, more correctly, (1.44) should be called "Ricci identity".

Remark 1.8 In global notation, for the $(0, 4)$-Riemann curvature tensor equation (1.44) becomes

$$\begin{aligned} &\mathrm{Riem}(X, Y, Z, W) + \mathrm{Riem}(X, Z, W, Y) + \mathrm{Riem}(X, W, Y, Z) = 0 \\ &\quad \text{for each } X, Y, Z, W \in \mathfrak{X}(M). \end{aligned}$$

An important consequence of (1.42) and (1.44) (see also [102]) is the symmetry

$$R^i_{jkt} = R^k_{tij} \qquad \text{(equivalently: } R_{ijkt} = R_{ktij}\text{)}. \tag{1.45}$$

Indeed, this is a consequence of the chain of equalities

$$\begin{aligned} R^i_{jkt} &= -R^j_{ikt} = R^j_{kti} + R^j_{tik} = -R^k_{jti} + R^t_{jki} \\ &= (R^k_{tij} + R^k_{ijt}) + R^t_{jki} = R^k_{tij} + R^k_{ijt} + (-R^t_{kij} - R^t_{ijk}) \\ &= 2R^k_{tij} + R^k_{ijt} - R^t_{ijk} = 2R^k_{tij} + R^i_{ktj} + R^i_{tjk} = 2R^k_{tij} - R^i_{jkt}. \end{aligned}$$

Remark 1.9 The symmetries of the Riemann curvature tensor show that

$$\mathrm{Riem} = R_{ijkt}\theta^i \otimes \theta^j \otimes \theta^k \otimes \theta^t = R_{ijkt}\theta^t \otimes \theta^k \otimes \theta^j \otimes \theta^i = R_{ijkt}\theta^k \otimes \theta^t \otimes \theta^i \otimes \theta^j, \tag{1.46}$$

that is, in global (Koszul) notation,

$$\mathrm{Riem}\,(X, Y, Z, W) = \mathrm{Riem}\,(W, Z, Y, X) = \mathrm{Riem}\,(Z, W, X, Y) \quad \forall\, X, Y, Z, W \in \mathfrak{X}(M). \tag{1.47}$$

In order to obtain what is called the *second Bianchi identity*, which is deduced differentiating the second structure equation, we first observe that, according to the general rule for covariant differentiation of tensor fields, the coefficients $R^i_{jkt,s}$ of the covariant derivative of the $(1, 3)$ curvature tensor $R^i_{jkt}\theta^k \otimes \theta^t \otimes \theta^j \otimes e_i$ are given by

$$R^i_{jkt,s}\theta^s = dR^i_{jkt} + R^l_{jkt}\theta^i_l - R^i_{lkt}\theta^l_j - R^i_{jlt}\theta^l_k - R^i_{jkl}\theta^l_t. \tag{1.48}$$

Note that the symmetries (1.35), (1.36), (1.44) and (1.45) hold for $R^i_{jkt,s}$, for instance

$$R^i_{jkt,s} = -R^i_{jtk,s}. \tag{1.49}$$

Using (1.34) we now rewrite the second structure equations (1.32) in the form

$$d\theta^i_j = -\theta^i_l \wedge \theta^l_j + \frac{1}{2}R^i_{jkt}\theta^k \wedge \theta^t. \tag{1.50}$$

We differentiate this equation and use the structure equations and (1.34) again to obtain

$$\begin{aligned} 0 &= d\theta^i_l \wedge \theta^l_j - \theta^i_l \wedge d\theta^l_j - \frac{1}{2}dR^i_{jkt} \wedge \theta^k \wedge \theta^t - \frac{1}{2}R^i_{jkt}d\theta^k \wedge \theta^t + \frac{1}{2}R^i_{jkt}\theta^k \wedge d\theta^t \\ &= (-\theta^i_s \wedge \theta^s_l + \Theta^i_l) \wedge \theta^l_j - \theta^i_l \wedge (-\theta^l_s \wedge \theta^s_j + \Theta^l_j) - \frac{1}{2}dR^i_{jkt} \wedge \theta^k \wedge \theta^t \\ &+ \frac{1}{2}R^i_{jkt}\theta^k_l \wedge \theta^l \wedge \theta^t - \frac{1}{2}R^i_{jkt}\theta^k \wedge \theta^t_l \wedge \theta^l \\ &= -\frac{1}{2}(dR^l_{jkt} + R^i_{jkt}\theta^i_l - R^i_{lkt}\theta^l_j - R^i_{jlt}\theta^l_k - R^i_{jkl}\theta^l_t) \wedge \theta^k \wedge \theta^t, \end{aligned}$$

that is,

$$R^i_{jkt,s}\theta^s \wedge \theta^k \wedge \theta^t = R^i_{jkt,s}\theta^k \wedge \theta^t \wedge \theta^s = 0.$$

Therefore, skew-symmetrizing,

$$\frac{1}{6}(R^i_{jkt,s} - R^i_{jks,t} + R^i_{jsk,t} - R^i_{jst,k} + R^i_{jts,k} - R^i_{jtk,s})\theta^k \wedge \theta^t \wedge \theta^s = 0,$$

from which, using the symmetries (1.49), we deduce the second Bianchi identity in its classical form

$$R^i_{jkt,l} + R^i_{jtl,k} + R^i_{jlk,t} = 0 \qquad \text{(equivalently: } R_{ijkt,l} + R_{ijtl,k} + R_{ijlk,t} = 0). \tag{1.51}$$

Remark 1.10 In global notation, for the $(0, 4)$-Riemann curvature tensor equation (1.51) becomes

$$\nabla \operatorname{Riem}(X, Y, Z, W; T) + \nabla \operatorname{Riem}(X, Y, W, T; Z) + \nabla \operatorname{Riem}(X, Y, T, Z; W) = 0$$

for each $X, Y, Z, W, T \in \mathfrak{X}(M)$.

Remark 1.11 Using the matrix notation we have an equivalent form of the second Bianchi identity:

$$\begin{aligned} d\Theta &= d(d\gamma + \gamma \wedge \gamma) = d\gamma \wedge \gamma - \gamma \wedge d\gamma = \\ &= (-\gamma \wedge \gamma + \Theta) \wedge \gamma - \gamma \wedge (-\gamma \wedge \gamma + \Theta) = \Theta \wedge \gamma - \gamma \wedge \Theta, \end{aligned}$$

which in components is

$$d\Theta^i_j = \Theta^i_k \wedge \theta^k_j - \theta^i_k \wedge \Theta^k_j.$$

The *Ricci tensor* Ric is obtained from (1.40) by tracing either with respect to i and k or, equivalently, due to the symmetries of the curvature tensor, with respect to j and t. Thus

$$\operatorname{Ric} = R_{ij}\theta^i \otimes \theta^j \tag{1.52}$$

with

$$R_{ij} = R_{kikj} = R_{ikjk}. \tag{1.53}$$

Note that, again because of the symmetries of the curvature tensor, $R_{ij} = R_{ji}$. Indeed,

$$R_{ij} = R_{kikj} = R_{kjki} = R_{ji}.$$

Thus Ric is a symmetric $(0,2)$-tensor field.

The *scalar curvature* S is defined as the trace of Ric, that is,

$$S = R_{ii} = R_{kiki}. \tag{1.54}$$

For the sake of clarity, when needed, we shall also use the notation ${}^M\mathrm{Ric}$ and ${}^MS$ to identify the underlying manifold M.

We now come to the *sectional curvature* $K_p(\Pi)$ of the 2-plane $\Pi \subset T_pM$ spanned by the vectors u and v. It is defined by

$$K_p(\Pi) = \frac{\mathrm{Riem}(u,v,u,v)}{\langle u,u\rangle\langle v,v\rangle - \langle u,v\rangle^2} \in \mathbb{R}. \tag{1.55}$$

It is not difficult to verify that the right-hand side of the above formula is in fact independent of the chosen basis of Π. Clearly, if $\{u,v\}$ is an orthonormal basis of Π, then

$$K_p(\Pi) = \mathrm{Riem}(u,v,u,v).$$

We note that a common notation, also used in the sequel, for the sectional curvature of the plane Π spanned by u and v is

$$K_p(\Pi) = K(u \wedge v).$$

Again, when needed, we shall use also the notation ${}^MK_p(\Pi)$ and ${}^MK(u \wedge v)$ to identify the manifold. We shall now show that the sectional curvatures $K_p(\Pi)$ defined in (1.55) completely determine the curvature tensor Riem_p.

First of all we note that, by its very definition, Riem satisfy the symmetry relations (1.35), (1.36), (1.44), (1.45) and (1.49). Considering Riem_p as a quadrilinear map $R = \mathrm{Riem}_p : T_pM \times T_pM \times T_pM \times T_pM \to \mathbb{R}$, (1.35), (1.44) and (1.33) rewrites in the form: for each $u, v, z, w \in T_pM$

$$R(u,v,z,w) + R(u,v,w,z) = 0, \tag{1.56}$$

$$R(u,v,z,w) + R(u,z,w,v) + R(u,w,v,z) = 0, \tag{1.57}$$

and

$$R(u,v,z,w) + R(v,u,z,w) = 0. \tag{1.58}$$

Thus letting V be any real vector space, and considering two quadrilinear maps $R, T : V \times V \times V \times V \to \mathbb{R}$ which satisfy (1.56)–(1.58), we claim that if for each $u, v \in V$

$$R(u,v,u,v) = T(u,v,u,v) \tag{1.59}$$

then $R \equiv T$. Thus in particular the sectional curvatures $K_p(\Pi)$, $\Pi \subset T_pM$, determine the entire tensor Riem_p. The proof of this claim can be found, for instance, in Lemma 3.3 of do Carmo's book [102].

We set

$$|u \wedge v|^2 = \langle u, u\rangle\langle v, v\rangle - \langle u, v\rangle^2$$

and for u and v linearly dependent set $K(u \wedge v) = 0$. Then for any pair of vectors u, v using (1.55) we have

$$\mathrm{Riem}(u, v, u, v) = |u \wedge v|^2 K(u \wedge v). \tag{1.60}$$

Since the quadrilinear map Riem is determined by its values $\mathrm{Riem}(u, v, u, v)$ on pairs of vectors, we expect the validity of a "polarization" formula. Indeed, one can check the validity of the following:

$$\begin{aligned}
\mathrm{Riem}(w, z, u, v) = \frac{1}{6}\{&K((u + w) \wedge (v + z))|(u + w) \wedge (v + z)|^2 \\
&-K((v + w) \wedge (u + z))|(v + w) \wedge (u + z)|^2 \\
&-K(u \wedge (v + z))|u \wedge (v + z)|^2 \\
&-K(v \wedge (u + w))|v \wedge (u + w)|^2 \\
&-K(z \wedge (u + w))|z \wedge (u + w)|^2 \\
&-K(w \wedge (v + z))|w \wedge (v + z)|^2 \\
&+K(u \wedge (v + w))|u \wedge (v + w)|^2 \\
&+K(v \wedge (z + w))|v \wedge (z + w)|^2 \\
&+K(z \wedge (v + w))|z \wedge (v + w)|^2 \\
&+K(w \wedge (u + z))|w \wedge (u + z)|^2 \\
&+K(u \wedge z)|u \wedge z|^2 + K(v \wedge w)|v \wedge w|^2 \\
&-K(u \wedge v)|u \wedge v|^2 - K(v \wedge z)|v \wedge z|^2\}
\end{aligned} \tag{1.61}$$

In particular, if for each 2-plane Π of T_pM, $K_p(\Pi) = C$ for some constant C, from the above formula one deduces

$$\mathrm{Riem}(u, v, z, w) = C\{\langle u, z\rangle\langle v, w\rangle - \langle u, w\rangle\langle v, z\rangle\}. \tag{1.62}$$

An alternative way to prove (1.62) is to define $R_1(u, v, z, w)$ as in the right-hand side of (1.62); observing that R_1 satisfies (1.56)–(1.58), the validity of (1.62) follows by showing that for each $u, v \in T_pM$,

$$\mathrm{Riem}(u, v, u, v) = R_1(u, v, u, v) = C|u \wedge v|^2,$$

which is exactly the definition of $K_p(\Pi) = C$ for each 2-plane Π.

The manifold $(M, \langle\, ,\, \rangle)$ is said to have *constant sectional curvature* C if $K_p(\Pi) = C$ for each $p \in M$ and for each 2-plane $\Pi \subset T_pM$. This is equivalent, by (1.62), to say that in any orthonormal coframe

$$R_{ijkt} = C\{\delta_{ik}\delta_{jt} - \delta_{it}\delta_{jk}\}. \tag{1.63}$$

We observe that if $m = \dim M \geq 3$, then $(M, \langle\, ,\, \rangle)$ has constant sectional curvature under the milder requirement that $K_p(\Pi)$ depends possibly only on p. This can be easily seen. Indeed, for each $p \in M$ we have (1.63) for some function $C = C(p)$ of class C^∞. Taking covariant derivatives and using $\delta_{ij,s} = 0$ we obtain

$$R_{ijkt,s} = C_s\{\delta_{ik}\delta_{jt} - \delta_{it}\delta_{jk}\}.$$

Using the second Bianchi identity (1.51) in its equivalent form

$$R_{ijkt,s} + R_{ijts,k} + R_{ijsk,t} = 0, \tag{1.64}$$

we then have

$$0 = C_s\{\delta_{ik}\delta_{jt} - \delta_{it}\delta_{jk}\} + C_k\{\delta_{it}\delta_{js} - \delta_{is}\delta_{jt}\} + C_t\{\delta_{is}\delta_{jk} - \delta_{ik}\delta_{js}\}.$$

Hence, for $s = j$ and $k \neq t \neq j \neq k$ (and the latter is possible because $m \geq 3$), from the above we obtain

$$C_k\delta_{it} - C_t\delta_{ik} = 0.$$

But i is still arbitrary, thus choosing $i = t$ we deduce

$$C_k = 0.$$

Since this can be done for each k, we conclude that $C = C(p)$ is a constant function as desired.

In case $\dim M = 2$ the result is of course false. In this case the Riemann curvature always expresses in the form

$$R_{ijkt} = K(p)\{\delta_{ik}\delta_{jt} - \delta_{it}\delta_{jk}\},$$

where $K(p)$ is the Gaussian curvature of the surface, which in general is nonconstant. Note that, in this case,

$$K = R_{1212} = R_{11} = R_{22} = \frac{1}{2}S.$$

We observe that the previous result often goes under the name of *Schur's theorem*.

We are now going to show a similar fact that, in the recent literature, also goes under the same name. First we recall that the manifold $(M, \langle\, ,\, \rangle)$, $m = \dim M \geq 2$, is said to be *Einstein* if

$$\mathrm{Ric} = \lambda \langle\, ,\, \rangle \tag{1.65}$$

for some $\lambda \in \mathbb{R}$. We observe that if $m \geq 3$ and (1.65) holds for some function $\lambda = \lambda(p)$ of class C^{∞}, then λ is constant. Indeed, tracing equation (1.65) we obtain

$$\lambda = \frac{S}{m}. \tag{1.66}$$

Next we trace the second Bianchi identity (1.64) with respect to the indices i and s to get

$$R_{ijkt,i} + R_{ijti,k} + R_{ijik,t} = 0.$$

Since covariant derivatives commute with tracing

$$R_{ijkt,i} = R_{jt,k} - R_{jk,t}, \tag{1.67}$$

R_{jt} being the components of the Ricci tensor. Whence contracting again, this time with respect to j and k, we obtain

$$R_{ikkt,i} = R_{kt,k} - R_{kk,t},$$

that is

$$2R_{kt,k} = S_t \tag{1.68}$$

(this equation is sometimes called *Schur's identity*). Now because of (1.65) and (1.66) we have

$$R_{kt} = \frac{S}{m}\delta_{kt}$$

and using again the fact that the metric tensor is parallel, we deduce

$$R_{kt,l} = \frac{1}{m} S_l \delta_{kt}.$$

Now tracing with respect to k and l we get

$$R_{kt,k} = \frac{1}{m} S_{,t}. \tag{1.69}$$

Substituting into (1.68) yields

$$\left(\frac{2}{m} - 1\right) S_{,t} = 0,$$

and we conclude that, if $m \geq 3$, the scalar curvature, and therefore λ, is constant.

The above result in particular enables us to draw the following conclusion: if $m \geq 3$, then $(M, \langle\,,\,\rangle)$ is Einstein if and only if the symmetric, $(0, 2)$-tensor called the *traceless Ricci tensor*

$$T = \mathrm{Ric} - \frac{S}{m} \langle, \rangle, \tag{1.70}$$

with components

$$T_{ij} = R_{ij} - \frac{S}{m} \delta_{ij}, \tag{1.71}$$

is identically zero.

1.4 Decompositions of the Curvature Tensor

In this section we give three decomposition of the Riemann curvature tensor that shall be useful in the next chapters.

Let $(M, \langle\,,\,\rangle)$ be a Riemannian manifold and consider a *pointwise conformal deformation of the metric* $\langle\,,\,\rangle$, that is, a new metric on M of the form

$$\widetilde{\langle, \rangle} = \varphi^2 \langle\,,\,\rangle, \tag{1.72}$$

for some strictly positive smooth function φ on M. Denoting by $\widetilde{\mathrm{Riem}}$ the curvature tensor of the metric $\widetilde{\langle\,,\,\rangle}$ and with Riem that of the metric $\langle\,,\,\rangle$, we want to determine their relationship. Let $\{\theta^i\}$ be a local orthonormal coframe on $(M, \langle\,,\,\rangle)$

with corresponding Levi-Civita connection forms $\{\theta^i_j\}$. In the new metric $\widetilde{\langle\, ,\, \rangle}$

$$\widetilde{\theta}^i = \varphi\theta^i \tag{1.73}$$

is a local orthonormal coframe on $(M, \widetilde{\langle\, ,\, \rangle})$. To determine the associated connection forms one can use Proposition 1.1, but it is immediate to see directly that, if $d\varphi = \varphi_t\theta^t$, the 1-forms

$$\widetilde{\theta}^i_j = \theta^i_j + \frac{\varphi_j}{\varphi}\theta^i - \frac{\varphi_i}{\varphi}\theta^j \tag{1.74}$$

are skew-symmetric and satisfy the first structure equations, thus they are the desired connection forms relative to the coframe defined in (1.73). In order to determine the curvature forms, we use the structure equations and the expression for the components of the Hessian of φ (that is, the covariant derivative of the 1-form $d\varphi$); according to the general rule given in Sect. 1.5, if $\nabla d\varphi = \varphi_{ij}\theta^i \otimes \theta^j$ then the components φ_{ij} are given by

$$\varphi_{ij}\theta^j = d\varphi_i - \varphi_t\theta^t_i. \tag{1.75}$$

Observe that

$$\varphi_{ij} = \varphi_{ji}. \tag{1.76}$$

This can be easily seen as follows: we differentiate the equation $d\varphi = \varphi_i\theta^i$ and use the first structure equations to get

$$\begin{aligned} 0 = d\varphi_i \wedge \theta^i + \varphi_i d\theta^i &= (\varphi_{ij}\theta^j + \varphi_k\theta^k_i) \wedge \theta^i - \varphi_i\theta^i_k \wedge \theta^k \\ &= \varphi_{ij}\theta^j \wedge \theta^i \\ &= \frac{1}{2}(\varphi_{ij} - \varphi_{ji})\theta^j \wedge \theta^i, \end{aligned}$$

hence the validity of (1.76).

Going back to the curvature forms $\widetilde{\Theta}^i_j$ we have

$$\begin{aligned} \widetilde{\Theta}^i_j &= d\widetilde{\theta}^i_j + \widetilde{\theta}^i_k \wedge \widetilde{\theta}^k_j \\ &= d\theta^i_j + d\left(\frac{\varphi_j}{\varphi}\right) \wedge \theta^i + \left(\frac{\varphi_j}{\varphi}\right) d\theta^i - d\left(\frac{\varphi_i}{\varphi}\right) \wedge \theta^j - \left(\frac{\varphi_i}{\varphi}\right) d\theta^j + \widetilde{\theta}^i_k \wedge \widetilde{\theta}^k_j \\ &= -\theta^i_k \wedge \theta^k_j + \Theta^i_j + \left(\frac{1}{\varphi}d\varphi_j - \frac{1}{\varphi^2}\varphi_k\varphi_j\theta^k\right) \wedge \theta^i - \frac{1}{\varphi}\varphi_j\theta^i_k \wedge \theta^k \end{aligned}$$

$$
\begin{aligned}
&-\left(\frac{1}{\varphi}d\varphi_i-\frac{1}{\varphi^2}\varphi_k\varphi_i\theta^k\right)\wedge\theta^j+\frac{1}{\varphi}\varphi_i\theta^j_k\wedge\theta^k\\
&+\left(\theta^i_k+\frac{\varphi_k}{\varphi}\theta^i-\frac{\varphi_i}{\varphi}\theta^k\right)\wedge\left(\theta^k_j+\frac{\varphi_j}{\varphi}\theta^k-\frac{\varphi_k}{\varphi}\theta^j\right)\\
=\;&\Theta^i_j+\left(\frac{\varphi_{jk}}{\varphi}-2\frac{\varphi_j\varphi_k}{\varphi^2}\right)\theta^k\wedge\theta^i-\left(\frac{\varphi_{ik}}{\varphi}-2\frac{\varphi_i\varphi_k}{\varphi^2}\right)\theta^k\wedge\theta^j-\frac{\varphi_k\varphi_k}{\varphi^2}\theta^i\wedge\theta^j,
\end{aligned}
$$

that is,

$$
\widetilde{\Theta}^i_j=\Theta^i_j+\left(\frac{\varphi_{jk}}{\varphi}-2\frac{\varphi_j\varphi_k}{\varphi^2}\right)\delta^i_t\theta^k\wedge\theta^t-\left(\frac{\varphi_{ik}}{\varphi}-2\frac{\varphi_i\varphi_k}{\varphi^2}\right)\delta^j_t\theta^k\wedge\theta^t-\frac{\varphi_l\varphi_l}{\varphi^2}\delta^i_k\delta^j_t\theta^k\wedge\theta^t.
$$

Hence, skew-symmetrizing the coefficients and recalling the definition of the curvature tensor, we obtain

$$
\begin{aligned}
\varphi^2\widetilde{R}^i_{jkt}=R^i_{jkt}&+\left(\frac{\varphi_{jk}}{\varphi}-2\frac{\varphi_j\varphi_k}{\varphi^2}\right)\delta^i_t-\left(\frac{\varphi_{jt}}{\varphi}-2\frac{\varphi_j\varphi_t}{\varphi^2}\right)\delta^i_k\\
&-\left(\frac{\varphi_{ik}}{\varphi}-2\frac{\varphi_i\varphi_k}{\varphi^2}\right)\delta^j_t+\left(\frac{\varphi_{it}}{\varphi}-2\frac{\varphi_i\varphi_t}{\varphi^2}\right)\delta^j_k\\
&-\frac{\varphi_l\varphi_l}{\varphi^2}(\delta^i_k\delta^j_t-\delta^i_t\delta^j_k).
\end{aligned}
\tag{1.77}
$$

To get the relation between the two Ricci tensors, we trace the above with respect to i and k. We have

$$
\varphi^2\widetilde{R}_{jt}=R_{jt}-(m-2)\frac{\varphi_{jt}}{\varphi}+2(m-2)\frac{\varphi_j\varphi_t}{\varphi^2}-(m-3)\frac{\varphi_l\varphi_l}{\varphi^2}\delta^j_t-\frac{\varphi_{kk}}{\varphi^2}\delta^j_t. \tag{1.78}
$$

Finally, a further tracing of (1.78) with respect to j and t yields

$$
\varphi^2\widetilde{S}=S-2(m-1)\frac{\Delta\varphi}{\varphi}-(m-1)(m-4)\frac{|\nabla\varphi|^2}{\varphi^2}, \tag{1.79}
$$

where $\Delta\varphi=\varphi_{kk}$ is the Laplacian of the function φ (see Sect. 1.5). Note that, for $m=2$, we have

$$
\varphi^2\widetilde{S}=S-2\Delta\log\varphi,
$$

that is, the well known formula relating the Gaussian curvatures of the two metrics

$$
\widetilde{K}=\frac{1}{\varphi^2}K-\frac{1}{\varphi^2}\Delta\log\varphi \tag{1.80}
$$

(see [214]). In the general case, that is for $m \geq 3$, using (1.77) and (1.78) we are able to detect a part of the curvature tensor which is naturally invariant with respect to a pointwise conformal change of the metric. Indeed, from (1.78) we have

$$(m-2)\left(\frac{\varphi_{jt}}{\varphi} - 2\frac{\varphi_j\varphi_t}{\varphi^2}\right) = R_{jt} - \varphi^2\tilde{R}_{jt} - \left((m-3)\frac{\varphi_l\varphi_l}{\varphi^2} + \frac{\varphi_{kk}}{\varphi}\right)\delta^j_t,$$

and inserting into (1.77) gives

$$\begin{aligned}
&\varphi^2\left(\tilde{R}^i_{jkt} - \frac{1}{m-2}\left(\tilde{R}_{ik}\delta^j_t - \tilde{R}_{jk}\delta^i_t + \tilde{R}_{jt}\delta^i_k - \tilde{R}_{it}\delta^j_k\right)\right)\\
&= R^i_{jkt} - \frac{1}{m-2}\left(R_{ik}\delta^j_t - R_{jk}\delta^i_t + R_{jt}\delta^i_k - R_{it}\delta^j_k\right)\\
&+\frac{1}{m-2}\left(2\frac{\Delta\varphi}{\varphi} + (m-4)\frac{|\nabla\varphi|^2}{\varphi^2}\right)(\delta^i_k\delta^j_t - \delta^i_t\delta^j_k).
\end{aligned}$$

On the other hand, by (1.79)

$$2\frac{\Delta\varphi}{\varphi} + (m-4)\frac{|\nabla\varphi|^2}{\varphi^2} = -\frac{1}{m-1}(\varphi^2\tilde{S} - S)$$

and we obtain

$$\begin{aligned}
&\varphi^2\left(\tilde{R}^i_{jkt} - \frac{1}{m-2}\left(\tilde{R}_{ik}\delta^j_t - \tilde{R}_{jk}\delta^i_t + \tilde{R}_{jt}\delta^i_k - \tilde{R}_{it}\delta^j_k\right) + \frac{\tilde{S}}{(m-1)(m-2)}(\delta^i_k\delta^j_t - \delta^i_t\delta^j_k)\right)\\
&= R^i_{jkt} - \frac{1}{m-2}\left(R_{ik}\delta^j_t - R_{jk}\delta^i_t + R_{jt}\delta^i_k - R_{it}\delta^j_k\right) + \frac{S}{(m-1)(m-2)}(\delta^i_k\delta^j_t - \delta^i_t\delta^j_k).
\end{aligned}$$

It follows, since $\widetilde{e_i} = \frac{1}{\varphi}e_i$ is the dual of $\widetilde{\theta^i}$, that the $(1,3)$-tensor W called the *Weyl tensor* and defined by

$$W = W^i_{jkt}\theta^k \otimes \theta^t \otimes \theta^j \otimes e_i,$$

with components

$$W^i_{jkt} = R^i_{jkt} - \frac{1}{m-2}\left(R_{ik}\delta^j_t - R_{jk}\delta^i_t + R_{jt}\delta^i_k - R_{it}\delta^j_k\right) + \frac{S}{(m-1)(m-2)}(\delta^i_k\delta^j_t - \delta^i_t\delta^j_k),$$

is invariant under a conformal change of the metric. It is worth to note that the corresponding $(0,4)$-version of W, with (local) components $W_{ijkt} = W^i_{jkt}$, is not conformally invariant.

As it can be seen by direct inspection, W^i_{jkt} has the same symmetries as R^i_{jkt}; that is,

$$W^i_{jkt} = -W^j_{ikt} = -W^i_{jtk} \qquad \text{(equivalently: } W_{ijkt} = -W_{jikt} = -W_{ijtk}) \tag{1.81}$$

and

$$W^i_{jkt} = W^k_{tij} \qquad \text{(equivalently: } W_{ijkt} = W_{ktij}). \tag{1.82}$$

Furthermore, it satisfies the first Bianchi identity

$$W^i_{jkt} + W^i_{ktj} + W^i_{tjk} = 0 \qquad \text{(equivalently: } W_{ijkt} + W_{iktj} + W_{itjk} = 0) \tag{1.83}$$

and, by inspection, we deduce that any of its traces is identically zero.

We have thus obtained a first *decomposition of the curvature tensor*, the one using its totally trace-free part (i.e. the Weyl tensor), its "Ricci part" and its "scalar curvature part", that is (in $(0,4)$ form)

$$R_{ijkt} = W_{ijkt} + \frac{1}{m-2}\left(R_{ik}\delta_{jt} - R_{jk}\delta_{it} + R_{jt}\delta_{ik} - R_{it}\delta_{jk}\right) - \frac{S}{(m-1)(m-2)}(\delta_{ik}\delta_{jt} - \delta_{it}\delta_{jk}). \tag{1.84}$$

To write (1.84) in a global way we introduce the *Kulkarni-Nomizu product* between two symmetric $(0,2)$-tensors η and κ, that we shall denote by $\eta \owedge \kappa$. The latter is the covariant $(0,4)$-tensor of components

$$(\eta \owedge \kappa)_{ijkt} = \eta_{ik}\kappa_{jt} - \eta_{it}\kappa_{jk} + \eta_{jt}\kappa_{ik} - \eta_{jk}\kappa_{it}. \tag{1.85}$$

Using (1.85) it is easy to see that (1.84) is equivalent to

$$\mathrm{Riem} = W + \frac{1}{m-2}\,\mathrm{Ric} \owedge g - \frac{S}{2(m-1)(m-2)} g \owedge g, \tag{1.86}$$

where we have indicated the $(0,4)$-version of the Weyl tensor with the same letter W.

We observe that, for $m = 3$, $W \equiv 0$: in fact in this case, because of (1.81) and (1.82), the only possibly nonzero coefficients have to be of the type

$$W^i_{kkt} \text{ (no sum over } k)$$

for $i \neq k \neq t$. From (1.84) we have

$$W^i_{kkt} = R^i_{kkt} + R_{it} \text{ (no sum over } k).$$

However, since $m = 3$ and $i \neq t$, $R^i_{kkt} = -R_{it}$ (no sum over k). Thus $W^i_{kkt} = 0$ (no sum over k).

Taking covariant derivatives of (1.84) we obtain

$$W^i_{jks,t} = R^i_{jks,t} - \frac{1}{m-2}\left(R_{ik,t}\delta_{js} - R_{is,t}\delta_{jk} + R_{js,t}\delta_{ik} - R_{jk,t}\delta_{is}\right) + \frac{S_t}{(m-1)(m-2)}(\delta_{ik}\delta_{js} - \delta_{is}\delta_{jk}).$$

Thus taking the divergence with respect to the first index, that is, $W^t_{jks,t}$, using (1.67) and (1.68) we get

$$\begin{aligned} W^t_{jks,t} &= R^t_{jks,t} - \frac{1}{m-2}R_{tk,t}\delta_{js} + \frac{1}{m-2}R_{ts,t}\delta_{jk} - \frac{1}{m-2}R_{js,k} \\ &\quad + \frac{1}{m-2}R_{jk,s} + \frac{S_k}{(m-1)(m-2)}\delta_{sj} - \frac{S_s}{(m-1)(m-2)}\delta_{jk} \\ &= -R_{jk,s} + R_{js,k} - \frac{1}{m-2}R_{jk,s} - \frac{1}{m-2}R_{js,k} \\ &\quad + \frac{1}{m-2}(\frac{1}{m-1} - \frac{1}{2})S_k\delta_{js} - \frac{1}{m-2}(\frac{1}{m-1} - \frac{1}{2})S_s\delta_{jk} \\ &= \frac{3-m}{m-2}R_{jk,s} + \frac{m-3}{m-2}R_{js,k} + \frac{1}{2}\frac{3-m}{m-2}\frac{S_k}{m-1}\delta_{js} + \frac{1}{2}\frac{m-3}{m-2}\frac{S_s}{m-1}\delta_{jk}, \end{aligned}$$

and we can write

$$W^t_{jks,t} = \left(\frac{m-3}{m-2}\right)C_{jsk} \tag{1.87}$$

where C_{jsk} are the components of the *Cotton tensor* C, that is,

$$C_{jsk} = R_{js,k} - R_{jk,s} + \frac{1}{2(m-1)}(S_s\delta_{jk} - S_k\delta_{js}). \tag{1.88}$$

Note that from (1.87) and the symmetries of the Weyl tensor, we deduce that any of the traces of C is zero,

$$C_{jsk} = -C_{jks} \quad \text{and} \quad C_{jsk} + C_{skj} + C_{kjs} = 0.$$

As far as the analogue of the second Bianchi identity for W is concerned, we have the following

Lemma 1.2 (The Fake Second Bianchi Identity for W)

$$W_{ijkt,l} + W_{ijlk,t} + W_{ijtl,k} = \frac{1}{m-2}\left(C_{itl}\delta_{jk} + C_{ilk}\delta_{jt} + C_{ikt}\delta_{jl} - C_{jtl}\delta_{ik} - C_{jlk}\delta_{it} - C_{jkt}\delta_{il}\right). \tag{1.89}$$

Proof We start taking the covariant derivative of (1.84):

$$R_{ijkt,l} = W_{ijkt,l} + \frac{1}{m-2}\big(R_{ik,l}\delta_{jt} - R_{it,l}\delta_{jk} + R_{jt,l}\delta_{ik} - R_{jk,l}\delta_{it}\big)$$
$$- \frac{S_l}{(m-1)(m-2)}\big(\delta_{ik}\delta_{jt} - \delta_{it}\delta_{jk}\big). \tag{1.90}$$

Permuting cyclically the last three indices, summing up and using (1.44) we deduce

$$-\big(W_{ijkt,l} + W_{ijlk,t} + W_{ijtl,k}\big)$$
$$= \frac{1}{m-2}\big[(R_{ik,l} - R_{il,k})\delta_{jt} + (R_{il,t} - R_{it,l})\delta_{jk} + (R_{it,k} - R_{ik,t})\delta_{jl}\big]$$
$$- \frac{1}{m-2}\big[\big(R_{jk,l} - R_{jl,k}\big)\delta_{it} + \big(R_{jl,t} - R_{jt,l}\big)\delta_{ik} + \big(R_{jt,k} - R_{jk,t}\big)\delta_{il}\big]$$
$$- \frac{1}{(m-1)(m-2)}\big[S_l\big(\delta_{ik}\delta_{jt} - \delta_{it}\delta_{jk}\big) + S_t\big(\delta_{il}\delta_{jk} - \delta_{ik}\delta_{jl}\big) + S_k\big(\delta_{it}\delta_{jl} - \delta_{il}\delta_{jt}\big)\big].$$

Working with the identity $R_{ij,k} - R_{ik,j} = C_{ijk} + \frac{1}{2(m-1)}\big(S_k\delta_{ij} - S_j\delta_{ik}\big)$, after some manipulation we get (1.89). □

The importance of the Weyl and the Cotton tensors is pointed out by a classical result. First recall that a Riemannian manifold $(M, \langle\,,\,\rangle)$ of dimension $m \geq 2$ is said to be *locally conformally flat* if, for each $p \in M$ there exist an open set $U \ni p$ and a function $\varphi \in C^\infty(U)$, $\varphi > 0$ on U such that the manifold $(U, \varphi^2\langle\,,\,\rangle)$ is flat.

We note that by a result of Korn [164] and Lichenstein [175], every 2-dimensional Riemannian manifold is locally conformally flat. Therefore, the above definition has full meaning only for $m \geq 3$. We have

Theorem 1.1 *Let $(M, \langle\,,\,\rangle)$ be a Riemannian manifold,* $\dim M = m \geq 3$. *A necessary and sufficient condition for M to be locally conformally flat is that*

$$C \equiv 0 \ \textit{if}\ m = 3$$
$$W \equiv 0 \ \textit{if}\ m > 3$$

This result is originally due to Weyl and Schouten; for a proof see [109].

Another way to interpret the Cotton tensor is as follows. Let

$$A = \operatorname{Ric} - \frac{S}{2(m-1)}\langle\,,\,\rangle$$

be the *Schouten tensor* of components

$$A_{ij} = R_{ij} - \frac{S}{2(m-1)}\delta_{ij}.$$

Clearly A is symmetric; hence taking covariant derivatives

$$A_{ij,k} = A_{ji,k}$$

but for the last two indices one immediately verifies that

$$A_{ij,k} - A_{ik,j} = C_{ijk}.$$

Hence we can think of the Cotton tensor as the obstruction for the Schouten tensor to be a Codazzi tensor. Quite often the Schouten tensor is used to write the decomposition of the Riemann tensor in a nice way; indeed, using (1.84) and (1.85), one easily deduces a second decomposition of Riem, the one using its totally trace-free part (i.e. the Weyl tensor) and its "Schouten part", that is (in $(0,4)$ form)

$$\mathrm{Riem} = W + \frac{1}{m-2} A \owedge g, \tag{1.91}$$

and componentwise

$$R_{ijkt} = W_{ijkt} + \frac{1}{m-2}\left(A_{ik}\delta_{jt} - A_{jk}\delta_{it} + A_{jt}\delta_{ik} - A_{it}\delta_{jk}\right). \tag{1.92}$$

In what follows we shall not use the Schouten tensor; we thus refer the interested reader to the treatise [41] for further information and results.

The third and final decomposition that we want to describe exploit the traceless Ricci tensor T: using (1.71) in (1.84) we deduce

$$R_{ijkt} = W_{ijkt} + \frac{1}{m-2}\left(T_{ik}\delta_{jt} - T_{jk}\delta_{it} + T_{jt}\delta_{ik} - T_{it}\delta_{jk}\right) + \frac{S}{m(m-1)}(\delta_{ik}\delta_{jt} - \delta_{it}\delta_{jk}). \tag{1.93}$$

Using the notation for instance of Huisken [150], the previous equation can be written in global form as

$$\mathrm{Riem} = W + U + V, \tag{1.94}$$

where the $(0,4)$-tensors U and V have components, respectively,

$$U_{ijkt} = \frac{S}{m(m-1)}(\delta_{ik}\delta_{jt} - \delta_{it}\delta_{jk}). \tag{1.95}$$

and

$$V_{ijkt} = \frac{1}{m-2}\left(T_{ik}\delta_{jt} - T_{jk}\delta_{it} + T_{jt}\delta_{ik} - T_{it}\delta_{jk}\right). \tag{1.96}$$

A simple check shows that W, U and V are mutually orthogonal:

$$W \perp U \perp V, \tag{1.97}$$

that is, in components,

$$W_{ijkt}U_{ijkt} = W_{ijkt}V_{ijkt} = U_{ijkt}V_{ijkt} = 0. \tag{1.98}$$

An easy computation shows that for V we have

$$|V|^2 = \frac{4}{m-2}\left(|\mathrm{Ric}|^2 - \frac{S^2}{m}\right) = \frac{4}{m-2}|T|^2 \tag{1.99}$$

and

$$V_{itjt} = T_{ij}, \tag{1.100}$$

while for U we have

$$|U|^2 = \frac{2}{m(m-1)}S^2 \tag{1.101}$$

and

$$U_{itjt} = \frac{S}{m}\delta_{ij}; \tag{1.102}$$

the previous relations imply that

$$|\mathrm{Riem}|^2 = |W|^2 + |U|^2 + |V|^2 = |W|^2 + \frac{4}{m-2}|\mathrm{Ric}|^2 - \frac{2}{(m-1)(m-2)}S^2. \tag{1.103}$$

Remark 1.12 Every $(0,4)$-tensor having the same symmetries of the Riemann curvature tensor can be decomposed in three orthogonal parts as in (1.94): for instance, if B is a $(0,4)$-tensor such that its components B_{ijkt} satisfy

$$B_{ijkt} = -B_{jikt} = -B_{ijtk} = B_{ktij},$$

we can write

$$B = B_1 + B_2 + B_3,$$

where $B_1 \perp B_2 \perp B_3$ and B_1 is the "scalar" part, B_2 is the "traceless Ricci" part and B_3 is the "Weyl" part (that is, the totally trace-free part). The explicit expressions for B_1 and B_2 are respectively, in components,

$$(B_1)_{ijkt} = \frac{B_{lsls}}{m(m-1)}\left(\delta_{ik}\delta_{jt} - \delta_{it}\delta_{jk}\right) \tag{1.104}$$

and

$$(B_2)_{ijkt} = \frac{1}{m-2}\left(b_{ik}\delta_{jt} - b_{jk}\delta_{it} + b_{jt}\delta_{ik} - b_{it}\delta_{jk}\right), \quad b_{ik} = B_{isks} - \frac{B_{lsls}}{m}\delta_{ik}, \tag{1.105}$$

while for B_3 we have $B_3 = B - B_1 - B_2$.

We will use the third decomposition (and also Remark 1.12) in Chap. 8 to prove a useful inequality (see Proposition 8.8).

We conclude this section by introducing another curvature tensor, the *projective curvature tensor* P. In a local orthonormal coframe its components (in the $(1, 3)$ version) are given by

$$P^i_{jkt} = R^i_{jkt} - \frac{1}{m-1}(R^i_k\delta_{jt} - R_{jk}\delta^i_t). \tag{1.106}$$

This tensor is invariant under *projective transformations*, that is, diffeomorphisms of M onto M leaving geodesics invariant; with this we mean that if $\langle\, , \rangle$ and $\widetilde{\langle\, , \rangle}$ are metrics whose Levi-Civita connection are projectively related (see for instance [128, pp. 121–122]) then the two tensors P and $\tilde{P}$ coincide. A simple computation shows that if $m = \dim M \geq 3$ then $P \equiv 0$ if and only if $(M, \langle\, , \rangle)$ has constant sectional curvature.

1.5 Commutation Rules

The aim of this section is to provide a number of commutation rules, also generically called Ricci identities, for covariant derivatives. We will describe two cases: functions and the curvature tensor. In doing so we will also implicitly describe the general procedure to obtain them. We begin with the case that we have briefly described in Sect. 1.4 for the function φ, the stretching factor of two conformally related metrics, $\widetilde{\langle\, , \rangle} = \varphi^2\langle\, , \rangle$. Thus, let $u \in C^\infty(M)$; if

$$du = u_i\theta^i \tag{1.107}$$

for some smooth coefficients u_i, the *Hessian* of u is defined as the $(0, 2)$ tensor field $\mathrm{Hess}(u) = \nabla du$ of components u_{ij} given by

$$u_{ij}\theta^j = du_i - u_k\theta^k_i, \tag{1.108}$$

that is,

$$\mathrm{Hess}(u) = u_{ij}\theta^j \otimes \theta^i. \tag{1.109}$$

As we have already proved (for $u = \varphi$ in (1.76))

$$u_{ij} = u_{ji}, \tag{1.110}$$

so that $\mathrm{Hess}(u)$ is a symmetric tensor. In global notation we have, for all $X, Y \in \mathfrak{X}(M)$,

$$\mathrm{Hess}(u)(X,Y) = (\nabla du)(X,Y) = Y(X(u)) - (\nabla_Y X)(u) = X(Y(u)) - (\nabla_X Y)(u); \tag{1.111}$$

using (1.30) it is also possible to show that, equivalently,

$$\mathrm{Hess}(u)(X,Y) = \frac{1}{2}(\mathscr{L}_{\nabla u}\langle\, ,\, \rangle)(X,Y). \tag{1.112}$$

The *Laplacian* of u is, by definition, the trace of the Hessian, (more precisely, of the $(1,1)$ version of the Hessian, see Sect. 1.9.1), that is,

$$\Delta u = \mathrm{Tr}(\mathrm{Hess}(u)) = u_{ii}. \tag{1.113}$$

The Laplacian of the function u can be defined, equivalently, as the divergence of its gradient, that is

$$\Delta u = \mathrm{div}\,(\nabla u).$$

The third derivatives of u are defined, according to the general rule for the derivative of the tensor $\mathrm{Hess}(u)$, by

$$u_{ijk}\theta^k = du_{ij} - u_{kj}\theta^k_i - u_{ik}\theta^k_j. \tag{1.114}$$

Remark 1.13 Note that, in case of functions, we use the notation u_{ijk} instead of $u_{ij,k}$ (and analogously for higher order derivatives).

Note that taking covariant derivative of (1.110) we have

$$u_{ijk} = u_{jik}. \tag{1.115}$$

To obtain the commutation rule of the last two indices we proceed as follows. We differentiate (1.108) and we use the structure equations to get

$$\begin{aligned} du_{ik} \wedge \theta^k - u_{ij}\theta^j_k \wedge \theta^k &= -du_t \wedge \theta^t_i + u_k\theta^k_t \wedge \theta^t_i - u_k\Theta^k_i \\ &= -(u_{tk}\theta^k + u_k\theta^k_t) \wedge \theta^t_i + u_k\theta^k_t \wedge \theta^t_i - \frac{1}{2}u_k R^k_{ijt}\theta^j \wedge \theta^t. \end{aligned}$$

Thus,

$$(du_{ik} - u_{tk}\theta_i^t - u_{it}\theta_k^t) \wedge \theta^k = -\frac{1}{2}u_t R_{ijk}^t \theta^j \wedge \theta^k,$$

and, by (1.114),

$$u_{ikj}\theta^j \wedge \theta^k = -\frac{1}{2}u_t R_{ijk}^t \theta^j \wedge \theta^k.$$

Skew-symmetrizing we obtain

$$\frac{1}{2}(u_{ikj} - u_{ijk})\theta^j \wedge \theta^k = -\frac{1}{2}u_t R_{ijk}^t \theta^j \wedge \theta^k,$$

thus

$$u_{ijk} = u_{ikj} + u_t R_{ijk}^t = u_{ikj} + u_t R_{tijk}. \tag{1.116}$$

Let us now consider the fourth order derivative of u. It is defined by

$$u_{ijkt}\theta^t = du_{ijk} - u_{tjk}\theta_i^t - u_{itk}\theta_j^t - u_{ijt}\theta_k^t. \tag{1.117}$$

By (1.115), taking covariant derivative, we deduce

$$u_{ijkt} = u_{jikt}. \tag{1.118}$$

Similarly, taking covariant derivative of (1.116)

$$u_{ijkt} = u_{ikjt} + u_{st}R_{sijk} + u_s R_{sijk,t}. \tag{1.119}$$

To obtain the commutation rule of the last two indices we differentiate both sides of (1.114). We use the structure equations and (1.114) itself to arrive at

$$u_{ijkt}\theta^t \wedge \theta^k = -\frac{1}{2}(u_{lj}R_{litk} + u_{il}R_{ljtk})\theta^t \wedge \theta^k.$$

Skew-symmetrizing we then deduce

$$u_{ijkt} = u_{ijtk} + u_{lj}R_{likt} + u_{il}R_{ljkt}. \tag{1.120}$$

We now determine some commutation relations for the second covariant derivatives of the curvature tensor that we shall use later on. Recall that the coefficients of the second covariant derivative of the (0, 4) curvature tensor $R_{ijkt,ls}$ are given by

$$R_{ijkt,ls}\theta^s = dR_{ijkt,l} - R_{sjkt,l}\theta_i^s - R_{iskt,l}\theta_j^s - R_{ijst,l}\theta_k^s - R_{ijks,l}\theta_t^s - R_{ijkt,s}\theta_l^s. \tag{1.121}$$

Of course these coefficients satisfy the symmetry relations obtained by covariantly derive those satisfied by the $R_{ijkt,l}$'s; what we need to determine here is the relation between the $R_{ijkt,ls}$ and the $R_{ijkt,sl}$. Towards this aim we rewrite (1.48) in the $(0,4)$ form, that is,

$$0 = R_{ijkt,l}\theta^l - dR_{ijkt} + R_{ljkt}\theta^l_i + R_{ilkt}\theta^l_j + R_{ijlt}\theta^l_k + R_{ijkl}\theta^l_t,$$

and we differentiate it. Using the first and second structure equations together with (1.34), we obtain

$$\begin{aligned}
0 &= dR_{ijkt,l} \wedge \theta^l - R_{ijkt,l}\theta^l_s \wedge \theta^s \\
&\quad + dR_{ljkt} \wedge \theta^l_i - R_{ljkt}\theta^l_s \wedge \theta^s_i + R_{ljkt}\Theta^l_i \\
&\quad + dR_{ilkt} \wedge \theta^l_j - R_{ilkt}\theta^l_s \wedge \theta^s_j + R_{ilkt}\Theta^l_j \\
&\quad + dR_{ijlt} \wedge \theta^l_k - R_{ijlt}\theta^l_s \wedge \theta^s_k + R_{ijlt}\Theta^l_k \\
&\quad + dR_{ijkl} \wedge \theta^l_t - R_{ijkl}\theta^l_s \wedge \theta^s_t + R_{ijkl}\Theta^l_t \\
&= (dR_{ijkt,l} - R_{ijkt,s}\theta^s_l) \wedge \theta^l \\
&\quad + (R_{ljkt,s}\theta^s + R_{sjkt}\theta^s_l + R_{lskt}\theta^s_j + R_{ljst}\theta^s_k + R_{ljks}\theta^s_t) \wedge \theta^l_i \\
&\quad - R_{ljkt}\theta^l_s \wedge \theta^s_i + \frac{1}{2}R_{ljkt}R_{ligv}\theta^g \wedge \theta^v \\
&\quad + (R_{ilkt,s}\theta^s + R_{slkt}\theta^s_i + R_{iskt}\theta^s_l + R_{ilst}\theta^s_k + R_{ilks}\theta^s_t) \wedge \theta^l_j \\
&\quad - R_{ilkt}\theta^l_s \wedge \theta^s_j + \frac{1}{2}R_{ilkt}R_{ljgv}\theta^g \wedge \theta^v \\
&\quad + (R_{ijlt,s}\theta^s + R_{sjlt}\theta^s_i + R_{islt}\theta^s_j + R_{ijst}\theta^s_l + R_{ijls}\theta^s_t) \wedge \theta^l_k \\
&\quad - R_{ijlt}\theta^l_s \wedge \theta^s_k + \frac{1}{2}R_{ijlt}R_{lkgv}\theta^g \wedge \theta^v \\
&\quad + (R_{ijkl,s}\theta^s + R_{sjkl}\theta^s_i + R_{iskl}\theta^s_j + R_{ijsl}\theta^s_k + R_{ijks}\theta^s_l) \wedge \theta^l_t \\
&\quad - R_{ijkl}\theta^l_s \wedge \theta^s_t + \frac{1}{2}R_{ijkl}R_{ltgv}\theta^g \wedge \theta^v \\
&= (dR_{ijkt,l} - R_{ijkt,s}\theta^s_l - R_{sjkt,l}\theta^s_i - R_{iskt,l}\theta^s_j - R_{ijst,l}\theta^s_k - R_{ijks,l}\theta^s_t) \wedge \theta^l \\
&\quad + \frac{1}{2}(R_{ljkt}R_{ligv} + R_{ilkt}R_{ljgv} + R_{ijlt}R_{lkgv} + R_{ijkl}R_{ltgv})\theta^g \wedge \theta^v.
\end{aligned}$$

Hence, using (1.121), we have

$$R_{ijkt,vg}\theta^g \wedge \theta^v = -\frac{1}{2}(R_{ljkt}R_{ligv} + R_{ilkt}R_{ljgv} + R_{ijlt}R_{lkgv} + R_{ijkl}R_{ltgv})\theta^g \wedge \theta^v.$$

Skew-symmetrizing the left-hand side, we thus obtain

$$R_{ijkt,vg} - R_{ijkt,gv} = R_{ljkt}R_{livg} + R_{ilkt}R_{ljvg} + R_{ijlt}R_{lkvg} + R_{ijkl}R_{ltvg}. \tag{1.122}$$

Contracting with respect to i and k we obtain the corresponding commutation rules for the second covariant derivative of the Ricci tensor

$$R_{jt,vg} - R_{jt,gv} = R_{ltvg}R_{jl} + R_{ljvg}R_{lt}. \tag{1.123}$$

It should now be clear how to proceed in the general case to determine commutation relations when needed (note that some others for vector fields are given in Sect. 8.1 of Chap. 8). For other commutation rules we refer the interested reader to [70].

1.6 Some Formulas for Immersed Submanifolds

Let $(N, \langle\,,\,\rangle_N)$ and M be respectively a Riemannian manifold and a manifold of dimensions n and m, with $m \le n$. Let $f : M \to N$ be an *immersion* and let $\langle\,,\,\rangle = f^*\langle\,,\,\rangle_N$ be the metric induced on M by f, where f^* denotes the pullback. If $\langle\,,\,\rangle_M$ is a given Riemannian metric on M and $f : M \to N$ is an immersion we will say that f is an *isometric immersion* if $\langle\,,\,\rangle_M = \langle\,,\,\rangle = f^*\langle\,,\,\rangle_N$.

We fix the following indices convention:

$$1 \le i, j, k, \ldots \le m, \quad m+1 \le \alpha, \beta, \gamma, \ldots \le n, \quad 1 \le a, b, c, \ldots \le n.$$

Let $V \subset N$ be an open set, and let $p \in f^{-1}(V)$; up to reducing V, we can assume that the connected component U of $f^{-1}(V)$ containing p is an embedded submanifold in the domain of a local flat chart. Using the Gram-Schmidt procedure, we can construct an orthonormal frame $\{E_a\}$ in a neighbourhood of $f(U)$ such that $\{E_i\}$ is a basis for $f_*(TU)$ (here f_* denotes the pushforward by the map f). We call this frame a *Darboux frame along* f, and we write $\{e_i\}$ for the basis of the tangent space at U such that $f_*e_i = E_i$ (where f_*e_i is the pushforward of e_i by the map f). The dual $\{\theta^a\}$ of a Darboux coframe is called a *Darboux coframe along* f. Note that the definition of a Darboux (co)frame is equivalent to say that the vectors $\{E_i\}$ (locally) span f_*TM, the image of TM through f in TN, while the vectors $\{E_\alpha\}$ are orthogonal to f_*TM and span in fact the *normal bundle* $TM^\perp$ (sometimes denoted by NM), that is the set of (local) vector fields in N that are orthogonal to f_*TM. A simple but fundamental consequence of the choice of a Darboux frame is that

$$f^*\theta^\alpha = 0, \tag{1.124}$$

where $f^*\theta^\alpha$ is the pullback of θ^α by the map f. Indeed, for every i, $(f^*\theta^\alpha)(e_i) = \theta^\alpha(f_*e_i) = \theta^\alpha(E_i) = 0$.

Let now $\{\theta^a_b\}$ be the Levi-Civita connection forms of N relative to $\{\theta^a\}$. Pulling-back on M the first structure equation of N, and using the properties of the pullback, we have

$$f^*(d\theta^a) = d\big(f^*\theta^a\big) = -f^*(\theta^a_b \wedge \theta^b) = -(f^*\theta^a_b) \wedge (f^*\theta^b).$$

Using (1.124) we obtain in particular that

$$d\big(f^*\theta^i\big) = -\big(f^*\theta^i_j\big) \wedge \big(f^*\theta^j\big); \tag{1.125}$$

moreover, we obviously have

$$f^*\big(\theta^i_j\big) + f^*(\theta^j_i) = 0,$$

thus by the uniqueness part in Proposition 1.1, we deduce that the $f^*\theta^i_j$'s are the Levi-Civita connection forms of M.

To simplify the notation, from now on we shall omit the pullback, being clear from the context where forms or tensors are considered. With such a convention equation (1.124) becomes

$$\theta^\alpha = 0 \text{ on } M \tag{1.126}$$

and for a Darboux coframe along f we have

$$\langle \, , \, \rangle = \sum_i^m (\theta^i)^2. \tag{1.127}$$

Moreover,

$$\theta^i_j + \theta^j_i = 0 \text{ on } M \tag{1.128}$$

and

$$d\theta^i = -\theta^i_j \wedge \theta^j. \tag{1.129}$$

To obtain further information we differentiate (1.126), use (1.129) and (1.126) again to obtain

$$0 = d\theta^\alpha = -\theta^\alpha_i \wedge \theta^i - \theta^\alpha_\beta \wedge \theta^\beta = -\theta^\alpha_i \wedge \theta^i. \tag{1.130}$$

Hence, from (1.130) and Cartan's Lemma 1.1 there exist (locally defined) smooth functions h^α_{ij} such that

$$\theta^\alpha_i = h^\alpha_{ij}\theta^j \tag{1.131}$$

and

$$h^\alpha_{ij} = h^\alpha_{ji}. \tag{1.132}$$

We claim that the h^α_{ij}'s are the coefficients of the *second fundamental tensor* $\mathrm{II} : TM \times TM \to TM^\perp$ of the immersion. II is a $(1,2)$-tensor *along f* (equivalently, a section of $T^*M \otimes T^*M \otimes TM^\perp$, viewing $TM^\perp$ as a subset of the *pullback bundle* f^*TN; see e.g. [232]) which in the present setting is defined by

$$\mathrm{II} = h^\alpha_{ij}\theta^i \otimes \theta^j \otimes E_\alpha. \tag{1.133}$$

Indeed, recall that, if $\nabla, \overline{\nabla}$ are the Levi-Civita connection respectively on M and N, by definition

$$\mathrm{II}(e_i, e_j) = \overline{\nabla} E_j(E_i) - \nabla e_j(e_i), \tag{1.134}$$

therefore

$$\mathrm{II}(e_i, e_j) = \theta^a_j(E_i)E_a - \theta^k_j(e_i)E_k = \theta^k_j(E_i)E_k + \theta^\alpha_j(E_i)E_\alpha - \theta^k_j(e_i)E_k = \theta^\alpha_j(E_i)E_\alpha$$

(note that, following the convention introduced before, the pullback is omitted, and $f_*e_i = E_i$). From (1.134) we deduce

$$\mathrm{II}(e_i, e_j) = h^\alpha_{jk}\theta^k(e_i)E_\alpha = h^\alpha_{ji}E_\alpha = h^\alpha_{ij}E_\alpha,$$

and the claim is proved. One can also verify that II is globally defined, and symmetric by (1.132). The *mean curvature vector field* is given by its normalized trace, that is

$$\mathbf{H} = \frac{1}{m}\mathrm{Tr}(\mathrm{II}) = \frac{1}{m}h^\alpha_{ii}E_\alpha.$$

From now on, to simplify the writing, we shall use the notation $\mathbf{H} = \frac{1}{m}h^\alpha_{ii}e_\alpha$.

We have the following general definitions:

(1) if $\mathrm{II}_p \equiv 0$ for $p \in M$ then the immersion is said to be *geodesic at p*, and *totally geodesic* if $\mathrm{II} \equiv 0$ on M. We recall that the immersion is geodesic at p if and only if every geodesic γ of M starting at p is a geodesic of N at p, that is, $\frac{D}{dt}(\dot\gamma)(0) = 0$, where $p = \gamma(0)$ and $\frac{D}{dt}$ is the covariant differentiation along a curve (see for instance [102], Proposition 2.9 for details).
(2) an *umbilic point p* is a point of M where $\mathrm{II}_p - \langle\, , \rangle_p \otimes \mathbf{H}_p = 0$, and the immersion is said to be *totally umbilical* if $\mathrm{II} - \langle\, , \rangle \otimes \mathbf{H} \equiv 0$ on M. Thus, if N is a space of constant sectional curvature and the only eigenvalue is constant, M lies in some $(m+1)$-dimensional totally geodesic submanifold of N.

(3) if $\mathbf{H} = 0$ on M then the immersion is said to be *minimal*. This terminology comes from the fact that such an immersion minimizes the volume in the induced metric. More precisely, if $f : M \to N$ is minimal and Ω is a sufficiently small domain with smooth boundary $\partial\Omega$, then the volume of Ω in the induced metric is less than or equal to the volume of any other submanifolds of M with the same boundary.

If ν is a globally defined unit normal vector field, the *mean curvature in the direction of* ν is

$$h^\nu = \langle \mathbf{H}, \nu \rangle_N .$$

If $m + 1 = n$ and both the hypersurface M and N are orientable, we can choose Darboux frames along f preserving orientations, that is, such that $\theta^1 \wedge \cdots \wedge \theta^{m+1}$ and $\theta^1 \wedge \cdots \wedge \theta^m$ give the correct orientations, respectively, of N and M. In this case the vector field E_{m+1} dual to θ^{m+1} on N is, when restricted to M, a global normal vector field on M that we shall indicate with ν. The mean curvature in the direction of ν is called the *mean curvature of the immersed hypersurface* and denoted by H. In this latter case, with $A = A_\nu : TM \to TM$ we shall indicate the *Weingarten operator*, sometimes called *shape operator*, defined, for each $X, Y \in T_pM$, by

$$\langle AX, Y \rangle = \langle \mathrm{II}(X, Y), \nu \rangle_N ; \tag{1.135}$$

componentwise this means that

$$A = h^{m+1}_{ij} \theta^i \otimes e_j .$$

When there is no ambiguity, to simplify the notation we shall write h_{ij} instead of h^{m+1}_{ij}. In fact, often we shall not distinguish between A and the second fundamental tensor in the direction of ν, that is, the map $\langle \mathrm{II}(\ ,\ ), \nu \rangle_N : TM \times TM \to \mathbb{R}$.

With this notation, the *k-th mean curvatures* of the hypersurface (in the direction of ν) are given by

$$H_k = \binom{m}{k}^{-1} S_k , \tag{1.136}$$

where $S_0 = 1$ and, for $1 \leq k \leq m$, S_k is the *k-th elementary symmetric function* of the eigenvalues of A (called also the *principal curvatures* of the hypersurface).

Remark 1.14 H_k for k even is well defined also in case M is not orientable.

In particular $H_1 = H$ is the mean curvature, H_m is the *Gauss-Kronecker curvature* and H_2 is strictly related to the scalar curvature of M; indeed, this can be seen by tracing Gauss equations that we are now going to introduce for general isometrically immersed submanifolds $f : M \to N$ (see Eqs. (1.139) and (1.142)).

On M we consider the second structure equations

$$d\theta^i_j = -\theta^i_k \wedge^k_j + \Omega^i_j \tag{1.137}$$

with Ω^i_j the curvature forms of M

$$\Omega^i_j = \frac{1}{2} {}^M R^i_{jkl} \theta^k \wedge \theta^l. \tag{1.138}$$

We now relate the curvature of M with that of N; towards this aim let

$$\Theta^a_b = \frac{1}{2} {}^N R^a_{bcd} \theta^c \wedge \theta^d$$

be the curvature forms of N. Pulling back the second structure equations of N to M and using (1.126), (1.138) and (1.131) we obtain

$$\begin{aligned} d\theta^i_j &= -\theta^i_k \wedge \theta^k_j - \theta^i_\alpha \wedge \theta^\alpha_j + \Theta^i_j \\ &= -\theta^i_k \wedge \theta^k_j + h^\alpha_{ik} h^\alpha_{jl} \theta^k \wedge \theta^l + \frac{1}{2} {}^N R^i_{jkl} \theta^k \wedge \theta^l. \end{aligned}$$

Therefore, skew-symmetrizing in k and l

$$\Omega^i_j = \frac{1}{2}(h^\alpha_{ik} h^\alpha_{jl} - h^\alpha_{il} h^\alpha_{jk} + {}^N R^i_{jkl}) \theta^k \wedge \theta^l$$

and we deduce the *Gauss equations*

$${}^M R^i_{jkl} = {}^N R^i_{jkl} + h^\alpha_{ik} h^\alpha_{jl} - h^\alpha_{il} h^\alpha_{jk}; \tag{1.139}$$

in global notation we have, for each $X, Y, Z, W \in \mathfrak{X}(M)$,

$$\langle {}^M R(X,Y)Z, W \rangle = \langle {}^N R(X,Y)Z, W \rangle_N - \langle \mathrm{II}(X,Z), \mathrm{II}(Y,W) \rangle_N + \langle \mathrm{II}(X,W), \mathrm{II}(Y,Z) \rangle_N,$$

or, equivalently,

$$\begin{aligned} {}^M\mathrm{Riem}\,(X,Y,Z,W) = {}^N\mathrm{Riem}\,(X,Y,Z,W) &+ \langle \mathrm{II}(X,Z), \mathrm{II}(Y,W) \rangle_N \\ &- \langle \mathrm{II}(X,W), \mathrm{II}(Y,Z) \rangle_N. \end{aligned}$$

For a hypersurface, if ν is a local unit normal and A is the Weingarten operator in the direction of ν, the above rewrites as, for each $X, Y, Z, W \in \mathfrak{X}(M)$,

$$\langle {}^M R(X,Y)Z, W \rangle = \langle {}^N R(X,Y)Z, W \rangle_N - \langle AX, Z \rangle \langle AY, W \rangle + \langle AX, W \rangle \langle AY, Z \rangle. \tag{1.140}$$

Tracing we have

$$ {}^{M}S = {}^{N}S - 2\,{}^{N}\mathrm{Ric}\,(\nu,\nu) + m^2H^2 - |A|^2, \tag{1.141} $$

and this can be rewritten as

$$ {}^{M}S = {}^{N}S - 2\,{}^{N}\mathrm{Ric}\,(\nu,\nu) + m(m-1)H_2. \tag{1.142} $$

We now need to extend covariant differentiation to tensors along f. We do this for II, a section of $T^*M \otimes T^*M \otimes TM^{\perp}$, but analogous definitions can be promptly given (and will be when needed) in different cases. Setting ∇II for the covariant derivative of II, a section of $T^*M \otimes T^*M \otimes T^*M \otimes TM^{\perp}$, its coefficients, $h^{\alpha}_{ij,k}$, are given by

$$ h^{\alpha}_{ij,k}\theta^k = dh^{\alpha}_{ij} - h^{\alpha}_{tj}\theta^t_i - h^{\alpha}_{it}\theta^t_j + h^{\beta}_{ij}\theta^{\alpha}_{\beta} \tag{1.143} $$

(as we shall see below, the θ^{α}_{β}'s are the connection forms of the Van der Waerden-Bortolotti covariant derivative on $TM^{\perp}$). Thus we have, locally,

$$ \nabla \mathrm{II} = h^{\alpha}_{ij,k}\theta^k \otimes \theta^i \otimes \theta^j \otimes e_{\alpha}; $$

note in particular the position of the new index, which is in the *first* position. As before, from the symmetry relation (1.132) we deduce

$$ h^{\alpha}_{ij,k} = h^{\alpha}_{ji,k}. \tag{1.144} $$

To determine the commutation relations in the last two indices, we differentiate (1.131), use (1.129) and the structure equations to obtain

$$ \begin{aligned} 0 &= d\theta^{\alpha}_i - d(h^{\alpha}_{ij}\theta^j) \\ &= -\theta^{\alpha}_j \wedge \theta^j_i - \theta^{\alpha}_{\gamma} \wedge \theta^{\gamma}_i + \Theta^{\alpha}_i - dh^{\alpha}_{ij} \wedge \theta^j + h^{\alpha}_{ij}\theta^j_k \wedge \theta^k + h^{\alpha}_{ij}\theta^j_{\gamma} \wedge \theta^{\gamma} \\ &= h^{\alpha}_{jk}\theta^j_i \wedge \theta^k - h^{\gamma}_{ik}\theta^{\alpha}_{\gamma} \wedge \theta^k - dh^{\alpha}_{ij} \wedge \theta^j + h^{\alpha}_{ij}\theta^j_k \wedge \theta^k + \frac{1}{2}{}^{N}R^{\alpha}_{ijk}\theta^j \wedge \theta^k. \end{aligned} $$

Thus, using (1.143), the above rewrites us

$$ (h^{\alpha}_{ij,k} + \frac{1}{2}{}^{N}R^{\alpha}_{ijk})\theta^j \wedge \theta^k = 0, $$

and skew-symmetrizing we obtain

$$ h^{\alpha}_{ij,k} = h^{\alpha}_{ik,j} - {}^{N}R^{\alpha}_{ijk}. \tag{1.145} $$

These commutation rules are known as the *Codazzi equations*; in global notation they become, for each $X, Y, Z \in \mathfrak{X}(M)$ and for all section η of $TM^{\perp}$,

$$\langle \nabla \mathrm{II}(Y, X, Z), \eta \rangle_N = \langle \nabla \mathrm{II}(Z, X, Y), \eta \rangle_N - \langle {}^N R(Y, Z)X, \eta \rangle_N.$$

We now briefly describe the *Van der Waerden-Bortolotti covariant derivative* in the normal bundle $TM^{\perp}$ in the above formalism.

Given the immersion $f : M \to (N, \langle\, , \rangle_N)$ we have a well defined bundle on M, the normal bundle $TM^{\perp}$, that pointwise is the orthogonal complement of $f_* T_p M$ in $T_p N$. Given a Darboux coframe along f, we locally define a covariant derivative by setting

$$De_\alpha = \theta_\alpha^\beta \otimes e_\beta.$$

$\{\theta_\beta^\alpha\}$ are called the connection forms and one verifies that this definition is meaningful globally.

We let the curvature forms Φ_β^α be defined via the second structure equations as follows:

$$d\theta_\beta^\alpha = -\theta_\gamma^\alpha \wedge \theta_\beta^\gamma + \Phi_\beta^\alpha \tag{1.146}$$

and we set

$$\Phi_\beta^\alpha = \frac{1}{2} {}^{\perp}R_{\beta ij}^\alpha \theta^i \wedge \theta^j.$$

The ${}^{\perp}R_{\beta ij}^\alpha$'s are the components of the *normal curvature tensor*. Comparing (1.146) with the pull back of the second structure equations of N, that is,

$$d\theta_\beta^\alpha = -\theta_\gamma^\alpha \wedge \theta_\beta^\gamma - \theta_i^\alpha \wedge \theta_\beta^i + \Theta_\beta^\alpha,$$

we deduce

$$\Phi_\beta^\alpha = \theta_i^\alpha \wedge \theta_i^\beta + \Theta_\beta^\alpha.$$

A simple computation similar to those presented above gives

$${}^{\perp}R_{\beta ij}^\alpha = h_{ki}^\alpha h_{kj}^\beta - h_{kj}^\alpha h_{ki}^\beta + {}^N R_{\beta ij}^\alpha. \tag{1.147}$$

These equations are often called the *Ricci equations*.

Next formula (1.148), known as Simons' formula, will be used in Chap. 6 (in the special case of hypersurfaces, see Eq. (1.149)).

Proposition 1.4 *Let* $f : M \to (N, \langle\, , \rangle_N)$*, with* $\dim M = m$ *and* $\dim N = n$*,* $m \leq n$*, be an isometric immersion, with second fundamental tensor* II. *Then the following* *formula holds:*

$$\frac{1}{2}\Delta|\mathrm{II}|^2 = |\nabla \mathrm{II}|^2 + h^\alpha_{ij}h^\alpha_{kk,ij} - 2h^\alpha_{ij}{}^N\!R_{i\alpha,j} + h^\alpha_{ij}{}^N\!R_{ij,\alpha} - h^\alpha_{ij}{}^N\!R_{i\alpha j\beta,\beta} \tag{1.148}$$
$$- h^\alpha_{ij}h^\alpha_{tk}{}^N\!R_{itjk} - h^\alpha_{ij}h^\beta_{ij}h^\alpha_{tk}h^\beta_{tk} + h^\alpha_{it}h^\alpha_{ij}{}^N\!R_{tj} - h^\alpha_{ij}h^\alpha_{it}{}^N\!R_{\beta t\beta j} + h^\alpha_{ij}h^\alpha_{it}h^\beta_{tj}h^\beta_{kk}$$
$$+ 2h^\alpha_{ij}h^\alpha_{tk}h^\beta_{tj}h^\beta_{ik} - 2h^\alpha_{ij}h^\alpha_{tj}h^\beta_{ik}h^\beta_{tk} - h^\alpha_{ij}h^\beta_{ik}{}^N\!R_{\alpha\beta jk} + h^\alpha_{ij}{}^N\!R_{\alpha\beta i\beta,j}.$$

If M is a hypersurface, so that $n = m + 1$, the previous equation becomes

$$\frac{1}{2}\Delta|\mathrm{II}|^2 = |\nabla \mathrm{II}|^2 + mh_{ij}H_{ij} - 2h_{ij}{}^N\!R_{i(m+1),j} + h_{ij}{}^N\!R_{ij,m+1} \tag{1.149}$$
$$- h_{ij}{}^N\!R_{i(m+1)j(m+1),m+1} - h_{ij}h_{tk}{}^N\!R_{itjk} - |\mathrm{II}|^4 + h_{it}h_{ij}{}^N\!R_{tj}$$
$$- h_{ij}h_{it}{}^N\!R_{(m+1)t(m+1)j} + mHh_{ij}h_{jt}h_{ti}.$$

Proof Since

$$|\mathrm{II}|^2 = h^\alpha_{ij}h^\alpha_{ij}$$

and

$$\left(|\mathrm{II}|^2\right)_k = 2h^\alpha_{ij}h^\alpha_{ij,k}$$

we have that

$$\frac{1}{2}\Delta|\mathrm{II}|^2 = \left(|\mathrm{II}|^2\right)_{kk} = h^\alpha_{ij,k}h^\alpha_{ij,k} + h^\alpha_{ij}h^\alpha_{ij,kk} = |\nabla \mathrm{II}|^2 + h^\alpha_{ij}h^\alpha_{ij,kk} \tag{1.150}$$

(see also Lemma 8.4 in Chap. 8). We need to compute $h^\alpha_{ij,kk}$. To this purpose, we first observe that, by definition of covariant derivative,

$$h^\alpha_{ij,kl}\theta^l = dh^\alpha_{ij,k} - h^\alpha_{tj,k}\theta^t_i - h^\alpha_{it,k}\theta^t_j - h^\alpha_{ij,t}\theta^t_k + h^\beta_{ij,k}\theta^\alpha_\beta. \tag{1.151}$$

Differentiating Eq. (1.143) we deduce

$$dh^\alpha_{ij,k}\wedge\theta^k - h^\alpha_{ij,k}\theta^k_t\wedge\theta^t = -dh^\alpha_{kj}\wedge\theta^k_i - h^\alpha_{kj}d\theta^k_i - dh^\alpha_{ik}\wedge\theta^k_j - h^\alpha_{ik}d\theta^k_j + dh^\beta_{ij}\wedge\theta^\alpha_\beta + h^\beta_{ij}d\theta^\alpha_\beta.$$

Next we use the second structure equation, Eq. (1.151) and again (1.143) in the previous relation, obtaining, after some manipulations,

$$h^\alpha_{ij,kl}\theta^l\wedge\theta^k = \left[-\frac{1}{2}\left(h^\alpha_{tj}{}^M\!R^t_{ilk} + h^\alpha_{it}{}^M\!R^t_{jlk}\right) + h^\beta_{ij}h^\beta_{kt}h^\alpha_{tl} + \frac{1}{2}h^\beta_{ij}{}^N\!R^\alpha_{\beta lk}\right]\theta^l\wedge\theta^k. \tag{1.152}$$

We now skew-symmetrize equation (1.152) and use Gauss equations (1.139), deducing the commutation relation for the second covariant derivative of the second fundamental tensor:

$$\begin{aligned} h^{\alpha}_{ij,kl} &= h^{\alpha}_{ij,lk} + h^{\alpha}_{tj}{}^{N}R^{t}_{ikl} + h^{\alpha}_{it}{}^{N}R^{t}_{jkl} - h^{\beta}_{ij}{}^{N}R^{\alpha}_{\beta kl} \\ &\quad + h^{\alpha}_{tj}\left(h^{\beta}_{tk}h^{\beta}_{il} - h^{\beta}_{tl}h^{\beta}_{ik}\right) + h^{\alpha}_{it}\left(h^{\beta}_{tk}h^{\beta}_{jl} - h^{\beta}_{tl}h^{\beta}_{jk}\right) \\ &\quad + h^{\beta}_{ij}\left(h^{\beta}_{tk}h^{\alpha}_{tl} - h^{\beta}_{tl}h^{\alpha}_{tk}\right). \end{aligned} \tag{1.153}$$

Renaming indices we can rewrite Eq. (1.153) in the form

$$\begin{aligned} h^{\alpha}_{ik,jl} &= h^{\alpha}_{ik,lj} + h^{\alpha}_{tk}{}^{N}R^{t}_{ijl} + h^{\alpha}_{it}{}^{N}R^{t}_{kjl} - h^{\beta}_{ik}{}^{N}R^{\alpha}_{\beta jl} \\ &\quad + h^{\alpha}_{tk}\left(h^{\beta}_{tj}h^{\beta}_{il} - h^{\beta}_{tl}h^{\beta}_{ij}\right) + h^{\alpha}_{it}\left(h^{\beta}_{tj}h^{\beta}_{kl} - h^{\beta}_{tl}h^{\beta}_{kj}\right) \\ &\quad + h^{\beta}_{ik}\left(h^{\beta}_{tj}h^{\alpha}_{tl} - h^{\beta}_{tl}h^{\alpha}_{tj}\right). \end{aligned} \tag{1.154}$$

From Codazzi equations (1.145) we deduce that

$$h^{\alpha}_{ij,kl} = h^{\alpha}_{ik,jl} - {}^{N}R^{\alpha}_{ijk,l}. \tag{1.155}$$

Next we use (1.154), (1.155) and the symmetry $h^{\alpha}_{ik,lj} = h^{\alpha}_{ki,lj}$ in the previous relation, obtaining the further commutation relation

$$\begin{aligned} h^{\alpha}_{ij,kl} &= h^{\alpha}_{kl,ij} + + h^{\alpha}_{tk}{}^{N}R^{t}_{ijl} + h^{\alpha}_{it}{}^{N}R^{t}_{kjl} - h^{\beta}_{ik}{}^{N}R^{\alpha}_{\beta jl} \\ &\quad + h^{\alpha}_{tk}\left(h^{\beta}_{tj}h^{\beta}_{il} - h^{\beta}_{tl}h^{\beta}_{ij}\right) + h^{\alpha}_{it}\left(h^{\beta}_{tj}h^{\beta}_{kl} - h^{\beta}_{tl}h^{\beta}_{kj}\right) \\ &\quad + h^{\beta}_{ik}\left(h^{\beta}_{tj}h^{\alpha}_{tl} - h^{\beta}_{tl}h^{\alpha}_{tj}\right) - {}^{N}R^{\alpha}_{ijk,l} - {}^{N}R^{\alpha}_{kil,j}. \end{aligned} \tag{1.156}$$

Tracing equation (1.156) with respect to k and l we deduce

$$\begin{aligned} h^{\alpha}_{ij,kk} &= h^{\alpha}_{kk,ij} + + h^{\alpha}_{tk}{}^{N}R^{t}_{ijk} + h^{\alpha}_{it}{}^{N}R^{t}_{kjk} - h^{\beta}_{ik}{}^{N}R^{\alpha}_{\beta jk} \\ &\quad + h^{\alpha}_{tk}\left(h^{\beta}_{tj}h^{\beta}_{ik} - h^{\beta}_{tk}h^{\beta}_{ij}\right) + h^{\alpha}_{it}\left(h^{\beta}_{tj}h^{\beta}_{kk} - h^{\beta}_{tk}h^{\beta}_{kj}\right) \\ &\quad + h^{\beta}_{ik}\left(h^{\beta}_{tj}h^{\alpha}_{tk} - h^{\beta}_{tk}h^{\alpha}_{tj}\right) - {}^{N}R^{\alpha}_{ijk,k} - {}^{N}R^{\alpha}_{kik,j}. \end{aligned} \tag{1.157}$$

We need now to analyze the three terms ${}^{N}R^{t}_{kjk}$, ${}^{N}R^{\alpha}_{ijk,k}$ and ${}^{N}R^{\alpha}_{kik,j}$. For the first we have

$${}^{N}R^{t}_{kjk} = {}^{N}R_{tkjk} = {}^{N}R_{taja} - {}^{N}R_{t\beta j\beta} = {}^{N}R_{tj} - {}^{N}R_{t\beta j\beta};$$

with the same reasoning for the third we obtain

$$ {}^N R^{\alpha}_{kik,j} = {}^N R_{\alpha kik,j} = {}^N R_{\alpha i,j} - {}^N R_{\alpha\beta i\beta,j} $$

and for the second, using also the second Bianchi identity,

$$ {}^N R^{\alpha}_{ijk,k} = {}^N R_{\alpha ijk,k} = {}^N R_{i\alpha,j} - {}^N R_{ij,\alpha} - {}^N R_{k\alpha ji,k} - {}^N R_{\beta\alpha ji,\beta} + {}^N R_{\beta j\alpha i,\beta}. $$

Inserting the latter three relations in (1.157) and contracting with h^{α}_{ij} we finally deduce

$$ \begin{aligned} h^{\alpha}_{ij} h^{\alpha}_{ij,kk} = {} & h^{\alpha}_{ij} h^{\alpha}_{kk,ij} - 2h^{\alpha}_{ij} {}^N R_{i\alpha,j} + h^{\alpha}_{ij} {}^N R_{ij,\alpha} - h^{\alpha}_{ij} {}^N R_{i\alpha j\beta,\beta} \\ & - h^{\alpha}_{ij} h^{\alpha}_{tk} {}^N R_{itjk} - h^{\alpha}_{ij} h^{\beta}_{ij} h^{\alpha}_{tk} h^{\beta}_{tk} + h^{\alpha}_{it} h^{\alpha}_{ij} {}^N R_{tj} - h^{\alpha}_{ij} h^{\alpha}_{it} {}^N R_{\beta t\beta j} + h^{\alpha}_{ij} h^{\alpha}_{it} h^{\beta}_{tj} h^{\beta}_{kk} \\ & + 2h^{\alpha}_{ij} h^{\alpha}_{tk} h^{\beta}_{tj} h^{\beta}_{ik} - 2h^{\alpha}_{ij} h^{\alpha}_{tj} h^{\beta}_{ik} h^{\beta}_{tk} - h^{\alpha}_{ij} h^{\beta}_{ik} {}^N R_{\alpha\beta jk} + h^{\alpha}_{ij} {}^N R_{\alpha\beta i\beta,j}, \end{aligned} \tag{1.158} $$

which implies, together with (1.150), Eq. (1.148). To deduce Eq. (1.149) it is sufficient to use the definition of the mean curvature H and to note that, in the case of a hypersurface, the last four terms of Eq. (1.158) vanish. □

1.7 The Geometry of Smooth Maps

In this section we briefly describe the geometry of smooth maps between Riemannian manifolds. The results we present will be used in particular in Chap. 5.

We let $(M, \langle\, , \rangle_M)$ and $(N, \langle\, , \rangle_N)$ be Riemannian manifolds of dimensions, respectively, m and n. We fix the indices convention $1 \leq i, j, k, \ldots \leq m$ and $1 \leq a, b, c, \ldots \leq n$. With $\{\theta^i\}$, $\{\theta^i_j\}$ and $\{\omega^a\}$, $\{\omega^a_b\}$ we shall respectively denote local orthonormal coframes and corresponding Levi-Civita connection forms on the open sets $U \subset M$ and $V \subset N$. Let $\varphi : M \to N$ be a smooth map and suppose, from now on, to have chosen the local coframes (frames) so that $\varphi^{-1}(V) \subset U$. We set

$$ \varphi^* \omega^a = \varphi^a_i \theta^i \tag{1.159} $$

so that the *differential of* φ, $d\varphi$, a section of the bundle $T^*M \otimes \varphi^{-1} TN$, can be written as

$$ d\varphi = \varphi^a_i \theta^i \otimes E_a, \tag{1.160} $$

with $\{E_a\}$ the frame dual to the coframe ω^a.

The *energy density*, $e(\varphi) : M \to \mathbb{R}$, of the map φ is then defined as $\frac{1}{2}$ the square of the Hilbert-Schmidt norm of $d\varphi$, that is

$$e(\varphi) = \frac{1}{2}|d\varphi|^2 = \frac{1}{2}\varphi_i^a\varphi_i^a, \tag{1.161}$$

where the two sums run over the appropriate indices. Note that we also have

$$e(\varphi) = \frac{1}{2}\mathrm{Tr}_{\langle\,,\,\rangle_M}\varphi^*(\langle\,,\,\rangle_N), \tag{1.162}$$

as immediately verified. The covariant derivative $\nabla d\varphi$ of $d\varphi$ is called the *generalized second fundamental tensor*; locally we have

$$\nabla d\varphi = \varphi_{ij}^a\theta^j \otimes \theta^i \otimes E_a, \tag{1.163}$$

where the coefficients φ_{ij}^a are defined according to the rule

$$\varphi_{ij}^a\theta^j = d\varphi_i^a - \varphi_k^a\theta_i^k + \varphi_i^b\omega_b^a. \tag{1.164}$$

Here, and from now on, in the last term we have omitted the pullback notation. Note that

$$\varphi_{ij}^a = \varphi_{ji}^a \tag{1.165}$$

so that the tensor field (along φ) $\nabla d\varphi$ is symmetric. The validity of (1.165) can be easily seen as it was done for (1.76) in the case of $\varphi : M \to \mathbb{R}$. Nevertheless, for the sake of completeness, to obtain it simply take exterior differentiation of (1.159) and use the structure equations of M and N to arrive at

$$(d\varphi_i^a - \varphi_k^a\theta_i^k + \varphi_i^b\omega_b^a) \wedge \theta^i = 0.$$

Thus from (1.164)

$$\varphi_{ij}^a\theta^j \wedge \theta^i = 0;$$

Skew-symmetrizing (or using Cartan's lemma) we deduce the validity of (1.165).

The *tension field* $\tau(\varphi)$ of φ is defined by

$$\tau(\varphi) = \mathrm{Tr}_{\langle\,,\,\rangle_M}\nabla d\varphi = \varphi_{ii}^a E_a. \tag{1.166}$$

Let $\Omega \subset M$ be a relatively compact domain and let $E_\Omega(\varphi)$ be the *energy functional* on Ω, that is,

$$E_\Omega(\varphi) = \int_\Omega e(\varphi). \tag{1.167}$$

We recall that a smooth map $\varphi : (M, \langle\,,\,\rangle_M) \to (N, \langle\,,\,\rangle_N)$ is *harmonic* if for each relatively compact domain $\Omega \subset M$ it is a stationary point of the energy functional $E_\Omega : C^\infty(M,N) \to \mathbb{R}$ with respect to variations preserving φ on $\partial\Omega$. It can be verified that φ is harmonic if and only if $\tau(\varphi) = 0$; for details we refer to [107].

Observe that, in case $\varphi = f$ is an isometric immersion, it is immediate to see that

$$e(f) = \frac{m}{2}, \tag{1.168}$$

$$\nabla df = \mathrm{II}, \tag{1.169}$$

$$\tau(f) = m\mathbf{H}, \tag{1.170}$$

$$E_\Omega(f) = \frac{m}{2}\mathrm{vol}(\Omega). \tag{1.171}$$

In particular, in this case f is harmonic if and only if f is a minimal immersion. This little observation points out that the geometry of smooth maps generalizes that of isometric immersions.

We shall be interested in the Bochner-Weitzenböck formula for the Laplacian of $|d\varphi|^2$. To derive it we need the commutation relation for the covariant derivative of $\nabla d\varphi$ that generalizes (1.116). First of all let φ^a_{ijk} be the coefficients of $\nabla(\nabla d\varphi)$, defined according to the rule

$$\varphi^a_{ijk}\theta^k = d\varphi^a_{ij} - \varphi^a_{kj}\theta^k_i - \varphi^a_{ik}\theta^k_j + \varphi^b_{ij}\omega^a_b. \tag{1.172}$$

Because of (1.165) we have

$$\varphi^a_{ijk} = \varphi^a_{jik}. \tag{1.173}$$

We want now to relate φ^a_{ijk} with φ^a_{ikj}. Towards this aim we compute the exterior derivative of (1.164) and use the structure equations on M and N to arrive at

$$\varphi^a_{ijk}\theta^k \wedge \theta^j = \frac{1}{2}({}^N R^a_{bcd}\varphi^b_i\varphi^c_k\varphi^d_j - {}^M R^t_{ikj}\varphi^a_t)\theta^k \wedge \theta^j.$$

Hence, skew-symmetrizing we obtain the desired commutation rule, that is

$$\varphi^a_{ijk} = \varphi^a_{ikj} + {}^N R^a_{bcd}\varphi^b_i\varphi^c_k\varphi^d_j - {}^M R^t_{ikj}\varphi^a_t. \tag{1.174}$$

We can now prove the next

Proposition 1.5 (Bochner-Weitzenböck Formula) *In the above setting and with the above notations*

$$\frac{1}{2}\Delta|d\varphi|^2 = |\nabla d\varphi|^2 + \varphi^a_i\varphi^a_{kki} + \varphi^a_i{}^N R^a_{bcd}\varphi^b_k\varphi^c_k\varphi^d_i + \varphi^a_i{}^M R_{ti}\varphi^a_t. \tag{1.175}$$

Remark 1.15 For those that better like a more modern global notation

$$\frac{1}{2}\Delta|d\varphi|^2 = |\nabla d\varphi|^2 + \langle\nabla\tau(\varphi), d\varphi\rangle_N + \sum_{i=1}^{m}\left\langle d\varphi({}^M\mathrm{Ric}(e_i, \,)^\sharp), d\varphi(e_i)\right\rangle_N$$

$$+ \sum_{i,j=1}^{m}\langle{}^NR(d\varphi(e_i), d\varphi(e_j))d\varphi(e_i), d\varphi(e_j)\rangle_N,$$

where $\{e_i\}$ is a local orthonormal frame on M, $\sharp : T^*M \to TM$ is the musical isomorphism (see Remark 1.2) and ${}^NR$ is the curvature tensor of N of type $(1, 3)$ according to Koszul definition.

Proof By definition $|d\varphi|^2 = \varphi_i^a\varphi_i^a$. Hence taking covariant derivatives

$$(\varphi_i^a\varphi_i^a)_j = 2\varphi_i^a\varphi_{ij}^a$$

and

$$(\varphi_i^a\varphi_i^a)_{jk} = 2\varphi_{ik}^a\varphi_{ij}^a + 2\varphi_i^a\varphi_{ijk}^a.$$

Tracing with respect to j and k (that is, in the metric $\langle\, ,\, \rangle_M$) yields

$$\frac{1}{2}\Delta|d\varphi|^2 = |\nabla d\varphi|^2 + \varphi_i^a\varphi_{ikk}^a.$$

Using the commutation rules (1.173) and (1.174) in the above gives formula (1.175). □

Remark 1.16 Note that if $u \in C^3(M)$, (1.175) gives the usual *Bochner formula*

$$\frac{1}{2}\Delta|\nabla u|^2 = |\,\mathrm{Hess}(u)|^2 + \langle\nabla\Delta u, \nabla u\rangle + {}^M\mathrm{Ric}(\nabla u, \nabla u). \tag{1.176}$$

Suppose now that

$$(M, \langle\, ,\, \rangle_M) \xrightarrow{\varphi} (N, \langle\, ,\, \rangle_N) \xrightarrow{\psi} (P, \langle\, ,\, \rangle_P)$$

are smooth maps between Riemannian manifolds and set $\xi = \psi \circ \varphi$. Then, with the previous formalism, the following formula is easily verified

$$\nabla d\xi = \nabla d\psi(d\varphi, d\varphi) + d\psi(\nabla d\varphi). \tag{1.177}$$

Indeed, let $\{\theta^i\}, \{e_i\}, \{\omega^a\}, \{E_a\}, \{\eta^\alpha\}, \{\varepsilon_\alpha\}$ be local orthonormal coframes and frames on M, N and P respectively (with $i, j, \ldots = 1, \ldots, \dim M$, $a, b, \ldots =$

$1, \ldots \dim N$, $\alpha, \beta, \ldots = 1, \ldots \dim P$). Then we can write

$$d\varphi = \varphi_i^a \theta^i \otimes E_a, \quad d\psi = \psi_b^\alpha \omega^b \otimes \varepsilon_\alpha, \quad d\xi = \xi_j^\alpha \theta^j \otimes \varepsilon_\alpha$$

and

$$\nabla d\varphi = \varphi_{ij}^a \theta^j \otimes \theta^i \otimes E_a, \quad \nabla d\psi = \psi_{bc}^\alpha \omega^c \otimes \omega^b \otimes \varepsilon_\alpha, \quad \nabla d\xi = \xi_{ij}^\alpha \theta^j \otimes \theta^i \otimes \varepsilon_\alpha.$$

Now we compute

$$\begin{aligned}
\xi^* \eta^\alpha &= \xi_i^\alpha \theta^i \\
&= (\psi \circ \varphi)^* \eta^\alpha = \varphi^* (\psi^* \eta^\alpha) = \varphi^* (\psi_b^\alpha \omega^b) \\
&= (\varphi^* \psi_b^\alpha) \varphi^* \omega^b = (\varphi^* \psi_b^\alpha) \varphi_i^b \theta^i = (\psi_b^\alpha \circ \varphi) \varphi_i^b \theta^i,
\end{aligned}$$

so that

$$\xi_i^\alpha = (\varphi^* \psi_b^\alpha) \varphi_i^b = (\psi_b^\alpha \circ \varphi) \varphi_i^b. \tag{1.178}$$

By definition of covariant derivative and using (1.178) we have

$$\begin{aligned}
\xi_{ij}^\alpha \theta^j &= d\xi_i^\alpha - \xi_k^\alpha \theta_i^k + \xi_i^\beta \left(\xi^* \eta_\beta^\alpha \right) \\
&= d[(\varphi^* \psi_b^\alpha) \varphi_i^b] - (\varphi^* \psi_b^\alpha) \varphi_k^b \theta_i^k + \left(\varphi^* \psi_b^\beta \right) \varphi_i^b \varphi^* (\psi^* \eta_b^\alpha) \\
&= \varphi^* (d\psi_b^\alpha) \varphi_i^b + (\varphi^* \psi_b^\alpha) d\varphi_i^b - (\varphi^* \psi_b^\alpha) \varphi_k^b \theta_i^k + \left(\varphi^* \psi_b^\beta \right) \varphi_i^b \varphi^* (\psi^* \eta_b^\alpha) \\
&= \varphi^* \left(\psi_{bc}^\alpha \omega^c + \psi_c^\alpha \omega_b^c - \psi_b^\gamma \left(\psi^* \eta_\gamma^\alpha \right) \right) \varphi_i^b + (\varphi^* \psi_b^\alpha) (\varphi_{ij}^b \theta^j + \varphi_k^b \theta_i^k - \eta_i^c (\varphi^* \omega_c^b)) \\
&\quad - (\varphi^* \psi_b^\alpha) \varphi_k^b \theta_i^k + \left(\varphi^* \psi_b^\beta \right) \varphi_i^b \varphi^* (\psi^* \eta_b^\alpha) \\
&= \left[\varphi^* (\psi_{bc}^\alpha) \varphi_j^c \theta^j + (\varphi^* \psi_c^\alpha)(\varphi^* \omega_b^c) - (\varphi^* \psi_b^\gamma) \left(\varphi^* \psi^* \eta_\gamma^\alpha \right) \right] \varphi_i^b \\
&\quad + (\varphi^* \psi_b^\alpha) (\varphi_{ij}^b \theta^j + \varphi_k^b \theta_i^k - \eta_i^c (\varphi^* \omega_c^b)) - (\varphi^* \psi_b^\alpha) \varphi_k^b \theta_i^k + \left(\varphi^* \psi_b^\beta \right) \varphi_i^b \varphi^* (\psi^* \eta_b^\alpha) \\
&= \varphi^* (\psi_{bc}^\alpha) \varphi_j^c \varphi_i^b \theta^j + \varphi^* (\psi_b^\alpha) \varphi_{ij}^b \theta^j
\end{aligned}$$

from which we deduce

$$\xi_{ij}^\alpha = \varphi^* (\psi_{bc}^\alpha) \varphi_j^c \varphi_i^b + \varphi^* (\psi_b^\alpha) \varphi_{ij}^b = (\psi_{bc}^\alpha \circ \varphi) \varphi_j^c \varphi_i^b + (\psi_b^\alpha \circ \varphi) \varphi_{ij}^b \tag{1.179}$$

and (1.177) now follows immediately.

Equation (1.177) in particular shows that

$$\tau(\xi) = \nabla d\psi(d\varphi(e_i), d\varphi(e_i)) + d\psi(\tau(\varphi)). \tag{1.180}$$

In case $P = \mathbb{R}$, that is $\xi : M \xrightarrow{\varphi} N \xrightarrow{\psi} \mathbb{R}$, (1.180) yields

$$\Delta\xi = \mathrm{Hess}(\psi)(d\varphi(e_i), d\varphi(e_i)) + \langle \nabla\psi, \tau(\varphi)\rangle_N. \tag{1.181}$$

Also observe that in the special case $N = P = \mathbb{R}$, so that $\psi : \mathbb{R} \to \mathbb{R}$, (1.177) becomes the well-known formula

$$\mathrm{Hess}(\psi \circ \varphi) = \psi'(\varphi)\,\mathrm{Hess}(\varphi) + \psi''(\varphi)d\varphi \otimes d\varphi. \tag{1.182}$$

Similarly, from (1.180) (or tracing (1.182)) we have

$$\Delta(\psi \circ \varphi) = \psi'(\varphi)\Delta\varphi + \psi''(\varphi)|\nabla\varphi|^2. \tag{1.183}$$

We will use (1.177) in the next section.

1.8 Warped Products

We now describe some of the geometry of *warped product spaces* and of their immersed hypersurfaces. This material will be used mainly in Chap. 7 (and Chap. 6). Towards the end of the section we shall also give some examples and we shall introduce model manifolds in the sense of Greene and Wu [129]; as we will see, the latter are strict relatives of warped products and they will be repeatedly used all over the book.

Let $N = I \times_\rho \mathbb{P}$ denote the ρ-warped product of the real interval $I \subseteq \mathbb{R}$, with $0 \in I$, and the Riemannian manifold $(\mathbb{P}, \langle\,,\,\rangle_{\mathbb{P}})$ of dimension m. Thus N is the $(m+1)$-dimensional manifold $I \times \mathbb{P}$ endowed with the metric

$$\langle\,,\,\rangle = \langle\,,\,\rangle_N = \pi_I^*\left(dt^2\right) + ((\rho \circ \pi_I)(t))^2 \pi_{\mathbb{P}}^*(\langle\,,\,\rangle_{\mathbb{P}}), \tag{1.184}$$

where t is a global parameter on I, $\rho : I \to \mathbb{R}^+$ is a smooth function and π_I and $\pi_{\mathbb{P}}$ are the projections on the two factors of the product. Since there will not be any possibility of misunderstanding we will indicate the above metric with the customary notation

$$\langle\,,\,\rangle = dt^2 + \rho(t)^2\langle\,,\,\rangle_{\mathbb{P}}. \tag{1.185}$$

We fix the indices convention $1 \le i, j, \ldots \le m$, $1 \le a, b, \ldots \le m+1$. We let $\{\theta^i\}$ be a local orthonormal coframe on $\mathbb{P}$ with corresponding Levi-Civita connection forms $\{\theta^i_j\}$ and curvature forms $\{\Theta^i_j\}$, so that

$$\Theta^i_j = \frac{1}{2}\,{}^{\mathbb{P}}R^i_{jkt}\theta^k \wedge \theta^t \tag{1.186}$$

define the components of the curvature tensor of $(\mathbb{P}, \langle\, ,\, \rangle_{\mathbb{P}})$. We introduce a local orthonormal coframe $\{\varphi^a\}$ on N by setting

$$\varphi^i = \rho(t)\theta^i, \quad \varphi^{m+1} = dt. \tag{1.187}$$

The corresponding connection and curvature forms are denoted respectively with φ^a_b and Φ^a_b. Note that

$$\Phi^a_b = \frac{1}{2}\,{}^N R^a_{bcd}\varphi^c \wedge \varphi^d. \tag{1.188}$$

A repeated use of exterior differentiation and of the structure equations of $\mathbb{P}$, together with the characterization of the Levi-Civita connection forms, gives

$$\begin{cases} \varphi^k_j = \theta^k_j \\ \varphi^k_{m+1} = \mathscr{H}\varphi^k = -\varphi^{m+1}_k, \end{cases} \tag{1.189}$$

where $\mathscr{H} = \mathscr{H}(t) = \frac{\rho'(t)}{\rho(t)}$. Consequently,

$$\begin{cases} \Phi^k_j = -\mathscr{H}^2\varphi^k \wedge \varphi^j + \Theta^k_j \\ \Phi^{m+1}_k = \left(\mathscr{H}^2 + \mathscr{H}'\right)\varphi^k \wedge \varphi^{m+1} = \frac{\rho''}{\rho}\varphi^k \wedge \varphi^{m+1} = -\Phi^k_{m+1}. \end{cases} \tag{1.190}$$

From here and (1.187) we immediately deduce

$$\begin{cases} {}^N R_{kj(m+1)t} = 0 \\ {}^N R_{(m+1)j(m+1)t} = -\frac{\rho''}{\rho}\delta_{jt} \\ {}^N R_{kjst} = \frac{1}{\rho^2}\,{}^{\mathbb{P}}R_{kjst} - \mathscr{H}^2(\delta_{ks}\delta_{jt} - \delta_{kt}\delta_{js}), \end{cases} \tag{1.191}$$

the remaining components being determined by the symmetries of the curvature tensor. Thus, the components of the Ricci tensor are

$$\begin{cases} {}^N R_{jt} = \frac{1}{\rho^2}\,{}^{\mathbb{P}}R_{jt} - \left((m-1)\mathscr{H}^2 + \frac{\rho''}{\rho}\right)\delta_{jt} \\ {}^N R_{(m+1)t} = 0 \\ {}^N R_{(m+1)(m+1)} = -m\frac{\rho''}{\rho}. \end{cases} \tag{1.192}$$

Therefore, using (1.187),

$$\begin{aligned}{}^{N}\mathrm{Ric} &= \frac{1}{\rho^2}{}^{\mathbb{P}}R_{jt}\varphi^j \otimes \varphi^t - \left[(m-1)\mathcal{H}^2 + \frac{\rho''}{\rho}\right]\delta_{jt}\varphi^j \otimes \varphi^t - m\frac{\rho''}{\rho}\varphi^{m+1} \otimes \varphi^{m+1} \\ &= {}^{\mathbb{P}}\mathrm{Ric} - \left[(m-1)\mathcal{H}^2 + \frac{\rho''}{\rho}\right]\rho^2\langle\, ,\,\rangle_{\mathbb{P}} - m\frac{\rho''}{\rho}dt \otimes dt,\end{aligned}$$

that is,

$${}^{N}\mathrm{Ric} = {}^{\mathbb{P}}\mathrm{Ric} - \left[(m-1)\left(\rho'\right)^2 + \rho''\rho\right]\langle\, ,\,\rangle_{\mathbb{P}} - m\frac{\rho''}{\rho}dt \otimes dt. \tag{1.193}$$

In light of these relations we have that N is Einstein with ${}^{N}\mathrm{Ric} = -m\mathcal{E}\langle\, ,\,\rangle$ and $\mathcal{E} \in \mathbb{R}$, if and only if

$$\begin{cases} {}^{\mathbb{P}}R_{kt} = \left((m-1)\mathcal{H}^2 + \frac{\rho''}{\rho} - m\mathcal{E}\right)\rho^2\delta_{kt} \\ \rho'' = \mathcal{E}\rho. \end{cases}$$

Because of the second equation we can rewrite the first as

$${}^{\mathbb{P}}R_{kt} = (m-1)(\rho'^2 - \mathcal{E}\rho^2)\delta_{kt}. \tag{1.194}$$

We note that the general solution of $\rho'' = \mathcal{E}\rho$ is explicitly given by

$$\rho(t) = \rho'(0)\mathrm{sn}_{-\mathcal{E}}(t) + \rho(0)\mathrm{cn}_{-\mathcal{E}}(t), \tag{1.195}$$

where

$$\mathrm{sn}_\kappa(t) = \begin{cases} \frac{1}{\sqrt{-\kappa}}\sinh(\sqrt{-\kappa}t) & \text{if } \kappa < 0, \\ t & \text{if } \kappa = 0, \\ \frac{1}{\sqrt{\kappa}}\sin(\sqrt{\kappa}t) & \text{if } \kappa > 0, \end{cases}$$

and

$$\mathrm{cn}_\kappa(t) = \mathrm{sn}'_\kappa(t).$$

Inserting (1.195) into (1.194) we obtain the next

Lemma 1.3 *Let* $(\mathbb{P}, \langle\, ,\, \rangle_{\mathbb{P}})$ *be a Riemannian manifold of dimension* m. *Consider the warped product manifold* $N = I \times_\rho \mathbb{P}$ *where* $0 \in I \subseteq \mathbb{R}$ *and* $\rho : I \to \mathbb{R}^+$ *is a smooth function. Then,* N *is Einstein with*

$$^N\mathrm{Ric} = -m\mathscr{E}\langle\, ,\, \rangle, \quad \mathscr{E} \in \mathbb{R},$$

if and only if

$$\rho(t) = \rho'(0)\mathrm{sn}_{-\mathscr{E}}(t) + \rho(0)\mathrm{cn}_{-\mathscr{E}}(t), \tag{1.196}$$

and $(\mathbb{P}, \langle\, ,\, \rangle_{\mathbb{P}})$ *is Einstein with*

$$^{\mathbb{P}}\mathrm{Ric} = (m-1)(\rho'(0)^2 - \mathscr{E}\rho(0)^2)\langle\, ,\, \rangle_{\mathbb{P}}.$$

There is a natural *foliation* $t \in I \to \mathbb{P}_t = \{t\} \times \mathbb{P}$ of N; the leaf $\mathbb{P}_t$ here will be called a *slice*. As a unit normal to $\mathbb{P}_t$ we take $\mathscr{T} = \frac{\partial}{\partial t}$ (note that we are identifying $\frac{\partial}{\partial t}$ on I with its lift on $I \times \mathbb{P}$). Then the local orthonormal coframe $\{\varphi^a\}$ when restricted to $\mathbb{P}_t$ satisfies

$$\varphi^{m+1} = 0 \quad \text{on } \mathbb{P}_t$$

and it is therefore a Darboux coframe along the *inclusion map* $i : \mathbb{P}_t \hookrightarrow N$. We compute the second fundamental tensor of the isometric immersion, being $\sum_{i=1}^{m}(\varphi^i)^2$ the metric $\langle\, ,\, \rangle_{\mathbb{P}_t}$ on $\mathbb{P}_t$. Using (1.189) we have

$$0 = d\varphi^{m+1} = -\varphi_i^{m+1} \wedge \varphi^i = \mathscr{H}(t)\delta_{ik}\varphi^k \wedge \varphi^i.$$

Thus the second fundamental tensor in the direction of $-\mathscr{T}$ is given by

$$A_t = \mathscr{H}(t)\delta_{ik}\varphi^i \otimes \varphi^k. \tag{1.197}$$

Hence the inclusion $i : \mathbb{P}_t \hookrightarrow N$ is totally umbilical (and totally geodesic if $\mathscr{H}(t) = 0$) with constant mean curvature, in the direction of $-\mathscr{T}$, given by

$$\mathscr{H}(t) = H = \frac{\rho'(t)}{\rho(t)}.$$

Since the *k-th mean curvature* (in the direction of $-\mathscr{T}$) is defined as $\binom{m}{k}^{-1}$ times the *k-th elementary symmetric function* in the eigenvalues of A_t, we have

$$H_k = \mathscr{H}^k(t) = \left(\frac{\rho'(t)}{\rho(t)}\right)^k, \quad 0 \leq k \leq m. \tag{1.198}$$

Of course the curvature tensor of $\mathbb{P}_t$ is "the same" as that of $\left(\mathbb{P}, \rho(t)^2\langle\, ,\, \rangle_{\mathbb{P}}\right)$. This can also be checked via Gauss equation: we have

$$ {}^{\mathbb{P}_t}R_{ijks} = {}^{N}R_{ijks} + \mathscr{H}^2(t)(\delta_{ik}\delta_{js} - \delta_{is}\delta_{jk}) = \frac{1}{\rho(t)^2}\, {}^{\mathbb{P}}R_{ijks}, $$

that is

$$ \rho^2(t)\, {}^{\mathbb{P}_t}R_{ijks} = {}^{\mathbb{P}}R_{ijks}, \tag{1.199} $$

from which we immediately deduce

$$ \rho^2(t)\, {}^{\mathbb{P}_t}R_{jl} = {}^{\mathbb{P}}R_{jl} \tag{1.200} $$

and

$$ \rho^2(t)\, {}^{\mathbb{P}_t}S = {}^{\mathbb{P}}S, \tag{1.201} $$

for the Ricci tensors and the scalar curvatures, respectively.

Note that the vector field $\mathscr{T}$ satisfies

$$ \mathscr{T}^i_j = \frac{\rho'(t)}{\rho(t)}\delta^i_j, \quad \mathscr{T}^{m+1}_i = \mathscr{T}^i_{m+1} = \mathscr{T}^{m+1}_{m+1} = 0, $$

as one can immediately compute by using the orthonormal coframe $\{\varphi^a\}$ in (1.187) and the relative connection forms in (1.189); note that we can put the above in the compact form

$$ \mathscr{T}^a_b = \frac{\rho'(t)}{\rho(t)}\left(\delta^a_b - \delta^{m+1}_b\delta^a_{m+1}\right). $$

It follows immediately that

$$ (\rho\mathscr{T}^a)_b = \rho'\left(\delta^{m+1}_b\mathscr{T}^a + \delta^a_b - \delta^{m+1}_b\delta^a_{m+1}\right) = \rho'\delta^a_b, $$

so that, for each vector field X on $I \times_\rho \mathbb{P}$,

$$ \nabla_X(\rho\mathscr{T}) = \rho' X. $$

In other words $\xi = \rho\mathscr{T}$ satisfies

$$ \nabla_X\xi = \psi_\xi X, \tag{1.202} $$

for some smooth function ψ_ξ and for each vector field X on $I \times_\rho \mathbb{P}$.

We recall that a vector field ξ satisfying Eq. (1.202) on a Riemannian manifold $(M, \langle\, ,\, \rangle)$ is called a *closed conformal vector field*. This terminology is justified by the following two observations:

(1) ξ is trivially a *conformal vector field*, that is the local flow it generates preserves the metric up to a multiplicative conformal factor. This is equivalent to say that the Lie derivative of the metric in the direction of ξ is a multiple of the metric itself, as it can be immediately checked using definition (1.202) and formula (1.30). Note also that the conformal factor is given by $\frac{2}{m}\operatorname{div}\xi$.

(2) $\xi^\sharp$, the 1-form metrically dual to ξ, is closed, that is, $d\xi^\sharp = 0$. To see this fix a local orthonormal coframe $\{\theta^a\}$ on M with corresponding Levi-Civita connection forms $\{\theta^a_b\}$ and dual frame $\{e_a\}$. If $\xi = \xi^a e_a$, since we are working in a orthonormal coframe we have $\xi^\sharp = \xi^a\theta^a$, that is $\left(\xi^\sharp\right)_a = \xi^a = \xi_a$. By using the first structure equation we get

$$d\xi^\sharp = \xi_{ab}\theta^a \wedge \theta^b.$$

Now Eq. (1.202) reads

$$\xi^a_b = \psi_\xi \delta^a_b;$$

substitution into the above yields $d\xi^\sharp = 0$.

Closed conformal vector fields are the key to understand warped structures: indeed, as observed by Montiel, if M is a Riemannian manifold with a nontrivial closed conformal field, then it is locally isometric to a warped product with a 1-dimensional factor; furthermore, the isometry is global if M is complete and simply connected (see [194] for details).

Let now $f : \Sigma \to N$ be an isometrically immersed hypersurface. On $N = I \times_\rho \mathbb{P}$ we have the projection map $\pi_I : N \to I$ and we can consider the composition $h = \pi_I \circ f$, often called the *height function* of the immersion. Later on we will be interested in $\operatorname{Hess}(h)$, that we are now going to compute using the present formalism and formula (1.177) that now reads in the form

$$\operatorname{Hess}(h) = \operatorname{Hess}(\pi_I)(df, df) + \langle \mathcal{T}, \nu \rangle A, \tag{1.203}$$

where $\mathcal{T} = \frac{\partial}{\partial t}$ as before, ν is a local unit normal vector to Σ and A is the second fundamental tensor in the direction of ν. Towards this aim let $\{\varphi^a\}$ be a local orthonormal coframe on N as above and fix a local orthonormal coframe $\{\omega^s\}$ on Σ. Then

$$\pi_I^* dt = \delta_a^{m+1}\varphi^a,$$

hence

$$(\pi_I)_a = \delta_a^{m+1}.$$

Thus

$$(\pi_I)_{ab}\varphi^b = d\delta_a^{m+1} - \delta_b^{m+1}\varphi_a^b = -\varphi_a^{m+1},$$

so that, using (1.189),

$$(\pi_I)_{jk} = \frac{\rho'}{\rho}\delta_{jk}, \tag{1.204}$$

the remaining coefficients being zero. Now

$$f^*\varphi^a = f_s^a\omega^s$$

and since the metric on Σ expresses as $\omega^t \otimes \omega^t$ we have

$$\delta_{st} = f_s^k f_t^k + f_s^{m+1} f_t^{m+1}; \tag{1.205}$$

furthermore,

$$A = \langle \nabla df, \nu \rangle = f_{sk}^{m+1}\omega^s \otimes \omega^k.$$

Using (1.203)–(1.205) we have

$$\begin{aligned}\mathrm{Hess}(h) &= \frac{\rho'}{\rho}(h)\big(\delta_{jk}f_s^j f_t^k\big)\omega^s \otimes \omega^t + \langle \mathscr{T}, \nu \rangle A \\ &= \frac{\rho'}{\rho}(h)\big(\delta_{st} - f_s^{m+1}f_t^{m+1}\big)\omega^s \otimes \omega^t + \langle \mathscr{T}, \nu \rangle A.\end{aligned}$$

Observing that $dh = h^*(dt) = f^*(\pi^* dt) = f_s^{m+1}\omega^s$, this can be written as

$$\mathrm{Hess}(h) = \frac{\rho'}{\rho}(h)(\langle\, ,\, \rangle_\Sigma - dh \otimes dh) + \langle \mathscr{T}, \nu \rangle A. \tag{1.206}$$

Introducing $\Theta = \langle \mathscr{T}, \nu \rangle$, a similar reasoning shows that Θ and h are related by the formula

$$\nabla h = \mathscr{T} - \Theta \nu, \tag{1.207}$$

which implies

$$|\nabla h|^2 = 1 - \Theta^2; \tag{1.208}$$

these formulas will be repeatedly used, e.g., in the study of the geometry of hypersurfaces in warped products (see Chap. 7).

Many classical spaces can be described as warped products. Let us consider for instance *pseudohyperbolic manifolds* (see Tashiro [263]): they are obtained as warped product spaces of the form $\mathbb{R} \times_\rho \mathbb{P}$, where the warping function ρ is a positive solution, for some $c < 0$, of the ordinary differential equation $\rho'' + c\rho = 0$ on $\mathbb{R}$. Thus, either $\rho(t) = \cosh\left(\sqrt{-c}t\right)$ or $\rho(t) = e^{\sqrt{-c}t}$ (note that if $\mathbb{P}$ is Ricci flat then $\mathbb{R} \times_\rho \mathbb{P}$ is Einstein with negative Ricci curvature, and if $\mathbb{P}$ is flat then $\mathbb{R} \times_\rho \mathbb{P}$ is a negatively curved space form). Tashiro terminology is due to the fact that with suitable choices of the fiber we obtain representatives of the hyperbolic space; to realize this (and for more details we refer to Montiel [194]), we look at the hyperbolic space $\mathbb{H}^{m+1}$ of constant sectional curvature -1 as a hypersphere in the Lorentz-Minkowski space (see Sect. 9.2 in Chap. 9), precisely as a connected component of the hyperquadric

$$\left\{x \in \mathbb{R}_1^{m+2},\ \langle x, x\rangle_L = -1\right\},$$

where $\langle\,,\,\rangle_L$ is the standard Lorentzian product in $\mathbb{R}^{m+2}$. If we fix $a \in \mathbb{R}^{m+2}$ and consider the closed conformal vector field on $\mathbb{H}^{m+1}$

$$T_x = a + \langle a, x\rangle_L x.$$

Depending on the causal character of a we have different foliations of $\mathbb{H}^{m+1}$, and hence different descriptions of it (or part of it) as a warped product: namely, if a is lightlike the hyperbolic space is foliated by horospheres and it can be viewed as $\mathbb{R} \times_{e^t} \mathbb{R}^m$; if a is spacelike the vector field T generates a foliation of $\mathbb{H}^{m+1}$ by means of totally geodesic hyperplanes and it can be represented as the warped product $\mathbb{R} \times_{\cosh t} \mathbb{H}^m$. In the last case, that is, when a is timelike, the hyperbolic space minus a point (say o) is foliated by spheres and $\mathbb{H}^{m+1} \setminus \{o\}$ can be described as the warped product $\mathbb{R}^+ \times_{\sinh t} \mathbb{S}^m$, of course with metric

$$\langle\,,\,\rangle = dt^2 + \sinh^2 t d\theta^2 \qquad \text{on } \mathbb{H}^{m+1} \setminus \{o\},$$

where $d\theta^2$ is the standard metric on $\mathbb{S}^m$ of constant sectional curvature 1. Due to the properties of the function $\sinh t$ at the origin, that is, $(\sinh t)^{(2k)}(0) = 0$ for $k = 1, 2, \ldots$ and $\sinh 0 = 0$, $(\sinh)'(0) = 1$, the metric above can be smoothly extended also to the point o, giving rise to the usual metric of the hyperbolic space of constant negative sectional curvature -1.

This latter structure is in fact a model in the sense of Greene and Wu [129], according to the following

Definition 1.1 A *model* M_g is a Riemannian manifold of dimension $m \geq 2$ with a pole o such that its metric $\langle\,,\,\rangle$ can be represented on $M_g \setminus \{o\} = (0, a) \times \mathbb{S}^{m-1}$, for some $a \in (0, +\infty]$, in the form

$$\langle\,,\,\rangle = dr^2 + g(r)^2 d\theta^2, \tag{1.209}$$

where, as above, $d\theta^2$ is the standard metric on $\mathbb{S}^{m-1}$ and $g \in C^\infty([0,a))$ satisfies $g > 0$ on $(0,a)$, $g(0) = 0$, $g'(0) = 1$ and $g^{(2k)}(0) = 0$ for $k = 1, 2, \ldots$.

Note that the metric extends smoothly to M_g and it is complete if and only if $a = +\infty$.

Thus, for instance, $\mathbb{R}^m$ can be described as the model M_g with $a = +\infty$, $g(r) = r$ while the hyperbolic space $\mathbb{H}^m$ can be viewed as a model as we did above. Let now M_g be given by $M_g \setminus \{0\} = (0, \pi) \times \mathbb{S}^{m-1}$ and metric $\langle\, , \rangle = dr^2 + \sin^2 r d\theta^2$; this represents the standard punctured sphere , for instance at the North pole, as a model.

Clearly the geometry of M_g outside the pole o is described as that of the corresponding warped product, and the description can be easily extended to the pole. We shall however only be interested in the following special formulas whose proof is left to the interested reader. On the model M_g we have

$$\mathrm{Hess}(r) = \frac{g'(r)}{g(r)}\{\langle\, , \rangle - dr \otimes dr\} \quad \text{on } M_g \setminus \{o\} \tag{1.210}$$

from which, tracing, we obtain

$$\Delta r = (m-1)\frac{g'(r)}{g(r)} \quad \text{on } M_g \setminus \{o\}. \tag{1.211}$$

Sometimes we will use also that

$$^{M_g}K_{rad} = -\frac{g''}{g}, \tag{1.212}$$

$$^{M_g}\mathrm{Ric}\,(\nabla r, \nabla r) = -(m-1)\frac{g''}{g}, \tag{1.213}$$

$$\mathrm{vol}\,(\partial B_R) = \omega_m g^{m-1}(R), \tag{1.214}$$

$$\mathrm{vol}(B_R) = \omega_m \int_0^R g(t)^{m-1}\, dt, \tag{1.215}$$

where ω_m is the volume of the unit sphere in $\mathbb{R}^m$ and where, from now on, $B_R = B_R(o)$ denotes the geodesic ball of radius R centered at the chosen origin o of the manifold, and $\partial B_R = \partial B_R(o)$ its boundary.

Occasionally we will consider on a model also less regular (that is, nonsmooth) metrics, for instance those obtained by requiring $g \in C^2([0,a))$, $g > 0$ on $(0,a)$, $g(0) = 0$, $g'(0) = 1$.

The real usefulness of models consists basically in two of their aspects, which are indeed interrelated. The first is that models, due to their structure, are very manageable to provide simple examples and counterexamples. The second is that, because of (1.212), given a function $G(r)$ that we can think as a lower or an upper bound for the sectional curvatures or for the Ricci tensor of a manifold, we can easily construct a model having that curvature by solving the Cauchy problem

$$\begin{cases} g'' - G(r)g = 0 \\ g(0) = 0, g'(0) = 1 \end{cases}$$

and guaranteeing the positivity of its solution at least on an interval $(0, a)$ depending, of course, on the behaviour of $G(r)$. We will go back to this in Sect. 1.9.1, where we shall deal with comparison results.

1.9 Comparison Results

We recall a few facts on the cut locus and the Riemannian distance function that will be repeatedly used in the sequel, referring to Chavel's book ([71] or [44] for proofs and further details).

Let o be a point in the complete manifolds $(M, \langle\, , \,\rangle)$, and let γ be a geodesic issuing from o. It is known that γ is locally minimizing. A point q in the image of γ is said to be a *cut point* for o along γ if γ minimizes the distance from o to q, but ceases to be minimizing beyond q. The set of cut points of o along geodesic emanating from o is the *cut locus* of o, and is denoted by $\mathrm{cut}(o)$. It turns out that $\mathrm{cut}(o)$ is a closed set of measure zero with respect to the Riemannian measure, and that the set $D_o = M \setminus \mathrm{cut}(o)$ is an open starshaped domain, which is in fact the maximal domain of the normal geodesic coordinates centered at o. At the tangent space level, we say that v is in the tangent cut locus of o, $\mathrm{Cut}(o)$, if the geodesic γ_v with initial velocity v minimizes distances for $t \in [0, 1]$ and does not minimize distances for $t > 1$. Thus $\mathrm{cut}(o)$ is the image of $\mathrm{Cut}(o)$ under the exponential map $\exp_o$, the set $E_o = \{tv \in T_oM : v \in \mathrm{Cut}(o), 0 \leq t < 1\}$, is the maximal starshaped domain with respect to o on which $\exp_o$ is a diffeomorphism, and finally $D_o = \exp_o(E_o)$. Moreover if $r(x)$ denotes the Riemannian distance function from o, namely, $r(x) = \mathrm{dist}_M(x, o) = |\exp^{-1}(x)|$, then $r(x)$ is smooth on $D_o \setminus \{o\}$.

Following Bishop [48] we say that q is an *ordinary* cut point for o if there are two or more minimizing geodesics joining o and q. Cut points which are not ordinary are said to be *singular*.

Bishop proves that ordinary cut points are dense in $\mathrm{cut}(o)$ ([48], Main Theorem). Since it is easily verified that the distance function $r(x)$ is not C^1 at ordinary cut points (see [48], Proposition), we deduce that if $r(x)$ is smooth on the punctured ball $B_R(o) \setminus \{o\}$, then $B_R(o) \cap \mathrm{cut}(o) = \emptyset$.

We recall that, given $p \in M$, the *injectivity radius of* p in M , $\mathrm{inj}_M(p)$, is given by $\mathrm{dist}\,(p, \mathrm{cut}(p))$. Clearly in $B_{\mathrm{inj}_M(p)}(p) \setminus \{p\}$ the distance function $r(x) = \mathrm{dist}\,(x, p)$ is smooth. Later on we shall occasionally use *regular balls*: with this terminology we mean a geodesic ball $B_R(p)$ such that $B_R(p) \cap \mathrm{cut}(p) = \emptyset$ and for which $\max\left\{0, \sup_{B_R(p)} K\right\}^{\frac{1}{2}} < \frac{\pi}{2R}$, where K are the sectional curvatures of M at points of $B_R(p)$. For more details we refer to [44].

1.9.1 The Laplacian Comparison Theorem

Now we show how (1.116) is the starting point to derive the classical Laplacian comparison theorem without using Jacobi fields. Fix a reference point o in $(M, \langle\ ,\ \rangle)$, and let γ be a minimizing geodesic parameterized by arclength issuing from o; we adopt the standard notation $\dot\gamma$ to denote the tangent vector of γ. Note that, since γ is a geodesic, we have $\nabla_{\dot\gamma}\dot\gamma = 0$. We define a unit vector field $Y \perp \dot\gamma$ along γ by parallel translation (see e.g. [170]); note that $\gamma(t)$ is an integral curve of ∇r, that is, $\dot\gamma(t) = (\nabla r)(\gamma(t))$. To perform calculations we let $\{\theta^i\}$ be a local orthonormal coframe and $\{e_i\}$ its dual frame. Then

$$dr = r_i\theta^i \qquad \text{and} \qquad Y = Y^j e_j.$$

By Gauss lemma (see for instance [102]) $|\nabla r|^2 = r_i r_i \equiv 1$ and covariantly differentiating we obtain

$$r_i r_{ij} = 0, \qquad j = 1, \dots m. \tag{1.216}$$

Therefore

$$Y^j r_{ij} r_i = 0.$$

Differentiating again the latter equation and using the fact that Y is parallel yields

$$r_i Y^j r_{ijk} + r_{ij} r_{ik} Y^j = 0;$$

hence, if $\dot\gamma = \dot\gamma^k e_k$, since $r_{ijk} = r_{jik}$,

$$r_{ist} Y^i \dot\gamma^s Y^t = -r_{ij} r_{ik} Y^j Y^k. \tag{1.217}$$

Now in formula (1.116) we take $u(x) = r(x)$ to deduce

$$r_{ijk}\dot\gamma^k Y^j Y^i - r_{ist} Y^i \dot\gamma^s Y^t = -R_{ijkt} Y^i \dot\gamma^j Y^k \dot\gamma^t. \tag{1.218}$$

Thus, inserting (1.217) into (1.218), we get

$$r_{ijk}\dot\gamma^k Y^j Y^i + r_{ij} r_{ik} Y^j Y^k = -R_{ijkt} Y^i \dot\gamma^j Y^k \dot\gamma^t. \tag{1.219}$$

Now we define the $(1, 1)$-*version of the Hessian*, hess, as the tensor field of type $(1, 1)$ such that, if u is a sufficiently smooth function and X and Y are smooth vector fields,

$$\langle \text{hess}\,(u)(X), Y\rangle = \text{Hess}\,(u)(X, Y).$$

Note that we can also write $\mathrm{hess}(u)(X) = \mathrm{Hess}(u)(X,\ )^\sharp$, see Remark 1.3. Thus we have

$$\mathrm{hess}(r)(Y) = \nabla_Y \nabla r,$$

so that

$$\mathrm{Hess}\,(r)(Y,X) = \langle \mathrm{hess}(r)(Y), X\rangle .$$

Having set

$$\mathrm{hess}^2(r)(Y) = \mathrm{hess}(r)(\mathrm{hess}(r)(Y)),$$

we define

$$\mathrm{Hess}^2(r)(Y,X) = \left\langle \mathrm{hess}^2(r)(Y), X\right\rangle .$$

Then, since $\nabla_{\dot\gamma} Y = 0$, (1.219) can be reinterpreted in the form

$$\frac{d}{dt}(\mathrm{Hess}\,(r)(\gamma)(Y,Y)) + \mathrm{Hess}^2(r)(\gamma)(Y,Y) = -K_\gamma(Y \wedge \dot\gamma). \tag{1.220}$$

Note that (1.216) rewrites as

$$\mathrm{hess}(r)(\nabla r) \equiv 0. \tag{1.221}$$

We sum (1.220) over an orthonormal basis $\{Y^i\}$ $(i = 2, \dots, m)$ of $\dot\gamma^\perp$ (where $\dot\gamma^\perp$ is the orthogonal complement of $\dot\gamma$) and use (1.221) to get

$$\frac{d}{dt}(\Delta r)(\gamma) + |\mathrm{Hess}\,(r)|^2(\gamma) = -\operatorname{Ric}(\nabla r, \nabla r)(\gamma). \tag{1.222}$$

Thus, using Newton's inequality

$$|\mathrm{Hess}\,(r)|^2 \geq \frac{(\Delta r)^2}{m-1},$$

we obtain

$$\frac{d}{dt}(\Delta r \circ \gamma) + \frac{(\Delta r \circ \gamma)^2}{m-1} \leq -\operatorname{Ric}(\nabla r \circ \gamma, \nabla r \circ \gamma). \tag{1.223}$$

In the literature, $\operatorname{Ric}(\nabla r, \nabla r)$ is called the *radial Ricci curvature*.

It follows that, assuming

$$\operatorname{Ric}(\nabla r, \nabla r) \geq -(m-1)G(r) \tag{1.224}$$

for some function $G \in C^0([0, +\infty))$,

$$\frac{d}{dt}(\Delta r \circ \gamma) + \frac{(\Delta r \circ \gamma)^2}{m-1} \leq (m-1)G(t). \tag{1.225}$$

Now we recall that $\Delta r = (\sqrt{g}(r,u))^{-1}\frac{\partial\sqrt{g}}{\partial r}$ (see for instance [71]), where $\sqrt{g}$ is the square root of the determinant of the metric in polar geodesic coordinates (r, u) centered at o. Also, $\sqrt{g} = \det\mathscr{G}(r, u)$ where $\mathscr{G}(r, u)$ is the matrix solution of the differential equation in $u^\perp \subset T_oM$

$$\mathscr{G}''(r,u) + \mathscr{R}(r,u)\mathscr{G}(r,u) = 0,$$

satisfying the initial conditions $\mathscr{G}(0, u) = 0$, $\mathscr{G}'(0, u) = Id$, and $\mathscr{R}(r, u)$ is the composition of the curvature operator at $\exp_o(ru)$ with parallel translation along the geodesic $\gamma_u(t) = \exp_o(tu)$ (see again [71, p. 114]). Thus

$$\mathscr{G}(r,u) = r\,\mathrm{Id} + O(r^2) \quad \text{and} \quad \mathscr{G}'(r,u) = \mathrm{Id} + O(r)$$

and we conclude that

$$\Delta r = \log(\det\mathscr{G})' = \mathrm{Tr}(\mathscr{G}'\mathscr{G}^{-1}) = \frac{m-1}{r} + O(r). \tag{1.226}$$

Hence, having set $\varphi(t) = \Delta r \circ \gamma$, using (1.225) and (1.226) and again the fact that γ is parameterized by arclength we deduce that, under assumption (1.224),

$$\begin{cases} \varphi'(t) + \dfrac{\varphi(t)^2}{m-1} \leq (m-1)G(t), \\ \varphi(t) = \dfrac{m-1}{t} + o(1) \quad \text{as } t \to 0^+. \end{cases} \tag{1.227}$$

Of course, in order to make sense from the analytical point of view, (1.227) has to be interpreted with the image of γ inside of the domain D_o of the normal geodesic coordinates centered at o, or, in other words, outside the cut locus of o. To analyze (1.227) we now need two simple calculus lemmas.

Lemma 1.4 *Let $G \in C^0([0, +\infty))$ and let $\varphi, \psi \in C^2((0, +\infty)) \cap C^1([0, +\infty))$ be solutions of the problems:*

$$(i)\begin{cases} \varphi'' - G\varphi \leq 0 \\ \varphi(0) = 0 \end{cases} \quad ; \quad (ii)\begin{cases} \psi'' - G\psi \geq 0 \\ \psi(0) = 0,\ \psi'(0) > 0. \end{cases} \tag{1.228}$$

If $\varphi(r) > 0$ *for* $r \in (0, T)$ *and* $\psi'(0) \geq \varphi'(0)$, *then* $\psi(r) > 0$ *in* $(0, T)$ *and*

$$\frac{\varphi'}{\varphi} \leq \frac{\psi'}{\psi}, \quad \psi \geq \varphi \text{ on } (0, T). \tag{1.229}$$

Proof Since $\psi'(0) > 0$, $\psi > 0$ in a neighborhood of 0. We observe in passing that if G is assumed to be nonnegative, then, integrating (1.228) (ii), we have

$$\psi'(r) \geq \psi'(0) + \int_0^r G(s)\psi(s)\, ds,$$

so that ψ' is positive in the interval where $\psi \geq 0$, and we conclude that, in fact, $\psi > 0$ on $(0, +\infty)$. In the general case, where no assumption is made on the sign of G, we let

$$\beta = \sup\{t : \psi > 0 \text{ in } (0, t)\};$$
$$\tau = \min\{\beta, T\}.$$

The function $\psi'\varphi - \psi\varphi' \in C^0([0, +\infty))$ vanishes in $r = 0$, and it satisfies

$$(\psi'\varphi - \psi\varphi')' = \psi''\varphi - \psi\varphi'' \geq 0$$

in $(0, \tau)$. Thus, $\psi'\varphi - \psi\varphi' \geq 0$ on $[0, \tau)$, and, dividing through by $\varphi\psi$, we deduce that

$$\frac{\psi'}{\psi} \geq \frac{\varphi'}{\varphi} \text{ in } (0, \tau).$$

Integrating between ε and r, with $0 < \varepsilon < r < \tau$, yields

$$\varphi(r) \leq \frac{\varphi(\varepsilon)}{\psi(\varepsilon)}\psi(r)$$

and, since

$$\lim_{\varepsilon \to 0+} \frac{\varphi(\varepsilon)}{\psi(\varepsilon)} = \frac{\varphi'(0)}{\psi'(0)} \leq 1,$$

we conclude that in fact

$$\varphi(r) \leq \psi(r) \text{ in } [0, \tau).$$

Since $\varphi > 0$ in $(0, T)$ by assumption, this in turn forces $\tau = T$, for, otherwise, $\tau = \beta < T$ and we would have $\varphi(\beta) > 0$, while, by continuity, $\psi(\beta) = 0$, a contradiction. □

Lemma 1.5 *Let $G \in C^0(\mathbb{R}_0^+)$ and let $g_i \in C^1((0, T_i))$, $i = 1, 2$ be solutions of the Riccati differential inequalities*

$$(i)\ g_1' + \frac{g_1^2}{\alpha} - \alpha G \le 0; \qquad (ii)\ g_2' + \frac{g_2^2}{\alpha} - \alpha G \ge 0 \tag{1.230}$$

satisfying the condition

$$g_i(t) = \frac{\alpha}{t} + O(1) \ \text{ as } t \to 0^+ \tag{1.231}$$

for some $\alpha > 0$. Then $T_1 \le T_2$ and $g_1(t) \le g_2(t)$ in $(0, T_1)$.

Proof Since $\widetilde{g_i} = \alpha^{-1} g_i$ satisfy the conditions in the statement with $\alpha = 1$, without loss of generality we assume $\alpha = 1$. Observe that the functions $g_i(s) - \frac{1}{s}$ are bounded and integrable in a neighborhood of $s = 0$, thus we define $\varphi_i \in C^2((0, T_i)) \cap C^1([0, T_i))$ on $[0, T_i)$, by setting

$$\varphi_i(t) = t e^{\int_0^t (g_i(s) - \frac{1}{s})\, ds}.$$

Then $\varphi_i(0) = 0$, $\varphi_i > 0$ on $(0, T_i)$ and straightforward computations show that

$$\varphi_i'(t) = g_i(t)\varphi_i(t), \qquad \varphi_i'(0) = 1$$

and

$$\begin{aligned} \varphi_1'' &\le G\varphi_1 \qquad \text{on } (0, T_1); \\ \varphi_2'' &\ge G\varphi_2 \qquad \text{on } (0, T_2). \end{aligned}$$

An application of Lemma 1.4 shows that $T_1 \le T_2$ and $g_1 = \frac{\varphi_1'}{\varphi_1} \le \frac{\varphi_2'}{\varphi_2} = g_2$ on $(0, T_1)$, as required. □

We are now ready to prove the next *Laplacian comparison theorem*, which is a simplified (but sufficient for our purposes) version of that appearing in [183]:

Theorem 1.2 *Let $(M, \langle\, , \rangle)$ be a complete manifold of dimension $m \ge 2$. Having fixed a reference point $o \in M$, let $r(x) = \mathrm{dist}_M(x, o)$. Assume that the radial Ricci curvature* $\mathrm{Ric}(\nabla r, \nabla r)$ *of M satisfies*

$$\mathrm{Ric}(\nabla r, \nabla r) \ge -(m-1)G(r) \tag{1.232}$$

for some nonnegative function $G \in C^0(\mathbb{R}_0^+)$. Let $h \in C^2(\mathbb{R}_0^+)$ be a solution of the problem

$$\begin{cases} h'' - Gh \ge 0 \\ h(0) = 0, \quad h'(0) = 1. \end{cases} \tag{1.233}$$

Then the inequality

$$\Delta r(x) \le (m-1)\frac{h'(r(x))}{h(r(x))} \tag{1.234}$$

holds pointwise on $M\setminus(\{o\}\cup \mathrm{cut}(o))$ *and weakly on all of* M.

Proof Fix any $x \in M\setminus(\{o\}\cup \mathrm{cut}(o))$ and let $\gamma : [0,l] \to M$ be a minimizing geodesic from o to x parameterized by arclength. We then arrive to (1.227), where the differential inequality is in $(0,l]$. Since $g = (m-1)\frac{h'}{h}$ satisfies

$$g'(t) + \frac{g(t)^2}{m-1} \ge (m-1)G(t) \ \text{ on } \mathbb{R}^+ \tag{1.235}$$

and (1.231) with $\alpha = m-1$, an application of Lemma 1.5 to (1.227) and (1.235) gives

$$\varphi(t) \le (m-1)\frac{h'(t)}{h(t)} \ \text{ in } (0,l].$$

Thus, in particular, since $\gamma(l) = x$ and $r(x) = l$,

$$\Delta r(x) \le (m-1)\frac{h'(r(x))}{h(r(x))},$$

showing the validity of (1.234) pointwise within the cut locus. It remains to show the validity of (1.234) weakly in all of M, which is guaranteed by the following Lemma. □

Lemma 1.6 *Set* $D_o = M\setminus \mathrm{cut}(o)$ *and suppose that*

$$\Delta r \le \alpha(r) \ \textit{ pointwise on } D_o\setminus\{o\}, \tag{1.236}$$

for some $\alpha \in C^0((0,+\infty))$. *Let* $v \in C^2(\mathbb{R})$ *be nonnegative and set* $u(x) = v(r(x))$ *on* M. *Suppose either*

$$(i)\ v' \le 0 \qquad \textit{or} \qquad (ii)\ v' \ge 0. \tag{1.237}$$

Then we respectively have

$$(i)\ \Delta u \ge v''(r) + \alpha(r)v'(r); \qquad (ii)\ \Delta u \le v''(r) + \alpha(r)v'(r) \tag{1.238}$$

weakly on M.

Proof Let E_o be the maximal star-shaped domain in T_oM on which $\exp_o$ is a diffeomorphism onto its image D_o, so that we have $\mathrm{cut}(o) = \partial(\exp_o(E_o))$. Since

E_o is a star-shaped domain, we can exhaust E_o by a family $\{E_o^n\}$ of relatively compact, star-shaped domains with smooth boundary such that $\overline{E}_o^n \subset E_o^{n+1}$. We set $D_o^n = \exp_o(E_o^n)$ so that

$$\overline{D}_o^n \subset D_o^{n+1} \text{ and } \bigcup_n D_o^n = D_o.$$

The fact that each E_o^n is star-shaped implies

$$\langle \nabla r, \nu_n \rangle > 0 \text{ on } \partial D_o^n, \tag{1.239}$$

where ν_n denotes the outward unit normal to ∂D_o^n. Now we assume the validity of (1.237) (i). Since $r \in C^\infty(D_o^n \backslash \{o\})$, computing we get

$$\Delta u \geq v'' + \alpha(r)v' \qquad \text{pointwise on } D_o^n \backslash \{o\}. \tag{1.240}$$

Let $0 \leq \varphi \in C_c^\infty(M)$, where $C_c^\infty(M)$ denotes the set of smooth function with compact support on M. We claim that, for each n,

$$\int_{D_o^n} u\Delta\varphi \geq \int_{D_o^n} (v'' + \alpha(r)v')\varphi + \varepsilon_n,$$

where $\varepsilon_n \to 0$ as $n \to +\infty$. Since $M = D_o \cup \mathrm{cut}(o)$ and $\mathrm{cut}(o)$ has measure 0, inequality (1.238) (i) will follow by letting $n \to +\infty$. To prove the claim we fix $\delta > 0$ small and we apply the second Green formula (see e.g. [71]) on $\overline{D_o^n} \backslash B_\delta(o)$ to obtain

$$\int_{D_o^n \backslash B_\delta(o)} u\Delta\varphi = \int_{D_o^n \backslash B_\delta(o)} \varphi\Delta u - \int_{\partial D_o^n \backslash \partial B_\delta(o)} (\varphi\langle \nabla u, \nu_n \rangle - u\langle \nabla\varphi, \nu_n \rangle), \tag{1.241}$$

where ν_n is the outward unit normal to $\partial D_o^n \backslash \partial B_\delta(o)$. We note that, according to (1.237) (i) and (1.239),

$$\langle \nabla u, \nu_n \rangle = v'(r)\langle \nabla r, \nu_n \rangle \leq 0 \quad \text{on } \partial D_o^n.$$

Using this, (1.239) and (1.241) we deduce

$$\int_{D_o^n} u\Delta\varphi \geq \int_{D_o^n} (v'' + \alpha(r)v')\varphi + \varepsilon_n + I_\delta,$$

with

$$\varepsilon_n = \int_{\partial D_o^n} u\langle \nabla\varphi, \nu_n\rangle,$$

$$I_\delta = \int_{B_\delta(o)} \left[u\Delta\varphi - \left(v'' + \alpha(r)v'\right)\varphi\right] - \int_{\partial B_\delta(o)} [u\langle\nabla\varphi, \nu_n\rangle - \varphi\langle\nabla u, \nu_n\rangle].$$

Clearly, $I_\delta \to 0$ as $\delta \downarrow 0^+$; on the other hand, since $\varphi \in C_c^\infty(M)$ and cut(o) has measure 0, using the divergence and Lebesgue theorems we see that, as $n \to +\infty$,

$$\varepsilon_n = \int_{D_o^n} \mathrm{div}(u\nabla\varphi) \to \int_{D_o} \mathrm{div}(u\nabla\varphi) = \int_M \mathrm{div}(u\nabla\varphi) = 0.$$

This proves the claim and the validity of (1.238) (i). The case (1.237) (ii) and (1.238) (ii) can be dealt with in a similar way. □

Remark 1.17 We note that, for the above proofs to work, it is not necessary that (1.232) holds on the entire M: instead, for instance, if (1.232) is valid on $B_R(o)$, then (1.234) holds on $B_R(o)\backslash(\{o\} \cup \mathrm{cut}(o))$ and weakly on $B_R(o)$.

We derive here another consequence of the differential inequality (1.223).

Let $D_o = M \setminus \mathrm{cut}\,(o)$ and $x \in D_o \setminus \{o\}$. We set $\varphi = \Delta r \circ \gamma$, where $\gamma : [0, r(x)] \to M$ is a unit speed minimizing geodesic from o to x. Then inequality (1.223) can be rewritten as

$$\varphi'(t) + \frac{\varphi(t)^2}{m-1} + \mathrm{Ric}\,(\dot\gamma, \dot\gamma)(t) \le 0 \quad \text{on } [0, r(x)]. \tag{1.242}$$

Furthermore we know from (1.227) that

$$\frac{1}{m-1}\varphi(t) = \frac{1}{t} + o(1) \quad \text{as } t \to 0^+. \tag{1.243}$$

Defining

$$u(t) = te^{\int_0^t \left(\frac{\varphi(s)}{m-1} - \frac{1}{s}\right) ds} \tag{1.244}$$

on $[0, r(x)]$, u is well defined because of (1.243) and a simple computation using (1.242) gives

$$u'' + \frac{\mathrm{Ric}\,(\dot\gamma, \dot\gamma)}{m-1} u \le 0 \quad \text{on } [0, r(x)]. \tag{1.245}$$

Next, we let $h \in C^1([0, r(x)])$ be such that $h(0) = 0 = h(r(x))$. Since $u > 0$ on $(0, r(x)]$, the function $h^2 \frac{u'}{u}$ is well defined on $(0, r(x)]$. Differentiating, using (1.245) and Young's inequality we get

$$\begin{aligned}\left(h^2\frac{u'}{u}\right)' &\leq -\frac{\operatorname{Ric}(\dot\gamma,\dot\gamma)}{m-1}h^2 - \left(\frac{u'}{u}\right)^2 h^2 + 2hh'\frac{u'}{u} \\ &\leq -\frac{\operatorname{Ric}(\dot\gamma,\dot\gamma)}{m-1}h^2 + (h')^2.\end{aligned}$$

Fix $\varepsilon > 0$ sufficiently small; integration of the above inequality on $[\varepsilon, r(x)]$ gives

$$-h^2(\varepsilon)\frac{u'(\varepsilon)}{u(\varepsilon)} \leq \int_\varepsilon^{r(x)} (h')^2 - \frac{\operatorname{Ric}(\dot\gamma,\dot\gamma)}{m-1}h^2.$$

Since $h(\varepsilon) = A\varepsilon + o(\varepsilon)$ as $\varepsilon \to 0^+$ for some $A \in \mathbb{R}$, letting $\varepsilon \to 0^+$ we obtain

$$\int_0^{r(x)} (h')^2 - \frac{\operatorname{Ric}(\dot\gamma,\dot\gamma)}{m-1}h^2 \geq 0, \tag{1.246}$$

that is, minimizing geodesics are stable.

Note that the above inequality can be extended to any $x \in M$ using "Calabi trick" (see Lemma 2.1 in Chap. 2). Indeed, suppose that $x \in \operatorname{cut}(o)$; translating the origin to $o_\varepsilon = \gamma(\varepsilon)$ so that $x \notin \operatorname{cut}(o)\varepsilon)$, using the triangle inequality and, finally, taking the limit as $\varepsilon \to 0$, one checks that (1.246) holds also in this case.

Inequality (1.246) will be repeatedly used in Chap. 8.

1.9.2 The Bishop-Gromov Comparison Theorem

We now show how to get from the previous results a (somewhat generalized) version of what is known in the literature as the *Bishop-Gromov comparison theorem* (see also [44]). We recall that $\operatorname{vol} B_R(o)$ and $\operatorname{vol} \partial B_R(o)$ denote the volume of the geodesic ball $B_R(o)$ and of its boundary $\partial B_R(o)$, respectively.

Theorem 1.3 *Let $(M, \langle\, ,\, \rangle)$ be a complete, m-dimensional Riemannian manifold satisfying*

$$\operatorname{Ric}(\nabla r, \nabla r) \geq -(m-1)G(r) \qquad \text{on } M \tag{1.247}$$

for some $G \in C^0(\mathbb{R}_0^+)$, $G \geq 0$, *where* $r(x) = \operatorname{dist}(x, o)$. *Let* $h \in C^2(\mathbb{R}_0^+)$ *be the nonnegative solution of the problem*

$$\begin{cases} h'' - G(t)h = 0 \\ h(0) = 0,\ h'(0) = 1. \end{cases} \tag{1.248}$$

Then, for almost every $R > 0$, *the function*

$$R \mapsto \frac{\operatorname{vol} \partial B_R(o)}{h(R)^{m-1}} \tag{1.249}$$

is nonincreasing, and

$$\operatorname{vol} \partial B_R(o) \leq \omega_m h(R)^{m-1}, \tag{1.250}$$

where ω_m *is the volume of the unit sphere in* $\mathbb{R}^m$. *Moreover,*

$$R \mapsto \frac{\operatorname{vol} B_R(o)}{\int_0^R h(t)^{m-1}\, dt} \tag{1.251}$$

is a nonincreasing function on $\mathbb{R}^+$.

Since it will be used in the proof of Theorem 1.3, and also in the next chapters, we first recall the useful *coarea formula.*

We denote by $W^{1,1}(M)$ the Sobolev space consisting of functions in $L^1(M)$ with (weak) gradient in $L^1(M)$. We also denote by $\partial\Omega_t^u$ the t-level set ($t \in \mathbb{R}$) of a function u on M, i.e. $\partial\Omega_t^u = \{x \in M | u(x) = t\}$. Following Schoen and Yau (see [252, p. 89]) we state the following

Proposition 1.6 *Let* M *be a compact Riemannian manifold with boundary and* $u \in W^{1,1}(M)$. *For any nonnegative measurable function* v *on* M *the following formula holds:*

$$\int_M v = \int_{-\infty}^{+\infty} \left(\int_{\partial\Omega_t^u} \frac{v}{|\nabla u|}\, d\sigma_u \right) dt, \tag{1.252}$$

where $d\sigma_u$ *is the* $(m-1)$*-dimensional Hausdorff measure of* $\partial\Omega_t^u$.

For a proof see the classical [117]. Note that, in particular, if $u(x) = r(x) = \operatorname{dist}_M(x, o)$, Eq. (1.252) becomes Fubini's formula

$$\int_M v = \int_0^D \left(\int_{\partial B_t(o)} v \right) dt, \tag{1.253}$$

where $D = \sup_M r(x)$.

Proof (of Theorem 1.3) In case o is a pole of M (see [129]) one integrates the divergence of the radial vector field

$$X = h(r(x))^{-m+1}\,\nabla r$$

on concentric balls $B_R(o)$, and uses the divergence and Laplacian comparison theorems. However, in general, objects are nonsmooth and inequalities are interpreted in the sense of distributions. Therefore, some extra care is needed. The Laplacian comparison theorem asserts that

$$\Delta r(x) \le (m-1)\frac{h'(r(x))}{h(r(x))} \tag{1.254}$$

pointwise on the open, star-shaped, full measured set $M\setminus \mathrm{cut}(o)$ and weakly on all of M. Thus, for each $0 \le \varphi \in \mathrm{Lip}_c(M)$,

$$-\int \langle \nabla r, \nabla \varphi\rangle \le (m-1)\int \frac{h'(r(x))}{h(r(x))}\varphi. \tag{1.255}$$

For $\varepsilon > 0$ fixed, consider the radial cut-off function

$$\varphi_\varepsilon(x) = \rho_\varepsilon(r(x))h(r(x))^{-m+1}, \tag{1.256}$$

where, for $0 < s < R$, ρ_ε is the piecewise linear function

$$\rho_\varepsilon(t) = \begin{cases} 0, & \text{if } t \in [0, s) \\ \frac{t-s}{\varepsilon}, & \text{if } t \in [s, s+\varepsilon) \\ 1, & \text{if } t \in [s+\varepsilon, R-\varepsilon) \\ \frac{R-t}{\varepsilon}, & \text{if } t \in [R-\varepsilon, R) \\ 0, & \text{if } t \in [R, +\infty). \end{cases} \tag{1.257}$$

Note that

$$\nabla \varphi_\varepsilon = \left\{ -\frac{\chi_{R-\varepsilon,R}}{\varepsilon} + \frac{\chi_{s,s+\varepsilon}}{\varepsilon} - (m-1)\frac{h'(r(x))}{h(r(x))}\rho_\varepsilon \right\} h(r(x))^{-m+1}\nabla r,$$

for almost all $x \in M$, where $\chi_{s,t}$ is the characteristic function of the annulus $B_t(o)\setminus B_s(o)$. Therefore, using φ_ε into (1.255) and simplifying, we get

$$\frac{1}{\varepsilon}\int_{B_R(o)\setminus B_{R-\varepsilon}(o)} h(r(x))^{-m+1} \le \frac{1}{\varepsilon}\int_{B_{s+\varepsilon}(o)\setminus B_s(o)} h(r(x))^{-m+1}.$$

Using the coarea formula (1.252) we deduce that

$$\frac{1}{\varepsilon}\int_{R-\varepsilon}^{R} \mathrm{vol}(\partial B_t(o))\, h(t)^{-m+1} dt \le \frac{1}{\varepsilon}\int_{s}^{s+\varepsilon} \mathrm{vol}(\partial B_t(o))\, h(t)^{-m+1} dt$$

and, letting $\varepsilon \downarrow 0$,

$$\frac{\mathrm{vol}(\partial B_R(o))}{h(R)^{m-1}} \le \frac{\mathrm{vol}(\partial B_s(o))}{h(s)^{m-1}} \tag{1.258}$$

for almost all $0 < s < R$. Letting $s \to 0$ and recalling that $h(s) \sim s$ and $\mathrm{vol}(\partial B_s) \sim \omega_m s^{m-1}$ as $s \to 0$ (which can be deduced, for instance, integrating Eq. (1.226) on a geodesic ball and using the divergence theorem and Gauss lemma), we conclude that, for almost any $R > 0$,

$$\mathrm{vol}\,\partial B_R(o) \le \omega_m h(R)^{m-1}.$$

To prove the second statement we note that, as observed in [74], for general real valued functions $f(t) \ge 0$, $g(t) > 0$, *if* $t \mapsto \frac{f(t)}{g(t)}$ *is decreasing, then* $t \mapsto \frac{\int_0^t f}{\int_0^t g}$ *is decreasing.* Indeed, since f/g is decreasing, if $0 < s < R$

$$\int_0^s f \int_s^R g = \int_0^s g\frac{f}{g} \int_s^R g \ge \frac{f(s)}{g(s)} \int_0^s g \int_s^R g \ge \int_0^s g \int_s^R g\frac{f}{g} = \int_0^s g \int_s^R f$$

whence

$$\int_0^s f \int_0^R g = \int_0^s f \int_0^s g + \int_0^s f \int_s^R g \ge \int_0^s f \int_0^s g + \int_0^s g \int_s^R f = \int_0^s g \int_0^R f.$$

In particular, applying this observation to (1.258) and using the coarea formula (1.252) we deduce that

$$s \mapsto \frac{\mathrm{vol}\, B_s(o)}{\int_0^s h(t)^{m-1}\, dt}$$

is decreasing, concluding the proof. □

Remark 1.18 The same argument will be applied in the proof of Proposition 8.10 on solitons.

To have a more precise idea of the estimates on $\mathrm{vol}\,\partial B_R(o)$ and $\mathrm{vol}\, B_R(o)$ that we can get via Theorem 1.3 we conclude with the following analytical result whose proof can be found in [44]:

Proposition 1.7 *Assume h is a solution of*

$$\begin{cases} h'' - B^2(1+r^2)^{\delta/2}h = 0 \\ h(0) = 0, \quad h'(0) = 1 \end{cases}$$

for some constants $B > 0$ and $\delta \geq -2$. Set

$$B' = \begin{cases} B, & \text{if } \delta > -2 \\ \frac{1}{2}(1+\sqrt{1+4B^2}), & \text{if } \delta = -2. \end{cases}$$

Then,

$$\frac{h'}{h}(r) \leq B' r^{\delta/2}(1+o(1)) \quad \text{as } r \to +\infty.$$

Moreover, there exists a constant $C > 0$ such that for $r > 1$

$$h(r) \leq C \begin{cases} \exp\left(\frac{2B'}{2+\delta}(1+r)^{1+\delta/2}\right) & \text{if } \delta \geq 0 \\ r^{-\delta/4}\exp\left(\frac{2B'}{2+\delta}r^{1+\delta/2}\right) & \text{if } -2 < \delta < 0 \\ r^{B'} & \text{if } \delta = -2. \end{cases}$$

1.9.3 The Hessian Comparison Theorem

For the sake of completeness we recall here the following *Hessian comparison theorem*; our discussion follows closely the one in [44]. Recall that the *radial sectional curvature* K_{rad} of a manifold is the sectional curvature of a 2-plane containing ∇r.

Theorem 1.4 *Let $(M, \langle\, , \rangle)$ be a complete manifold of dimension m. Having fixed a reference point $o \in M$, let $r(x) = \mathrm{dist}_M(x, o)$, and let $D_o = M \setminus \mathrm{cut}(o)$ be the domain of the normal geodesic coordinates centered at o. Given $G \in C^0(\mathbb{R}_0^+)$, let h be the solution of the Cauchy problem*

$$(i) \begin{cases} h'' - Gh \geq 0 \\ h(0) = 0, \quad h'(0) = 1, \end{cases} \quad \text{or} \quad (ii) \begin{cases} h'' - Gh \leq 0 \\ h(0) = 0, \quad h'(0) = 1, \end{cases} \tag{1.259}$$

and let $I = (0, R_0) \subseteq \mathbb{R}^+$ be the maximal interval where h is positive. If the radial sectional curvature of M satisfies

$$K_{rad} \geq -G(r(x)) \qquad \text{on } B_{R_0}(o), \tag{1.260}$$

then

$$\mathrm{Hess}\,(r)(x) \le \frac{h'(r(x))}{h(r(x))}\{\langle\,,\,\rangle - dr \otimes dr\} \tag{1.261}$$

on $(D_o \setminus \{o\}) \cap B_R(o)$ *in the sense of quadratic forms, where* h *solves (i). On the other hand, if*

$$K_{rad} \le -G(r(x)) \qquad \text{on } B_{R_0}(o), \tag{1.262}$$

then

$$\mathrm{Hess}\,(r)(x) \ge \frac{h'(r(x))}{h(r(x))}\{\langle\,,\,\rangle - dr \otimes dr\}, \tag{1.263}$$

on $(D_o \setminus \{o\}) \cap B_R(o)$ *in the sense of quadratic forms, where* h *solves (ii).*

Remark 1.19 By taking traces in Theorem 1.4 we immediately obtain the corresponding estimates for Δr. However, as we have seen in Theorem 1.2, the estimate from above for the Laplacian of the distance function holds under the weaker assumption that the radial Ricci curvature (and not the full radial sectional curvature) is bounded from below by $-(m-1)G(r(x))$. Furthermore the estimate in this latter case can be extended, in weak form, to the entire manifold. This is not the case for the above estimates on $\mathrm{Hess}\,(r)$.

To prove Theorem 1.4 we first need some results concerning comparison theory for Riccati equations in the matrix-valued setting.

Let E be a finite dimensional vector space endowed with an inner product $\langle\,,\,\rangle$ and induced norm $||\cdot||$, and let $S(E)$ be the space of self-adjoint linear endomorphism of E. We say that $A \in S(E)$ satisfies $A \ge 0$ if A is positive semi-definite; analogously, we say that $A \le B$ if $B - A$ is positive semi-definite. We denote with $I \in S(E)$ the identity transformation. The following comparison result is due to Eschenburg and Heintze [114].

Theorem 1.5 *Let* $R_i : \mathbb{R}_0^+ \to S(E)$, $i = 1, 2$, *be smooth curves, and assume that* $R_1 \le R_2$. *For each i, let* $B_i : (0, s_i) \to S(E)$ *be a maximally defined solution of the matrix Riccati equation*

$$B_i' + B_i^2 = R_i.$$

Suppose that $U = B_2 - B_1$ *can be continuously extended at* $s = 0$ *and* $U(0^+) \ge 0$. *Then*

$$s_1 \le s_2 \qquad \text{and } B_1 \le B_2 \quad \text{on } (0, s_1).$$

Furthermore, $d(s) = \dim\ker U(s)$ *is nonincreasing on* $(0, s_1)$. *In particular, if* $B_1(\widetilde{s}) = B_2(\widetilde{s})$, *then* $B_1 \equiv B_2$ *on* $(0, \widetilde{s})$.

Proof Set $s_0 = \min\{s_1, s_2\}$ and observe that , on $(0, s_0)$, $U = B_2 - B_1$ satisfies

$$U' = UX + XU + S, \qquad \text{where} \begin{cases} S &= R_2 - R_1 \geq 0 \\ X &= -\frac{1}{2}(B_2 + B_1). \end{cases} \tag{1.264}$$

We claim that X is bounded from above near $s = 0$. Indeed, by the Riccati equation $B_i' \leq R_i$, hence for every unit vector $x \in E$ the function $\eta_i(s) = \langle B_i(s)x, x\rangle$ satisfies $\eta_i' \leq \langle R_i(s)x, x\rangle \leq ||R_i(s)|| \leq C$, where the last inequality follows since R_i is bounded on $[0, s_0]$. Integrating on some $[s, \widetilde{s}] \subset (0, s_0)$,

$$\eta_i(s) \geq -C(\widetilde{s} - s) + \eta_i(\widetilde{s}) \geq -C\widetilde{s} - ||B_i(\widetilde{s})||$$

independently on x. Therefore, each B_i is bounded from below as $s \to 0$, and thus there exists $a > 0$ such that $X \leq aI$ near $s = 0$, as claimed. The solution U of (1.264) can be computed via the method of the variation of constants. First, fix $\widetilde{s} \in (0, s_0)$ and consider the solution of the Cauchy problem

$$\begin{cases} g' = Xg \\ g(\widetilde{s}) = I, \end{cases}$$

where $I \in S(E)$ is the identity. Then, g is nonsingular on $(0, s_0)$: indeed, its inverse is given by the function $\bar{g}$ satisfying $\bar{g}' = -\bar{g}X$, $\bar{g}(\widetilde{s}) = I$. The general solution U of (1.264) is thus

$$U = gV\,{}^{T}g, \tag{1.265}$$

where $V : (0, s_0) \to S(E)$ is the general solution of

$$V' = g^{-1}S\,{}^{T}(g^{-1}).$$

Since $S \geq 0$, we deduce $V' \geq 0$. Hence, for every fixed $x \in E$, $\langle V(s)x, x\rangle : (0, s_0) \to \mathbb{R}$ is nondecreasing. This shows that the pointwise limit $\langle V(0)x, x\rangle$ exists, possibly infinite. We claim that $\langle V(0)x, x\rangle$ is finite, hence $V(0)$ can be defined by polarization. Furthermore, we shall show that $V(0) \geq 0$. Towards this aim, from (1.265) and setting, for notational convenience, $h = {}^{T}(g^{-1})$,

$$\langle Vx, x\rangle = \left\langle g^{-1}U\left({}^{T}g\right)^{-1}x, x\right\rangle = \left\langle U\,{}^{T}(g^{-1})x, {}^{T}(g^{-1})x\right\rangle = \langle Uhx, hx\rangle, \tag{1.266}$$

so that

$$|\langle Vx, x\rangle| \leq ||U|| \cdot ||hx||^2.$$

Since, by assumption, $||U||$ is bounded as $s \to 0$, to prove that $|\langle Vx, x\rangle|$ is bounded in a neighbourhood of zero we shall show that so is the function $f(s) = ||h(s)x||^2$. Note that, by its very definition and the properties of g, $h' = -Xh$. Hence,

$$f'(s) = 2\big\langle h'(s)x, h(s)x\big\rangle = -2\langle Xh(s)x, h(s)x\rangle \geq -2af.$$

By Gronwall lemma, f cannot diverge as $s \to 0^+$, as required. As a consequence, for every $s_k \to 0$ the set $\{y_k\} = \{h(s_k)x\} \subset E$ is bounded. By compactness, up to a subsequence $y_k \to y$, for some $y \in E$. Therefore, by (1.266)

$$\langle V(0)x, x\rangle = \lim_k \langle V(s_k)x, x\rangle = \lim_k \langle U(s_k)y_k, y_k\rangle = \langle U(0)y, y\rangle \geq 0,$$

hence $V(0) \geq 0$. From $V' \geq 0$, we deduce $V \geq 0$, thus by (1.265) $U \geq 0$, as desired. Since V is nonnegative and nondecreasing, so is $\dim\ker U(s) = d(s)$, and this conclude the proof. □

Now, using the notation of Sect. 1.9, let $p \in D_o$ and let $\gamma : [0, r(x)] \to M$ be the minimizing geodesic from o to p, so that $r(\gamma(s)) = s$ and $\nabla r \circ \gamma = \dot\gamma$ for every s. Fix a local orthonormal frame $\{e_i\}$ around p, with dual coframe $\{\theta^i\}$; then $\dot\gamma = \nabla r = r_i e_i$, $dr = r_i\theta^i$ and differentiating $|\nabla r|^2 = r_i r_i = 1$ we obtain (see Eq. (1.216))

$$r_{ij}r_i = 0, \qquad \text{that is} \qquad \mathrm{Hess}(r)(\nabla r, \ ) = 0. \tag{1.267}$$

A further covariant differentiation of (1.267) gives

$$r_{ijk}r_i + r_{ij}r_{ik} = 0,$$

which can be rewritten using (1.115) and (1.116) as

$$0 = r_{ijk}r_i + r_{ij}r_{ik} = r_{jik}r_i + r_{ij}r_{ik} = r_{jki}r_i + r_t R_{tjik} + r_{ij}r_{ik}.$$

Contracting the above equation with two parallel vector fields X and Y along γ and perpendicular to ∇r we get

$$0 = r_{jki}X^jY^kr_i + X^jY^kr_tr_iR_{tjik} + r_{ij}r_{ik}X^jY^k;$$

in Koszul notation, and using the symmetries of the curvature tensor (see (1.45)), the above relation reads

$$\begin{aligned} 0 &= \langle \nabla\,\mathrm{hess}(r)(\nabla r, X, Y)\rangle + \langle \mathrm{hess}(r)(X), \mathrm{hess}(r)(Y)\rangle + \langle R(\nabla r, X)Y, \nabla r\rangle \\ &= \langle \nabla\,\mathrm{hess}(r)(\nabla r, X, Y)\rangle + \langle \mathrm{hess}(r)(X), \mathrm{hess}(r)(Y)\rangle + \langle R(X, \nabla r)\nabla r, Y\rangle. \end{aligned} \tag{1.268}$$

(compare with Eqs. (1.219) and (1.220)). Since hess(r) is self-adjoint, denoting with R_γ the self-adjoint map

$$X \mapsto R_\gamma(X) = R(X, \nabla r)\nabla r, \tag{1.269}$$

and with a prime the covariant differentiation along γ, (1.268) becomes

$$0 = \left\langle \left((\mathrm{hess}(r))' + (\mathrm{hess}(r))^2 + R_\gamma\right)(X), Y \right\rangle = 0 \tag{1.270}$$

for each $X, Y \in \nabla r^\perp$, parallel. Note that, by (1.267) and the properties of the curvature tensor, both hess(r) and R_γ can be thought as endomorphisms of $\nabla r^\perp$. Furthermore, for every unit vector $X \in \nabla r^\perp$,

$$\left\langle R_\gamma(X), X \right\rangle = K(X \wedge \nabla r) = K_{rad}(X), \tag{1.271}$$

that is, the sectional curvature of $X \wedge \nabla r$. Since X and Y are arbitrary, we have

$$(\mathrm{hess}(r))' + (\mathrm{hess}(r))^2 + R_\gamma = 0 \tag{1.272}$$

as a section of End $\left(\nabla r^\perp\right)$ along γ. By parallel translation, we can identify the fibers of the vector bundle $\nabla r^\perp$; indeed, if we consider an orthonormal basis $\{E_i\} \subset \nabla r^\perp$ of parallel vector fields along γ, and we denote with $B = (r_{ij})$, $R_\gamma = \left((R_\gamma)_{ij}\right)$ the representation of $\mathrm{hess}(r)_{|\nabla r^\perp}$ and R_γ in the basis $\{E_i\}$, (1.272) becomes the matrix Riccati equation

$$B' + B^2 + R_\gamma = 0. \tag{1.273}$$

Taking into account the asymptotic relation for $K = \{o\}$ (see [219]),

$$\mathrm{Hess}(r) = \frac{1}{s}(\langle\, , \rangle - dr \otimes dr) + o(1) \quad \text{as } s \to 0^+,$$

and B satisfies

$$\begin{cases} B' + B^2 + R_\gamma = 0 & \text{on } (0, r(x)] \\ B(s) = s^{-1}I + o(1) & \text{as } s \to 0^+. \end{cases} \tag{1.274}$$

Now, assume either

$$(i)\ K_{rad} \geq -G(r) \qquad \text{or} \qquad (ii)\ K_{rad} \leq -G(r),$$

for some $G(r) \in C^0\left(\mathbb{R}_0^+\right)$. Henceforth, (i) (resp. (ii)) means that the inequality

$$K(\Pi)(x) \geq -G(r(x))$$

(resp. $\leq$) holds for every 2-plane Π containing ∇r. Then, by (1.271), respectively

$$(i)\ R_\gamma \geq -G(s)I \qquad (ii)\ R_\gamma \leq -G(s)I,$$

and by (1.273) this yields the following matrix Riccati inequalities:

$$\text{case }(i):\ \begin{cases} B' + B^2 \leq GI, \\ B(s) = s^{-1}I + o(1) \quad \text{as } s \to 0^+; \end{cases} \tag{1.275}$$

$$\text{case }(ii):\ \begin{cases} B' + B^2 \geq GI, \\ B(s) = s^{-1}I + o(1) \quad \text{as } s \to 0^+. \end{cases} \tag{1.276}$$

Now, consider a solution h to

$$\begin{cases} h'' - Gh \geq 0 \\ h(0) = 0,\ h'(0) = 1 \end{cases} \quad \text{for } (i),$$

$$\begin{cases} h'' - Gh \leq 0 \\ h(0) = 0,\ h'(0) = 1 \end{cases} \quad \text{for } (ii),$$

and assume that h is positive on some maximal interval $I = (0, R_0)$. Setting $B_h = (h'/h)I$ we have that

$$\text{case }(i):\ \begin{cases} B_h' + B_h^2 \geq GI, \\ B_h(s) = s^{-1}I + o(1) \quad \text{as } s \to 0^+; \end{cases} \tag{1.277}$$

$$\text{case }(ii):\ \begin{cases} B_h' + B_h^2 \leq GI, \\ B_h(s) = s^{-1}I + o(1) \quad \text{as } s \to 0^+. \end{cases} \tag{1.278}$$

By the matrix Riccati comparison Theorem 1.5, $B \leq B_h$ when (i) holds, and $B \geq B_h$ under assumption (ii). This yields the proof of Theorem 1.4.

Chapter 2
The Omori-Yau Maximum Principle

The aim of this chapter is to introduce the Omori-Yau and the weak maximum principles. We begin with some analytical motivations of a general nature and we then proceed to introduce the various concepts, results and related discussions. In this process we follow a perspective quite different from the original approach of Omori and Yau. Indeed, we introduce a function theoretic formulation of the principle that does not tie it to curvature assumptions as in the pioneering works of Omori [210] and Yau [279] (see for instance the statement of Theorem 2.4 below). This formulation reaches a great advantage in applications as it will become crystal clear from the geometric and analytic results contained in the subsequent chapters. We then relax the original statement of the principle to obtain what we call the weak maximum principle. This simple minded procedure, originally justified by some geometric applications, leads to an unexpected bridge: the weak maximum principle (for the Laplace-Beltrami operator Δ) on a possibly nongeodesically complete manifold $(M, \langle\, ,\, \rangle)$ is equivalent to stochastic completeness of the Brownian motion (associated to Δ). This fact, beside the many applications, has a deep theoretical value which however we do not exploit here. The introduction of the weak maximum principle enables us also to shed light on the notion of parabolicity that we present as a stronger version of the former. This well explain the often apparently strange similarities between various phenomena linked, respectively, to stochastic completeness and recurrence of the Brownian motion.

In Theorem 2.5 we give a sufficient condition for the validity of the Omori-Yau maximum principle in terms of curvature conditions, thus involving the distance function r from a fixed reference point $o \in M$. We do this with the intent to introduce, in this most simple case of the Hessian and the Laplace-Beltrami operators, a technique to deal with the cut locus and with solutions in the weak sense that we will address later on, when considering the very general family of operators introduced in Chap. 3.

L.J. Alías et al., *Maximum Principles and Geometric Applications*,
Springer Monographs in Mathematics, DOI 10.1007/978-3-319-24337-5_2

2.1 Some Preliminary Considerations

Let $u : [a, b] \to \mathbb{R}$ be a continuous function. Then u attains its maximum u^* at some point $x_0 \in [a, b]$. If $x_0 \in (a, b)$ and u has continuous second derivative in a neighborhood of x_0, then

$$\text{(i) } u'(x_0) = 0 \text{ and (ii) } u''(x_0) \leq 0. \tag{2.1}$$

It follows easily that, if u satisfies a differential inequality of the type

$$u''(x) + g(x)u'(x) > 0 \tag{2.2}$$

on the open interval (a, b), where g is any bounded function, then either $x_0 = a$ or $x_0 = b$. Otherwise one would get

$$0 < u''(x_0) + g(x_0)u'(x_0) \leq 0.$$

Note, however, that if we relax (2.2) to the nonstrict inequality

$$u''(x) + g(x)u'(x) \geq 0 \tag{2.3}$$

on (a, b), then the constant solutions $u \equiv c$ are admitted, and for such a solution the maximum is attained at any point in $[a, b]$. The content of the usual maximum principle is the fact that this exception is the only possible, and it is stated in the following form.

Theorem 2.1 *Let $u : [a, b] \to \mathbb{R}$ be a twice continuously differentiable function satisfying*

$$u''(x) + g(x)u'(x) \geq 0$$

on (a, b), where g is any bounded function. Then, u cannot have an interior maximum in (a, b), unless u is constant.

The argument, due to Hopf, to prove Theorem 2.1 is a tricky way to pass from the nonstrict inequality (2.3) to the strict inequality (2.2) for a new function v properly related to u. Then one concludes with the aid of the previous discussion (see, for instance, Theorem 1 on page 2 in the classical book of Protter and Weinberger [233]). Thus, the core of the maximum principle indeed relies on $u(x_0) = u^*$ and conditions (i) and (ii) in (2.1).

Substituting $[a, b] \subset \mathbb{R}$ with a compact Riemannian manifold M without boundary, for any given function $u \in C^2(M)$, there exists a point $x_0 \in M$ such that

$$\text{(i) } u(x_0) = u^*, \;\; \text{(ii) } |\nabla u(x_0)| = 0, \;\; \text{and (iii) } \Delta u(x_0) \leq 0, \tag{2.4}$$

where $u^* = \sup_M u < +\infty$, or, more generally,

$$\text{(i) } u(x_0) = u^*, \text{ (ii) } |\nabla u(x_0)| = 0, \text{ and (iii)' Hess }(u)(x_0) \leq 0, \tag{2.5}$$

in the sense that

$$\text{Hess }(u)(x_0)(v, v) \leq 0 \quad \text{for all } v \in T_{x_0}M.$$

As we know from Chap. 1, ∇, Δ and Hess stand here, respectively, for the gradient, the Laplacian and the Hessian operators on the Riemannian manifold $(M, \langle\, , \rangle)$. Following Yau, the validity of either (2.4) or (2.5) on M is called the *usual maximum principle* (equivalently, the *finite maximum principle*). To immediately point out its importance let us recall the following typical application in the context of classical surface theory, that is, the proof that every compact surface in $\mathbb{R}^3$ has an *elliptic point*, in other words, a point where the Gaussian curvature is positive (see Corollary 5.1 and Proposition 5.1). In particular, no compact Riemannian surface with nonpositive Gaussian curvature, for instance a minimal surface, can be isometrically immersed into $\mathbb{R}^3$.

Obviously, when M is not compact it is not always possible, given a continuous function $u : M \to \mathbb{R}$ with $u^* = \sup_M u < +\infty$, to find a point $x_0 \in M$ such that $u(x_0) = u^*$. Nevertheless, if $u : \mathbb{R} \to \mathbb{R}$ is a twice continuously differentiable function with $u^* < +\infty$, then it is not difficult to realize the existence of a sequence $\{x_k\}_{k\in\mathbb{N}} \subset \mathbb{R}$ with the following properties:

$$\text{(i) } u(x_k) > u^* - \frac{1}{k}, \text{ (ii) } |u'(x_k)| < \frac{1}{k}, \text{ and (iii) } u''(x_k) < \frac{1}{k}$$

for each $k \in \mathbb{N}$. More generally, given a twice continuously differentiable function $u : \mathbb{R}^m \to \mathbb{R}$, $m \geq 1$, with $u^* < +\infty$, there exists a sequence $\{x_k\}_{k\in\mathbb{N}} \subset \mathbb{R}^m$ such that

$$\text{(i) } u(x_k) > u^* - \frac{1}{k}, \text{ (ii) } |\nabla u(x_k)| < \frac{1}{k}, \text{ and (iii) } \Delta u(x_k) < \frac{1}{k} \tag{2.6}$$

for each $k \in \mathbb{N}$. The main idea to prove this result goes back to Ahlfors [1] and even earlier, and it consists in considering a family of functions each of which attains a maximum at some point of $\mathbb{R}^m$ and then apply the usual maximum principle. For instance, to prove (2.6) we fix a sequence $\{\varepsilon_i\} \searrow 0^+$ and define

$$u_i(x) = u(x) - \varepsilon_i |x|^2.$$

Clearly, u_i takes its absolute maximum at some point $x_i \in \mathbb{R}^m$, where

$$\nabla u_i(x_i) = 0 \text{ and } \Delta u_i(x_i) \leq 0.$$

Since in $\mathbb{R}^m$ we have $\nabla|x|^2 = 2x$ and $\Delta|x|^2 = 2m$, we obtain

$$\nabla u(x_i) = 2\varepsilon_i x_i \tag{2.7}$$

and

$$\Delta u(x_i) \leq 2m\varepsilon_i. \tag{2.8}$$

On the other hand,

$$u(x_i) - \varepsilon_i|x_i|^2 = u_i(x_i) \geq u_i(0) = u(0),$$

and therefore

$$\varepsilon_i|x_i|^2 \leq u(x_i) - u(0) \leq u^* - u(0) \leq C$$

for some positive constant C. It then follows that

$$|x_i| \leq \sqrt{\frac{C}{\varepsilon_i}},$$

which jointly with (2.7) implies

$$|\nabla u(x_i)| \leq 2\sqrt{C\varepsilon_i}. \tag{2.9}$$

To conclude, fix $k \in \mathbb{N}$. Then, there exists a point $y_k \in \mathbb{R}^m$ such that

$$u(y_k) > u^* - \frac{1}{2k}.$$

For every $i \in \mathbb{N}$ we have

$$u_i(x_i) = u(x_i) - \varepsilon_i|x_i|^2 \geq u_i(y_k) = u(y_k) - \varepsilon_i|y_k|^2 > u^* - \frac{1}{2k} - \varepsilon_i|y_k|^2,$$

that is,

$$u(x_i) > u^* - \frac{1}{2k} - \varepsilon_i|y_k|^2 + \varepsilon_i|x_i|^2 \geq u^* - \frac{1}{2k} - \varepsilon_i|y_k|^2 \tag{2.10}$$

for every $i \in \mathbb{N}$. Choosing now $i = i_k$ sufficiently large such that

$$\varepsilon_{i_k}|y_k|^2 < \frac{1}{2k}, \quad 2\sqrt{C\varepsilon_{i_k}} < \frac{1}{k} \quad \text{and } 2m\varepsilon_{i_k} < \frac{1}{k},$$

from (2.10), (2.9) and (2.8) it follows, respectively, that

$$u(x_{i_k}) > u^* - \frac{1}{k}, \quad |\nabla u(x_{i_k})| < \frac{1}{k} \quad \text{and } \Delta u(x_{i_k}) < \frac{1}{k}.$$

Therefore, the choice $x_k = x_{i_k}$ completes the proof.

In the previous argument there are two important facts that need to be stressed. The first is the equality

$$\Delta|x|^2 = 2m,$$

which is tightly related to the geometry of $\mathbb{R}^m$. The second is the linearity of the Laplacian operator for which we have been able to perform the following computation:

$$\Delta u_k = \Delta u - \varepsilon_k \Delta|x|^2.$$

Of course, it is possible to reformulate (2.6) on an m-dimensional Riemannian manifold M. In this general context, it is not difficult to see that if the manifold is, for instance, complete then for any smooth function $u \in C^2(M)$ with $u^* < +\infty$ one can always find a sequence of points $\{x_k\}_{k\in\mathbb{N}} \subset M$ satisfying (i) and (ii) in (2.6). This is a direct consequence of the following general fact.

Proposition 2.1 *Let M be a Riemannian manifold and let $u \in C^2(M)$ be such that $u^* < +\infty$. Given $\varepsilon > 0$, let $y \in M$ satisfy $u(y) > u^* - \varepsilon^2$ and suppose that the closed ball $\overline{B_\varepsilon(y)}$ is compact. Then, there exists a point $x \in \overline{B_\varepsilon(y)}$ with the following properties*

$$\textit{(i)}\ u(x) \geq u(y), \ \textit{and (ii)}\ |\nabla u(x)| \leq \varepsilon.$$

For a geometric differential proof and the need of compactness of the closed ball $\overline{B_\varepsilon(y)}$ see [227, Proposition 1.7]. Here we will provide a different argument.

In Proposition 2.1 the alert reader has certainly recognized a form of the *Ekeland quasi-minimum variational principle* (of course written in the form of a quasi-maximum principle). We give here a simple proof due to Crandall, as reported in [110, p. 444].

Theorem 2.2 *Let (M, d) be a complete metric space and $u : M \to \mathbb{R}$ an upper semicontinuous function with $u^* = \sup_M u < +\infty$. Fix $\varepsilon, \delta > 0$ and let $y \in M$ satisfy*

$$u(y) \geq u^* - \varepsilon. \tag{2.11}$$

Then, there exists $x \in M$ such that

(i) $d(x, y) \leq \delta$,
(ii) $u(x) \geq u(y)$, and
(iii) for every $z \neq x$, $u(z) < u(x) + \frac{\varepsilon}{\delta} d(x, z)$.

Proof We define a sequence $\{x_n\} \subset M$ as follows. We set $x_0 = y$ and let us suppose to have chosen x_n. Then either

$$\text{for each } z \neq x_n, u(z) < u(x_n) + \frac{\varepsilon}{\delta} d(x_n, z), \tag{2.12}$$

and in this case we set $x_{n+1} = x_n$ or

$$\text{there exists } z \neq x_n \text{ such that } u(z) \geq u(x_n) + \frac{\varepsilon}{\delta} d(x_n, z). \tag{2.13}$$

In this latter case we define

$$S_n = \{z \neq x : u(z) \geq u(x_n) + \frac{\varepsilon}{\delta} d(x_n, z)\}.$$

Note that

$$u(x_n) < \sup_{S_n} u \leq u^* < +\infty$$

and therefore

$$u(x_n) - \sup_{S_n} u < 0.$$

We then choose $x_{n+1} \in S_n$ such that

$$u(x_{n+1}) \geq \sup_{S_n} u + \frac{1}{2}(u(x_n) - \sup_{S_n} u) = \frac{1}{2} u(x_n) + \frac{1}{2} \sup_{S_n} u. \tag{2.14}$$

We now show that the sequence $\{x_n\}$ is a Cauchy sequence. First we observe that if (2.12) holds for a certain n_0, then $x_n = x_{n_0}$ for each $n \geq n_0$ and the sequence is clearly Cauchy. If this is not the case then (2.13) holds for each n. Then, according to (2.13), we have

$$\frac{\varepsilon}{\delta} d(x_n, x_{n+1}) \leq u(x_{n+1}) - u(x_n) \text{ for each } n. \tag{2.15}$$

Let $p \geq n$. Summing up, using (2.14) and the triangle inequality, we get

$$\frac{\varepsilon}{\delta} d(x_n, x_p) \leq \frac{\varepsilon}{\delta} d(x_n, x_{n+1}) + \cdots + \frac{\varepsilon}{\delta} d(x_{p-1}, x_p) \leq u(x_p) - u(x_n). \tag{2.16}$$

Thus the sequence $\{u(x_n)\} \subset \mathbb{R}$ is nondecreasing and bounded above by u^*. It follows that it converges and (2.16) yields that $\{x_n\} \subset M$ is a Cauchy sequence.

Completeness of (M, d) implies that $x_n \to x \in M$ as $n \to +\infty$. We show that x satisfies items (i)–(iii) in the statement of the theorem. Since u is upper semicontinuous and $u(x_n)$ is nondecreasing, we have

$$u(x) \geq \limsup_{n\to+\infty} u(x_n) = \lim_{n\to+\infty} u(x_n) \geq u(x_0) = u(y),$$

proving (ii).

To prove (i), we chose $n = 0$ in (2.16) and we use (2.11) to get

$$\frac{\varepsilon}{\delta} d(y, x_p) \leq u(x_p) - u(y) \leq u^* - u(y) \leq \varepsilon.$$

Therefore

$$d(y, x_p) \leq \delta$$

and letting $p \to +\infty$ we deduce the validity of (i). Next, if (iii) were false there would exist $z \neq x$ such that

$$u(z) \geq u(x) + \frac{\varepsilon}{\delta} d(x, z). \tag{2.17}$$

Letting $p \to +\infty$ into (2.16) we obtain

$$\frac{\varepsilon}{\delta} d(x_n, x) \leq u(x) - u(x_n) \tag{2.18}$$

and therefore, using (2.17) and (2.18) we obtain

$$u(z) \geq u(x) + \frac{\varepsilon}{\delta} d(x, z) \geq u(x_n) + \frac{\varepsilon}{\delta}(d(x_n, x) + d(x, z)) \geq u(x_n) + \frac{\varepsilon}{\delta} d(x_n, z),$$

that is, $z \in S_n$ for each n.

On the other hand, from (2.14)

$$2u(x_{n+1}) - u(x_n) \geq \sup_{S_n} u \geq u(z)$$

for each n, and, letting $n \to +\infty$,

$$u(x_n) \to \ell \quad \text{and} \quad \ell \geq u(z).$$

Since u is upper semicontinuous

$$u(x) \geq \ell \geq u(z)$$

contradicting (2.17). □

Remark 2.1 Often the conclusion of Theorem 2.2 is stated in the weaker form of the validity of

(jj) $u(x) \geq u(y)$, and
(jjj) for every $z \neq x$, $u(z) < u(x) + \varepsilon d(x, z)$.

It can be proved, see [126], that if this conclusion is true for each upper semicontinuous function $u : M \to \mathbb{R} \cup \{-\infty\}$, $u \not\equiv -\infty$, on a metric space (M, d), then the latter is necessarily complete.

We next provide, using Ekeland principle, that is Theorem 2.2, a proof of a stronger form of the claim preceding Proposition 2.1.

Proposition 2.2 *Let $(M, \langle\,,\,\rangle)$ be a complete manifold and $u : M \to \mathbb{R}$ a C^1 function such that $u^* = \sup_M u < +\infty$. Then, for every sequence $\{y_n\} \subset M$ such that $u(y_n) \to u^*$ as $n \to +\infty$ there exists a sequence $\{x_n\} \subset M$ with the properties*

(i) $u(x_n) \to u^*$,
(ii) $|\nabla u(x_n)| \to 0$ *and*
(iii) $d(x_n, y_n) \to 0$

as $n \to +\infty$.

Proof For each $n \in \mathbb{N}$, let $\varepsilon_n = u^* - u(y_n)$ and $\delta_n = \sqrt{\varepsilon_n}$. If $\varepsilon_n = 0$ we choose $x_n = y_n$, otherwise $\varepsilon_n > 0$ and by Theorem 2.2 there exists x_n such that

$$(i)\, u(y_n) \leq u(x_n); \quad (ii)\, d(x_n, y_n) \leq \varepsilon_n, \tag{2.19}$$

and

$$\text{for every } z \neq x_n, \quad u(z) < u(x_n) + \sqrt{\varepsilon} d(x_n, z). \tag{2.20}$$

Now fix $v \in T_{x_n}M$, $|v| = 1$, and let $\gamma : (-\alpha, \alpha) \to M$, $\alpha > 0$, be the unit speed geodesic such that $\gamma(0) = x_n$ and $\dot{\gamma}(0) = v$. We can assume to have chosen α so small that γ realizes the distance between $\gamma(0)$ and any other of its points and furthermore $\gamma(t) \neq \gamma(0)$ for every $t \in (-\alpha, \alpha)$, $t \neq 0$. Let $z = \gamma(t)$ so that from (2.20) we get

$$u(\gamma(t)) - u(\gamma(0)) < \sqrt{\varepsilon_n} d(\gamma(t), x_n) = \sqrt{\varepsilon_n} |t|.$$

Since $u \in C^1(M)$, from here it follows immediately that

$$|\langle \nabla u(x_n), v \rangle| \leq \sqrt{\varepsilon_n}$$

for every $v \in T_{x_n}M$, $|v| = 1$, and therefore

$$|\nabla u(x_n)| \leq \sqrt{\varepsilon_n}. \tag{2.21}$$

Now, letting $n \to +\infty$, $\varepsilon_n, \delta_n \to 0$ and (i)–(iii) follow from (2.19) and (2.21). □

Remark 2.2 Completeness of $(M, \langle\, ,\, \rangle)$ cannot be avoided. Indeed, for $M = \mathbb{R}^m \setminus \{0\}$ with the induced Euclidean metric, the function $u(x) = e^{-|x|}$ on M is such that $1 = \sup_M u = \lim_{|x| \to 0} e^{-|x|}$, while $\lim_{|x| \to 0} |\nabla u|(x) = 1$. See also Remark 2.1.

The following example shows that in general there might be no sequences satisfying all the three conditions in (2.6) at the same time, and points to the fact that some geometric conditions on M need to be imposed in order to obtain the validity of the whole (2.6). The choice of the dimension $m = 2$ is made to simplify the writing.

Example 2.1 Let M_g be the 2-dimensional model with metric given in polar coordinates, outside the origin o, by

$$dr^2 + g(r)^2 d\theta^2, \tag{2.22}$$

where $d\theta^2$ is the standard metric of $\mathbb{S}^1$ and $g \in C^\infty(\mathbb{R}_0^+)$ is such that $g(r) > 0$ for $r > 0$ and

$$g(r) = \begin{cases} r & \text{if } 0 \le r < 1, \\ r(\log r)^{1+\mu} e^{r^2 (\log r)^{1+\mu}} & \text{if } r > 3. \end{cases}$$

for some positive constant μ. As we observed in Sect. 1.7 of Chap. 1, the behaviour of g near 0 guarantees that the metric (2.22) can be smoothly defined on all of M_g. Furthermore, observe also that this metric is obviously complete. Let

$$\alpha(r) = \int_0^r \left(\frac{1}{g(t)} \int_0^t g(s) ds \right) dt$$

and consider the function given by

$$u(x) = \alpha(r(x)).$$

Then, $u \in C^2(M)$ and it satisfies

$$\Delta u = \alpha'(r) \Delta r + \alpha''(r) \equiv 1.$$

Therefore, in this case property (iii) in (2.6) cannot hold; however, since $\mu > 0$, an easy check shows that $u^* < +\infty$. It is worth pointing out that in this example the Gaussian curvature K and the volume growth of the geodesic ball $B_R = B_R(o)$ have the asymptotic behaviours

$$K(r) = -\frac{g''(r)}{g(r)} \sim -c^2 r^2 (\log r)^{2(1+\mu)} \quad \text{as } r \to +\infty$$

for some constant $c > 0$, and

$$\mathrm{vol}(B_R) \sim \frac{1}{2} e^{R^2 (\log R)^{(1+\mu)}} \quad \text{as } R \to +\infty.$$

Thus it seems reasonable to expect the failure of (2.6) in case of a fast divergence of the curvature to $-\infty$ or in case of a strong growth of the volume of the geodesic balls of exploding radius. The results of Chaps. 3 and 4, together with the subsequent geometric applications, will point out a more intricate and subtle situation.

2.2 The Generalized Omori-Yau Maximum Principle

In [210] Omori proved that if $(M, \langle\, ,\, \rangle)$ is a complete Riemannian manifold with *sectional curvature bounded from below*, then for any smooth function $u \in C^2(M)$ with $u^* < +\infty$ there exists a sequence of points $\{x_k\}_{k\in\mathbb{N}} \subset M$ satisfying

$$\text{(i) } u(x_k) > u^* - \frac{1}{k}, \quad \text{(ii) } |\nabla u(x_k)| < \frac{1}{k}, \quad \text{and (iii)' } \operatorname{Hess}(u)(x_k) < \frac{1}{k}\langle\, ,\, \rangle, \tag{2.23}$$

in the sense of quadratic forms, that is,

$$\operatorname{Hess}(u)(x_k)(v, v) < \frac{1}{k}|v|^2 \quad \text{for all } v \in T_{x_k}M, \; v \neq 0.$$

Later on, Yau [279] (see also Cheng and Yau [81]) gave a version of this result for complete Riemannian manifolds with *Ricci curvature bounded from below*, replacing condition (iii)' in (2.23) with condition (iii) in (2.6). For this reason, and following the terminology introduced by Pigola et al. in [227], we state the following definition.

Definition 2.1 Let $(M, \langle\, ,\, \rangle)$ be a (not necessarily complete) Riemannian manifold. The *Omori-Yau maximum principle for the Laplacian* is said to hold on M if for any function $u \in C^2(M)$ with $u^* = \sup_M u < +\infty$ there exists a sequence of points $\{x_k\}_{k\in\mathbb{N}} \subset M$ satisfying

$$\text{(i) } u(x_k) > u^* - \frac{1}{k}, \quad \text{(ii) } |\nabla u(x_k)| < \frac{1}{k}, \quad \text{and (iii) } \Delta u(x_k) < \frac{1}{k} \tag{2.24}$$

for each $k \in \mathbb{N}$. Equivalently, for any function $u \in C^2(M)$ with $u_* = \inf_M u > -\infty$ there exists a sequence of points $\{x_k\}_{k\in\mathbb{N}} \subset M$ with the properties

$$\text{(i) } u(x_k) < u_* + \frac{1}{k}, \quad \text{(ii) } |\nabla u(x_k)| < \frac{1}{k}, \quad \text{and (iii) } \Delta u(x_k) > -\frac{1}{k} \tag{2.25}$$

for each $k \in \mathbb{N}$. In the case where the stronger statement (iii)' in (2.23) concerning the Hessian is satisfied, we say that the *Omori-Yau maximum principle for the Hessian* holds on M.

With this terminology, the results given by Omori [210] and Yau [279] can be stated as follows.

Theorem 2.3

(i) The Omori-Yau maximum principle for the Hessian holds on every complete Riemannian manifold with sectional curvature bounded from below.
(ii) The Omori-Yau maximum principle for the Laplacian holds on every complete Riemannian manifold with Ricci curvature bounded from below.

More generally, as observed by Pigola, Rigoli and Setti in [227], the validity of the Omori-Yau maximum principle does not depend on curvature bounds as much as one would expect. Actually, a condition to guarantee the validity of (2.24) or (2.23) can be expressed in a function theoretic form. This is the content of a generalization of Theorem 2.3 due to Pigola et al. [227, Theorem 1.9]. See also [239] for the underlying ideas of the proof.

Recently, this latter result has been improved by Albanese et al. [5] to the following

Theorem 2.4 *The Omori-Yau maximum principle for the Laplacian holds on every Riemannian manifold $(M, \langle\,,\,\rangle)$ admitting a C^2 function $\gamma : M \to \mathbb{R}$ satisfying the following requirements:*

(i) $\gamma(x) \to +\infty$ *as* $x \to \infty$*;*
(ii) $|\nabla\gamma| \leq G(\gamma)$ *outside a compact subset of* M*;*
(iii) $\Delta\gamma \leq G(\gamma)$ *outside a compact subset of* M*,*

with $G \in C^1(\mathbb{R}^+)$*, positive near infinity and such that*

$$\frac{1}{G} \notin L^1(+\infty) \quad \text{and} \quad G'(t) \geq -A(\log t + 1),$$

for $t \gg 1$ *and some* $A \geq 0$*. An analogous statement holds for the case of the Omori-Yau maximum principle for the Hessian, by replacing assumption (iii) above with*

(iii)' $\operatorname{Hess}(\gamma) \leq G(\gamma)\langle\,,\,\rangle$ *(in the sense of quadratic forms) outside a compact subset of* M*.*

Remark 2.3 As observed in Theorem 3.5 of Chap. 3 the requirement $u^* < +\infty$ for the validity of the maximum principle can be relaxed to $u(x) = o(\gamma(x))$ as $x \to \infty$.

Remark 2.4 Especially significant examples of functions G satisfying the conditions in Theorem 2.4 are given by

$$G(t) = t \prod_{j=1}^{N} \log^{(j)}(t), \ t \gg 1,$$

where $\log^{(j)}$ stands for the j-th iterated logarithm.

Remark 2.5 It is also worth pointing out that although in the statement of Theorem 2.4 the manifold M is not required to be geodesically complete, the two assumptions (i) and (ii) imply it. See the proof of Theorem 3.2.

Remark 2.6 The proof of Theorem 2.4 shows that one needs γ to be C^2 only in a neighborhood of certain points in a set Z and that one also needs the validity of (ii) and (iii) or (iii)' there. In the important situation where γ is the composition of an appropriate function with the Riemannian distance from a fixed reference point o, this is the case if Z does not intersect the cut locus of o. Otherwise, elaborating on a trick of Calabi [55] one can solve the problem. We will consider this in Theorem 2.5 below. Note that in the case of the Laplacian, since we have an upper bound for Δr which holds in the weak sense on the entire manifold we can also use a second argument which is contained in the proof of Theorem 3.11 via the comparison Theorem 3.5 of [236] or Proposition 3.1 below.

2.2.1 Two Significant Examples

Of course we expect that the most natural examples of functions γ should be built via the Hessian and Laplacian comparison theorems through the distance function r to a fixed origin $o \in M$. However, in general r is only Lipschitz on M. Fortunately enough the result holds true also in this case as expressed in the next Theorem 2.5. Its proof also yields the validity of Theorem 2.4, at least for $A = 0$, while for $A > 0$ see Remark 3.2 of Chap. 3. We also observe that the technique we introduce here will reveal basic in extending the maximum principle to more general operators. We elaborate on an old idea of Calabi, [55], known as "Calabi trick", contained in the next

Lemma 2.1 *Let $r : M \to \mathbb{R}_0^+$ be the distance function from the point o in the complete manifold $(M, \langle\,,\,\rangle)$. Suppose that r is not differentiable at q and let $\sigma : [0, l] \to M$ be a unit speed geodesic such that $\sigma(0) = o$, $\sigma(l) = q$ and with $l = r(q)$. Fix $\varepsilon > 0$ sufficiently small and let $o_\varepsilon = \sigma(\varepsilon)$. Then $r_\varepsilon(x) = \mathrm{dist}\,(x, o_\varepsilon)$ is smooth at q.*

Proof The following argument is taken from Petersen's book [219, p. 284]. By contradiction suppose that $r_\varepsilon(x)$ is not smooth at q. Then it is well known (see for instance [219, Chap. 5]), that either

(i) there are (at least) two minimizing geodesics from o_ε to q
or
(ii) q is a critical value for $\exp_{o_\varepsilon}$, the exponential map at o_ε.

In case (i) we would have a nonsmooth curve of length l from p to q, which is not possible. Thus case (ii) must hold. To obtain a contradiction we show that this implies that $\exp_q$ has $o_\varepsilon = \sigma(\varepsilon)$ as a critical value. Since q is a critical value for $\exp_{o_\varepsilon}$ there exists a Jacobi field $J : [\varepsilon, l] \to TM$ along $\sigma|_{[\varepsilon,l]}$ such that $J(\varepsilon) = 0$, $J'(\varepsilon) \neq 0$ and $J(l) = 0$. Then, also $J'(l) \neq 0$ since J solves a second order linear equation. Running backwards from q to $o_\varepsilon = \sigma(\varepsilon)$ shows that $\exp_q$ is critical at $\sigma(\varepsilon)$. This contradicts the minimality of $\sigma : [0, l] \to M$. □

Theorem 2.5 *Let $(M, \langle\,,\,\rangle)$ be a complete, noncompact, Riemannian manifold of dimension m; let $o \in M$ be a reference point and denote by $r(x)$ the Riemannian distance function from o. Assume that the sectional curvature of M satisfies*

$$^{M}K(x) \geq -G^2(r(x)), \tag{2.26}$$

where $G \in C^1(\mathbb{R}_0^+)$ satisfies

$$(i)\, G(0) > 0, \quad (ii)\, G'(t) \geq 0, \quad (iii)\, \frac{1}{G(t)} \notin L^1(+\infty). \tag{2.27}$$

Then the Omori-Yau maximum principle for the Hessian holds on M.

If we only assume, instead of (2.26), that the Ricci curvature satisfies

$$\mathrm{Ric} \geq -(m-1)G^2(r)\langle\,,\,\rangle, \tag{2.28}$$

then the Omori-Yau maximum principle for the Laplacian holds on M.

Remark 2.7 As it will become apparent from the proof, in case $o \in M$ is a pole (that is, $\mathrm{cut}(o) = \emptyset$), (2.26) and (2.28) can be replaced, respectively, with

$$K_{\mathrm{rad}}(x) \geq -G^2(r(x)), \tag{2.29}$$

and

$$\mathrm{Ric}(\nabla r, \nabla r) \geq -(m-1)G^2(r). \tag{2.30}$$

Here K_{rad} is the radial sectional curvature of M, that is, the sectional curvature of 2-planes containing ∇r. Observe that the last part of the theorem holds in the only assumption (2.30) also in case $o \in M$ is not a pole; see Remark 2.8 below.

Proof (of Theorem 2.5) Let $D_o = M \setminus \mathrm{cut}(o)$ be the domain of normal geodesic coordinates centered at o. On D_o, from (2.26) and the general Hessian comparison theorem, Theorem 1.4, we have

$$\mathrm{Hess}\,(r) \leq \frac{g'(r)}{g(r)}\left(\langle\,,\,\rangle - dr \otimes dr\right) \tag{2.31}$$

where $g(t)$ is the solution on $\mathbb{R}_0^+$ of the Cauchy problem

$$\begin{cases} g''(t) - G^2(t)g(t) = 0 & \text{on } \mathbb{R}_0^+, \\ g(0) = 0, \quad g'(0) = 1. \end{cases} \tag{2.32}$$

Observe that $g > 0$ and $g' > 0$ on $(0, +\infty)$. Actually, since $g(0) = 0$ it suffices to prove that $g' > 0$ on $(0, +\infty)$. Suppose, to the contrary, that there exists a first $t_0 > 0$ such that $g'(t_0) = 0$. Thus on $(0, t_0)$ we have $g > 0$ and $g' > 0$. Then $g''(t) = G(t)^2 g(t) \geq 0$ on $(0, t_0)$, and

$$g'(t_0) - g'(0) = 0 - 1 = \int_0^{t_0} g''(t)dt \geq 0,$$

which is a contradiction. Letting

$$\psi(t) = \frac{1}{G(0)}\left(e^{\int_0^t G(s)ds} - 1\right) \tag{2.33}$$

we have $\psi(0) = 0$, $\psi'(0) = 1$ and

$$\psi'' - G^2(t)\psi = \frac{1}{G(0)}\left(G^2(t) + G'(t)e^{\int_0^t G(s)ds}\right) \geq 0,$$

that is, ψ is a subsolution of (2.32). By the Sturm comparison theorem

$$\frac{g'(t)}{g(t)} \leq \frac{\psi'(t)}{\psi(t)} = G(t)\frac{e^{\int_0^t G(s)ds}}{e^{\int_0^t G(s)ds} - 1}. \tag{2.34}$$

Thus, for every $v \in T_xM$, we have

$$\mathrm{Hess}\,(r)(x)(v, v) \leq G(r(x))\frac{e^{\int_0^{r(x)} G(s)ds}}{e^{\int_0^{r(x)} G(s)ds} - 1}\left(|v|^2 - \langle \nabla r(x), v\rangle^2\right).$$

Now $|v|^2 - \langle \nabla r(x), v\rangle^2 \geq 0$; hence since $G > 0$ and $G' \geq 0$

$$\mathrm{Hess}\,(r)(x)(v, v) \leq G(r(x) + 1)\frac{e^{\int_0^{r(x)} G(s)ds}}{e^{\int_1^{r(x)} G(s)ds} - 1}\left(|v|^2 - \langle \nabla r(x), v\rangle^2\right), \tag{2.35}$$

for, say, $r(x) \geq 2$. Define

$$\varphi(t) = \int_0^t \frac{ds}{G(s+1)} \tag{2.36}$$

so that

$$\varphi'(t) = \frac{1}{G(t+1)} \quad \text{and} \quad \varphi''(t) \leq 0. \tag{2.37}$$

Set

$$\gamma(x) = \varphi(r(x)) \quad \text{on } M \setminus \overline{B}_2$$

and note that

$$\gamma(x) \to +\infty \quad \text{as } x \to \infty \tag{2.38}$$

because $\varphi(t) \to +\infty$ as $t \to +\infty$ since $1/G \notin L^1(+\infty)$. We also observe that, from $G \notin L^1(+\infty)$, we have

$$0 \leq \sup_{t \geq 2} \frac{e^{\int_0^t G(s)ds}}{e^{\int_1^t G(s)ds} - 1} = \Lambda < +\infty. \tag{2.39}$$

Therefore, using (2.35), (2.37) and (2.39) we deduce that for each $x \in D_o \cap (M \setminus \overline{B}_2)$ and $v \in T_xM$

$$\begin{aligned}\operatorname{Hess}(\gamma)(x)(v,v) &= \varphi'(r(x)) \operatorname{Hess}(r)(x)(v,v) + \varphi''(r(x))\langle v, \nabla r(x)\rangle^2 \\ &\leq \Lambda \left(|v|^2 - \langle v, \nabla r(x)\rangle^2\right),\end{aligned}$$

that is,

$$\operatorname{Hess}(\gamma)(v,v) \leq \Lambda |v|^2. \tag{2.40}$$

Furthermore, observe that

$$|\nabla \gamma| = \frac{1}{G(r+1)} \leq \frac{1}{G(1)} \leq \Lambda, \tag{2.41}$$

up to choosing Λ in (2.40) sufficiently large. Let now $u \in C^2(M)$ with

$$u^* = \sup_M u < +\infty. \tag{2.42}$$

For a fixed $\eta > 0$ consider the sets

$$A_\eta = \{x \in M : u(x) > u^* - \eta\} \tag{2.43}$$

and

$$B_\eta = \{x \in A_\eta : |\nabla u(x)| < \eta\}. \tag{2.44}$$

Since $(M, \langle\, ,\, \rangle)$ is complete, from Ekeland quasi-minimum principle (precisely, Proposition 2.2), we deduce $B_\eta \neq \emptyset$. We have to show that

$$\inf_{B_\eta} \mathrm{Hess}\,(u)(x) \leq 0 \tag{2.45}$$

in the sense of symmetric bilinear forms. To prove (2.45) we reason by contradiction and we suppose that there exists $\sigma_0 > 0$ such that for each $x \in B_\eta$ there exists $\bar{v} \in T_xM$, $|\bar{v}| = 1$, such that

$$\mathrm{Hess}\,(u)(x)(\bar{v}, \bar{v}) \geq \sigma_0. \tag{2.46}$$

First we observe that u^* cannot be attained at any point $x_0 \in M$, for otherwise $x_0 \in B_\eta$ and since $\mathrm{Hess}(u)(x_0)$ has to be negative semi-definite we contradict (2.46). We set

$$\Omega_t = \{x \in M : \gamma(x) > t\}.$$

Then $\Omega_t^c = M \setminus \Omega_t$ is closed and hence compact by (2.38). Define

$$u_t^* = \max_{x \in \Omega_t^c} u(x).$$

Since u^* is not attained on M and $\{\Omega_t^c\}$ is a telescoping family exhausting M, there exists a divergent sequence $\{t_j\}_{j \in \mathbb{N}} \subset \mathbb{R}^+$ such that

$$u_{t_j}^* \to u^* \quad \text{as } j \to +\infty, \tag{2.47}$$

and $T_1 > 0$ sufficiently large that $u_{T_1}^* > u^* - \eta/2$ and $\Omega_{T_1} \subset M \setminus \overline{B}_2$. In particular (2.40) and (2.41) hold on $\Omega_{T_1} \cap D_o$. Choose α such that $u_{T_1}^* < \alpha < u^*$. Because of (2.47) we can find j sufficiently large such that $T_2 = t_j > T_1$ and $u_{T_2}^* > \alpha$. Then, we select $\delta > 0$ small enough to have

$$\alpha + \delta < u_{T_2}^*. \tag{2.48}$$

For $\sigma > 0$ define

$$\gamma_\sigma(x) = \alpha + \sigma(\gamma(x) - T_1).$$

Then

$$\gamma_\sigma(x) = \alpha \quad \text{for every } x \in \partial\Omega_{T_1},$$

and for σ sufficiently small, from (2.40), (2.41) we have

$$\text{Hess}\,(\gamma_\sigma)(x) = \sigma\,\text{Hess}\,(\gamma)(x) \le \sigma\Lambda < \sigma_0 \quad \text{on } D_o \cap \Omega_{T_1}, \tag{2.49}$$

$$|\nabla\gamma_\sigma| = \sigma|\nabla\gamma| \le \sigma\Lambda < \eta \quad \text{on } D_o \cap \Omega_{T_1}. \tag{2.50}$$

On $\Omega_{T_1} \setminus \Omega_{T_2}$

$$\alpha \le \gamma_\sigma(x) \le \alpha + \sigma(T_2 - T_1). \tag{2.51}$$

Thus, choosing $\sigma > 0$ sufficiently small that

$$\sigma(T_2 - T_1) < \delta, \tag{2.52}$$

we obtain

$$\alpha \le \gamma_\sigma(x) < \alpha + \delta \quad \text{on } \Omega_{T_1} \setminus \Omega_{T_2}.$$

For $x \in \partial\Omega_{T_1}$, $\gamma_\sigma(x) = \alpha > u^*_{T_1} \ge u(x)$. Hence

$$(u - \gamma_\sigma)(x) < 0 \quad \text{on } \partial\Omega_{T_1}. \tag{2.53}$$

Let $\bar{x} \in \Omega_{T_1} \setminus \Omega_{T_2}$ be such that $u(\bar{x}) = u^*_{T_2} > \alpha + \delta$. Using (2.52) and (2.51) we deduce

$$(u - \gamma_\sigma)(\bar{x}) \ge u^*_{T_2} - \alpha - \sigma(T_2 - T_1) > u^*_{T_2} - \alpha - \delta > 0.$$

Moreover, from (2.38) and $u^* < +\infty$, for $T_3 > T_2$ sufficiently large we have

$$(u - \gamma_\sigma)(x) < 0 \quad \text{on } \Omega_{T_3}. \tag{2.54}$$

Therefore,

$$\mu = \sup_{x \in \overline{\Omega}_1} (u - \gamma_\sigma)(x) > 0$$

is in fact a maximum attained at a point z_0 in the compact set $\overline{\Omega}_1 \setminus \Omega_{T_3}$. From (2.54) we know that $\gamma(z_0) > T_1$. Thus

$$u(z_0) = \gamma_\sigma(z_0) + \mu > \gamma_\sigma(z_0) > \alpha > u^*_{T_1} > u^* - \eta/2$$

and hence $z_0 \in A_\eta \cap \Omega_{T_1}$. Next we have to distinguish two cases according to $z_0 \in D_o$ or not. If $z_0 \in D_o$, since z_0 is a maximum for $u - \gamma_\sigma$, we get $\nabla(u - \gamma_\sigma)(z_0) = 0$. Using this fact we infer that $z_0 \in B_\eta$ since, by (2.50),

$$|\nabla u(z_0)| = |\nabla \gamma_\sigma(z_0)| < \sigma \Lambda < \eta.$$

Thus $z_0 \in B_\eta \cap \Omega_{T_1}$. Again since z_0 is a maximum for $u - \gamma_\sigma$, we have

$$\text{Hess}\,(u)(z_0) \leq \text{Hess}\,(\gamma_\sigma)(z_0)$$

and this, jointly with (2.49), yields

$$\text{Hess}\,(u)(z_0) < \sigma_0 \langle\, , \,\rangle$$

in the sense of symmetric bilinear forms, contradicting (2.46). This concludes the proof when $z_0 \in D_0$.

In case $z_0 \notin D_o$ we reason as follows. Fix $0 < \varepsilon < 1$ sufficiently small so that for the minimizing geodesic ξ parametrized by arclength and joining o with z_0, the point $o_\varepsilon = \xi(\varepsilon) \neq z_0$ and $z_0 \notin \text{cut}(o_\varepsilon)$. Thus, by Lemma 2.1, the function $r_\varepsilon(x) = \text{dist}(x, o_\varepsilon)$ is C^2 in a neighborhood of z_0. By the triangle inequality,

$$r(x) \leq r_\varepsilon(x) + \varepsilon, \tag{2.55}$$

equality holding at z_0. With φ defined in (2.36), set

$$\gamma^\varepsilon(x) = \varphi(r_\varepsilon(x) + \varepsilon).$$

Since φ is increasing

$$\gamma(x) = \varphi(r(x)) \leq \varphi(r_\varepsilon(x) + \varepsilon) = \gamma^\varepsilon(x) \tag{2.56}$$

and

$$\gamma(z_0) = \gamma^\varepsilon(z_0). \tag{2.57}$$

Next, consider the function

$$\gamma_\sigma^\varepsilon(x) = \alpha + \sigma(\gamma^\varepsilon(x) - T_1).$$

Because of (2.56) and (2.57), in a neighborhood of z_0 we have

$$u(x) - \gamma_\sigma^\varepsilon(x) \leq u(x) - \gamma_\sigma(x) \leq \mu,$$

and

$$u(z_0) - \gamma_\sigma^\varepsilon(z_0) \le u(z_0) - \gamma_\sigma(z_0) = \mu.$$

Hence z_0 is also a local maximum for $u(x) - \gamma_\sigma^\varepsilon(x)$. Therefore

$$\nabla u(z_0) = \nabla \gamma_\sigma^\varepsilon(z_0) \tag{2.58}$$

and

$$\operatorname{Hess}(u)(z_0) \le \operatorname{Hess}\left(\gamma_\sigma^\varepsilon\right)(z_0). \tag{2.59}$$

From (2.58) we deduce

$$\begin{aligned} |\nabla u(z_0)| &= \sigma|\nabla\gamma^\varepsilon(z_0)| = \sigma\varphi'(r_\varepsilon(z_0)+\varepsilon)|\nabla r_\varepsilon(z_0)| \\ &= \frac{\sigma}{G(r(z_0)+1)} \le \frac{\sigma}{G(1)} < \eta. \end{aligned}$$

Since we already know that $z_0 \in A_\eta$ we conclude that $z_0 \in B_\eta$. Now we analyze (2.59). Because of (2.31), (2.55) and $G' \ge 0$, we have

$${}^M K(x) \ge -G^2(r(x)) \ge -G^2(r_\varepsilon(x)+\varepsilon).$$

Set $G_\varepsilon(t) = G(t+\varepsilon)$ and consider the Cauchy problem (2.32) with G_ε instead of G. Again by the Hessian comparison theorem, on D_{o_ε} we have

$$\operatorname{Hess}(r_\varepsilon)(x) \le \frac{\psi_\varepsilon'(r_\varepsilon(x)}{\psi_\varepsilon(r_\varepsilon(x)}\left(\langle\, ,\,\rangle - dr_\varepsilon \otimes dr_\varepsilon\right),$$

where

$$\psi_\varepsilon(t) = \frac{1}{G_\varepsilon(0)}\left(e^{\int_0^t G_\varepsilon(s)ds} - 1\right).$$

Observing that $z_0 \in D_{o_\varepsilon}$, using (2.55) and (2.39), for $v \in T_{z_0}M$, $|v| = 1$, we obtain

$$\begin{aligned} \operatorname{Hess}(\gamma^\varepsilon)(z_0)(v,v) &\le \varphi'(r_\varepsilon(z_0)+\varepsilon)\operatorname{Hess}(r_\varepsilon)(z_0)(v,v) \\ &= \frac{1}{G(r_\varepsilon(z_0)+\varepsilon)+1}\operatorname{Hess}(r_\varepsilon)(z_0)(v,v) \\ &= \frac{1}{G(r(z_0))+1}\operatorname{Hess}(r_\varepsilon)(z_0)(v,v) \\ &\le \frac{1}{G(r(z_0))+1}\frac{\psi_\varepsilon'(r_\varepsilon(z_0)}{\psi_\varepsilon(r_\varepsilon(z_0)}\left(|v|^2 - \langle\nabla r_\varepsilon(z_0), v\rangle^2\right) \end{aligned}$$

$$
\begin{aligned}
&= \frac{1}{G(r(z_0))+1} G_\varepsilon(r_\varepsilon(z_0)) \\
&\quad \times \frac{e^{\int_0^{r_\varepsilon(z_0)} G(s+\varepsilon)ds}}{e^{\int_0^{r_\varepsilon(z_0)} G(s+\varepsilon)ds} - 1} \left(|v|^2 - \langle \nabla r_\varepsilon(z_0), v\rangle^2\right) \\
&= \frac{G(r_\varepsilon(z_0)+\varepsilon)}{G(r(z_0))+1} \frac{e^{\int_\varepsilon^{r_\varepsilon(z_0)+\varepsilon} G(s)ds}}{e^{\int_\varepsilon^{r_\varepsilon(z_0)+\varepsilon} G(s)ds} - 1} \left(|v|^2 - \langle \nabla r_\varepsilon(z_0), v\rangle^2\right) \\
&= \frac{G(r(z_0))}{G(r(z_0))+1} \frac{e^{\int_\varepsilon^{r(z_0)} G(s)ds}}{e^{\int_\varepsilon^{r(z_0)} G(s)ds} - 1} \left(|v|^2 - \langle \nabla r_\varepsilon(z_0), v\rangle^2\right) \\
&\le \frac{e^{\int_0^{r(z_0)} G(s)ds}}{e^{\int_1^{r(z_0)} G(s)ds} - 1} \left(|v|^2 - \langle \nabla r_\varepsilon(z_0), v\rangle^2\right) \\
&\le \Lambda |v|^2.
\end{aligned}
$$

Thus,

$$
\mathrm{Hess}\,(\gamma_\sigma^\varepsilon)(z_0)(v,v) = \sigma\,\mathrm{Hess}\,(\gamma^\varepsilon)(z_0)(v,v) \le \sigma\Lambda|v|^2 < \sigma_0|v|^2,
$$

contradicting (2.46). The second part of the theorem, dealing with the Laplace-Beltrami operator, can be proved in an analogous way under assumption (2.28). □

Remark 2.8 As observed in Remark 2.7, the final conclusion of the theorem, that is, the validity of the Omori-Yau maximum principle for the Laplacian, can be proved under the relaxed assumption (2.30), even if o is not a pole. The argument is based on a comparison procedure and it makes essential use of the validity of the differential inequality

$$
\Delta r \le (m-1)\frac{\psi'(r)}{\psi(r)} \tag{2.60}
$$

in the weak sense on all of M. Indeed, as in the proof above, we reason by contradiction and we suppose that

$$
\inf_{B_\eta} \Delta u \ge \sigma_0 \tag{2.61}
$$

for some $\sigma_0 > 0$. As above, we prove the existence of $z_0 \in A_\eta$ that is a point of maximum for $u - \gamma_\sigma$. Now note that if $z_0 \in \mathrm{cut}(o)$, then we can prove, via the trick of (2.58), that $|\nabla u(z_0)| < \eta$, so that $z_0 \in B_\eta$. Otherwise, if $z_0 \notin \mathrm{cut}(o)$ then $z_0 \in B_\eta$ trivially. Let

$$
Z = \{x \in M \setminus \overline{\Omega}_1 : (u - \gamma_\sigma)(x) = \mu\} \subset B_\eta.
$$

Since Z is compact and B_η is open, there exists an open neighborhood U_Z of Z contained in B_η. Pick any $y \in Z$, fix $\beta \in (0, \mu)$ and call $Z_{\beta,y}$ the connected component of the set

$$\{x \in M \setminus \overline{\Omega}_1 : (u - \gamma_\sigma)(x) > \beta\}$$

containing y. Since $\beta > 0, Z_{\beta,y} \subset M \setminus \overline{\Omega}_1$, and we can also choose β sufficiently near to μ so that $\overline{Z_{\beta,y}} \subset B_\eta$. Furthermore, $\overline{Z_{\beta,y}}$ is compact. Using (2.60), that presently plays the role of (2.40), we have

$$\Delta\gamma \leq \Lambda$$

in the weak sense on M. Using the latter and (2.61) we deduce

$$\Delta\gamma_\sigma \leq \sigma\Lambda < \sigma_0,$$

and therefore

$$\Delta u \geq \sigma_0 > \Delta\gamma_\sigma = \Delta(\gamma_\sigma + \beta) \tag{2.62}$$

in the weak sense on $Z_{\beta,y}$. Moreover, on $\partial Z_{\beta,y}$

$$u(x) = \gamma_\sigma(x) + \beta,$$

and hence by Proposition 3.1 and Remark 3.9 in Chap. 3, $u(x) \leq \gamma_\sigma(x) + \beta$ on $Z_{\beta,y}$. However, at $y \in Z_{\beta,y}$

$$u(y) = \gamma_\sigma(y) + \mu > \gamma_\sigma(y) + \beta = u(y),$$

which is a contradiction.

Remark 2.9 A key point in the previous proof is to guarantee the existence of $z_0 \in \overline{\Omega}_{T_1} \setminus \Omega_{T_3}$ where the function $u - \gamma_\sigma$ attains its positive maximum μ. Towards this end, inequality (2.54) is essential. However, we can guarantee the validity of the latter under the assumptions that $u(x) = o(\gamma(x))$ as $x \to \infty$, that is,

$$u(x) = o\left(\int_0^{r(x)+1} \frac{ds}{G(s)}\right) \quad \text{as } r(x) \to +\infty, \tag{2.63}$$

which is clearly weaker than $u^* < +\infty$. This observation will be used in geometric applications.

Before proving the next result we recall the following notation: let $f : M \to N$ be a map between two manifolds; then $f(x) \to \infty$ as $x \to \infty$ in M means that for each compact set $\Omega_N \subset N$ there exists a compact $\Omega_M \subset M$ such that, for each

$x \in M \setminus \Omega_M, f(x) \in N \setminus \Omega_N$. Similarly for $f : M \to \mathbb{R}$ and $f(x) \to +\infty$ as $x \to \infty$ in M.

Theorem 2.6 *Let $f : M \to N$ be an isometric immersion into a complete Riemannian manifold N with mean curvature vector field $\mathbf{H}$. Let $o_N \in N$ and assume that $f(M) \cap cut(o_N) = \emptyset$. Suppose that the radial sectional curvature of N with respect to o_N satisfies*

$$^{N}K_{\mathrm{rad}} \geq -G_N^2(\rho), \tag{2.64}$$

where ρ denotes the Riemannian distance function on N from the point o_N, and $G_N \in C^0\left(\mathbb{R}_0^+\right)$ is a positive function satisfying

$$\frac{1}{G_N} \notin L^1(+\infty).$$

Define

$$\varphi_N(t) = \int_0^t \frac{ds}{G_N(s+1)}.$$

If the immersion is proper and there exists a positive $G \in C^1\left(\mathbb{R}_0^+\right)$ such that $\frac{1}{G} \notin L^1(+\infty)$, $G' \geq 0$ and

$$|\mathbf{H}| \leq G(\varphi_N \circ \rho \circ f), \tag{2.65}$$

in the complement of a compact set in M, then the Omori-Yau maximum principle for the Laplacian holds on M.

Proof Clearly we can suppose that M is not compact, otherwise there is nothing to prove. Next, note that $\varphi_N \circ \rho \circ f$ is $C^2(M)$ because of the assumptions, and that, without loss of generality, we can suppose G_N nondecreasing with $G_N(0) > 0$. Now, since φ_N is defined as in (2.36), the corresponding of (2.37) holds. Set

$$\gamma(x) = (\varphi_N \circ \rho \circ f)(x) = \int_0^{\rho(f(x))} \frac{ds}{G_N(s+1)}.$$

Then, since f is proper we have that $f(x) \to \infty$ in N as $x \to \infty$ in M. Hence, using $\frac{1}{G_N} \notin L^1(+\infty)$, we deduce

$$\gamma(x) \to +\infty \quad \text{as } x \to \infty \text{ in } M.$$

Furthermore, with the aid of Gauss lemma,

$$|\nabla\gamma| = \left|\varphi_N'(\rho \circ f)\right| = \frac{1}{G_N(\rho \circ f + 1)} \leq \Lambda \quad \text{on } M$$

for some constant $\Lambda > 0$. Letting $m = \dim M$, using (1.180) and (1.170), and indicating with $\{e_i\}$ a local orthonormal frame on M, we have

$$
\begin{aligned}
{}^M\Delta\gamma &= \sum_{i=1}^{m} \text{Hess}\,(\varphi_N \circ \rho)(df(e_i), df(e_i)) + m\langle \nabla(\varphi_N \circ \rho), \mathbf{H}\rangle \\
&= \sum_{i=1}^{m} \left\{ \varphi_N'(\rho)\,\text{Hess}\,(\rho)(df(e_i), df(e_i)) + \varphi_N''(\rho)(d(\rho \circ f)(e_i))^2 \right\} \\
&\quad + m\varphi_N'(\rho)\langle \nabla\rho, \mathbf{H}\rangle.
\end{aligned}
$$

By the Hessian comparison Theorem 1.4 and Eq. (2.34), having defined ψ as in (2.33) with G replaced by G_N, we have,

$$
\text{Hess}(\rho) \le \frac{\psi'(\rho)}{\psi(\rho)}\{\langle\,,\,\rangle_N - d\rho \otimes d\rho\} = G_N(\rho)\frac{e^{\int_0^\rho G_N(s)\,ds}}{e^{\int_0^\rho G_N(s)\,ds} - 1}\{\langle\,,\,\rangle_N - d\rho \otimes d\rho\}.
$$

As remarked,

$$
f(x) \to \infty \quad \text{in } N
$$

as $x \to \infty$ in M. Hence, noting that $\varphi_N'' \le 0$, proceeding as in Theorem 2.5, we can choose Λ large enough and a compact set $K \subset M$ such that

$$
{}^M\Delta\gamma \le \Lambda + \frac{m}{G_N(0)}|\mathbf{H}| \quad \text{on } M \setminus K.
$$

Let now G be as in statement of the theorem; since

$$
|\mathbf{H}| < G(\gamma)
$$

the function γ satisfies the hypotheses of Theorem 2.4, therefore we have the validity of the Omori-Yau maximum principle for the Laplacian on M. □

Remark 2.10 Similar extrinsic sufficient conditions for the validity of the Omori-Yau maximum principle for the Laplacian are given in the proof of Theorem 5.9, item (ii), in Chap. 5, and in Theorem 7.1 of Chap. 7.

2.3 Stochastic Completeness and the Weak Maximum Principle

Let us recall that *stochastic completeness* is the property for a stochastic process to have infinite (intrinsic) life time. In other words, the total probability of the particle being found in the state space is constantly equal to 1. A classical analytic condition to express stochastic completeness is as follows.

Definition 2.2 A Riemannian manifold $(M, \langle\, , \rangle)$ is said to be stochastically complete if for some (and hence, any) $(x, t) \in M \times (0, +\infty)$

$$\int_M p(x, y, t) dy = 1, \tag{2.66}$$

where $p(x, y, t)$ is the (minimal) positive *heat kernel* of the Laplace-Beltrami operator Δ, that is, the smallest positive fundamental solution of the heat equation

$$\frac{\partial p}{\partial t} = \frac{1}{2}\Delta p \quad \text{on } M,$$

in the variables (x, t) (the point y is considered as fixed), with initial data

$$p(\cdot, y, t) \to \delta_y \quad \text{for } t \to 0^+,$$

where δ_y is the Dirac delta centered at y.

Observe that p is smooth in $(x, y, t) \in \mathbb{R}^+ \times M \times M$. Note also that in the above definition the Riemannian manifold M is not assumed to be geodesically complete. Indeed, following Dodziuk [104], one can construct a minimal heat kernel on an arbitrary Riemannian manifold as the supremum of the Dirichlet heat kernels on an exhausting sequence of relatively compact domains with smooth boundary. The analytic condition expressed in (2.66) is equivalent to a number of other properties. For instance, one has the following equivalent characterizations (for a proof, see [131, Theorem 6.2] and also Theorem 2.14 below).

Theorem 2.7 *Let $(M, \langle\, , \rangle)$ be a Riemannian manifold. Then the following are equivalent:*

(i) M is stochastically complete.
(ii) For every $\lambda > 0$, the only nonnegative bounded C^2 solution of $\Delta u \geq \lambda u$ on M is $u \equiv 0$.
(iii) For every $\lambda > 0$, the only nonnegative bounded C^2 solution of $\Delta u = \lambda u$ on M is $u \equiv 0$.
(iv) For every $T > 0$, the only bounded solution on $M \times (0, T)$ of the Cauchy problem

$$\begin{cases} \frac{\partial u}{\partial t} = \frac{1}{2}\Delta u \\ u|_{t=0^+} = 0 \quad \textit{in the } L^1_{loc}(M) \textit{ sense} \end{cases}$$

is $u \equiv 0$.

By way of example, on which we shall come back extensively in Sect. 2.5, recall that a Riemannian manifold is said to be *parabolic* if every subharmonic function on M which is bounded from above is constant, that is, $\Delta u \geq 0$ and $u^* = \sup_M u < +\infty$ implies that $u =$ constant. In particular, every parabolic Riemannian manifold clearly satisfies condition (ii) in Theorem 2.7 and hence it is stochastically complete.

In [225], Pigola et al. found the following characterization of stochastic completeness.

Theorem 2.8 *Let $(M, \langle\, ,\, \rangle)$ be a Riemannian manifold. Then the following are equivalent:*

(i) M is stochastically complete.
(ii) For every function $u \in C^2(M)$ with $u^* = \sup_M u < +\infty$, and for every $\varepsilon > 0$,

$$\inf_{\Omega_\varepsilon} \Delta u \leq 0$$

where $\Omega_\varepsilon = \{x \in M : u(x) > u^* - \varepsilon\}$.
(iii) For every function $u \in C^2(M)$ with $u^* = \sup_M u < +\infty$ there exists a sequence of points $\{x_k\} \subset M$ satisfying

$$(i)\ u(x_k) > u^* - \frac{1}{k}, \ \text{and (ii)}\ \Delta u(x_k) < \frac{1}{k}$$

for each $k \in \mathbb{N}$.
(iv) For every function $u \in C^2(M)$ with $u^* = \sup_M u < +\infty$ and every $f \in C^0(\mathbb{R})$, if $\Delta u \geq f(u)$ on the subset $\Omega_\varepsilon = \{x \in M : u(x) > u^* - \varepsilon\}$, for some $\varepsilon > 0$, then $f(u^*) \leq 0$.

Proof In an obvious way, (ii) implies (iii), simply by choosing $\varepsilon = 1/k$ for each $k \in \mathbb{N}$ and taking $x_k \in \Omega_{1/k}$ such that $\Delta u(x_k) < 1/k$, since $\inf_{\Omega_{1/k}} \Delta u < 1/k$. On the other hand, (iii) clearly implies (iv). Indeed, $x_k \in \Omega_\varepsilon$ if k is sufficiently large, so that

$$\frac{1}{k} > \Delta u(x_k) \geq f(u(x_k)),$$

and taking limits here yields $f(u^*) = \lim_{k\to+\infty} f(u(x_k)) \leq 0$. Furthermore, (iv) clearly implies condition (ii) in Theorem 2.7, and hence (i), simply by choosing $f(u) = \lambda u$.

Therefore, it only remains to prove that (i) implies (ii). To see this, we argue by contradiction, and assume that there exists a function $u \in C^2(M)$ with $u^* < +\infty$ and such that, for some $\varepsilon > 0$,

$$\inf_{\Omega_\varepsilon} \Delta u \geq 2c > 0.$$

We let $\Omega^* = \{x \in M : \Delta u(x) > c\}$, so that $\overline{\Omega_\varepsilon} \subset \Omega^*$. Having set $\lambda = c/\varepsilon$, at each $x \in \Omega^*$ we have

$$\Delta u(x) > c \geq c + \lambda(u(x) - u^*) = \lambda(u(x) + \varepsilon - u^*),$$

so that $u + \varepsilon - u^*$ is a C^2 subsolution of

$$Lu = \Delta u - \lambda u = 0 \tag{2.67}$$

on Ω^*. Since the constant function 0 is obviously a subsolution of equality (2.67) on M, we see that $u_\varepsilon = \max\{u + \varepsilon - u^*, 0\}$ is also a subsolution on M. Since u is C^2, u_ε belongs to $C^0(M) \cap W^{1,2}_{loc}(M)$. Furthermore, $u_\varepsilon \not\equiv 0$ and $0 \le u_\varepsilon \le \varepsilon < +\infty$. Noting that any positive constant is a supersolution of (2.67), choosing $u_+ > \varepsilon$, and applying the monotone iteration scheme (see [240, Proposition 2.4] for the formulation needed here) yields a smooth solution v of (2.67) on M such that $u_\varepsilon \le v \le u_+$. Now, since u_ε does not vanish identically, the same holds for v, and this contradicts condition (iii) in Theorem 2.7 and, equivalently, condition (i). □

Comparison with the Omori-Yau maximum principle for the Laplacian suggests the following

Definition 2.3 Let M be a (not necessarily complete) Riemannian manifold. The *weak maximum principle* is said to hold for the Laplacian on M if, for any function $u \in C^2(M)$ with $u^* = \sup_M u < +\infty$, there exists a sequence of points $\{x_k\}_{k\in\mathbb{N}} \subset M$ satisfying

$$\text{(i) } u(x_k) > u^* - \frac{1}{k}, \text{ and (ii) } \Delta u(x_k) < \frac{1}{k}.$$

Analogously, the weak maximum principle *for the Hessian* is said to hold on M if, for any function $u \in C^2(M)$ having $u^* = \sup_M u < +\infty$, there exists a sequence of points $\{x_k\}_{k\in\mathbb{N}} \subset M$ satisfying

$$\text{(i) } u(x_k) > u^* - \frac{1}{k}, \text{ and (ii) } \operatorname{Hess}(u)(x_k) < \frac{1}{k}\langle\, ,\, \rangle.$$

The chain of equivalences described in Theorem 2.8 shows that this seemingly simple minded definition is in fact surprisingly deep. First of all, the validity of the Omori-Yau maximum principle immediately implies stochastic completeness. Thus, for instance, by Theorem 2.5 and Remark 2.7, this is the case if

$$\operatorname{Ric}(\nabla r, \nabla r) \ge -(m-1)G^2(r),$$

where $G : \mathbb{R}^+_0 \to \mathbb{R}^+_0$ satisfies

$$G(r) \sim Cr(\log r)(\log\log r)\cdots \quad \text{as } r \to +\infty$$

for some constant $C > 0$. Indeed, since the condition on Ric is expressed as an inequality, we can always redefine G to satisfy also (i) and (ii) of (2.27) (note that this formulation of the Omori-Yau maximum principle greatly improves on [279]).

On the other hand, the function theoretic characterization of stochastic completeness given in Theorem 2.8 often enables one to analyze consequences of the latter in a simple way. This is the case, for instance, of the following straightforward proof of a sufficient condition for stochastic completeness due to Khas'minskii [159] (see [267] for a proof based on the standard argument). We remark that, as proved by Mari and Valtorta in [182], condition (2.68) below is also necessary.

Theorem 2.9 *Let $(M, \langle\, ,\, \rangle)$ be a Riemannian manifold. If M supports a C^2 function γ such that $\gamma(x) \to +\infty$ as $x \to \infty$ and, for some positive constant $\lambda > 0$,*

$$\Delta\gamma \leq \lambda\gamma \tag{2.68}$$

outside a compact subset of M, then M is stochastically complete.

Proof Note that by adding a constant to γ we may assume without loss of generality that γ is everywhere positive and that (2.68) holds on the whole M. We will prove that the weak maximum principle holds on M. To this end, let u be a C^2 function on M with $u^* < +\infty$, and assume by contradiction that condition (ii) in Theorem 2.8 does not hold. That is, there exists $\varepsilon > 0$ such that $\inf_{\Omega_\varepsilon} \Delta u > 0$, where $\Omega_\varepsilon = \{x \in M : u(x) > u^* - \varepsilon\}$. Therefore, choosing

$$\varepsilon' < \min\{\varepsilon, \inf_{\Omega_\varepsilon} \Delta u\},$$

we have

$$\Delta u > \varepsilon' \tag{2.69}$$

on the subset $\Omega_{\varepsilon'} = \{x \in M : u(x) > u^* - \varepsilon'\}$. Let $\eta \leq \min\{\varepsilon'/2, \varepsilon'/2\lambda\}$ and let $\hat{x} \in M$ be such that

$$u(\hat{x}) > u^* - \eta/2. \tag{2.70}$$

Choose $c > 0$ small enough that

$$c\gamma(\hat{x}) \leq \eta/2, \tag{2.71}$$

and consider the function $u - c\gamma$. Since γ tends to $+\infty$ as x goes to ∞ in M and $u^* < +\infty$, the function $u - c\gamma$ attains its absolute maximum at some point $x_0 \in M$. We claim that

$$u(x_0) > u^* - \varepsilon'/2 \text{ and } c\lambda\gamma(x_0) < \varepsilon'/2. \tag{2.72}$$

Indeed, by (2.70) and (2.71) we have

$$u(x_0) - c\gamma(x_0) \geq u(\hat{x}) - c\gamma(\hat{x}) > u^* - \eta.$$

Thus,

$$u(x_0) \geq u(x_0) - c\gamma(x_0) > u^* - \eta \geq u^* - \varepsilon'/2$$

and

$$c\gamma(x_0) < u(x_0) - u^* + \eta \leq \eta \leq \varepsilon'/2\lambda.$$

This proves (2.72). Therefore, $x_0 \in \Omega_{\varepsilon'}$ and (2.69) holds at x_0. But, recalling that $u - c\gamma$ attains its absolute maximum at x_0, and using (2.68) and (2.72), we have

$$0 \geq \Delta(u - c\gamma)(x_0) = \Delta u(x_0) - c\Delta\gamma(x_0) > \varepsilon' - c\lambda\gamma(x_0) > \varepsilon'/2,$$

which is a contradiction. □

Remark 2.11 One can indeed relax the regularity of γ to $\gamma \in C^0(M \setminus K) \cap W^{1,2}_{loc}(M \setminus K))$ for some compact set $K \subset M$. See Theorem A in [229]. This fact will be used in the proof of Theorem 2.12 below.

A minor modification of the above proof yields the following version of Theorem 2.9 for the Hessian.

Theorem 2.10 *Let $(M, \langle\, ,\, \rangle)$ be a Riemannian manifold. If M supports a C^2 function γ such that $\gamma(x) \to +\infty$ as $x \to \infty$ and, for some positive constant $\lambda > 0$, it satisfies the differential inequality*

$$\text{Hess}\,(\gamma) \leq \lambda\gamma\langle\, ,\, \rangle$$

outside a compact subset of M (in the sense of quadratic forms), then the weak maximum principle for the Hessian holds on M.

It is interesting to remark that the existence of a function γ satisfying the requirements in Theorem 2.9 does not force the manifold to be geodesically complete. This should be compared with the observation after Remark 2.5. Contrary to what happened there, in the present situation no conditions are imposed on the gradient of the function γ, and this allows one to find functions satisfying the due requirements even on noncomplete manifolds, as in the following example.

Example 2.2 Let M be the geodesically incomplete Riemannian manifold given by $\mathbb{R}^m \setminus \{0\}$, with the usual Euclidean metric and $m \geq 3$. On M we consider the function

$$\gamma(x) = |x|^2 + |x|^{2-m} = \frac{|x|^m + 1}{|x|^{m-2}}.$$

Clearly, $\gamma(x) \to +\infty$ as $x \to \infty$. Moreover, since for every $x \in \mathbb{R}^m \setminus \{0\}$

$$|\nabla |x||^2 = 1 \quad \text{and} \quad |x|\Delta|x| = m - 1,$$

it follows that

$$\begin{aligned}\Delta\gamma(x) &= 2|x|\Delta|x| + 2 + (2-m)|x|^{1-m}\Delta|x| + (2-m)(1-m)|x|^{-m}\\ &= 2m + (2-m)(m-1)|x|^{-m} - (2-m)(m-1)|x|^{-m},\end{aligned}$$

that is,

$$\Delta\gamma(x) = 2m$$

on $\mathbb{R}^m\backslash\{0\}$. Thus γ satisfies the conditions in Theorem 2.9 and this shows that $M = \mathbb{R}^m\backslash\{0\}$, $m \geq 3$, is stochastically complete.

However, contrary to Theorem 2.9, the conditions on γ in Theorem 2.10 imply that $(M, \langle\,,\,\rangle)$ is complete, although we have no restrictions on $\nabla\gamma$. This can be seen as follows: suppose that

$$\text{Hess}\,(\gamma) \leq \lambda\gamma\langle\,,\,\rangle \tag{2.73}$$

is satisfied outside some compact set $K \subset M$. Without loss of generality we can suppose $\lambda = 1$ and that $\gamma > 0$ on $M \setminus K$. Let $\xi : [0, l) \to M$ be a maximal geodesic path parameterized by arclength; we need to show that $l = +\infty$. Towards this aim note that ξ has to be a *divergent path*, that is, it eventually leaves each fixed compact set of M. Thus, there exists $t_0 > 0$ such that $\xi(t) \notin K$ for each $t \geq t_o$. Consider the unit speed geodesic $\Gamma : [0, l - t_0) \to M \setminus K$, $\Gamma(t) = \xi(t + t_0)$; set $\varphi = \gamma \circ \Gamma$. A computations using (2.73) shows that $\varphi(t)$ satisfies

$$\varphi''(t) \leq \varphi(t) \qquad \text{on } [0, l - t_0). \tag{2.74}$$

Furthermore,

$$\varphi((l - t_0)^-) = +\infty. \tag{2.75}$$

On the other hand, using the classical Sturm comparison argument (see [44]), (2.74) shows that the function

$$\sinh(t)\varphi'(t) - \cosh(t)\varphi(t)$$

is nonincreasing. As a consequence

$$\frac{\varphi'(t)}{\varphi(t)} \leq \coth(t),$$

which, integrated, implies that φ cannot explode in finite time, contradicting (2.75).

The Khas'minskii test in Theorem 2.9 may be used to deduce conditions that ensure the stochastic completeness of a Riemannian manifold. For instance, we may apply it to the case where M is a *radial model* in the sense of Greene and Wu [129] (see Definition 1.1 in Sect. 1.8).

Example 2.3 Let $M_g = \mathbb{R}^m$ be the rotationally symmetric manifold with metric given in polar coordinates on $(0, +\infty) \times \mathbb{S}^{m-1}$ by

$$dr^2 + g(r)^2 d\theta^2,$$

where $d\theta^2$ is the standard metric on the unit sphere $\mathbb{S}^{m-1} \subset \mathbb{R}^m$ and $g \in C^\infty(\mathbb{R}_0^+)$ is such that $g(r) > 0$ for $r > 0$, $g'(0) = 1$ and $g^{(2k)}(0) = 0$ for $k = 0, 1, 2, \ldots$. Then, as recalled in Sect. 1.8, denoting with B_R the geodesic ball of radius R centered at $o \in \mathbb{R}^m$,

$$\mathrm{vol}(\partial B_R) = \omega_m g^{m-1}(R),$$

where ω_m stands for the volume of the unit sphere $\mathbb{S}^{m-1}$ of $\mathbb{R}^m$, and

$$\mathrm{vol}(B_R) = \omega_m \int_0^R g^{m-1}(t)dt.$$

Define

$$\begin{aligned}\gamma(x) = \gamma(r(x)) &= \int_0^{r(x)} \frac{\mathrm{vol}(B_t)}{\mathrm{vol}(\partial B_t)} dt \\ &= \int_0^{r(x)} \left(\frac{1}{g(t)^{m-1}} \int_0^t g(s)^{m-1} ds \right) dt.\end{aligned}$$

Since for $r > 0$

$$\Delta r = (m-1)\frac{g'(r)}{g(r)},$$

it follows that

$$\begin{aligned}\Delta\gamma &= \gamma'(r)\Delta r + \gamma''(r) \\ &= (m-1)\frac{g'(r)}{g(r)^m} \int_0^r g(s)^{m-1} ds + 1 - (m-1)\frac{g'(r)}{g(r)^m} \int_0^r g(s)^{m-1} ds \\ &= 1,\end{aligned}$$

that is, $\Delta\gamma \equiv 1$ on M_g. Therefore, if

$$\frac{\mathrm{vol}(B_R)}{\mathrm{vol}(\partial B_R)} \notin L^1(+\infty),$$

from Theorem 2.9 we deduce that M_g is stochastically complete. On the other hand, if

$$\frac{\mathrm{vol}(B_R)}{\mathrm{vol}(\partial B_R)} \in L^1(+\infty),$$

then γ is a bounded C^2 function on M_g with $\Delta\gamma \equiv 1$, so that the weak maximum principle does not hold on M_g. In other words, M_g is not stochastically complete. We collect these observations in the following result.

Proposition 2.3 *A model manifold M_g is stochastically complete if and only if*

$$\frac{vol(B_R)}{vol(\partial B_R)} \notin L^1(+\infty). \tag{2.76}$$

It has been conjectured (see [131, 227]) that (2.76) is a sufficient condition for a general complete manifold M to be stochastically complete. This conjecture has been recently proved to be false by Bär and Bessa [36]. To date, the best volume growth sufficient condition for stochastic completeness of a general complete Riemannian manifold is due to Grigor'yan [130] (see also [131, Theorem 9.1]), and it is expressed in the next.

Theorem 2.11 *Let $(M, \langle\,,\,\rangle)$ be a complete Riemannian manifold. If, for some reference point,*

$$\frac{R}{\log vol(B_R)} \notin L^1(+\infty), \tag{2.77}$$

then M is stochastically complete.

Observe that condition (2.77) implies (2.76), the converse being generally false, so that Grigor'yan condition is slightly stronger than the necessary and sufficient condition for the stochastic completeness of model manifolds; to see this refer, for instance, to Lemma 2.6 at the very end of the chapter. We also note that (2.77) is implied, via the Bishop comparison theorem, by a lower bound on the radial Ricci curvature of the type

$$\mathrm{Ric}(\nabla r, \nabla r) \geq -(m-1)G^2(r),$$

where $G : \mathbb{R}_0^+ \to \mathbb{R}_0^+$ is a nondecreasing function satisfying

$$\frac{1}{G(t)} \notin L^1(+\infty),$$

a typical example of such G being $G(t) = A\sqrt{1+t^2}$ (see Theorem 1.3 together with Proposition 1.7). Theorem 2.11 can be seen also as a consequence of Theorem 2.15

below, recently proved in [227, Proposition 3.17]. In some sense the situation is similar to what happen for parabolicity, see Remark 2.19 below. A clarification in the present case would certainly be most welcome.

On the other hand, in order to prove stochastic completeness one can also use comparison with a suitable model and the following theorem. For a version of this result extended to a large class of operators on M we refer to [229].

Theorem 2.12 *Let $(M, \langle\, ,\, \rangle)$ be a complete Riemannian manifold of dimension m, let $o \in M$ be a reference point and denote by $r(x)$ the Riemannian distance function from o. Let $g \in C^{\infty}(\mathbb{R}_0^+)$ be such that $g(t) > 0$ for $t > 0$, $g(0) = 0$, $g'(0) = 1$ and $g^{(2k)}(0) = 0$ for $k = 1, 2, \ldots$, and consider the corresponding model manifold M_g of the same dimension m. Assume that*

$$\Delta r(x) \leq (m-1)\frac{g'(r(x))}{g(r(x))}$$

holds on $M \backslash (\{o\} \cup cut(o) \cup K)$ for some compact set $K \subset M$. If M_g is stochastically complete, then M is also stochastically complete.

Proof From Proposition 2.3 we know that M_g is stochastically complete if and only if (2.76) holds and the latter is clearly equivalent to

$$g(t)^{1-m}\int_1^t g(s)^{m-1}ds \notin L^1(+\infty). \tag{2.78}$$

Fix $\lambda > 0$; choose R sufficiently large that $K \subset B_R = B_R(o)$, set

$$\alpha(r) = \lambda\int_R^r \left(g(t)^{1-m}\int_R^t g(s)^{m-1}ds\right)dt \quad \text{on } [R, +\infty), \tag{2.79}$$

and define $\gamma(x) = \alpha(r(x))$ on $M \setminus B_R$. Note that since M_g is stochastically complete (2.78) holds and

$$\gamma(x) \to +\infty \qquad \text{as } x \to \infty.$$

Next, according to the assumptions of the theorem, on $M \backslash (\text{cut}(o) \cup B_R)$ we have

$$\Delta\gamma(x) = \alpha''(r(x)) + \Delta r(x)\alpha'(r(x)) \leq \alpha''(r(x)) + (m-1)\frac{g'(r(x))}{g(r(x))}\alpha'(r(x))$$

since $\alpha' \geq 0$. Using (2.79) we easily see that

$$\Delta\gamma(x) \leq \lambda\gamma(x) \quad \text{on } M \backslash (\text{cut}(o) \cup B_R). \tag{2.80}$$

Now we use a trick of Cheng and Yau [81] to show that (2.80) is satisfied in the weak sense on $M \setminus B_R$. By Theorem 2.9 and Remark 2.11 this will be enough to conclude that M is stochastically complete.

Towards this aim we consider an exhaustion $\{\Omega_n\}$ of $M \setminus \mathrm{cut}(o)$ by bounded domains with smooth boundaries, star-shaped with respect to o. Let ν be the outward unit normal to $\partial\Omega_n$. Denote by $\rho(x)$ the distance function from x to $\partial\Omega_n$, with the convention that $\rho(x) > 0$ if $x \in \Omega_n$ and $\rho(x) < 0$ if $x \notin \overline{\Omega_n}$. Thus ρ is the radial coordinate for the Fermi coordinates relative to $\partial\Omega_n$ (see for instance [72]). By Gauss lemma $|\nabla\rho| = 1$ and $\nabla\rho = -\nu$ on $\partial\Omega_n$. Let

$$\Omega_n^\varepsilon = \{x \in \Omega_n : \rho(x) > \varepsilon\}$$

for some $\varepsilon > 0$ sufficiently small and define the Lipschitz function

$$\psi_\varepsilon(x) = \begin{cases} 1, & \text{if } x \in \Omega_n^\varepsilon; \\ \rho(x)/\varepsilon, & x \in \Omega_n \setminus \Omega_n^\varepsilon; \\ 0, & x \in M \setminus \Omega_n. \end{cases} \tag{2.81}$$

Let $\varphi \in C_c^\infty(M \setminus B_R)$, $\varphi \geq 0$. Since γ satisfies (2.80) in $(M \setminus B_R) \cap \Omega_n$ and $\varphi\psi_\varepsilon \in W_0^{1,2}((M \setminus B_R) \cap \Omega_n)$ we have

$$\int_{\Omega_n} (\lambda\gamma)(\varphi\psi_\varepsilon) \geq -\int_{\Omega_n} \langle\nabla\gamma, \nabla(\varphi\psi_\varepsilon)\rangle = -\int_{\Omega_n} \langle\nabla\gamma, \nabla\varphi\rangle\psi_\varepsilon - \frac{1}{\varepsilon}\int_{\Omega_n\setminus\Omega_n^\varepsilon} \langle\nabla\gamma, \nabla\rho\rangle\varphi.$$

Therefore, by the coarea formula,

$$\int_{\Omega_n} (\lambda\gamma)(\varphi\psi_\varepsilon) \geq -\int_{\Omega_n} \langle\nabla\gamma, \nabla\varphi\rangle\psi_\varepsilon - \frac{1}{\varepsilon}\int_0^\varepsilon dt \int_{\partial\Omega_n^t} \langle\nabla\gamma, \nabla\rho\rangle\varphi,$$

where $\Omega_n^t = \{x \in \Omega_n : \rho(x) > t\}$. Letting $\varepsilon \to 0^+$, we get

$$\int_{\Omega_n} \lambda\gamma\varphi \geq -\int_{\Omega_n} \langle\nabla\gamma, \nabla\varphi\rangle + \int_{\partial\Omega_n} \alpha'(r)\langle\nabla r, \nu\rangle\varphi.$$

Since Ω_n is star-shaped with respect to o and $\alpha' \geq 0$ we deduce

$$\int_{\Omega_n} (\langle\nabla\gamma, \nabla\varphi\rangle + \lambda\gamma\varphi) \geq 0.$$

Now $\mathrm{cut}(o)$ has measure zero, and letting $n \to +\infty$ we finally obtain

$$\int_{M\setminus B_R} (\langle\nabla\gamma, \nabla\varphi\rangle + \lambda\gamma\varphi) \geq 0,$$

showing that (2.80) is satisfied in the weak sense on $M \setminus B_R$. □

Remark 2.12 Considering that $\gamma(x) = \alpha(r(x))$, with $\alpha' \geq 0$, a proof analogous to that of Theorem 2.5 applies here too.

As a consequence of Theorem 2.12 we have the following result extending Varopoulos [267] (see also [149, 202]), which detects the maximum amount of negative curvature that can be allowed without destroying stochastic completeness.

Theorem 2.13 *Let $(M, \langle\ ,\ \rangle)$ be a complete Riemannian manifold of dimension m, let $o \in M$ be a fixed origin and denote by $r(x)$ the Riemannian distance function from o. Assume that the radial Ricci curvature satisfies*

$$\operatorname{Ric}(\nabla r, \nabla r) \geq -(m-1)G^2(r),$$

for some positive nondecreasing continuous function G with

$$\frac{1}{G} \notin L^1(+\infty). \tag{2.82}$$

Then M is stochastically complete.

Proof Note that without loss of generality we can further suppose that $G \in C^\infty(\mathbb{R}_0^+)$ and $G^{(2k+1)}(0) = 0$ for $k = 0, 1, 2, \ldots$. We let $g(t)$ be the positive solution of the Cauchy problem given by

$$\begin{cases} g'' - G^2(t)g = 0 & \text{on } \mathbb{R}_0^+, \\ g(0) = 0, \quad g'(0) = 1. \end{cases} \tag{2.83}$$

Observe that our assumptions on G imply that $g^{(2k)}(0) = 0$ for each $k = 1, 2, \ldots$. By the Laplacian comparison theorem (see Theorem 1.2), we have that

$$\Delta r(x) \leq (m-1)\frac{g'(r(x))}{g(r(x))}$$

on $M \backslash (\{o\} \cup \operatorname{cut}(o))$. Therefore, by Theorem 2.12, it is enough to show that the model M_g is stochastically complete. From Proposition 2.3, this is equivalent to show that

$$\frac{\operatorname{vol}(B_R)}{\operatorname{vol}(\partial B_R)} \notin L^1(+\infty), \tag{2.84}$$

with

$$\frac{\operatorname{vol}(B_t)}{\operatorname{vol}(\partial B_t)} = \frac{\int_0^t g^{m-1}(s)ds}{g^{m-1}(t)}.$$

Similarly to what we did in Theorem 2.5, Eq. (2.33), we define the function $h(t)$ on $\mathbb{R}_0^+$ setting

$$h(t) = \frac{1}{G(0)}\left(e^{\int_0^t G(s)ds} - 1\right).$$

Then h is a subsolution to the Cauchy problem (2.83), and by Sturm comparison theorem we have

$$\frac{g'(t)}{g(t)} \le \frac{h'(t)}{h(t)}.$$

Since

$$\frac{h'(t)}{h(t)} = G(t)\frac{e^{\int_0^t G(s)}ds}{e^{\int_0^t G(s)}ds - 1} \sim CG(t), \qquad \text{as } t \to +\infty,$$

for some constant $C > 0$, we conclude that

$$\frac{g'(t)}{g(t)} \le CG(t)$$

for some (other) constant $C > 0$, whenever t is sufficiently large. Note that since $G(t) \ge 0$, the function $g(t)$ diverges to infinity as $t \to +\infty$. Then, one shows (2.84) using (2.82) and de l'Hospital's rule. □

It is clear that Theorem 2.8 also gives useful information on the study of bounded above solutions of differential inequalities of the form $\Delta u \ge f(u)$. Indeed the statement

(v) *For every function $f \in C^0(\mathbb{R})$ and every $u \in C^2(M)$ with $u^* = \sup_M u < +\infty$ solving the differential inequality $\Delta u \ge f(u)$, we have $f(u^*) \le 0$*

is equivalent to any of the statements (i) to (iv) of that theorem. To see this simply observe that (v) is clearly implied by (iii) and it implies (ii).

We are now going to extend our investigation to a more general class of differential inequalities which includes those of the type $\Delta u \ge b(x)f(u)$. This in part justifies the study of the class of operators that we introduce next. For a definitely more compelling reason, see Chap. 8.

Let A, b, V be smooth functions on M with $A, b > 0$ and $V \ge 0$. We consider the elliptic operator defined by

$$Lu = \frac{1}{b}\left(\operatorname{div}(A\nabla u) - Vu\right), u \in C^2(M). \tag{2.85}$$

Then $-L$ is a positive, symmetric operator on $C_c^\infty(M) \subset L^2(M, b(x)dx)$, where dx is the Riemannian volume element. For ease of notation, we use the same symbol L

to denote the Friedrichs extension of L (note that L may fail to be essentially self-adjoint on $C_c^\infty(M)$, but it is so for instance in the important case where $V \equiv 0$). The following result is proved in Sect. 3 of [227, Theorem 3.11]. We refer to this paper for the proof and a discussion on it and related results.

Theorem 2.14 *Let $(M, \langle\, ,\, \rangle)$ be a Riemannian manifold and let $L = b^{-1}\,(\mathrm{div}(A\nabla) - V)$ where A, b, V are smooth functions with $A, b > 0$ and $V \geq 0$. Consider the following statements:*

(i) If $u \in C^2(M)$ is such that $u \geq 0$, $u^* < +\infty$ and $Lu \geq \lambda u$ for some $\lambda > 0$, then u vanishes identically.
(ii) If $u \in C^2(M)$ is such that $u \geq 0$, $u^* < +\infty$ and $Lu = \lambda u$ for some $\lambda > 0$, then u vanishes identically.
(iii) For every $u \in C^2(M)$ with $u^* < +\infty$ and every $\varepsilon > 0$,

$$\inf_{\Omega_\varepsilon} Lu \leq 0$$

where $\Omega_\varepsilon = \{x \in M : u(x) > u^* - \varepsilon\}$.
(iv) For every $u \in C^2(M)$ with $u^* < +\infty$ there exists a sequence $\{x_k\} \subset M$ such that

$$u(x_k) > u^* - \frac{1}{k}, \quad Lu(x_k) < \frac{1}{k}$$

for every $k \in \mathbb{N}$.
(v) Any nonnegative bounded solution $u(x,t)$ of

$$\begin{cases} Lu \geq \frac{\partial u}{\partial t}, & \text{on } M \times (0,T); \\ u(x, 0^+) = 0, & \text{in the } L^1_{loc}(M, b(x)dx) \text{ sense} \end{cases}$$

is identically zero.
(vi) Any nonnegative bounded solution $u(x,t)$ of

$$\begin{cases} Lu = \frac{\partial u}{\partial t}, & \text{on } M \times (0,T); \\ u(x, 0^+) = 0, & \text{in the } L^1_{loc}(M, b(x)dx) \text{ sense} \end{cases}$$

is identically zero.

Then, the following chain of implications holds under the additional assumption specified on the corresponding dashed implication arrow

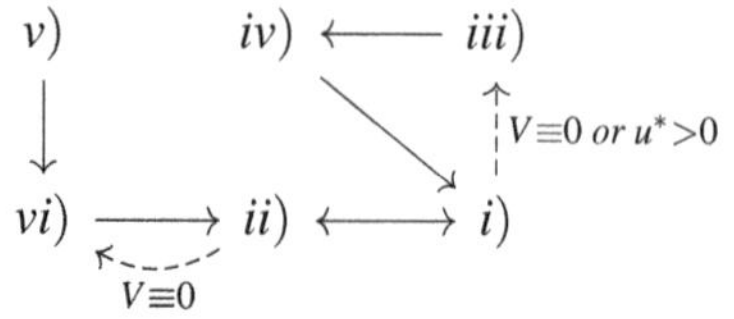

$$\begin{array}{ccccc} v) & & iv) & \longleftarrow & iii) \\ \downarrow & & & \searrow & \uparrow{\scriptstyle V\equiv 0 \text{ or } u^*>0} \\ vi) & \longrightarrow & ii) & \longleftrightarrow & i) \\ & \underset{V\equiv 0}{\dashleftarrow} & & & \end{array}$$

In particular, if $V \equiv 0$ all of the previous items but (v) are equivalent.

We note that actually the proof shows that in (i) and (ii) we may replace the condition that u is C^2 with the weaker assumption that $u \in W^{1,2}_{loc}(M)$ and that the (in)equality $Lu = \lambda u$ ($\geq \lambda u$) holds in the weak sense.

We generalize the definition of the weak maximum principle for the class of operators $L = b^{-1}(\mathrm{div}(A\nabla) - V)$ as follows.

Definition 2.4 Let $(M, \langle\, ,\, \rangle)$ be a Riemannian manifold. We say that L satisfies the weak maximum principle on M if for every $u \in C^2(M)$ with $u^* < +\infty$ there exists a sequence $\{x_k\} \subset M$ such that

$$u(x_k) > u^* - \frac{1}{k}, \quad Lu(x_k) < \frac{1}{k}$$

for every $k \in \mathbb{N}$.

In case $V \equiv 0$ it follows easily from Theorem 2.14 that the weak maximum principle for the operator L is equivalent to the validity of the following condition:

(vii) *For every $f \in C^0(\mathbb{R})$ and every $u \in C^2(M)$ with $u^* < +\infty$ solving the differential inequality* $\mathrm{div}(A\nabla u) \geq b(x)f(u)$ *we have* $f(u^*) \leq 0$.

One can find sufficient conditions for the validity of the weak maximum principle for the operator L. For instance we have

Theorem 2.15 *Let $(M, \langle\, ,\, \rangle)$ be a complete Riemannian manifold, and assume that, for some reference point,*

$$\frac{r^{1-\mu}}{\log \mathrm{vol}\, B_r} \notin L^1(+\infty), \tag{2.86}$$

for some $\mu \in \mathbb{R}$. Then, for every $u \in C^2(M)$ with $u^* = \sup_M u < +\infty$, and for every $\varepsilon > 0$, it holds

$$\inf_{\Omega_\varepsilon}(1 + r)^\mu \Delta u \leq 0,$$

where $\Omega_\varepsilon = \{x \in M : u(x) > u^* - \varepsilon\}$.

Remark 2.13 Condition (2.86) implies $\mu \leq 2$. In case $\mu = 2$, and as an application of [260], (2.86) can be improved to

$$\frac{\log r}{r \log \mathrm{vol}\, B_r} \notin L^1(+\infty). \tag{2.87}$$

The proof of Theorem 2.15 elaborates on some ideas of Grigor'yan and, in particular, uses heat equation techniques. Since this approach is different from that we shall follow in the sequel, we report here the argument taken from [227]. In fact we shall prove the next more general

Theorem 2.16 *Let $(M, \langle\ ,\ \rangle)$ be a complete Riemannian manifold and let A, b, V be smooth, $A, b > 0$, $V \geq 0$ and such that*

$$b(x) \geq \frac{H}{r(x)^{\mu}}; \quad A(x) \leq Kr(x)^{\gamma} \tag{2.88}$$

on $M \setminus B_{R_0}$ for some R_0, H, $K > 0$, γ, $\mu \in \mathbb{R}$. Let $u(x,t) \in C^0(M \times (0,T])$ for some $T > 0$ be a nonnegative C^2 in the space variable x and C^1 in t solution of the problem

$$\begin{cases} \operatorname{div}(A\nabla u) - Vu \geq b(x)\frac{\partial u}{\partial t} & \text{on } M \times (0,T] \\ u|_{t\to 0+} = 0, \end{cases} \tag{2.89}$$

where the initial data is considered in the $L^2_{loc}(M, b(x)dx)$ sense. Assume that, for every $R \gg 1$,

$$\int_0^T \int_{B_{2R}\setminus B_{\frac{3}{2}R}} A(x)u(x,t)^2\, dxdt \leq e^{f(2R)}, \tag{2.90}$$

where f is a positive function defined for $r \gg 1$ and such that

$$\frac{r^{1-\gamma-\mu}}{f(r)} \notin L^1(+\infty). \tag{2.91}$$

Then u vanishes identically in $M \times (0,T]$.

Remark 2.14 If u is bounded and $u|_{t\to 0+} = 0$ in the $L^1_{loc}(M, b(x)dx)$ sense, then the equality also holds in the $L^2_{loc}(M, b(x)dx)$ sense.

Remark 2.15 Theorem 2.16 generalizes results of Grigor'yan, see for instance [131, Theorem 9.2].

The proof of Theorem 2.16 will follow immediately by combining the next two lemmas.

Lemma 2.2 *Let $A, b > 0$ on M and let $f(t)$ be a positive nondecreasing function defined for $t \gg 1$. Suppose that*

$$b(x) \geq \frac{H}{r(x)^{\mu}} \quad \text{for } r(x) \gg 1 \quad \text{and some } H > 0, \mu \in \mathbb{R}, \tag{2.92}$$

$$A(x) \leq Kr(x)^{\gamma} \quad \text{for } r(x) \gg 1 \quad \text{and some } K > 0, \gamma \in \mathbb{R}. \tag{2.93}$$

Let $u \in C^0(M \times (0,T])$, $T > 0$ *be such that* $u|_{t\to 0^+} = 0$ *in the* $L^2_{loc}(M, b(x)dx)$ *sense, and assume that, for every* $0 < \tau < T$, $R \gg 1$ *and* δ *satisfying*

$$0 < \delta \leq \min\left\{\tau, \left(\inf_{\overline{B}_{2R}\setminus B_R} \frac{b(x)}{A(x)}\right)\frac{R^2}{16f(2R)}\right\}, \tag{2.94}$$

we have

$$\int_{B_R} u(x,\tau)^2 b(x) \leq \int_{B_{2R}} u(x,\tau-\delta)^2 b(x) + \frac{C^2}{R^2}, \tag{2.95}$$

for some absolute constant $C > 0$. *If*

$$\frac{r^{1-\gamma-\mu}}{f(r)} \notin L^1(+\infty), \tag{2.96}$$

then $u \equiv 0$ *in* $M \times (0,T]$.

Proof We fix $R \gg 1$ and $\tau \in (0,T]$ and for each $k \in \mathbb{N}$ we define

$$R_k = 2^k R \qquad \text{and } \delta_k \in (0,\tau]$$

such that

$$\begin{aligned}\delta_k &\leq \left(\inf_{\overline{B}_{2R_k}\setminus B_{R_k}} \frac{b(x)}{A(x)}\right)\frac{R_k^2}{16f(2R_k)} \\ &= \left(\inf_{\overline{B}_{R_{k+1}}\setminus B_{R_k}} \frac{b(x)}{A(x)}\right)\frac{R_{k+1}^2}{64f(R_{k+1})}.\end{aligned} \tag{2.97}$$

We also define inductively a decreasing sequence $\{\tau_k\}$ setting

$$\tau_0 = \tau, \quad \tau_{k+1} = \tau_k - \delta_k. \tag{2.98}$$

If τ_k and τ_{k+1} are positive, assumption (2.95) implies that

$$\int_{B_{R_k}} b(x)u(x,\tau_k)^2 \leq \int_{B_{R_{k+1}}} b(x)u(x,\tau_{k+1})^2 + \frac{C^2}{R_k^2}. \tag{2.99}$$

Indeed, choosing $\delta = \tau_k - \tau_{k+1}$ we have

$$\tau_k - \delta = \tau_{k+1}$$

and $\delta = \delta_k < \min\left(\tau, \left(\inf_{\overline{B}_{2R_k}\setminus B_{R_k}} \frac{b(x)}{A(x)}\right)\frac{R_k^2}{16f(2R_k)}\right)$ as required for the validity of (2.95). Since $u|_{t=0^+} = 0$ in the $L^2_{loc}(M, b(x)dx)$ sense, the same inequality continues to hold, passing to the limit, even if $\tau_{k+1} = 0$, in which case we have

$$\int_{B_{R_{k+1}}} b(x)u\big(x, 0^+\big)^2 = 0. \tag{2.100}$$

Thus, if we can show that the sequence τ_k can be chosen in such a way that $\tau_{\bar{k}} = 0$ for some finite $\bar{k}$, iterating (2.99) and using (2.100) we obtain

$$\int_{B_R} b(x)u(x,\tau)^2 \leq C^2 \sum_{k=0}^{\bar{k}} \frac{1}{R_k^2} < \frac{C^2}{2R^2}.$$

Letting $R \to +\infty$, we deduce that $u(\cdot, \tau) \equiv 0$ and this holds for each $\tau \in (0, T]$; then $u \equiv 0$ in $M \times (0, T]$.

Having fixed $\tau \in (0, T)$, the sequence $\{\tau_k\}$ in (2.98) will reach 0 for some finite index $\bar{k}$ if

$$\tau = \delta_0 + \delta_1 + \ldots + \delta_{\bar{k}}. \tag{2.101}$$

Towards this end, note that if

$$\sum_{k=1}^{+\infty} \left(\inf_{\overline{B}_{R_k}\setminus B_{R_{k-1}}} \frac{b(x)}{A(x)}\right)\frac{R_k^2}{f(R_k)} = +\infty \tag{2.102}$$

then we may choose the sequence δ_k in such a way that (2.97) holds for every k and

$$\sum_{k=0}^{+\infty} \delta_k = +\infty.$$

Thus, by possibly making some δ_k smaller, we can find $\bar{k}$ in such a way that (2.101) holds.

Hence, it remains to prove that (2.102) is satisfied. Taking into account (2.92) and (2.93), this amounts to showing that

$$\sum_{k=0}^{+\infty} \frac{R_k^{2-\gamma-\mu}}{f(R_k)} = +\infty,$$

which in turn follows easily from (2.96) and the fact that f is nondecreasing. □

In the next result we see how to guarantee condition (2.95).

Lemma 2.3 *Let A, b and $(M, \langle\, ,\, \rangle)$ be as in the previous Lemma. Let $u(x,t) \in C^0(M \times (0,T])$, for some $T > 0$, be C^2 in the space variable x and C^1 in t, and assume that it is a solution of*

$$u \operatorname{div}(A\nabla u) \geq b(x)u\frac{\partial u}{\partial t} \quad \text{on } M \times (0,T]. \tag{2.103}$$

Assume also that

$$\int_0^T \int_{B_{2R}\setminus B_{\frac{3}{2}R}} A(x)u(x,t)^2\, dxdt \leq e^{f(2R)} \quad \text{for } R \gg 1, \tag{2.104}$$

where f is a positive function defined for $r \gg 1$. Then, for every $0 < \tau < T$ and for every δ satisfying

$$0 < \delta \leq \min\left(\tau, \left(\inf_{\overline{B}_{2R}\setminus B_R} \frac{b(x)}{A(x)}\right)\frac{R^2}{16f(2R)}\right), \tag{2.105}$$

we have

$$\int_{B_R} u(x,\tau)^2 b(x) \leq \int_{B_{2R}} u(x,\tau-\delta)^2 b(x) + \frac{C^2}{R^2}, \tag{2.106}$$

for each $R \gg 1$ and for some absolute constant $C > 0$.

Proof Let $R \gg 1$ be chosen so that (2.104) holds, and let η be a smooth cutoff function satisfying

$$\operatorname{supp}\eta \subseteq B_{2R}, \quad \eta \equiv 1 \text{ on } B_{\frac{3}{2}R}, \quad 0 \leq \eta \leq 1, \quad |\nabla\eta| \leq \frac{C}{R},$$

for a constant $C > 0$ independent of R. Let $\xi(x,t)$ be a Lipschitz function in the x variable for each $t \in [0,T]$ to be specified later. Consider the nonnegative function $\eta^2 e^{\xi} \in \mathrm{Lip}_c(M)$; interpreting (2.103) in the weak sense we have

$$\int_{B_{2R}} \eta^2 e^{\xi} bu\frac{\partial u}{\partial t} \leq -\left(\int_{B_{2R}} 2\eta u e^{\xi} A\langle\nabla u, \nabla\eta\rangle + \eta^2 u e^{\xi} A\langle\nabla\xi, \nabla u\rangle + \eta^2 e^{\xi} A|\nabla u|^2\right),$$

from which, using the Cauchy-Schwarz inequality we deduce

$$\int_{B_{2R}} \eta^2 e^{\xi} bu\frac{\partial u}{\partial t} \leq 2\int_{B_{2R}} \eta|u|e^{\xi}A|\nabla u||\nabla\eta| + \eta^2|u|e^{\xi}A|\nabla\xi||\nabla u| - \eta^2 e^{\xi}A|\nabla u|^2.$$

We now apply the elementary inequality

$$ab \leq \frac{\varepsilon^2}{2}a^2 + \frac{b^2}{2\varepsilon^2}, \quad a, b > 0, \varepsilon > 0$$

to the first two integrand in the right-hand side to obtain, after some manipulations,

$$\int_{B_{2R}} \eta^2 e^\xi bu \frac{\partial u}{\partial t} \leq 2 \int_{B_{2R}} u^2 e^\xi A|\nabla \eta|^2 + \frac{1}{2} \int_{B_{2R}} \eta^2 u^2 e^\xi A|\nabla \xi|^2.$$

Therefore, integrating the above inequality over $[\tau - \delta, \tau]$ and then integrating by parts the left-hand side and rearranging we obtain

$$\frac{1}{2} \int_{B_{2R}} \eta^2 e^\xi bu^2 \Big|_{\tau-\delta}^{\tau} \leq \int_{\tau-\delta}^{\tau} \int_{B_{2R}} \left(\frac{1}{2} u^2 \eta^2 e^\xi \left[b \frac{\partial \xi}{\partial t} + A|\nabla \xi|^2 \right] + 2u^2 e^\xi A|\nabla \eta|^2 \right). \tag{2.107}$$

We now specify the function ξ. Let

$$\rho(x) = \begin{cases} 0 & \text{if } x \in B_R \\ \text{dist}\,(x, o) - R & \text{otherwise}, \end{cases} \tag{2.108}$$

and note that $|\nabla \rho| = 1$ on $M \setminus B_R$. To simplify notation set

$$a_R = \inf_{\overline{B}_{2R} \setminus B_R} \frac{b(x)}{A(x)}, \tag{2.109}$$

and observe that, since $A, b > 0$ on M, $a_R > 0$. Finally define

$$\xi(x, t) = \frac{-a_R \rho(x)^2}{2(\tau + \delta - t)} \quad \text{on } M \times [\tau - \delta, \tau],$$

so that

$$\xi(x, t) \leq 0 \quad \text{on } M \times [\tau - \delta, \tau] \tag{2.110}$$

$$\xi(x, t) = 0 \quad \text{on } B_R \times [\tau - \delta, \tau]. \tag{2.111}$$

A simple computation that uses (2.109) shows that, on $\overline{B}_{2R} \setminus B_R$,

$$b \frac{\partial \xi}{\partial t} + A|\nabla \xi|^2 \leq 0 \quad \forall\, t \in [\tau - \delta, \tau], \tag{2.112}$$

and this holds on B_R as well, since ξ vanishes identically there. Since $\rho(x) \geq \frac{R}{2}$ on $B_{2R} \setminus B_{\frac{3}{2}R}$ and $\tau + \delta - t \leq 2\delta$ for $t \in [\tau - \delta, \tau]$, we also have

$$\xi(x,t) \leq -\frac{a_R R^2}{16\delta} \quad \text{on} \ \left(B_{2R} \setminus B_{\frac{3}{2}R}\right) \times [\tau - \delta, \tau]. \tag{2.113}$$

Inserting (2.112) into (2.110), using the properties of η, (2.113), (2.110) and (2.111) we obtain

$$\frac{1}{2}\int_{B_R} b(x)u(x,\tau)^2 \leq \frac{1}{2}\int_{B_{2R}} b(x)u(x,\tau-\delta)^2 + \frac{C^2}{R^2}\int_{\tau-\delta}^{\tau}\int_{B_{2R}\setminus B_{\frac{3}{2}R}} A(x)u^2 e^{-\frac{a_R R^2}{16\delta}}. \tag{2.114}$$

According to assumption (2.104), the second integral on the right-hand side is bounded above by

$$\int_{B_{2R}} b(x)u(x,\tau-\delta)^2 + \frac{C^2}{R^2}$$

and the required conclusion follows. $\square$

For the sake of completeness, we end the section with some observations on the difference of applicability between the weak maximum principles for the Laplacian and for the full Hessian operator (for instance, see Theorem 5.7 and Corollary 5.6 of Chap. 5). A first striking difference is pointed out by Proposition 2.4 below (see Proposition 40 of [228]), which states that every Riemannian manifold satisfying the weak maximum principle for the Hessian must be *nonextendible* (that is, non-isometric to any proper open subset of another connected Riemannian manifold $(N, (\ ,\ ))$). For example, for every Riemannian manifold M and $p \in M$, $M \setminus \{p\}$ does not satisfy the weak maximum principle for the Hessian.

Proposition 2.4 *Suppose that $(M, \langle\ ,\ \rangle)$ satisfies the weak maximum principle for the Hessian. Then $(M, \langle\ ,\ \rangle)$ is nonextendible.*

Proof By contradiction, suppose the contrary and let $p \in \partial M$, the boundary of M in N. Define $r(x) = \mathrm{dist}_N(x, p)$. Next, fix $0 < R < \mathrm{inj}_N(p)$ and let $u \in C^\infty(N \setminus \{p\}) \cup C^0(N)$ be a radial nonincreasing function such that

$$u(x) = \begin{cases} e^{-r(x)} & \text{if } r(x) < \frac{R}{2} \\ 0 & \text{if } r(x) > R. \end{cases}$$

Clearly $u \in C^\infty(M)$ is bounded from above with

$$u^* = u(p) = L.$$

A computation shows that

$$\mathrm{Hess}(u)(\nabla r, \nabla r) \geq e^{-r} \geq e^{-R/2}$$

on ${}^N B_{R/2}(p) \setminus \{p\}$. Since any sequence $\{x_k\} \subset M$ along which u attains its supremum must be eventually contained in ${}^N B_{R/2}(p) \setminus \{p\}$, we conclude that the weak maximum principle for the Hessian is not satisfied on M, which is a contradiction. □

Since, by Theorem 2.5, geodesic completeness and a well-behaved sectional curvature imply the full Omori-Yau maximum principle for the Hessian, one might ask if, keeping a well-behaved sectional curvature and relaxing geodesic completeness to the property of nonextendibility, one could prove the validity of the weak maximum principle for the Hessian. This is false, as the following simple counterexample shows. Consider the standard cone in the Euclidean space $\mathbb{R}^3$ given by

$$M = \{x = (x_1, x_2, x_3) \neq (0,0,0) : x_3 = \sqrt{x_1^2 + x_2^2}\}.$$

In polar coordinates (r, θ), where $r \in \mathbb{R}^+$ and $\theta \in [0, 2\pi)$, the cone can be parameterized as $x_1 = r\cos\theta$, $x_2 = r\sin\theta$, $x_3 = r$. Therefore, the induced metric reads

$$ds^2 = 2dr^2 + r^2 d\theta^2;$$

this shows that the cone is trivially nonextendible as a Riemannian manifold (every such extension N must contain only one point not in M, but the metric is singular in $r = 0$). However, since M is a flat embedded hypersurface trivially contained into a nondegenerate cone, because of Theorem 5.7 of Chap. 5 the weak maximum principle for the Hessian necessarily fails. Nevertheless, M is stochastically complete; indeed, from the form of the metric we deduce that the normal projection onto the hyperplane $x_3 = 0$ gives a quasi-isometry between M and $\mathbb{R}^2 \setminus \{(0,0)\}$, preserving divergent sequences and such that the derivatives of the metric on M are controlled by those of $\mathbb{R}^2 \setminus \{(0,0)\}$. Therefore, stochastic completeness follows applying a slight modification of Proposition 3.4 in [227]; see also Proposition 4.1 of Chap. 4.

2.4 Two Applications of Stochastic Completeness

The aim of this section is to give and idea of the use of stochastic completeness via the weak maximum principle, by showing the validity of some geometric results. We begin by proving Theorem 2.17 below, see [187], that shall be used (but only in the compact case originally due to Tachibana, [261]) in Remark 8.6 and in the proof of Theorems 8.8 and 8.9 of Chap. 8.

In order to state the result we need to introduce some terminology. In what follows, the Riemannian manifold $(M, \langle\, , \rangle)$ is always supposed to have dimension $m \geq 3$. Let

$$\mathrm{Riem} = R_{ijk\ell}\theta^i \otimes \theta^j \otimes \theta^k \otimes \theta^\ell \tag{2.115}$$

denote the $(0, 4)$-type Riemann curvature tensor, with respect to a local orthonormal coframe $\{\theta^i\}$. With $\mathfrak{R} : \Lambda^2(M) \to \Lambda^2(M)$ we denote the symmetric endomorphism determined by Riem, that is, if $\theta^i \wedge \theta^j$, $1 \leq i < j \leq m$, is a local basis of $\Lambda^2(M)$

$$\mathfrak{R}(\theta^i \wedge \theta^j) = \sum_{k<\ell} R_{ijk\ell}\theta^k \wedge \theta^\ell = \frac{1}{2}R_{ijk\ell}\theta^k \wedge \theta^\ell. \tag{2.116}$$

Then M is said to have a *positive curvature operator* if there exists a constant $\Lambda > 0$ such that all the eigenvalues of $\mathfrak{R}$ are bounded below by Λ. In other words, for any $\omega \in \Lambda^2(M)$

$$\langle \mathfrak{R}\omega, \omega\rangle \geq \Lambda|\omega|^2. \tag{2.117}$$

It is worth to recall here some formulas and facts presented in Chap. 1. In conformally invariant form, the components of the Weyl and Cotton tensors are respectively given by

$$\begin{aligned} W^i_{jk\ell} = R^i_{jk\ell} &- \frac{1}{m-2}(R_{ik}\delta^j_\ell - R_{jk}\delta^i_\ell + R_{j\ell}\delta^i_k - R_{i\ell}\delta^j_k) \\ &- \frac{S}{(m-1)(m-2)}(\delta^i_k\delta^j_\ell - \delta^i_\ell\delta^j_k) \end{aligned} \tag{2.118}$$

and

$$C_{jsk} = R_{js,k} - R_{jk,s} + \frac{1}{2(m-1)}(S_s\delta_{jk} - S_k\delta_{js}). \tag{2.119}$$

Here S is the scalar curvature and $R_{js,k}$ are the components of the covariant derivative of the Ricci tensor. We recall that for $m = 3$, $W \equiv 0$ always, while, by Theorem 1.1, $W \equiv 0$ for $m \geq 4$ and $C \equiv 0$ for $m = 3$ are equivalent to conformal flatness of the manifold. Recall also that the $(0, 4)$-projective curvature tensor P (see Eq. (1.106) for the $(1, 3)$-version), whose components in the local orthonormal coframe $\{\theta^i\}$ are given by

$$P_{ijkt} = R_{ijkt} - \frac{1}{m-1}(R_{ik}\delta_{jt} - R_{jk}\delta_{it}), \tag{2.120}$$

is zero if and only if the manifold has constant sectional curvature. A simple computation shows that

$$|P|^2 = |\operatorname{Riem}|^2 - \frac{2}{m-1}|\operatorname{Ric}|^2. \tag{2.121}$$

The next result is due to Lichnerowicz [175, p.10]; a general formulation can be found in [53]. We present here a simple computational proof which is a reorganization of the original one by Lichnerowicz. In the next Lemma we denote by div Riem the 3-covariant tensor whose components are given by

$$(\operatorname{div}\operatorname{Riem})_{jkt} = R_{ijkt,i}.$$

Lemma 2.4 *With the above notations*

$$\frac{1}{4}\Delta|\operatorname{Riem}|^2 = K + \frac{1}{2}|\nabla\operatorname{Riem}|^2 - |\operatorname{div}\operatorname{Riem}|^2 + (R_{ijkt}R_{rjkt,r})_i \tag{2.122}$$

where K is the scalar function defined by

$$K = R_{ri}R_{ijkt}R_{rjkt} - \frac{1}{2}R_{ijkt}R_{ijrs}R_{rskt} - 2R_{ijkt}R_{irks}R_{jrts}. \tag{2.123}$$

Proof First we observe that

$$\frac{1}{4}\Delta|\operatorname{Riem}|^2 = \frac{1}{4}(|\operatorname{Riem}|^2)_{tt} = \frac{1}{2}R_{ijk\ell}R_{ijk\ell,tt} + \frac{1}{2}|\nabla\operatorname{Riem}|^2. \tag{2.124}$$

We now consider the commutation relations for the second covariant derivative of the curvature tensor given in (1.122)

$$R_{ijk\ell,st} - R_{ijk\ell,ts} = R_{rjkl}R_{rist} + R_{irk\ell}R_{rjst} + R_{ijr\ell}R_{rkst} + R_{ijkr}R_{r\ell st}. \tag{2.125}$$

Tracing with respect to i and t we get

$$R_{ijk\ell,si} - R_{ijk\ell,is} = R_{rjkl}R_{rs} + R_{irk\ell}R_{rjsi} + R_{ijr\ell}R_{rksi} + R_{ijkr}R_{r\ell si};$$

multiplying both sides by R_{ijkt} and renaming the indices we deduce

$$R_{ijkt}R_{rjkt,ir} = R_{ijkt}R_{rjkt,ri} + \tilde{K} \tag{2.126}$$

where $\tilde{K}$ is the scalar function defined by

$$\tilde{K} = R_{ijkt}R_{ri}R_{rjkt} + R_{ijkt}R_{srkt}R_{rjis} + R_{ijkt}R_{sjrt}R_{rkis} + R_{ijkt}R_{sjkr}R_{rtis}. \tag{2.127}$$

Now we observe that, using the symmetries of the curvature tensor and the second Bianchi identity, we can write

$$R_{ijkt}R_{rjkt,ir} = \frac{1}{2}R_{ijkt}(R_{rjkt,i} - R_{rikt,j})_r = \frac{1}{2}R_{ijkt}(R_{ktrj,i} - R_{ktri,j})_r = \frac{1}{2}R_{ijkt}R_{ijkt,rr}. \tag{2.128}$$

Moreover,

$$R_{ijkt}R_{rjkt,ri} = (R_{ijkt}R_{rjkt,r})_i - R_{ijkt,i}R_{rjkt,r} = (R_{ijkt}R_{rjkt,r})_i - |\operatorname{div}\operatorname{Riem}|^2. \tag{2.129}$$

Inserting (2.128) and (2.129) into (2.126) and using (2.124) we get

$$\frac{1}{4}\Delta|\operatorname{Riem}|^2 = \tilde{K} + \frac{1}{2}|\nabla\operatorname{Riem}|^2 - |\operatorname{div}\operatorname{Riem}|^2 + (R_{ijkt}R_{rjkt,r})_i. \tag{2.130}$$

Next, we show that $\tilde{K} = K$, so that (2.130) proves the validity of (2.122). Note that the first terms in (2.123) and (2.127) are equal. For the remaining terms, we use the first Bianchi identity, the symmetries of the Riemann curvature tensor and we rename some of the indices to get

$$-\frac{1}{2}R_{ijkt}R_{ijrs}R_{rskt} = \frac{1}{2}(R_{ijkt}R_{rijs}R_{rskt} - R_{ijkt}R_{rjis}R_{rskt}) = R_{ijkt}R_{srkt}R_{rjis}$$

and

$$R_{ijkt}R_{sjrt}R_{rkis} = R_{ijkt}R_{sjrt}R_{isrk} = -R_{ijkt}R_{rjst}R_{irks} = -R_{ijkt}R_{jrts}R_{irks},$$

and finally

$$R_{ijkt}R_{sjkr}R_{rtis} = R_{ijkt}R_{rjks}R_{irst} = -RijtkR_{rjts}R_{irks} = -R_{ijkt}R_{jrts}R_{irks},$$

so that

$$R_{ijkt}R_{sjrt}R_{rkis} + R_{ijkt}R_{sjkr}R_{rtis} = -2R_{ijkt}R_{irks}R_{jrts},$$

which implies $K = \tilde{K}$. □

Tachibana [261] has shown the validity of the following

Lemma 2.5 *Let M have positive curvature operator $\mathfrak{R}$, that is,*

$$\langle\mathfrak{R}\omega, \omega\rangle \geq \Lambda|\omega|^2$$

for some constant $\Lambda > 0$ and each $\omega \in \Lambda^2(M)$. Let K be the function defined in (2.123) and P the projective curvature tensor. Then

$$K \geq \frac{\Lambda}{2}(m-1)|P|^2. \tag{2.131}$$

Remark 2.16 Note that Lemma 2.5 also holds if Λ is a positive function on M.

We are now ready to prove the next

Theorem 2.17 *Let $(M, \langle\,,\,\rangle)$ be a stochastically complete Riemannian manifold of dimension $m \geq 3$ with positive curvature operator and scalar curvature S. Assume that either one of the following conditions is satisfied:*

(i) Ric *is parallel.*
(ii) M is locally conformally flat and S is constant.

Then either $|\operatorname{Riem}|^* = \sup_M |\operatorname{Riem}| = +\infty$ or M has positive constant sectional curvature. In particular, this is the case if M is Einstein.

Remark 2.17 Note that if $(M, \langle\,,\,\rangle)$ is also geodesically complete and $|\operatorname{Riem}|^* < +\infty$, then M is compact by Myers theorem [203]. Observe also that conditions (i) and (ii) are necessary for M to have constant sectional curvature.

Proof First of all we observe that under anyone of the conditions (i) or (ii), we have $\operatorname{div}\operatorname{Riem} \equiv 0$. For (i) use (1.67); while for (ii) we note that, from (1.67) and (2.119),

$$R_{ijtk,i} = C_{jkt} + \frac{1}{2(m-1)}\left(S_t\delta_{jk} - S_k\delta_{jt}\right),$$

and from the constancy of S and the local conformal flatness we have $R_{ijkt,i} = 0$. Furthermore, in case (i) $|\operatorname{Ric}|^2$ is obviously constant, while in case (ii) by the usual decomposition (2.118) of the Riemann curvature tensor we immediately deduce

$$|\operatorname{Ric}|^2 = \frac{m-2}{4}|\operatorname{Riem}|^2 + \frac{1}{2(m-1)}S^2. \tag{2.132}$$

Equation (2.122), together with (2.131), (2.121) and $\operatorname{div}\operatorname{Riem} \equiv 0$, yields

$$\frac{1}{4}\Delta|\operatorname{Riem}|^2 \geq \frac{\Lambda}{2}(m-1)|P|^2 = \frac{\Lambda}{2}(m-1)\left(|\operatorname{Riem}|^2 - \frac{2}{m-1}|\operatorname{Ric}|^2\right). \tag{2.133}$$

In case (i), since $|\operatorname{Ric}|^2$ is constant, if $|\operatorname{Riem}|^* < +\infty$, applying the weak maximum principle gives

$$|\operatorname{Riem}|^2 \leq \frac{2}{m-1}|\operatorname{Ric}|^2 \quad \text{on } M,$$

and (2.121) implies $|P| \equiv 0$. In case (ii), using (1.103),

$$\frac{1}{4}\Delta|\,\mathrm{Riem}\,|^2 \geq \frac{\Lambda}{2}(m-1)|P|^2 = \frac{\Lambda}{2}(m-1)\left(\frac{m}{2(m-1)}|\,\mathrm{Riem}\,|^2 - \frac{S^2}{(m-1)^2}\right) \tag{2.134}$$

and again, if $|\,\mathrm{Riem}\,|^* < +\infty$, since S is constant, applying the weak maximum principle we have

$$|\,\mathrm{Riem}\,|^2 \leq \frac{2}{m(m-1)}S^2$$

which, from (2.121) implies $|P| \equiv 0$. □

We give here a second application of stochastic completeness. Let L be a symmetric diffusion operator, of the type $Lu = A^{-1}\operatorname{div}(A\nabla u)$ for some $A \in C^2(M)$, $A > 0$. We are interested in the following problem: if $u \in C^2(M)$,

$$Lu \geq 0 \quad \text{on } M,$$

and $u \in L^1(M, Adx)$, is it true that u is constant? More generally, one could ask for $u \in L^p(M, Adx)$ for some $p \geq 1$. Sometimes positive results in this direction are called *L^p-type Liouville results*. The case $p = +\infty$ corresponds to the usual notion of parabolicity that we will consider in the next Sect. 2.5.

While in case $u \in L^p(M, Adx)$, with $p > 1$, and $u(x_0) > 0$ for some $x_0 \in M$ we have a positive answer (see for instance [243] for a result on general operators that does not cover the present case, but whose proof can be adapted to the purpose), the case $p = 1$ requires some "extra conditions" as shown by the following example.

Example 2.4 Consider the complete manifold given by an m-dimensional model M_g with polar coordinates (r, θ) on $\mathbb{R}^+ \times \mathbb{S}^{m-1}$ as in Definition 1.1 of Chap. 1. We choose

$$g(r) = r \quad \text{on } [0, 1].$$

Define

$$\alpha(r) = \int_0^r g(t)^{1-m} \int_0^t g(s)^{m-1} ds dt$$

and set

$$u(x) = \alpha(r(x)). \tag{2.135}$$

Then $u(x)$ is nonconstant and the computation we did in Example 2.1 shows that u satisfies

$$\Delta u \equiv 1 \quad \text{on } M_g. \tag{2.136}$$

Let $A(x) = \tilde{A}(r(x))$ for some positive function $\tilde{A} \in C^1(\mathbb{R}_0^+)$ with $\tilde{A}'(r) \geq 0$. Since $g \geq 0$ on $\mathbb{R}_0^+$, using (2.136) we have

$$Lu = A^{-1}\operatorname{div}(A\nabla u) = 1 + \frac{\tilde{A}'(r)}{\tilde{A}(r)} \frac{\int_0^r g^{m-1}(s)ds}{g^{m-1}(r)} > 0$$

on M_g. Let now $\varepsilon > 0$ and choose g such that

$$g(r) = \left(\frac{1}{r\log^{1+\varepsilon}(r)e^{r\log^{1+\varepsilon}(r)}}\right)^{1/(m-1)}$$

for $r \gg 1$. With this choice, from the definition of $\alpha(r)$, we have

$$\alpha(r) \sim Ce^{r\log^{1+\varepsilon}(r)} \quad \text{as } r \to +\infty,$$

and

$$\int_{\partial B_R} u \sim \frac{C}{R\log^{1+\varepsilon}(R)} \quad \text{as } R \to +\infty, \tag{2.137}$$

for some constant $C > 0$. Thus, if we require that

$$\tilde{A}(r) \to \Lambda > 0 \quad \text{as } r \to +\infty,$$

then

$$\int_{\partial B_R} uA \sim \frac{C\Lambda}{R\log^{1+\varepsilon}(R)} \quad \text{as } R \to +\infty,$$

and being $\varepsilon > 0$, $u \in L^1(M, Adx)$.

In the next result the role of the "extra condition" is played by L-stochastic completeness.

Theorem 2.18 *Let $(M, \langle\, ,\, \rangle)$ be a complete, L-stochastically complete manifold with $L = A^{-1}\operatorname{div}(A\nabla\,)$ for some $A \in C^2(M)$, $A > 0$. Let $u \in C^2(M) \cap L^1(M, Adx)$ and suppose that u is a nonnegative, L-superharmonic function on M. Then u is constant.*

Proof We reason by contradiction and we assume the existence of a nonconstant $u \geq 0$, $u \in C^2(M) \cap L^1(M, Adx)$, such that $Lu \leq 0$. By the usual maximum principle it follows that, since $u \not\equiv 0$, $u > 0$ on M. We denote by G^L the Green kernel of L. Fix $y_0 \in M$ and observe that $G^L_{y_0}(x) = G^L(x, y_0)$ is L-harmonic on $M \setminus \{y_0\}$. We now show that there exists a compact set K and a constant $C > 0$ such that $y_0 \in K$ and

$$G^L_{y_0} \leq Cu \quad \text{on } M \setminus K. \tag{2.138}$$

Since the singularity of $G^L_{y_0}$ in y_0 is Adx-integrable and $u \in L^1(M, Adx)$, the above inequality yields

$$G^L_{y_0} \in L^1(M, Adx). \tag{2.139}$$

To show (2.138) we fix an exhaustion $\{\Omega_k\}$ of M by relatively compact domains with smooth boundaries and with the property that $B_\varepsilon(y_0) \subset \Omega_1$ for some $\varepsilon > 0$ sufficiently small. Let $\mathcal{G}^L_k$ be the Green kernel in Ω_k, so that $\mathcal{G}^L_k$ satisfies

$$\begin{cases} L\mathcal{G}^L_{k,y_0} = 0 & \text{on } \Omega_k \\ \mathcal{G}^L_{k,y_0} = 0 & \text{on } \partial\Omega_k \end{cases} \tag{2.140}$$

and recall that

$$G^L_{y_0}(x) = \lim_{k \to +\infty} \mathcal{G}^L_{k,y_0}(x), \quad x \neq y_0, \tag{2.141}$$

locally uniformly (see [71]). Fix $C > \sup_{\partial B_\varepsilon(y_0)} G^L_{y_0}$. Then, up to choosing k sufficiently large

$$C > \mathcal{G}^L_{k,y_0}(x) \quad \text{for } x \in \partial B_\varepsilon(y_0).$$

Thus, there exists a constant $\lambda > 0$ sufficiently small such that, for $k \gg 1$,

$$\lambda \mathcal{G}^L_{k,y_0}(x) \leq u(x) \quad \text{for } x \in \partial B_\varepsilon(y_0).$$

Note that this is possible since $u > 0$ on M. Because of (2.140) we also have

$$\lambda \mathcal{G}^L_{k,y_0}(x) \leq u(x) \quad \text{on } \partial\Omega_k.$$

By the usual maximum principle we then deduce

$$\lambda \mathcal{G}^L_{k,y_0}(x) \leq u(x) \quad \text{on } \Omega_k \setminus B_\varepsilon(y_0).$$

Taking $K = \overline{\Omega}_{k_0}$ for some fixed k_0 sufficiently large and using (2.141) we then get (2.138).

Since $y_0 \in M$ was fixed arbitrarily, we have obtained

$$G^L_y(x) \in L^1(M, Adx) \quad \text{for each } y \in M.$$

Using this fact we define

$$v(y) = \int_M G_y^L(x)A(x)dx.$$

Clearly, $v \geq 0$ and $Lv = -1$. Thus the weak maximum principle for the operator L cannot hold on M, and therefore M is not L-stochastically complete. Contradiction. □

Is there any other "natural" condition that could replace L-stochastic completeness in the above result? See [206, 243] for some results in this direction.

2.5 Parabolicity

The aim of this section is to show that the usual notion of parabolicity can be thought as a stronger version of the weak maximum principle. We recall that, according to the well known "Liouville-type" property, we have

Definition 2.5 A manifold $(M, \langle\ ,\ \rangle)$ is *parabolic* if there are no nonconstant bounded above, $C^2(M)$, subharmonic functions on it.

Let us for the moment enlarge the above definition to functions $u \in C^0(M) \cap W^{1,2}_{loc}(M)$. In this case, if $\alpha \in \mathbb{R}$ is any constant, then

$$w(x) = \max\{u(x), \alpha\}$$

is still subharmonic, in the weak sense, anytime u is so. This can be easily seen, but in any case we refer to Proposition 4.3 of Chap. 4 where we prove this fact in a greater generality. Thus, in the enlarged class of functions $C^0(M) \cap W^{1,2}_{loc}(M)$, the above definition is equivalent to:

$$\text{for each } u \in C^0(M) \cap W^{1,2}_{loc}(M), \text{ the properties } \Delta u \geq 0 \text{ and } 0 \leq u \leq u^* < +\infty$$
$$\text{imply that } u \text{ is constant.} \tag{2.142}$$

Now suppose that the property in Definition 2.5 holds for $u \in C^2(M)$ and assume by contradiction that for some $v \in C^0(M) \cap W^{1,2}_{loc}(M)$ we have that v is a nonconstant bounded above solution of $\Delta v \geq 0$ on M. By adding a constant we can in fact suppose that

$$v_* = \inf_M v < 0 < \sup_M v = v^*.$$

It follows that, if $G \geq 0$, $G \not\equiv 0$ is a smooth function with compact support contained in the set $\{x \in M : v(x) \leq 0\}$, then $\gamma_- = \max\{v, 0\}$ is a nonnegative, nonconstant,

$C^0(M) \cap W^{1,2}_{loc}(M)$ solution of

$$\Delta\gamma_- \geq G(x)\gamma_-.$$

Now a sufficiently large positive constant γ_+ satisfies $\gamma_- \leq \gamma_+$ and

$$\Delta\gamma_+ \leq G(x)\gamma_+.$$

Hence by the sub-supersolution method (see for instance [31, 240, 250]) we conclude that the equation

$$\Delta u = G(x)u$$

has a $C^2(M)$ (even smooth) solution u satisfying $0 \leq \gamma_- \leq u \leq \gamma_+$ so that u is a bounded above $C^2(M)$ solution of $\Delta u \geq 0$ which is nonconstant, contradicting the validity of Definition 2.5. Thus we have proven that if the property of Definition 2.5 holds for $C^2(M)$ functions then it holds for $C^0(M) \cap W^{1,2}_{loc}(M)$ functions, the converse being of course trivially true.

Now note that if (2.142) holds in the class $C^2(M)$ and not in the class $C^0(M) \cap W^{1,2}_{loc}(M)$, then similarly to what we have done above we can construct a nonconstant $C^2(M)$ function v satisfying (2.142), contradiction. In other words, (2.142) is equivalent to

$$\text{for each } u \in C^2(M) \text{ the properties } \Delta u \geq 0 \text{ and } 0 \leq u \leq u^* < +\infty$$
$$\text{imply that } u \text{ is constant.}$$

Clearly the latter is similar to the requirement of stochastic completeness expressed in Theorem 2.7 (ii); in fact it is formally the same for $\lambda = 0$. This suggests that (2.142) is equivalent to any one of the following properties:

for each $u \in C^2(M)$, $u^* < +\infty$, u nonconstant on M and for each $\gamma < u^*$,

$$\inf_{\Omega_\gamma} \Delta u < 0, \tag{2.143}$$

where $\Omega_\gamma = \{x \in M : u(x) > \gamma\}$;

$$\text{for each } u \in C^2(M),\ u^* < +\infty,\ u \text{ nonconstant on } M, \text{ there exists } \{x_k\} \subset M$$
$$\text{such that } u(x_k) > u^* - \frac{1}{k}, \quad \Delta u(x_k) < 0 \quad \forall\, k \in \mathbb{N}. \tag{2.144}$$

It is clear that (2.143) and (2.144) are equivalent, but while obviously (2.144) cannot be expressed in a weak form, when $u \in C^0(M) \cap W^{1,2}_{loc}(M)$, (2.143) can be interpreted in a weak sense as

$$\text{there exist } \varepsilon > 0 \text{ and } \psi \in C^\infty_c(M), \psi \geq 0, \psi \not\equiv 0, \text{ such that}$$
$$-\int_M \langle \nabla u, \nabla \psi \rangle \leq -\varepsilon \int_M \psi.$$

Furthermore, it is not hard to show that (2.143) is equivalent to its weak formulation as we described above, that we call (2.143)-weak. Indeed, suppose the validity of (2.143) and that, by contradiction, there exists $u \in C^0(M) \cap W^{1,2}_{loc}(M)$, $u^* < +\infty$, u nonconstant such that for each $\varepsilon > 0$ and for each $\psi \in C^\infty_c(M)$, $\psi \geq 0, \psi \not\equiv 0$,

$$-\int_M \langle \nabla u, \nabla \psi \rangle > -\varepsilon \int_M \psi.$$

Letting $\varepsilon \downarrow 0^+$ this means

$$\Delta u \geq 0 \qquad \text{on } M$$

in the weak sense. Thus, as above we can construct a nonconstant $v \in C^2(M)$, $v^* < +\infty$, such that $\Delta v \geq 0$ on M; in particular, for any $\gamma < v^*$,

$$\inf_{\Omega_\gamma} \Delta v \geq 0,$$

contradicting the validity of (2.143). The other implication is trivial.

Our next step is contained in

Theorem 2.19 *Properties* (2.142) *and* (2.143) *are equivalent in the functional class* $C^0(M) \cap W^{1,2}_{loc}(M)$, *and therefore in the class* $C^2(M)$.

Proof Clearly (2.143)-weak implies (2.142). Viceversa assume (2.142) and let $u \in C^0(M) \cap W^{1,2}_{loc}(M)$ satisfy $u^* < +\infty$, u nonconstant and by contradiction suppose that, for some $\tilde{\gamma} < u^*$,

$$\Delta u \geq 0 \qquad \text{on } \Omega_{\tilde{\gamma}}$$

in the weak sense. Pick $0 < \varepsilon < u^* - \tilde{\gamma}$ and define

$$v(x) = \max\left\{u(x), \tilde{\gamma} + \frac{\varepsilon}{2}\right\}.$$

Then $v \in C^0(M) \cap W^{1,2}_{loc}(M)$ is a bounded above subharmonic function on M and thus by (2.142), which is equivalent to the property in Definition 2.5 on $C^0(M) \cap$

$W^{1,2}_{loc}(M)$, we have that v is constant on M. Since $\Omega_{\tilde{\gamma}+\frac{\varepsilon}{2}} \neq \emptyset$ and $v = u$ on $\Omega_{\tilde{\gamma}+\frac{\varepsilon}{2}}$ we conclude that u is constant on $\Omega_{\tilde{\gamma}+\frac{\varepsilon}{2}}$, contradiction. □

There is a further characterization of parabolicity which expresses in the form of a (classical weak) maximum principle (see for instance Theorem 8.1 in [125]). The original result for surfaces is due to Ahlfors (see Theorem 6 C in [2]).

Theorem 2.20 *The manifold $(M, \langle\ ,\ \rangle)$ is parabolic if and only if for each open set $\Omega \subset M$ with $\partial\Omega \neq \emptyset$ and for each $v \in C^0(\bar{\Omega}) \cap W^{1,2}_{loc}(\Omega)$ satisfying*

$$\begin{cases} \Delta v \geq 0 & \text{on } \Omega \\ \sup_{\Omega} v < +\infty \end{cases} \tag{2.145}$$

we have

$$\sup_{\Omega} v = \sup_{\partial\Omega} v. \tag{2.146}$$

Proof First assume that $(M, \langle\ ,\ \rangle)$ is parabolic and by contradiction suppose that there exist $\Omega \subset M$, $\partial\Omega \neq \emptyset$ open and $v \in C^0(\bar{\Omega}) \cap W^{1,2}_{loc}(\Omega)$ satisfying (2.145) but for which

$$\sup_{\Omega} v > \sup_{\partial\Omega} v.$$

Choose $\varepsilon > 0$ sufficiently small that

$$\sup_{\Omega} v > \sup_{\partial\Omega} v + \varepsilon,$$

and consider the open set

$$\Omega_\varepsilon = \left\{ x \in \Omega : v(x) > \sup_{\Omega} v - \varepsilon \right\} \neq \emptyset.$$

Then $\overline{\Omega_\varepsilon} \subset \Omega$ and therefore

$$u(x) = \begin{cases} \max\left\{ v(x), \sup_{\Omega} v - \varepsilon \right\} & \text{on } \Omega \\ \sup_{\Omega} v - \varepsilon \quad \text{on } M \setminus \Omega \end{cases}$$

defines a $C^0(M) \cap W^{1,2}_{loc}(M)$ solution of $\Delta u \geq 0$ on M. Furthermore, $\sup_M u = \sup_\Omega v < +\infty$. Since $(M, \langle\ ,\ \rangle)$ is parabolic we have that u is constant; since $\Omega_\varepsilon \neq \emptyset$, $u = \sup_\Omega v - \varepsilon$ on Ω_ε contradicting the definition of the latter.

Viceversa, assume the validity of (2.146) and by contradiction suppose that $(M, \langle\, ,\, \rangle)$ is not parabolic. Then there exists a nonconstant function $u \in C^0(M) \cap W^{1,2}_{loc}(M)$ satisfying

$$\begin{cases} \Delta u \geq 0 & \text{on } M \\ u^* = \sup_M u < +\infty. \end{cases} \tag{2.147}$$

Choose $\gamma < u^*$ and let Ω_γ as in (2.143). Since u is nonconstant, then, up to choosing γ sufficiently close to u^*, $\partial\Omega_\gamma \neq \emptyset$. Because of the validity of (2.145) with $v = u_{|\overline{\Omega}_\gamma}$ on $\Omega = \Omega_\gamma$, from (2.146) we deduce

$$u^* = \sup_{\Omega_\gamma} v = \sup_{\partial\Omega_\gamma} v = \gamma,$$

contradiction. □

Remark 2.18 The first part of the proof of the previous theorem is based on the important fact that, for the Laplace-Beltrami operator Δ, the supremum of two subsolutions or at least of a subsolution with a constant is still a subsolution. As far as we know, this fact does not generalize to the entire class of operators we shall consider later on; hence the need for a proof based on a different argument. Let us go back to the reasoning in the first part of the proof and observe that, without loss of generality, by adding a positive constant we can suppose $\sup_\Omega v > 0$. We now choose $\varepsilon > 0$ small enough that $\overline{\Omega_{2\varepsilon}} \subset \Omega$. Clearly

$$\overline{\Omega_\varepsilon} \subset \Omega_{2\varepsilon}.$$

Let $\varphi \in C^\infty(M)$ be a cut-off function such that

$$\varphi \equiv 1 \quad \text{on } \overline{\Omega_\varepsilon}, \quad \varphi \equiv 0 \quad \text{on } \Omega \setminus \Omega_{2\varepsilon}$$

and define

$$u(x) = \begin{cases} \varphi(x)v(x) & \text{on } \Omega \\ 0 & \text{on } M \setminus \Omega. \end{cases}$$

Then $u \in C^0(M) \cap W^{1,2}_{loc}(M)$ and $u^* = \sup_\Omega v < +\infty$. Furthermore, on the upper level set $\Omega_\varepsilon \subset \Omega$ we have

$$\Delta u = \Delta v \geq 0.$$

To obtain a contradiction using (2.143) for functions $u \in C^0(M) \cap W^{1,2}_{loc}(M)$ which are not constant we have to show that Ω_ε is a upper level set for u. Towards this aim we observe that we can choose $\varepsilon > 0$ sufficiently small so that

$$\gamma = \sup_{\Omega} v - \varepsilon > 0.$$

We let

$$\Omega_\gamma = \{x \in M : u(x) > \gamma\};$$

we claim that $\Omega_\gamma = \Omega_\varepsilon$. Indeed, let $x \in \Omega_\varepsilon$; then $v(x) > \gamma > 0$ and since $\varphi(x) = 1$ on $\overline{\Omega_\varepsilon}$, $u(x) = \varphi(x)v(x) = v(x) > \gamma$. Thus $x \in \Omega_\gamma$, or, in other words, $\Omega_\varepsilon \subset \Omega_\gamma$. Suppose now that $x \in \Omega_\gamma$; since $\gamma > 0$, by the definition of u we deduce that $x \in \Omega$ and $v(x) > 0$. Thus

$$v(x) \geq \varphi(x)v(x) = u(x) > \gamma = \sup_{\Omega} v - \varepsilon;$$

in other words, $x \in \Omega_\varepsilon$. Thus

$$\Delta u \geq 0 \qquad \text{on } \Omega_\gamma,$$

contradicting (2.143).

This argument will be used in Chap. 4, where we shall deal with the notion of "parabolicity" for a very general class of operators. Furthermore, in view of the results we shall present there, it is worth to recall, in the present particularly simple setting, at least one sufficient condition for parabolicity of the Laplacian on a complete, noncompact manifold. Towards this aim we follow the classical path used in potential theory of relating parabolicity with capacity. For a wealth of information see the survey article by Grigor'yan [131]. Note however that this approach does not extend in general to nonlinear operators (see Sect. 4.4 in Chap. 4), with the exception of operators strictly related to the p-Laplacian. For this latter case we refer the interested reader to the comprehensive monograph [142].

First we recall the following

Definition 2.6 Let Ω and K be respectively an open and a compact set in M such that $K \subset \Omega$. Define the capacity of K in Ω, $\operatorname{cap}(K, \Omega)$, by setting

$$\operatorname{cap}(K, \Omega) = \inf_{\phi \in D(K,\Omega)} \int_{\Omega} |\nabla \phi|^2,$$

where

$$D(K, \Omega) = \{\phi \in C_c^\infty(\Omega) : 0 \leq \phi \leq 1 \text{ and } \phi \equiv 1 \text{ in a neighborhood of } K\}.$$

If K is relatively compact in Ω we set

$$\operatorname{cap}(K, \Omega) = \operatorname{cap}(\overline{K}, \Omega),$$

and finally, in case $\Omega = M$ we simply set

$$\operatorname{cap}(K) = \operatorname{cap}(K, M).$$

In fact, where needed, we can substitute the space $D(K, \Omega)$ in the definition above with

$$L(K, \Omega) = \{\phi \in \operatorname{Lip}_{loc}(M) : \operatorname{supp}\phi \subset \overline{\Omega}, 0 \leq \phi \leq 1 \text{ and } \phi \equiv 1 \text{ on } K\}. \tag{2.148}$$

We have the following result (see Theorem 5.1 in [131]).

Theorem 2.21 *Let $(M, \langle , \rangle)$ be a Riemannian manifold. Then Δ is parabolic on M if and only if for each compact set K, $\operatorname{cap}(K) = 0$.*

Given for granted the proof of Theorem 2.21 we establish the following estimates that will enable us to provide a "volume growth"-type sufficient condition for the parabolicity of Δ.

Theorem 2.22 *Let $(M, \langle , \rangle)$ be a complete Riemannian manifold and $0 < s < R$. Then we have the following estimates*

$$\frac{1}{\operatorname{cap}(B_s, B_R)} \geq \frac{1}{2}\int_s^R \frac{\rho - s}{\operatorname{vol}(B_\rho) - \operatorname{vol}(B_s)} d\rho, \tag{2.149}$$

and

$$\frac{1}{\operatorname{cap}(B_s, B_R)} \geq \int_s^R \frac{d\rho}{\operatorname{vol}(\partial B_\rho)}. \tag{2.150}$$

In particular,

$$\frac{1}{\operatorname{cap}(B_s)} \geq \frac{1}{2}\int_s^{+\infty} \frac{\rho - s}{\operatorname{vol}(B_\rho) - \operatorname{vol}(B_s)} d\rho, \tag{2.151}$$

and

$$\frac{1}{\operatorname{cap}(B_s)} \geq \int_s^{+\infty} \frac{d\rho}{\operatorname{vol}(\partial B_\rho)}. \tag{2.152}$$

Proof Using the definition of capacity

$$\operatorname{cap}(B_s, B_R) = \inf_{u \in L(\overline{B}_s, B_R)} \int_{B_R} |\nabla u|^2,$$

where $L(\overline{B_s}, B_R)$ is defined in (2.148). To prove (2.149) we fix $\varepsilon > 0$ and we show that we can find $u \in L(\overline{B_s}, B_R)$ such that

$$\int_{B_R} |\nabla u|^2 \leq 2 \left(\int_s^R \frac{t-s}{\mathrm{vol}(B_t) - \mathrm{vol}(B_r) + \varepsilon} dt \right)^{-1}. \tag{2.153}$$

Towards this aim we let $g \in \mathrm{Lip}(\mathbb{R}_0^+)$ be such that

$$g \equiv 1 \text{ on } [0, s) \text{ and } g \equiv 0 \text{ on } (R, +\infty). \tag{2.154}$$

We set $\rho(y) = \mathrm{dist}_M(y, o)$, $o \in M$ a fixed origin, and we define $u = g \circ \rho$. From Gauss lemma $|\nabla u| = |g'(\rho)|$, and therefore

$$\int_{B_R} |\nabla u|^2 = \int_{B_R} |g'(\rho)|^2 = \int_s^R |g'(\rho)|^2 \mathrm{vol}(\partial B_\rho) d\rho. \tag{2.155}$$

We now choose

$$g(\rho) = a \int_\rho^R \frac{t-s}{\mathrm{vol}(B_t) - \mathrm{vol}(B_s) + \varepsilon} dt \quad \text{on } [s, R]$$

with

$$a = \left(\int_s^R \frac{t-s}{\mathrm{vol}(B_t) - \mathrm{vol}(B_s) + \varepsilon} dt \right)^{-1}. \tag{2.156}$$

Observe that $g(s) = 1$ and $g(R) = 0$, so that (2.155) can be verified. Furthermore,

$$g'(\rho) = -a \frac{\rho - s}{\mathrm{vol}(B_\rho) - \mathrm{vol}(B_s) + \varepsilon} \text{ on } [s, R].$$

Substituting into (2.154) and using $(\mathrm{vol}(B_\rho))' = \mathrm{vol}(\partial B_\rho)$ by the coarea formula, together with (2.156), we have

$$\begin{aligned}
\int_{B_R} |\nabla u|^2 &= a^2 \int_s^R \frac{(\rho - s)^2 \mathrm{vol}(\partial B_\rho)}{(\mathrm{vol}(B_\rho) - \mathrm{vol}(B_s) + \varepsilon)^2} d\rho \\
&= -a^2 \int_s^R (\rho - s)^2 \left(\frac{1}{\mathrm{vol}(B_\rho) - \mathrm{vol}(B_s) + \varepsilon} \right)' d\rho \\
&= -a^2 \frac{(\rho - s)^2}{\mathrm{vol}(B_\rho) - \mathrm{vol}(B_s) + \varepsilon} \Big|_s^R + 2a^2 \int_s^R \frac{\rho - s}{\mathrm{vol}(B_\rho) - \mathrm{vol}(B_s) + \varepsilon} d\rho \leq 2a.
\end{aligned}$$

This gives (2.153).

We now prove (2.150). Towards this aim we start from (2.155) but we choose g in a different way. Letting (2.154) to hold, we set

$$g(\rho) = a \int_{\rho}^{R} \frac{dt}{\mathrm{vol}(\partial B_t) + \varepsilon}$$

with

$$a = \left(\int_{s}^{R} \frac{dt}{\mathrm{vol}(\partial B_t) + \varepsilon} \right)^{-1}, \tag{2.157}$$

so that $g(s) = 1$ and $g(R) = 0$. Since

$$g'(\rho) = -\frac{a}{\mathrm{vol}(\partial B_\rho) + \varepsilon},$$

substituting into (2.155) and using (2.157) we obtain

$$\int_{B_R} |\nabla u|^2 = a^2 \int_{s}^{R} \frac{\mathrm{vol}(\partial B_\rho)}{(\mathrm{vol}(\partial B_\rho) + \varepsilon)^2} d\rho \le a^2 \int_{s}^{R} \frac{d\rho}{\mathrm{vol}(\partial B_\rho) + \varepsilon} = a,$$

and (2.150) follows at once. □

Estimates (2.151) and (2.152) together with Theorem 2.21 can be used to obtain the next sufficient conditions for parabolicity.

Theorem 2.23 *Let $(M, \langle , \rangle)$ be a complete Riemannian manifold and suppose that, for some fixed origin o, either*

$$\frac{R}{\mathrm{vol}(B_R)} \notin L^1(+\infty) \tag{2.158}$$

or

$$\frac{1}{\mathrm{vol}(\partial B_R)} \notin L^1(+\infty). \tag{2.159}$$

Then Δ is parabolic on M.

Proof Assume (2.158) holds. Then from (2.151)

$$\mathrm{cap}(B_s) \le 2 \left(\int_{s}^{+\infty} \frac{\rho - s}{\mathrm{vol}(B_\rho) - \mathrm{vol}(B_s)} d\rho \right)^{-1} = 0.$$

Now if K is any compact set, by the Hopf-Rinow theorem there exists $s > 0$ sufficiently large such that $K \subset B_s$. But clearly, from the definition of capacity, $\mathrm{cap}(K) \le \mathrm{cap}(B_s) = 0$. Applying Theorem 2.21 we get the desired conclusion. Similarly for (2.159). □

Remark 2.19 Condition (2.159) does not imply, in general, condition (2.158); see for instance [148]. While (2.159) is necessary and sufficient for a model manifold M_g to be parabolic, in some instances (2.158) is also necessary. This is the case, by a result of Varopoulos [266], when $\mathrm{Ric} \geq 0$ on the complete manifold $(M, \langle , \rangle)$.

On the other hand (2.158) implies (2.159). This is an immediate consequence of the following

Lemma 2.6 *Let $(M, \langle , \rangle)$ be a complete manifold, $h \in C^0(M)$, $h \geq 0$, and set*

$$v(t) = \int_{B_t} h,$$

so that

$$v'(t) = \int_{\partial B_t} h.$$

Fix $s > 0$ and let $R > s$. Then, for each $\delta > 0$,

$$\int_s^R \left(\frac{t-s}{v(t)}\right)^{1/\delta} dt \leq C \int_s^R \left(\frac{1}{v'(t)}\right)^{1/\delta} dt, \tag{2.160}$$

for some constant $C > 0$ independent of R. In particular

$$\left(\frac{t}{v(t)}\right)^{1/\delta} \notin L^1(+\infty) \text{ implies } \left(\frac{1}{v'(t)}\right)^{1/\delta} \notin L^1(+\infty). \tag{2.161}$$

Proof Fix $\varepsilon > 0$ and set

$$v_\varepsilon(t) = \int_{B_t} (h + \varepsilon).$$

From the coarea formula

$$v_\varepsilon'(t) = \int_{\partial B_t} (h + \varepsilon).$$

Applying Hölder's inequalities with conjugate exponents $1 + \delta$ and $1 + 1/\delta$ we obtain

$$\int_s^R \left(\frac{t-s}{v_\varepsilon(t)}\right)^{1/\delta} dt \leq C \left(\int_s^R \left(\frac{t-s}{v_\varepsilon(t)}\right)^{1+1/\delta} v_\varepsilon'(t) dt\right)^{1/(1+\delta)} \left(\int_s^R \frac{dt}{(v_\varepsilon'(t))^{1/\delta}}\right)^{\delta/(1+\delta)} \tag{2.162}$$

Integrating by parts the first integral in the right-hand side of the above inequality we have

$$\int_s^R \left(\frac{t-s}{v_\varepsilon(t)}\right)^{1+1/\delta} v_\varepsilon'(t)dt = -\delta \frac{(R-s)^{1+1/\delta}}{v_\varepsilon(R)^{1/\delta}} + (1+\delta)\int_s^R \left(\frac{t-s}{v_\varepsilon(t)}\right)^{1+1/\delta} dt$$
$$\leq (1+\delta)\int_s^R \left(\frac{t-s}{v_\varepsilon(t)}\right)^{1+1/\delta} dt,$$

and therefore, substituting into (2.162),

$$\int_s^R \left(\frac{t-s}{v_\varepsilon(t)}\right)^{1/\delta} dt \leq (1+\delta)^{1/\delta} \int_s^R \frac{dt}{(v_\varepsilon'(t))^{1/\delta}}. \tag{2.163}$$

By Lebesgue theorem as $\varepsilon \to 0$, v_ε and v_ε' decrease, respectively, to v and v'. Inequality (2.160) then follows by applying the monotone convergence theorem to both members of (2.163). Since

$$\left(\frac{t-s}{v(t)}\right)^{1/\delta} \geq \frac{1}{2^{1/\delta}}\left(\frac{t}{v(t)}\right)^{1/\delta} \quad \text{for } t \geq 2s,$$

it is clear that (2.161) follows from (2.160). □

The following result will be used in the 2-dimensional case in Chap. 9.

Theorem 2.24 *Let M be a Riemannian manifold of dimension m and let $B_R(p)$ be relatively compact in M. Let $u \in C^2(B_R(p))$ satisfy*

$$u\Delta u \geq 0 \tag{2.164}$$

on $B_R(p)$. Then, for $r \in [0, R)$,

$$\int_{B_R(p)} u\Delta u \leq \frac{4}{\int_r^R \frac{dt}{\mathrm{vol}\,(\partial B_t)}} \sup_{B_R(p)} u^2. \tag{2.165}$$

Furthermore, if $\rho(x) = \mathrm{dist}\,(x, p)$ and

$$\mathrm{Ric}\,(\nabla\rho, \nabla\rho) \geq -(m-1)G(\rho) \tag{2.166}$$

on $B_R(p)$ then (2.165) yields, for $r \in [0, R)$,

$$\int_{B_R(p)} u\Delta u \leq \frac{4\omega_m}{\int_r^R \frac{1}{h(t)^{m-1}} dt} \sup_{B_R(p)} u^2, \tag{2.167}$$

where ω_m is the volume of the unit sphere in $\mathbb{R}^m$, provided that the solution h of the Cauchy problem

$$\begin{cases} h'' - G(t)h = 0 \quad on \ [0, R), \\ h(0) = 0, \ h'(0) = 1 \end{cases} \tag{2.168}$$

is positive on $(0, R)$.

Proof Let $\zeta \in D(B_r, B_R)$, where the latter is as in Definition 2.6. Integrating the divergence of the vector field $W = \zeta^2 u \nabla u$ and applying Cauchy-Schwarz and Young's inequalities we obtain

$$\int_{B_R(p)} \zeta^2 \Big(|\nabla u|^2 + u\Delta u \Big) \leq 2 \int_{B_R(p)} \zeta u \langle \nabla \zeta, \nabla u \rangle \leq \int_{B_R(p)} \zeta^2 |\nabla u|^2 + 4 \int_{B_R(p)} u^2 |\nabla \zeta|^2.$$

Hence, using $u\Delta u \geq 0$,

$$\int_{B_R(p)} u\Delta u \leq \int_{B_R(p)} \zeta^2 u\Delta u \leq 4 \sup_{B_R(p)} u^2 \int_{B_R(p)} |\nabla \zeta|^2. \tag{2.169}$$

Taking the infimum on $\zeta \in D(B_r, B_R)$ we deduce

$$\int_{B_R(p)} u\Delta u \leq 4\,\mathrm{cap}\,(B_R, B_r) \sup_{B_R(p)} u^2.$$

Thus, using inequality (2.150) from here we infer (2.165). As for (2.167) simply observe that, by the Bishop-Gromov comparison Theorem 1.3 we have

$$\mathrm{vol}\,(\partial B_t) \leq \omega_m h(t)^{m-1},$$

so that the latter follows at once from (2.165). □

Chapter 3
New Forms of the Maximum Principle

In the previous chapter we described the Omori-Yau maximum principle for the Laplace-Beltrami operator Δ, giving some analytical motivations, and later we introduced the weak maximum principle, illustrating its deep equivalence with stochastic completeness. Furthermore, to show the power and effectiveness of these tools when applied to some specific problem, we gave a few applications to geometry. The aim of the present chapter is to extend the investigation to a much more general class of differential operators containing those that naturally appear when dealing with the geometry of submanifolds or, more generally, in tackling some analytical problems on complete manifolds: for instance, the p-Laplacian, the (generalized) mean curvature operator, trace operators, and so on. In doing so we give sufficient conditions for the validity of two types of maximum principles corresponding, respectively, to the Omori-Yau and to the weak maximum principle. In this chapter we focus our attention on conditions that basically require the existence of a function, indicated throughout with γ, whose existence is, in many instances, guaranteed by the geometry of the problem. First we deal with the linear case, that presents less analytical difficulties, and we conclude our discussion by providing a first a priori estimate; again by way of example, we show its use in a geometric problem. Note that in the next chapter we will provide a second type of sufficient condition for the validity of the weak maximum principle when the operator is in divergence form, basically in terms of the volume growth of geodesic balls with a fixed center on M. Clearly, this kind of condition is very mild and immediately implied by suitable curvature assumptions. We then move to the nonlinear case, where the analytical difficulties that we have to face are definitely deeper; for this reason and for an intrinsic interest, we devote an entire subsection to a careful proof of a general form of some auxiliary analytical results that we shall need for our purposes. We finally prove our general nonlinear results in Theorems 3.11 and 3.13, concluding the chapter.

L.J. Alías et al., *Maximum Principles and Geometric Applications*,
Springer Monographs in Mathematics, DOI 10.1007/978-3-319-24337-5_3

3.1 New Forms of the Weak and Omori-Yau Maximum Principles

Motivated by the discussion and the examples in the previous chapter we now prove a weak maximum principle, Theorem 3.1, an Omori-Yau type maximum principle, Theorem 3.2, and further related results for a large class of linear differential operators of geometrical interest. We shall deal with nonlinear operators in Sect. 3.3.

To describe our first result let T be a symmetric positive semi-definite $(0,2)$-tensor field on M and X a vector field. We set $L = L_{T,X}$ to denote the differential operator acting, say, on $u \in C^2(M)$ by

$$Lu = \operatorname{div}(T(\nabla u,\ )^\sharp) - \langle X, \nabla u\rangle = \operatorname{Tr}(t \circ \operatorname{hess}(u)) + \operatorname{div} T(\nabla u) - \langle X, \nabla u\rangle \tag{3.1}$$

where $^\sharp$ is the musical isomorphism, Tr is the trace and t and $\operatorname{hess}(u)$ are the endomorphisms of TM corresponding, respectively, to T and $\operatorname{Hess}(u)$.

For instance if $T = \langle\ ,\ \rangle$ and X is a vector field on M for $u \in C^2(M)$ we have

$$Lu = \Delta u - \langle X, \nabla u\rangle \tag{3.2}$$

and L coincides with the X-*Laplacian*, denoted by Δ_X, used in the study of general soliton structures, see [188] and also Chap. 8; in particular if $X = \nabla f$ then $L = \Delta_f$ is the f-*Laplacian*, appearing also as the natural symmetric diffusion operator in the study of the *weighted Riemannian manifold* $(M, \langle\ ,\ \rangle, e^{-f}dx)$, [132] (see Chap. 8 for applications to solitons theory). If $T = p(x)\langle\ ,\ \rangle$ for some $p \in C^1(M)$, $p > 0$ on M, and $X \equiv 0$, then $q(x)L$ is (at least on the set where q is positive) a typical (nonsymmetric) diffusion operator. On the other hand, if T is as above and $X = (\operatorname{div} T)^\sharp$, then for $u \in C^2(M)$, Lu becomes the *trace operator*

$$Lu = \operatorname{Tr}(t \circ \operatorname{hess}(u)); \tag{3.3}$$

we will deal with trace operators especially in Chaps. 6 and 7 in a geometric context.

Theorem 3.1 *Let $(M, \langle\ ,\ \rangle)$ be a Riemannian manifold and L be as above. Let $q(x) \in C^0(M)$, $q(x) \geq 0$ and suppose that*

$$q(x) > 0 \text{ outside a compact set.} \tag{3.4}$$

Let $\gamma \in C^2(M)$ be such that

$$\begin{cases} (i) & \gamma(x) \to +\infty \quad \text{as } x \to \infty, \\ (ii) & q(x)L\gamma(x) \leq B \quad \text{outside a compact set} \end{cases} \tag{Γ}$$

for some constant $B > 0$. *If* $u \in C^2(M)$ *and* $u^* < +\infty$, *then there exists a sequence* $\{x_k\} \subset M$ *with the properties*

$$(i)\, u(x_k) > u^* - \frac{1}{k}, \quad \textit{and} \quad (ii)\, q(x_k)Lu(x_k) < \frac{1}{k} \tag{3.5}$$

for each $k \in \mathbb{N}$.

If the conclusion of the theorem holds on M we shall say that the *q-weak maximum principle for the operator L holds on* $(M, \langle\ ,\ \rangle)$. Clearly if $q \equiv 1$, or more generally q is a positive constant, we shall say that the weak maximum principle for the operator L holds on $(M, \langle\ ,\ \rangle)$. Obviously, if the q-weak maximum principle holds for L and $0 \le \hat{q}(x) \le q(x)$, $\hat{q}(x)$ satisfying (3.4), then the $\hat{q}$-weak maximum principle for the operator L also holds.

Remark 3.1 We underline that when $q(x)$ is bounded between two positive constants the validity of the weak maximum principle is equivalent to that of the q-weak maximum principle. In fact it is easy to see that when q is bounded from below by a positive constant, then the q-weak maximum principle implies the weak maximum principle, while the converse occurs when $q(x)$ is bounded from above.

Remark 3.2 We stress that the Riemannian manifold M is not assumed to be geodesically complete. This matches with the fact that for $L = \Delta$ and $q(x) \equiv 1$, conditions (Γ) (i), (ii) are exactly the Khas'minski conditions that we have considered before in Sect. 2.3 of Chap. 2. In fact, as we shall show below in the next subsection, condition (ii) in (Γ) can be substituted, for instance, by

$$\text{(ii)}' \quad q(x)L\gamma(x) \le G(\gamma(x)) \quad \text{outside a compact subset of } M \tag{Γ}$$

where $G \in C^1(\mathbb{R}^+)$ is nonnegative and satisfies

$$\text{(i)} \quad \tfrac{1}{G} \notin L^1(+\infty); \;\; \text{(ii)} \quad G'(t) \ge -A(\log t + 1), \tag{3.6}$$

for $t \gg 1$ and some constant $A \ge 0$. For instance, the functions $G(t) = t$, $G(t) = t \log t$, $t >> 1$, $G(t) = t \log t \log\log t$, $t \gg 1$, and so on, satisfy (i) and (ii) in (3.6) with $A = 0$.

The "Omori-Yau" type version of Theorem 3.1 is as follows.

Theorem 3.2 *Let* $(M, \langle\ ,\ \rangle)$ *be a Riemannian manifold and* L *be as above. Let* $q(x) \in C^0(M)$, $q(x) \ge 0$ *and suppose*

$$q(x) > 0 \quad \textit{outside a compact set.} \tag{3.7}$$

Let $\gamma \in C^2(M)$ be such that

$$\begin{cases} (i) & \gamma(x) \to +\infty \quad \text{as } x \to \infty, \\ (ii) & q(x)L\gamma \le B \quad \text{outside a compact subset of } M, \\ (iii) & |\nabla \gamma| \le B \quad \text{outside a compact subset of } M \end{cases} \tag{Γ_B}$$

for some constant $B > 0$. If $u \in C^2(M)$ and $u^* < +\infty$ then there exists a sequence $\{x_k\} \subset M$ with the properties

$$(i)\ u(x_k) > u^* - \frac{1}{k}, \quad (ii)\ q(x_k)Lu(x_k) < \frac{1}{k}, \quad \text{and } (iii)\ |\nabla u(x_k)| < \frac{1}{k} \tag{3.8}$$

for each $k \in \mathbb{N}$.

If the conclusion of the theorem holds we shall say that the *q-Omori-Yau maximum principle for the operator L holds on $(M, \langle\ ,\ \rangle)$.*

Remark 3.3 Also in this case conditions (ii) and (iii) in (Γ_B) can be replaced by the apparently weaker requirement

$$\begin{cases} (\text{ii})' & q(x)L\gamma \le G(\gamma) \\ (\text{iii})' & |\nabla \gamma| \le G(\gamma) \end{cases} \tag{Γ_B}$$

outside a compact subset of M, where $G \in C^1(R_0^+)$ is a positive function satisfying (3.6) (*i*), (*ii*).

We observe that when $(M, \langle\ ,\ \rangle)$ is a complete, noncompact Riemannian manifold a special candidate for γ, in both Theorems 3.1 and 3.2, is some composition of an at least C^2 function with the distance $r(x)$ from a fixed origin $o \in M$. Of course $r(x)$ is smooth only outside $\{o\} \cup \text{cut}(o)$, where cut(o) is the cut locus of o, but, as we have seen in Theorem 2.5, this problem can be bypassed by elaborating on the old trick of Calabi [55]. In fact, a proof similar to that of Theorem 2.5 holds true. A different way is to understand the differential inequality involving the considered composition with $r(x)$ and the operator L only in the weak-Lip sense, and apply Theorem 5.3 of [236] or Theorem 3.9 together with Remark 3.10 below, instead of Proposition 3.1 in Remark 2.8. We underline that the same arguments, via the comparison principle of Theorem 5.3 in [236], also shows that if $\gamma \in C^1(M)$ satisfies (Γ_B) (i), (iii), and is a classical weak solution of (Γ_B) (ii), then Theorem 3.2 is still valid. The same, of course, applies to Theorem 3.1 (and to the regularity of u), where however we need to make the further requirement $\frac{1}{q} \in L^1_{loc}(M)$ [see the proof of Theorem 3.11, inequality (3.174)].

As a further step, given T and X as above, we introduce the operator $H = H_{T,X}$ acting on $u \in C^2(M)$ by

$$Hu = H_{T,X}u = T(\text{hess}(u)\cdot, \cdot) + (\text{div}T - X^\flat) \otimes du,$$

where $^\flat : TM \to TM^*$ is the inverse of the musical isomorphism $^\sharp$. Observe that $Lu = \mathrm{Tr}(Hu)$. The above theorems admit then the following general versions.

Theorem 3.3 *Let $(M, \langle\ ,\ \rangle)$ be a Riemannian manifold and $H = H_{T,X}$ be as above. Let $q(x) \in C^0(M)$, $q(x) \geq 0$ and suppose that*

$$q(x) > 0 \text{ outside a compact set.} \tag{3.9}$$

Let $\gamma \in C^2(M)$ be such that

$$\begin{cases} (i) & \gamma(x) \to +\infty \quad \text{as } x \to \infty, \\ (ii) & q(x)H\gamma(x)(v,v) \leq B|v|^2 \end{cases} \tag{Γ_C}$$

for some constant $B > 0$ and for every $x \in M \setminus K$, for some compact $K \subset M$, and for every $v \in T_xM$. If $u \in C^2(M)$ and $u^* < +\infty$, then there exists a sequence $\{x_k\} \subset M$ with the properties

$$(i)\, u(x_k) > u^* - \frac{1}{k}, \quad \text{and} \quad (ii)\ q(x_k)Hu(x_k)(v,v) < \frac{1}{k}|v|^2 \tag{3.10}$$

for each $k \in \mathbb{N}$ and every $v \in T_{x_k}M$, $v \neq 0$.

Theorem 3.4 *Let $(M, \langle\ ,\ \rangle)$ be a Riemannian manifold and $H = H_{T,X}$ be as above. Let $q(x) \in C^0(M)$, $q(x) \geq 0$ and suppose that*

$$q(x) > 0 \text{ outside a compact set.} \tag{3.11}$$

Let $\gamma \in C^2(M)$ be such that

$$\begin{cases} (i) & \gamma(x) \to +\infty \quad \text{as } x \to \infty, \\ (ii) & q(x)H\gamma(x)(v,v) \leq B|v|^2, \\ (iii) & |\nabla\gamma(x)| \leq B \end{cases} \tag{Γ_D}$$

for some constant $B > 0$, for every $x \in M \setminus K$, for some compact $K \subset M$, and for every $v \in T_xM$. If $u \in C^2(M)$ and $u^* < +\infty$, then there exists a sequence $\{x_k\} \subset M$ with the properties

$$(i)\ u(x_k) > u^* - \frac{1}{k}, \quad (ii)\ q(x_k)Hu(x_k)(v,v) < \frac{1}{k}|v|^2, \quad \text{and } |\nabla u(x_k)| < \frac{1}{k} \tag{3.12}$$

for each $k \in \mathbb{N}$ and every $v \in T_{x_k}M$, $v \neq 0$.

Similar to what happens in Theorems 3.1 and 3.2, condition (ii) in (Γ_C) and conditions (ii) and (iii) in (Γ_D) can be substituted, respectively, by

$$\text{(ii)}' \quad q(x)H\gamma(x)(v,v) \le G(\gamma)|v|^2 \qquad (\Gamma_C)$$

and

$$\begin{cases} \text{(ii)}' & q(x)H\gamma(x)(v,v) \le G(\gamma)|v|^2 \\ \text{(iii)}' & |\nabla\gamma| \le G(\gamma) \end{cases} \qquad (\Gamma_D)$$

outside a compact subset of M, where $G \in C^1(R_0^+)$ is a positive function satisfying (3.6).

Observe now that Theorem 2.4 is just a particular case of Theorems 3.2 and 3.4 with $q(x) \equiv 1$ and $L = \Delta$.

3.1.1 Proof of Theorem 3.1 and Related Results

In this section we give a proof of Theorem 3.1 and of some companion results.

Proof (of Theorem 3.1) We fix $\eta > 0$ and let

$$A_\eta = \{x \in M : \ u(x) > u^* - \eta\}. \tag{3.13}$$

We claim that

$$\inf_{A_\eta}\{q(x)Lu(x)\} \le 0. \tag{3.14}$$

Note that (3.14) is equivalent to conclusion (3.5) of the theorem.

We reason by contradiction and we suppose that

$$q(x)Lu(x) \ge \sigma_0 > 0 \quad \text{on } A_\eta. \tag{3.15}$$

First we observe that u^* cannot be attained at any point $x_0 \in M$, for otherwise $x_0 \in A_\eta$, $\nabla u(x_0) = 0$, and $Lu(x_0)$ reduces to $Lu(x_0) = \mathrm{Tr}(t \circ \mathrm{hess}(u))(x_0)$, so that, since T is positive semi-definite, $q(x_0)Lu(x_0) \le 0$ contradicting (3.15).

Next we let

$$\Omega_t = \{x \in M : \ \gamma(x) > t\}, \tag{3.16}$$

and define

$$u_t^* = \sup_{x \in \Omega_t^c} u(x). \tag{3.17}$$

Clearly Ω_t^c is closed; we show that it is also compact. In fact, by (Γ) (i) there exists a compact set K_t such that $\gamma(x) > t$ for every $x \notin K_t$. In other words, $\Omega_t^c \subset K_t$ and hence it is compact. In particular, $u_t^* = \max_{x\in\Omega_t^c} u(x)$.

Since u^* is not attained in M and $\{\Omega_t^c\}$ is a telescoping exhaustion of M, we find a divergent sequence $\{t_j\} \subset \mathbb{R}_0^+$ such that

$$u_{t_j}^* \to u^* \quad \text{as } j \to +\infty, \tag{3.18}$$

and we can choose $T_1 > 0$ sufficiently large in such a way that

$$u_{T_1}^* > u^* - \frac{\eta}{2}. \tag{3.19}$$

Furthermore we can also suppose to have chosen T_1 sufficiently large that $q(x) > 0$ and (Γ) (ii) holds on Ω_{T_1}. We now choose α such that $u_{T_1}^* < \alpha < u^*$. Because of (3.18) we can find j sufficiently large that

$$T_2 = t_j > T_1 \quad \text{and} \quad u_{T_2}^* > \alpha. \tag{3.20}$$

We select $\overline{\eta} > 0$ small enough that

$$\alpha + \overline{\eta} < u_{T_2}^*. \tag{3.21}$$

For $\sigma \in (0, \sigma_0)$ we define

$$\gamma_\sigma(x) = \alpha + \sigma(\gamma - T_1). \tag{3.22}$$

We note that

$$\gamma_\sigma(x) = \alpha \quad \text{for every } x \in \partial\Omega_{T_1}, \tag{3.23}$$

and

$$q(x)L\gamma_\sigma(x) = \sigma q(x)L\gamma(x) \leq \sigma B < \sigma_0 \quad \text{on } \Omega_{T_1}, \tag{3.24}$$

up to have chosen σ sufficiently small.

Since on $\Omega_{T_1} \setminus \Omega_{T_2}$ we have

$$\alpha < \gamma_\sigma(x) \leq \alpha + \sigma(T_2 - T_1) \tag{3.25}$$

we can choose $\sigma \in (0, \sigma_0)$ sufficiently small, so that

$$\sigma(T_2 - T_1) < \overline{\eta} \tag{3.26}$$

and then

$$\alpha \le \gamma_\sigma(x) < \alpha + \overline{\eta} \quad \text{on} \quad \Omega_{T_1} \setminus \Omega_{T_2}. \tag{3.27}$$

For any such σ, on $\partial\Omega_{T_1}$ we have

$$\gamma_\sigma(x) = \alpha > u^*_{T_1} \ge u(x), \tag{3.28}$$

so that

$$(u - \gamma_\sigma)(x) < 0 \quad \text{on } \partial\Omega_{T_1}. \tag{3.29}$$

Furthermore, if $\overline{x} \in \Omega_{T_1} \setminus \Omega_{T_2}$ is such that

$$u(\overline{x}) = u^*_{T_2} > \alpha + \overline{\eta}$$

then

$$(u - \gamma_\sigma)(\overline{x}) \ge u^*_{T_2} - \alpha - \sigma(T_2 - T_1) > u^*_{T_2} - \alpha - \overline{\eta} > 0$$

by (3.21) and (3.26). Finally, (Γ) (i) and the fact that $u^* < +\infty$ imply

$$(u - \gamma_\sigma)(x) < 0 \quad \text{on } \Omega_{T_3} \tag{3.30}$$

for $T_3 > T_2$ sufficiently large. Therefore,

$$\mu = \sup_{x \in \overline{\Omega}_{T_1}} (u - \gamma_\sigma)(x) > 0,$$

and it is in fact a positive maximum attained at a certain point z_0 in the compact set $\overline{\Omega}_{T_1} \setminus \Omega_{T_3}$. In particular, $\nabla(u - \gamma_\sigma)(z_0) = 0$ and $L(u - \gamma_\sigma)(z_0)$ reduces to $\mathrm{Tr}(t \circ \mathrm{hess}(u - \gamma_\sigma))(z_0)$. Therefore, since T is positive semi-definite we have that $Lu(z_0) \le L\gamma_\sigma(z_0)$.

By (3.29) we know that $\gamma(z_0) > T_1$. Therefore, at z_0 we have

$$u(z_0) = \gamma_\sigma(z_0) + \mu > \gamma_\sigma(z_0) > \alpha > u^*_{T_1} > u^* - \frac{\eta}{2}, \tag{3.31}$$

and hence $z_0 \in A_\eta \cap \Omega_{T_1}$. In particular $q(z_0) > 0$ and (Γ) (ii) holds at z_0. From (3.15) we obtain

$$0 < \sigma_0 \le q(z_0)Lu(z) \le q(z_0)L\gamma_\sigma(z_0) \le \sigma B < \sigma_0, \tag{3.32}$$

that is, the desired contradiction. □

We observe that we can relax the assumption in Theorem 3.1 on the boundedness of the function u from above to a control of u at infinity via the function γ. This is the content of the next result.

Theorem 3.5 *Let $(M, \langle\, ,\, \rangle)$ be a Riemannian manifold and $L = L_{T,X}$ be as above. Let $q(x) \in C^0(M)$, $q(x) \geq 0$ and suppose that*

$$q(x) > 0 \text{ outside a compact set.} \tag{3.33}$$

Let $\gamma \in C^2(M)$ be such that

$$\begin{cases} (i) & \gamma(x) \to +\infty \quad \text{as } x \to \infty, \\ (ii) & q(x)L\gamma(x) \leq B \quad \text{outside a compact set} \end{cases} \tag{Γ}$$

for some constant $B > 0$. If $u \in C^2(M)$ and

$$u(x) = o(\gamma(x)) \quad \text{as } x \to \infty, \tag{3.34}$$

then for each τ such that

$$\Omega_\tau = \{x \in M : u(x) > \tau\} \neq \emptyset$$

we have

$$\inf_{\Omega_\tau}\{q(x)Lu(x)\} \leq 0.$$

Proof Of course we consider here the case $u^* = +\infty$. We reason by contradiction as in the proof of Theorem 3.1 and we suppose the validity of (3.15) on Ω_τ. Next we proceed as in the above proof (obviously in this case u^* is not attained on M) to arrive to (3.18) that now takes the form

$$u^*_{t_j} \to +\infty \text{ as } j \to \infty, \tag{3.35}$$

and we choose $T_1 > 0$ sufficiently large in such a way that (3.19) now becomes

$$u^*_{T_1} > 2\tau. \tag{3.36}$$

Furthermore we can suppose to have chosen T_1 sufficiently large that $q(x) > 0$ and (Γ) (ii) holds on Ω_{T_1}. We choose α such that $\alpha > u^*_{T_1}$. Because of (3.35) we can find j sufficiently large that

$$T_2 = t_j > T_1 \quad \text{and} \quad u^*_{T_2} > \alpha. \tag{3.37}$$

We then proceed up to (3.30) which is now true on Ω_{T_3} for T_3 sufficiently large since, due to condition (3.34), the function

$$(u - \gamma_\sigma)(x) = \gamma_\sigma \left(\frac{u}{\gamma_\sigma} - 1 \right)(x);$$

becomes negative on Ω_{T_3}, for T_3 sufficiently large.

The rest of the proof is as Theorem 3.1. □

We now show the validity of Remark 3.2. Thus we assume (Γ) $(ii)'$ with G and $A \geq 0$ as in (3.6). We set

$$\varphi(t) = \int_{t_0}^{t} \frac{ds}{G(s) + A\, s \log s} \tag{3.38}$$

on $[t_0, +\infty)$ for some $t_0 > 0$. Note that, by (3.6) (i), $\varphi(t) \to +\infty$ as $t \to +\infty$. Thus, defining $\widehat{\gamma} = \varphi(\gamma)$, (Γ) (i) implies that

$$\widehat{\gamma}(x) \to +\infty \quad \text{as } x \to \infty. \tag{3.39}$$

Next, using that

$$L(\varphi(u)) = \varphi'(u)Lu + \varphi''(u)T(\nabla u, \nabla u),$$

a computation gives

$$\begin{aligned} q(x)L\widehat{\gamma}(x) = &\frac{q(x)L\gamma(x)}{G(\gamma(x)) + A\gamma(x)\log\gamma(x)} \\ &- \frac{G'(\gamma(x)) + A(1 + \log\gamma(x))}{(G(\gamma(x)) + A\gamma(x)\log\gamma(x))^2} q(x)T(\nabla\gamma(x), \nabla\gamma(x)) \end{aligned}$$

outside a sufficiently large compact set. Since $T(\nabla\gamma, \nabla\gamma) \geq 0$, $q(x) \geq 0$ and (3.6) (ii) holds, we deduce

$$q(x)L\widehat{\gamma}(x) \leq \frac{q(x)L\gamma(x)}{G(\gamma(x)) + A\gamma(x)\log\gamma(x)} \tag{3.40}$$

if $\gamma(x)$ is sufficiently large. Thus, from (Γ) $(ii)'$ and $G \geq 0$ we finally obtain

$$q(x)L\widehat{\gamma}(x) \leq B \tag{3.41}$$

outside a compact set. Then (3.39) and (3.41) show the validity of (Γ) (i), (ii) for the function $\widehat{\gamma}$. This finishes the proof of Remark 3.2 and also points out further possible extensions of condition (3.6) (ii).

Regarding Theorem 3.5, if we substitute (Γ) (ii) with (Γ) $(ii)'$, G satisfying (3.6), then condition (3.34) has to be replaced by

$$u(x) = o\left(\int_0^{\gamma(x)} \frac{ds}{G(s) + As\log s}\right) \quad \text{as } x \to \infty. \tag{3.42}$$

Thus for instance if $G(t) = t$, so that we can choose $A = 0$, (Γ) (ii)$'$ is $q(x)L\gamma(x) \leq \gamma(x)$ but (3.42) becomes $u(x) = o(\log\gamma(x))$ as $x \to \infty$, showing a balancing effect between the two conditions.

Proof (of Theorem 3.3) For a proof of Theorem 3.3 we proceed as in the proof of Theorem 3.1 letting

$$A_\eta = \{x \in M : u(x) > u^* - \eta\}. \tag{3.43}$$

We claim that for every $\varepsilon > 0$ there exists $x \in A_\eta$ such that

$$q(x)Hu(x)(v, v) < \varepsilon$$

for each $v \in T_xM$ with $|v| = 1$. By contradiction, suppose that there exists $\sigma_0 > 0$ such that, for every $x \in A_\eta$ there exists $\bar{v} \in T_xM$, $|\bar{v}| = 1$, such that

$$q(x)Hu(x)(\bar{v}, \bar{v}) \geq \sigma_0. \tag{3.44}$$

Now we follow the argument of the proof of Theorem 3.1 up to Eq. (3.24), which is now replaced by

$$q(x)H\gamma_\sigma(x)(\bar{v}, \bar{v}) = \sigma q(x)H\gamma(x)(\bar{v}, \bar{v}) \leq \sigma B < \sigma_0 \quad \text{on } \Omega_{T_1}, \tag{3.45}$$

up to have chosen σ sufficiently small. We then proceed up to the existence of a certain point z_0 in the compact set $\overline{\Omega}_{T_1} \setminus \Omega_{T_3}$ where the function $u - \gamma_\sigma$ attains its positive maximum. In particular, $\nabla(u - \gamma_\sigma)(z_0) = 0$ and $H(u - \gamma_\sigma)(z_0)$ reduces to

$$H(u - \gamma_\sigma)(z_0)(v, v) = T(\text{hess}(u - \gamma_\sigma)(z_0)v, v) \quad \text{for every } v \in T_{z_0}M.$$

Therefore, since T is positive semi-definite we have

$$Hu(z_0)(v, v) \leq H\gamma_\sigma(z_0)(v, v)$$

for every $v \in T_{z_0}M$.

Proceeding as in the proof of Theorem 3.1, we deduce that $z_0 \in A_\eta \cap \Omega_{T_1}$. In particular $q(z_0) > 0$ and (Γ) $(ii)'$ holds at z_0. On the other hand, from (3.44) we have

$$0 < \sigma_0 \leq q(z_0)Hu(z_0)(\bar{v}, \bar{v}) \leq q(z_0)H\gamma_\sigma(z_0)(\bar{v}, \bar{v}) \leq \sigma B < \sigma_0, \tag{3.46}$$

giving the desired contradiction. □

3.1.2 Proof of Theorem 3.2 and Some Related Results

We follow the notation of the previous section to give a proof of Theorem 3.2.

Proof (of Theorem 3.2) We first observe that, although it is not required in the statement of the theorem, the two assumptions (Γ_B) *(i)* and *(iii)* imply that the manifold M is geodesically complete. To see this, let $\varsigma : [0, \ell) \to M$ be any divergent path parameterized by arc-length, that is, as in the previous chapter, a path that eventually lies outside any compact subset of M. From (Γ_B) *(iii)* we have that $|\nabla\gamma| \leq B$ outside a compact subset K of M. We set $h(t) = \gamma(\varsigma(t))$ on $[t_0, \ell)$, where t_0 has been chosen so that $\varsigma(t) \notin K$ for all $t_0 \leq t < \ell$. Then, for every $t \in [t_0, \ell)$ we have

$$|h(t) - h(t_0)| = \left| \int_{t_0}^{t} h'(s)ds \right| \leq \int_{t_0}^{t} |\nabla\gamma(\varsigma(s))|ds \leq B(t - t_0).$$

Since ς is divergent, then $\varsigma(t) \to \infty$ as $t \to \ell^-$, so that $h(t) \to +\infty$ as $t \to \ell^-$ because of assumption (Γ_B) *(i)*. Therefore, letting $t \to \ell^-$ in the inequality above, we conclude that $\ell = +\infty$. This shows that divergent paths in M have infinite length. In other words, the metric on M is complete.

As in the proof of Theorem 3.1 we fix $\eta > 0$ but, instead of the set A_η of (3.13), we now consider the set

$$B_\eta = \{x \in M : u(x) > u^* - \eta \text{ and } |\nabla u(x)| < \eta\}. \tag{3.47}$$

Since the manifold is complete, by applying Ekeland quasi-minimum principle (see Proposition 2.2) we deduce that $B_\eta \neq \emptyset$. We claim that

$$\inf_{B_\eta} \{q(x)Lu(x)\} \leq 0. \tag{3.48}$$

Note that (3.48) is equivalent to conclusion (3.8) of Theorem 3.2. We reason by contradiction and suppose that

$$q(x)Lu(x) \geq \sigma_0 > 0 \quad \text{on } B_\eta. \tag{3.49}$$

Now the proof follows the pattern of that of Theorem 3.1 with the choice of T_1, such that also (Γ) *(iii)* holds on Ω_{T_1}, with Ω_t as in (3.16). We observe that in this case

$$\gamma_\sigma(x) = \alpha \quad \text{for every } x \in \partial\Omega_{T_1}, \tag{3.50}$$

$$q(x)L\gamma_\sigma(x) = \sigma q(x)L\gamma(x) \leq \sigma B < \sigma_0 \quad \text{on } \Omega_{T_1}, \tag{3.51}$$

and

$$|\nabla\gamma_\sigma(x)| = \sigma|\nabla\gamma(x)| \leq \sigma B < \eta \quad \text{on } \Omega_{T_1}, \tag{3.52}$$

up to have chosen σ sufficiently small.

Therefore, we find a point $z_0 \in \overline{\Omega_{T_1}} \setminus \Omega_{T_3}$ where $u - \gamma_\sigma$ attains a positive absolute maximum μ. As in the proof of Theorem 3.1, $z_0 \in \Omega_{T_1}$ and at z_0 we have

$$u(z_0) > \gamma_\sigma(z_0) > \alpha > u^*_{T_1} > u^* - \frac{\eta}{2} > u^* - \eta; \tag{3.53}$$

furthermore

$$|\nabla u(z_0)| = |\nabla \gamma_\sigma(z_0)| = \sigma |\nabla \gamma(z_0)| \leq \sigma B < \eta, \tag{3.54}$$

by our choice of σ. Thus $z_0 \in B_\eta \cap \Omega_{T_1}$ and a contradiction is achieved as at the end of the proof of Theorem 3.1. □

We note that the validity of Remark 3.3 is immediate. Indeed defining $\widehat{\gamma} = \varphi(\gamma)$ as in the previous subsection, conditions (Γ_B) (i), (ii) are satisfied for $\widehat{\gamma}$; as for condition (Γ_B) (iii), using (Γ_B) $(iii)'$ and $G \geq 0$, we have

$$|\nabla \widehat{\gamma}| = \frac{|\nabla \gamma|}{G(\gamma) + A\gamma \log \gamma} \leq \frac{G(\gamma)}{G(\gamma) + A\gamma \log \gamma} \leq 1 \tag{3.55}$$

outside a compact set. Thus, we also have the validity of (Γ_B) (iii) for $\widehat{\gamma}$.

As already pointed out in Theorem 2.5, on a complete manifold $(M, \langle\ ,\ \rangle)$ a naturale candidate for γ is some composition of the distance function $r(x)$ from a fixed origin o with an appropriate real function say, φ, under some curvature conditions. As we know the technical difficulty arising from this choice is related to the lack of smoothness; this forces us to introduce a reasoning in some way similar to approaching the problem via viscosity solutions.

We omit the details of the proof of Theorem 3.4, which follows similarly from the proof of Theorem 3.2.

3.2 An A Priori Estimate

A typical application of Theorem 3.2 is the following a priori estimate. Note that condition (3.59) below coincides (for $f = F$) with the Keller-Osserman condition for the Laplace-Beltrami operator (see [183]) showing that in this type of results what really matters is the structure, in this case linear, of the differential operator. We observe that we shall also give an a priori estimate in the nonlinear case, but the latter is definitely more complicated to prove (see Sect. 4.1).

Theorem 3.6 *Assume on $(M, \langle\ ,\ \rangle)$ the validity of the q-maximum principle for the operator $L = L_{T,X}$ and suppose that*

$$q(x) T(\ ,\ ) \leq C \langle\ ,\ \rangle \tag{3.56}$$

for some $C > 0$. Let $u \in C^2(M)$ be a solution of the differential inequality

$$q(x)Lu \geq \phi(u, |\nabla u|) \tag{3.57}$$

with $\phi(t, y)$ continuous in t, C^2 in y and such that

$$\frac{\partial^2 \phi}{\partial y^2}(t, y) \geq 0. \tag{3.58}$$

Set $f(t) = \phi(t, 0)$. Then a sufficient condition to guarantee

$$u^* = \sup_M u < +\infty$$

is the existence of a continuous function F positive on $[a, +\infty)$ for some $a \in \mathbb{R}$, satisfying the following

$$\left(\int_a^t F(s)ds\right)^{-1/2} \in L^1(+\infty), \tag{3.59}$$

$$\limsup_{t \to +\infty} \frac{\int_a^t F(s)ds}{tF(t)} < +\infty, \tag{3.60}$$

$$\liminf_{t \to +\infty} \frac{f(t)}{F(t)} > 0 \tag{3.61}$$

and

$$\liminf_{t \to +\infty} \frac{\left(\int_a^t F(s)ds\right)^{-1/2}}{F(t)} \frac{\partial \phi}{\partial y}(t, 0) > -\infty. \tag{3.62}$$

Furthermore, in this case, we have

$$f(u^*) \leq 0. \tag{3.63}$$

Proof Following the proof of Theorem 1.31 in [227] we choose $g \in C^2(\mathbb{R})$ to be increasing from 1 to 2 on $(-\infty, a+1)$ and defined by

$$g(t) = \int_{a+1}^t \frac{ds}{\left(\int_a^s F(r)dr\right)^{1/2}} + 2 \quad on \quad [a+1, +\infty).$$

Observe that

$$g'(t) = \frac{t}{\left(\int_a^t F(s)ds\right)^{1/2}} \quad and \quad g''(t) = -\frac{F(t)}{2} g'(t)^3 < 0 \tag{3.64}$$

on $(a+1,+\infty)$. We reason by contradiction and assume that $u^* = +\infty$. Since g is increasing,

$$\inf_M \frac{1}{g(u)} = \frac{1}{g(u^*)} = \frac{1}{g(+\infty)} > 0.$$

By applying the q-maximum principle for L to $1/g$, there exists a sequence $\{x_k\} \subset M$ such that

$$\lim_{k\to+\infty} \frac{1}{g(u(x_k))} = \frac{1}{g(+\infty)}, \tag{3.65}$$

or equivalently

$$\lim_{k\to+\infty} u(x_k) = +\infty, \tag{3.66}$$

$$\left|\nabla \frac{1}{g(u)}(x_k)\right| = \frac{g'(u(x_k))}{g(u(x_k))^2}|\nabla u(x_k)| < \frac{1}{k} \tag{3.67}$$

and finally

$$\begin{aligned}-\frac{1}{k} < q(x_k)L(\frac{1}{g(u)})(x_k) = q(x_k)\Bigg\{ &-\frac{g'(u(x_k))}{g(u(x_k))^2}Lu(x_k) + \\ &+ \left(\frac{2g'(u(x_k))^2}{g(u(x_k))^3} - \frac{g''(u(x_k))}{g(u(x_k))^2}\right)T(\nabla u(x_k), \nabla u(x_k))\Bigg\}\end{aligned} \tag{3.68}$$

for each $k \in \mathbb{N}$. Because of (3.66), we can suppose that the sequence $\{x_k\}$ satisfies $u(x_k) > a+1$, so that (3.64) holds along the sequence $u(x_k)$. Multiplying (3.68) by

$$\frac{g'(u(x_k))^2}{-g(u(x_k))^2 g''(u(x_k))} > 0$$

and using (3.57), we obtain

$$\begin{aligned}\frac{g'(u(x_k))^3}{g(u(x_k))^4|g''(u(x_k))|}\phi(u(x_k), |\nabla u(x_k)|) \le \frac{1}{k}\frac{g'(u(x_k))^2}{g(u(x_k))^2|g''(u(x_k))|} + \\ + \left(\frac{2g'(u(x_k))^4}{g(u(x_k))^5|g''(u(x_k))|} + \frac{g'(u(x_k))^2}{g(u(x_k))^4}\right) q(x_k)T(\nabla u(x_k), \nabla u(x_k)).\end{aligned} \tag{3.69}$$

Since $g \ge 1$, then $1/g^2 \le 1/g$ and

$$\frac{g'(u(x_k))^2}{g(u(x_k))^2|g''(u(x_k))|} \le \frac{g'(u(x_k))^2}{g(u(x_k))|g''(u(x_k))|}.$$

On the other hand, by (3.56) we also have

$$q(x_k)T(\nabla u(x_k), \nabla u(x_k)) \leq C|\nabla u(x_k)|^2.$$

Using these two facts in (3.69), jointly with (3.67), yields

$$\frac{g'(u(x_k))^3}{g(u(x_k))^4|g''(u(x_k))|}\phi(u(x_k), |\nabla u(x_k)|) \leq \frac{g'(u(x_k))^2}{g(u(x_k))|g''(u(x_k))|}\left(\frac{1}{k} + \frac{2C}{k^2}\right) + \frac{C}{k^2}.$$

Next, we use Taylor formula with respect to y centered at $(u(x_k), 0)$ and (3.58) to deduce

$$\varphi(u(x_k), |\nabla u(x_k)|) \geq f(u(x_k)) + \frac{\partial \phi}{\partial y}(u(x_k), 0)|\nabla u(x_k)|,$$

so that

$$\frac{g'(u(x_k))^3 f(u(x_k))}{g(u(x_k))^4|g''(u(x_k))|} + A_k \leq \frac{g'(u(x_k))^2}{g(u(x_k))|g''(u(x_k))|}\left(\frac{1}{k} + \frac{2C}{k^2}\right) + \frac{C}{k^2}, \tag{3.70}$$

where

$$A_k := \min\left\{0, \frac{1}{k}\frac{\partial \phi}{\partial y}(u(x_k), 0)\frac{g'(u(x_k))^2}{g(u(x_k))^2|g''(u(x_k))|}\right\}.$$

In what follows, we always assume that t is taken sufficiently large. Observe that we have

$$\frac{g'(t)^2}{g(t)|g''(t)|} = 2\frac{(\int_a^t F(s)ds)^{1/2}}{g(t)F(t)} = 2\frac{\int_a^t F(s)ds}{g(t)(\int_a^t F(s)ds)^{1/2}F(t)},$$

and

$$g(t) \geq \frac{t-a-1}{(\int_a^t F(s)ds)^{1/2}},$$

so that

$$\frac{g'(t)^2}{g(t)|g''(t)|} \leq C\frac{\int_a^t F(s)ds}{tF(t)}, \quad t \gg 1,$$

for some positive constant C. Therefore, using (3.60) we deduce

$$\limsup_{k\to+\infty} \frac{g'(u(x_k))^2}{g(u(x_k))|g''(u(x_k))|} < +\infty,$$

and then

$$\limsup_{k\to+\infty} \frac{g'(u(x_k))^2}{g(u(x_k))|g''(u(x_k))|}\left(\frac{1}{k}+\frac{2C}{k^2}\right)+\frac{C}{k^2}=0. \tag{3.71}$$

On the other hand,

$$\frac{g'(t)^3 f(t)}{g(t)^4|g''(t)|}=\frac{2f(t)}{g(t)^4F(t)}\geq c\frac{f(t)}{F(t)}$$

for some $c>0$, since $\sup_M g<+\infty$ by (3.59). Therefore, using (3.61) we have

$$\liminf_{k\to+\infty} \frac{g'(u(x_k))^3 f(u(x_k))}{g(u(x_k))^4|g''(u(x_k))|}>0. \tag{3.72}$$

Finally, observe that

$$\frac{\partial\phi}{\partial y}(t,0)\frac{g'(t)^2}{g(t)^2|g''(t)|}=\frac{1}{g(t)^2}\left(\frac{\partial\phi}{\partial y}(t,0)\frac{(\int_a^t F(s)ds)^{1/2}}{F(t)}\right)$$

whence, using $\sup_M g<+\infty$ and (3.62), we get

$$\liminf_{t\to+\infty}\left(\frac{\partial\phi}{\partial y}(t,0)\frac{g'(t)^2}{g(t)^2|g''(t)|}\right)>-\infty.$$

Thus,

$$\liminf_{k\to+\infty} A_k=0. \tag{3.73}$$

Therefore, taking $k\to+\infty$ in (3.70) and using (3.71)–(3.73) we obtain the desired contradiction.

As for the conclusion $f(u^*)\leq 0$, we note that if ϕ were continuous in both variables, then to reach the desired conclusion it would be enough to apply the q-maximum principle to u to get a sequence $\{y_k\}$ with $\lim u(y_k)=u^*$, $\lim|\nabla u(y_k)|=0$ and

$$\frac{1}{k}>q(y_k)Lu(y_k)\geq\phi(u(y_k),|\nabla u(y_k)|).$$

Thus, taking the limit as $k\to+\infty$ we would get $f(u^*)\leq 0$. Otherwise, in our more general assumptions, we can argue in the following way. We re-define the function $g(t)$ at the very beginning of the proof in such a way that it changes concavity only once at the point $T=\min\{u^*,a\}-1$. We emphasize that with this choice $g''<0$ on $(T,+\infty)$. We now proceed as in the proof of the first part of the Theorem, applying the q-maximum principle to the function $1/g(u)$, and get the existence of a sequence $\{x_k\}$ as before, with $g''(u(x_k)) < 0$ if k is sufficiently large. That is all we need to arrive at (3.70). Taking the limit in the latter for $k \to +\infty$ and using $\lim_{k\to+\infty} u(x_k) = u^* < +\infty$, we conclude that $f(u^*) \leq 0$. □

As an application, we shall now combine Theorems 3.6 and 2.5 to deal with the following problem.

Let M be an m-dimensional manifold with $m \geq 3$ and let h be a given symmetric $(0,2)$-tensor field on M. Can h be realized as the Ricci tensor of some metric $\langle\ ,\ \rangle$ on M?

Of course there are natural obstructions to the existence of $\langle\ ,\ \rangle$ solving

$$\mathrm{Ric}_{\langle\,,\rangle} = h. \tag{3.74}$$

For instance, if M is compact and $\mathrm{Ric}_{\langle\,,\rangle}$ is positive definite then the first Betti number of M has to be zero as proved by Bochner [49]. Again by a result of Myers [203], if the lowest eigenvalue of $\mathrm{Ric}_{\langle\,,\rangle}$ is bounded below by a positive constant and $(M, \langle\ ,\ \rangle)$ is complete, then it is compact and has finite fundamental group. A similar obstruction exists also when $\mathrm{Ric}_{\langle\,,\rangle}$ is possibly negative; for details see [44]. Hamilton [133] has proved that any compact 3-dimensional manifold with positive Ricci curvature is diffeomorphic to a 3-manifolds with constant positive sectional curvature. Schoen and Yau [251] have proved that a complete, noncompact, 3-dimensional manifold with positive Ricci tensor is diffeomorphic to $\mathbb{R}^3$.

In case h is positive definite, and therefore gives rise to a metric on M, we are going to present an obstruction to the existence of $\langle\ ,\ \rangle$ satisfying (3.74) which is obtained via a special harmonic map.

Lemma 3.1 *Let $(M, \langle\ ,\ \rangle)$ be a Riemannian manifold and let $(\cdot,\cdot)$ be a second metric on M such that*

$$\mathrm{Ric}_{\langle\,,\rangle} = (\ ,\ ). \tag{3.75}$$

Let $\varphi : (M\langle\ ,\ \rangle)) \to (M, (\ ,\ ))$ be the identity map. Then φ is harmonic.

Proof In the notation of Sect. 1.7 of Chap. 1 we let $\{\theta^i\}$, $\{\theta^i_j\}$ and $\{\omega^i\}$, $\{\omega^i_j\}$ be local orthonormal coframes with corresponding Levi-Civita connections forms, respectively, on $(M, \langle\ ,\ \rangle)$ and $(M, (\ ,\ ))$. Let

$$\omega^i = \varphi^i_j \theta^j. \tag{3.76}$$

Then, using (3.75), we deduce

$$R_{ij} = \varphi^t_i \varphi^t_j. \tag{3.77}$$

Taking covariant derivatives

$$R_{ij,k} = \varphi^t_{ik}\varphi^t_j + \varphi^t_i\varphi^t_{jk},$$

from which we also deduce

$$R_{jk,i} = \varphi^t_{ji}\varphi^t_k + \varphi^t_j\varphi^t_{ki}$$

and

$$R_{ik,j} = \varphi^t_{ij}\varphi^t_k + \varphi^t_i\varphi^t_{kj}.$$

Using the symmetry relations $\varphi^t_{ij} = \varphi^t_{ji}$, we immediately obtain

$$R_{ij,k} - R_{jk,i} + R_{ik,j} = 2\varphi^t_i\varphi^t_{jk}.$$

We now trace in the metric $\langle\ ,\ \rangle$ with respect to the indices j and k and recall Schur's identities (1.68)

$$2R_{ik,k} = S_i,$$

S the scalar curvature of $\langle\ ,\ \rangle$, to deduce

$$\varphi^t_i\varphi^t_{kk} = 0.$$

But φ is a diffeomorphism and thus (φ^t_i) is an invertible matrix, from which we infer $\varphi^t_{kk} = 0$, that is, φ is harmonic. □

Note that, from the proof of Lemma 3.1, precisely from Eq. (3.77), we also have

$$S = |d\varphi|^2. \tag{3.78}$$

In particular, if Δ stands for the Laplacian operator of $(M, \langle\ ,\ \rangle)$, then ΔS can be obtained via the Bochner-Weitzenböck formula (1.175) that, since φ is harmonic, reads

$$\frac{1}{2}\Delta S = |\nabla d\varphi|^2 + \widetilde{R}^a_{bcd}\varphi^a_i\varphi^b_k\varphi^c_k\varphi^d_i + R_{ti}\varphi^a_i\varphi^a_t,$$

where R and $\widetilde{R}$ denote, respectively, the curvature tensors in the metrics $\langle\ ,\ \rangle$ and $(\ ,\ )$ on M. Again, because of (3.77), from the above we deduce

$$\frac{1}{2}\Delta S = |\nabla d\varphi|^2 + \widetilde{R}^a_{bcd}\varphi^a_i\varphi^b_k\varphi^c_k\varphi^d_i + |\operatorname{Ric}_{\langle,\rangle}|^2_{\langle,\rangle}. \tag{3.79}$$

To interpret the middle term in the above formula, and similarly to what has been done at the beginning of Sect. 2.4, we will introduce a second curvature operator, that we shall indicate with $\mathfrak{R}$, now acting on symmetric $(0,2)$-tensors. Let

$$\widetilde{\operatorname{Riem}} = \widetilde{R}_{ijk\ell}\,\omega^i \otimes \omega^j \otimes \omega^k \otimes \omega^\ell$$

denote the $(0, 4)$-type Riemann curvature tensor of $(M, (\, , \,))$, with respect to the local orthonormal coframe $\{\omega^i\}$. Let $\alpha \in S^2(M)$ be a symmetric $(0, 2)$-tensor

$$\alpha = \alpha_{ij}\omega^i \otimes \omega^j \tag{3.80}$$

with $\alpha_{ij} = \alpha_{ji}$. Then

$$\mathfrak{R}(\alpha) = \widetilde{R}_{ijk\ell}\alpha_{j\ell}\omega^i \otimes \omega^k. \tag{3.81}$$

It is immediate to verify that $\mathfrak{R}$ is well defined. Furthermore, since $\widetilde{R}$ satisfies the symmetry relations $\widetilde{R}_{ijk\ell} = \widetilde{R}_{k\ell ij}$, $\mathfrak{R}(\alpha) \in S^2(M)$ and we have an endomorphism

$$\mathfrak{R} : S^2(M) \to S^2(M).$$

Even more, since $\widetilde{R}$ also satisfies $\widetilde{R}_{ijk\ell} = -\widetilde{R}_{jik\ell}$, given to $S^2(M)$ the obvious inner product induced by $(\, , \,)$, that we shall indicate with the same notation, we have that $\mathfrak{R}$ is self-adjoint. Indeed, for $\alpha, \beta \in S^2(M)$ we have

$$\begin{aligned}(\alpha, \mathfrak{R}(\beta)) &= \alpha_{j\ell}\widetilde{R}_{ji\ell k}\beta_{ik} = -\alpha_{j\ell}\widetilde{R}_{ij\ell k}\beta_{ik} = -\alpha_{jl}\widetilde{R}_{\ell kij}\beta_{ik} \\ &= \alpha_{j\ell}\widetilde{R}_{k\ell ij}\beta_{ik} = \alpha_{j\ell}\widetilde{R}_{ijk\ell}\beta_{ik} = (\mathfrak{R}(\alpha), \beta).\end{aligned}$$

In particular, $\mathfrak{R}$ is diagonalizable on $S^2(M)$.

To simplify the writing let $g = \langle \, , \, \rangle$. Setting $(\ell^s_t) = (\varphi^i_j)^{-1}$ with φ^i_j as in (3.76), we have

$$g = \langle \, , \, \rangle = \ell^s_i\ell^s_j\omega^i \otimes \omega^j.$$

We define a new symmetric $(0, 2)$-tensor g^{-1} as the tensor whose coefficients in the local orthonormal basis $\{\omega^i\}$ are given by the coefficient of the inverse of the matrix $(\ell^s_i\ell^s_j)$. It is immediate to verify that the latter is the matrix $(\varphi^k_s\varphi^t_s)$. Thus

$$g^{-1} = \varphi^k_s\varphi^t_s\omega^k \otimes \omega^t.$$

It follows that

$$-(\mathfrak{R}(g^{-1}), g^{-1}) = \widetilde{R}_{abdc}\varphi^b_k\varphi^d_k\varphi^a_s\varphi^c_s,$$

thus (3.79) can be written as

$$\frac{1}{2}\Delta S = |\nabla d\varphi|^2 - (\mathfrak{R}(g^{-1}), g^{-1}) + |\operatorname{Ric}_{\langle , \rangle}|^2_{\langle , \rangle}. \tag{3.82}$$

Hence, if $\Lambda(x)$ is the maximum of the eigenvalues of $\mathfrak{R}$ at x, we have

$$\begin{aligned}\frac{1}{2}\Delta S &\geq |\nabla d\varphi|^2 - \Lambda(x)(g^{-1}, g^{-1}) + |\operatorname{Ric}_{\langle , \rangle}|^2_{\langle , \rangle} \\ &= |\nabla d\varphi|^2 + (1 - \Lambda(x))|\operatorname{Ric}_{\langle , \rangle}|^2_{\langle , \rangle}.\end{aligned}$$

On the other hand,

$$|\operatorname{Ric}_{\langle\,,\,\rangle}|^2_{\langle\,,\,\rangle} \geq \frac{S^2}{m},$$

so that, if $\Lambda(x) \leq 1$ on M, we finally arrive to the differential inequality

$$\frac{1}{2}\Delta S \geq |\nabla d\varphi|^2 + \frac{1}{m}(1 - \Lambda(x))S^2. \tag{3.83}$$

We are now ready to prove the following

Theorem 3.7 *Let $(\,,\,)$ be a Riemannian metric on M and let $\Lambda(x)$ be the largest eigenvalue of the curvature operator of $(\,,\,)$ acting on symmetric $(0, 2)$-tensors at $x \in M$. Assume that*

$$\sup_M \Lambda(x) < 1. \tag{3.84}$$

Then there is no complete metric $\langle\,,\,\rangle$ on M such that

$$\operatorname{Ric}_{\langle\,,\,\rangle} = (\,,\,).$$

Proof Suppose by contradiction the existence of a complete metric $\langle\,,\,\rangle$ on M satisfying the above requirements. Then, from (3.84) and (3.83) there exists $C > 0$ such that

$$\Delta S \geq CS^2 \quad \text{on } M. \tag{3.85}$$

By the completeness of the metric $\langle\,,\,\rangle$ and the positivity of its Ricci tensor, we have the validity of the Omori-Yau maximum principle for Δ on $(M, \langle\,,\,\rangle)$. Therefore by Theorem 3.6 we conclude that $S \equiv 0$; this contradicts the fact that $S = \operatorname{Tr}_{\langle\,,\,\rangle}(\,,\,) > 0$ on M. □

Remark 3.4 Theorem 3.7 improves on DeTurck and Koiso [100] and Delanoë [99].

3.3 The Nonlinear Case

In this section we will introduce an extension of Theorems 3.1 and 3.2 to the nonlinear case. Since solutions of PDE's involving the type of operators we shall consider are not, in general, even for constant coefficients, of class C^2, it will be more appropriate to work, from the very beginning, in the weak setting (think for instance of the p-Laplace operator with $p \neq 2, p > 1$).

We let $A : \mathbb{R}^+ \to \mathbb{R}$ and we define $\varphi(t) = tA(t)$. The next assumptions will be crucial to apply the version of Theorems 5.2 and 5.4 of [236] that we present below in Theorems 3.8 and 3.10:

(A1) $A \in C^1(\mathbb{R}^+)$.
(A2) (i) $\varphi'(t) > 0$ on $\mathbb{R}^+$, (ii) $\varphi(t) \to 0$ as $t \to 0^+$.
(T1) T is a positive definite, symmetric, 2-covariant tensor field on M.
(T2) For every $x \in M$ and for every $\xi \in T_x M$, $\xi \neq 0$, the bilinear form

$$\frac{A'(|\xi|)}{|\xi|}\langle \xi, \ \rangle \odot T(\xi, \ ) + A(|\xi|)T(\ , \ )$$

is symmetric and positive definite. Here $\odot$ denotes the symmetric tensor product.

Note that the above requirements are not mutually independent. Indeed the bilinear form in (T2) is automatically symmetric when T does. Furthermore, if we write it in terms of φ, being positive definite means that for every $x \in M$ and for every $\xi, v \in T_x M$, $\xi, v \neq 0$,

$$\frac{1}{|\xi|^2}\left(\varphi'(|\xi|) - \frac{\varphi(|\xi|)}{|\xi|}\right)\langle \xi, v\rangle T(\xi, v) + \frac{\varphi(|\xi|)}{|\xi|}T(v, v) > 0.$$

In particular, the choice $v = \xi$ shows that

$$\varphi'(t) > 0 \quad \text{on} \quad \mathbb{R}^+,$$

that is, requirement (i) in (A2). Requirement (T2) is in fact equivalent to (i) in (A2) in case $T = t(x)\langle \ , \ \rangle$ is a "pointwise conformal" deformation of the metric for some smooth function $t(x) > 0$ on M. Indeed, in this case (T2) reduces to

$$\frac{1}{|\xi|^2}\varphi'(|\xi|)t(x)\langle \xi, v\rangle^2 + \frac{\varphi(|\xi|)}{|\xi|^3}t(x)\left(|v|^2|\xi|^2 - \langle \xi, v\rangle^2\right) > 0$$

for every $x \in M$ and for every $\xi, v \in T_x M$, $\xi, v \neq 0$.

Having fixed a vector field X on M, we define the operator $L = L_{A,T,X}$

$$Lu = \operatorname{div}\left(A(|\nabla u|)T(\nabla u, \cdot)^\sharp\right) - \langle X, \nabla u\rangle \tag{3.86}$$

acting on $C^1(M)$, where $^\sharp : T^*M \to TM$ denotes the musical isomorphism. Of course, the above operator L has to be understood in the appropriate weak sense.

L gives rise to various familiar operators. For instance, choosing $T = \langle \ , \ \rangle$ and $X = 0$ we have

1. For $\varphi(t) = t^{p-1}$, $p > 1$,

$$Lu = \operatorname{div}\left(|\nabla u|^{p-2}\nabla u\right)$$

is the usual *p-Laplacian*. Of course the case $p = 2$ yields the usual Laplace-Beltrami operator.

2. For $\varphi(t) = t/\sqrt{1+t^2}$ the operator

$$Lu = \operatorname{div}\left(\frac{\nabla u}{\sqrt{1+|\nabla u|^2}}\right)$$

is the usual *mean curvature operator.*

We let, as in the linear case, $q(x) \in C^0(M)$, $q(x) \geq 0$, be such that, for some compact $K \subset M$, $q(x) > 0$ on $M \setminus K$. However, since our setting now is that of solutions in the weak sense, for technical reasons (see for instance (3.173) in the proof of Theorem 3.11 below) we need the local integrability of $1/q$ also inside K. Thus, when needed, we will also assume

$$\frac{1}{q} \in L^1_{loc}(M). \tag{Q}$$

This fact was already pointed out after Remark 3.3 of the linear case whenever we deal with functions u on M which are merely of class C^1.

3.3.1 Analytic Preliminaries

The aim of this section is to prove the comparison and the strong maximum principles that we will need later for C^1 or even Lip_{loc} solutions. However, instead of proving them just as needed, we present these two results in a more general form involving a function f satisfying some, accordingly to the results we are presenting, of the following conditions:

(F1) $f \in C^0(\mathbb{R}_0^+)$;
(F2) f is positive on some interval $(0, \delta)$ with $0 < \delta \leq +\infty$;
(F3) $f(0) = 0$ and f is nondecreasing on some interval $(0, \delta)$ with $0 < \delta \leq +\infty$.

This choice is motivated mainly by two reasons: first, the results in this general form are useful in many different applications; second, in the maximum principle, when $f \not\equiv 0$, it appears a condition on f which is somehow dual to the Keller-Osserman condition given in the linear case in (3.59). We shall briefly comment on this later in Sect. 4.1 of Chap. 4.

We begin by proving an auxiliary lemma; recall that, given a smooth curve $c : [0, 1] \to M$, a vector field X_t *along* c is smooth map $X : [0, 1] \to TM$ such that $X_t \in T_{c(t)}M$.

Lemma 3.2 *Assume (A1) and let T be a $(0,2)$-tensor field on M. Let $\nabla u, \nabla v \in T_xM$, for some $x \in M$, be such that $X_t = t\nabla u + (1-t)\nabla v \neq 0$ for each $t \in [0,1]$. Then at x we have*

$$\langle A(|\nabla u|)T(\nabla u, \ )^\sharp - A(|\nabla v|)T(\nabla v, \ )^\sharp, \nabla u - \nabla v\rangle =$$

$$\int_0^1 \left(\frac{A'(|X_t|)}{|X_t|}\langle X_t, \nabla u - \nabla v\rangle T(X_t, \nabla u - \nabla v) + A(|X_t|)T(\nabla u - \nabla v, \nabla u - \nabla v)\right) dt.$$

Proof Let $c : [0,1] \to M$ be the constant curve $c(t) = x$ for all $t \in [0,1]$, and consider the vector field X_t along c given by $X_t = t\nabla u + (1-t)\nabla v \neq 0$. To simplify notations we set $Y = \nabla u - \nabla v$. Let $\{e_i\}$ be a local orthonormal frame at x satisfying $\nabla_{e_j} e_i(x) = 0$ for all $i, j = 1, \ldots, m$. Using the latter, jointly with the properties of covariant differentiation D/dt along the curve, the fact that $\dot{c} \equiv 0$ on $[0,1]$, and $X_t \neq 0$ on $[0,1]$ by assumption, we have

$$\begin{aligned}
\frac{d}{dt}\langle A(|X_t|)T(X_t, \ )^\sharp, Y\rangle &= \langle \frac{D}{dt}A(|X_t|)T(X_t, \ )^\sharp, Y\rangle = \langle \frac{D}{dt}A(|X_t|)T(X_t, e_i)e_i, Y\rangle \\
&= \frac{d}{dt}\left(A(|X_t|)T(X_t, e_i)\right)\langle e_i, Y\rangle \\
&= T(X_t, e_i)\frac{A'(|X_t|)}{|X_t|}\langle \frac{D}{dt}X_t, X_t\rangle\langle e_i, Y\rangle \\
&\quad + A(|X_t|)\frac{d}{dt}\left(T(X_t, e_i)\right)\langle e_i, Y\rangle \\
&= T(X_t, e_i)\frac{A'(|X_t|)}{|X_t|}\langle X_t, Y\rangle\langle e_i, Y\rangle \\
&\quad + A(|X_t|)\langle e_i, Y\rangle\left((\nabla_{\dot{c}(t)}T)(X_t, e_i) + T(\frac{D}{d}X_t, e_i)\right) \\
&= T(X_t, Y)\langle X_t, Y\rangle\frac{A'(|X_t|)}{|X_t|} + A(|X_t|)T(Y, Y).
\end{aligned}$$

Then the result follows immediately by integration. □

We are now ready to prove the following

Theorem 3.8 *Assume (A1), (T1), (T2), (F1) and (F3). Let X be a vector field on M and $\Omega \subset M$ be a relatively compact domain. Let $u, v \in C^0(\overline{\Omega}) \cap C^1(\Omega)$ be weak solutions of*

$$\operatorname{div}\left(A(|\nabla u|)T(\nabla u, \ )^\sharp\right) - \langle X, \nabla u\rangle - f(u) \leq 0 \quad \text{in } \Omega, \tag{3.87}$$

$$\operatorname{div}\left(A(|\nabla v|)T(\nabla v, \ )^\sharp\right) - \langle X, \nabla v\rangle - f(v) \geq 0 \quad \text{in } \Omega, \tag{3.88}$$

respectively, with $v < \delta$ *for* δ *as in (F3). Assume that*

$$|\nabla u| + |\nabla v| > 0 \quad \text{on } \overline{\Omega}, \tag{3.89}$$

and

$$\text{either } |\nabla u| < b \text{ or } |\nabla v| < b \quad \text{in } \Omega \tag{3.90}$$

for some $b > 0$. *If*

$$u \geq v \quad \text{on } \partial\Omega \tag{3.91}$$

then $u \geq v$ *on* $\overline{\Omega}$.

Remark 3.5 We underline the essential requirement (F3).

Proof We reason by contradiction and, setting $w = u - v$, we suppose that

$$\bar{\varepsilon} = -\inf_{\Omega} w > 0. \tag{3.92}$$

Next, for $a \in [\bar{\varepsilon}/2, \bar{\varepsilon})$ we let $w_a = w + a$ and set

$$\Sigma_a = \{x \in \Omega : w_a(x) < 0\}.$$

Of course $\overline{\Sigma_a} \subset \Omega$ and therefore Σ_a is relatively compact. Next there exists $0 < d < 2b$ such that

$$|\nabla u| + |\nabla v| \geq 4d \quad \text{on } \Sigma_{\bar{\varepsilon}/2} \supset \Sigma_a. \tag{3.93}$$

Indeed, $\overline{\Sigma_{\bar{\varepsilon}/2}} \subset \Omega$ and $|\nabla u| + |\nabla v| > 0$ by assumption (3.89). We now claim that we can choose a sufficiently close to $\bar{\varepsilon}$ so that for each $t \in [0, 1]$

$$|t\nabla u + (1-t)\nabla v| \geq d \quad \text{on } \Sigma_a, \tag{3.94}$$

and

$$|\nabla u|, |\nabla v| \leq b \quad \text{on } \Sigma_a. \tag{3.95}$$

To prove the claim, observe that the set

$$E = \{x \in \Omega : w(x) = -\bar{\varepsilon}\} \subset \Sigma_a$$

since $a \in [\bar{\varepsilon}/2, \bar{\varepsilon})$; furthermore $E \neq \emptyset$ because of (3.91). The points of E are absolute minima for w and thus

$$\nabla u = \nabla v \quad \text{on } E.$$

We observe that, because of (3.90), $w(x) \geq -\bar{\varepsilon}$ on Ω. Hence, for $x \in \Sigma_a$

$$-\bar{\varepsilon} \leq w(x) < -a,$$

and choosing a sufficiently close to $\bar{\varepsilon}$, by continuity,

$$|\nabla u - \nabla v| < d \quad \text{on } \Sigma_a.$$

In particular, for such values of a, since by (3.93)

$$\max\{|\nabla u|, |\nabla v|\} \geq 2d \quad \text{on } \Sigma_a,$$

for all $t \in [0, 1]$ we have

$$|t\nabla u + (1-t)\nabla v| \geq \max\{|\nabla u|, |\nabla v|\} - |\nabla u - \nabla v| \geq d \quad \text{on } \Sigma_a,$$

that is, (3.94). To prove (3.95) consider, without loss of generality, the case $|\nabla v| < b$ on Ω in (3.90). Define

$$\bar{b} = \sup_{\Sigma_{\bar{\varepsilon}/2}} |\nabla v|.$$

Since $\overline{\Sigma}_{\bar{\varepsilon}/2} \subset \Omega$ and $|\nabla v| < b$ on Ω, we have $\bar{b} < b$, and if we choose a sufficiently close to $\bar{\varepsilon}$, then also $|\nabla u - \nabla v| < b - \bar{b}$ in Σ_a. It follows that

$$|\nabla u| \leq |\nabla v| + |\nabla u - \nabla v| < b \quad \text{in } \Sigma_a,$$

that is, (3.95).

Hence, setting $X_t = t\nabla u + (1-t)\nabla v$, then for all $a \in [\bar{a}, \bar{\varepsilon})$ with $\bar{a}$ sufficiently close to $\bar{\varepsilon}$, we have

$$d \leq |X_t| \leq 2b \quad \text{on } \Sigma_a.$$

This fact, (T2) and the compactness of $\overline{\Sigma}_{\bar{a}}$ imply the existence of a constant $\lambda > 0$, independent of a and t, such that

$$\frac{A'(|X_t|)}{|X_t|}\langle X_t, \nabla u - \nabla v\rangle T(X_t, \nabla u - \nabla v) + A(|X_t|)T(\nabla u - \nabla v, \nabla u - \nabla v) \geq \lambda|\nabla u - \nabla v|^2 \tag{3.96}$$

on Σ_a for all $a \in [\bar{a}, \bar{\varepsilon})$ and for all $t \in [0, 1]$.

We now extend w_a to be 0 outside Σ_a and we use this nonpositive function as a test function. We have

$$\int_{\Sigma_a} \langle A(|\nabla u|)T(\nabla u,\cdot)^\sharp - A(|\nabla v|)T(\nabla v,\cdot)^\sharp, \nabla w_a\rangle$$
$$\leq \int_{\Sigma_a} \langle X, \nabla v - \nabla u\rangle w_a + \int_{\Sigma_a} (f(v) - f(u))w_a. \tag{3.97}$$

Using (3.96), the fact that since $w_a \leq 0$ on Σ_a both u and v are strictly less than δ of (F3), Lemma 3.2 and $\nabla w_a = \nabla u - \nabla v$ on Σ_a, we have

$$\lambda \int_{\Sigma_a} |\nabla u - \nabla v|^2 \leq \int_{\Sigma_a} \langle X, \nabla v - \nabla u\rangle w_a \leq \sup_{\overline{\Omega}} |X| \int_{\Sigma_a} |\nabla w_a||w_a|,$$

that is,

$$\lambda \int_{\Sigma_a} |\nabla w_a|^2 \leq \eta \int_{\Sigma_a} |\nabla w_a||w_a|, \tag{3.98}$$

with λ and η positive constants independent of $a \in [\bar{a}, \bar{\varepsilon})$. We define

$$\Gamma_a = \{x \in \Omega : a - \bar{\varepsilon} < w_a(x) < 0\} \subset \Sigma_a$$

and observe that

$$\Sigma_a \setminus \Gamma_a = E.$$

Hence $\nabla w_a = 0$ in $\Sigma_a \setminus \Gamma_a$. From (3.98) we then deduce

$$\lambda \int_{\Gamma_a} |\nabla w_a|^2 \leq \eta \int_{\Gamma_a} |\nabla w_a||w_a|.$$

Applying Hölder's inequality to the right-hand side of the above, we obtain

$$\lambda \int_{\Gamma_a} |\nabla w_a|^2 \leq \left(\frac{\eta}{\lambda}\right)^2 \int_{\Gamma_a} |w_a|^2. \tag{3.99}$$

Note that this is possible since $\lambda \int_{\Gamma_a} |\nabla w_a|^2 \neq 0$ for each $a \in [\bar{a}, \bar{\varepsilon})$. Indeed, as we have already observed

$$\lambda \int_{\Gamma_a} |\nabla w_a|^2 = \lambda \int_{\Sigma_a} |\nabla w_a|^2.$$

Now consider the isoperimetric constant

$$0 < S_{\Sigma_a} = \inf_{\varphi \in H_0^1(\Sigma_a), \varphi \not\equiv 0} \frac{\int_{\Sigma_a} |\nabla \varphi|}{\|\varphi\|_{L^{m'}(\Sigma_a)}} \tag{3.100}$$

with m' the Hölder conjugate of m (see [141]). Note that $w_a \in W_0^{1,2}(\Sigma_a)$ and $w_a \not\equiv 0$. Therefore, using again Hölder's inequality

$$0 < \left(\int_{\Sigma_a} |\nabla w_a| \right)^2 \leq \operatorname{vol}(\Sigma_a) \int_{\Sigma_a} |\nabla w_a|^2.$$

To finish the proof, first we consider the case $m \geq 3$. We apply (3.99), and Hölder and Sobolev inequalities to obtain

$$\begin{aligned}
\left(\frac{\eta}{\lambda}\right)^2 \operatorname{vol}(\Gamma_a)^{2/m} \left(\int_{\Gamma_a} |w_a|^{\frac{2m}{m-2}} \right)^{\frac{m-2}{m}} &\geq \left(\frac{\eta}{\lambda}\right)^2 \int_{\Gamma_a} |w_a|^2 \\
&\geq \int_{\Gamma_a} |\nabla w_a|^2 = \int_{\Sigma_a} |\nabla w_a|^2 \\
&\geq \left(\frac{m-2}{2} S_{\Sigma_a} \right)^2 \left(\int_{\Sigma_a} |w_a|^{\frac{2m}{m-2}} \right)^{\frac{m-2}{m}} \\
&\geq \left(\frac{m-2}{2} S_{\Sigma_a} \right)^2 \left(\int_{\Gamma_a} |w_a|^{\frac{2m}{m-2}} \right)^{\frac{m-2}{m}}.
\end{aligned}$$

Since $|w_a| \neq 0$ on Γ_a, for $a \in [\bar{a}, \bar{\varepsilon})$ we have

$$\left(\frac{\eta}{\lambda}\right)^2 \operatorname{vol}(\Gamma_a)^{2/m} \geq \left(\frac{m-2}{2} S_{\Sigma_a} \right)^2 \geq \left(\frac{m-2}{2} S_{\Sigma_{\bar{a}}} \right)^2 > 0.$$

Letting $a \to \bar{a}$ and noting that $\Gamma_a \to \emptyset$, from the above we obtain the desired contradiction.

When $m = 2$ we proceed from (3.99) as above with $\frac{m}{m-2}$ replaced by any fixed exponent $q > 1$. □

Remark 3.6

(i) Of course (3.98) is true also in case $\sup_{\overline{\Omega}} |X| \equiv 0$ and the proof follows. However, in this case, the argument simplifies; indeed, we have

$$\lambda \int_{\Sigma_a} |\nabla u - \nabla v|^2 \leq \int_{\Sigma_a} (f(v) - f(u)) w_a \leq 0,$$

and therefore $\nabla w \equiv 0$ on Σ_a. Let $y \in E$ and let U_y be the connected component of Σ_a containing y. Note that $w = -a$ on ∂U_y ($\neq \emptyset$). On the other hand $w(y) = -\bar{\varepsilon}$ and w is constant on U_y, contradicting the fact that $a \in [\bar{\varepsilon}/2, \bar{\varepsilon})$.

(ii) A further observation is that we can substitute the term $\langle X, \nabla u\rangle + f(u)$ in (3.87) with a general $B(x, u, \nabla u)$ with the property that

$$B(x, u, \xi) \leq k\varphi(|\xi|) + f(u) \tag{3.101}$$

for $x \in M$, $u \in \mathbb{R}_0^+$ and $|\xi| \leq 1$, for some $k \geq 0$ and f satisfying (F1), (F3), provided (A2) (i) holds. This is essential for the very general form of Theorem 3.10 below; to see this note that the right-hand side of inequality (3.97) now becomes

$$\int_{\Sigma_a} \{k[\varphi(|\nabla v|) - \varphi(|\nabla u|)] + f(v) - f(u)\}w_a$$

and

$$\varphi(|\nabla u|) - \varphi(|\nabla v|) = \int_0^1 \varphi'(X_t)(|\nabla u| - |\nabla v|) \leq \eta|\nabla u - \nabla v|$$

with $\eta = \max_{[b,2d]} \varphi' > 0$. This allows us to obtain (3.98) again.

Remark 3.7 The reasoning in the proof of the theorem above shows that Ω can be any domain, that is, not necessarily relatively compact, if we add to (3.91) the further requirement

$$\limsup_{x\in\Omega, r(x)\to+\infty} (u(x) - v(x)) \geq 0 \tag{3.102}$$

and the condition $\sup_\Omega |X| < +\infty$. The above condition (3.102) on u and v will be also considered in Proposition 3.1 that follows.

For our needs we will use a simplified form of the comparison principle like that expressed in the next result and in Theorem 3.9 below.

Proposition 3.1 *Assume (A1), (A2). Let $\Omega \subset M$ be a domain and suppose that $u, v \in C^0(\overline{\Omega}) \cap C^1(\Omega)$ satisfy*

$$\begin{cases} (i)\ \operatorname{div}(A(|\nabla u|)\nabla u) \geq \operatorname{div}(A(|\nabla v|)\nabla v) & \textit{weakly on } \Omega \\ (ii)\ u \leq v \textit{ on } \partial\Omega \textit{ and } \limsup_{x\in\Omega, r(x)\to+\infty}(u(x) - v(x)) \leq 0, \end{cases} \tag{3.103}$$

where the last condition only appears in case Ω is unbounded. Then $u \leq v$ on $\overline{\Omega}$.

Remark 3.8 The proof of the proposition is much simpler and direct than that of Theorem 3.8. This is due to the fact that inequality (3.96), requiring $X_t \neq 0 \ \forall\, t \in [0,1]$, can now be avoided because the tensor T coincides with the metric. This will be clear in the argument below.

Proof Set $w = v - u$ in $\overline{\Omega}$ and by contradiction assume that there exists $y \in \Omega$ such that $w(y) < 0$. Fix $\varepsilon > 0$ sufficiently small so that $w(y) + \varepsilon < 0$. By assumption (3.103) (ii), $w \geq 0$ on $\partial\Omega$ and "at infinity"; it follows that $w_\varepsilon = \min\{w + \varepsilon, 0\}$ is a nonpositive Lipschitz function with compact support in Ω. By the meaning of weak solution of (3.103) (i), taking $-w_\varepsilon$ as a test function we get

$$\int_\Omega h \leq 0, \tag{3.104}$$

where we have set

$$\begin{aligned} h &= \langle A(|\nabla v|)\nabla v - A(|\nabla u|)\nabla u, \nabla w_\varepsilon \rangle \\ &= \left\langle |\nabla v|^{-1}\varphi(|\nabla v|)\nabla v - |\nabla u|^{-1}\varphi(|\nabla u|)\nabla u, \nabla w_\varepsilon \right\rangle. \end{aligned}$$

Clearly $h = 0$ on the set $\{x \in \Omega : w(x) + \varepsilon = 0\}$. On the other hand, on $\{x \in \Omega : w(x) + \varepsilon < 0\}$,

$$\begin{aligned} h = {} & [\varphi(|\nabla v|) - \varphi(|\nabla u|)][|\nabla v| - |\nabla u|] \\ & + \left[|\nabla v|^{-1}\varphi(|\nabla v|) + |\nabla u|^{-1}\varphi(|\nabla u|)\right][|\nabla u||\nabla v| - \langle \nabla u, \nabla v \rangle]. \end{aligned}$$

Whence, using Cauchy-Schwarz inequality and (A2), that also implies $\varphi(t) > 0$ for $t > 0$, we obtain

$$h \geq 0 \quad \text{a. e. in } \Omega.$$

From this and (3.104) it follows that $h \equiv 0$ a. e. in Ω, which in turn forces $\nabla w_\varepsilon = 0$ a. e. in Ω. This shows that $w_\varepsilon = w + \varepsilon = w(y) + \varepsilon < 0$ so that $v - u = w < -\varepsilon$ in Ω contradicting (3.103) (ii). □

Remark 3.9 The same proof works also in case u and v, solutions of (3.103) (i), are in $C^0(\overline{\Omega}) \cap \mathrm{Lip}_{loc}(\Omega)$.

We now give a version of Proposition 3.1 in case T is not the metric. The proof introduces a further point of view resting on the distributional divergence of a vector field. Since this approach will also be used later on, for instance in the proofs of Theorems 4.4, 4.1 and 4.2, it seems rewarding to introduce it here. See however Remark 3.10 below.

To prove our result we need a second version of Lemma 3.2, but first let us introduce the next function. Fix $x \in M$, $\xi \in T_xM$ and consider

$$g_{x,\xi} : (T_xM \setminus \{0\}) \to \mathbb{R}_0^+$$

defined by

$$g_{x,\xi}(v) = T\left(\frac{\varphi(|v|)}{|v|}v, \xi\right).$$

Since

$$\left|g_{x,\xi}(v)\right| \leq |T|_x \varphi(|v|)|\xi|,$$

the validity of (A2) (ii) implies that

$$g_{x,\xi}(v) \to 0 \quad \text{as } v \to 0.$$

This allows us to define $g_{x,\xi} : T_xM \to \mathbb{R}_0^+$ continuously by setting $g_{x,\xi}(0) = 0$.

Next we strengthen (A2) to

(A2)' (i) $\varphi'(t) > 0$ on $\mathbb{R}^+$, (ii) $\varphi(t) \to 0$ as $t \to 0^+$, (iii) $\frac{\varphi(t)}{t} \in L^1(0^+)$.

Observe that (i) and (ii) imply $\varphi'(t) \in L^1(0^+)$. We are now ready to state

Lemma 3.3 *Assume the validity of (A1), (A2)' and define $g_{x,\xi}$ for $x \in M$, $\xi \in T_xM$ as above. Let ∇u, $\nabla v \in T_xM$ and set $X_t = t\nabla u + (1-t)\nabla v$ for $t \in [0,1]$. Suppose that $|\nabla u| + |\nabla v| > 0$ and let T be a $(0,2)$-tensor field on M. Then at x we have*

$$\begin{aligned} h(x) &= g_{x,\nabla u - \nabla v}(\nabla u) - g_{x,\nabla u - \nabla v}(\nabla v) \\ &= \int_0^1 \left\{ \frac{\varphi(|X_t|)}{|X_t|} T(\nabla u - \nabla v, \nabla u - \nabla v) \right. \\ &\quad \left. + \frac{1}{|X_t|^2}\left(\varphi'(|X_t|) - \frac{\varphi(|X_t|)}{|X_t|}\right)\langle X_t, \nabla u - \nabla v\rangle T(X_t, \nabla u - \nabla v) \right\} dt. \end{aligned} \tag{3.105}$$

Furthermore, if (T1) and (T2) hold, then $h(x) \geq 0$ and $h(x) = 0$ if and only if $\nabla u = \nabla v$.

Proof If $X_t \neq 0$ on $[0,1]$ the result coincides with Lemma 3.2. Thus suppose there exists $t_0 \in [0,1]$ with $X_{t_0} = 0$. Note that t_0 is unique. The cases $t_0 = 0$ and $t_0 = 1$ are simpler, so let us assume $t_0 \in (0,1)$. Let I be the integrand in (3.105). For $\varepsilon > 0$ sufficiently small, integrating on the intervals $[0, t_0 - \varepsilon]$ and $[t_0 + \varepsilon, 1]$ we get

$$\int_0^{t_0-\varepsilon} I\,dt + \int_{t_0+\varepsilon}^1 I\,dt = g_{x,\nabla u - \nabla v}(X_{t_0-\varepsilon}) - g_{x,\nabla u - \nabla v}(\nabla v) + g_{x,\nabla u - \nabla v}(\nabla u) - g_{x,\nabla u - \nabla v}(X_{t_0+\varepsilon}).$$

By the continuity of $g_{x,\xi}$, its linearity in ξ and since $X_{t_0} = 0$, the right-hand side of the above converges to $h(x)$ as $\varepsilon \to 0$ and the left-hand side converges to $\int_0^1 I$ because of (A2)'.

Under the validity of (T1) and (T2), the fact that $h(x) \geq 0$ and $h(x) = 0$ if and only if $\nabla u = \nabla v$ follows immediately from (3.105). □

Theorem 3.9 *Assume (A1), (A2)', (T1), (T2) and let $\Omega \subset M$ be a relatively compact domain. Let $u, v \in C^0(\overline{\Omega}) \cap C^1(\Omega)$ satisfy*

$$\begin{cases} \operatorname{div}\left(A(|\nabla u|)T(\nabla u, \,)^\sharp\right) \geq \operatorname{div}\left(A(|\nabla v|)T(\nabla v, \,)^\sharp\right) & \text{weakly on } \Omega \\ u \leq v \ \text{ on } \partial\Omega. \end{cases} \tag{3.106}$$

Then $u \leq v$ on $\overline{\Omega}$.

Proof Clearly it suffices to prove that for each $\varepsilon > 0$ we have

$$u \leq v + \varepsilon \quad \text{on } \Omega. \tag{3.107}$$

Towards this aim fix $\varepsilon > 0$ and let $\tilde{\Omega}$ be an open set with smooth boundary such that

$$\Theta = \{x \in \Omega : u(x) > v(x) + \varepsilon\} \subset\subset \tilde{\Omega} \subset\subset \Omega.$$

Note that to construct $\tilde{\Omega}$ we can choose a smooth nonnegative function z such that $z \equiv 1$ on Θ and $z \equiv 0$ on $M \setminus \Omega$. If $c \in \left(\frac{1}{4}, \frac{3}{4}\right)$ is a regular value of z (which exists by Sard's theorem) we may set $\tilde{\Omega} = \{x \in \Omega : z(x) > c\}$. Let $\alpha \in C^1(\mathbb{R})$ be such that $\alpha(t) = 0$ if $t \leq \varepsilon$ and $\alpha'(t) > 0$ if $t > \varepsilon$, so that $\alpha(t) > 0$ for $t > \varepsilon$. Let W be the vector field defined by

$$W = \alpha(u - v)\Big[|\nabla u|^{-1}\varphi(|\nabla u|)T(\nabla u, \,)^\sharp - |\nabla v|^{-1}\varphi(|\nabla v|)T(\nabla v, \,)^\sharp\Big]; \tag{3.108}$$

note that W continuously extends to all of Ω whenever ∇u or ∇v are zero. Furthermore, note that, by the definition of α on $\tilde{\Omega}$, $W \equiv 0$ in a neighbourhood of $\tilde{\Omega}$. Using (3.106) we have

$$\begin{aligned} \operatorname{div} W = {} & \alpha(u - v)\Big\{\operatorname{div}\left(|\nabla u|^{-1}\varphi(|\nabla u|)T(\nabla u, \,)^\sharp\right) - \operatorname{div}\left(|\nabla v|^{-1}\varphi(|\nabla v|)T(\nabla v, \,)^\sharp\right)\Big\} \\ & + \alpha'(u - v)T\Big(|\nabla u|^{-1}\varphi(|\nabla u|)\nabla u - |\nabla v|^{-1}\varphi(|\nabla v|)\nabla v, \nabla u - \nabla v\Big), \end{aligned}$$

so that

$$\operatorname{div} W \geq \alpha'(u - v)T\Big(|\nabla u|^{-1}\varphi(|\nabla u|)\nabla u - |\nabla v|^{-1}\varphi(|\nabla v|)\nabla v, \nabla u - \nabla v\Big) \quad \text{on } \Omega. \tag{3.109}$$

Denote by ρ the distance function from $\partial\tilde{\Omega}$, with the convention that $\rho(x) > 0$ if $x \in \tilde{\Omega}$ and $\rho(x) < 0$ if $x \notin \overline{\tilde{\Omega}}$, so that ρ is the radial coordinate in the Fermi coordinates (see also Chap. 2, Sect. 2.3) with respect to $\partial\tilde{\Omega}$. By Gauss lemma, $|\nabla\rho| = 1$. Let

$$\tilde{\Omega}_\gamma = \{x \in \tilde{\Omega} : \rho(x) > \gamma\}$$

and let ψ_γ be the Lipschitz function defined by

$$\psi_\gamma(x) \begin{cases} 1 & \text{if } x \in \tilde{\Omega}_\gamma \\ \frac{1}{\gamma}\rho(x) & \text{if } x \in \tilde{\Omega} \setminus \tilde{\Omega}_\gamma \\ 0 & \text{if } x \notin \tilde{\Omega}. \end{cases}$$

Note that, since W vanishes in a neighbourhood of $\partial\tilde{\Omega}$, for each $\gamma > 0$ sufficiently small W vanishes off $\tilde{\Omega}_\gamma$ and by definition of weak divergence we have

$$\int_{\tilde{\Omega}} \psi_\gamma \operatorname{div} W = -\int_{\tilde{\Omega}} \langle W, \nabla\psi_\gamma\rangle = -\frac{1}{\gamma}\int_{\tilde{\Omega}\setminus\tilde{\Omega}_\gamma} \langle W, \nabla\rho\rangle = 0. \tag{3.110}$$

Thus, using (3.109),

$$\int_{\tilde{\Omega}} \alpha'(u-v)T\Big(|\nabla u|^{-1}\varphi(|\nabla u|)\nabla u - |\nabla v|^{-1}\varphi(|\nabla v|)\nabla v, \nabla u - \nabla v\Big) \le 0.$$

By Lemma 3.3

$$h = T\Big(|\nabla u|^{-1}\varphi(|\nabla u|)\nabla u - |\nabla v|^{-1}\varphi(|\nabla v|)\nabla v, \nabla u - \nabla v\Big) \ge 0.$$

Thus

$$\int_{\Theta} \alpha'(u-v)h \le 0.$$

Since $\alpha'(u-v) > 0$ on Θ and $h > 0$ if $\nabla u \ne \nabla v$ we deduce that $\nabla u \equiv \nabla v$ on Θ; but then $u - v$ is constant on each connected component of Θ. Since $u = v + \varepsilon$ on $\partial\Omega$ this contradicts the definition of Θ. □

Remark 3.10 The above proof extends to the case where u and v are only Lipschitz in Ω. Note that by Lemma 1.16 in [142] the conclusion $\nabla u = \nabla v$ a.e. implies $u \equiv v$ on every connected component of Θ.

We now consider a solution $u \ge 0$ of the differential inequality

$$\operatorname{div}\Big(A(|\nabla u|)T(\nabla u,)^\sharp\Big) - f(u) \le 0, \tag{3.111}$$

where f satisfies (F1), while A and T are as before.

We say that the *strong maximum principle holds for* (3.111) *in a relatively compact domain* Ω if for any solution $u \in C^1(\Omega)$, $u \geq 0$, the existence of $x_0 \in \Omega$ such that $u(x_0) = 0$ implies $u \equiv 0$ in Ω. Before stating the theorem we define on $\mathbb{R}_0^+$ the function

$$H(t) = t\varphi(t) - \int_0^t \varphi(s)\, ds. \tag{3.112}$$

Note that H is strictly increasing on $\mathbb{R}_0^+$; indeed, let $0 \leq t_0 < t_1$, then using (A2) we have

$$t_1\varphi(t_1) - t_0\varphi(t_0) > (t_1 - t_0)\varphi(t_1) > \int_{t_0}^{t_1} \varphi(s)\, ds.$$

In particular $H(t) > 0$ on $\mathbb{R}^+$ and $H(+\infty) \leq +\infty$. Note that the case $H(+\infty) < +\infty$ can indeed happen (for instance with $\varphi(t) = \frac{t}{\sqrt{1+t^2}}$). With f satisfying (F2) on $(0, \varepsilon)$ we define

$$F(t) = \int_0^s f(s)\, ds; \tag{3.113}$$

thus $F(t)$ is positive for $t > 0$ sufficiently small and $F(0) = 0$. Hence, for $t \in \mathbb{R}^+$ sufficiently small the function $H^{-1}(F(t))$ is well defined and the requirement

$$\frac{1}{H^{-1}(F(t))} \notin L^1(0^+) \tag{3.114}$$

is meaningful.

Although we will use Theorem 3.10 below only for $f \equiv 0$, we consider here the more general case for its relation to condition (3.114) as we briefly explain.

Recall that, given the differential inequality

$$\begin{cases} \operatorname{div}\left(A(|\nabla u|)T(\nabla u, \ )^\sharp\right) \geq f(u) \\ u \geq 0 \end{cases}$$

in $\Omega = M \setminus K$ for some compact set K of M, we say that the *compact support principle* (CSP for short) holds for it if the condition $u(x) \to 0$ as $x \to \infty$ implies that u has compact support in Ω.

As proved in [236], a necessary condition for this to happen is the validity, in the above notation, of

$$\frac{1}{H^{-1}(F(t))} \in L^1(0^+). \tag{3.115}$$

This requirement can be, in some sense, thought as "dual" to the Keller-Osserman condition

$$\frac{1}{H^{-1}(F(t))} \in L^1(+\infty).$$

Loosely speaking, and as already pointed out in Theorem 3.6, the failure of the latter is strictly related to the existence of unbounded positive solutions on M of the differential inequality

$$\operatorname{div}\left(A(|\nabla u|)T(\nabla u, \ )^\sharp\right) \geq f(u).$$

Similarly, the failure of (3.115), in other words, the validity of (3.114), yields the validity of the maximum principle in Theorem 3.10 below, which in turn allows us to construct counterexamples (see [236]) to the CSP. A full clarification of the mutual relation between these conditions is still open, although some progress has been made in [45].

Theorem 3.10 *Assume (A1), (A2), (T1) and (T2). Let $\Omega \subset M$ be a relatively compact domain and let $u \in C^1(\Omega)$ be a solution of*

$$\begin{cases} \operatorname{div}\left(A(|\nabla u|)T(\nabla u, \ )^\sharp\right) - B(x, u, \nabla u) \leq 0 \ \ in \ \ \Omega, \\ u \geq 0 \ \ in \ \ \Omega \end{cases} \tag{3.116}$$

with $B(x, u, \nabla u) \leq \kappa\varphi(|\nabla u|) + f(u)$ for some $\kappa \geq 0$. Then, for the strong maximum principle to hold, it is sufficient that either $f \equiv 0$ on $[0, \mu)$ for some $\mu > 0$ or that f satisfies (F3) and (3.114)*.*

Remark 3.11 If f satisfies (F3) but $f \not\equiv 0$ on $[0, \mu]$ for some $\mu > 0$, then $f > 0$ on $(0, \mu]$ for some $\mu > 0$ because of (F3), hence (F2) holds and (3.114) is meaningful.

Remark 3.12 If $\kappa = 0$ then (A2) is not needed.

The proof of Theorem 3.10 is based on the original idea of Hopf [144] to compare the solution u of (3.116) with an appropriate function v to obtain a contradiction in case u violates the strong maximum principle in Ω. However, since we are dealing with nonlinear operators of a very general type, the construction of v is quite delicate. Thus, in order to prove the theorem, we first need to establish a number of auxiliary results. We begin with the next

Lemma 3.4 *Assume (F1), (F3). If $\tau \geq 1$ and* (3.114) *holds then*

$$\frac{1}{H^{-1}(\tau F(t))} \notin L^1(0^+). \tag{3.117}$$

Proof Let $\sigma \in [0, 1]$; since, by (F3), f is nondecreasing for $t \in [0, \delta]$, we have

$$\sigma f(\sigma t) \leq \sigma f(t).$$

It follows that

$$\int_0^{\sigma t} f(x)\,dx = \sigma \int_0^t f(\sigma s)\,ds \le \sigma \int_0^t f(s)\,ds,$$

that is,

$$F(\sigma t) \le \sigma F(t). \tag{3.118}$$

Choose $\sigma = \frac{1}{\tau}$; then

$$\tau F\Big(\frac{t}{\tau}\Big) \le F(t), \qquad 0 < t \ll 1.$$

But H^{-1} is increasing and thus

$$\frac{1}{H^{-1}\big(\tau F\big(\frac{t}{\tau}\big)\big)} \ge \frac{1}{H^{-1}(F(t))}.$$

Fix $\varepsilon > 0$ small and $a > \varepsilon$ sufficiently small. Then

$$\tau \int_{\frac{\varepsilon}{\tau}}^{\frac{a}{\tau}} \frac{dx}{H^{-1}(\tau F(x))} = \int_\varepsilon^a \frac{dt}{H^{-1}\big(\tau F\big(\frac{t}{\tau}\big)\big)} \ge \int_\varepsilon^a \frac{dt}{H^{-1}(F(t))}.$$

Letting $\varepsilon \downarrow 0^+$ we obtain (3.117). □

The following result will reveal essential in what follows because it guarantees $w'(T) > 0$ for a $C^1([0,T])$-solution of problem (3.120) below.

Note that, under assumption (A2) (ii), for technical reasons we extend the definition of φ to a continuous function on $\mathbb{R}$, still called φ, by setting $\varphi(t) = -\varphi(-t)$ for $t < 0$. Clearly $\varphi(0) = 0$. From (A2) (i) we also have $t\varphi(t) > 0$ for $t \neq 0$.

Lemma 3.5 *Let $T > 0$, assume (A2) and let φ be as above. Let*

$$q \in C^0((0,T)), \quad q > 0 \ \ in \ \ (0,T). \tag{3.119}$$

Then any C^1-weak solution $w = w(t)$ of

$$\begin{cases} [\operatorname{sgn} w(t)][q(t)\varphi(w'(t))]' \ge 0 \ \ in \ (0,T), \\ w(0) = 0, \ \ w(T) = m > 0, \end{cases} \tag{3.120}$$

where sgn *is the* signum function, *that is,*

$$\operatorname{sgn} t = \begin{cases} \frac{t}{|t|} & if \ t \neq 0 \\ 0 & if \ t = 0, \end{cases}$$

is such that

$$w \geq 0, \;\; w' \geq 0 \;\; in \;\; (0,T). \tag{3.121}$$

Even more, there exists $t_0 \in [0,T)$ *with the property that*

$$w \equiv 0 \quad in \;\; [0,t_0], \quad w > 0, \;\; w' > 0 \;\; in \;\; (t_0,T) \tag{3.122}$$

and if $w \in C^1((0,T])$ *then*

$$w'(T) > 0. \tag{3.123}$$

Proof We claim $w \geq 0$ in $[0,T]$. Otherwise there exist $0 \leq t_0 < t_1 < T$ such that $w(t_0) = w_(t_1) = 0$ and $w < 0$ on (t_0,t_1). We use

$$\psi(t) = \begin{cases} w(t) & \text{if } \; t \in [t_0,t_1], \\ 0 & \text{otherwise} \end{cases}$$

as a test function. Since $\psi(t) \leq 0$, (3.120) gives

$$\int_{t_0}^{t} q(t)\varphi\big(w'(t)\big)w'(t)\,dt \leq 0.$$

Now, for $\rho \neq 0$, $\varphi(\rho)\rho > 0$, thus from (3.119) we deduce that the integrand is nonnegative. Therefore, necessarily $w' \equiv 0$ on $[t_0,t_1]$. It follows that $w \equiv 0$ on $[t_0,t_1]$. This contradiction proves the claim.

Now let

$$J = \big\{t \in (0,T) : w'(t) > 0\big\}.$$

Obviously, $J \neq \emptyset$ since $w(0) = 0$ and $w(T) > 0$, and J is open in $(0,T)$ since $w \in C^1((0,T))$. Let $t_0 = \inf J \in [0,T)$; then $w \equiv 0$ in $[0,t_0]$, since we already know that $w \geq 0$ in $[0,T]$. Next, for any fixed $t \in (t_0,T)$ necessarily there exists $t_1 \in (t_0,t)$ such that $w'(t_1) > 0$. Integrating (3.120) on $[t_1,t]$, and recalling that $w \geq 0$ on $(0,T)$, because of (3.119) and (A2) (i), that is, φ increasing on $\mathbb{R}^+$, we get

$$q(t)\varphi\big(w'(t)\big) \geq q(t_1)\varphi\big(w'(t_1)\big) > 0,$$

so that $w' > 0$ on (t_0,T) and (3.123) holds in case $w \in C^1((0,T))$. Now by integration $w > 0$ in (t_0,T), completing the proof of (3.122). □

Our next step is to solve the following singular two-point boundary value problem:

$$\begin{cases} [q(t)\varphi(w'(t))]' - q(t)f(w(t)) = 0 \;\; \text{in} \;\; (0,T), \\ w(0) = 0, \;\; w(T) = m > 0. \end{cases} \tag{3.124}$$

Here we assume

$$q \in C^1([0,T]), \ q > 0 \text{ on } [0,T] \tag{3.125}$$

and we set

$$q_0 = \min_{[0,T]} q(t), \quad q_1 = \max_{[0,T]} q(t).$$

We have

Proposition 3.2 *Let* (3.125), *(F1), (F3),* (A1), (A2) *hold.*

(i) Let $\varphi(+\infty) = +\infty$. *Then problem* (3.124) *admits a* C^1*-weak solution with the properties*

$$w \in C^1([0,T]), \ \varphi(w') \in C^1([0,T]), \ w' \geq 0. \tag{3.126}$$

Moreover we have

$$w'(T) > 0 \tag{3.127}$$

and

$$\|w'\|_\infty \leq \varphi^{-1}\left(\frac{q_1}{q_0}\left[T\bar{f}(m) + \varphi\left(\frac{m}{T}\right)\right]\right), \tag{3.128}$$

where $\bar{f}(m) = \max_{[0,m]} f(t)$. *In particular*

$$w' \leq 1 \tag{3.129}$$

if $m > 0$ *and* $T > 0$ *are sufficiently small.*

(ii) If $\varphi(+\infty) = \omega < +\infty$ *let* $m \in (0, \bar{\delta})$, $0 < T \ll 1$ *be such that*

$$\frac{q_1}{q_0}\left[T\bar{f}(m) + \varphi\left(\frac{m}{T}\right)\right] < \omega. \tag{3.130}$$

Then the conclusion of part (i) continues to hold.

Remark 3.13 Note that we could relax (F3) to f nonnegative on $[0,\delta)$ for some $\delta > 0$.

Proof Note that, for $\rho < 0$, $\varphi(\rho) = -\varphi(-\rho)$; this definition does not affect the generality, since the ultimate solution w satisfies $w' \geq 0$. It is also convenient to redefine f so that $f(t) = f(m)$ for $t \geq m$ and to set $f(t) = 0$ for $t \leq 0$. Again this will not affect the conclusion of the proposition, since clearly any ultimate solution with $w' > 0$ satisfies $0 \leq w \leq m$.

Case (i). Let

$$\mu_1 = q_1\Big[T\bar{f}(m) + \varphi\Big(\frac{m}{T}\Big)\Big] \tag{3.131}$$

and

$$I = [0, \mu_1].$$

To show existence we shall use Browder's version of the Leray-Schauder Theorem (see Theorem 11.6 of [125]). Towards this aim we let X be the Banach space $X = C^0([0, T])$ with the sup norm $\| \|_\infty$. Let $\mathscr{F} : X \to X$ be defined by

$$\mathscr{F}(w)(t) = m - \int_t^T \varphi^{-1}\Big(\frac{1}{q(s)}\Big[\mu - \int_s^T q(\tau)f(w(\tau))\,d\tau\Big]\Big)\,ds, \tag{3.132}$$

$t \in [0, T]$, where $\mu = \mu(w) \in I$ is chosen in such a way that

$$\mathscr{F}(w)(0) = 0. \tag{3.133}$$

Such a choice of μ is possible and in fact unique; indeed, for any fixed $w \in X$ and $\mu \in I$ we have

$$-\frac{\bar{f}(m)}{q_0}\int_0^T q(t)\,dt \le \frac{1}{q(s)}\Big[\mu - \int_s^T q(\tau)f(w(\tau))\,d\tau\Big] \le \frac{\mu_1}{q_0}; \tag{3.134}$$

hence $\mathscr{F}(w)$ is well defined for each $\mu \in I$. Moreover, for $\mu = 0$ we see that, for all $w \in X$,

$$\mathscr{F}(w)(0) \ge m.$$

On the other hand, for $\mu = \mu_1$, for all $w \in X$ we find

$$\begin{aligned}\mathscr{F}(w)(0) = m \\ &- \int_t^T \varphi^{-1}\Big(\frac{q_1}{q(s)}\varphi\Big(\frac{m}{T}\Big) + \frac{1}{q(s)}\Big[q_1 T\bar{f}(m) - \int_s^T q(\tau)f(w(\tau))\,d\tau\Big]\Big)\,ds \\ &\le m - \int_0^T \varphi^{-1}\Big(\varphi\Big(\frac{m}{T}\Big)\Big)\,ds = 0,\end{aligned}$$

since φ^{-1} is increasing and $\Big[q_1 T\bar{f}(m) - \int_s^T q(\tau)f(w(\tau))\,d\tau\Big] \ge 0$. Now, the integral on the right-hand side of (3.132) is a strictly increasing function of μ for w fixed; it is therefore clear that there exists a unique $\mu \in I$ such that (3.133) holds true. Next we define the homotopy $\mathscr{H} : X \times [0,1] \to X$ by setting

$$\mathscr{H}(w,\sigma)(t) = \sigma m - \int_t^T \varphi^{-1}\left(\frac{1}{q(s)}\left[\mu_\sigma - \sigma \int_s^T q(\tau)f(w(\tau))\,d\tau\right]\right) ds, \quad (3.135)$$

where $\mu_\sigma = \mu(w,\sigma) \in I$ is chosen in such a way that

$$\mathscr{H}(w,\sigma)(0) = 0. \quad (3.136)$$

Proceeding as above we see that μ_σ exists and is unique, and the mapping $\mathscr{H}$ is well defined. By construction any fixed point $w_\sigma = \mathscr{H}(w_\sigma,\sigma)$ is of class $C^1([0,T])$, has the property that $\varphi(w') \in C^1([0,T])$ and it is a C^1-weak solution of the problem

$$\begin{cases} \left[q(t)\varphi\big(w'_\sigma(t)\big)\right]' - \sigma q(t)f(w_\sigma(t)) = 0 \ \text{ in } [0,T], \\ w_\sigma(0) = 0, \ \ w_\sigma(T) = \sigma m. \end{cases} \quad (3.137)$$

Note that with our new definition of f, w_σ satisfies the differential inequality in (3.120). Thus, by Lemma 3.5, a fixed point $w = \mathscr{H}(w,1)$ satisfies $w, w' \geq 0$ and so is a solution of (3.124) satisfying (3.126) and (3.127).

It remains to show that such a fixed point $w = w_1$ exists. We begin by verifying the first step to apply Leray-Schauder's Theorem. When $\sigma = 0$, then clearly $\mu_\sigma = 0$ and therefore, for all $w \in X$, $\mathscr{H}(w,0)(t) = 0$, that is, $\mathscr{H}(\cdot,0)$ maps X into the single point $0 \in X = C^0([0,T])$. We now show that $\mathscr{H} : X \times [0,1] \to X$ is compact and continuous. Let (w_k,σ_k) be a bounded sequence in $X \times [0,1]$. Clearly $\mu_{\sigma_k} \in I$; therefore, using the fact that $0 \leq f(t) \leq \bar{f}(m)$ for every $t \geq 0$ together with (3.134) we deduce that

$$\|\mathscr{H}'(w_k,\sigma_k)\|_\infty \leq C',$$

where

$$C' = \max\left\{\frac{\bar{f}(m)}{q_0}\int_0^T q(t)\,dt, \varphi^{-1}\left(\frac{\mu_1}{q_0}\right)\right\}. \quad (3.138)$$

Since C' is independent of k, by Ascoli-Arzelà theorem $\mathscr{H}$ maps bounded sequences into relatively compact sequences in X.

To show that $\mathscr{H}$ is continuous on $X \times [0,1]$ let $\{(w_j,\sigma_j)\} \subset X \times [0,1]$ be such that $w_j \to w$, $\sigma_j \to \sigma$. Then, in (3.135), $\sigma_j f(w_j) \to \sigma f(w)$, $\sigma_j h \to \sigma h$ since the modified function f is continuous on $\mathbb{R}$ (here we use the condition $f(0) = 0$). We need to show that $\mu\big(w_j,\sigma_j\big) \to \mu(w,\sigma)$. By contradiction suppose this is not the case; then, for some subsequence still called $\{(w_j,\sigma_j)\}$, we have

$$\mu_j = \mu\big(w_j,\sigma_j\big) \to \tilde{\mu} \neq \mu = \mu(w,\sigma).$$

From (3.136) we deduce

$$0 = (\sigma_j - \sigma)m - \int_0^T \varphi^{-1}\left(\frac{1}{q}\left[\mu_j - \sigma_j \int_s^T q(\tau)f(w_j(\tau))\,d\tau\right]\right) ds$$
$$+ \int_0^T \varphi^{-1}\left(\frac{1}{q}\left[\tilde{\mu} - \sigma \int_s^T q(\tau)f(w(\tau))\,d\tau\right]\right) ds,$$

and letting $j \to +\infty$,

$$\begin{aligned} 0 \; &= \int_0^T \Big\{\varphi^{-1}\Big(\tfrac{1}{q}\Big[\tilde{\mu} - \sigma \int_s^T q(\tau)f(w(\tau))\,d\tau\Big]\Big) \\ &\quad -\varphi^{-1}\Big(\tfrac{1}{q}\Big[\mu - \sigma \int_s^T q(\tau)f(w(\tau))\,d\tau\Big]\Big)\Big\}\, ds. \end{aligned} \tag{3.139}$$

But φ^{-1} is monotone increasing, so the integrand is either everywhere positive or negative contradicting (3.139).

To apply the Leray-Schauder Theorem it is now enough to show that there is a constant $\lambda > 0$ such that

$$\|w\|_\infty \le \Lambda \quad \text{for every } (w,\sigma) \in X \times [0,1] \quad \text{with } \mathscr{H}(w,\sigma) = w. \tag{3.140}$$

Towards this aim let (w,σ) be as in (3.140). As observed above, since $w' \ge 0$, $\|w\|_\infty = w(T) = \sigma m \le m$ and we can choose $\Lambda = m$.

The Leray-Schauder Theorem can therefore be applied and the mapping $\mathscr{I}(w) = \mathscr{H}(w,1)$ has a fixed point $w \in X$ which is the required solution of (3.124). That (3.126), (3.127) hold has already been noted. To prove (3.128) consider (3.132) evaluated at a fixed point w. From (3.134) and $\mu \in I$ we have

$$w'(t) = \mathscr{F}'(w)(t) = \varphi^{-1}\left(\frac{1}{q(t)}\left[\mu - \int_s^T q(\tau)f(w(\tau))\,d\tau\right]\right)$$
$$\le \varphi^{-1}\left(\frac{\mu_1}{q_0}\right) = \varphi^{-1}\left[\frac{q_1}{q_0}\left(T\bar{f}(m) + \varphi\left(\frac{m}{T}\right)\right)\right]$$

and (3.128) follows at once.

Case (ii). The argument is the same as before, except that in (3.134) the right-hand side, $\frac{\mu_1}{q_0}$, is now less than w because of (3.130). Thus, $\mathscr{F}$ is well defined on X and the rest of the proof remains unchanged. □

Proposition 3.3 *Assume* (3.125), *(F1)*, (A1), (A2) *and suppose additionally that either* $f(t) \equiv 0$ *on* $(0,\mu)$ *for some* $\mu > 0$, *or that (F2) and* (3.114) *hold. Then the solution* w *of the problem* (3.124) *given in Proposition 3.2 has the further properties*

$$w(t) > 0 \quad in \;\; (0,T]; \qquad w'(t) > 0 \quad in \;\; [0,T]. \tag{3.141}$$

Proof **Case (i)**. Let $f \equiv 0$ on $(0, \mu)$ for some $\mu > 0$. Then, from (3.124),

$$\left[q(t)\varphi\left(w'(t)\right)\right]' \equiv 0$$

at least for t near 0. Hence for t sufficiently small

$$q(t)\varphi\left(w'(t)\right) = q(0)\varphi\left(w'(0)\right)$$

is constant. We claim that $w'(0) > 0$. Indeed suppose the contrary; then, since φ^{-1} is increasing and $\varphi^{-1}(0) = 0$, we would have $w'(t) \equiv 0$ for t sufficiently small and then, by continuation, for $t \in [0, T]$, contradicting the boundary condition $w(T) = m > 0$. Hence $w'(0) > 0$ so that, in the second part of Lemma 3.5, $t_0 = 0$ and it therefore follows that $w'(t) > 0$ in $[0, T]$ and $w > 0$ in $(0, T]$ as required.

Case (ii). Let (3.114) hold. Because of (3.126) $\varphi(w') \in C^1([0, T])$. We also already know that $w'(0) \geq 0$ and $0 \leq w \leq m$. If we show that $w'(0) > 0$ then the conclusion follows as before. Hence, let w be a solution of

$$\begin{cases} [q(t)\varphi(w'(t))]' - q(t)f(w(t)) \leq 0 \text{ on } (0, T), \\ w(0) = 0, \quad w(T) = m > 0, \quad w' \geq 0 \end{cases} \tag{3.142}$$

with $\varphi(w') \in C^1([0, T])$. Suppose that $f(u) > 0$ for $u > 0$. If $w'(0) = 0$ we claim that

$$\frac{1}{H^{-1}(F(s))} \in L^1\left(0^+\right). \tag{3.143}$$

Proving this yields a contradiction with (3.114), so that $w'(0) > 0$ and the proposition is proved. Towards this aim we need the following auxiliary

Lemma 3.6 *Assume the validity of (A1), (A2), (F1), (F3) and that*

$$q \in C^1([0, T]), \quad q > 0 \quad on \ [0, T] \tag{3.144}$$

holds. Then, for each weak solution $w \in C^1([0, T])$ of

$$\begin{cases} [q(t)\varphi(w'(t))]' - q(t)f(w(t)) \leq 0 \ on \ (0, T), \\ w(0) = 0, \ \ 0 \leq w(T) \leq \mu, \quad w' \geq 0 \ on \ (0, T) \end{cases} \tag{3.145}$$

for which $w'(0) = 0$ and $\varphi(w') \in C^1((0, T))$, for F as in (3.113) *we have*

$$H\left(w'(t)\right) \leq B(t)F(w(t)) \quad on \ (0, T), \tag{3.146}$$

where

$$B(t) = 1 + \sup_{s\in[0,t]} \left(-\frac{q'(s)}{q(s)^2} \int_0^s q(\tau)\,d\tau \right)_+. \tag{3.147}$$

In particular, if $q' \geq 0$, (3.146) *implies*

$$H\big(w'(t)\big) \leq F(w(t)). \tag{3.148}$$

Proof Note that (3.148) follows from (3.146) since by definition $B(t) \equiv 1$ if $q' \geq 0$. Denote by E the energy function associated to H, that is,

$$E(t) = H\big(w'(t)\big) - F(w(t)). \tag{3.149}$$

Since by assumption $\varphi(w') \in C^1((0,T))$, we claim that $H(w') \in C^1((0,T))$ and we have

$$\big(H\big(w'\big)\big)' = w'(t)\big(\varphi\big(w'(t)\big)\big)' \quad \text{on } (0,T). \tag{3.150}$$

Indeed, note that by Stieltjes formula (see [257]) H can be written as

$$H(\rho) = \int_0^\rho s\varphi'(s)\,ds = \int_0^\rho s d\varphi(s) = \int_0^{\varphi(\rho)} \varphi^{-1}(\tau)\,d\tau$$

so that

$$H\big(w'(t)\big) = \int_0^{\varphi(w'(t))} \varphi^{-1}(\tau)\,d\tau.$$

Thus $H(w') \in C^1((0,T))$ and (3.150) follows by differentiation. Therefore from (3.150) and (3.145) one deduces

$$E'(t) = w'\Big[\big(\varphi\big(w'(t)\big)\big)' - f(w)\Big] \leq -\frac{q'(t)}{q(t)}\varphi\big(w'\big)w' \quad \text{on } (0,T)$$

since by assumption $w' \geq 0$ on $(0,T)$ and $q > 0$ on $(0,T)$. Next we let $0 < t < T$ and we integrate on $(0,t)$. Using $w(0) = w'(0) = 0$ we deduce

$$H\big(w'(t)\big) \leq F(w(t)) - \int_0^t \frac{q'(s)}{q(s)}\varphi\big(w'\big)w'\,ds. \tag{3.151}$$

We claim that

$$\varphi\big(w'(t)\big) \leq \frac{f(w(t))}{q(t)} \int_0^t q(s)\,ds. \tag{3.152}$$

Indeed, integrating the differential inequality in (3.145) on $[0, t]$ and using again $w(0) = w'(0) = 0$ we get

$$\varphi\big(w'(t)\big) \le \frac{1}{q(t)} \int_0^t q(s)f(w(s))\,ds.$$

Now, since $w(t) \in [0, \delta]$, $w' \ge 0$ and f is nondecreasing on $[0, \delta]$ by (F3), we have $f(w(s)) \le f(w(t))$ for $s \le t$ that immediately yields (3.152). Using (3.151) we thus deduce

$$\begin{aligned} H\big(w'(t)\big) &\le F(w(t)) - \int_0^t \frac{q'(s)}{q(s)}\varphi\big(w'\big)w'\,ds \\ &\le F(w(t)) + \int_0^t \left(-\frac{q'(s)}{q(s)}\varphi\big(w'\big)\right)_+ w'\,ds \\ &\le F(w(t)) + \int_0^t \left(-\frac{q'(s)}{q(s)}\int_0^s q(\tau)\,d\tau\right)_+ f(w)w'\,ds \\ &= B(t)F(w(t)), \end{aligned}$$

proving (3.146). □

We now go back to the proof of (3.143). From the second line of (3.142) we infer the existence of $t_0 \in [0, T)$ such that $w(t) \equiv 0$ on $[0, t_0]$, while $w > 0$ on $(t_0, T]$. If $t_0 = 0$ then $w'(t_0) = 0$ by our assumption, otherwise if $t_0 > 0$ then $w(t_0) = w'(t_0) = 0$ since w is C^1. Let $t_2 \in (t_0, T]$ and set

$$B = B(t_2) = 1 + \sup_{s\in[0,t_2]} \left(-\frac{q'(s)}{q(s)}\int_0^s q(\tau)\,d\tau\right)_+;$$

let δ be as in (F3). Then, there exists $t_1 \in (t_0, t_2)$ such that $n_1 = w(t_1) > 0$ satisfies

$$n_1 \le \frac{\delta}{B} \quad \text{and} \quad F(Bn_1) < H(+\infty). \tag{3.153}$$

We now apply Lemma 3.6 on the interval $[t_0, t_1]$ with the new function defined on $[t_0, t_1]$,

$$\overline{B}(t) = 1 + \sup_{s\in[t_0,t]} \left(-\frac{q'(s)}{q(s)}\int_{t_0}^s q(\tau)\,d\tau\right)_+.$$

Since clearly $\overline{B}(t) \le B(t)$ we obtain

$$H\big(w'(t)\big) \le \overline{B}(t)F(w(t)) \le B(t)F(w(t)) \le BF(w(t)).$$

Observe that $B \geq 1$, so that applying (3.118) of Lemma 3.4 with $\sigma = \frac{1}{B}$ we get

$$BF(w(t)) \leq F(Bw(t)),$$

hence

$$H\big(w'(t)\big) \leq F(Bw(t)) \quad \text{on} \quad (t_0, t_1),$$

therefore, using (3.153),

$$w'(t) \leq H^{-1}(F(Bw(t))) \quad \text{on} \quad (t_0, t_1).$$

Recalling that $f(t) > 0$ for $t \in (0, \delta)$ and thus $F(t) > 0$ for $t \in (0, \delta)$, integration yields

$$\int_0^{Bn_1} \frac{ds}{H^{-1}(F(s))} = B \int_0^{n_1} \frac{d\tau}{H^{-1}(F(B\tau))} = B \int_0^{t_1} \frac{w'(t)\, dt}{H^{-1}(F(Bw(t)))} \leq B(t_1 - t_0) < +\infty$$

as required to show the validity of (3.143). □

We are now ready for the

Proof (of Theorem 3.10) Here we extend the definition of φ to a continuous function on $\mathbb{R}$ by setting $\varphi(0) = 0$ and $\varphi(t) = -\varphi(-t)$ for $t < 0$. Of course continuity is due to (A2). We begin by constructing an auxiliary function. Towards this aim we fix an origin $o \in \Omega$ and $R > 0$ sufficiently small so that

$$E_R = \overline{B_R(o)} \setminus B_{\frac{R}{2}}(o) \subset \Omega \setminus \mathrm{cut}(o).$$

The choice of o will be done at the end of the proof. Since E_R is compact, there exists $\Gamma > 0$ such that $K_{rad} \geq -\Gamma^2$ on E_R and by $(T1)$ there exist constants $\alpha \geq 0$, $0 < \lambda < \Lambda$ such that

$$\lambda \leq T(X,X) \leq \Lambda, \quad |(\operatorname{div} T)(X)| \leq \alpha \quad \text{for every}\, x \in E_R,\ X \in T_xM,\ |X| = 1. \tag{3.154}$$

Letting $\mathrm{hess}(r)$ and $t : TM \to TM$ be defined as in Chap. 1, that is

$$\mathrm{hess}(r)(Y) = \mathrm{Hess}(r)(Y, \cdot)^\sharp, \quad t(Y) = T(Y, \cdot)^\sharp,$$

a computation shows that if $v = w(r)$ is radial, then for the operator in (3.116) we have

$$\begin{aligned} \operatorname{div}\Big(A(|\nabla v|)T(\nabla v, \cdot)^\sharp\Big) - \kappa\varphi(|\nabla v|) - f(v) = {} & T(\nabla r, \nabla r)\big[\varphi(w')\big]' \\ & + [(\operatorname{div} T)(\nabla r) + \mathrm{Tr}\,(t \circ \mathrm{hess}(r))]\varphi(w') \\ & - \kappa\varphi\big(|w'|\big) - f(w). \end{aligned} \tag{3.155}$$

Letting $g(r) = \Gamma \sinh(\Gamma r)$, using $K_{rad} \geq -\Gamma^2$ and the Hessian comparison theorem, that is Theorem 1.4, we immediately see that at x

$$\mathrm{Tr}\,(t \circ \mathrm{hess}(r)) \leq (m-1)\frac{g'(r)}{g(r)} \max_k \lambda_k \leq (m-1)\frac{g'(r)}{g(r)}\Lambda, \tag{3.156}$$

where the λ_k are the eigenvalues of T. Now we use the extended definition of φ on $\mathbb{R}$ to observe that, if $w' \leq 0$, then $\varphi(w') \leq 0$; hence using (3.154) we obtain

$$(\mathrm{div}\, T)(\nabla r)\varphi(w') \geq \alpha\varphi(w'). \tag{3.157}$$

Putting together (3.156) and (3.157) and the fact that $-\kappa\varphi(|w'|) = \kappa\varphi(w')$ we get

$$-\kappa\varphi\big(|w'|\big)+[(\mathrm{div}\, T)(\nabla r) + \mathrm{Tr}\,(t \circ \mathrm{hess}(r))]\varphi(w') \geq \left[\kappa + \alpha + (m-1)\frac{g'(r)}{g(r)}\Lambda\right]\varphi(w'). \tag{3.158}$$

Using (3.158) into (3.155) we obtain

$$\begin{aligned}\mathrm{div}\left(A(|\nabla v|)T(\nabla v,\cdot)^\sharp\right) - \langle X, \nabla v\rangle - f(v) \geq T(\nabla r, \nabla r)\big[\varphi(w')\big]' \\ + \left[\kappa + \alpha + (m-1)\frac{g'(r)}{g(r)}\Lambda\right]\varphi(w') - f(w).\end{aligned} \tag{3.159}$$

Thus, using (3.154) and (F3), the right-hand side is nonnegative if w satisfies

$$\begin{cases} \lambda[\varphi(w')]' + \left(\kappa + \alpha + (m-1)\frac{g'}{g}\Lambda\right)\varphi(w') - f(w) \geq 0 \text{ in } \left[\frac{R}{2}, R\right], \\ 0 \leq w < \delta, \;\; w' \leq 0 \;\text{ in } \left[\frac{R}{2}, R\right]. \end{cases} \tag{3.160}$$

We set

$$\ell(r) = \exp\left(\frac{1}{\lambda(m-1)}\int_{\frac{R}{3}}^{r} [\kappa + \alpha + (m-1)\frac{g'}{g}(t)\Lambda]dt\right);$$

then (3.160) can be written as

$$\begin{cases} \lambda\big[\ell^{m-1}(r)\varphi(w')\big]' - \ell^{m-1}(r)f(w) \geq 0 \text{ in } \left[\frac{R}{2}, R\right], \\ 0 \leq w < \delta, \;\; w' \leq 0 \;\text{ in } \left[\frac{R}{2}, R\right]. \end{cases} \tag{3.161}$$

Thus, if w solves (3.161), then $v(r) = w(r(x))$ solves

$$\begin{cases} \mathrm{div}\left(A(|\nabla v|)T(\nabla v,\cdot)^\sharp\right) - \kappa\varphi(|\nabla v|) - f(v) \geq 0 \;\text{ in } E_R \\ 0 \leq v < \delta, \;\; \langle \nabla v, \nabla r\rangle \leq 0 \;\text{ in } E_R. \end{cases} \tag{3.162}$$

In fact, for technical reasons that shall be apparent below, we also need

$$-1 \leq \langle \nabla v, \nabla r \rangle < 0 \ \text{ in } E_R, \tag{3.163}$$

which is a strengthening of the second in (3.162) and that corresponds to

$$-1 \leq w' < 0 \ \text{ in } \left[\frac{R}{2}, R\right]. \tag{3.164}$$

To solve (3.161), (3.163) we set

$$q(t) = \ell^{m-1}((R-t)), \quad t \in [0, T], \ T = \frac{R}{2}. \tag{3.165}$$

We observe that, without loss of generality, we can suppose $\lambda < 1$ so that $\frac{1}{\lambda} > 1$ and from (3.114) and Lemma 3.4 we deduce

$$\frac{1}{H^{-1}\left(\frac{1}{\lambda}F(t)\right)} \notin L^1(0^+).$$

We can therefore apply Proposition 3.3 to guarantee the existence of a solution $z = z(t)$ of the problem

$$\begin{cases} \lambda[q(t)\varphi(z')]' - q(t)f(z) = 0 \text{ in } [0, T], \\ z(0) = 0, \ \ 0 < z(T) = a < \delta, \\ z > 0 \text{ in } (0, T], \ z'(t) > 0 \text{ in } [0, T], \end{cases}$$

where a and $T = \frac{R}{2}$ are chosen so small that

$$\rho = \frac{\max_{[0,T]} q}{\min_{[0,T]} q}\left[\frac{T}{\lambda}\max_{[0,a]} f(s) + \varphi\left(\frac{a}{T}\right)\right] < \varphi(+\infty).$$

Set $r = R - t$ and define $w(r) = z(t)$. Then w satisfies (3.161) with $w(R) = 0$, $w\left(\frac{R}{2}\right) = a$, $w(r) > 0$ in $(\frac{R}{2}, R]$, $w'(r) < 0$ in $[\frac{R}{2}, R]$. In particular $0 \leq w \leq a < \delta$ since $a \in (0, \delta)$. Furthermore, by (3.128) of Proposition 3.2,

$$\|w'\|_\infty \leq \varphi^{-1}(\rho).$$

Thus, up to choosing $a \in (0, \delta)$ and R sufficiently small we can suppose that

$$\|w'\|_\infty \leq 1$$

so that w satisfies (3.164). It follows that $v(x) = w(r(x))$ is a C^1-solution of (3.162), (3.163) and moreover

$$v(x) = a \quad \text{on } \partial B_{\frac{R}{2}}(o), \; v(x) = 0 \quad \text{on } \partial B_R(o). \tag{3.166}$$

To finish the proof of the theorem we now reason by contradiction and we suppose the existence of a C^1-solution in Ω of (3.116) and of $x_0 \in \Omega$ such that $u(x_0) = 0$ but $u \not\equiv 0$ in Ω. Let

$$\Omega^+ = \{x \in \Omega : u(x) > 0\};$$

then $x_0 \in \partial\Omega^+ \cap \Omega \neq \emptyset$. We let $x_1 \in \Omega^+$ be such that

$$\operatorname{dist}\left(x_1, \partial\Omega^+\right) < \operatorname{dist}(x_1, \partial\Omega).$$

Let $B(x_1)$ be the largest geodesic ball centered at x_1 and contained in Ω^+; then $u > 0$ in $B(x_1)$, while $u(\bar{x}) = 0$ for some $\bar{x} \in \partial B(x_1) \cap \Omega$. Let ν be the exterior unit normal to $\partial B(x_1)$ at $\bar{x}$; since $\bar{x}$ is an absolute minimum for u in Ω we have

$$\langle \nabla u, \nu \rangle = 0. \tag{3.167}$$

We shall contradict (3.167). Towards this aim we fix $y \in B(x_1)$ and $R < \operatorname{inj}_M(y)$ sufficiently small that $B_R(y) \subset B(x_1)$, $u < \delta$ on $B_R(y)$ and $\bar{x} \in \partial B_R(y)$. Note that, since $R < \operatorname{inj}_M(y)$, the distance function from y is smooth outside y in $B_R(y)$. We construct v to solve (3.162), (3.163) in $E_R \subset \overline{B_R(y)}$ by choosing $a \in (0, \delta)$ in the above construction so small that also

$$v(x) \leq u(x) \quad \text{on } \partial B_{\frac{R}{2}}(y).$$

Note that $v \equiv 0$ on $\partial B_R(y)$ while $u \geq 0$ on $\partial B_R(y)$. Next we note that $|\nabla u| + |\nabla v| > 0$ and $|\nabla v| \leq 1$ in E_R. Since u satisfies (3.116) in E_R, from Theorem 3.8 we have

$$u \geq v \quad \text{on } E_R. \tag{3.168}$$

Now $\nu(\bar{x}) = \nabla r(\bar{x})$, with $r() = \operatorname{dist}(\,, y)$, so that $\langle \nabla(u - v), \nu \rangle(\bar{x}) \leq 0$ and by (3.163)

$$\langle \nabla u, \nu \rangle(\bar{x}) \leq \langle \nabla v, \nu \rangle(\bar{x}) < 0, \tag{3.169}$$

contradicting (3.167). □

Remark 3.14 To obtain (3.168) we used Theorem 3.8 in its generality, since on E_R we had to compare u and v, $u \geq v$ on ∂E_R, solutions respectively of

$$\operatorname{div}\left(A(|\nabla u|)T(\nabla u, \cdot)^\sharp\right) \leq B(x, u, \nabla u) \leq \kappa\varphi(|\nabla u|) + f(u)$$

and

$$\operatorname{div}\left(A(|\nabla v|)T(\nabla v,\cdot)^\sharp\right) \geq \kappa\varphi(|\nabla v|)+f(v).$$

However, if $\kappa = 0$ and (A2)' and (F3) hold, the above inequalities can be used as follows. By contradiction suppose (3.168) false so that

$$D = \{x \in E_R : u(x) < v(x)\} \neq \emptyset.$$

Then $\overline{D} \subseteq \overline{E}_R$ and $u = v$ on ∂D. Furthermore, because of (F3),

$$\operatorname{div}\left(A(|\nabla u|)T(\nabla u,\cdot)^\sharp\right) \leq f(u) \leq f(v) \leq \operatorname{div} A\Big(|\nabla v|T(\nabla v,\cdot)^\sharp\Big)$$

on D. By Theorem 3.9 we have $u \geq v$ on D, contradiction. Thus the validity of (3.168).

Note that, in this case, by Remark 3.10, we can assume $u \in \mathrm{Lip}_{loc}(M)$, definitely relaxing the regularity assumption on u in Theorem 3.10.

3.3.2 The Maximum Principle

In what follows we recall that $q(x) \in C^0(M)$, $q(x) \geq 0$ and we suppose that $q(x) > 0$ outside a compact set K. The further assumption

$$\frac{1}{q} \in L^1_{loc}(M) \qquad \text{(Q)}$$

is in force throughout this section. The differential operator L is the one given in (3.86).

Next, we introduce the following Khas'minskiĭ type condition.

Definition 3.1 We say that the *(q-SK) condition* holds if there exists a telescoping exhaustion of relatively compact open sets $\{\Sigma_j\}_{j\in\mathbb{N}}$ such that $K \subset \Sigma_1$, $\overline{\Sigma}_j \subset \Sigma_{j+1}$ for every j and, for any pair $\Omega_1 = \Sigma_{j_1}$, $\Omega_2 = \Sigma_{j_2}$, with $j_1 < j_2$, and for each $\varepsilon > 0$, there exists $\gamma \in C^0(M \setminus \Omega_1) \cap C^1(M \setminus \overline{\Omega}_1)$ with the following properties:

(i) $\gamma \equiv 0$ on $\partial\Omega_1$,
(ii) $\gamma > 0$ on $M \setminus \Omega_1$,
(iii) $\gamma \leq \varepsilon$ on $\Omega_2 \setminus \Omega_1$,
(iv) $\gamma(x)\to+\infty$ when $x\to\infty$,
(v) $q(x)L\gamma \leq \varepsilon$ on $M \setminus \overline{\Omega}_1$.

Since property (v) has to be interpreted in the weak sense we mean that

$$L\gamma \le \frac{\varepsilon}{q(x)} \text{ weakly on } M \setminus \overline{\Omega}_1,$$

that is, for all $\psi \in C_c^\infty(M \setminus \overline{\Omega}_1)$, $\psi \ge 0$,

$$\int_{M\setminus\overline{\Omega}_1} \left(A(|\nabla\gamma|)T(\nabla\gamma, \nabla\psi) + \langle X, \nabla\gamma\rangle\psi + \frac{\varepsilon}{q}\psi \right) \ge 0.$$

Of course we expect the (q-SK) condition in Definition 3.1 to be equivalent in the linear case to the weak form of (Γ) of Theorem 3.1, which obviously reads as follows:

Definition 3.2 We say that the *(q-KL) condition* holds if there exist a compact set $H \supset K$ and a function $\tilde{\gamma} \in C^1(M)$ with the following properties:

(j) $\tilde{\gamma}(x) \to +\infty$ when $x \to \infty$,
(jj) $q(x)L\tilde{\gamma} \le B$ on $M \setminus H$ for some constant B, in the weak sense.

Obviously, the (q-SK) condition implies the (q-KL) condition simply by choosing $H = \overline{\Omega}_2$, setting $\tilde{\gamma} = \gamma$ on $M \setminus \Omega_2$ and extending it on Ω_2 to be of class C^1 on M.

We shall prove the equivalence of the two conditions in the linear case after the proof of Theorem 3.11. Note that the (q-SK) Khas'minskiĭ type condition is not only sufficient for the validity of the q-weak maximum principle but indeed equivalent in some cases (see [182]).

Before stating Theorem 3.11 we recall that for an operator L, a function $q(x) > 0$ on an open set $\Omega \subset M$ and $u \in C^1(\Omega)$ the inequality

$$\inf_{\Omega}\{q(x)Lu(x)\} \le 0 \tag{3.170}$$

holds in the weak sense if for each $\varepsilon > 0$

$$-\int_\Omega (A(|\nabla u|)T(\nabla u, \nabla\psi) + \langle X, \nabla u\rangle\psi) \le \int_\Omega \frac{\varepsilon}{q(x)}\psi$$

for some $\psi \in C_c^\infty(\Omega)$, $\psi \ge 0$, $\psi \not\equiv 0$.

We are now ready to state the nonlinear version of Theorem 3.1.

Theorem 3.11 *Let $(M, \langle\, ,\, \rangle)$ be a Riemannian manifold, let L be as above (that is, as in* (3.86)*), with (A1), (A2)', (T1), (T2) holding. Let $q(x) \in C^0(M)$, $q(x) \ge 0$, suppose that $q(x) > 0$ outside some compact set $K \subset M$ and that q satisfies (Q). Assume the validity of (q-SK). If $u \in C^1(M)$ and $u^* = \sup_M u < +\infty$ then for each $\eta > 0$*

$$\inf_{A_\eta}\{q(x)Lu(x)\} \le 0 \tag{3.171}$$

holds in the weak sense, where

$$A_\eta = \{x \in M : u(x) > u^* - \eta\}. \tag{3.172}$$

Remark 3.15 Here $u \in \mathrm{Lip}_{loc}(M)$ suffices, as noted in the course of the proof.

Similarly to what we did in the linear case, if the conclusion of the theorem holds we shall say that the *q-weak maximum principle for the operator L holds on* $(M, \langle\, , \rangle)$.

Proof We argue by contradiction and we suppose that for some $\eta > 0$ there exists $\varepsilon_0 > 0$ such that

$$Lu \geq \frac{\varepsilon_0}{q(x)}$$

holds weakly on A_η, that is, for each $\psi \in C_c^\infty(A_\eta)$, $\psi \geq 0$,

$$\int_{A_\eta} \left(A(|\nabla u|)T(\nabla u, \nabla\psi) + \langle X, \nabla u\rangle\psi + \frac{\varepsilon_0}{q}\psi \right) \leq 0. \tag{3.173}$$

Note that since in general $A_\eta \not\subset M \setminus K$ assumption (Q) is here essential.

First we observe that u^* cannot be attained at any point $x_0 \in M$. Otherwise $x_0 \in A_\eta$ and, because of (3.173), on the open set A_η it holds weakly

$$Lu \geq \frac{\varepsilon_0}{q(x)}. \tag{3.174}$$

Now, at x_0, $\nabla u(x_0) = 0$. Choose a small geodesic ball $B_R(x_0) \subset A_\eta$ such that $-\frac{\varepsilon_0}{2q_0} \leq \langle X, \nabla u\rangle \leq \frac{\varepsilon_0}{2q_0}$ on $B_R(x_0)$, where $q_0 = \sup_{B_R(x_0)} q > 0$. From (3.174), on $B_R(x_0)$ we have

$$\operatorname{div}\left(A(|\nabla u|)T(\nabla u, \cdot)^\sharp\right) \geq \langle X, \nabla u\rangle + \frac{\varepsilon_0}{q_0} \geq \frac{\varepsilon_0}{2q_0} \geq 0.$$

By Remark 3.14 and Theorem 3.10 with $f \equiv 0$, $\kappa \equiv 0$ and (so that u may be only $\mathrm{Lip}_{loc}(M)$) we deduce that $u \equiv u^*$ on $B_R(x_0)$, contradicting (3.174).

Next we let Σ_j be the telescoping sequence of relatively compact open domains of condition (q-SK) in Definition 3.1. Given $u^* - \frac{\eta}{2}$, there exists Σ_{j_1} such that

$$u^*_{j_1} = \sup_{\overline{\Sigma}_{j_1}} u > u^* - \frac{\eta}{2}.$$

We set $\Omega_1 = \Sigma_{j_1}$ and define

$$u^*_1 = u^*_{j_1}.$$

Note that, since u^* is not attained on M

$$u^* - \frac{\eta}{2} < u_1^* < u^*. \tag{3.175}$$

We can therefore fix α so that

$$u_1^* < \alpha < u^*. \tag{3.176}$$

Since $\alpha > u_1^*$, there exists Σ_{j_2} with $j_2 > j_1$ such that, setting $\Omega_2 = \Sigma_{j_2}$, $u_2^* = \sup_{\Omega_2} u = \max_{\bar{\Omega}_2} u$, we have

$$\overline{\Omega}_1 \subset \Omega_2$$

and furthermore

$$u_1^* < \alpha < u_2^* < u^*. \tag{3.177}$$

We fix $\bar{\eta} > 0$ so small that

$$\alpha + \bar{\eta} < u_2^* \tag{3.178}$$

and

$$\bar{\eta} < \varepsilon_0. \tag{3.179}$$

We apply the (q-SK) condition with the choice $\varepsilon = \bar{\eta}$ and Ω_1 and Ω_2 as above to obtain the existence of $\gamma \in C^0(M \setminus \Omega_1) \cap C^1(M \setminus \overline{\Omega}_1)$ satisfying the properties listed in Definition 3.1. We introduce the function

$$\sigma(x) = \alpha + \gamma(x). \tag{3.180}$$

Then

$$\sigma(x) = \alpha \text{ on } \partial\Omega_1, \tag{3.181}$$

$$\alpha < \sigma(x) \leq \alpha + \bar{\eta} \text{ on } \Omega_2 \setminus \bar{\Omega}_1, \tag{3.182}$$

$$\sigma(x) \to +\infty \text{ as } x \to \infty, \tag{3.183}$$

and, since $\nabla\sigma = \nabla\gamma$, $L\sigma = L\gamma$ and by (v) of Definition 3.1

$$q(x)L\sigma \leq \bar{\eta} \text{ in the weak sense on } M \setminus \bar{\Omega}_1. \tag{3.184}$$

Next, we consider the function $u - \sigma$. Because of (3.181) and (3.176), we have for every $x \in \partial\Omega_1$

$$(u - \sigma)(x) = u(x) - \alpha \leq u_1^* - \alpha < 0. \tag{3.185}$$

Since $u_2^* = \max_{\bar{\Omega}_2} u$ and $\bar{\Omega}_2$ is compact, u_2^* is attained at some $\bar{x} \in \bar{\Omega}_2$. Note that $\bar{x} \notin \bar{\Omega}_1$ because otherwise

$$u_1^* \geq u(\bar{x}) = u_2^*,$$

contradicting (3.177). Thus $\bar{x} \in \bar{\Omega}_2 \setminus \bar{\Omega}_1$. By (3.178) we have

$$u(\bar{x}) > \alpha + \bar{\eta}.$$

Thus, by (3.182) and (3.178), we deduce

$$(u - \sigma)(\bar{x}) = u_2^* - \sigma(\bar{x}) \geq u_2^* - \alpha - \bar{\eta} > 0. \tag{3.186}$$

Finally, (3.183) implies the existence of Σ_ℓ, $\ell > j_2$, such that

$$(u - \sigma)(x) < 0 \text{ on } M \setminus \Sigma_\ell. \tag{3.187}$$

Because of (3.185)–(3.187) the function $u - \sigma$ attains an absolute maximum $\mu > 0$ at a certain point $z_0 \in \Sigma_\ell \setminus \bar{\Omega}_1 \subset M \setminus \bar{\Omega}_1$. At z_0, by (3.176) and (3.175), we have

$$u(z_0) = \sigma(z_0) + \mu > \sigma(z_0) = \alpha + \gamma(z_0) \geq \alpha > u_1^* > u^* - \frac{\eta}{2},$$

and hence $z_0 \in A_\eta$. It follows that

$$\Xi = \{x \in M \setminus \bar{\Omega}_1 : (u - \sigma)(x) = \mu\} \subset A_\eta. \tag{3.188}$$

Since A_η is open there exists a neighborhood U_Ξ of Ξ contained in A_η. Pick any $y \in \Xi$, fix $\beta \in (0, \mu)$ and call $\Xi_{\beta,y}$ the connected component of the set

$$\{x \in M \setminus \bar{\Omega}_1 : (u - \sigma)(x) > \beta\}$$

containing y. Since $\beta > 0$,

$$\Xi_{\beta,y} \subset \bar{\Sigma}_\ell \setminus \bar{\Omega}_1 \subset M \setminus \bar{\Omega}_1,$$

and we can also choose β sufficiently near to μ so that $\bar{\Xi}_{\beta,y} \subset A_\eta$. Furthermore, $\bar{\Xi}_{\beta,y}$ is compact. Because of (3.184), (3.179) and (3.173), on $\Xi_{\beta,y}$ we have

$$q(x)Lu(x) \geq \varepsilon_0 > q(x)L\gamma(x)$$

in the weak sense. Furthermore,

$$u(x) = \sigma(x) + \beta \quad \text{on } \partial\Xi_{\beta,y}.$$

Hence by Theorem 3.9 (so that here it suffices $u \in \mathrm{Lip}_{loc}(M)$)

$$u(x) \le \sigma(x) + \beta \quad \text{on } \Xi_{\beta,y}.$$

This contradicts the fact that $y \in \Xi_{\beta,y}$, indeed,

$$u(y) = \sigma(y) + \mu > \sigma(y) + \beta$$

since $\mu > \beta$. This completes the proof of Theorem 3.11. □

We now prove that for a linear operator the (q-KL) condition implies the (q-SK) condition.

Towards this aim observe that since L is linear, $A(t) = 1$ or equivalently $\varphi(t) = t$. Thus, once (T1) is satisfied, assumptions (A1), (A2) and (T2) are also satisfied. Now assume (q-KL) and fix a strictly increasing divergent sequence $\{T_j\} \nearrow +\infty$. With the notation in Definition 3.2, let

$$\Sigma_j = \{x \in M : \tilde{\gamma}(x) < T_j\}.$$

Obviously, each Σ_j is open and because of (j) in (q-KL) one immediately verifies that $\bar{\Sigma}_j = \{x \in M : \tilde{\gamma}(x) \le T_j\}$ is compact. For the same reason we can suppose to have chosen T_1 sufficiently large that $K \subset H \subset \Sigma_1$. Furthermore $\bar{\Sigma}_j \subset \Sigma_{j+1}$ and again by (j) in (q-KL), $\{\Sigma_j\}$ is a telescoping exhaustion. Consider any pair

$$\Omega_1 = \Sigma_{j_1} = \{x \in M : \tilde{\gamma}(x) < T_{j_1}\}$$

and

$$\Omega_2 = \Sigma_{j_2} = \{x \in M : \tilde{\gamma}(x) < T_{j_2}\}$$

with $j_2 > j_1$, and choose $\varepsilon > 0$. Let $\sigma \in (0, \sigma_0)$ and define $\gamma : M \setminus \Omega_1 \to \mathbb{R}_0^+$ by setting

$$\gamma(x) = \sigma(\tilde{\gamma}(x) - T_{j_1}).$$

Then

(i) $\gamma(x) = 0$ for every $x \in \partial\Omega_1$,
(ii) $\gamma(x) > 0$ if $x \in M \setminus \overline{\Omega_1} = \{x \in M : \tilde{\gamma}(x) > T_{j_1}\}$,
(iii) on $\Omega_2 \setminus \Omega_1 = \{x \in M : T_{j_1} \le \tilde{\gamma}(x) < T_{j_2}\}$ we have $\gamma(x) < \sigma(T_{j_2} - T_{j_1})$ and hence, up to have chosen σ_0 sufficiently small, $\gamma(x) \le \varepsilon$ on $\Omega_2 \setminus \Omega_1$,
(iv) $\gamma(x) \to +\infty$ when $x \to \infty$, because of (j),

(v) on $M \setminus \overline{\Omega}_1$, by the linearity of L and (jj),

$$q(x)L\gamma = q(x)L(\sigma(\tilde{\gamma} - T_{j_1})) = q(x)\sigma L\tilde{\gamma} \leq \sigma B \leq \varepsilon$$

and up to have chosen σ_0 sufficiently small.

It is worth giving some examples where the (q-SK) condition is satisfied. For the sake of simplicity we limit ourselves to the case $T = \langle\, , \rangle$ and $X \equiv 0$. Let $(M, \langle\, , \rangle)$ be a complete, noncompact Riemannian manifold of dimension $m \geq 2$. Let $o \in M$ be a fixed reference point, denote by $r(x)$ the Riemannian distance from o and suppose that

$$\mathrm{Ric}(\nabla r, \nabla r) \geq -(m-1)G(r)^2 \tag{3.189}$$

for some positive nondecreasing function $G(r) \in C^0(\mathbb{R}_0^+)$, $G(r) > 0$, with $1/G \notin L^1(+\infty)$. Similarly to what has been done in Sect. 2.2.1 and for the same ψ defined there in (2.33), by the Laplacian comparison theorem we have

$$\Delta r \leq (m-1)\frac{\psi'}{\psi}(r) \tag{3.190}$$

weakly on M for $r \geq R_0 > 0$ sufficiently large. Let $A(t)$ and the corresponding $\varphi(t) = tA(t)$ satisfy (A1), (A2) and assume that

$$\varphi(t) \leq Ct^\delta \quad \text{on } \mathbb{R}_0^+ \quad \text{for some } C, \delta > 0. \tag{A3}$$

Note that (A3) implies (A2)' (iii).

Suppose now that the function $q(x) \in C^0(M)$, $q(x) \geq 0$, satisfies

$$q(x) \leq \Theta(r(x)) \tag{3.191}$$

outside a compact set $K \subset M$, for some nonincreasing continuous function $\Theta : \mathbb{R}_0^+ \to \mathbb{R}^+$ with the property that

$$\Theta(t) \leq BG^{\delta-1}(t) \tag{3.192}$$

for $t \gg 1$, some constant $B > 0$ and δ as in (A3). Note that if $\delta \geq 1$, (3.192) is automatically satisfied.

Fix $\sigma > 0$ and $R \geq R_0$ such that $K \subset B_R$, the geodesic ball of radius R centered at o. On $[R, +\infty)$ define the function

$$\chi_\sigma(r) = \int_R^r \varphi^{-1}(\sigma h(t))\, dt, \tag{3.193}$$

where

$$h(t) = \psi^{1-m}(t) \int_R^t \frac{\psi^{m-1}(s)}{\Theta(s)} ds.$$

Note that, since $\varphi : \mathbb{R}_0^+ \to [0, \varphi(+\infty)) = I \subseteq \mathbb{R}_0^+$ increasingly, the inverse function $\varphi^{-1} : I \to \mathbb{R}_0^+$ does indeed exist. Furthermore, in order that χ_σ be well defined when $\varphi(+\infty) < +\infty$, we need that for every $t \in [R, +\infty)$

$$\sigma h(t) \in I. \tag{3.194}$$

Towards this end we note that

$$\frac{\psi'}{\psi}(t) = G(t) \frac{e^{\int_0^t G(s)ds}}{e^{\int_0^t G(s)ds} - 1} \sim CG(t) \quad \text{as} \quad t \to +\infty \tag{3.195}$$

for some constant $C > 0$. Then

$$h(t) \le \frac{1}{\Theta(t)} \psi^{1-m}(t) \int_R^t \psi^{m-1}(s)ds \le \frac{C}{\Theta(t)G(t)} \tag{3.196}$$

for $t \gg 1$ and some $C > 0$. The assumption

$$\limsup_{r \to +\infty} \frac{1}{\Theta(r)G(r)} < +\infty$$

is therefore enough to guarantee that $h(t)$ is bounded above. By choosing σ sufficiently small, say $0 < \sigma \le \sigma_0$, we obtain the validity of (3.194) so that (3.193) is well defined on $[R, +\infty)$.

We now set $\gamma(x) = \chi_\sigma(r(x))$ for $x \in M \setminus B_R$ and note that

(i) $\gamma \equiv 0$ on ∂B_R,
(ii) $\gamma > 0$ on $M \setminus \overline{B_R}$,

Moreover, having fixed $\varepsilon > 0$ and a second geodesic ball $B_{\tilde{R}}$ with $\tilde{R} > R$, since $\varphi^{-1}(t) \to 0$ as $t \to 0^+$, up to choosing $\sigma > 0$ sufficiently small we also have $\chi_\sigma(r) \le \varepsilon$ if $R \le r < \tilde{R}$, so that

(iii) $\gamma \le \varepsilon$ on $B_{\tilde{R}} \setminus B_R$,

On the other hand, since $1/G \notin L^1(+\infty)$, to prove that

(iv) $\gamma(x) \to +\infty$ when $x \to \infty$

it suffices to show that

$$\varphi^{-1}(\sigma h(t)) \ge \frac{\tilde{C}}{G(t)} \qquad \text{for } t \gg 1$$

for some constant $\tilde{C} > 0$. Equivalently, that there exists a constant $\tilde{C} > 0$ such that

$$\frac{h(t)}{\varphi\left(\frac{\tilde{C}}{G(t)}\right)} \geq \frac{1}{\sigma} \quad \text{for } t \gg 1. \tag{3.197}$$

Without loss of generality we can suppose $G(t) \to +\infty$ as $t \to +\infty$. By the structural condition (A3) on φ we have

$$\varphi\left(\frac{\tilde{C}}{G(t)}\right) \leq C\frac{\tilde{C}^{\delta}}{G(t)^{\delta}},$$

so that

$$\frac{h(t)}{\varphi\left(\frac{\tilde{C}}{G(t)}\right)} \geq \frac{A(t)}{B(t)}$$

with

$$A(t) = G(t)^{\delta}\int_R^t \frac{\psi^{m-1}(s)}{\Theta(s)}ds$$

and

$$B(t) = C\tilde{C}^{\delta}\psi^{m-1}(t).$$

Note that both $A(t)$ and $B(t)$ diverge to $+\infty$ as $t \to +\infty$. Hence,

$$\liminf_{t\to+\infty}\frac{A(t)}{B(t)} \geq \liminf_{t\to+\infty}\frac{A'(t)}{B'(t)}.$$

A computation that uses $G' \geq 0$, $\Theta > 0$ and (3.192) shows that

$$\frac{A'(t)}{B'(t)} \geq \frac{G(t)}{BC\tilde{C}^{\delta}(m-1)\frac{\psi'(t)}{\psi(t)}}, \quad t \gg 1,$$

and since $\psi'(t)/\psi(t) \sim G(t)$ as $t \to +\infty$, we can choose $\tilde{C} > 0$ sufficiently small that

$$\liminf_{t\to+\infty}\frac{A'(t)}{B'(t)} \geq \frac{1}{\sigma},$$

proving the validity of (3.197)

Clearly, by definition, $\chi_\sigma(t)$ is nondecreasing and satisfies $\chi'_\sigma(t) = \varphi^{-1}(\sigma h(t))$, that is, $\varphi(\chi'_\sigma(t)) = \sigma h(t)$. Therefore

$$\nabla\gamma = \chi'_\sigma(r)\nabla r, \quad |\nabla\gamma| = \chi'_\sigma(r) \quad \text{and} \quad \varphi(|\nabla\gamma|) = \sigma h(r).$$

Since

$$h'(t) = \frac{1}{\Theta(t)} - (m-1)\frac{\psi'}{\psi}(t)h(t),$$

a computation using (3.190) and (3.191) gives

$$\begin{aligned} L\gamma &= \operatorname{div}\left(|\nabla\gamma|^{-1}\varphi(|\nabla\gamma|)\nabla\gamma\right) = \operatorname{div}(\sigma h(r)\nabla r) = \sigma h'(r)|\nabla r|^2 + \sigma h(r)\Delta r \\ &= \frac{\sigma}{\Theta(r)} + \sigma h(r)\left(\Delta r - (m-1)\frac{\psi'}{\psi}(r)\right) \le \frac{\sigma}{\Theta(r)} \le \frac{\sigma}{q(x)} \end{aligned} \tag{3.198}$$

if $r \ge R$. That is,

(v) $q(x)L\gamma \le \sigma$ on $M \setminus \overline{B_R}$

outside the cut locus and weakly on all of $M \setminus \overline{B_R}$ as it can be easily proved.

It is now clear how to satisfy the requirements of the (q-SK) condition in Definition 3.1 by choosing a telescoping exhaustion $\{B_{R+j}\}_{j\in\mathbb{N}}$.

Summarizing we have proved the following

Theorem 3.12 *Let $(M, \langle\, ,\, \rangle)$ be a complete Riemannian manifold satisfying* (3.189) *for some $G \in C^0(\mathbb{R}_0^+)$, $G > 0$ on $\mathbb{R}_0^+$, $\frac{1}{G} \notin L^1(+\infty)$. Let $A(t)$ and $\varphi(t) = tA(t)$ satisfy (A1), (A2), (A3), $q \in C^0(M)$, $q \ge 0$, $q(x) > 0$ outside a compact set K. Furthermore assume that, for some nonincreasing function $\Theta : \mathbb{R}_0^+ \to \mathbb{R}_0^+$, $q(x) \le \Theta(r(x))$ on $M \setminus K$ and, if $\delta < 1$, $\Theta(t) \le BG(t)^{\delta-1}$ for some $B > 0$, with δ the coefficient appearing in (A3). Finally suppose* $\limsup_{r\to+\infty} \frac{1}{\Theta(r)G(r)} < +\infty$. *Let L be the operator acting on $u \in C^1(M)$ by*

$$Lu = \operatorname{div} A(|\nabla u|)\nabla u.$$

If $u^* < +\infty$ then for each $\eta > 0$

$$\inf_{A_\eta} q(x)Lu \le 0$$

holds in the weak sense, where A_η is as in (3.172).

Here we introduce another example where the (q-SK) condition is satisfied with $T = \langle\, ,\, \rangle$ and arbitrary X. Let $(M, \langle\, ,\, \rangle)$, $o \in M$ be as above and follow the same notation. Suppose, as in the previous example, that

$$\operatorname{Ric}(\nabla r, \nabla r) \ge -(m-1)G(r)^2 \tag{3.199}$$

with G satisfying the requirements in (3.189). We know that, for the same function ψ of (2.33) in Sect. 2.2.1,

$$\Delta r \leq (m-1)\frac{\psi'}{\psi}(r) \leq CG(r) \tag{3.200}$$

weakly on M for $r \geq R_0 > 0$ sufficiently large and some $C > 0$.

Suppose now that the function $q(x) \in C^0(M)$, $q(x) \geq 0$, satisfies

$$q(x) \leq \frac{1}{G(r(x)) + |X(x)|} \tag{3.201}$$

outside a compact set $K \subset M$. We fix $\sigma > 0$ and $R \geq R_0$ such that $K \subset B_R$, then we define the function

$$\gamma(x) = \sigma(r(x) - R) \quad \text{for } x \in M \setminus B_R. \tag{3.202}$$

Obviously,

(i) $\gamma \equiv 0$ on ∂B_R,
(ii) $\gamma > 0$ on $M \setminus \overline{B_R}$,

Then, having fixed $\varepsilon > 0$ and a second geodesic ball $B_{\tilde{R}}$ with $\tilde{R} > R$, up to choosing $\sigma > 0$ sufficiently small we also have

(iii) $\gamma \leq \varepsilon$ on $B_{\hat{R}} \setminus B_R$.

Moreover, since M is complete

(iv) $\gamma(x) \to +\infty$ when $x \to \infty$.

Finally, a direct computation using (3.200) and (3.201) gives

$$\begin{aligned} L\gamma &= \operatorname{div}\left(|\nabla\gamma|^{-1}\varphi(|\nabla\gamma|)\nabla\gamma\right) - \langle X, \nabla\gamma\rangle = \operatorname{div}(\varphi(\sigma)\nabla r) - \sigma\langle X, \nabla r\rangle \\ &= \varphi(\sigma)\Delta r - \sigma\langle X, \nabla r\rangle \leq \varphi(\sigma)CG(r) + \sigma|X| \\ &\leq \varepsilon(G(r) + |X|) \leq \frac{\varepsilon}{q(x)} \end{aligned}$$

if $r \geq R$, up to choosing $\sigma > 0$ sufficiently small, since $\varphi(\sigma) \to 0$ as $\sigma \to 0^+$; in other words,

(v) $q(x)L\gamma \leq \varepsilon$ on $M \setminus \overline{B_R}$

outside the cut locus cut(o) and weakly on all of $M \setminus \overline{B_R}$ as it can be easily proved. It is now clear how to satisfy the requirements of the (q-SK) condition in Definition 3.1 by choosing a telescoping exhaustion $\{B_{R+j}\}_{j\in\mathbb{N}}$.

For the next result we define the *(q-SK∇) condition* as the (q-SK) condition with the added requirement

(vi) $|\nabla\gamma| < \varepsilon$ on $M \setminus \Omega_1$.

Theorem 3.13 *Let* $(M, \langle\, ,\, \rangle)$ *be a Riemannian manifold and let* L *be as in* (3.86), *with (A1), (A2)', (T1), (T2) holding. Let* $q(x) \in C^0(M)$, $q(x) \geq 0$; *suppose* $q(x) > 0$ *outside some compact set* $K \subset M$ *and that* q *satisfy (Q). Assume the validity of (q-SK∇). If* $u \in C^1(M)$ *and* $u^* = \sup_M u < +\infty$ *then for each* $\eta > 0$

$$\inf_{B_\eta}\{q(x)Lu(x)\} \leq 0 \tag{3.203}$$

holds in the weak sense, where

$$B_\eta = \{x \in M : u(x) > u^* - \eta \quad and \quad |\nabla u(x)| < \eta\}.$$

Proof First of all note that the validity of (q-SK∇) implies, once we fix arbitrarily a pair $\Omega_1 \subset \Omega_2$, an $\varepsilon > 0$ and a corresponding γ, that the metric is geodesically complete. Indeed, let $\varsigma : [0, \ell) \to M$ be any divergent path parametrized by arc-length. Thus ς lies eventually outside any compact subset of M. From (vi), $|\nabla\gamma| \leq \varepsilon$ outside the compact subset $\bar{\Omega}_1$. We set $h(t) = \gamma(\varsigma(t))$ on $[t_0, \ell)$, where t_0 has been chosen so that $\varsigma(t) \notin \bar{\Omega}_1$ for all $t_0 \leq t < \ell$. Then, for every $t \in [t_0, \ell)$ we have

$$|h(t) - h(t_0)| = \left|\int_{t_0}^t h'(s)ds\right| \leq \int_{t_0}^t |\nabla\gamma(\varsigma(s))|ds \leq \varepsilon(t - t_0).$$

Since ς is divergent, then $\varsigma(t) \to \infty$ as $t \to \ell^-$, so that $h(t) \to +\infty$ as $t \to \ell^-$ because of (iv). Therefore, letting $t \to \ell^-$ in the inequality above, we conclude that $\ell = +\infty$. This shows that divergent paths in M have infinite length and in other words, that the metric is complete.

Since the metric is complete, we can apply Ekeland quasi-minimum principle to deduce that $B_\eta \neq \emptyset$ and therefore that the infimum in (3.203) is meaningful.

Now we proceed as in the proof of Theorem 3.11 substituting, as in the linear case, the subset A_η with the smaller open set B_η. We need to show that the compact set Ξ defined in (3.188) satisfies $\Xi \subset B_\eta$. Because of (3.178) it is enough to prove that for every $z \in \Xi$,

$$|\nabla u(z)| < \eta. \tag{3.204}$$

But z is a point of absolute maximum for $(u - \sigma)$ and $z \in M \setminus \bar{\Omega}_1$, hence using (vi) of (q-SK∇),

$$|\nabla u(z)| = |\nabla\sigma(z)| = |\nabla\gamma(z)| < \varepsilon.$$

thus $\Xi \subset B_\eta$ and the rest of the proof is now exactly as at the end of Theorem 3.11. This finishes the proof of Theorem 3.13. □

Remark 3.16 Here the assumption $u \in C^1(M)$ enables us to express B_η in a easy form. Compare with the remark after the statement of Theorem 3.11, where we can suppose u only in $\mathrm{Lip}_{loc}(M)$.

Suppose now that L is linear; we have an analog condition (q-KL), that is, (q-KL∇), adding

(jjj) $|\nabla\tilde{\gamma}| \leq B$ on $M \setminus H$, for some constant $B > 0$ and $H \subset M$ compact.

It is immediate to show that this condition and linearity of L imply (q-KS∇).

Chapter 4
Sufficient Conditions for the Validity of the Weak Maximum Principle

As anticipated in the final part of the introduction to Chap. 3, the aim of this chapter is to prove the validity of the weak maximum principle for a large class of operators under the sole assumption of a controlled volume growth of geodesic balls related to the structure of the operator "at infinity". In doing so, we provide a second a priori estimate for solutions of certain differential inequalities, that, as an application given at the end of the chapter, enables us to generalize a Liouville-type result due, for the case of the Laplacian, to Dancer and Du [97] (see also the previous work by Aronson and Weinberger [32]). We also localize the principle to the family of the open sets with nonempty boundary of the manifold. This new formulation reminds of the (weak form of the) maximum principle as it appears, for instance, in the classical books by Protter and Weinberger [233], Gilbarg and Trudinger [125], or in the more recent work by Pucci and Serrin [235]. We underline its importance by giving an analytical application to the uniqueness problem for the positive solutions of certain Lichnerowicz-type equations.

Section 4.2 is devoted to the proof of a controlled growth weak maximum principle, that is, we allow the function u to be not necessarily bounded above but with a certain growth controlled by a power of the distance function from a fixed origin.

Finally, the observations and the discussion in Chap. 2 on parabolicity suggest us to introduce a new notion, that we call *strong parabolicity*, for which we give some sufficient conditions for its validity. We note that, for a large class of operators, the usual notion of parabolicity, in the sense of a Liouville-type result, is equivalent to strong parabolicity. Effectiveness of the latter will appear, for instance, when dealing with generic Ricci solitons in Chap. 8. A word of warning: in the proofs of various results that follow we define vector fields (typically called W) that are only continuous, and we then apply the divergence theorem. This procedure is intended to better explain the underlying argument of the proof; having done this it then becomes an easy matter to provide a proof for the general continuous case either

L.J. Alías et al., *Maximum Principles and Geometric Applications*,
Springer Monographs in Mathematics, DOI 10.1007/978-3-319-24337-5_4

using the weak formulation of the divergence theorem or simply following the weak formulation of the problem from the very beginning (see for instance [243]).

4.1 Volume Growth Conditions and Another A Priori Estimate

In this section we prove the validity of the weak maximum principle for a large class of operators under the sole assumption of a controlled volume growth for geodesic balls related to the structure of the operator "at infinity" as in condition (4.2) (iii) below.

Let T be a $(0, 2)$ symmetric tensor field on M. Assume that T satisfies

$$T_-(r) \leq T(X,X) \leq T_+(r) \tag{4.1}$$

for each $X \in T_xM$, $|X| = 1$, $x \in \partial B_r$ (where B_r is, as usual, the geodesic ball of radius r centered at a fixed origin $o \in M$) and some $T_\pm \in C^0(\mathbb{R}_0^+)$. Let φ : $M \times \mathbb{R}_0^+ \to \mathbb{R}_0^+$ be such that $\varphi(\ ,t) \in C^0(M)$ for each $t \in \mathbb{R}_0^+$, $\varphi(x,\ ) \in C^0(\mathbb{R}_0^+) \cap C^1(\mathbb{R}^+)$ for each $x \in M$, and

$$\begin{cases} (i)\ \varphi(x,0) = 0, & \text{for each } x \in M; \\ (ii)\ \varphi(x,t) > 0 \ \text{ on } \mathbb{R}^+, & \text{for each } x \in M; \\ (iii)\ \varphi(x,t) \leq A(x)t^\delta, & \text{on } M \times \mathbb{R}^+ \end{cases} \tag{4.2}$$

for some $\delta > 0$ and $A(x) \in C^0(M)$, $A(x) > 0$ on M.

Set

$$T_\delta(r) = \begin{cases} T_+(r), & \text{if } 0 < \delta \leq 1; \\ T_-(r)^{\frac{1-\delta}{2}} T_+(r)^{\frac{1+\delta}{2}}, & \text{if } \delta > 1. \end{cases} \tag{4.3}$$

and

$$\Theta(r) = \max_{[0,r]} T_\delta(s). \tag{4.4}$$

Define the operator $L = L_{\varphi,T}$ by setting, for each $u \in C^2(M)$,

$$Lu = \frac{1}{A(x)} \operatorname{div}\left(|\nabla u|^{-1}\varphi(x, |\nabla u|)T(\nabla u,\ )^\sharp\right). \tag{4.5}$$

We are now ready to prove the next

Theorem 4.1 *Let $(M\langle\,,\,\rangle)$ be a complete Riemannian manifold and φ, T be as above. Assume $T_-(r) > 0$ on $\mathbb{R}^+$ and let Θ be as in (4.4). Let $b(x) \in C^0(M)$ satisfy*

$$b(x) \geq \frac{1}{Q(r(x))} \tag{4.6}$$

where $Q : \mathbb{R}_0^+ \to \mathbb{R}^+$ is continuous and nondecreasing. Given $f \in C^0(\mathbb{R})$, assume that $u \in C^1(M)$ satisfies $u^* < +\infty$ and

$$Lu \geq b(x)f(u) \tag{4.7}$$

on the upper level set

$$\Omega_\gamma = \Omega_\gamma^u = \{x \in M : u(x) > \gamma\} \tag{4.8}$$

for some $\gamma < u^*$. If

$$\lim_{r\to+\infty} \frac{\Theta(r)Q(r)}{r^{1+\delta}} = 0 \tag{4.9}$$

and

$$\liminf_{r\to+\infty} \frac{\Theta(r)Q(r)}{r^{1+\delta}} \log\left(\int_{B_r} A(x)\right) < +\infty, \tag{4.10}$$

then $f(u^*) \leq 0$.

Remark 4.1 As it will be clear from the proof below, if $0 < \delta \leq 1$ we can relax the assumption $T_-(r) > 0$ to $T_-(r) \geq 0$.

Proof First of all, we note that if (4.7) holds on Ω_γ then it holds on $\Omega_{\gamma'}$ for each $\gamma \leq \gamma' < u^*$. Next, we assume by contradiction that $f(u^*) > 0$. By continuity of f and by increasing γ if necessary, we may suppose that $f(u) \geq C > 0$ on Ω_γ, and that u satisfies

$$Lu \geq \frac{B}{Q(r(x))} \quad \text{on } \Omega_\gamma$$

for some $B > 0$ that, without loss of generality we can suppose to be 1. Fix $0 < \eta < 1$. By choosing γ sufficiently close to u^*, we may also suppose that

$$\Gamma = \gamma - u^* + \eta \geq \frac{\eta}{2} > 0,$$

thus, having defined $v = u - u^* + \eta$, we have

$$v^* = \sup v = \eta, \quad \Omega_\Gamma^v = \Omega_\gamma^u,$$

with the obvious meaning of the notation. Furthermore,

$$Lv \geq \frac{1}{Q(r(x))} \quad \text{on } \Omega_\Gamma^v. \tag{4.11}$$

Choose $R > 0$ large enough that, for $r \geq R$, $B_r \cap \Omega_\Gamma^v \neq \emptyset$; fix $\zeta > 1$ to be determined later, and let $\psi : M \to [0, 1]$ be a smooth cut-off function such that

$$\begin{cases} (i)\ \psi \equiv 1, & \text{on } B_r; \\ (ii)\ \psi \equiv 0, & \text{on } M \setminus B_{2r}; \\ (iii)\ |\nabla \psi| \leq \frac{C_0}{r} \psi^{1/\zeta}, \end{cases} \tag{4.12}$$

for some constant $C_0 = C_0(\zeta) > 0$ and $r \geq R$. Note that the latter requirement (iii) is possible because $\zeta > 1$. Next, let $\lambda : \mathbb{R} \to \mathbb{R}_0^+$ be a C^1 function such that

$$\lambda(t) = 0 \quad \text{on } (-\infty, \Gamma], \quad \lambda'(t) \geq 0 \quad \text{on } \mathbb{R}. \tag{4.13}$$

Fix $\alpha > 2$ to be determined later, and consider the vector field W defined by

$$W = \psi^{2\alpha} \lambda(v) v^{\alpha-1} |\nabla v|^{-1} \varphi(x, |\nabla v|) T(\nabla v, \ )^\sharp \quad \text{on } \Omega_\Gamma^v \tag{4.14}$$

and $W \equiv 0$ outside. Note that in fact $W \equiv 0$ off $B_{2r} \cap \Omega_\Gamma^v$. For the ease of notation we set

$$T_v = \frac{T(\nabla v, \nabla v)}{|\nabla v|^2}.$$

From (4.1) and the assumptions of the theorem

$$0 < T_-(r) \leq T_v. \tag{4.15}$$

Furthermore,

$$|T(\nabla v, \nabla \psi)| \leq \sqrt{\frac{T(\nabla v, \nabla v)}{|\nabla v|^2}} |\nabla v| \sqrt{\frac{T(\nabla \psi, \nabla \psi)}{|\nabla \psi|^2}} |\nabla \psi| \leq T_v^{1/2} T_+^{1/2}(r) |\nabla v| |\nabla \psi|,$$

that is,

$$|T(\nabla v, \nabla \psi)| \leq T_v^{1/2} T_+^{1/2}(r) |\nabla v| |\nabla \psi| \tag{4.16}$$

and

$$|\nabla v|\varphi(x,|\nabla v|)\geq A(x)^{-1/\delta}\varphi(x,|\nabla v|)^{1+1/\delta}. \tag{4.17}$$

Using these facts, $\lambda'\geq 0$ and inequality (4.11) we now compute $\operatorname{div} W$. We have

$$\begin{aligned}\operatorname{div} W \geq{}& \psi^{2\alpha}\lambda(v)v^{\alpha-1}\frac{A(x)}{Q(r)}-2\alpha\psi^{2\alpha-1}\lambda(v)v^{\alpha-1}\varphi(x,|\nabla v|)|\nabla\psi|T_v^{1/2}T_+^{1/2}\\&+\frac{\alpha-1}{A(x)^{1/\delta}}\psi^{2\alpha}\lambda(v)v^{\alpha-2}\varphi(x,|\nabla v|)^{1+1/\delta}T_v.\end{aligned}$$

Since W is compactly supported, integrating and applying the divergence theorem, we obtain

$$\begin{aligned}\int\psi^{2\alpha}\lambda(v)v^{\alpha-1}\frac{A(x)}{Q(r)}\leq{}&-(\alpha-1)\int A(x)^{-1/\delta}\psi^{2\alpha}\lambda(v)v^{\alpha-2}\varphi(x,|\nabla v|)^{1+1/\delta}T_v\\&+2\alpha\int\psi^{2\alpha-1}\lambda(v)v^{\alpha-1}\varphi(x,|\nabla v|)|\nabla\psi|T_v^{1/2}T_+^{1/2}.\end{aligned}$$

We apply to the second integral on the right-hand side the inequality

$$ab\leq\sigma^p\frac{a^p}{p}+\frac{b^q}{q\sigma^q},\quad a,b\geq 0,$$

with $p=1+1/\delta$, $q=1+\delta$ and $\sigma>0$ chosen in such a way that the first integral on the right-hand side cancels out. Indeed, we have

$$\begin{aligned}&\int\psi^{2\alpha}\lambda(v)v^{\alpha-1}\frac{A(x)}{Q(r)}\leq\\&\frac{1}{(1+\delta)\sigma^{1+\delta}}\int A(x)\psi^{2\alpha-1-\delta}\lambda(v)v^{\alpha-1+\delta}|\nabla\psi|^{1+\delta}T_v^{(1-\delta)/2}T_+^{(1+\delta)/2}\end{aligned} \tag{4.18}$$

with σ satisfying

$$(2\alpha\sigma)^{1+1/\delta}=\frac{(1+\delta)(\alpha-1)}{\delta},$$

so that

$$\frac{1}{(1+\delta)\sigma^{1+\delta}}=\frac{2^{\delta+1}\delta^\delta\alpha^\delta}{(\alpha-1)^\delta(1+\delta)^{(1+\delta)}}\alpha. \tag{4.19}$$

Now, since ψ is supported on B_{2r} and Q is nondecreasing, $Q(r(x)) \leq Q(2r)$ on the support of ψ and the left-hand side of (4.18) is bounded from below by

$$Q(2r)^{-1} \int \psi^{2\alpha} \lambda(v) v^{\alpha-1} A(x). \tag{4.20}$$

On the other hand, since

$$\left(\frac{\alpha}{\alpha-1}\right)^{\delta} \leq 2^{\delta} \text{ for } \alpha \geq 2,$$

from (4.19) we have the estimate

$$\frac{1}{(1+\delta)\sigma^{1+\delta}} \leq C(\delta)\alpha$$

with

$$C(\delta) = \frac{2^{2\delta+1}\delta^{\delta}}{(1+\delta)^{(1+\delta)}} > 0$$

independent of $\alpha \geq 2$. Furthermore, using (4.12) (iii), we may write

$$\psi^{2\alpha-1-\delta}|\nabla\psi|^{1+\delta} = \psi^{2\alpha-(1+\delta)(1-1/\zeta)}(\psi^{-1/\zeta}|\nabla\psi|)^{1+\delta} \leq \psi^{2\alpha-(1+\delta)(1-1/\zeta)} \frac{C_0}{r^{1+\delta}}.$$

Finally, recalling that $T_-(r(x)) \leq T_v(x) \leq T_+(r(x))$, we see that

$$T_v(x)^{(1-\delta)/2} \leq \begin{cases} T_+^{(1-\delta)/2}(r(x)), \text{ if } 0 < \delta \leq 1; \\ T_-^{(1-\delta)/2}(r(x)), \text{ if } \delta > 1. \end{cases}$$

and therefore,

$$T_v^{(1-\delta)/2}(x) T_+^{(1+\delta)/2}(r(x)) \leq T_\delta(r(x)) \leq \Theta(2r) \quad \text{on } B_{2r}. \tag{4.21}$$

Thus, the right-hand side of (4.18) is estimated from above by

$$\frac{C_0 C(\delta)}{r^{1+\delta}} \alpha \Theta(2r) \int A(x) \psi^{2\alpha-(1+\delta)(1-1/\zeta)}(x) v^{\alpha-1+\delta}(x).$$

Now, we choose $\zeta > 1$ close enough to 1 that $2 - (1+\delta)(1-1/\zeta) > 0$, and we apply Hölder's inequality with conjugate exponents $\alpha/(\alpha-1)$ and α to estimate from above this last expression with

$$\frac{C_0 C(\delta)}{r^{1+\delta}} \alpha \Theta(2r) \left(\int A(x) \psi^{2\alpha} v^{\alpha-1} \lambda(v)\right)^{\frac{\alpha-1}{\alpha}} \left(\int A(x) \psi^{2\alpha-(1+\delta)(1-1/\zeta)\alpha} v^{\alpha-1+\delta\alpha} \lambda(v)\right)^{\frac{1}{\alpha}}. \tag{4.22}$$

Using (4.20) and (4.22) into (4.18) we have

$$Q(2r)^{-1}\int A(x)\psi^{2\alpha}v^{\alpha-1}\lambda(v)\leq$$

$$\frac{C_0C(\delta)}{r^{1+\delta}}\alpha\Theta(2r)\left(\int A(x)\psi^{2\alpha}v^{\alpha-1}\lambda(v)\right)^{\frac{\alpha-1}{\alpha}}\left(\int A(x)\psi^{2\alpha-(1+\delta)(1-1/\zeta)\alpha}v^{\alpha-1+\delta\alpha}\lambda(v)\right)^{\frac{1}{\alpha}},$$

that is,

$$\int A(x)\psi^{2\alpha}v^{\alpha-1}\lambda(v)\leq\left(\frac{C_0C(\delta)}{r^{1+\delta}}\alpha\Theta(2r)Q(2r)\right)^{\alpha}\int A(x)\psi^{2\alpha-(1+\delta)(1-1/\zeta)\alpha}v^{\alpha-1+\delta\alpha}\lambda(v).$$

Recalling that $\psi\equiv 1$ on B_r, $\psi\equiv 0$ on $M\setminus B_{2r}$ and that $\eta/2\leq v\leq\eta$ on Ω_Γ^v when $\lambda(v)>0$, we deduce that

$$\int_{B_r}A(x)\lambda(v)\leq\left(2^{(\alpha-1)/\alpha}\eta^{(1-\alpha)/\alpha}\frac{C_0C(\delta)}{r^{1+\delta}}\alpha\Theta(2r)Q(2r)\eta^{(\alpha-1)/\alpha+\delta}\right)^{\alpha}\int_{B_{2r}}A(x)\lambda(v).$$

Hence

$$\int_{B_r}A(x)\lambda(v)\leq\left(\frac{C_1}{r^{1+\delta}}\alpha\Theta(2r)Q(2r)\eta^{\delta}\right)^{\alpha}\int_{B_{2r}}A(x)\lambda(v),\tag{4.23}$$

with

$$C_1=2C_0C(\delta).$$

We now set

$$\alpha=\alpha(r)=\frac{r^{1+\delta}}{4C_1\Theta(2r)Q(2r)\eta^{\delta}}.$$

Note that, for r sufficiently large, $\alpha=\alpha(r)\geq 2$, so that we can rewrite (4.23) as

$$\int_{B_r}A(x)\lambda(v)\leq\left(\frac{1}{2}\right)^{\frac{1}{4C_1}\eta^{-\delta}\frac{r^{1+\delta}}{\Theta(2r)Q(2r)}}\int_{B_{2r}}A(x)\lambda(v),\tag{4.24}$$

for each $r\geq R$. Note that C_1 is independent of r and η. We now need the following

Lemma 4.1 *Let $G,F:[R,+\infty)\to\mathbb{R}_0^+$ be nondecreasing functions such that for some constants $0<\Lambda<1$ and $B,\theta>0$,*

$$G(r)\leq\Lambda^{B\frac{r^{\theta}}{F(2r)}}G(2r),\ \textit{for each } r\geq R.\tag{4.25}$$

Then there exists a constant $S = S(\theta) > 0$ such that for each $r \geq 2R$

$$\frac{F(r)}{r^\theta}\log G(r) \geq \frac{F(r)}{r^\theta}\log G(R) + SB\log(\frac{1}{\Lambda}). \tag{4.26}$$

Proof Let $r_0 = R$ and $r_k = 2^k r_0$. Then, for each $r \leq 2r_0$ there exists k such that $r_k \leq r \leq r_{k+1}$. Applying inequality (4.25) k-times, we obtain

$$G(r_0) \leq \Lambda^{B\sum_{j=0}^{k-1}\frac{r_j^\theta}{F(2r_j)}} G(r_k). \tag{4.27}$$

Using the definition of r_j and the fact that F is nondecreasing, we estimate

$$\sum_{j=0}^{k-1}\frac{r_j^\theta}{F(2r_j)} \geq \frac{r_0^\theta}{F(2r_{k-1})}\sum_{j=0}^{k-1}2^{j\theta} = \frac{r_{k+1}^\theta}{F(2r_{k-1})}\frac{1-2^{-k\theta}}{2^\theta-1}2^{-\theta} \geq S\frac{r^\theta}{F(r)}$$

with $S = 2^{-\theta}/(2^\theta - 1)$. Substituting into (4.27), recalling that $0 < \Lambda < 1$ and that G is nondecreasing, we conclude that

$$G(r_0) \leq \Lambda^{BS\frac{r^\theta}{F(r)}} G(r).$$

Hence (4.26) follows by taking logarithms. □

We apply Lemma 4.1 with $G(r) = \int_{B_r} A(x)\lambda(v)$, $\theta = 1+\delta$, $\Lambda = 1/2$, $B = \frac{1}{4C_1}\eta^{-\delta}$, $F(r) = Q(r)\Theta(r)$ to deduce the existence of a constant $S = S(\delta) > 0$ such that for each $r \geq 2R$

$$\frac{Q(r)\Theta(r)}{r^{1+\delta}}\log\int_{B_r} A(x)\lambda(v) \geq \frac{Q(r)\Theta(r)}{r^{1+\delta}}\log\int_{B_R} A(x)\lambda(v) + \frac{S}{4C_1}\eta^{-\delta}\log 2. \tag{4.28}$$

Now we choose λ in such a way that $\sup\lambda = 1$. Letting $r \to +\infty$ in (4.28) and using (4.9) we obtain

$$\liminf_{r\to+\infty}\frac{Q(r)\Theta(r)}{r^{1+\delta}}\log\int_{B_r} A(x) \geq \frac{S}{4C_1}\eta^{-\delta}\log 2,$$

with S and C_1 independent of η. Letting $\eta \to 0^+$ we contradict (4.10). This completes the proof of the theorem. □

Remark 4.2 We note that a minor change of the above argument allows us to replace assumption (4.10) with

$$\liminf_{r\to+\infty}\frac{Q(r)\Theta(r)}{r^{1+\delta}}\int_{B_r} A(x)|u|^p < +\infty \tag{4.29}$$

for some $p > 0$.

We now give an a priori estimate similar, in some sense, to that given in Theorem 3.6. In other words, we prove that, under appropriate assumptions, solutions of $Lu \geq b(x)f(u)$ are necessarily bounded above. Again, as in Theorem 3.6, a key role is played by an assumption implying the Keller-Osserman condition for these general operators. We shall discuss this after the proof of Theorem 4.3.

Remark 4.3 Let A, φ, f, b, Q, T and Θ be as in Theorem 4.1 and assume that $u \in C^1(M)$ is such that $u_* = \inf_M u > -\infty$ and it satisfies $Lu \leq -b(x)f(u)$ on the set $\Omega_\gamma^- = \{x \in M : u(x) < \gamma\}$ for some $\gamma > u_*$. If (4.9) and (4.10) [or (4.9) and (4.29)] hold, then $f(u_*) \leq 0$. Indeed, it suffices to note that the function $v = -u$ is bounded above, $v^* = -u_*$ and v satisfies $Lv \geq b(x)g(v)$ with $g(t) = f(-t)$. In the assumptions of Theorem 4.1, $g(v^*) = f(u_*) \leq 0$.

Theorem 4.2 *Let φ, b, Q, A, T and Θ be as in Theorem 4.1 and assume that $u \in C^1(M)$ satisfies*

$$Lu \geq b(x)f(u) \tag{4.30}$$

on the set $\Omega_\gamma = \{x \in M : u(x) > \gamma\}$ for some $\gamma < u^* \leq +\infty$, where f is a continuous function on $\mathbb{R}$ such that

$$\liminf_{t\to+\infty} \frac{f(t)}{t^\sigma} > 0 \tag{4.31}$$

for some $\sigma > \delta$. If (4.10) [or (4.29)] holds, then u is bounded above.

Proof Assume by contradiction that u is not bounded above, so that the set

$$\Omega_\gamma = \{x \in M : u(x) > \gamma\}$$

is nonempty for each $\gamma > 0$. By increasing γ, if necessary, we may assume that $f(t) \geq Bt^\sigma$ for some $B > 0$ if $t \geq \gamma$. For the ease of notation, we let $B = 1$ so that on Ω_γ

$$Lu = L_{\varphi,T}u = \frac{1}{A(x)} \operatorname{div}\left(|\nabla u|^{-1}\varphi(x, |\nabla u|)T(\nabla u, \ )^\sharp\right) \geq b(x)u^\sigma,$$

weakly, that is, for each $\psi \in C_c^\infty(\Omega_\gamma)$, $\psi \geq 0$,

$$\int_{\Omega_\gamma} \left(|\nabla u|^{-1}\varphi(x, |\nabla u|)T(\nabla u, \nabla\psi) + b(x)u^\sigma A(x)\psi\right) \leq 0.$$

Clearly we may also assume that $b(x)$ is bounded above. Let $R > 0$ be large enough that $\Omega_\gamma \cap B_R \neq \emptyset$. Let $\lambda : \mathbb{R} \to \mathbb{R}_0^+$ be a C^1, nondecreasing function such that

$\lambda(t) = 0$ for $t \leq \gamma$; fix $\xi > 1$ satisfying

$$1 - \frac{1+\delta}{\sigma - \delta}\left(1 - \frac{1}{\xi}\right) > 0 \tag{4.32}$$

and, as in the proof of Theorem 4.1, choose a C^∞ cut-off function $\psi = \psi_r : M \to [0, 1]$ such that, for $r \geq R$

$$(i)\psi \equiv 1 \text{ on } B_R; \;\; (ii)\psi \equiv 0 \text{ on } M \setminus B_{2r}; \;\; (iii)|\nabla \psi| \leq \frac{C_0}{r}\psi^{1/\xi} \tag{4.33}$$

for some constant $C_0 = C_0(\xi) > 0$. Finally, fix $\alpha > \max\{1+\delta, 2\sigma\}$ and $\beta > 0$ to be determined later. Consider the vector field W defined by

$$W = \psi^\alpha \lambda(u) u^\beta |\nabla u|^{-1} \varphi(x, |\nabla u|) T(\nabla u, \,)^\sharp$$

on Ω_γ and $W \equiv 0$ everywhere else. Note that the properties of λ and ψ imply that W vanishes off $B_{2r} \cap \Omega_\gamma$. Proceeding as in the proof of Theorem 4.1, we estimate

$$\begin{aligned} \operatorname{div} W \geq\; & \psi^\alpha \lambda(u) u^{\sigma+\beta} b(x) A(x) + \frac{\beta}{A(x)^{1/\delta}} \psi^\alpha \lambda(u) u^{\beta-1} \varphi(x, |\nabla u|)^{1+1/\delta} T_u \\ & - \alpha \psi^{\alpha-1} \lambda(u) u^\beta \varphi(x, |\nabla u|) |\nabla \psi| T_u^{1/2} T_+^{1/2}, \end{aligned}$$

where $T_\pm$ are defined in (4.1) and

$$T_u = \frac{T(\nabla u, \nabla u)}{|\nabla u|^2}.$$

Next we apply to the second term of the right-hand side of the inequality above the following inequality

$$ab \leq \varepsilon^p \frac{a^p}{p} + \frac{b^q}{q\varepsilon^q}, \quad a, b \geq 0,$$

with $p = 1+\delta$, $q = (1+\delta)/\delta$, and with $\varepsilon > 0$ chosen in such a way that the last term of the right-hand side cancels out, that is,

$$\varepsilon^{1+1/\delta} = \frac{\delta}{1+\delta} \frac{\alpha}{\beta}.$$

Inserting the resulting inequality in the above estimate, we obtain

$$\begin{aligned} \operatorname{div} W \geq\; & \psi^\alpha \lambda(u) u^{\sigma+\beta} b(x) A(x) \\ & - \frac{\delta^\delta}{(1+\delta)^{1+\delta}} \left(\frac{\alpha}{\beta}\right)^\delta \alpha A(x) \psi^{\alpha-1-\delta} \lambda(u) u^{\beta+\delta} T_u^{(1-\delta)/2} T_+^{(1+\delta)/2} |\nabla \psi|^{1+\delta}. \end{aligned}$$

Next, we integrate, apply the divergence theorem, and recall that W has compact support to obtain

$$\int \psi^{\alpha}\lambda(u)b(x)u^{\sigma+\beta}A(x)$$
$$\leq \frac{\delta^{\delta}}{(1+\delta)^{1+\delta}}\left(\frac{\alpha}{\beta}\right)^{\delta}\alpha\int \psi^{\alpha-1-\delta}\lambda(u)u^{\beta+\delta}T_u^{(1-\delta)/2}T_+^{(1+\delta)/2}|\nabla\psi|^{1+\delta}A(x). \tag{4.34}$$

Multiplying and dividing by $b(x)^{1/p}$ in the integral on the right-hand side, and applying Hölder's inequality with conjugate exponents p and q, yields

$$\int \psi^{\alpha-1-\delta}\lambda(u)u^{\beta+\delta}T_u^{(1-\delta)/2}T_+^{(1+\delta)/2}|\nabla\psi|^{1+\delta}A(x) \leq$$
$$\left(\int \psi^{\alpha}b(x)\lambda(u)u^{(\delta+\delta}A(x)\right)^{1/p}$$
$$\times\left(\int \psi^{\alpha-(1+\delta)(1-1/\xi)q}\lambda(u)b(x)^{1-q}T_u^{(1-\delta)q/2}T_+^{(1+\delta)q/2}\left(\frac{|\nabla\psi|}{\psi^{1/\xi}}\right)^{(1+\delta)q}A(x)\right)^{1/q},$$

provided

$$\alpha-(1+\delta)(1-1/\xi)q>0. \tag{4.35}$$

Choosing $p=\dfrac{\beta+\sigma}{\beta+\delta}>1$ since $\sigma>\delta$, the first integral on the right-hand side of the above inequality is equal to the integral on the left-hand side of (4.34). Thus, inserting into the latter and simplifying, we obtain

$$\int \psi^{\alpha}\lambda(u)b(x)u^{\sigma+\beta}A(x)\leq\left(\frac{\delta}{(1+\delta)^2}\frac{\alpha^{1+\delta}}{\beta}\right)^{q} \tag{4.36}$$
$$\times\int \psi^{\alpha-(1+\delta)(1-1/\xi)q}\lambda(u)b(x)^{1-q}T_u^{(1-\delta)q/2}T_+^{(1+\delta)q/2}\left(\frac{|\nabla\psi|}{\psi^{1/\xi}}\right)^{(1+\delta)q}A(x).$$

Since $u>\gamma$ on Ω_{γ} and $\psi\equiv 1$ on B_r,

$$\gamma^{\beta+\sigma}\int_{B_r} b(x)\lambda(u)A(x)\leq\int \psi^{\alpha}\lambda(u)b(x)u^{\sigma+\beta}A(x).$$

On the other hand, using (4.33) (ii), (iii) and the fact that ψ is supported on B_{2r}, we have

$$\left(\frac{\delta^{\delta}}{(1+\delta)^{1+\delta}}\left(\frac{\alpha}{\beta}\right)^{\delta}\alpha\right)^{q}\times$$

$$\int \psi^{\alpha-(1+\delta)(1-1/\xi)q}\lambda(u)b(x)^{1-q}T_{u}^{(1-\delta)q/2}T_{+}^{(1+\delta)q/2}\left(\frac{|\nabla\psi|}{\psi^{1/\xi}}\right)^{(1+\delta)q}A(x)$$

$$\leq\left(\frac{\delta^{\delta}}{(1+\delta)^{1+\delta}}\left(\frac{\alpha}{\beta}\right)^{\delta}\alpha\frac{C_0^{1+\delta}}{r^{1+\delta}}\sup_{B_{2r}}\frac{T_u^{\frac{1-\delta}{2}}T_+^{\frac{1+\delta}{2}}}{b(x)}\right)^{q}\int_{B_{2r}}b(x)\lambda(u)A(x).$$

We insert these two latter inequalities into (4.36); we use $b(x)\geq Q(r(x))^{-1}$ with Q nondecreasing and the validity of (similar to (4.21) in the proof of Theorem 4.1)

$$T_u^{\frac{1-\delta}{2}}(x)T_+^{\frac{1+\delta}{2}}(r(x))\leq T_\delta(r(x))\leq\Theta(2r) \tag{4.37}$$

on B_{2r}, and

$$q=\frac{\beta+\sigma}{\sigma-\delta}$$

to obtain

$$\int_{B_r}b(x)\lambda(u)A(x)\leq\left(\frac{C}{\gamma^{\sigma-\delta}}\frac{\Theta(2r)Q(2r)}{r^{1+\delta}}\left(\frac{\alpha}{\beta}\right)^{\delta}\alpha\right)^{\frac{\beta+\sigma}{\sigma-\delta}}\int_{B_{2r}}b(x)\lambda(u)A(x), \tag{4.38}$$

with $C=C(\delta,C_0)>0$. Now we choose

$$\alpha=\beta+\sigma=\frac{1}{4C}\gamma^{\sigma-\delta}\frac{r^{1+\delta}}{\Theta(2r)Q(2r)}$$

so that (4.32) implies that (4.35) holds. Moreover, because of (4.9), $\alpha\to+\infty$ as $r\to+\infty$. Hence, for r sufficiently large $\frac{\alpha}{\beta}\leq 2$. It follows that, for such values of r, (4.38) gives

$$\int_{B_r}b(x)\lambda(u)A(x)\leq\left(\frac{1}{2}\right)^{\frac{\gamma^{\sigma-\delta}}{4C(\sigma-\delta)}\frac{r^{1+\delta}}{\Theta(2r)Q(2r)}}\int_{B_{2r}}b(x)\lambda(u)A(x). \tag{4.39}$$

We let

$$G(r)=\int_{B_r}b(x)\lambda(u)A(x)$$

and

$$F(r) = \Theta(r)Q(r)$$

defined on $[R, +\infty)$ for some R sufficiently large such that (4.39) holds for $r \geq R$. Then

$$G(r) \leq \left(\frac{1}{2}\right)^{\Gamma \frac{r^{1+\delta}}{F(2r)}} G(2r)$$

with

$$\Gamma = \frac{\gamma^{\sigma-\delta}}{4C(\sigma-\delta)} > 0. \tag{4.40}$$

Then by Lemma 4.1, there exists a constant $S > 0$ such that, for each $r \geq 2R$

$$\frac{Q(r)\Theta(r)}{r^{1+\delta}} \log \int_{B_r} b(x)\lambda(u)A(x) \geq \frac{Q(r)\Theta(r)}{r^{1+\delta}} \log \int_{B_R} b(x)\lambda(u)A(x) + S\Gamma \log 2,$$

To reach the desired contradiction, we choose λ satisfying $\sup \lambda = \frac{1}{\sup_M b} > 0$ so that $b(x)\lambda(u) \leq 1$. Taking r going to $+\infty$ in the above and using (4.9) we deduce

$$\liminf_{r\to+\infty} \frac{Q(r)\Theta(r)}{r^{1+\delta}} \log \int_{B_r} A(x) \geq SB \log 2.$$

This contradicts (4.10) by choosing γ sufficiently large in the expression (4.40) of Γ. □

As a simple application of Theorems 4.1 and 4.2 we have

Theorem 4.3 *Let $(M, \langle\ ,\ \rangle)$ be a complete Riemannian manifold of dimension $m \geq 3$ with nonnegative scalar curvature $S(x)$. Fix an origin o and let $r(x) = \operatorname{dist}(x, o)$. Let $K(x) \in C^{\infty}(M)$ be nonpositive and such that*

$$K(x) \leq -\frac{C}{r(x)^{\mu}} \quad \textit{for } r(x) \gg 1 \tag{4.41}$$

and some constants $C > 0$, $\mu \in \mathbb{R}$. Assume

$$\liminf_{r\to+\infty} \frac{\log \operatorname{vol} B_r}{r^{2-\mu}} < +\infty. \tag{4.42}$$

Then there are no conformal deformations of the metric to a new metric with scalar curvature $K(x)$.

Proof By contradiction suppose it is possible to find a conformal deformations of the metric to a new metric with scalar curvature $K(x)$; then (and here we use $m \geq 3$), setting $\varphi = u^{\frac{2}{m-2}}$ in Eq. (1.79), there exists $u > 0$, smooth solution on M of the (standard) *Yamabe equation*

$$c_m \Delta u - S(x)u + K(x)u^{\frac{m+2}{m-2}} = 0.$$

Here $c_m = 4\frac{m-1}{m-2}$ and the pointwise conformal deformation $\widetilde{\langle\, ,\, \rangle}$ of $\langle\, ,\, \rangle$ having scalar curvature $K(x)$ is $\widetilde{\langle\, ,\, \rangle} = u^{\frac{4}{m-2}} \langle\, ,\, \rangle$. By Proposition 3.10 of [189] and the subsequent remark, there exist $R > 0$ sufficiently large, $\tilde{K}(x) \in C^{\infty}(M)$, a constant $C_1 > 0$ and a $C^{\infty}(M)$ positive function v such that

$$(i)\ \tilde{K}(x) < 0 \quad \text{on } M, \quad (ii)\ \tilde{K}(x) = C_1 K(x) \quad \text{on } M \setminus B_R$$

and v solves

$$c_m \Delta v - S(x)v + \tilde{K}(x)v^{\frac{m+2}{m-2}} = 0 \quad \text{on } M.$$

Hence, since $S(x) \geq 0$,

$$c_m \Delta v \geq -\tilde{K}(x)v^{\frac{m+2}{m-2}} \quad \text{on } M.$$

Because of (i), (ii) and assumption (4.41),

$$-\tilde{K}(x) \geq \frac{C_2}{(1 + r(x))^{\mu}} \quad \text{on } M$$

for some constant $C_2 > 0$. It follows that

$$\Delta v \geq \frac{C_3}{(1 + r(x))^{\mu}} v^{\frac{m+2}{m-2}} \quad \text{on } M.$$

Since (4.42) holds and $\frac{m+2}{m-2} > 1$, we can apply Theorem 4.2 to deduce $v^* < +\infty$. By Theorem 4.1 we then have $v \equiv 0$, contradiction. □

We briefly comment on condition (4.31) and its relation with the Keller-Osserman condition . The latter historically was introduced independently by Keller [157] and Osserman [213] analyzing the differential inequality

$$\Delta u \geq f(u) \quad \text{on } \mathbb{R}^m. \tag{4.43}$$

Letting, for $f(t) > 0$ on $\mathbb{R}^+, f(0) = 0$,

$$F(t) = \int_0^t f(s)\, ds, \tag{4.44}$$

the *Keller-Osserman condition* for (4.43) expresses as

$$\frac{1}{\sqrt{F(t)}} \in L^1(+\infty). \tag{4.45}$$

It is well known that if (4.45) is satisfied, then there are no nonnegative solutions of (4.43) on $\mathbb{R}^m$ besides the trivial $u \equiv 0$. On the contrary, if (4.45) fails then (4.43) admits positive solutions on $\mathbb{R}^m$ exploding at infinity. We note that (4.45) coincides with (3.59) (for $F = f$ in the notation there). Indeed, (4.45) implies, by Theorem 3.6, boundedness of u and then the weak maximum principle for Δ on $\mathbb{R}^m$ yields $f(u^*) \leq 0$, so that $u \equiv 0$.

The Keller-Osserman condition can be generalized to other operators, and for instance, for those considered in Sect. 3.3.1, it becomes

$$\frac{1}{H^{-1}(F(t))} \in L^1(+\infty), \tag{4.46}$$

where H is the function defined in (3.112) for the operator

$$\operatorname{div}(A(|\nabla u|)\nabla u) = \operatorname{div}\left(|\nabla u|^{-1}\varphi(|\nabla u|)\nabla u\right).$$

For the operator L in (4.30) of Theorem 4.2, condition (4.31) implies the corresponding Keller-Osserman condition of the type (4.46) as explained in detail in [183].

While in the linear case of Theorem 3.6 we have been able to use directly the Keller-Osserman condition to obtain an a priori upper bound, in the nonlinear case the matter becomes quite complicate and it remains an open problem to replace a condition like (4.31) with the corresponding Keller-Osserman condition that it implies.

4.2 A Controlled Growth Weak Maximum Principle

The aim of this section is to prove a weak maximum principle type result when the function u is not necessary bounded above.

Theorem 4.4 *Let $(M, \langle\,,\rangle)$ be a complete Riemannian manifold, let o be a reference point in M, and let $r(x)$ be the distance function from o. Let T be a symmetric $(0, 2)$ tensor field. Assume that, for some positive continuous functions T_- and T_+ defined on $\mathbb{R}_0^+$, the tensor T satisfies the following bound*

$$0 < T_-(r) \leq T(X, X) \leq T_+(r) \tag{4.47}$$

for every $X \in T_xM$, $|X| = 1$, and every $x \in \partial B_r$, where B_r denotes the geodesic ball of radius r centered at o. Let $\varphi : M \times \mathbb{R}_0^+ \to \mathbb{R}_0^+$ be such that $\varphi(\,, t) \in C^0(M)$ for *every* $t \in \mathbb{R}_0^+$, $\varphi(x, \ ) \in C^0(\mathbb{R}_0^+) \cap C^1(\mathbb{R}^+)$ *for every* $x \in M$, *and*

$$(i)\ \varphi(x, 0) = 0$$

for every $x \in M$,

$$(ii)\ \varphi(x, t) > 0 \quad \text{on } \mathbb{R}^+$$

for every $x \in M$, *and*

$$(iii)\ \varphi(x, t) \leq A(x)t^{\delta}$$

on $M \times \mathbb{R}^+$ *for some* $\delta > 0$ *and* $A(x) \in C^0(M)$, $A(x) > 0$ *on* M. *Furthermore, assume that*

$$\inf_M \frac{T_-(r(x))}{T_+(r(x))} \frac{1}{A(x)^{1/\delta}} = \frac{1}{\Sigma^{1/\delta}} \tag{4.48}$$

for some $\Sigma > 0$. *Given* $\sigma, \mu \in \mathbb{R}$ *we let*

$$\eta = \mu + (\sigma - 1)(1 + \delta) \tag{4.49}$$

and we assume that

$$\sigma \geq 0, \quad \sigma - \eta > 0. \tag{4.50}$$

Let $u \in C^1(M)$ *be a function such that*

$$\hat{u} = \limsup_{r(x) \to +\infty} \frac{u(x)}{r(x)^{\sigma}} < +\infty. \tag{4.51}$$

Suppose that for some function $f \in C^{\infty}(M)$

$$\liminf_{R \to +\infty} \frac{\log \int_{B_R} T_+(r) e^{-f}}{R^{\sigma - \eta}} = d_0 < +\infty. \tag{4.52}$$

Define

$$Lu = L_{\varphi, T, f} = e^f \operatorname{div}(e^{-f} |\nabla u|^{-1} \varphi(x, |\nabla u|) T(\nabla u, \ )^{\sharp}), \tag{4.53}$$

where $^\sharp$ *denotes the musical isomorphism. For* $\gamma \in \mathbb{R}$ *suppose that the set*

$$\Omega_\gamma = \{x \in M : u(x) > \gamma\} \tag{4.54}$$

is nonempty. Then

$$\inf_{\Omega_\gamma} \frac{(1+r(x))^\mu}{T_+(r(x))} Lu(x) \leq \begin{cases} 0, & \text{if } \sigma = 0; \\ \Sigma d_0 \max\{\hat{u}, 0\}^\delta (\sigma - \eta)^{1+\delta}, & \text{if } \sigma > 0 \text{ and } \eta < 0; \\ \Sigma d_0 \max\{\hat{u}, 0\}^\delta \sigma^\delta (\sigma - \eta), & \text{if } \sigma > 0 \text{ and } \eta \geq 0; \end{cases} \tag{4.55}$$

Proof We begin observing that if a is any constant and $u_a = u + a$, then

$$Lu_a = Lu$$

and

$$\Omega_\gamma = \{x \in M : u_a(x) > \gamma + a\}$$

Furthermore, if $\sigma > 0$ then $\hat{u}_a = \hat{u}$, and if $\sigma = 0$ then $\hat{u}_a = \hat{u} + a$. So in order to estimate

$$\inf_{\Omega_\gamma} \frac{(1+r(x))^\mu}{h_+(r(x))} Lu(x)$$

we may replace u with a suitable translate u_a. Next, fix $b > \max\{\hat{u}, 0\}$. It is easy to see that there exists a constant a such that

$$\frac{u_a(x)}{(1+r(x))^\sigma} < b \quad \text{on} \quad M \tag{4.56}$$

and $u_a(x_0) > 0$ for some $x_0 \in M$. This is obvious if u is bounded above and in particular, due to (4.51) if $\sigma = 0$. On the other hand, if u is not bounded above, and therefore $\sigma > 0$, then by (4.51) there exists $\bar{R} > 0$ such that

$$\frac{u(x)}{(1+r(x))^\sigma} < b \quad \text{on } M \backslash B_{\bar{R}},$$

and it is clear that there exists $a \in \mathbb{R}$ such that $u_a(x_0) > 0$ for some $x_0 \in \bar{B}_{\bar{R}}$ and $\frac{u_a(x)}{(1+r(x))^\sigma} < b$ on $\bar{B}_{\bar{R}}$ and on all of M. We will assume that a constant a has been selected in such a way that (4.56) holds. In accordance to the observation made above, we are going to replace u with u_a and, for the ease of notation, we suppress the subscript a. Furthermore, if $\gamma_1 \leq \gamma$ and

$$\inf_{\Omega_\gamma} \frac{(1+r(x))^\mu}{T_+(r(x))} Lu(x) \leq D$$

then

$$\inf_{\Omega_{\gamma_1}} \frac{(1+r(x))^\mu}{T_+(r(x))} Lu(x) \leq D$$

so that, without loss of generality, we may suppose $\gamma \geq 0$. Next, let

$$K = \inf_{\Omega_\gamma} \frac{(1+r(x))^\mu}{T_+(r(x))} Lu(x)$$

and suppose $K > 0$, otherwise there is nothing to prove. In this case u is nonconstant on any component of Ω_γ and

$$\frac{(1+r(x))^\mu}{T_+(r(x))} Lu(x) \geq K > 0 \quad \text{on } \Omega_\gamma \tag{4.57}$$

We fix $\theta \in (1/2, 1)$ and we choose $R_0 > 0$ large enough that $B_{R_0} \cap \Omega_\gamma \neq \emptyset$ and $|\nabla u| \not\equiv 0$ on it. Given $R > R_0$ we let $\psi \in C^\infty(M)$ be a cutoff function such that

$$0 \leq \psi \leq 1, \psi \equiv 1 \text{ on } B_{\theta R},\ \psi \equiv 0 \text{ on } M \backslash B_R,\ |\nabla \psi| \leq \frac{C}{R(1-\theta)} \tag{4.58}$$

for some constant $C > 0$. Let also $\lambda \in C^1(\mathbb{R})$ and $F(v, r) \in C^1(\mathbb{R}^2)$ be such that

$$0 \leq \lambda \leq 1,\ \ \lambda = 0 \text{ on } (-\infty, \gamma],\ \ \lambda > 0,\ \ \lambda' \geq 0 \text{ on } (\gamma, +\infty) \tag{4.59}$$

and

$$F(v, r) > 0,\ \ \frac{\partial F}{\partial v}(v, r) < 0 \text{ on } \mathbb{R}_0^+ \times \mathbb{R}_0^+. \tag{4.60}$$

Finally we let W be the vector field defined on Ω_γ by

$$W = \psi^{1+\delta} \lambda(u) F(v, r) e^{-f} |\nabla u|^{-1} \varphi(x, |\nabla u|) T(\nabla u, \ )^\sharp, \tag{4.61}$$

where v is given by

$$v = \alpha(1+r)^\sigma - u \tag{4.62}$$

and $\alpha > b$ is a constant so that $v > 0$ on Ω_γ. Indeed, according to (4.56) and the assumption $\gamma \geq 0$, so that $u > 0$ on Ω_γ, we have

$$(\alpha - b)(1+r)^\sigma \leq v \leq \alpha(1+r)^\sigma \quad \text{on } \Omega_\gamma. \tag{4.63}$$

Note that W vanishes on $\partial(\Omega_\gamma \cap B_R)$ and it extends to a continuous vector field on the whole of M by defining it to be zero in the complement of $\Omega_\gamma \cap B_R$.

We now compute the divergence of W. Note that, from (iii), we have

$$t\varphi(x, t) \geq A(x)^{-1/\delta} \varphi(x, t)^{1+1/\delta} \quad \text{on } M \times \mathbb{R}_0^+. \tag{4.64}$$

Furthermore, from the properties (4.47) of T,

$$|T(\nabla u, \nabla v)| \leq \sqrt{T(\nabla u, \nabla u)}\sqrt{T(\nabla v, \nabla v)} \leq T_+(r(x))|\nabla u||\nabla v|. \tag{4.65}$$

We compute

$$\begin{aligned}
e^f \operatorname{div} W &= \psi^{1+\delta}\lambda(u)F(v,r)Lu \\
&\quad +(1+\delta)\psi^\delta\lambda(u)F(v,r)|\nabla u|^{-1}\varphi(x,|\nabla u|)T(\nabla u,\nabla\psi) \\
&\quad +\psi^{1+\delta}\lambda'(u)F(v,r)|\nabla u|^{-1}\varphi(x,|\nabla u|)T(\nabla u,\nabla u) \\
&\quad +\psi^{1+\delta}\lambda(u)\frac{\partial F}{\partial v}(v,r)|\nabla u|^{-1}\varphi(x,|\nabla u|)T(\nabla u,\nabla v) \\
&\quad +\psi^{1+\delta}\lambda(u)\frac{\partial F}{\partial r}(v,r)|\nabla u|^{-1}\varphi(x,|\nabla u|)T(\nabla u,\nabla r) \\
&\geq \psi^{1+\delta}\lambda(u)F(v,r)K(1+r)^{-\mu}T_+(r) \\
&\quad -(1+\delta)\psi^\delta\lambda(u)F(v,r)\varphi(x,|\nabla u|)T_+(r)|\nabla\psi| \\
&\quad +\psi^{1+\delta}\lambda(u)\frac{\partial F}{\partial v}(v,r)|\nabla u|^{-1}\varphi(x,|\nabla u|)T(\nabla u,\alpha\sigma(1+r)^{\sigma-1}\nabla r-\nabla u) \\
&\quad +\psi^{1+\delta}\lambda(u)\frac{\partial F}{\partial r}(v,r)|\nabla u|^{-1}\varphi(x,|\nabla u|)T(\nabla u,\nabla r),
\end{aligned}$$

where to obtain the last inequality we have used (4.47), (4.57), (4.59), (4.62) and (4.65). Using now (4.47), (4.60) and (4.64) we obtain

$$\begin{aligned}
e^f \operatorname{div} W &\geq -(1+\delta)\psi^\delta\lambda(u)F(v,r)\varphi(x,|\nabla u|)T_+(r)|\nabla\psi| \\
&\quad +\psi^{1+\delta}\lambda(u)F(v,r)K(1+r)^{-\mu}T_+(r) \\
&\quad -\psi^{1+\delta}\lambda(u)\frac{\partial F}{\partial v}(v,r)\varphi(x,|\nabla u|)|\nabla u|T_-(r) \\
&\quad +\alpha\sigma(1+r)^{\sigma-1}\psi^{1+\delta}\lambda(u)\frac{\partial F}{\partial v}(v,r)|\nabla u|^{-1}\varphi(x,|\nabla u|)T(\nabla u,\nabla r) \\
&\quad +\psi^{1+\delta}\lambda(u)\frac{\partial F}{\partial r}(v,r)|\nabla u|^{-1}\varphi(x,|\nabla u|)T(\nabla u,\nabla r) \\
&\geq -(1+\delta)\psi^\delta\lambda(u)F(v,r)\varphi(x,|\nabla u|)T_+(r)|\nabla\psi| \\
&\quad +\psi^{1+\delta}\lambda(u)F(v,r)K(1+r)^{-\mu}T_+(r) \\
&\quad +\psi^{1+\delta}\lambda(u)\left|\frac{\partial F}{\partial v}(v,r)\right|\frac{T_-(r)}{A(x)^{1/\delta}}\varphi(x,|\nabla u|)^{1+1/\delta}
\end{aligned}$$

$$-\psi^{1+\delta}\lambda(u)\left|\frac{\partial F}{\partial v}(v,r)\right|\alpha\sigma(1+r)^{\sigma-1}|\nabla u|^{-1}\varphi(x,|\nabla u|)T(\nabla u,\nabla r)$$

$$+\psi^{1+\delta}\lambda(u)\left|\frac{\partial F}{\partial v}(v,r)\right|\frac{\frac{\partial F}{\partial r}(v,r)}{\left|\frac{\partial F}{\partial v}(v,r)\right|}|\nabla u|^{-1}\varphi(x,|\nabla u|)T(\nabla u,\nabla r).$$

Therefore,

$$\begin{aligned} e^{f}\operatorname{div}W \geq &-(1+\delta)\psi^{\delta}\lambda(u)F(v,r)\varphi(x,|\nabla u|)T_{+}(r)|\nabla\psi| \\ &+\psi^{1+\delta}\lambda(u)\left|\frac{\partial F}{\partial v}(v,r)\right|B(x,\nabla u,r), \end{aligned} \tag{4.66}$$

where

$$\begin{aligned} B(x,\nabla u,r) = &\frac{T_{-}(r)}{A(x)^{1/\delta}}\varphi(x,|\nabla u|)^{1+1/\delta} + \frac{F(v,r)}{\left|\frac{\partial F}{\partial v}(v,r)\right|}K(1+r)^{-\mu}T_{+}(r) \\ &+\left(\frac{\frac{\partial F}{\partial r}(v,r)}{\left|\frac{\partial F}{\partial v}(v,r)\right|}-\alpha\sigma(1+r)^{\sigma-1}\right)|\nabla u|^{-1}\varphi(x,|\nabla u|)T(\nabla u,\nabla r). \end{aligned} \tag{4.67}$$

Using (4.48) we obtain

$$\begin{aligned} B(x,\nabla u,r) \geq &\left(\frac{1}{\Sigma^{1/\delta}}\varphi(x,|\nabla u|)^{1+1/\delta} + \frac{F(v,r)}{\left|\frac{\partial F}{\partial v}(v,r)\right|}K(1+r)^{-\mu}\right)T_{+}(r) \\ &+\left(\frac{\frac{\partial F}{\partial r}(v,r)}{\left|\frac{\partial F}{\partial v}(v,r)\right|}-\alpha\sigma(1+r)^{\sigma-1}\right)|\nabla u|^{-1}\varphi(x,|\nabla u|)T(\nabla u,\nabla r). \end{aligned} \tag{4.68}$$

Next, we consider different cases.

Case I: $\eta < 0$. We choose

$$F(v,r) = e^{-qv(1+r)^{-\eta}}$$

where $q > 0$ is a constant that will be specified later. From (4.62), (4.63), (4.50) and $\alpha > 0$ we obtain

$$0 \geq \frac{\frac{\partial F}{\partial r}(v,r)}{\left|\frac{\partial F}{\partial v}(v,r)\right|} - \alpha\sigma(1+r)^{\sigma-1} \geq -\alpha(\sigma-\eta)(1+r)^{\sigma-1} \text{ on } \Omega_{\gamma} \tag{4.69}$$

and

$$\frac{F(v,r)}{\left|\frac{\partial F}{\partial v}(v,r)\right|} = \frac{1}{q}(1+r)^{\eta}. \tag{4.70}$$

We also note that

$$|\nabla u|^{-1}\varphi(x,|\nabla u|)T(\nabla u,\nabla r)\leq\varphi(x,|\nabla u|)T_{+}(r).$$

Inserting (4.69) and (4.70) into (4.68) and using (4.49) we deduce

$$\begin{aligned}B(x,\nabla u,r)\geq\Big[&\frac{1}{\Sigma^{1/\delta}}\varphi(x,|\nabla u|)^{1+1/\delta}+\frac{K}{q}(1+r)^{(\sigma-1)(1+\delta)}\\&-\alpha(\sigma-\eta)(1+r)^{\sigma-1}\varphi(x,|\nabla u|)\Big]T_{+}(r).\end{aligned}\tag{4.71}$$

At this point we need to estimate the right-hand side of (4.71) so to have

$$B(x,\nabla u,r)\geq\Lambda\varphi(x,|\nabla u|)^{1+1/\delta}T_{+}(r)\tag{4.72}$$

for some positive constant Λ independent of ∇u, r and x. For this purpose we use the next lemma whose proof is a calculus exercise.

Lemma 4.2 *Let $\delta,\varrho,\beta,\omega$ be positive constants and let $\hat{f}$ be the function defined on $\mathbb{R}_0^+$ by $\hat{f}(s)=\omega s^{1+1/\delta}-\beta s+\varrho$. Then the inequality $\hat{f}(s)\geq\Lambda s^{1+1/\delta}$ holds on $\mathbb{R}_0^+$ provided*

$$\Lambda\leq\omega-\frac{\delta\beta^{1+1/\delta}}{(1+\delta)^{1+1/\delta}\varrho^{1/\delta}}.$$

Applying Lemma 4.2 with $s=\varphi(x,|\nabla u|)$ and x fixed, it is easy to verify that (4.71) holds independently of x if we can choose a positive Λ such that

$$\Lambda\leq\frac{1}{\Sigma^{1/\delta}}-\frac{q^{1/\delta}\delta(\alpha(\sigma-\eta))^{1+1/\delta}}{(1+\delta)^{1+1/\delta}K^{1/\delta}}.\tag{4.73}$$

Note that the above is independent of $r=r(x)$. Thus, if $\tau\in(0,1)$ and we choose

$$q=\frac{\tau^{\delta}K(1+\delta)^{1+\delta}}{\Sigma\delta^{\delta}(\alpha(\sigma-\eta))^{1+\delta}}\quad\text{and}\quad\Lambda=\frac{1-\tau}{\Sigma^{1/\delta}}\tag{4.74}$$

then $\Lambda>0$ and it satisfies (4.73).

We insert (4.72) and the expression for $\partial F/\partial v$ into (4.66) to obtain

$$\begin{aligned}e^{f}\operatorname{div}W\geq&-(1+\delta)\psi^{\delta}\lambda(u)F(v,r)\varphi(x,|\nabla u|)T_{+}(r)|\nabla\psi|\\&+q\Lambda\psi^{1+\delta}\lambda(u)(1+r)^{-\eta}F(v,r)\varphi(x,|\nabla u|)^{1+1/\delta}T_{+}(r).\end{aligned}$$

We integrate this inequality on $\Omega_\gamma \cap B_R$, apply the divergence theorem and recall that W vanishes on $\partial(\Omega_\gamma \cap B_R)$ to obtain

$$\frac{q\Lambda}{1+\delta}\int_{\Omega_\gamma\cap B_R} \psi^{1+\delta}\lambda(u)(1+r)^{-\eta}F(v,r)\varphi(x,|\nabla u|)^{1+1/\delta}T_+(r)e^{-f} \leq \int_{\Omega_\gamma\cap B_R} \psi^{\delta}\lambda(u)F(v,r)\varphi(x,|\nabla u|)T_+(r)|\nabla\psi|e^{-f}.$$

Write

$$\psi^{\delta}\lambda(u)F(v,r)\varphi(x,|\nabla u|)T_+(r)|\nabla\psi|e^{-f} = g_1g_2$$

with

$$g_1 = (\lambda(u)F(v,r)T_+(r))^{\frac{1}{1+\delta}}\,|\nabla\psi|(1+r)^{\frac{\eta\delta}{1+\delta}}e^{-\frac{1}{1+\delta}f}$$

and

$$g_2 = (\lambda(u)F(v,r)T_+(r))^{\frac{\delta}{1+\delta}}\,\psi^{\delta}\varphi(x,|\nabla u|))(1+r)^{-\frac{\eta\delta}{1+\delta}}e^{-\frac{\delta}{1+\delta}f}.$$

Applying Hölder's inequality with conjugate exponents $1+\delta$ and $1+1/\delta$ to the integral on the right-hand side we obtain

$$\int_{\Omega_\gamma\cap B_R} \psi^{\delta}\lambda(u)F(v,r)\varphi(x,|\nabla u|)T_+(r)|\nabla\psi|e^{-f}$$
$$\leq \left(\int_{\Omega_\gamma\cap B_R} \lambda(u)F(v,r)T_+(r)|\nabla\psi|^{1+\delta}(1+r)^{\eta\delta}e^{-f}\right)^{\frac{1}{1+\delta}} \times$$
$$\left(\int_{\Omega_\gamma\cap B_R} \lambda(u)F(v,r)T_+(r)\psi^{1+\delta}\varphi(x,|\nabla u|)^{\frac{1+\delta}{\delta}}(1+r)^{-\eta}e^{-f}\right)^{\frac{\delta}{1+\delta}},$$

and, after some simplification, from the above we get

$$\left(\frac{q\Lambda}{1+\delta}\right)^{1+\delta}\int_{\Omega_\gamma\cap B_R} \psi^{1+\delta}\lambda(u)(1+r)^{-\eta}F(v,r)\varphi(x,|\nabla u|)^{1+1/\delta}T_+(r)e^{-f} \leq \int_{\Omega_\gamma\cap B_R} \lambda(u)F(v,r)(1+r)^{\eta\delta}|\nabla\psi|^{1+\delta}T_+(r)e^{-f}.$$

Let $R > 2R_0$; then $\theta R > R/2 > R_0$ and using the properties of λ and ψ we deduce

$$E = \left(\frac{q\Lambda}{1+\delta}\right)^{1+\delta} \int_{\Omega_\gamma \cap B_{R_0}} \lambda(u)F(v,r)\varphi(x,|\nabla u|)^{1+1/\delta}T_+(r)e^{-f} \tag{4.75}$$

$$\leq C^{1+\delta}(1+\theta R)^{\eta\delta}[(1-\theta)R]^{-(1+\delta)} \int_{\Omega_\gamma \cap (B_R \setminus B_{\theta R})} F(v,r)T_+(r)e^{-f}.$$

Using (4.63) for v and the expression of F on $\Omega_\gamma \cap (B_R \setminus B_{\theta R})$

$$F(v,r) \leq e^{-q(\alpha-b)(1+\theta R)^{\sigma-\eta}},$$

thus from (4.75)

$$E \leq \hat{C}R^{\delta\eta-1-\delta}e^{-q(\alpha-b)(1+\theta R)^{\sigma-\eta}} \int_{B_R} T_+(r)e^{-f}$$

for some constant $\hat{C} > 0$. Now observe that, since $|\nabla u| \not\equiv 0$ on $\Omega_\gamma \cap B_{R_0}$, $E > 0$. From assumption (4.52), for every fixed $d > d_0$ there exists a strictly increasing sequence $R_k \nearrow +\infty$ with $R_1 > 2R_0$ and such that

$$\log \int_{B_{R_k}} T_+(r)e^{-f} \leq dR_k^{\sigma-\eta}; \tag{4.76}$$

and from the above inequality with $R = R_k$ we obtain

$$0 < E \leq \hat{C}R_k^{\delta\eta-1-\delta}e^{-q(\alpha-b)(1+\theta R_k)^{\sigma-\eta}} \int_{B_{R_k}} T_+(r)e^{-f}$$

$$\leq \hat{C}R_k^{\delta\eta-1-\delta}e^{dR_k^{\sigma-\eta}-q(\alpha-b)(1+\theta R_k)^{\sigma-\eta}},$$

where the constant $\hat{C} > 0$ is independent of k. In order for this inequality to hold for every k, we must have

$$d \geq (\alpha-b)q\theta^{\sigma-\eta},$$

whence, letting $\theta \to 1$,

$$d \geq (\alpha-b)q.$$

We set $\alpha = tb$, with $t > 1$, and we insert the choice (4.74) of q in the above inequality, solve with respect to K, and let $\tau \nearrow 1$ to obtain

$$K \leq \Sigma d b^\delta(\sigma-\eta)^{1+\delta}\frac{\delta^\delta}{(1+\delta)^{1+\delta}}\frac{t^{1+\delta}}{t-1}.$$

Therefore, minimizing with respect to $t > 1$ and letting $d \to d_0$, $b \to \max\{\hat{u}, 0\}$, we have

$$K \leq \Sigma d_0 \max\{\hat{u}, 0\}^{\delta}(\sigma - \eta)^{1+\delta}.$$

In other words,

$$\inf_{\Omega_\gamma} \frac{(1+r)^{\mu}}{T_+(r)} Lu \leq \Sigma d_0 \max\{\hat{u}, 0\}^{\delta}(\sigma - \eta)^{1+\delta}. \tag{4.77}$$

This finishes the proof when $\sigma > 0$ and $\eta < 0$.

For $\sigma = 0$ [and necessarily $\eta < 0$ by (4.50)] we can improve the above estimate as follows. We apply (4.77) to the function $u - \hat{u}$ on the set

$$\{x \in M : u(x) - \hat{u} > \gamma - \hat{u}\} = \Omega_\gamma,$$

observing that $\widehat{u - \hat{u}} = 0$ and that $Lu = L(u - \hat{u})$, to obtain

$$\inf_{\Omega_\gamma} \frac{(1+r)^{\mu}}{T_+(r)} Lu \leq 0.$$

Case II: $\eta \geq 0$ [and necessarily $\sigma > 0$ by (4.50)]. We choose

$$F(v, r) = F(v) = e^{-qv^{(\sigma-\eta)/\sigma}}$$

where $q > 0$ is a constant to be specified later. Noting that the exponent of v is positive by (4.50), a computation yields

$$\frac{\partial F}{\partial v}(v, r) = -q\frac{\sigma - \eta}{\sigma} v^{-\eta/\sigma} F(v) < 0,$$

while clearly, $\partial F/\partial r \equiv 0$.

Using estimate (4.68), recalling that by (4.63) $v \geq (\alpha - b)(1+r)^{\sigma}$ and proceeding as in Case I, we estimate

$$B(x, \nabla u, r) \geq \left[\frac{1}{\Sigma^{1/\delta}}\varphi(x, |\nabla u|)^{1+1/\delta} - \alpha\sigma(1+r)^{\sigma-1}\varphi(x, |\nabla u|) + \frac{\sigma}{q(\sigma - \eta)}(\alpha - b)^{\eta/\sigma} K(1+r)^{(\sigma-1)(1+\delta)}\right] T_+(r). \tag{4.78}$$

According to Lemma 4.2, for every $r \geq 0$ fixed, the right-hand side of the above inequality is bounded from below by $\Lambda\varphi(x, |\nabla u|)^{1+1/\delta} T_+(r)$ provided

$$\Lambda \leq \frac{1}{\Sigma^{1/\delta}} - \frac{q^{1/\delta}\delta(\alpha\sigma)^{1+1/\delta}(\sigma - \eta)^{1/\delta}}{(1+\delta)^{1+1/\delta}(K\sigma)^{1/\delta}(\alpha - b)^{\eta/(\delta\sigma)}}. \tag{4.79}$$

Since the right-hand side of the above inequality is independent of r, for every such Λ we have $B(x, \nabla u, r) \geq \Lambda\varphi(x, |\nabla u|)^{1+1/\delta}T_+(r)$. In particular, if $\tau \in (0, 1)$ and we choose

$$q = \frac{\tau^\delta K\sigma(1+\delta)^{1+\delta}(\alpha - b)^{\eta/\sigma}}{\Sigma\delta^\delta(\alpha\sigma)^{1+\delta}(\sigma - \eta)} \text{ and } \Lambda = \frac{1-\tau}{\Sigma^{1/\delta}} \tag{4.80}$$

then $\Lambda > 0$ and it satisfies (4.79). Substituting into (4.66), and using the expression for $\partial F/\partial v$, we deduce that

$$\begin{aligned} e^f \operatorname{div} W \geq &-(1+\delta)\psi^\delta\lambda(u)F(v)\varphi(x, |\nabla u|)T_+(r)|\nabla\psi| \\ &+q\frac{\sigma-\eta}{\sigma}\Lambda\psi^{1+\delta}\lambda(u)v^{\eta/\sigma}F(v)\varphi(x, |\nabla u|)^{1+1/\delta}T_+(r). \end{aligned}$$

We now proceed as in Case I, repeating, with minor adaptations, the arguments that lead to (4.75), to conclude instead that

$$\begin{aligned} 0 < E = &\int_{\Omega_\gamma \cap B_{R_0}} \lambda(u)F(v)\varphi(x, |\nabla u|)^{1+1/\delta}T_+(r)e^{-f} \qquad (4.81) \\ \leq \hat{C}(1+\theta R_k)^{\eta\delta}[(1-\theta)R_k]^{-(1+\delta)} &\int_{\Omega_\gamma \cap (B_{R_k} \setminus B_{\theta R_k})} F(v)T_+(r)e^{-f}, \end{aligned}$$

where $\hat{C}$ is a constant independent of k and θ. Using the inequality

$$F(v) \leq e^{-q(\alpha-b)^{(\sigma-\eta)/\sigma}(1+\theta R_k)^{\sigma-\eta}}$$

valid on $\Omega_\gamma \cap (B_{R_k} \setminus B_{\theta R_k})$, and (4.76), we conclude that for every k we have

$$0 < E \leq \hat{C}R_k^{\delta\eta-1-\delta}e^{dR_k^{\sigma-\eta}-q(\alpha-b)^{(\sigma-\eta)/\sigma}(1+\theta R_k)^{\sigma-\eta}}.$$

Again, this forces

$$d \geq q(\alpha - b)^{(\sigma-\eta)/\sigma}\theta^{\sigma-\eta}.$$

Therefore, setting $\alpha = tb$, with $t > 1$, letting $\theta \nearrow 1$, inserting the value of q given by (4.80), solving with respect to K and letting $\tau \nearrow 1$, $d \searrow d_0$, $b \searrow \max\{\hat{u}, 0\}$, we obtain

$$K \leq \Sigma d_0 \max\{\hat{u}, 0\}^\delta \frac{(\delta\sigma)^\delta(\sigma-\eta)}{(1+\delta)^{1+\delta}}\frac{t^{1+\delta}}{t-1};$$

whence, again minimizing with respect to $t > 1$, we conclude that

$$K \leq \Sigma d_0 \max\{\hat{u}, 0\}^\delta\sigma^\delta(\sigma - \eta).$$

In other words,

$$\inf_{\Omega_\gamma} \frac{(1+r)^\mu}{T_+(r)} Lu \le \Sigma d_0 \max\{\hat{u}, 0\}^\delta \sigma^\delta (\sigma - \eta).$$

This finishes the proof when $\eta \ge 0$. □

To see how one can use Theorem 4.4 let us consider the following geometric setting.

Let $(N, \langle\, ,\, \rangle)$ be a $(m+1)$-dimensional Riemannian manifold endowed with a nonsingular Killing vector field Y with complete flow lines such that the orthogonal distribution $D : N \to TN$, that is

$$D : x \to D_x = \{v \in T_xN : \langle Y_x, v \rangle = 0\} \subset T_xN$$

is integrable. It is not difficult to verify that the (maximal) integral leaves of D are totally geodesic hypersurfaces in N [95]. In particular, if N is complete, then any leaf is complete. We fix an integral leaf M; the flow $\Phi : \mathbb{R} \times M \to N$ generated by Y takes isometrically $M = M_0$ to the leaf $M_s = \phi_s(M_0)$ for any $s \in \mathbb{R}$, where $\phi_s = \Phi(s, \cdot)$. Then, given $u \in C^\infty(M)$, the *Killing graph* Γ_u associated to u is the hypersurface

$$\Gamma_u : M \to N$$

given by

$$\Gamma_u : x \to \Phi(u(x), x).$$

One can show (see for instance [95]) that the Killing graph Γ_u has mean curvature H if and only if u satisfies the equation

$$\operatorname{div}\left(\frac{\nabla u}{W}\right) - \left\langle \frac{\nabla \gamma}{2\gamma}, \frac{\nabla u}{W} \right\rangle = mH, \tag{4.82}$$

where

$$\gamma(x) = \frac{1}{|Y(x)|^2}, \tag{4.83}$$

$$W(x) = \sqrt{\gamma(x) + |\nabla u(x)|^2} \tag{4.84}$$

and the mean curvature H is computed with respect to the orientation given by the normal

$$\nu = \frac{1}{W}(\gamma Y - \Phi_*(\nabla u)). \tag{4.85}$$

Note that the operators div and ∇ are on M with the metric induced by the inclusion $M = M_0 \hookrightarrow N$.

In the special case of a product $N = \mathbb{R} \times M$, with $(M, \langle\, , \rangle)$ a Riemannian manifold, indicating with s the (global) coordinate on $\mathbb{R}$ we can choose $Y = \frac{\partial}{\partial s}$ so that $|Y| \equiv 1$ and

$$\Phi(s, x) = (s, x).$$

In this case, for the Killing graph

$$\Gamma_u : x \to (u(x), x),$$

Eq. (4.82) reduces to the well-known *mean curvature equation*

$$\operatorname{div}\left(\frac{\nabla u}{\sqrt{1 + |\nabla u|^2}}\right) = mH. \tag{4.86}$$

When N is complete, we fix an origin $o \in M = M_0$ and set $r(x) = \operatorname{dist}_M(x, o)$. We have (see [24])

Theorem 4.5 *Let N be a complete Riemannian manifold endowed with a complete nonsingular Killing field Y and let M be an integral leaf of the Killing foliation. Assume that*

$$\sup_M |Y| < +\infty \tag{4.87}$$

and

$$\liminf_{R \to +\infty} \frac{\log \int_{B_R} |Y|}{R^{2-\sigma}} = 0 \tag{4.88}$$

for some $0 \le \sigma < 2$. Then, any constant mean curvature Killing graph $\Gamma_u(x) = \Phi(u(x), x)$, $x \in M$, lying between the graphs $\Gamma_{\sigma,\pm\beta}(x) = \Phi(\pm\beta r(x)^\sigma, x)$ outside a compact set of M, for some $\beta > 0$, is minimal.

Proof We note that for the Killing graph Γ_u we have the validity of (4.82) for some constant H. Passing to $-u$ if necessary, we may assume that $H > 0$. Now observe that since N is complete then M is complete. To apply Theorem 4.4 we choose T to be the metric on M so that T_- and T_+ are both identically equal to 1, we let $f = \log \sqrt{\gamma} = -\log |Y|$ and we define

$$\varphi(x, t) = \frac{t}{\sqrt{\gamma(x) + t^2}}.$$

Then φ clearly satisfies (i) and (ii) of Theorem 4.4 and since $\gamma(x) > 0$ for each x, $\varphi(x, \ ) \in C^0(\mathbb{R}_0^+) \cap C^1(\mathbb{R}^+)$ for each $x \in M$. Furthermore, since

$$\varphi(x,t) \le \frac{1}{\sqrt{\gamma(x)}} t,$$

it satisfies (iii) with the choices $\delta = 1$, $A(x) = |Y(x)|$. Thus assumption (4.87) guarantees the validity of (4.48). Since Γ_u lies between the graphs $\Gamma_{\sigma,\pm\beta}$ we have

$$\hat{u} = \limsup_{r(x)\to+\infty} \frac{u(x)}{r(x)^\sigma} \le \beta < +\infty.$$

We now let $\eta = 2(\sigma - 1)$ and observe that

$$\sigma \ge 0, \qquad \sigma - \eta = 2 - \sigma > 0.$$

Furthermore, (4.88) corresponds to (4.52) with $d_0 = 0$. Next we choose any ζ such that

$$\Omega_\zeta = \{x \in M : u(x) > \zeta\} \neq \emptyset.$$

By applying Theorem 4.4 we have

$$0 \ge \inf_{\Omega_\zeta} Lu = \inf_{\Omega_\zeta} \left\{ \operatorname{div}\left(\frac{\nabla u}{W}\right) - \left\langle \frac{\nabla\gamma}{2\gamma}, \frac{\nabla u}{W} \right\rangle \right\} = H$$

so that $H = 0$. □

We note that condition (4.88) cannot be relaxed. Indeed, consider the case $N = \mathbb{R} \times \mathbb{H}^m$, $Y = \frac{\partial}{\partial s}$, s the coordinate on $\mathbb{R}$ and $M = \mathbb{H}^m$ the hyperbolic space of constant sectional curvature -1. Then, realizing the metric of $\mathbb{H}^m$ in polar coordinates $(r, \theta) \in \mathbb{R}^+ \times \mathbb{S}^{m-1}$ as

$$\langle\, , \,\rangle = dr^2 + \sinh^2 r d\theta^2,$$

where $d\theta^2$ is the canonical metric on $\mathbb{S}^{m-1}$, we have

$$\operatorname{vol} B_R \sim Ce^{(m-1)R} \quad \text{as } R \to +\infty$$

for some constant $C > 0$. Since $|Y| \equiv 1$ we have

$$\liminf_{R\to+\infty} \frac{\log \int_{B_R} |Y|}{R^{2-\sigma}} = \begin{cases} +\infty & \text{if } \sigma \in (1,2), \\ C(m-1) & \text{if } \sigma = 1, \\ 0 & \text{if } \sigma \in [0,1). \end{cases}$$

Now, for any $H \in \left(0, \frac{m-1}{m}\right]$, the smooth function

$$u(x) = \int_0^{r(x)} \frac{\sinh^{1-m}(t) \int_0^t mH \sinh^{m-1}(s)\, ds}{\left\{1 - \sinh^{2(1-m)}(t)\left(\int_0^t mH \sinh^{m-1}(s)\, ds\right)^2\right\}^{\frac{1}{2}}}\, dt$$

defines an entire graph on $\mathbb{H}^m$ with constant mean curvature H. Furthermore, $u(x) \sim r(x)$ as $r(x) \to +\infty$, which means that the graph lies between the graphs $\psi_{1,\pm 2}(x) = (\pm 2r(x), x)$ outside a compact set of $\mathbb{H}^m$.

Remark 4.4 The problem of the existence of a Killing graph with nonzero constant mean curvature H is related to the value of the (appropriately weighted) Cheeger's constant of the leaf M of the Killing foliation. This is explained in detail in [24].

4.3 An Equivalent Open Form of the Weak Maximum Principle

The aim of this section, partially based on the recent [28], is to present another form of the weak maximum principle which turns out to be very useful in geometric applications. We focus our attention on the general class of operators that have been defined in Sect. 4.2 (see Eq. (4.53), with $f = 0$) and that we consider, for instance, in [5, 24]. For the sake of completeness, and for the ease of reading, we recall the definition once more.

We let T be a symmetric, 2-covariant tensor field on a Riemannian manifold $(M, \langle\ ,\ \rangle)$. Assume that, for some continuous functions T_- and T_+ on $\mathbb{R}_0^+$, the tensor T satisfies the following bounds

$$0 < T_-(r) \le T(Y, Y) \le T_+(r) \tag{4.89}$$

for each $Y \in T_xM$, $|Y| = 1$, and every $x \in \partial B_r$, where B_r denotes the geodesic ball of radius r centered at an origin o. Let $\varphi : M \times \mathbb{R}_0^+ \to \mathbb{R}_0^+$ be such that $\varphi(\ , t) \in C^0(M)$ for each $t \in \mathbb{R}_0^+$, $\varphi(x,\ ) \in C^0(\mathbb{R}_0^+) \cap C^1(\mathbb{R}^+)$ for each $x \in M$ and

$$\begin{cases} \text{(i) } \varphi(x, 0) = 0 \text{ for every } x \in M; \\ \text{(ii) } \varphi(x, t) > 0 \text{ on } M \times \mathbb{R}^+; \\ \text{(iii) } \varphi(x, t) \le A(x)t^\delta \text{ on } M \times \mathbb{R}^+ \end{cases} \tag{4.90}$$

for some $\delta > 0$ and $A(x) \in C^0(M)$, $A(x) > 0$. Let X be a vector field on M. For $u \in C^1(M)$ we define

$$Lu = L_{\varphi,T,X}u = \operatorname{div}\left(|\nabla u|^{-1}\varphi(x, |\nabla u|)T(\nabla u, \ )^\sharp\right) - \langle X, \nabla u\rangle \tag{4.91}$$

in the weak sense, where $^\sharp : T^*M \to TM$ denotes the musical isomorphism.

Remark 4.5 Note that the left-hand side inequality in (4.89) and requirement (ii) in (4.90) are ellipticity conditions for the operator L. As a matter of fact properties (4.89) and (4.90) will not be used in proving the equivalence in Theorem 4.6 below. On the other hand, they are basic in looking for sufficient conditions to guarantee that the property expressed in Definition 4.1 below holds on the manifold we are considering. In fact, when $\varphi(x,t) = t$ it is enough to consider $u \in \mathrm{Lip}_{loc}(M)$; the more restrictive $u \in C^1(M)$ enables us to deal with the nonlinear case. Furthermore, for those theorems giving sufficient conditions in terms of the volume growth of geodesic balls we can enlarge the class of admissible solutions to $C^0(M) \cap W^{1,1+\delta}_{loc}(M)$. This is due to the fact that the argument of proof for these results is based only on the notion of weak solution.

In what follows we recall the next concept introduced immediately after the statement of Theorem 3.1 in Chap. 3. Let $q(x) \in C^0(M)$, $q(x) > 0$.

Definition 4.1 We say that the *q-WMP* (the *q-weak maximum principle*) holds on M for the operator L in (4.91) if, for each $u \in C^1(M)$ with $u^* = \sup_M u < +\infty$ and for each $\gamma \in \mathbb{R}$ with $\gamma < u^*$, we have

$$\inf_{\Omega_\gamma}\{q(x)Lu\} \le 0 \tag{4.92}$$

in the weak sense, where

$$\Omega_\gamma = \{x \in M : u(x) > \gamma\}. \tag{4.93}$$

Note that (4.92) in the weak sense expresses as follows: for every $\varepsilon > 0$

$$-\int_{\Omega_\gamma}\left(|\nabla u|^{-1}\varphi(x, |\nabla u|)T(\nabla u, \nabla\psi) + \langle X, \nabla u\rangle\psi\right) \le \int_{\Omega_\gamma}\frac{\varepsilon}{q(x)}\psi, \tag{4.94}$$

for some $\psi \in C^\infty_c(\Omega_\gamma)$, $\psi \ge 0$, $\psi \not\equiv 0$. In case $q(x)$ is a positive constant we will simply say that L satisfies the WMP (the weak maximum principle).

Of course Remark 3.1 applies.

We observe that, from an analytic point of view, the usual form of the maximum principle loosely states that, for a certain operator, say L, if u satisfies for instance

$$Lu \ge 0$$

on a region Ω, then

$$\sup_{\Omega} u = \sup_{\partial\Omega} u$$

(see Chap. 10 in [125] for a good reference). In the previous chapters we took on Yau's point of view based on the observation that if a C^2-function u attains its maximum at x_0, then

$$\nabla u(x_0) = 0, \qquad \Delta u(x_0) \leq 0,$$

and we formulated our form of the maximum principle accordingly to the Omori-Yau philosophy; see for instance Theorems 2.4 and 3.2. As we will see in a shortwhile, with the next result we basically go back to the original point of view, at least for the weak maximum principle.

We remark that the various assumptions on φ and T given in the definition of $L_{\varphi,T,X}$ will be very marginally used in some of the forthcoming results. In fact in the next theorem we will only use the property $Lu = L(u + a)$ for any constant $a \in \mathbb{R}$.

Theorem 4.6 *The q-WMP holds on M for the operator L if and only if the* open q-WMP *holds on M, that is, for each $f \in C^0(\mathbb{R})$, for each open set $\Omega \subset M$ with $\partial\Omega \neq \emptyset$, and for each $v \in C^0(\overline{\Omega}) \cap C^1(\Omega)$ satisfying*

$$\begin{cases} (i)\ q(x)Lv \geq f(v) \text{ on } \Omega; \\ (ii)\ \sup_{\Omega} v < +\infty, \end{cases} \tag{4.95}$$

we have that either

$$\sup_{\Omega} v = \sup_{\partial\Omega} v \tag{4.96}$$

or

$$f(\sup_{\Omega} v) \leq 0. \tag{4.97}$$

Remark 4.6 Observe that the q-WMP on M for the operator L is also equivalent to the following dual statement: The q-WMP holds on M for the operator L if and only if for each $f \in C^0(\mathbb{R})$, for each open set $\Omega \subset M$ with $\partial\Omega \neq \emptyset$, and for each $v \in C^0(\overline{\Omega}) \cap C^1(\Omega)$ satisfying

$$\begin{cases} \text{(i) } q(x)Lv \leq f(v) \text{ on } \Omega; \\ \text{(ii) } \inf_{\Omega} v > -\infty, \end{cases} \tag{4.98}$$

we have that either

$$\inf_{\Omega} v = \inf_{\partial\Omega} v \tag{4.99}$$

or

$$f(\inf_{\Omega} v) \geq 0. \tag{4.100}$$

Proof (of Theorem 4.6) Assume that the q-WMP holds for the operator L on M and let f, v and Ω be as in the statement of the theorem. Suppose that (4.96) is not satisfied, that is

$$\sup_{\Omega} v > \sup_{\partial\Omega} v. \tag{4.101}$$

Fix $\varepsilon > 0$ sufficiently small that

$$\sup_{\Omega} v - 2\varepsilon > \sup_{\partial\Omega} v + 2\varepsilon \tag{4.102}$$

and define

$$U_{2\varepsilon} = \{x \in \Omega : v(x) > \sup_{\Omega} v - 2\varepsilon\}. \tag{4.103}$$

Note that $U_{2\varepsilon} \neq \emptyset$. Moreover, for every $x \in \overline{U}_{2\varepsilon}$ from (4.102) one has

$$v(x) \geq \sup_{\Omega} v - 2\varepsilon > \sup_{\partial\Omega} v + 2\varepsilon > \sup_{\partial\Omega} v,$$

so that $x \in \Omega$. That is, $\overline{U}_{2\varepsilon} \subset \Omega$, and therefore

$$\overline{U}_{\varepsilon} \subset U_{2\varepsilon} \subset \overline{U}_{2\varepsilon} \subset \Omega,$$

where U_{ε} obviously is defined in a way similar to (4.103).

By adding, if necessary, a positive constant to v, we can suppose that $\sup_{\Omega} v > 2\varepsilon$ and we let $\gamma = \sup_{\Omega} v - \varepsilon > 0$. Next we choose a smooth cut-off function $\psi : M \to [0, 1]$ such that

$$\psi \equiv 1 \text{ on } U_{\varepsilon} \quad \text{and} \quad \psi \equiv 0 \text{ on } M \setminus U_{2\varepsilon}$$

and we define

$$u(x) = \begin{cases} \psi(x)v(x) \text{ on } \Omega, \\ 0 \text{ on } M \setminus \Omega. \end{cases} \tag{4.104}$$

Then $u \in C^1(M)$, $u^* < +\infty$ and

$$Lu = Lv \quad \text{on } U_\varepsilon. \tag{4.105}$$

We claim that

$$\Omega_\gamma = \{x \in M : u(x) > \gamma\} = U_\varepsilon = \{x \in \Omega : v(x) > \gamma = \sup_\Omega v - \varepsilon\}. \tag{4.106}$$

Clearly it suffices to show that $\Omega_\gamma \subset U_\varepsilon$. For every $x \in \Omega_\gamma$ one has $u(x) > \gamma > 0$. In particular, by (4.104), it follows that $x \in \Omega$ and $v(x) > 0$, so that

$$v(x) \geq \psi(x)v(x) = u(x) > \gamma = \sup_\Omega v - \varepsilon.$$

Since $x \in \Omega$, this means that $x \in U_\varepsilon$.

Now for any constant $a \in \mathbb{R}$, $L(v + a) = Lv$, thus using (4.105) and (4.95) we deduce

$$Lu = L(v + a) = Lv \geq \frac{1}{q(x)} f(v) \quad \text{on } \Omega_\gamma.$$

In other words

$$q(x)Lu \geq f(v) \quad \text{on } \Omega_\gamma.$$

Applying the q-WMP to u we infer

$$0 \geq \inf_{\Omega_\gamma}\{q(x)Lu\} \geq \inf_{\Omega_\gamma} f(v).$$

But $\Omega_\gamma = U_\varepsilon$ and thus, letting $\varepsilon \to 0^+$ and using continuity of f we obtain (4.97).

For the converse, assume the validity of the open q-WMP for L. We reason by contradiction and we suppose that the q-WMP is false. Then, there exists $u \in C^1(M)$ with $u^* < +\infty$, and $\gamma < u^*$ such that

$$\beta = \inf_{\Omega_\gamma}\{q(x)Lu\} > 0. \tag{4.107}$$

This implies that u is nonconstant and therefore, since β is increasing with γ, up to choosing γ sufficiently near to u^*, we can suppose that

$$\partial\Omega_\gamma = \{x \in M : u(x) = \gamma\} \neq \emptyset.$$

Set $\Omega = \Omega_\gamma$ and $v = u|_{\overline{\Omega}}$. Because of (4.107) and $u^* < +\infty$ we have

$$\begin{cases} q(x)Lv \geq \beta \text{ on } \Omega, \\ \sup_\Omega v = u^* < +\infty. \end{cases} \tag{4.108}$$

Since $f(v) \equiv \beta > 0$, alternative (4.97) cannot occur. However alternative (4.96) cannot occur either because

$$\sup_\Omega v = u^* > \gamma = \sup_{\partial\Omega} v.$$

This yields the desired contradiction. □

Remark 4.7 Note that the above proof works for any of the choices of the functional classes of the solutions that we have been considering in Remark 4.5. Of course in Definition 4.1 we have to enlarge the functional class accordingly.

A careful reading of the above proof yields the validity of the following form of the theorem useful in applications.

Theorem 4.7 *The q-WMP holds on M for the operator L if and only if for each $\beta \in \mathbb{R}^+$, for each open set $\Omega \subset M$ with $\partial\Omega \neq \emptyset$, and for each $v \in C^0(\overline{\Omega}) \cap C^1(\Omega)$ satisfying*

$$\begin{cases} (i)\ q(x)Lv \geq \beta \text{ on } \Omega; \\ (ii)\ \sup_\Omega v < +\infty, \end{cases} \tag{4.109}$$

we have

$$\sup_\Omega v = \sup_{\partial\Omega} v. \tag{4.110}$$

The following fact seems worth mentioning; it extends Proposition 3.4 in [225] to general operators.

Proposition 4.1 *Let $(M, \langle\, ,\, \rangle_M)$ and $(N, \langle\, ,\, \rangle_N)$ be noncompact Riemannian manifolds, and assume that there exist compact sets $A \subset M$ and $B \subset N$ and a Riemannian isometry $f : M \setminus A \to N \setminus B$ which preserves divergent sequences in the ambient spaces, that is, $\{x_k\}$ diverges in M if and only if $\{f(x_k)\}$ diverges in N. Let X be a vector field on M, T a symmetric $(0, 2)$ tensor field on M satisfying (4.89) and φ as in (4.90) that define the differential operator $L_{\varphi,T,X}$ on M; let Y, S, ψ be with the same properties on N and define the differential operator $L_{\psi,S,Y}$ on N. Assume that*

$$Y = f_*X, \quad S = f_*T, \quad \psi(y, t) = \varphi(f^{-1}(y), t)$$

on $N \setminus B$. Then the WMP holds on M for the operator $L_{\varphi,T,X}$ if and only if the WMP holds on N for the operator $L_{\psi,S,Y}$.

Observe that the condition that $\{x_k\}$ diverges in M if and only if $\{f(x_k)\}$ diverges in N makes sense for any divergent sequence in M even if f is not globally defined on M because the sequence eventually leaves the compact set A.

Proof Suppose that the WMP holds on M for the operator $L_{\varphi,T,X}$. Let $v \in C^1(N)$ with $v^* < +\infty$. Without loss of generality we may assume that v^* is not attained and strictly positive. Consider K_1, K_2 be two relatively compact domains in M such that $A \subseteq K_1 \subseteq \overline{K}_1 \subseteq K_2$. Choose a smooth cutoff function $\lambda : M \to [0, 1]$ satisfying $\lambda \equiv 0$ on K_1, $\lambda \equiv 1$ on $M \setminus K_2$, and define a function $u \in C^1(M)$ by

$$u = \begin{cases} \lambda(v \circ f), & \text{on } M \setminus A; \\ 0, & \text{on } A. \end{cases} \tag{4.111}$$

We claim that $v^* = u^*$ and that u^* is not attained. By construction $v^* \leq u^*$. On the other hand, let $\{\overline{y}_k\}$ be a sequence in N such that $v(\overline{y}_k) \nearrow v^*$. Since v does not attain v^* the sequence $\{\overline{y}_k\}$ is divergent, therefore for k sufficiently large $\overline{y}_k$ lies outside B. By the assumption on f, $\{\overline{x}_k\} = \{f^{-1}(\overline{y}_k)\}$ is a divergent sequence in M. Thus $u(\overline{x}_k) = \lambda(\overline{x}_k)(v \circ f(\overline{x}_k)) = v(\overline{y}_k)$ for k sufficiently large, showing that $u(\overline{x}_k) \nearrow v^*$, and $v^* = u^*$. Furthermore, u^* is not attained, indeed, $u(x) = 0$ on A, and $u(x) \leq v(f(x)) < v^* = u^*$ on $M \setminus A$, hence u does not attain u^*, as claimed. Therefore, we can fix $\gamma < v^*$ sufficiently close to v^* such that

$$\Sigma_\gamma = \{y \in N : v(y) > \gamma\} \subset N \setminus (B \cup f(\overline{K}_2 \setminus A))$$

and consider $f^{-1}(\Sigma_\gamma) = \{x \in M \setminus A : (v \circ f)(x) > \gamma\}$.

Since $v^* = u^* > 0$ we can suppose that $\gamma > 0$ and it follows that

$$\Omega_\gamma = \{x \in M : u(x) > \gamma\} = \{x \in M \setminus A : \lambda(x)(v \circ f)(x) > \gamma\}.$$

In particular $(v \circ f)(x) > \gamma$ so that $\Omega_\gamma \subseteq f^{-1}(\Sigma_\gamma)$.

The validity of the WMP on M, yields that, for each $\varepsilon > 0$ there exist some $\tilde{\psi} \in C_c^\infty(\Omega_\gamma)$, $\tilde{\psi} \geq 0$, $\tilde{\psi} \not\equiv 0$ such that

$$\begin{aligned}
\int_{F^{-1}(\Sigma_\gamma)} \varepsilon\tilde{\psi} = \int_{\Omega_\gamma} \varepsilon\tilde{\psi} &\geq -\int_{\Omega_\gamma} \Big(|\nabla u|^{-1}\varphi(x, |\nabla u|)T(\nabla u, \nabla\tilde{\psi}) + \langle X, \nabla u\rangle\tilde{\psi}\Big) \\
&= -\int_{f^{-1}(\Sigma_\gamma)} \Big(|\nabla u|^{-1}\varphi(x, |\nabla u|)T(\nabla u, \nabla\tilde{\psi}) + \langle X, \nabla u\rangle\tilde{\psi}\Big) \\
&= -\int_{\Sigma_\gamma} \Big\{|\nabla(u \circ f^{-1})|^{-1}\varphi(f^{-1}(y), |\nabla(u \circ f^{-1})|)T(\nabla(u \circ f^{-1}), \nabla(\tilde{\psi} \circ f^{-1}))\Big\} \\
&\quad - \int_{\Sigma_\gamma} \Big\{\langle X \circ f^{-1}, \nabla(u \circ f^{-1})\rangle\tilde{\psi} \circ f^{-1}\Big\}.
\end{aligned}$$

But for $y \in \Sigma_\gamma, f^{-1}(y) \in M \setminus K_2$ hence

$$u(f^{-1}(y)) = (v \circ f)(f^{-1}(y)) = v(y);$$

$$\tilde{\varphi} = \tilde{\psi} \circ f^{-1} \in C_c^\infty(\Sigma_\gamma), \tilde{\varphi} \geq 0, \tilde{\varphi} \not\equiv 0;$$

$$X \circ f^{-1} = Y \text{ and } T(\nabla(u \circ f^{-1}), \nabla\tilde{\varphi}) = S(\nabla v, \nabla\tilde{\varphi}).$$

Therefore, being f an isometry,

$$\int_{\Sigma_\gamma} \varepsilon\tilde{\varphi} \geq -\int_{\Sigma_\gamma} \left(|\nabla v|^{-1}\psi(y, |\nabla v|)S(\nabla v, \nabla\tilde{\varphi}) + \langle Y, \nabla v\rangle\tilde{\varphi}\right).$$

This proves that the WMP holds for the operator $L_{\psi,S,Y}$ on N.

Repeating the same argument with M and N interchanged shows that if the WMP holds in N, so it holds in M (note that $f^{-1} : N \setminus B \to M \setminus A$ is a Riemannian isometry which maps divergent sequences to divergent sequences). □

4.3.1 A First Application to PDE's

The open form of the weak maximum principle will be applied in the sequel in some geometric context. However, it seems interesting to present here a uniqueness result for positive solutions of certain PDE's obtained via its use.

Towards this aim we recall that a *Lichnerowicz-type equation* is a PDE of the form

$$\Delta u + a(x)u - b(x)u^\sigma + c(x)u^\tau = 0 \tag{4.112}$$

for some $a(x)$, $b(x)$, $c(x)$ continuous on the Riemannian manifold $(M, \langle\, ,\, \rangle)$. Here $\sigma > 1$ and $\tau < 1$, so that the latter can be negative too. For the sake of simplicity we consider positive C^2-solutions of (4.112) on an open set $\Omega \subset M$, possibly with boundary.

Equations of the type of (4.112) arise in the analysis of the Einstein field equations in General Relativity, in the initial data set for the nonlinear wave system, and the coefficients $a(x)$, $b(x)$, $c(x)$ have a precise physical meaning. In particular, in some models $b(x)$ and $c(x)$ are, respectively, positive and nonnegative. This fact will somehow justify our assumptions in Theorem 4.8 below. For details we refer to [4, 90, 174] and the references therein.

Our uniqueness result will be an immediate consequence of the following comparison theorem (see [4]). We recall that, given a vector field X on M, Δ_X, the X-Laplacian, is the operator $\Delta_X u = \Delta u - \langle X, \nabla u\rangle$, with, say, $u \in C^2(M)$; see also Sect. 3.1.

Theorem 4.8 *Let $(M, \langle\,,\,\rangle)$ be a complete manifold, $a(x)$, $b(x)$, $c(x) \in C^0(M)$, X a vector field on M, $\sigma, \tau \in \mathbb{R}$ be such that $\sigma > 1$ and $\tau < 1$. Let Ω be a relatively compact open set in M. Assume*

$$(i)\, b(x) > 0 \quad on\ M \setminus \Omega; \qquad (ii)\, c(x) \geq 0 \quad on\ M \setminus \Omega; \tag{4.113}$$

$$(i)\ \sup_M \frac{a_-(x)}{b(x)} < +\infty; \qquad (ii)\ \sup_M \frac{c(x)}{b(x)} < +\infty, \tag{4.114}$$

where a_- denotes the negative part of a. Let $u, v \in C^2(M \setminus \overline{\Omega}) \cap C^0(M \setminus \Omega)$ be positive solutions of

$$\begin{cases} \Delta_X u + a(x)u - b(x)u^\sigma + c(x)u^\tau \geq 0 \\ \Delta_X v + a(x)v - b(x)v^\sigma + c(x)v^\tau \leq 0 \end{cases} \tag{4.115}$$

on $M \setminus \overline{\Omega}$ satisfying

$$(i)\ \liminf_{x\to\infty} v(x) > 0, \qquad (ii)\ \limsup_{x\to\infty} u(x) < +\infty \tag{4.116}$$

and

$$0 < \inf_{\partial\Omega} u \leq u(x) \leq v(x) \qquad on\ \partial\Omega. \tag{4.117}$$

Assume the validity of the $\frac{1}{b}$-WMP for the operator Δ_X on $M \setminus \overline{\Omega}$. Then

$$u(x) \leq v(x) \qquad on\ M \setminus \Omega. \tag{4.118}$$

Remark 4.8 As it will be clear from the proof, in case $0 \leq \tau < 1$ assumption (4.114) (ii) can be dropped. Furthermore, if $\Omega = \emptyset$ assumption (4.117) is empty.

Proof To simplify the writing set L for $\Delta_X = \Delta - \langle X, \nabla\rangle$; furthermore, without loss of generality observe that we can suppose that $M \setminus \overline{\Omega}$ is connected. From positivity of v, (4.116) (i), (ii) and (4.117) there exist constants $C_1, C_2 > 0$ such that

$$v(x) \geq C_1, \qquad u(x) \leq C_2 \quad \text{on } M \setminus \Omega. \tag{4.119}$$

We set

$$\xi = \sup_{M\setminus\overline{\Omega}} \left(\frac{u}{v}\right).$$

From the assumptions on u, v and (4.119) it follows that ξ satisfies

$$0 < \xi < +\infty. \tag{4.120}$$

Clearly if $\xi \leq 1$ then $u \leq v$ on $M \setminus \Omega$. By contradiction assume that $\xi > 1$ and define

$$w = u - \xi v;$$

then $w \leq 0$ on $M \setminus \Omega$. It is a simple matter to realize, using (4.120) and the definition of ξ, that

$$\sup_{M \setminus \Omega} w = 0. \tag{4.121}$$

We now use (4.115) to compute

$$\begin{aligned} Lw \geq & -a(x)w + b(x)[u^\sigma - (\xi v)^\sigma] - c(x)[u^\tau - (\xi v)^\tau] \\ & + b(x)\xi v\Big[(\xi v)^{\sigma-1} - v^{\sigma-1}\Big] + c(x)\xi v\Big[v^{\tau-1} - (\xi v)^{\tau-1}\Big]. \end{aligned} \tag{4.122}$$

We let

$$h(x) = \begin{cases} \sigma u^{\sigma-1}(x) & \text{if } u(x) = \xi v(x) \\ \frac{\sigma}{u(x) - \xi v(x)} \int_{\xi v(x)}^{u(x)} t^{\sigma-1}\, dt & \text{if } u(x) < \xi v(x) \end{cases}$$

and similarly, for $\tau \neq 0$,

$$j(x) = \begin{cases} -\tau u^{\tau-1}(x) & \text{if } u(x) = \xi v(x) \\ \frac{\tau}{\xi v(x) - u(x)} \int_{\xi v(x)}^{u(x)} t^{\tau-1}\, dt & \text{if } u(x) < \xi v(x). \end{cases}$$

In case $\tau = 0$, choose $j(x) \equiv 0$. Observe that h and j are continuous on $M \setminus \Omega$ and h is nonnegative. Using h and j and observing that $-a(x)w \geq a_-(x)w$, from (4.122) we obtain

$$\begin{aligned} Lw \geq & [a_-(x) + b(x)h(x) + c(x)j(x)]w \\ & + b(x)\xi v\Big[(\xi v)^{\sigma-1} - v^{\sigma-1}\Big] + c(x)\xi v\Big[v^{\tau-1} - (\xi v)^{\tau-1}\Big]. \end{aligned} \tag{4.123}$$

Let

$$\Omega_{-1} = \big\{x \in M \setminus \overline{\Omega} : w(x) > -1\big\}.$$

Since u is bounded above on $M \setminus \Omega$, there exists a constant $C > 0$ such that

$$v(x) = \frac{1}{\xi}(u(x) - w(x)) \leq \frac{1}{\xi}(C + 1) \tag{4.124}$$

on Ω_{-1}. Using the definition of h and j, from the mean value theorem for integrals we deduce

$$h(x) = \sigma y_h^{\sigma-1}, \quad j(x) = -\tau y_j^{\tau-1}$$

for some $y_h = y_h(x)$ and $y_j = y_j(x)$ in the range $[u(x), \xi v(x)]$. Since $u(x)$ and $v(x)$ are bounded above on Ω_{-1},

$$\max\{h(x), j(x)\} \le C \tag{4.125}$$

on Ω_{-1} for some constant $C > 0$. Next we recall that $b(x) > 0$ on $M \setminus \Omega$ to rewrite (4.123) in the form

$$\begin{aligned}\frac{1}{b(x)}Lw \ge &\left[\frac{a_-(x)}{b(x)} + h(x) + \frac{c(x)}{b(x)}j_+(x)\right]w \\ &+ \xi v\Big[(\xi v)^{\sigma-1} - v^{\sigma-1}\Big] + \frac{c(x)}{b(x)}\xi v\Big[v^{\tau-1} - (\xi v)^{\tau-1}\Big].\end{aligned}$$

Since $w \le 0$, (4.113), (4.114) and (4.125) imply

$$\left[\frac{a_-(x)}{b(x)} + h(x) + \frac{c(x)}{b(x)}j_+(x)\right]w \ge Cw$$

for some constant $C > 0$ on Ω_{-1}. For further use we observe here that when $0 \le \tau < 1, j_+(x) \equiv 0$ so that in this case assumption (4.114) (ii) is not needed to obtain this last inequality. Thus

$$\frac{1}{b(x)}Lw \ge Cw + \xi v\Big[(\xi v)^{\sigma-1} - v^{\sigma-1}\Big] + \frac{c(x)}{b(x)}\xi v\Big[v^{\tau-1} - (\xi v)^{\tau-1}\Big]$$

on Ω_{-1}. Recalling the elementary inequalities

$$\begin{cases} a^s - b^s \ge sb^{s-1}(a-b) & \text{for } s < 0 \text{ and } s > 1 \\ a^s - b^s \ge sa^{s-1}(a-b) & \text{for } 0 \le s \le 1, \end{cases} \tag{4.126}$$

with $a, b > 0$, coming from the mean value theorem for integrals, we conclude that

$$\frac{1}{b(x)}Lw \ge Cw + (\sigma-1)\xi^{\min\{1,\sigma-1\}}(\xi-1)v^\sigma + (1-\tau)\frac{c(x)}{b(x)}\frac{\xi-1}{\xi^{1-\tau}}v^\tau$$

on Ω_{-1}. Now we use the fact that $\tau < 1$, v is bounded from below by a positive constant, (4.113), (4.114) (ii) to get [again if $0 \le \tau < 1$ we do not need (4.114) (ii)]

$$\frac{1}{b(x)}Lw \ge Cw + B \quad \text{on } \Omega_{-1},$$

for some constants $B, C > 0$. Finally we choose $0 < \varepsilon < 1$ sufficiently small that

$$Cw > -\frac{1}{2}B$$

on

$$\Omega_{-\varepsilon} = \left\{x \in M \setminus \overline{\Omega} : w(x) > -\varepsilon\right\} \subset \Omega_{-1},$$

and $\partial\Omega_{-\varepsilon} \subset M \setminus \Omega$. Therefore

$$\frac{1}{b(x)}Lw \geq \frac{1}{2}B > 0 \quad \text{on } \Omega_{-\varepsilon}. \tag{4.127}$$

Furthermore, note that

$$w(x) \leq \min\{-\varepsilon, (1-\xi)\min_{\partial\Omega} v\} < 0$$

on $\Omega_{-\varepsilon}$. As a consequence $\sup_{\partial\Omega_{-\varepsilon}} w < 0$, while $\sup_{\Omega_{-\varepsilon}} w = 0$. Applying the open form of the $\frac{1}{b}$-weak maximum principle to (4.127) we obtain the desired contradiction. □

As an immediate consequence we obtain the following uniqueness result.

Corollary 4.1 *In the assumptions of Theorem 4.8 the equation*

$$\Delta_X u + a(x)u - b(x)u^\sigma + c(x)u^\tau = 0 \quad \text{on } M \setminus \overline{\Omega}$$

admits at most a unique positive solution $u \in C^2\left(M \setminus \overline{\Omega}\right) \cap C^0(M \setminus \Omega)$ *with assigned boundary data on* $\partial\Omega$ *and satisfying*

$$C_1 \leq u(x) \leq C_2 \quad \text{on } M \setminus \overline{\Omega} \tag{4.128}$$

for some constants C_1, $C_2 > 0$ *provided that the* $\frac{1}{b}$*-weak maximum principle holds on* M *for the operator* Δ_X.

4.4 Strong Parabolicity

In Sect. 2.5 of Chap. 2 we briefly discussed parabolicity for the Laplace-Beltrami operator Δ showing that it is equivalent to a stronger form of the weak maximum principle for Δ. Motivated by this observation, here we introduce a stronger notion of parabolicity and indicate when this is equivalent to the usual one that we specify in Definition 4.3 below. In doing so we follow [28]. We begin with the next

Definition 4.2 We say that the operator $L = L_{\varphi,T,X}$ defined in (4.91) of Sect. 4.3 is *strongly parabolic* on M if for each nonconstant $u \in C^1(M)$ with $u^* < +\infty$ and for each $\gamma \in \mathbb{R}$ with $\gamma < u^*$ we have

$$\inf_{\Omega_\gamma}\{Lu\} < 0 \tag{4.129}$$

in the weak sense, where, as usual, $\Omega_\gamma = \{x \in M : u(x) > \gamma\}$.

Recall that the strict inequality (4.129) in the weak sense means that for some $\varepsilon > 0$

$$-\int_{\Omega_\gamma} \left(|\nabla u|^{-1}\varphi(x, |\nabla u|)T(\nabla u, \nabla\psi) + \langle X, \nabla u\rangle\psi\right) \le -\int_{\Omega_\gamma} \varepsilon\psi, \tag{4.130}$$

for some $\psi \in C_c^\infty(\Omega_\gamma)$, $\psi \ge 0$, $\psi \not\equiv 0$.

It is immediate to compare this definition with the more familiar

Definition 4.3 We say that the operator $L = L_{\varphi,T,X}$ is *parabolic* on M if each $u \in C^1(M)$ with $u^* < +\infty$ and satisfying $Lu \ge 0$ on M is constant.

It is clear that strong parabolicity of L implies parabolicity. The converse is also true if we enlarge the functional class of u to $\mathrm{Lip}_{loc}(M)$ or $C^0(M) \cap W^{1,1+\delta}_{loc}(M)$ and we assume the validity of the following proposition:

Proposition 4.2 *For every open set $\Omega \subseteq M$, if $u \in \mathrm{Lip}_{loc}(\Omega)$ or $u \in C^0(\overline{\Omega}) \cap W^{1,1+\delta}_{loc}(\Omega)$ satisfies $Lu \ge 0$ on Ω then, for each fixed $\alpha \in \mathbb{R}$, the function $v(x) = \max\{u(x), \alpha\}$ satisfies $Lv \ge 0$ on Ω.*

Indeed, we have

Theorem 4.9 *Let $(M, \langle\, ,\, \rangle)$ be a Riemannian manifold and $L = L_{\varphi,T,X}$ be an operator as in* (4.91)*. Assume the validity of Proposition 4.2. Then L is strongly parabolic on M if and only if L is parabolic.*

Proof We only have to show that parabolicity implies strong parabolicity. We reason by contradiction and we assume the existence of a nonconstant u with $u^* < +\infty$ and $\gamma \in \mathbb{R}$, $\gamma < u^*$ such that

$$Lu \ge 0$$

on Ω_γ.

Up to increasing γ we may assume $\partial\Omega_\gamma \ne \emptyset$, because otherwise $\Omega_\gamma = M$ and the result is immediate. Consider the function

$$v(x) = \begin{cases} \max\{u(x), \gamma + \frac{u^*-\gamma}{2}\} \text{ on } \Omega_\gamma, \\ \\ \gamma + \frac{u^*-\gamma}{2} \text{ on } M \setminus \Omega_\gamma. \end{cases}$$

Then $v^* = u^* < +\infty$ and, because of Proposition 4.2 (on Ω_γ)

$$Lv \geq 0 \quad \text{on} \quad M.$$

By Definition 4.3 v is the constant $\gamma + \frac{u*-\gamma}{2} < u^* = v^*$, contradiction. □

A large class of operators satisfies Proposition 4.2. For instance

Proposition 4.3 *Let $(M, \langle\, ,\, \rangle)$ be a Riemannian manifold and $L = L_{\varphi,T,X}$ be an operator as in (4.91), with T a symmetric, positive semi-definite $(0,2)$-tensor field on M. Define $A(x,t) = t^{-1}\varphi(x,t)$ on $M \times \mathbb{R}^+$ and suppose that, for each $x \in M$, $A(x,\ )$ is nondecreasing on $\mathbb{R}^+$. Then Proposition 4.2 holds.*

Proof Since, for any $\beta \in \mathbb{R}$, $u + \beta$ is still a solution of $Lv \geq 0$ if u is so, in Proposition 4.2 we can suppose $\alpha = 0$. In this case $v(x) = \max\{u(x), 0\} = u_+(x)$, so that it remains to show that $Lu_+ \geq 0$ on M. Towards this end we fix $\psi \in C_c^\infty(M)$, $\psi \geq 0$, and we recall that $Lu \geq 0$ yields

$$\int_M (A(x, |\nabla u|)T(\nabla u, \nabla\psi) + \langle X, \nabla u\rangle\psi) \leq 0. \tag{4.131}$$

We let $\varepsilon > 0$ and we set

$$u_\varepsilon = \sqrt{u^2 + \varepsilon}, \quad \psi_\varepsilon = \frac{1}{2}\left(1 + \frac{u}{u_\varepsilon}\right)\psi.$$

Observe that ψ_ε is still an admissible test function for $Lu \geq 0$. Furthermore, since $\nabla u_\varepsilon = \dfrac{u}{u_\varepsilon}\nabla u$, we have

$$u_\varepsilon \to |u|, \quad \nabla u_\varepsilon \to \text{sign}(u)\nabla u, \quad \text{and } \psi_\varepsilon \to \frac{1}{2}\left(1 + \text{sign}u\right)\psi$$

as $\varepsilon \to 0^+$. A computation shows that

$$T(\nabla u_\varepsilon, \nabla\psi) = T(\nabla u, \nabla(\frac{u}{u_\varepsilon}\psi)) - \frac{\psi}{u_\varepsilon^3}(u_\varepsilon^2 - u^2)T(\nabla u, \nabla u),$$

so that, since T is positive semi-definite,

$$T((\nabla u_\varepsilon, \nabla\psi) \leq T(\nabla u, \nabla(\frac{u}{u_\varepsilon}\psi)).$$

From this last inequality it follows immediately

$$T\left(\nabla(\frac{u_\varepsilon + u}{2}), \nabla\psi\right) \leq T(\nabla u, \nabla\psi_\varepsilon);$$

on the other hand

$$\left|\nabla(\frac{u_\varepsilon + u}{2})\right| = \frac{1}{2}(1 + \frac{u}{u_\varepsilon})|\nabla u| \leq |\nabla u|$$

so that, using the above and the fact that $A(x, \ )$ is nondecreasing on $\mathbb{R}^+$ we deduce

$$A\left(x, |\nabla(\frac{u_\varepsilon + u}{2})|\right)T\left(\nabla(\frac{u_\varepsilon + u}{2}), \nabla\psi\right) \leq A(x, |\nabla u|)T(\nabla u, \nabla\psi_\varepsilon). \tag{4.132}$$

Now , by the definition of subsolution for u, we have

$$\int_M (A(x, |\nabla u|)T(\nabla u, \nabla\psi_\varepsilon) + \langle X, \nabla u\rangle\psi_\varepsilon) \leq 0,$$

and therefore, using (4.132)

$$\int_M \left(A(x, |\nabla(\frac{u_\varepsilon + u}{2})|)T(\nabla(\frac{u_\varepsilon + u}{2}), \nabla\psi) + \langle X, \nabla u\rangle\psi_\varepsilon\right) \leq 0.$$

Letting $\varepsilon \to 0^+$ and using Fatou's lemma we deduce

$$\int_M (A(x, |\nabla u_+|)T(\nabla u_+, \nabla\psi) + \langle X, \nabla u_+\rangle\psi) \leq 0.$$

that is, u_+ is also a subsolution. □

In particular, for the trace operator

$$Lu = \mathrm{Tr}(t \circ \mathrm{hess}(u)) = \mathrm{div}(T(\nabla u, \cdot)^\sharp) - \langle \mathrm{div}T, \nabla u\rangle$$

the assumptions of Proposition 4.3 are satisfied. Thus, for trace operators strong parabolicity and parabolicity coincide.

Note that the function $A(t) = t^{p-2}$ is decreasing for $1 < p < 2$, however Proposition 4.2 still holds; in fact, it holds for the more general class of operators

$$L_{p,f}u = e^f \mathrm{div}(e^{-f}|\nabla u|^{p-2}\nabla u) = \mathrm{div}\big(|\nabla u|^{p-2}\nabla u\big) - \langle \nabla f, \nabla u\rangle,$$

with $p \in (1, +\infty)$ and $f \in C^\infty(M)$ a potential function. This is proved, with a nontrivial argument, in Lemma 7.1 of [46].

The same happens for the mean curvature operator

$$Lu = \mathrm{div}\left(\frac{\nabla u}{\sqrt{1 + |\nabla u|^2}}\right).$$

Further results in this direction are contained in Lemma 1.3 of [229], in Lemma 3.1 of [6] and in the original work of Le [169].

As expected we have the following open version of strong parabolicity for the operator L. Since the proof is very similar to that of Theorem 4.6, we leave it to the interested reader.

Theorem 4.10 *The strong parabolicity of the operator L as in Definition 4.2 is equivalent to the following* open strong parabolicity*: for each $f \in C^0(\mathbb{R})$, for each open set $\Omega \subset M$ with $\partial\Omega \neq \emptyset$ and for each $v \in C^0(\overline{\Omega}) \cap C^1(\Omega)$, nonconstant and satisfying*

$$\begin{cases} Lv \geq f(v) \text{ on } \Omega, \\ \sup_\Omega v < +\infty \end{cases} \tag{4.133}$$

we have that either

$$\sup_\Omega v = \sup_{\partial\Omega} v \tag{4.134}$$

or, for each $\varepsilon > 0$

$$\inf_{U_\varepsilon} f(v) < 0 \tag{4.135}$$

where

$$U_\varepsilon = \{x \in \Omega : v(x) > \sup_\Omega v - \varepsilon\}.$$

Remark 4.9 Note the minor, but essential, difference between conclusion (4.135) of Theorem 4.10 and (4.97) of Theorem 4.6.

As a consequence of Theorem 4.10 we deduce that, if the operator L is strongly parabolic on M, then for each open set $\Omega \subset M$ with $\partial\Omega \neq \emptyset$ and for each nonconstant $v \in C^0(\overline{\Omega}) \cap C^1(\Omega)$, satisfying

$$\begin{cases} Lv \geq 0 \text{ on } \Omega, \\ \sup_\Omega v < +\infty \end{cases} \tag{4.136}$$

we have

$$\sup_\Omega v = \sup_{\partial\Omega} v. \tag{4.137}$$

Interestingly enough, also the converse is true; that is, calling this latter property *Ahlfors parabolicity* (see Ahlfors-Sario [2], Theorem 6C), in strict analogy with Theorem 4.7 we have

Theorem 4.11 *The operator L is strongly parabolic on M if and only if it is Ahlfors parabolic.*

Proof We only need to prove that Ahlfors parabolicity implies strong parabolicity. We reason by contradiction and we suppose the existence of a nonconstant $u \in C^1(M)$ with $u^* < +\infty$ and of $\gamma \in \mathbb{R}$, $\gamma < u^*$ such that $\inf_{\Omega_\gamma} Lu \geq 0$, that is,

$$Lu \geq 0 \quad \text{on } \Omega_\gamma.$$

Since u is nonconstant, by possibly increasing γ we can suppose $\partial\Omega_\gamma \neq \emptyset$. Let $v = u|_{\overline{\Omega}_\gamma}$ so that, for $\Omega = \Omega_\gamma$, $v \in C^0(\overline{\Omega}) \cap C^1(\Omega)$, v is nonconstant on Ω and it satisfies (4.136). Then, by (4.137),

$$u^* = \sup_{\Omega} v = \sup_{\partial\Omega} v = \gamma$$

contradiction. □

We now need sufficient conditions to guarantee both strong parabolicity and parabolicity for the operator L on $(M, \langle\, , \,\rangle)$. We begin by considering the linear case; in this case the result is obtained by a minor modification of the proof of Theorem 3.1.

Theorem 4.12 *Let $(M, \langle\, , \,\rangle)$ be a Riemannian manifold and let $L = L_{T,X}$ be the operator*

$$Lu = \operatorname{div}(T(\nabla u, \,)^\sharp) - \langle X, \nabla u\rangle.$$

Assume the existence of $\gamma \in C^2(M)$ such that

$$\begin{cases} \gamma(x) \to +\infty \text{ as } x \to \infty, \\ L\gamma \leq 0 \text{ if } X \equiv 0 \text{ or } L\gamma < 0 \text{ if } X \not\equiv 0 \end{cases}$$

outside a compact set. Then L is strongly parabolic in $\mathrm{Lip}_{loc}(M)$ *if $X \equiv 0$, in $C^2(M)$ if $X \not\equiv 0$.*

Proof Let $X \equiv 0$. We reason by contradiction and we assume the existence of a nonconstant $u \in \mathrm{Lip}_{loc}(M)$ with $u^* < +\infty$ and of $\eta > 0$ such that

$$Lu \geq 0 \quad \text{on } \Omega_\eta = \{x \in M : u(x) > u^* - \eta\}. \tag{4.138}$$

First we observe that u^* cannot be attained at any point $x_0 \in M$, for otherwise $x_0 \in \Omega_\eta$ and by the strong maximum principle for the linear operator $L = L_{T,X}$, Theorem 3.10, and the final observation in Remark 3.14, we have that u is constantly equal to u^* on the connected component of Ω_η containing x_0. From this and the connectedness of M it follows easily that u is constant on M, contradiction. The rest of the proof proceeds similarly to that of Theorem 3.1, up to inequality (3.30), having only to substitute (3.24) with

$$L\gamma_\sigma = \sigma L\gamma \leq 0 \quad \text{on } \Omega_{T_1}, \tag{4.139}$$

due to the assumptions on γ. To finish the proof we now argue as follows. We let

$$\mu = \sup_{x \in \overline{\Omega}_{T_1}} (u - \gamma_\sigma)(x) > 0; \tag{4.140}$$

μ is in fact a positive maximum attained at some point z_0 in the compact set $\overline{\Omega}_{T_1} \setminus \Omega_{T_3}$. Thus

$$\Sigma = \{x \in \Omega_{T_1} : (u - \gamma_\sigma)(x) = \mu\} \neq \emptyset.$$

Furthermore, for $y \in \Sigma$,

$$u(y) = \gamma_\sigma(y) + \mu > \gamma_\sigma(y) = \alpha + \sigma(\gamma(y) - T_1) > \alpha > u^*_{T_1} > u^* - \frac{\eta}{2},$$

so that

$$\Sigma \subset \Omega_\eta$$

and, by (4.140), $\Sigma \subset \overline{\Omega}_{T_1} \setminus \Omega_{T_3}$ and therefore Σ is compact. Hence, there exists an open neighbourhood Σ_U of Σ, such that $\Sigma_U \subset \Omega_\eta$. Fix $y \in \Sigma$ and $\beta \in (0, \mu)$ and call $\Sigma_{\beta,y}$ the connected component of the set

$$\{x \in \Omega_{T_1} : (u - \gamma_\sigma)(x) > \beta\}$$

containing y. We can choose β sufficiently close to μ so that $\overline{\Sigma}_{\beta,y} \subset \Omega_\eta \cap \Omega_{T_1}$. Note that, since $\beta > 0$, $\overline{\Sigma}_{\beta,y}$ is compact. Because of (4.139) and (4.138),

$$Lu \geq 0 \geq \sigma L\gamma = L\gamma_\sigma \quad \text{on } \Sigma_{\beta,y}$$

in the weak sense. Furthermore, $u(x) = \gamma_\sigma(x) + \beta$ on $\partial\Sigma_{\beta,y}$. By Proposition 3.1, $u(x) \leq \gamma_\sigma(x) + \beta$ on $\overline{\Sigma}_{\beta,y}$. However, $y \in \Sigma_{\beta,y}$ and we have

$$u(y) = \gamma_\sigma(y) + \mu > \gamma_\sigma(y) + \beta$$

by our choice of β, contradiction.

If $X \not\equiv 0$ we suppose u in (4.138) to be of class $C^2(M)$. We still can claim that u^* cannot be attained at any point $x_0 \in M$. We do this again applying Theorem 3.10 because L is linear and $B(x, u, \nabla u) = \langle X, \nabla u\rangle \leq |X||\nabla u| \leq \sup|X||\nabla u|$, where we can always suppose $\sup|X| < +\infty$ up to restricting our reasoning on a small, therefore with compact closure, ball centered at x_0. Now in (4.139) we have

$$L\gamma_\sigma = \sigma L\gamma < 0 \quad \text{on } \Omega_{T_1}$$

and, having detected $z_0 \in \overline{\Omega}_{T_1} \setminus \Omega_{T_3}$ as above we have

$$0 \leq Lu(z_0) \leq L\gamma_\sigma(z_0) = \sigma L\gamma(z_0) < 0,$$

contradiction. □

Note that to apply Theorem 3.10 in case $X \not\equiv 0$ we need the crucial estimate for $\langle X, \nabla u\rangle$ with $\varphi(t) = t$. This is not possible if φ is nonlinear. This fact will limit us to the case $X \equiv 0$ in the nonlinear case.

To take care of the latter first we need to modify Definition 3.1 to the following new Khas'minskiĭ type condition. Here the operator L is

$$Lu = \operatorname{div}\left(|\nabla u|^{-1}\varphi(|\nabla u|)T(\nabla u, \,)^\sharp\right),$$

with the validity of (A1), (A2), (T1), (T2) (see Sect. 3.3 in Chap. 3).

Definition 4.4 We say that the *nonlinear strong parabolicity condition* holds if there exists a telescoping exhaustion of relatively compact open sets $\{\Sigma_j\}_{j\in\mathbb{N}}$ such that $\overline{\Sigma}_j \subset \Sigma_{j+1}$ for every j and, for any pair $\Omega_1 = \Sigma_{j_1}$, $\Omega_2 = \Sigma_{j_2}$, with $j_1 < j_2$, and for each $\varepsilon > 0$, there exists $\gamma \in C^0(M \setminus \Omega_1) \cap C^2(M \setminus \overline{\Omega}_1)$ if $X \not\equiv 0$ and $\mathrm{Lip}_{loc}(M \setminus \Omega_1)$ if $X \equiv 0$ with the following properties:

(i) $\gamma \equiv 0$ on $\partial\Omega_1$,
(ii) $\gamma > 0$ on $M \setminus \Omega_1$,
(iii) $\gamma \leq \varepsilon$ on $\Omega_2 \setminus \Omega_1$,
(iv) $\gamma(x) \to +\infty$ when $x \to \infty$,
(v) $L\gamma \leq 0$ on $M \setminus \overline{\Omega}_1$.

We are now ready to state

Theorem 4.13 *Let $(M, \langle , \rangle)$ be a Riemannian manifold and let L be as above acting on* $\mathrm{Lip}_{loc}(M)$. *Assume the validity of the nonlinear strong parabolicity condition. Then L is strongly parabolic on M in the class* $\mathrm{Lip}_{loc}(M)$.

Remark 4.10 The proof of Theorem 4.13 is a simple adaptation of the proof of Theorem 3.11. The only delicate points are

(i) to show that u^* cannot be attained at $x_0 \in M$,
(ii) the comparison between u and γ at the end of the proof.

Since $X \equiv 0$, for (i) we apply Theorem 3.10 with Remark 3.14, while for (ii) we apply Theorem 3.9.

The next is a sufficient condition, definitely more satisfactory, for parabolicity. However, also here we have limitations; indeed the vector field X is of the special form $X = \nabla f$. This is in order to be able to express the differential operator L basically in divergence form.

Theorem 4.14 *Let $(M, \langle\ ,\ \rangle)$ be a complete Riemannian manifold, $o \in M$ a fixed origin and $r(x) = dist_M(x, o)$. Let $L = L_{\varphi,T,f}$ be the operator defined, for $u \in C^1(M)$, by*

$$Lu = e^f \operatorname{div}\left(e^{-f}|\nabla u|^{-1}\varphi(x, |\nabla u|)T(\nabla u,\ )^\sharp\right)$$

and suppose that, for some $T_-, T_+ \in C^0(\mathbb{R}_0^+)$,

$$0 < T_-(r) \le T(Y, Y) \le T_+(r) \tag{4.141}$$

for every $Y \in T_xM$, $|Y| = 1$, and every $x \in \partial B_r$, where B_r denotes the geodesic ball of radius r centered at o. Let $\varphi : M \times \mathbb{R}_0^+ \to \mathbb{R}_0^+$ be such that $\varphi(\ , t) \in C^0(M)$ for every $t \in \mathbb{R}_0^+$, $\varphi(x,\ ) \in C^0(\mathbb{R}_0^+) \cap C^1(\mathbb{R}^+)$ for every $x \in M$, and

$$\begin{array}{lll} (i) & \varphi(x, 0) = 0, & \text{for every } x \in M; \\ (ii) & \varphi(x, t) > 0, & \text{on } M \times \mathbb{R}^+; \\ (iii) & \varphi(x, t) \le A(x)t^\delta, & \text{on } M \times \mathbb{R}^+, \end{array} \tag{4.142}$$

for some $\delta > 0$ and $A(x) \in C^0(M)$, $A(x) > 0$ on M. Furthermore, assume that

$$\inf_M \frac{T_-(r(x))}{T_+(r(x))} \frac{1}{A(x)^{1/\delta}} = \frac{1}{C_0^{1/\delta}} \tag{4.143}$$

for some $C_0 > 0$. If

$$\frac{1}{\left(T_+(t) \int_{\partial B_t} e^{-f}\right)^{\frac{1}{\delta}}} \notin L^1(+\infty) \tag{4.144}$$

then L is parabolic on M.

Remark 4.11 Note that in general Proposition 4.2 does not hold here, so that L may not be strongly parabolic on M.

Remark 4.12 Note that in case $T = \langle\ ,\ \rangle, f \equiv 0$ and $\varphi(x, t) = t$, Theorem 4.14 reduces to the second case of Theorem 2.23. However, the proof here is not based on a capacity argument, which is definitely nonapplicable because of the possible "strongly" nonlinear nature of the differential operator.

The proof is based on the following approach (see [243]):

Lemma 4.3 *In the assumptions of Theorem 4.14, let $\kappa \in C^0(\mathbb{R})$ and let u be a nonconstant $C^1(M)$ solution of the differential inequality*

$$Lu \geq |\nabla u|^{-1}\varphi(x, |\nabla u|)T(\nabla u, \nabla u)\kappa(u) \tag{4.145}$$

on M. Assume that there exist functions $\alpha \in C^1(I)$ and $\beta \in C^0(I)$ defined on an interval $I \supset u(M)$ such that

$$\alpha(u) \geq 0, \tag{4.146}$$

$$\alpha'(u) + \kappa(u)\alpha(u) \geq \beta(u) > 0 \tag{4.147}$$

on M. Then, there exist $R_0 > 0$ depending only on u and a constant $C > 0$ independent of α and β, such that, for any $r > R \geq R_0$

$$\left(\int_{B_r} \beta(u)\varphi(x, |\nabla u|)|\nabla u|T_-(r)e^{-f}\right)^{-1} \geq C\left(\int_R^r \left(\int_{\partial B_r} e^{-f}T_+(r)\frac{\alpha(u)^{1+\delta}}{\beta(u)^\delta}\right)^{-\frac{1}{\delta}}\right)^\delta. \tag{4.148}$$

Proof We consider the vector field

$$Z = \alpha(u)e^{-f}|\nabla u|^{-1}\varphi(x, |\nabla u|)T(\nabla u, \cdot)^\sharp. \tag{4.149}$$

We compute the distributional divergence of Z and we use our assumptions on α, β and (4.145) to obtain

$$\begin{aligned} \mathrm{divZ} &\geq \big(\alpha'(u) + \kappa(u)\alpha(u)\big)e^{-f}|\nabla u|^{-1}\varphi(x, |\nabla u|)T(\nabla u, \nabla u) \\ &\geq \beta(u)e^{-f}|\nabla u|^{-1}\varphi(x, |\nabla u|)T(\nabla u, \nabla u). \end{aligned}$$

Using (4.141) we immediately get

$$\mathrm{divZ} \geq \beta(u)e^{-f}|\nabla u|\varphi(x, |\nabla u|)T_-(r). \tag{4.150}$$

Integrating over B_t and applying the divergence theorem gives

$$\int_{\partial B_t} \langle Z, \nabla r\rangle e^{-f} \geq \int_{B_t} \beta(u)|\nabla u|\varphi(x, |\nabla u|)T_-(r)e^{-f}. \tag{4.151}$$

On the other hand, using Cauchy-Schwarz inequality and (4.141), we have

$$\int_{\partial B_t} \langle Z, \nabla r\rangle e^{-f} \leq T_+(t)\int_{\partial B_t} \alpha(u)\varphi(x, |\nabla u|)e^{-f}. \tag{4.152}$$

We observe that assumption (4.142) (*iii*) on φ implies

$$t\varphi(x,t) \geq A(x)^{-\frac{1}{\delta}}\varphi(x,t)^{1+\frac{1}{\delta}}. \tag{4.153}$$

Hence,

$$\begin{aligned}&\alpha(u)\varphi(x,|\nabla u|)T_+(r)e^{-f}\\ &\leq A(x)^{\frac{1}{1+\delta}}\frac{\alpha(u)}{\beta(u)^{\frac{\delta}{1+\delta}}}\frac{T_+(r)}{T_-(r)^{\frac{\delta}{1+\delta}}}e^{-f\frac{1}{1+\delta}}\left(|\nabla u|\varphi(x,|\nabla u|)\right)^{\frac{\delta}{1+\delta}}e^{-f\frac{\delta}{1+\delta}}\beta(u)^{\frac{\delta}{1+\delta}}T_-(r)^{\frac{\delta}{1+\delta}}.\end{aligned}$$

Thus, applying Hölder's inequality with conjugate exponents $p = 1+\delta$ and $q = 1+\frac{1}{\delta}$ we obtain

$$\begin{aligned}\int_{\partial B_t}\langle Z,\nabla r\rangle e^{-f} &\leq \int_{\partial B_t}\alpha(u)\varphi(x,|\nabla u|)T_+(t)e^{-f} \\ &\leq \left(\int_{\partial B_t}A(x)\frac{\alpha(u)^{1+\delta}}{\beta(u)^{\delta}}\frac{T_+(t)^{1+\delta}}{T_-(t)^{\delta}}e^{-f}\right)^{\frac{1}{1+\delta}}\times \\ &\quad\left(\int_{\partial B_t}|\nabla u|\varphi(x,|\nabla u|)\beta(u)T_-(t)e^{-f}\right)^{\frac{\delta}{1+\delta}}.\end{aligned} \tag{4.154}$$

We set

$$G(R) = \int_{B_R}\beta(u)T_-(r)|\nabla u|\varphi(x,|\nabla u|)e^{-f} \tag{4.155}$$

and we observe that, since u is nonconstant, there exists $R_0 > 0$ sufficiently large such that, for any $R \geq R_0$, it holds that $G(R) > 0$. Using the coarea formula and putting together (4.151) and (4.154) we obtain

$$G(R)^{\frac{1+\delta}{\delta}} \leq G'(R)\left(\int_{\partial B_R}A(x)\frac{\alpha(u)^{1+\delta}}{\beta(u)^{\delta}}\frac{T_+(R)^{\delta}}{T_-(R)^{\delta}}T_+(R)e^{-f}\right)^{\frac{1}{\delta}} \tag{4.156}$$

for $R \geq R_0$. In particular the term between parenthesis of the above inequality is positive and we can rewrite (4.156) in the form

$$\left(\int_{\partial B_R}A(x)\frac{T_+(R)^{\delta}}{T_-(R)^{\delta}}T_+(R)\frac{\alpha(u)^{1+\delta}}{\beta(u)^{\delta}}e^{-f}\right)^{-\frac{1}{\delta}} \leq \frac{G'(R)}{G(R)^{1+\frac{1}{\delta}}} \tag{4.157}$$

on $[R_0, +\infty)$. Hence, using (4.143),

$$C_0^{-\frac{1}{\delta}}\left(\int_{\partial B_R}\frac{\alpha(u)^{1+\delta}}{\beta(u)^{\delta}}T_+(R)e^{-f}\right)^{-\frac{1}{\delta}} \leq \frac{G'(R)}{G(R)^{1+\frac{1}{\delta}}}.$$

Thus, integrating on $[R, r]$ with $R_0 \le R < r$, we deduce

$$G(R)^{-\frac{1}{\delta}} \ge G(R)^{-\frac{1}{\delta}} - G(r)^{-\frac{1}{\delta}} \ge \frac{1}{\delta C_0^{-\frac{1}{\delta}}} \int_R^r \left(\frac{\alpha(u)^{1+\delta}}{\beta(u)^\delta} T_+(t) e^{-f} \right)^{-\frac{1}{\delta}} dt.$$

We then obtain (4.148) with $C = \left(\delta^\delta C_0\right)^{-1}$. □

Proof (of Theorem 4.14) Let $\zeta > 0$ and set $\alpha(t) = e^{\zeta t}$, $\beta(t) = \zeta e^{\zeta t}$. Hence (4.146), (4.147) are satisfied with $\kappa \equiv 0$. If u is a solution of $Lu \ge 0$ which is nonconstant and with $u^* = \sup_M u < +\infty$, applying (4.148) of Lemma 4.3 for $r > R \ge R_0$ we have

$$\frac{1}{\zeta \int_{B_R} e^{\zeta u} \varphi(x, |\nabla u|) |\nabla u| T_-(r) e^{-f}} \ge C \left(\int_R^r \frac{dt}{\left(\int_{\partial B_t} T_+(t) \frac{e^{u^*}}{\zeta} e^{-f} \right)^{\frac{1}{\delta}}} \right)^\delta. \tag{4.158}$$

Letting $r \to +\infty$ and using (4.144) we obtain the desired contradiction. □

Remark 4.13 By a simple modification of the above proof we see that we can replace assumption (4.144) of the theorem with

$$\frac{1}{\left(T_+(t) \int_{\partial B_t} u^q e^{-f}\right)^{1/\delta}} \notin L^1(+\infty),$$

provided u is nonnegative and $q > \delta$. The case $q = \delta$ requires extra care and some further assumption; for instance see Theorem C in [243].

A minor variation in the proof of Lemma 4.3 yields the following alternative statement of Theorem 4.14.

Theorem 4.15 *Let $(M, \langle\ ,\ \rangle)$ be a complete Riemannian manifold, $o \in M$ a fixed origin and $r(x) = dist_M(x, o)$. Let $L = L_{\varphi,Tf}$ be the operator defined, for $u \in C^1(M)$, by*

$$Lu = e^f \operatorname{div}\left(e^{-f} |\nabla u|^{-1} \varphi(x, |\nabla u|) T(\nabla u,\ )^\sharp\right)$$

and suppose that, for some $T_-, T_+ \in C^0\left(\mathbb{R}_0^+\right)$,

$$0 \le T_-(r) \le T(Y, Y) \le T_+(r) \tag{4.159}$$

for every $Y \in T_xM$, $|Y| = 1$, and every $x \in \partial B_r$, where B_r denotes the geodesic ball of radius r centered at o, and with the further assumption $T_-(r) > 0$ if $\delta > 1$, with δ as in (4.161). Define

$$T_\delta(r) = \begin{cases} T_+(r) & \text{if } 0 < \delta \leq 1, \\ T_-(r)^{(1-\delta)/2} T_+(r)^{(1+\delta)/2} & \text{if } \delta \geq 1. \end{cases} \tag{4.160}$$

Let $\varphi : M \times \mathbb{R}_0^+ \to \mathbb{R}_0^+$ be such that $\varphi(\,, t) \in C^0(M)$ for every $t \in \mathbb{R}_0^+$, $\varphi(x, \,) \in C^0(\mathbb{R}_0^+) \cap C^1(\mathbb{R}^+)$ for every $x \in M$, and

$$\begin{array}{ll} (i)\ \varphi(x,0) = 0, & \text{for every } x \in M; \\ (ii)\ \varphi(x,t) > 0, & \text{on } M \times \mathbb{R}^+; \\ (iii)\ \varphi(x,t) \leq A(x)t^\delta, & \text{on } M \times \mathbb{R}^+, \end{array} \tag{4.161}$$

for some $\delta > 0$ and $A(x) \in C^0(M)$, $A(x) > 0$ on M. If

$$\frac{1}{\left(T_\delta(t) \int_{\partial B_t} A(x) e^{-f}\right)^{1/\delta}} \notin L^1(+\infty) \tag{4.162}$$

then L is parabolic on M.

Observe that in this new formulation the infimum of $T_-(r)$ can be 0 and if $\delta \leq 1$ $T_-(r)$ could even be 0 for some $r \in \mathbb{R}_0^+$; in other words, L could even be semi-elliptic. Both cases were excluded by assumptions and (4.141), (4.143) in Theorem 4.14. On the other hand, the definition of $T_\delta(r)$ is not symmetric with respect to the choice of the parameter δ.

4.5 A Liouville-Type Theorem

The aim of this section is to provide a proof for the Liouville-type result given in Theorem 4.19. In doing so we comment on the various assumptions and we compare with a previous result of Dancer and Du [97] which, in turn, generalizes to the elliptic case a consequence in the pioneering work of Aronson and Weinberger [32]. In order to properly comment on the various assumptions we briefly sketch the above mentioned results.

Let us consider the semilinear diffusion equation

$$\frac{\partial u}{\partial t} = \Delta u + f(u) \qquad \text{on } \mathbb{R}_0^+ \times \mathbb{R}^m, \tag{4.163}$$

which arises in population biology and chemical reaction theory. In [32] Aronson and Weinberger showed that if f satisfies

$$f \in C^1(\mathbb{R}_0^+),\ f(0) = 0 = f(a), f(t) > 0 \text{ on } (0, a),\ f(t) \le 0 \text{ on } (a, +\infty) \tag{4.164}$$

for some $a > 0$ and

$$\liminf_{t \to 0+} \frac{f(t)}{t^{1+\frac{2}{m}}} > 0, \tag{4.165}$$

then a "hair trigger" effect takes place, and any nonidentically zero solution $u(x, t)$ of (4.163) with values in $[0, a]$ is such that

$$\lim_{t \to +\infty} u(x, t) = 0$$

uniformly in $x \in \mathbb{R}^m$. Moreover, the exponent of t in (4.165) is sharp in the sense that the hair trigger effect fails if $1 + \frac{2}{m}$ is replaced by any larger σ.

As a consequence one deduces a Liouville result for the elliptic problem associated to (4.163), that is,

$$\Delta u + f(u) = 0. \tag{4.166}$$

Precisely, any solution u of (4.166) on $\mathbb{R}^m$ with values in $[0, a]$ is constant and identically equal to 0 or a.

As for the sharpness of the exponent $1 + \frac{2}{m}$ in (4.165), in order that this type of Liouville result holds it was shown by Dancer [96] that, if $m > 2$ and $\sigma > \frac{m}{m-2}$, one can find a function $f \in C^1(\mathbb{R})$ satisfying (4.164) and $f(t) \ge ct^\sigma$ for $t \to 0^+$ such that (4.166) has a positive solution u with $0 < u < a$ which tends to zero at infinity. In a subsequent work, Du and Guo [105] analyzed the case of the p-Laplace operator and conjectured that, if $m > p$, then the sharp exponent should be given by Serrin's exponent $\sigma = \frac{m(p-1)}{m-p}$ (which reduces to $\sigma = \frac{m}{m-2}$ in the case of the Laplace-Beltrami operator).

The conjecture was proved correct by Dancer and Du [97], using results of Bidaut-Véron and Pohozaev [47] and Serrin and Zou [255]. Here is their result.

Theorem 4.16 (Dancer and Du) *Let $f \in C^0(\mathbb{R}_0^+)$ and locally quasi monotone. Assume that f satisfies* (4.164) *for some $a > 0$. Let $p > 1$ and, if $m \ge p$, assume furthermore that there exist $\varepsilon > 0$ and $C > 0$ such that*

$$f(t) \ge Ct^\xi \quad \text{on } (0, \varepsilon) \tag{4.167}$$

where

$$\xi \in \mathbb{R}^+ \text{ if } m = p \quad \text{and } \xi \in (0, \frac{m(p-1)}{m-p}] \text{ if } m > p. \tag{4.168}$$

Let $b(x) \in C^0(\mathbb{R}^m)$ satisfy $0 < C_1 \leq b(x) \leq C_2 < +\infty$ on $\mathbb{R}^m$. Then any solution of

$$\operatorname{div}\left(|\nabla u|^{p-2}\nabla u\right) + b(x)f(u) = 0 \quad on\ \mathbb{R}^m \tag{4.169}$$

satisfying $0 \leq u \leq a$ is constant (and identically equal to either 0 or a).

We recall that f is said to be *locally quasi monotone* on $\mathbb{R}_0^+$ if for any bounded interval $[\alpha, \beta] \subset \mathbb{R}_0^+$ there exists a continuous increasing function h such that $f(s) + h(s)$ is nondecreasing in $[\alpha, \beta]$.

As remarked in [97], the range of values of ξ in (4.168) for inequality (4.167) is sharp. Furthermore, it follows from the condition that $f(s) < 0$ for $s > a$ that any globally bounded nonnegative solution of (4.169) satisfies $0 \leq u \leq a$, and if in addition f satisfies a condition of the type

$$\liminf_{t\to+\infty} \frac{f(t)}{t^\sigma} > 0 \tag{4.170}$$

for some $\sigma > p - 1$, then any nonnegative solution of (4.169) is in fact globally bounded (see [105]).

The result we shall present below is an extension of these achievements in various directions, and to better compare with the above theorem we consider the following version that can be immediately obtained from Theorem 4.19 below. Here, as usual, $\varphi(t) = tA(t)$ and we suppose the validity of

(A1) $A \in C^1(\mathbb{R}^+)$
(A2) (i) $\varphi'(t) > 0$ on $\mathbb{R}^+$, (ii) $\varphi(t) \to 0$ as $t \to 0^+$.
(A3) $\varphi(t) \leq Ct^\delta$ on $\mathbb{R}^+$ for some $C, \delta > 0$.

In this setting we have

Theorem 4.17 *Let $(M, \langle\ ,\ \rangle)$ be a complete manifold, A and φ be as above and satisfying (A1)–(A3). Let $f \in C^0\left(\mathbb{R}_0^+\right)$ satisfy (4.164) for some $a > 0$ and (4.170) for some $\sigma > \max\{1, \delta\}$; let also $b(x) \in C^0(M)$ and suppose that*

$$b(x) \geq \frac{C}{(1 + r(x))^\mu} \quad on\ M \tag{4.171}$$

for some $C > 0$ and $0 \leq \mu < 1 + \delta$. Let u be a nonnegative solution of

$$\operatorname{div}(A(|\nabla u|)\nabla u) + b(x)f(u) = 0 \quad on\ M. \tag{4.172}$$

Assume that

$$\liminf_{R\to+\infty} \frac{\log \operatorname{vol} B_R}{R^{1+\delta-\mu}} < +\infty, \tag{4.173}$$

and if

$$\frac{1}{(\mathrm{vol}\, \partial B_R)^{1/\delta}} \in L^1(+\infty) \tag{4.174}$$

assume furthermore that

$$f(t) \geq Ct^{\xi}, \quad 0 < t \ll 1 \tag{4.175}$$

for some $\xi > 0$ *and* $C > 0$*. Finally, if*

$$\xi \geq \delta, \tag{4.176}$$

suppose also that

$$u(x) \geq Cr(x)^{-\theta}, \quad r(x) \gg 1 \tag{4.177}$$

for some $\theta \geq 0$, $C > 0$ *and that*

$$\liminf_{R \to +\infty} \frac{\log \mathrm{vol}\, B_R}{R^{1+\delta-\theta(\xi-\delta+\varepsilon)-\mu}} < +\infty \tag{4.178}$$

for some $\varepsilon > 0$*. Then* u *is constant and identically equal to* 0 *or* a*.*

We observe that, by way of example, it is not difficult to see that condition (4.173) may hold independently of the validity of (4.174). More elaborate arguments allow to construct model manifolds such that

$$\frac{1}{(\mathrm{vol}\, \partial B_R)^{1/\delta}} \notin L^1(+\infty), \tag{4.179}$$

and yet vol B_R grows arbitrarily fast, as recalled in Remark 2.19. In particular (4.179) does not imply neither (4.173) nor (4.178).

As for condition (4.177), it has no counterpart in the result of Dancer and Du, but in fact is automatically satisfied in the situation they consider; it is necessary in our more general setting. We will come back to this in a shortwhile.

Now suppose $M = \mathbb{R}^m$ and $A(t) = t^{p-2}$, $p > 1$, so that in our Theorem 4.17 we consider the case of the p-Laplacian on $\mathbb{R}^m$ as in Dancer and Du. Assumption (4.170) is common and needed to guarantee that $u \geq 0$ is bounded above and satisfies $0 \leq u \leq a$ on $\mathbb{R}^m$; thus we concentrate on the remaining assumptions. In the Euclidean space $\mathbb{R}^m$, (4.173) is automatically true since $0 \leq \mu < 1 + \delta$ (Dancer and Du case has $\mu = 0$). Since $\delta = p - 1$ for the p-Laplacian, (4.174) becomes

$$\frac{1}{R^{\frac{m-1}{p-1}}} \in L^1(+\infty)$$

which is the case for $m > p$. Hence for $m = p$ we do not require, contrary to Dancer and Du, the validity of (4.167). For $m > p$ in (4.175) we require (4.167) but, if $\xi < p-1$, no further assumption is needed. We have to see what happens for the range $\xi \in \left[p-1, \frac{m(p-1)}{m-p}\right]$ considered in [97]. Towards this aim we need to find a priori lower bounds for nonnegative solutions of $\Delta_p u = 0$. Let us consider here the more general case of nonnegative solutions of

$$\operatorname{div}\left(A(|\nabla u|)\nabla u\right) \leq 0. \tag{4.180}$$

As above we let $\varphi(t) = tA(t)$; we prove

Lemma 4.4 *Let $\varphi \in C^0\left(\mathbb{R}_0^+\right) \cap C^1\left(\mathbb{R}^+\right)$ satisfy*

$$(i)\ \varphi(0) = 0; \quad (ii)\ \varphi(t) > 0 \ \text{on} \ \mathbb{R}^+ \tag{4.181}$$

and assume that φ is strictly increasing in $[0,\varepsilon)$ for some $\varepsilon > 0$ and that

$$\varphi(t) \sim C_0 t^{\xi} \quad \text{as} \ t \to 0^+ \tag{4.182}$$

for some $C_0, \xi > 0$. Let $g \in C^1\left(\mathbb{R}_0^+\right)$ be such that $g(0) = 0$, $g(t) > 0$ on $\mathbb{R}^+$, $g'(t) > 0$ for $t \gg 1$ and suppose that, for some $m > 1$,

$$g(t)^{-\frac{m-1}{\xi}} \in L^1(+\infty). \tag{4.183}$$

Fix $R, H > 0$. Then there exists $B > 0$ such that, having denoted with $\psi : [0,\varphi(\varepsilon)) \to [0,\varepsilon)$ the local inverse of φ, the function α defined by

$$\alpha(r) = \int_r^{+\infty} \psi\left(Bg(t)^{1-m}\right) dt \tag{4.184}$$

is defined and C^2 on $[R,+\infty)$ and satisfies

$$\begin{cases} \varphi(|\alpha'|)' + (m-1)\frac{g'}{g}\varphi(|\alpha'|) = 0 \\ \alpha(r) \leq \alpha(R) = D < H,\ \alpha'(r) < 0 \ \text{for} \ r \geq R. \end{cases} \tag{4.185}$$

Furthermore,

$$\alpha(r) \sim \left(\frac{B}{C_0}\right)^{1/\xi} \int_r^{+\infty} g(t)^{-\frac{m-1}{\xi}} dt \ \text{ as } r \to +\infty. \tag{4.186}$$

In particular, if

$$\limsup_{r\to+\infty} \frac{g'}{g}(r) < +\infty,$$

then there exists $C > 0$ *such that*

$$\alpha(r) \geq Cg(r)^{-\frac{m-1}{\xi}} \quad \textit{for } r \geq R, \tag{4.187}$$

and if

$$\frac{g'}{g}(r) \ \textit{ is eventually decreasing },$$

$$\alpha(r) \geq C\frac{g(r)}{g'(r)}g(r)^{-\frac{m-1}{\xi}} \quad \textit{for } r \geq R. \tag{4.188}$$

Proof Note that, since g is eventually increasing and $g(t)^{-\frac{m-1}{\xi}}$ is integrable at infinity, $g(t) \to +\infty$ as $t \to +\infty$. In particular, for $B > 0$ sufficiently small, $Bg(t)^{-(m-1)} < \varphi(\varepsilon)$ for every $t \geq R$. Furthermore, it follows from (4.182) that $\psi(t) \sim \left(\frac{t}{C_0}\right)^{1/\xi}$ as $t \to 0^+$ so that

$$\psi\left(Bg(t)^{-(m-1)}\right) \sim \left(\frac{B}{C_0}\right)^{1/\xi} g(t)^{-\frac{m-1}{\xi}} \quad \text{as } t \to 0^+ \tag{4.189}$$

and the integral in (4.184) is well defined for each $r \geq R$. It is clear that α is C^2, decreasing, and, by choosing a smaller B if necessary, it can be arranged that $\alpha(r) < H$ on $[R, +\infty)$. A computation shows that α satisfies (4.185); it follows from (4.189) that α satisfies also (4.186). Finally, if $\frac{g'}{g} \leq \eta$ for $t \geq R$, the integrand in (4.186) is bounded from below by

$$\frac{1}{\eta}g(t)^{-\frac{m-1}{\xi}-1}g'(t),$$

and (4.187) follows by integration recalling that $g(t) \to +\infty$ as $t \to +\infty$. A similar reasoning proves that if $\frac{g'}{g}(t)$ is eventually decreasing then (4.188) holds. □

Proposition 4.4 *Let* φ *and* g *satisfy the conditions listed in Lemma 4.4, and assume*

$$\Delta r \leq (m-1)\frac{g'}{g}(r) \tag{4.190}$$

pointwise in the complement of the cut locus of the fixed origin o*. Let* u *be a nonnegative* C^1 *solution of*

$$\operatorname{div}(A(|\nabla u|)\nabla u) \leq 0. \tag{4.191}$$

Then, there exist constants C and $R > 0$ such that

$$u(x) \geq C \int_{r(x)}^{+\infty} g(t)^{-\frac{m-1}{\xi}} \, dt \quad on \ M \setminus B_R. \tag{4.192}$$

Furthermore, if

$$\limsup_{r \to +\infty} \frac{g'}{g}(r) < +\infty,$$

then there exists $C > 0$ such that

$$u(x) \geq C g(r(x))^{-\frac{m-1}{\xi}} \quad when \ x \in M \setminus B_R, \tag{4.193}$$

and if

$$\frac{g'}{g} \ \ is \ eventually \ decreasing$$

then

$$u(x) \geq C \frac{g(r(x))}{g'(r(x))} g(r(x))^{-\frac{m-1}{\xi}} \quad where \ x \in M \setminus B_R. \tag{4.194}$$

Proof Fix $R > 0$ so that $g'(t) > 0$ on $(R, +\infty)$ and choose $B > 0$ small enough that the function α defined in (4.184) satisfies the conditions of the statement of Lemma 4.4 with $H = \inf_{\partial B_R} u$. Set $v(x) = \alpha(r(x))$. It follows from (4.185) and (4.190) that the inequality

$$\begin{aligned} \operatorname{div}(A(|\nabla v|)\nabla v) &= -\varphi(|\alpha'|)' - \varphi(|\alpha'|)\Delta r \\ &\geq -\varphi(|\alpha'|)' - (m-1)\frac{g'}{g}\varphi(|\alpha'|) = 0 \end{aligned} \tag{4.195}$$

holds pointwise in the complement of the cut locus of o and, similarly to what we did for instance in Lemma 1.6 by adapting an argument of Yau [280], weakly on M. Thus

$$\begin{cases} \operatorname{div}(A(|\nabla v|)\nabla v) \geq \operatorname{div}(A(|\nabla u|)\nabla u) & \text{on } M \setminus B_R \\ v < u & \text{on } \partial B_R. \end{cases} \tag{4.196}$$

We claim that $u \geq v$ on $M \setminus B_R$. Indeed, otherwise, there would exist $\eta > 0$ and $x_0 \in M \setminus \overline{B}_R$ such that $u(x_0) < v(x_0) - \eta$. Thus the set

$$A_\eta = \{x \in M \setminus B_R : u(x) < v(x) - \eta\}$$

would be open, nonempty, and $x_0 \in A_\eta \subseteq \overline{A}_\eta \subseteq M \setminus \overline{B}_R$. Since $v(x) \to 0$ as $r(x) \to +\infty$ while u is positive on M, $\overline{A}_\eta$ is bounded; thus completeness of M implies that it is compact. Since $u = v - \eta$ on ∂A_η by Proposition 3.1, we have $u \geq v - \eta$ on A_η and therefore $u(x_0) \geq v(x_0) - \eta$, contradicting the definition of η and x_0. Now the required lower estimates follow from Lemma 4.4. □

Corollary 4.2 *Let $(M, \langle\, , \rangle)$ be a complete m-dimensional Riemannian manifold with a fixed origin o and $r(x) = \operatorname{dist}(x, o)$. Assume that the radial Ricci curvature satisfies*

$$\operatorname{Ric}(\nabla r, \nabla r) \geq -(m-1)G^2(r) \tag{4.197}$$

for some positive function $G \in C^1\left(\mathbb{R}_0^+\right)$ such that

$$\begin{cases} (i)\ \inf_{\mathbb{R}^+} \frac{G'}{G^2} > -\infty \\ (ii)\ \limsup_{r\to+\infty} G(r) < +\infty \\ (iii)\ G(r) \notin L^1(+\infty) \\ (iv)\ e^{-\frac{m-1}{\zeta}D_0 \int_0^r G(s)\,ds} \in L^1(+\infty) \end{cases} \tag{4.198}$$

with ζ as in Lemma 4.4, for some $D_0 > 0$. Let φ be as in Lemma 4.4 and for $\varphi(t) = tA(t)$ let u be a nonnegative, nonidentically zero solution of

$$\operatorname{div}(A(|\nabla u|)\nabla u) \leq 0 \quad \text{on } M.$$

Then, there exist constants $C > 0$ and $D \geq D_0$ such that

$$u(x) \geq Ce^{-\frac{m-1}{\zeta}D\int_0^{r(x)} G(s)\,ds}. \tag{4.199}$$

If G is assumed to be nonincreasing then

$$u(x) \geq CG^{-1}(r(x))e^{-\frac{m-1}{\zeta}D\int_0^{r(x)} G(s)\,ds}. \tag{4.200}$$

Proof We set

$$g(r) = \frac{1}{DG(0)}\left\{e^{D\int_0^r G(s)\,ds} - 1\right\}.$$

As we already know, see (2.33), using the Laplacian comparison Theorem, for $D > 0$ sufficiently large we have

$$\Delta r \leq (m-1)\frac{g'(r)}{g(r)}$$

pointwise in the complement of the cut locus of o and weakly on M. Note that (4.198) (iii) implies $g(r) \to +\infty$ as $r \to +\infty$, and

$$\frac{g'(r)}{g(r)} \sim DG(r) \quad \text{as } r \to +\infty.$$

We choose $D \geq D_0$ so that, by (4.198) (iv), condition (4.183) of Lemma 4.4 holds, and applying Proposition 4.4 we deduce that, for some $H > 0$,

$$u(r) \geq Hg(r(x))^{-\frac{m-1}{\xi}} \geq Ce^{-\frac{m-1}{\xi}D\int_0^{r(x)} G(s)\,ds},$$

which can be improved to

$$u(x) \geq CG(r(x))^{-\frac{1}{2}}e^{-\frac{m-1}{\xi}D\int_0^{r(x)} G(s)\,ds}$$

if G is nonincreasing. □

Note that if $(M, \langle\, ,\, \rangle)$ is $\mathbb{R}^m$ with its flat metric we have $\Delta r = \frac{m-1}{r}$ so that the inequality $\Delta r \leq (m-1)\frac{g'}{g}$ holds if $g(r) = r^D$ with $D \geq 1$ and we deduce that nonnegative solutions of

$$\Delta_p u \leq 0$$

satisfy the bound

$$u(x) \geq Cr^{1-\frac{m-1}{p-1}} \quad \text{if } m > p \tag{4.201}$$

for some $C > 0$, while, if $m \leq p$, for every $\eta > 0$ there exists $C = C(\eta) > 0$ such that

$$u(x) \geq Cr^{-\eta}. \tag{4.202}$$

Inserting (4.201) in the statement of Theorem 4.17 with $\delta = p - 1$ and $\mu = 0$ we see that condition (4.178) becomes

$$\liminf_{r\to+\infty} \frac{m\log r}{r^{p-\frac{m-p}{p-1}(\xi-p+1+\varepsilon)}} < +\infty$$

for some $\varepsilon > 0$. It follows that in this case Theorem 4.17 is applicable provided

$$0 < \xi < \frac{m(p-1)}{m-p},$$

which should be compared with the range

$$0 < \xi \leq \frac{m(p-1)}{m-p}$$

obtained by Dancer and Du.

Note that Corollary 4.2 can be applied to obtain lower bounds for $u(x)$ in other situations, for instance when $G^2(r) = \frac{B^2}{1+r^2}$ which corresponds to a geometric behaviour borderline between the Euclidean and the non-Euclidean case. For φ as in the statement of Lemma 4.4, having chosen $D > \frac{\xi}{m-1}$ one can show that any nonnegative solution of

$$\mathrm{div}\,(A(|\nabla u|)\nabla u) \leq 0$$

satisfies the bound

$$u(x) \geq Cr^{1-\frac{D(m-1)}{\xi}}$$

for some $C > 0$.

In Theorem 4.2 above we proved, under some assumptions, that if $u \in C^1(M)$ satisfies

$$Lu \geq b(x)f(u)$$

on a set $\Omega_\gamma = \{x \in M : u(x) > \gamma\} \neq \emptyset$, and

$$\liminf_{t \to -\infty} \frac{f(t)}{t^\xi} > 0$$

for some $\xi > \delta$ then $u^* < +\infty$. We are now going to look for an a priori lower bound.

Theorem 4.18 *Let φ, b, Q, T and Θ satisfy the assumptions of Theorem 4.1 with $A(x) = A$ a positive constant, φ independent of x and T satisfying (T2). Let $f \in C^0(\mathbb{R})$ and assume that $u \in C^1(M)$ is a nonnegative and nonidentically zero solution of*

$$Lu = \mathrm{div}\left(|\nabla u|^{-1}\varphi(|\nabla u|)T(\nabla u, \cdot)^\sharp\right) \leq -b(x)f(u) \tag{4.203}$$

on the set

$$\Omega_{\gamma_0} = \{x \in M : u(x) < \gamma_0\} \tag{4.204}$$

for some $\gamma_0 > u_* = \inf_M u$. *If*

$$f(t) \geq Ct^{\xi} \quad \text{as } t \to 0^+ \quad \text{for some } \xi < \delta, \tag{4.205}$$

and either

$$\liminf_{r\to+\infty} \frac{Q(r)\Theta(r)\log \operatorname{vol} B_r}{r^{1+\delta}} < +\infty \tag{4.206}$$

or

$$\liminf_{r\to+\infty} \frac{Q(r)\Theta(r)}{r^{1+\delta}} \int_{B_r} |u|^p < +\infty \tag{4.207}$$

holds for some $p > 0$, *then* $u_* > 0$.

Proof Observe that, by the strong maximum principle of Theorem 3.10, u is strictly positive on M. We assume by contradiction that $u_* = 0$, so that u satisfies (4.203) on Ω_γ for any $0 < \gamma < \gamma_0$. Observe that in this case by Ω_γ we denote the set $\Omega_\gamma = \{x \in M : u(x) < \gamma\}$. Fix such a γ in such a way that, for $t \in (0, \gamma)$,

$$f(t) \geq Bt^{\xi}$$

for some constant $B > 0$. It follows that

$$-Lu \geq \frac{B}{Q(r)} u^{\xi} \qquad \text{on } \Omega_\gamma.$$

For the ease of notation we may suppose that $B = 1$. Similarly to what we did in the proof of Theorem 4.1, we let $\lambda : \mathbb{R} \to \mathbb{R}$ be a C^1 function such that $\lambda(t) = 0$ if $t \geq \gamma$, $\lambda(t) > 0$ if $t < \gamma$ and $\lambda' \leq 0$. Choose $R > 0$ large enough that $B_R \cap \Omega_\gamma \neq \emptyset$ and, for $r \geq R$, let $\psi = \psi_r$ be a smooth cutoff function with $\psi = 1$ on B_r, $\psi = 0$ off B_{2r} and $|\nabla\psi| \leq \frac{c_0}{r}\psi^{\frac{1}{\Gamma}}$ for some c_0 and $\Gamma > 1$ independent of r. Finally we let W be the vector field defined by

$$W = -\psi^{\alpha}\lambda(u)u^{-\beta}|\nabla u|^{-1}\varphi(|\nabla u|)T(\nabla u, \,)^{\sharp}, \tag{4.208}$$

where $\alpha, \beta > 0$ are constants to be determined later. Using, as in the proof of Theorem 4.1, $\lambda' \leq 0$, $|\nabla u|^{-1}\varphi(|\nabla u|) \geq A^{-1/\delta}\varphi(|\nabla u|)^{1+\frac{1}{\delta}}$, $T_u = \frac{T(\nabla u, \nabla u)}{|\nabla u|^2} > 0$ and $|T(\nabla u, \nabla\psi)| \leq T_u^{1/2}T_+^{1/2}|\nabla u||\nabla\psi|$ we estimate

$$\begin{aligned}\operatorname{div} W \geq{}& \lambda(u)\psi^{\alpha}b(x)u^{\xi-\beta} + \frac{\beta}{A^{1/\delta}}\psi^{\alpha}u^{-\beta-1}\lambda(u)\varphi(|\nabla u|)^{1+1/\delta}T_u \\ &- \alpha\psi^{\alpha-1}\lambda(u)u^{-\beta}\varphi(|\nabla u|)|\nabla\psi|T_u^{1/2}T_+^{1/2}.\end{aligned} \tag{4.209}$$

Now we argue as in the proof of Theorem 4.1 and we estimate the last term on the right-hand side using the inequality

$$ab \leq \frac{\sigma^p a^p}{p} + \frac{b^q}{\sigma^q q}, \quad a, b \geq 0$$

with $p = 1 + 1/\delta$, $q = 1 + \delta$ and with

$$\sigma = \left[\beta(1+\delta)/\left(A^{1/\delta}\alpha\delta\right)\right]^{\frac{\delta}{1+\delta}}$$

chosen in such a way as to cancel the second term.

Indeed, using $|\nabla\psi| \leq \frac{C_0}{r}\psi^{1/\Gamma}$, we obtain

$$\begin{aligned}
&\alpha\psi^{\alpha-1}\lambda(u)u^{-\beta}\varphi(|\nabla u|)|\nabla\psi|T_u^{1/2}T_+^{1/2} \\
&= \left(\alpha^{\frac{1}{p}}\lambda(u)^{\frac{1}{p}}u^{-\frac{\beta}{p}-\frac{1}{p}}\psi^{\frac{\alpha}{p}}\varphi(|\nabla u|)T_u^{\frac{1}{p}}\right) \\
&\times\left(\alpha^{\frac{1}{q}}\lambda(u)^{\frac{1}{q}}u^{-\frac{\beta}{q}+\frac{1}{p}}\psi^{\frac{\alpha}{q}-1}T_u^{\frac{1}{2}-\frac{1}{p}}T_+^{\frac{1}{2}}|\nabla\psi|\right) \\
&\leq \frac{\beta}{A^{\frac{1}{\delta}}}\lambda(u)u^{-\beta-1}\psi^{\alpha}\varphi(|\nabla u|)^{1+\frac{1}{\delta}} \\
&+ A\left(\frac{\alpha}{\beta}\right)^{\delta}\frac{\delta^{\delta}}{(1+\delta)^{1+\delta}}\alpha\frac{C_0^{1+\delta}}{r^{1+\delta}}\lambda(u)u^{\delta-\beta}\psi^{\alpha-(1+\delta)\left(1-\frac{1}{\Gamma}\right)}T_u^{\frac{1-\delta}{2}}T_+^{\frac{1+\delta}{2}}.
\end{aligned}$$

Setting $C_1 = A\frac{\delta^{\delta}}{(1+\delta)^{1+\delta}}C_0^{1+\delta}$ and inserting in (4.209) we have

$$\operatorname{div} W \geq \lambda(u)\psi^{\alpha}b(x)u^{\xi-\beta} - C_1\left(\frac{\alpha}{\beta}\right)^{\delta}\frac{\alpha}{r^{1+\delta}}\lambda(u)u^{\delta-\beta}\psi^{\alpha-(1+\delta)\left(1-\frac{1}{\Gamma}\right)}T_u^{\frac{1-\delta}{2}}T_+^{\frac{1+\delta}{2}}.$$

Integrating this inequality, applying the divergence theorem and observing that W is compactly supported we deduce

$$\int_M \lambda(u)\psi^{\alpha}b(x)u^{\xi-\beta} \leq C_1\left(\frac{\alpha}{\beta}\right)^{\delta}\frac{\alpha}{r^{1+\delta}}\int_M \psi^{\alpha-(1+\delta)\left(1-\frac{1}{\Gamma}\right)}\lambda(u)u^{\delta-\beta}T_u^{\frac{1-\delta}{2}}T_+^{\frac{1+\delta}{2}},$$

provided $\alpha - (1+\delta)\left(1-\frac{1}{\Gamma}\right) \geq 0$. Furthermore note that the constant C_1 is independent of α, β, r. Since $Q(r)$ is nondecreasing and $b(x) \geq \frac{1}{Q(r(x))}$ from the above we obtain

$$Q(2r)^{-1}\int_M \lambda(u)\psi^{\alpha}u^{\xi-\beta} \leq C_1\left(\frac{\alpha}{\beta}\right)^{\delta}\frac{\alpha}{r^{1+\delta}}\int_M \psi^{\alpha-(1+\delta)\left(1-\frac{1}{\Gamma}\right)}\lambda(u)u^{\delta-\beta}T_u^{\frac{1-\delta}{2}}T_+^{\frac{1+\delta}{2}}. \tag{4.210}$$

Next we estimate the integral on the right-hand side of (4.210). We let p and q be conjugate exponents, so that

$$\int_M \psi^{\alpha-(1+\delta)(1-\frac{1}{\Gamma})}\lambda(u)u^{\delta-\beta}T_u^{\frac{1-\delta}{2}}T_+^{\frac{1+\delta}{2}}$$

$$\leq \int_M \psi^{\frac{\alpha}{p}}\lambda(u)^{\frac{1}{p}}u^{\delta-\beta}\psi^{\frac{\alpha}{q}-(1+\delta)(1-\frac{1}{\Gamma})}\lambda(u)^{\frac{1}{q}}T_u^{\frac{1-\delta}{2}}T_+^{\frac{1+\delta}{2}}$$

$$\leq \left\{\int_M \psi^{\alpha}\lambda(u)u^{(\delta-\beta)p}\right\}^{\frac{1}{p}}$$

$$\times\left\{\int_M \psi^{\alpha-(1+\delta)(1-\frac{1}{\Gamma})q}\lambda(u)T_u^{\frac{1-\delta}{2}q}T_+^{\frac{1+\delta}{2}q}\right\}^{\frac{1}{q}},$$

provided

$$\alpha-(1+\delta)\left(1-\frac{1}{\Gamma}\right)q \geq 0. \tag{4.211}$$

We choose $p=\frac{\beta-\xi}{\beta-\delta}>1$, by the assumption $\xi<\delta$. It follows that the first integral above is equal to the integral on the left-hand side of (4.210). Thus, we insert into (4.210), we simplify, we use the definition of Θ, (4.21) and the properties of ψ to obtain

$$\int_{B_r}\lambda(u)u^{\xi-\beta} \leq \left\{C_1\left(\frac{\alpha}{\beta}\right)^{\delta}\frac{\Theta(2r)Q(2r)}{r^{1+\delta}}\alpha\right\}^{q}\int_{B_{2r}}\lambda(u)$$

provided the validity of (4.211). Since

$$q=\frac{\beta-\xi}{\delta-\xi}$$

if we choose

$$\beta=\alpha+\xi$$

condition (4.211) becomes

$$1>(1+\delta)\left(1-\frac{1}{\Gamma}\right)/(\delta-\xi),$$

which holds provided Γ is sufficiently close to 1. Now, since $u < \gamma$ on Ω_γ, $u^{\xi-\beta} > \gamma^{\xi-\beta}$ for $\beta > \xi$; furthermore, $\frac{\alpha}{\beta} < 1$. We therefore deduce

$$\int_{B_r} \lambda(u) \le \left\{ C_1 \gamma^{\delta-\xi} \alpha \frac{\Theta(2r)Q(2r)}{r^{1+\delta}} \right\}^{\frac{\alpha}{\delta-\xi}} \int_{B_{2r}} \lambda(u),$$

hence choosing

$$\alpha = \frac{1}{2C_1} \gamma^{-(\delta-\xi)} \frac{r^{1+\delta}}{\Theta(2r)Q(2r)}$$

we get

$$\int_{B_r} \lambda(u) \le \left(\frac{1}{2}\right)^{\frac{1}{2C_1(\delta-\xi)\gamma^{\delta-\xi}} \frac{r^{1+\delta}}{\Theta(2r)Q(2r)}} \int_{B_{2r}} \lambda(u).$$

Now the proof proceeds as at the end of the proof of Theorem 4.1 in either one of the assumptions (4.206) or (4.207). □

We are now ready to state our Liouville-type result.

Theorem 4.19 *Let φ, T, T_δ and Θ be as in the statement of Theorem 4.18 and suppose that*

$$\Theta(r) \le Cr^{\nu} \tag{4.212}$$

for some $\nu > 0$. Let $u \in C^1(M)$ be a nonnegative solution of

$$L_{\varphi,T} u = -b(x) f(u) \quad \text{on } M, \tag{4.213}$$

where $b \in C^0(M)$ is such that

$$b(x) \ge \frac{C}{(1 + r(x))^{\mu}} \quad \text{on } M \tag{4.214}$$

for some $C > 0$ and $0 \le \mu < 1 + \delta$, and $f \in C^0(\mathbb{R}_0^+)$ satisfies $f(0) = f(a) = 0$, $f(t) > 0$ on $(0, a)$, $f(t) < 0$ in $(a, +\infty)$ for some $a > 0$, and

$$\liminf_{t\to+\infty} -\frac{f(t)}{t^{\sigma}} > 0 \tag{4.215}$$

for some $\sigma > \max\{1, \delta\}$. *Assume that*

$$\inf_{\mathbb{R}_0^+} T_- > 0, \tag{4.216}$$

$$\liminf_{r \to +\infty} \frac{\log \operatorname{vol} B_r}{r^{1+\delta-(\mu+\nu)}} < +\infty \tag{4.217}$$

and, if

$$(T_+(r) \operatorname{vol}(\partial B_r))^{-1/\delta} \in L^1(+\infty) \tag{4.218}$$

assume furthermore that

$$f(t) \geq Ct^{\xi}, \quad 0 < t \ll 1 \tag{4.219}$$

for some $\xi > 0$ *and* $C > 0$. *Finally, if*

$$\xi \geq \delta \tag{4.220}$$

suppose also that

$$u(x) \geq Cr(x)^{-\theta} \quad \textit{for } r(x) \gg 1 \tag{4.221}$$

for some $\theta \geq 0$, $C > 0$, *and that*

$$\liminf_{r \to +\infty} \frac{\log \operatorname{vol} B_r}{r^{1+\delta-\theta(\xi-\delta+\varepsilon)-(\mu+\nu)}} < +\infty \tag{4.222}$$

for some $\varepsilon > 0$. *Then* u *is constant and identically equal to* 0 *or* a.

Remark 4.14 Defining $T_\delta(r)$ as in (4.160), the theorem holds getting rid of (4.216) and substituting (4.218) with

$$(T_\delta(r) \operatorname{vol}(\partial B_r))^{-1/\delta} \in L^1(+\infty).$$

Proof (of Theorem 4.19) We set $u^* = \sup_M u$ and $u_* = \inf_M u$. Next we divide the argument into several steps.

Step 1. Assumption (4.215) gives

$$-f(t) \geq Ct^{\sigma}$$

for $t \gg 1$, some constant $C > 0$ and $\sigma > \max\{1, \delta\}$. Putting this together with (4.214), $0 \leq \mu < 1 + \delta$ and the volume growth condition (4.217) yields, by

Theorem 4.2, $u^* < +\infty$. Note that the same conclusion holds if we assume that condition (4.217) is replaced by condition (4.29).

Step 2. Since u is bounded above and (4.217) holds, Theorem 4.1 implies that $-f(u^*) \le 0$ so that, by the properties of f, $u^* \in [0, a]$ and $0 \le u \le a$ on M. It follows that

$$L_{\varphi,T} u \le 0 \quad \text{on } M. \tag{4.223}$$

Again, the same conclusion holds if we assume condition (4.29)instead of (4.217).

Step 3. If

$$(T_+(r) \operatorname{vol}(\partial B_r))^{-1/\delta} \notin L^1(+\infty),$$

using (4.216), by Theorem 4.14 the manifold $(M, \langle\, , \rangle)$ is $L_{\varphi,T}$-parabolic and therefore $u \ge 0$ together with (4.223) implies that u is constant. Since $b(x) > 0$ on M and f vanishes only in 0 and a, it follows from (4.213) that either $u \equiv 0$ or $u \equiv a$.

Step 4. If

$$(T_+(r) \operatorname{vol}(\partial B_r))^{-1/\delta} \in L^1(+\infty),$$

then the manifold $(M, \langle\, , \rangle)$ is not necessarily $L_{\varphi,T}$-parabolic and further analysis is required. First we note that $0 \le u_* \le a$; then by Theorem 4.1 and Remark 4.3 we have $f(u_*) \le 0$. Thus, u_* is either 0 or a. In the latter case we have $u_* = u^* = a$, so that $u \equiv a$; if $u_* = u^* = 0$ again u is constant, $u \equiv 0$. Thus the only case to consider is $u_* = 0$ and $0 < u^* \le a$. To show that this cannot happen it is enough to show that under our assumptions $u_* > 0$.

Now, since u satisfies (4.223) and it does not vanish identically by the strong maximum principle, Theorem 3.10, u is strictly positive on M. If (4.219) holds and $\xi < \delta$ we apply Theorem 4.18 to conclude that $u_* > 0$. Otherwise, that is, if $\xi \ge \delta$, we observe that u is a solution of

$$L_{\varphi,T} u = -\tilde{b}(x)\tilde{f}(u)$$

with

$$\tilde{f}(u) = f(u)u^{-(\xi-\delta+\varepsilon)} \quad \text{and } \tilde{b}(x) = b(x)u^{\xi-\delta+\varepsilon}.$$

According to (4.221) and (4.214) we have

$$\tilde{b}(x) \ge C(1 + r(x))^{-\mu-\theta(\xi-\delta+\varepsilon)}$$

and the required conclusion follows from (4.222) and a further application of Theorem 4.18. □

Remark 4.15 As pointed out in the proof, assumption (4.217) can be substituted with condition (4.29) of Remark 4.2.

Chapter 5
Miscellany Results for Submanifolds

This chapter is basically devoted to miscellany applications of the results presented in Chaps. 3 and 4. We show how, with the aid of various forms of the maximum principle, we can improve on some classical results. In fact we begin with some introductory considerations to motivate a nonimmersibility result for a manifold M into cones of the Euclidean space due to Omori [210] and of which we provide an improved version in Theorem 5.1. We recall that, in the cited work of Omori, we have the first appearance of what is now known in the literature as the Omori-Yau maximum principle. We then continue our investigation by establishing a quantitative estimate, according to the results presented in [181], for the width of the cone of $\mathbb{R}^n$ containing the image of M under a smooth map, see Theorem 5.2. Later, we elaborate on some old result of Jorge and Koutroufiotis [154], (see Theorem 5.6) and we provide a "quantitative" version for immersions into a cone with the aid of the WMP for the Hessian (see Theorem 5.7). With the help of this result and of the theory of flat bilinear forms we are able to consider also the case where M is a Kähler manifold in Corollary 5.7.

A good portion of the chapter deals with cylindrically bounded submanifolds that are strictly related to a famous conjecture of Calabi [56] of which we prove the validity under some very mild additional assumptions (for instance see Theorem 5.9 and Corollary 5.8). As it is well known, this conjecture in its original formulation is false, see [155, 205]. The chapter ends with some consequences on the geometry of the Gauss map for submanifolds of Euclidean space with parallel mean curvature vector, where we use the well known result of Ruh and Vilms on the harmonicity of the Gauss map. In particular we give a sufficient condition for a parallel mean curvature immersion of M in $\mathbb{R}^n$ to be minimal and we analyze the size of its image under the Gauss map respectively in Theorems 5.11 and 5.12.

In the very final section we deal with an application of the open form of the WMP.

L.J. Alías et al., *Maximum Principles and Geometric Applications*,
Springer Monographs in Mathematics, DOI 10.1007/978-3-319-24337-5_5

5.1 Immersions into Nondegenerate Cones in Euclidean Space

A typical application of the usual maximum principle for compact submanifolds in Euclidean space is the proof of the fact that every compact surface in $\mathbb{R}^3$ has an elliptic point, that is, a point where the Gaussian curvature is positive. As a consequence, we have the well known classical fact expressed in the next

Corollary 5.1 *No compact Riemannian surface with everywhere nonpositive Gaussian curvature can be isometrically immersed in* $\mathbb{R}^3$.

More generally, by applying the usual maximum principle we can also prove the following

Proposition 5.1 *Let $f : M \to \mathbb{R}^n$ be an isometric immersion of a compact manifold. Then there exists a point $x_0 \in M$ and a normal vector $\xi \in T_{x_0}M^{\perp}$ such that the second fundamental form at x_0 with respect to ξ is positive definite.*

Proof To see this, given f consider the smooth function $u : M \to \mathbb{R}$ defined by $u(x) = \frac{1}{2}|f(x)|^2$. With the notation of Sect. 1.6 of Chap. 1, supposing $\dim M = m$, we now compute $\mathrm{Hess}(u)$. Thus let $\{\theta^a\}$ be a Darboux frame along f, so that $\theta^\alpha = 0$ on M and let $\{\theta^a_b\}$ be the corresponding Levi-Civita connection forms. We have

$$du = \langle df,f\rangle = \langle e_i,f\rangle\theta^i,$$

so that

$$u_i = \langle e_i,f\rangle. \tag{5.1}$$

Then

$$\begin{aligned} u_{ij}\theta^j &= du_i - u_t\theta^t_j = \langle de_i,f\rangle + \langle e_i,df\rangle - \langle e_t,f\rangle\theta^t_j \\ &= \theta^k_i\langle e_k,f\rangle + h^\alpha_{ij}\langle e_\alpha,f\rangle\theta^j + \langle e_i,f_j\rangle\theta^j - \langle e_t,f\rangle\theta^t_j \\ &= \left(h^\alpha_{ij}\langle e_\alpha,f\rangle + \delta_{ij}\right)\theta^j. \end{aligned}$$

It follows that

$$\mathrm{Hess}(u) = \langle\,,\,\rangle_M + \langle \mathrm{II}(\,,\,),f\rangle = \langle\,,\,\rangle_M + \langle \mathrm{II}(\,,\,),f^{\perp}\rangle. \tag{5.2}$$

Since M is compact, there exists a point $x_0 \in M$ at which u attains its maximum $u^* > 0$, and by the usual maximum principle we have

$$\text{(i) } u(x_0) = u^* > 0, \;\; \text{(ii) } \nabla u(x_0) = 0, \;\; \text{and (iii)' } \mathrm{Hess}(u)(x_0) \le 0, \tag{5.3}$$

in the sense that

$$\text{Hess}(u)(x_0)(v, v) \leq 0 \quad \text{for all } v \in T_{x_0}M.$$

By (5.1), conditions (i) and (ii) mean that $f(x_0) = f(x_0)^\perp \neq 0$ is a normal vector at the point x_0. Then, by (5.2) and choosing $\xi = -f(x_0)^\perp$, condition (iii)$'$ yields

$$\langle \text{II}(v, v), \xi \rangle \geq |v|^2$$

for every $v \in T_{x_0}M$. This proves the Proposition. □

As a consequence, no compact Riemannian manifold M can be isometrically immersed as a minimal submanifold into an Euclidean space $\mathbb{R}^n$; in other words, there exists no minimal compact submanifold of Euclidean space.

Motivated by this fact, in [210], and as the first application of the Omori-Yau maximum principle, Omori proved that for every complete Riemannian manifold M with sectional curvature bounded from below for which there exists an isometric immersion $f : M \to \mathbb{R}^n$ with $f(M)$ contained into a nondegenerate cone of $\mathbb{R}^n$, there exists a point $x_0 \in M$ and a normal vector $\xi \in T_{x_0}M^\perp$ such that the second fundamental form at x_0 with respect to ξ is positive definite. In particular, no complete Riemannian manifold M with sectional curvature bounded from below can be isometrically immersed as a minimal submanifold into a nondegenerate cone of $\mathbb{R}^n$.

Here by a *nondegenerate cone of* $\mathbb{R}^n$ we mean the following. Fix an origin $o \in \mathbb{R}^n$ and a unit vector $\zeta \in \mathbb{S}^{n-1}$. We set $\mathscr{C}_{o,\zeta,\theta}$ (shortly, $\mathscr{C}$) to denote the nondegenerate cone of $\mathbb{R}^n$ with vertex o, direction ζ and width $\theta \in (0, \pi/2)$, that is,

$$\mathscr{C}_{o,\zeta,\theta} = \mathscr{C} = \left\{ p \in \mathbb{R}^n \backslash \{o\} : \left\langle \frac{p - o}{|p - o|}, \zeta \right\rangle \geq \cos\theta \right\}.$$

By *nondegenerate* we mean that it is strictly smaller than a half-space.

Following essentially the proof given by Omori, we can derive the following stronger result (for a weaker form using the Omori-Yau maximum principle see [227, Theorem 1.28]).

Theorem 5.1 *Let M be a Riemannian manifold of dimension m, which satisfies the weak maximum principle for the Laplacian. Then M does not admit an isometric, minimal immersion into any nondegenerate cone of some Euclidean space $\mathbb{R}^n$.*

Proof We reason by contradiction and assume that there exists an isometric, minimal immersion $f : M \to \mathbb{R}^n$ with $f(M)$ contained in a nondegenerate cone of $\mathbb{R}^n$. We may assume without loss of generality that the vertex of the cone is the origin $0 \in \mathbb{R}^n$, so that there exists $\zeta \in \mathbb{S}^{n-1}$ and $\theta \in (0, \pi/2)$ such that

$$\frac{\langle f(x), \zeta \rangle}{|f(x)|} \geq \cos\theta \;\text{ on } M. \tag{5.4}$$

For each $x \in M$, let $\hat{f}(x)$ denote the orthogonal projection of $f(x)$ onto the hyperplane orthogonal to ζ; that is,

$$\hat{f}(x) = f(x) - \langle f(x), \zeta \rangle \zeta,$$

so that

$$|\hat{f}(x)|^2 = |f(x)|^2 - \langle f(x), \zeta \rangle^2. \tag{5.5}$$

It follows that

$$\begin{aligned} \langle f(x), \zeta \rangle^2 - \cos^2\theta |\hat{f}(x)|^2 &= \langle f(x), \zeta \rangle^2 - \cos^2\theta |f(x)|^2 + \cos^2\theta \langle f(x), \zeta \rangle^2 \\ &\geq \langle f(x), \zeta \rangle^2 - \cos^2\theta |f(x)|^2 \geq 0. \end{aligned} \tag{5.6}$$

We let $0 < \varepsilon < \frac{1}{\sqrt{2}} \cos\theta$ and define

$$u_\varepsilon(x) = \sqrt{1 + \varepsilon^2 |\hat{f}(x)|^2} - \langle f(x), \zeta \rangle$$

for every $x \in M$. Then, independently of ε, one has

$$u_\varepsilon(x) \leq 1 \quad \text{on } M. \tag{5.7}$$

Indeed, (5.7) is equivalent to

$$\sqrt{1 + \varepsilon^2 |\hat{f}(x)|^2} \leq 1 + \langle f(x), \zeta \rangle,$$

that is,

$$\langle f(x), \zeta \rangle^2 - \varepsilon^2 |\hat{f}(x)|^2 + 2\langle f(x), \zeta \rangle \geq 0. \tag{5.8}$$

Now, since $\langle f(x), \zeta \rangle > 0$ and $\varepsilon^2 < \cos^2\theta$, by (5.6) we have

$$\langle f(x), \zeta \rangle^2 - \varepsilon^2 |\hat{f}(x)|^2 + 2\langle f(x), \zeta \rangle \geq \langle f(x), \zeta \rangle^2 - \cos^2\theta |\hat{f}(x)|^2 \geq 0,$$

hence the validity of (5.8).

Fix a point $x_0 \in M$ and define

$$\Omega_\varepsilon = \{x \in M : u_\varepsilon(x) \geq u_\varepsilon(x_0)\} \neq \emptyset,$$

so that on Ω_ε we have

$$\sqrt{1 + \varepsilon^2 |\hat{f}(x)|^2} \geq u_\varepsilon(x_0) + \langle f(x), \zeta \rangle$$

and

$$\langle f(x), \zeta \rangle \geq \cos\theta |f(x)|.$$

Using these inequalities we get

$$\begin{aligned}\sqrt{1+\varepsilon^2|f(x)|^2} &= \sqrt{1+\varepsilon^2\left(|\hat{f}(x)|^2 + \langle f(x), v\rangle^2\right)} \\ &\geq \sqrt{1+\varepsilon^2|\hat{f}(x)|^2} \geq u_\varepsilon(x_0) + \cos\theta|f(x)| \qquad (5.9)\end{aligned}$$

on Ω_ε. Next, we set

$$\Omega_\varepsilon^\pm = \{x \in \Omega_\varepsilon : u_\varepsilon(x_0) + \cos\theta|f(x)| \gtrless 0\},$$

so that for every $x \in \Omega_\varepsilon^-$ we have

$$|f(x)| < \frac{-u_\varepsilon(x_0)}{\cos\theta} = \frac{\langle f(x_0), \zeta\rangle - \sqrt{1+\varepsilon^2|\hat{f}(x_0)|^2}}{\cos\theta} \leq \frac{\langle f(x_0), \zeta\rangle}{\cos\theta},$$

independently of ε. On the other hand, for $x \in \Omega_\varepsilon^+$ and squaring (5.9) we see that

$$1+\varepsilon^2|f(x)|^2 \geq (u_\varepsilon(x_0) + \cos\theta|f(x)|)^2,$$

that is,

$$\left(\cos^2\theta - \varepsilon^2\right)|f(x)|^2 + 2\cos\theta u_\varepsilon(x_0)|f(x)| + u_\varepsilon^2(x_0) - 1 \leq 0.$$

Therefore, for $x \in \Omega_\varepsilon^+$,

$$\begin{aligned}|f(x)| &\leq \frac{-\cos\theta u_\varepsilon(x_0) + \sqrt{\cos^2\theta - \varepsilon^2 + \varepsilon^2 u_\varepsilon^2(x_0)}}{\cos^2\theta - \varepsilon^2} \\ &\leq 2\frac{|u_\varepsilon(x_0)| + \sqrt{1+u_\varepsilon^2(x_0)}}{\cos\theta}. \qquad (5.10)\end{aligned}$$

Since $0 < \varepsilon^2 < \frac{1}{2}\cos^2\theta$ and $\langle f(x_0), \zeta\rangle > 0$, we have

$$\begin{aligned}u_\varepsilon^2(x_0) &= \langle f(x_0), \zeta\rangle^2 + 1 + \varepsilon^2|\hat{f}(x_0)|^2 - 2\langle f(x_0), \zeta\rangle\sqrt{1+\varepsilon^2|\hat{f}(x_0)|^2} \\ &\leq \langle f(x_0), \zeta\rangle^2 + 1 + \frac{\cos^2\theta}{2}|f(x_0)|^2,\end{aligned}$$

which, jointly with (5.10), implies that for every $x \in \Omega_\varepsilon^+$

$$|f(x)| \leq C(x_0, \zeta, \theta), \tag{5.11}$$

for a certain constant $C(x_0, \zeta, \theta) > 0$ independent of ε. This shows that $|f|$ is bounded on Ω_ε, independently of $\varepsilon \in \left(0, \frac{1}{2}\cos\theta\right)$.

Let us now consider the function $u : M \to \mathbb{R}$ given by

$$u(x) = u_\varepsilon(x) - u_\varepsilon(x_0).$$

By (5.7) we have

$$u(x) \leq 1 - u_\varepsilon(x_0) = 1 + \langle f(x_0), \zeta\rangle - \sqrt{1 + \varepsilon^2|\hat{f}(x_0)|^2} \leq \langle f(x_0), \zeta\rangle$$

for every $x \in M$, so that u is bounded above on M, independently of ε. Observe that u is nonnegative exactly on the set Ω_ε and $u(x_0) = 0$. Using the formalism introduced in Sect. 1.6 we now compute Δu. From the very definition of u_ε, using (5.5) we have

$$u_\varepsilon(x) = \sqrt{1 + \varepsilon^2\left[|f(x)|^2 - \langle f(x), \zeta\rangle^2\right]} - \langle f(x), \zeta\rangle.$$

Now we let $\{\theta^a\}$ be a Darboux frame along f, so that $\theta^\alpha = 0$ on M, and let $\{\theta^a_b\}$ be the corresponding Levi-Civita connection forms. Then

$$\begin{aligned} du = du_\varepsilon &= -\langle df, \zeta\rangle + \frac{1}{2}\frac{\varepsilon^2\{2\langle df, f\rangle - 2\langle f, \zeta\rangle\langle df, \zeta\rangle\}}{\sqrt{1 + \varepsilon^2\left[|f|^2 - \langle f, \zeta\rangle^2\right]}} \\ &= -\langle e_i, \zeta\rangle\theta^i + \frac{\varepsilon^2}{\sqrt{1 + \varepsilon^2\left[|f|^2 - \langle f, \zeta\rangle^2\right]}}\{\langle e_i, f\rangle - \langle f, \zeta\rangle\langle e_i, \zeta\rangle\}\theta^i. \end{aligned}$$

Hence,

$$u_i = \frac{\varepsilon^2}{\sqrt{1 + \varepsilon^2\left[|f|^2 - \langle f, \zeta\rangle^2\right]}}\{\langle e_i, f\rangle - \langle f, \zeta\rangle\langle e_i, \zeta\rangle\} - \langle e_i, \zeta\rangle;$$

in particular,

$$\nabla u = \nabla u_\varepsilon = -\zeta^\top + \frac{\varepsilon^2}{\sqrt{1 + \varepsilon^2|\hat{f}|^2}}(f^\top - \langle f, \zeta\rangle\zeta^\top) = -\zeta^\top + \frac{\varepsilon^2}{\sqrt{1 + \varepsilon^2|\hat{f}|^2}}\hat{f}^\top. \tag{5.12}$$

It follows that

$$
\begin{aligned}
u_{ij}\theta^j &= du_i - u_k\theta_i^k \\
&= -\frac{\varepsilon^4\{\langle e_k,f\rangle - \langle f,\zeta\rangle\langle e_k,\zeta\rangle\}}{\sqrt{\left(1+\varepsilon^2\left[|f|^2 - \langle f,\zeta\rangle^2\right]\right)^3}}\{\langle e_i,f\rangle - \langle f,\zeta\rangle\langle e_i,\zeta\rangle\}\theta^k \\
&+ \frac{\varepsilon^2}{\sqrt{1+\varepsilon^2\left[|f|^2 - \langle f,\zeta\rangle^2\right]}} \\
&\quad \{\theta_i^k\langle e_k,f\rangle + \langle e_i,e_k\rangle\theta^k + \theta_i^\alpha\langle e_\alpha,f\rangle - \langle e_k,\zeta\rangle\langle e_i,\zeta\rangle\theta^k - \langle f,\zeta\rangle\langle\theta_i^k e_k,\zeta\rangle\} \\
&- \theta_i^k\langle e_k,\zeta\rangle - \theta_i^\alpha\langle e_\alpha,\zeta\rangle \\
&- \frac{\varepsilon^2}{\sqrt{1+\varepsilon^2\left[|f|^2 - \langle f,\zeta\rangle^2\right]}}\{\theta_i^k\langle e_k,f\rangle - \theta_i^k\langle f,\zeta\rangle\langle e_k,\zeta\rangle\} + \theta_i^k\langle e_k,\zeta\rangle \\
&= \left\{-\frac{\varepsilon^4}{\sqrt{\left(1+\varepsilon^2\left[|f|^2 - \langle f,\zeta\rangle^2\right]\right)^3}}(\langle e_j,f\rangle - \langle f,\zeta\rangle\langle e_j,\zeta\rangle)(\langle e_i,f\rangle - \langle f,\zeta\rangle\langle e_i,\zeta\rangle)\right\}\theta^j \\
&+ \left\{\frac{\varepsilon^2}{\sqrt{1+\varepsilon^2\left[|f|^2 - \langle f,\zeta\rangle^2\right]}}(\delta_{jk} - \langle e_j,\zeta\rangle\langle e_i,\zeta\rangle)\right\}\theta^j + [h_{ij}^\alpha\langle e_\alpha,f\rangle - h_{ij}^\alpha\langle e_\alpha,\zeta\rangle]\theta^j,
\end{aligned}
$$

thus

$$
\begin{aligned}
u_{ij} = &-\frac{\varepsilon^4}{\sqrt{\left(1+\varepsilon^2\left|\hat{f}\right|^2\right)^3}}(\langle e_j,f\rangle - \langle f,\zeta\rangle\langle e_j,\zeta\rangle)(\langle e_i,f\rangle - \langle f,\zeta\rangle\langle e_i,\zeta\rangle) \\
&+ \frac{\varepsilon^2}{\sqrt{1+\varepsilon^2\left|\hat{f}\right|^2}}(\delta_{jk} - \langle e_j,\zeta\rangle\langle e_i,\zeta\rangle) \\
&+ h_{ij}^\alpha\langle e_\alpha,f\rangle - h_{ij}^\alpha\langle e_\alpha,\zeta\rangle
\end{aligned}
$$

and

$$
\Delta u = -\frac{\varepsilon^4}{(1+\varepsilon^2|\hat{f}|^2)^{3/2}}\left|\hat{f}\right|^2 + \frac{\varepsilon^2}{\sqrt{1+\varepsilon^2|\hat{f}|^2}}(m - |\zeta^\top|^2) + m\langle\mathbf{H},\eta\rangle, \tag{5.13}
$$

with

$$\eta = -\zeta^{\perp} + \frac{\varepsilon^2}{\sqrt{1+\varepsilon^2|\hat{f}|^2}}\hat{f}^{\perp}.$$

Using (5.12) into (5.13) one gets

$$\begin{aligned}\Delta u = & \frac{m\varepsilon^2}{\sqrt{1+\varepsilon^2|\hat{f}|^2}} + \frac{(1-\varepsilon^2)|\zeta^{\top}|^2}{\sqrt{1+\varepsilon^2|\hat{f}|^2}} \\ & - \frac{|\nabla u|^2}{\sqrt{1+\varepsilon^2|\hat{f}|^2}} - \frac{2\varepsilon^2\langle\zeta^{\top},\hat{f}^{\top}\rangle}{(1+\varepsilon^2|\hat{f}|^2)} + m\langle \mathbf{H}, \eta\rangle.\end{aligned}$$

In particular, if the immersion is minimal then $\mathbf{H} \equiv 0$ and we have

$$\Delta u + \frac{|\nabla u|^2}{\sqrt{1+\varepsilon^2|\hat{f}|^2}} = \frac{m\varepsilon^2}{\sqrt{1+\varepsilon^2|\hat{f}|^2}} + \frac{(1-\varepsilon^2)|\zeta^{\top}|^2}{\sqrt{1+\varepsilon^2|\hat{f}|^2}} - \frac{2\varepsilon^2\langle\zeta^{\top},\hat{f}^{\top}\rangle}{(1+\varepsilon^2|\hat{f}|^2)}. \tag{5.14}$$

Observe that

$$\begin{aligned}& \frac{m\varepsilon^2}{\sqrt{1+\varepsilon^2|\hat{f}|^2}} + \frac{(1-\varepsilon^2)|\zeta^{\top}|^2}{\sqrt{1+\varepsilon^2|\hat{f}|^2}} - \frac{2\varepsilon^2\langle\zeta^{\top},\hat{f}^{\top}\rangle}{(1+\varepsilon^2|\hat{f}|^2)} \\ & \geq \frac{m\varepsilon^2}{\sqrt{1+\varepsilon^2|\hat{f}|^2}} + \frac{(1-\varepsilon^2)|\zeta^{\top}|^2}{(1+\varepsilon^2|\hat{f}|^2)} - \frac{2\varepsilon^2\langle\zeta^{\top},\hat{f}^{\top}\rangle}{(1+\varepsilon^2|\hat{f}|^2)} \\ & = \frac{m\varepsilon^2}{\sqrt{1+\varepsilon^2|\hat{f}|^2}} - \frac{\varepsilon^2|\zeta^{\top}|^2}{(1+\varepsilon^2|\hat{f}|^2)} + \frac{|\zeta^{\top}-\varepsilon^2\hat{f}^{\top}|^2}{(1+\varepsilon^2|\hat{f}|^2)} - \frac{\varepsilon^4|\hat{f}^{\top}|^2}{(1+\varepsilon^2|\hat{f}|^2)} \\ & \geq \varepsilon^2\left(\frac{m}{\sqrt{1+\varepsilon^2|\hat{f}|^2}} - 1\right) \geq \varepsilon^2\left(\frac{m}{\sqrt{1+\varepsilon^2|f|^2}} - 1\right),\end{aligned}$$

where we have used the facts that $|\zeta^{\top}|^2 \leq 1$, $|\hat{f}^{\top}|^2 \leq |\hat{f}|^2$ and $|\hat{f}|^2 \leq |f|^2$, which hold at every $x \in M$. Using this in (5.14), one obtains

$$\Delta u + |\nabla u|^2 \geq \Delta u + \frac{|\nabla u|^2}{\sqrt{1+\varepsilon^2|\hat{f}|^2}} \geq \varepsilon^2\left(\frac{m}{\sqrt{1+\varepsilon^2|f|^2}} - 1\right) \tag{5.15}$$

on M. Recall now that $|f|$ is bounded on Ω_ε, independently of ε, that is (5.11) holds; therefore, on Ω_ε one has

$$\varepsilon^2\left(\frac{m}{\sqrt{1+\varepsilon^2|f|^2}}-1\right)\geq\varepsilon^2\left(\frac{m}{\sqrt{1+\varepsilon^2C^2}}-1\right),$$

and choosing $\varepsilon<\sqrt{m^2-1}/C$ we have from here and from (5.15) that

$$\Delta u+|\nabla u|^2\geq\varepsilon^2\left(\frac{m}{\sqrt{1+\varepsilon^2C^2}}-1\right)>0$$

on Ω_ε. Thus, setting $w=e^u$, from the above we immediately obtain

$$\Delta w=w(\Delta u+|\nabla u|^2)\geq\varepsilon^2\left(\frac{m}{\sqrt{1+\varepsilon^2C^2}}-1\right)w \text{ on } \Omega_\varepsilon. \tag{5.16}$$

Since $u^*=\sup_M u<+\infty$, $w^*<+\infty$ and by the weak maximum principle, there exists a sequence of points $\{x_k\}\subset M$ such that

$$\text{(i) } w(x_k)>w^*-\frac{1}{k}, \text{ and (ii) } \Delta w(x_k)<\frac{1}{k}$$

for each $k\in\mathbb{N}$. Since $u(x_k)\to u^*$ and $u<0$ outside of Ω_ε, we can assume without loss of generality that $x_k\in\Omega_\varepsilon$. Hence, using (5.16) we get

$$\frac{1}{k}>\Delta w(x_k)\geq\varepsilon^2\left(\frac{m}{\sqrt{1+\varepsilon^2C^2}}-1\right)w(x_k)\geq\varepsilon^2\left(\frac{m}{\sqrt{1+\varepsilon^2C^2}}-1\right)>0.$$

Finally, letting $k\to+\infty$ in this inequality we obtain a contradiction. □

As an application of Theorems 5.1 and 2.6 we immediately obtain the following [227, Corollary 1.29].

Corollary 5.2 *A complete Riemannian manifold M does not admit a proper, isometric, minimal immersion into any nondegenerate cone of some Euclidean space $\mathbb{R}^n$.*

5.2 Maps into Nondegenerate Cones in Euclidean Space

Related to the results of the previous section, in the recent paper [181] Mari and Rigoli consider smooth maps $\varphi : M \to \mathbb{R}^n$ with image contained into a nondegenerate cone and, under quite general assumptions on M, they provide a lower bound for the width of the cone in terms of the energy, the tension of the map φ

and a metric parameter. As an application of their results, they recover and/or extend some well-known results about harmonic maps, minimal and isometric immersions.

From the previous section we recall that, given the Euclidean space $\mathbb{R}^n$ with its flat canonical metric $\langle\,,\,\rangle$, having fixed an origin $o \in \mathbb{R}^n$ and a unit vector $\zeta \in \mathbb{S}^{n-1}$, we set $\mathscr{C}_{o,\zeta,\theta}$, shortly $\mathscr{C}$, to denote the nondegenerate cone with vertex in o, direction ζ and width θ, $\theta \in (0, \pi/2)$, that is,

$$\mathscr{C} = \left\{ z \in \mathbb{R}^n \backslash \{o\} \, : \, \left\langle \frac{z-o}{|z-o|}, \zeta \right\rangle \geq \cos(\theta) \right\}. \tag{5.17}$$

Let $(M, (,))$ be a connected, m-dimensional, $m \geq 2$, Riemannian manifold, and let

$$\varphi \, : (M, (,)) \longrightarrow (\mathbb{R}^n, \langle\,,\,\rangle)$$

be a smooth map. We indicate with $|d\varphi|^2$ the square of the Hilbert-Schmidt norm of the differential $d\varphi$ (in other words, twice the energy density of φ) and with $\tau(\varphi)$ the tension field of φ. Recall, see Sect. 1.7, that in case φ is an isometric immersion, $|d\varphi|^2 = m$ and $\tau(\varphi) = m\mathbf{H}$, where $\mathbf{H}$ is the mean curvature vector. We fix an origin $q \in M$ and we consider the distance function from q, $r(x) = d(x, q)$. We set B_R for the geodesic ball with radius R centered at q.

To state the next theorems, given $\eta > 0$, we define

$$A_\eta = \sup_{(\xi,\alpha)\in\Lambda} \left\{ \xi\alpha^2\sqrt{1-\alpha^2} \right\}, \tag{5.18}$$

where $\Lambda = \{\xi, \alpha) \in \mathbb{R}^2 : 0 < \xi < 1, 0 < \alpha < \min\{1, \eta\sqrt{1-\xi}\}\}$.

The constant A_η can be easily computed, but the actual value is irrelevant for our purposes. Note also that A_η is a nondecreasing function of η.

Theorem 5.2 *Let M be a connected, noncompact m-dimensional Riemannian manifold, and let*

$$\varphi \, : (M, (,)) \longrightarrow (\mathbb{R}^n, \langle\,,\,\rangle)$$

be a map of class C^2 such that $|d\varphi(x)|^2 > 0$ on M. Consider the elliptic operator $L = |d\varphi|^{-2}\Delta$, and assume that M is L-stochastically complete. Let $\mathscr{C} = \mathscr{C}_{o,\zeta,\theta}$ be a cone with vertex at $o \in \mathbb{R}^n \backslash \varphi(M)$, let π_ζ be the hyperplane orthogonal to ζ passing through o and let $d(\pi_\zeta, \varphi(M))$ be the Euclidean distance between this hyperplane and $\varphi(M)$.

If $\varphi(M)$ is contained in $\mathscr{C}$, then

$$\cos\theta \leq \sqrt{\frac{1}{A_1} d(\pi_\zeta, \varphi(M)) \sup_M \left[\frac{|\tau(\varphi)|}{|d\varphi|^2} \right]}. \tag{5.19}$$

In case φ is an isometric immersion, we can replace A_1 with A_m in (5.19) obtaining a sharper estimate.

In the next remarks we comment on the content of the theorem.

Remark 5.1 Note that, in case

$$\sup_M \left[\frac{|\tau(\varphi)|}{|d\varphi|^2} \right] = +\infty$$

and $d(\pi_\zeta, \varphi(M)) = 0$, that is, $\varphi(M)$ "gathers around the origin o", as we shall see in the proof, we have no restriction on θ.

Remark 5.2 For the condition that $L = |d\varphi|^{-2}\Delta$ generates a conservative diffusion, that is, L is stochastically complete, as in the case of Δ-stochastic completeness, no geodesic completeness of M is required. On the other hand, if M is complete, L-stochastic completeness has been analyzed in Chap. 3, for instance in Theorem 2.15. See also previous work of Grigor'yan [131], Sturm [260] and Pigola et al. [227]. In particular, by Theorem 2.15, if there exist $C > 0$, $\mu \in \mathbb{R}$ such that

$$|d\varphi(x)|^2 \geq \frac{C}{(1 + r(x))^\mu} \quad \text{on } M \tag{5.20}$$

and

$$\frac{r^{1-\mu}}{\log(\mathrm{Vol}(B_r))} \notin L^1(+\infty), \tag{5.21}$$

then the weak maximum principle holds for $L = |d\varphi|^{-2}\Delta$. It is worth to observe that (5.21) implies $\mu \leq 2$, but no restriction on nonnegativity of μ is needed. As already observed in Remark 2.13 in case $\mu = 2$, an application of [260] leads to slightly improving (5.21) to

$$\frac{\log r}{r \log(\mathrm{Vol}(B_r))} \notin L^1(+\infty).$$

Remark 5.3 Due to the form of (5.19), we cannot expect the result to be significant when $\varphi(M)$ is far from π_ζ, in the following sense: for every M, $\mathscr{C}$ and φ satisfying the assumptions of Theorem 5.2, and for every $k \geq 0$, we can consider the map $\varphi_k = \varphi + k\zeta$. Then $d(\pi_\zeta, \varphi_k(M)) = d(\pi_\zeta, \varphi(M)) + k$, while the other parameters in the right-hand side of (5.19) remain unchanged. Therefore, for k sufficiently large inequality (5.19) becomes meaningless unless $\tau(\varphi) \equiv 0$. On the contrary, we show with a simple example that, when $d(\pi_\zeta, \varphi(M))$ is very small, (5.19) is sharp in the following sense: for every fixed hyperplane π_ζ, and for every origin $o \in \pi_\zeta$, there exists a family of maps φ_d, $d > 0$ representing $d(\pi_\zeta, \varphi_d(M))$, such that, if we denote by θ_d the width of the nondegenerate tangent cone containing $\varphi_d(M)$,

$$\frac{\cos^2\theta_d}{d} \geq C \qquad \text{when} \quad d \to 0^+,$$

for some constant $C > 0$. Indeed, for every fixed $d > 0$ consider the hypersurface $\varphi_d : \mathbb{R}^m \to \mathbb{R}^{m+1}$ given by the graph $\varphi(x) = (x, |x|^2 + d)$, with the induced metric. Indicating with π_ζ the hyperplane $x_{m+1} = 0$, we have by standard calculations

$$|\tau(\varphi_d)| = \frac{2m + 8(m-1)|x|^2}{(1 + 4|x|^2)^{3/2}} \quad \text{and} \quad |d\varphi_d|^2 = m$$

Therefore $\sup_M |\tau(\varphi_d)|/|d\varphi_d|^2 = 2$. Moreover, for the tangent cone passing through the origin

$$\cos^2\theta_d = \frac{4d}{1+4d},$$

thus, since $d \equiv d(\pi_\zeta, \varphi_d(M))$, we reach the desired conclusion.

Proof (of Theorem 5.2) First of all we observe that

$$d(\pi_\zeta, \varphi(M)) = \inf_{x_o \in M} \langle \varphi(x_o), \zeta \rangle,$$

and that the right-hand side of (5.19) is invariant under homothetic transformations of $\mathbb{R}^n$. We choose o as the origin of global coordinates, and for the ease of notation we set

$$b = \cos\theta \qquad b \in (0, 1).$$

Furthermore, for future use, note that $\varphi(M) \subseteq \mathscr{C}$ implies

$$\langle \varphi(x), \zeta \rangle \geq b|\varphi(x)| > 0 \qquad \text{for every } x \in M. \tag{5.22}$$

Next, we reason by contradiction and we suppose that (5.19) is false. Therefore, there exists $x_o \in M$ such that

$$\langle \varphi(x_o), \zeta \rangle \sup_{x \in M} \left[\frac{|\tau(\varphi(x))|}{|d\varphi(x)|^2} \right] < A_1 b^2.$$

By definition, and the fact that the inequality is strict, we can find

$$\xi \in (0, 1) \quad , \quad \alpha \in \left(0, \sqrt{1-\xi}\right)$$

such that

$$\langle \varphi(x_o), \zeta \rangle \sup_{x \in M} \left[\frac{|\tau(\varphi(x))|}{|d\varphi(x)|^2} \right] < \left(\xi \alpha^2 \sqrt{1-\alpha^2} \right) b^2,$$

thus

$$\langle \varphi(x_o), \zeta \rangle |\tau(\varphi(x))| < \left(\xi \alpha^2 \sqrt{1-\alpha^2} \right) b^2 |d\varphi(x)|^2 \qquad \text{for each } x \in M. \tag{5.23}$$

For the ease of notation we set $T = \langle \varphi(x_o), \zeta \rangle > 0$ and $a = b\alpha$; the last relation becomes

$$T|\tau(\varphi(x))| < \frac{\xi a^2 \sqrt{b^2 - a^2}}{b} |d\varphi(x)|^2 \qquad \text{for each } \ x \in M. \tag{5.24}$$

Note also that

$$a \in (0, b\sqrt{1-\xi}) \subseteq (0, b). \tag{5.25}$$

Now, we define the following function:

$$u(x) = \sqrt{T^2 + a^2 |\varphi(x)|^2} - \langle \varphi(x), \zeta \rangle, \tag{5.26}$$

and we note that, by construction, $u(x_o) > 0$. We first claim that

$$u < T \qquad \text{on } M. \tag{5.27}$$

Indeed, an algebraic manipulation shows that (5.27) is equivalent to

$$\langle \varphi(x), \zeta \rangle^2 + 2T \langle \varphi(x), \zeta \rangle - a^2 |\varphi(x)|^2 > 0 \qquad \text{on } M.$$

On the other hand, using (5.22), since $a < b$ the left-hand side of the above inequality is bounded from below by $(b^2 - a^2)|\varphi(x)|^2 > 0$ and the claim is proved.

We now consider the closed nonempty set:

$$\Omega_o = \{ x \in M \ : \ u(x) \geq u(x_o) \}.$$

Using (5.22) and the definition of Ω_o we deduce:

$$\sqrt{T^2 + a^2 |\varphi(x)|^2} \geq b|\varphi(x)| + u(x_o). \tag{5.28}$$

Since $u(x_o) > 0$ by construction, we can square inequality (5.28) to obtain

$$(b^2 - a^2)|\varphi(x)|^2 + 2bu(x_o)|\varphi(x)| + u(x_o)^2 - T^2 \leq 0. \tag{5.29}$$

Since $(b^2 - a^2) > 0$, the left-hand side of the above inequality is a quadratic polynomial in $|\varphi(x)|$ with two distinct roots $\alpha_- < 0 < \alpha_+$ [use Cartesio rule and (5.27)], where the roots $\alpha_\pm$ are given by

$$\alpha_\pm = \left[b^2 - a^2\right]^{-1}\left\{\pm\sqrt{(b^2 - a^2)T^2 + a^2u(x_o)^2} - bu(x_o)\right\};$$

therefore, (5.29) implies

$$|\varphi(x)| \le \left[b^2 - a^2\right]^{-1}\left\{\sqrt{(b^2 - a^2)T^2 + a^2u(x_o)^2} - bu(x_o)\right\} \qquad \text{on } \Omega_o. \tag{5.30}$$

We then use the elementary inequality $\sqrt{1+t^2} \le 1 + t$ on $\mathbb{R}_0^+$ to deduce

$$\begin{aligned}
&\left[b^2 - a^2\right]^{-1}\left\{\sqrt{(b^2 - a^2)T^2 + a^2u(x_o)^2} - bu(x_o)\right\} \\
&= \frac{au(x_o)}{b^2 - a^2}\sqrt{1 + \frac{(b^2 - a^2)T^2}{a^2u(x_o)^2}} - \frac{bu(x_o)}{b^2 - a^2} \\
&\le \frac{au(x_o)}{b^2 - a^2}\left(1 + \frac{T\sqrt{b^2 - a^2}}{au(x_o)}\right) - \frac{bu(x_o)}{b^2 - a^2} \\
&= \frac{T}{\sqrt{b^2 - a^2}} - \frac{u(x_o)}{b + a}
\end{aligned}$$

and thus (5.30) together with $u(x_o) > 0$ yields

$$|\varphi(x)| \le \frac{T}{\sqrt{b^2 - a^2}} = \varphi_{max} \qquad \text{on } \Omega_o. \tag{5.31}$$

To compute Δu, we fix a local orthonormal frame $\{e_i\}$ and its dual coframe $\{\theta^i\}$. Then, writing $du = u_i\theta^i$, a simple computation shows that

$$u_i = \frac{a^2\langle d\varphi(e_i), \varphi\rangle}{\sqrt{T^2 + a^2|\varphi|^2}} - \langle d\varphi(e_i), \zeta\rangle, \tag{5.32}$$

and taking the covariant derivative we have $\nabla du = u_{ij}\theta^i \otimes \theta^j$, where

$$\begin{aligned}
u_{ij} = &-\frac{a^4\langle d\varphi(e_i), \varphi\rangle\langle d\varphi(e_j), \varphi\rangle}{\left(T^2 + a^2|\varphi|^2\right)^{3/2}} - \langle \nabla d\varphi(e_i, e_j), \zeta\rangle \\
&+ \frac{a^2\langle \nabla d\varphi(e_i, e_j), \varphi\rangle + a^2\langle d\varphi(e_i), d\varphi(e_j)\rangle}{\sqrt{T^2 + a^2|\varphi|^2}}.
\end{aligned}$$

Tracing the above expression we get

$$\Delta u = \langle \frac{S}{|\varphi|}\varphi - \zeta, \tau(\varphi)\rangle + S\frac{|d\varphi|^2}{|\varphi|} - \frac{1}{|\varphi|^2}\frac{S^2}{\sqrt{T^2 + a^2|\varphi|^2}}\sum_{i=1}^{m}\langle\varphi, d\varphi(e_i)\rangle^2 \quad (5.33)$$

on M, where we have defined

$$S = S(x) = \frac{a^2|\varphi(x)|}{\sqrt{T^2 + a^2|\varphi(x)|^2}}. \quad (5.34)$$

Note that, by (5.22),

$$\left|\frac{S}{|\varphi|}\varphi - \zeta\right|^2 \leq S^2 - 2bS + 1, \quad (5.35)$$

and that

$$\sum_{i=1}^{m}\langle\varphi, d\varphi(e_i)\rangle^2 \leq \begin{cases} |\varphi|^2\sum_{i=1}^{m}|d\varphi(e_i)|^2 = |\varphi|^2|d\varphi|^2; \\ |\varphi|^2 = \frac{1}{m}|\varphi|^2|d\varphi|^2 \qquad \text{if } \varphi \text{ is isometric.} \end{cases} \quad (5.36)$$

The possibility, for the isometric case, of substituting A_1 with A_m in (5.19) depends only on the above difference. Since the next passages are the same, we carry on with the general case. Substituting (5.35), (5.36) in (5.33) it follows that

$$\Delta u \geq -|\tau(\varphi)|\sqrt{S^2 - 2bS + 1} + S\frac{|d\varphi|^2}{|\varphi|} - \frac{S^2}{\sqrt{T^2 + a^2|\varphi|^2}}|d\varphi|^2. \quad (5.37)$$

We now restrict our estimates on the right-hand side of (5.37) on Ω_o. Then, (5.31) holds and from (5.24) we obtain

$$\frac{|\tau(\varphi)|}{|d\varphi|^2} < \frac{\xi a^2\sqrt{b^2 - a^2}}{Tb} = \frac{\xi a^2}{\sqrt{T^2 + a^2\varphi_{max}^2}} \leq \frac{\xi a^2}{\sqrt{T^2 + a^2|\varphi|^2}} = \frac{\xi S}{|\varphi|}.$$

Inserting this inequality into (5.37) we have

$$\Delta u \geq \frac{a^2|d\varphi|^2}{\sqrt{T^2 + a^2|\varphi|^2}}\left[1 - \xi\sqrt{S^2 - 2bS + 1} - \frac{S^2}{a^2}\right]. \quad (5.38)$$

We want to find a strictly positive lower bound for $(1 - \xi\sqrt{S^2 - 2bS + 1} - S^2/a^2)$ on Ω_o. Since $1 - 2bS + S^2$ represents a convex parabola and since S is increasing in the variable $|\varphi|$ on $[0, \varphi_{max}]$, its maximum is attained either in 0 or in φ_{max}. Since

$S(0) = 0$, $S(\varphi_{max}) = a^2/b > 0$ we have

$$S(\varphi_{max})^2 - 2bS(\varphi_{max}) + 1 = 1 + a^2\Big(\frac{a^2}{b^2} - 2\Big) < 1 = S(0)^2 - 2bS(0) + 1,$$

thus we can roughly bound as follows:

$$1 - \xi\sqrt{S^2 - 2bS + 1} - \frac{S^2}{a^2} \geq 1 - \xi - \frac{a^2}{b^2}$$

and the right-hand side of the above inequality is strictly positive since $a \in (0, b\sqrt{1-\xi})$. Therefore, (5.38) together with (5.31) yield

$$Lu = |d\varphi|^{-2}\Delta u \geq \frac{a^2}{\sqrt{T^2 + a^2|\varphi|^2}}\Big[1 - \xi - \frac{a^2}{b^2}\Big] \geq \delta \qquad \text{on } \Omega_o, \tag{5.39}$$

for some $\delta > 0$.

There are now two possibilities:

(i) x_o is an absolute maximum for u on M. By assumption $|d\varphi(x_o)|^2 > 0$, and the finite form of the maximum principle yields $\Delta u(x_o) \leq 0$, so that $Lu(x_o) \leq 0$. Since $x_o \in \Omega_o$ (5.39) immediately gives a contradiction.

(ii) $\mathrm{Int}(\Omega_o) = \{x \in M : u(x) > u(x_o)\} \neq \emptyset$. In this case, since $u(x)$ is bounded above on M, it is enough to evaluate inequality (5.39) along a sequence $\{x_k\}$ realizing the weak maximum principle for L, that is $u(x_k) > u^* - 1/k$, $Lu(x_k) < 1/k$. Note that this sequence eventually lies in $\mathrm{Int}(\Omega_o)$.

□

As an immediate consequence of Theorem 5.2, we recover Atsuji's result [33]:

Corollary 5.3 *Let $\varphi : (M, (\,,\,)) \to \mathbb{R}^n$ be harmonic and such that $|d\varphi|^2 \geq C$ for some positive constant C. If M is stochastically complete, then $\varphi(M)$ cannot be contained in any nondegenerate cone of $\mathbb{R}^n$. In particular, a stochastically complete manifold cannot be minimally immersed into a nondegenerate cone of $\mathbb{R}^n$.*

Proof If M is stochastically complete and $|d\varphi|^2 \geq C$, then, as we have already observed in Remark 3.1, it is straightforward to deduce that M is L-stochastically complete, where $L = |d\varphi|^{-2}\Delta$. Indeed, for every $u \in C^2(M)$ with $u^* < +\infty$, along the sequence $\{x_k\}$ realizing the weak maximum principle for Δ we have also

$$Lu(x_k) = |d\varphi(x_k)|^{-2}\Delta u(x_k) \leq \frac{1}{Ck}.$$

The result follows setting $\tau(\varphi) \equiv 0$ in Theorem 5.2. □

Note that even the statement of [33] in its full generality requires $|d\varphi|^2 \geq C > 0$, an assumption that can be overcome by the weighted requirements (5.20), (5.21).

Furthermore, in case $\mu = 0$ we can replace stochastic completeness and the uniform control from below in (5.20) with the properness of φ.

Corollary 5.4 *Let $(M, (\,,\,))$ be a Riemannian manifold. Then, there does not exist any proper harmonic map $\varphi : M \to \mathbb{R}^n$, such that $|d\varphi(x)| > 0$ on M and $\varphi(M)$ is contained into a nondegenerate cone of $\mathbb{R}^n$.*

Proof From (5.31) in the proof of Theorem 5.2 we deduce that $\varphi(\Omega_o)$ is bounded, hence $\overline{\varphi(\Omega_o)}$ is compact. The properness assumption implies that $\varphi^{-1}(\overline{\varphi(\Omega_o)})$ is compact, thus Ω_o is compact. Therefore, it is enough to use the finite form of the maximum principle in (5.39). □

Remark 5.4 It is a well known open problem to deal with the case $\theta = \pi/2$, that is, when the cone degenerates to a half-space and the dimension m is greater than 2. When $m = 2$, $n = 3$, by Hoffman-Meeks' half-space theorem [146] the only properly embedded minimal surfaces in a half-space are affine planes. On the contrary, if $m \geq 3$ there exist properly embedded minimal hypersurfaces even contained between two parallel hyperplanes (the so called generalized catenoids). It is still an open problem to find sufficient conditions on M, φ in order to have a Hoffman-Meeks' type result, and it seems quite difficult to adapt the methods of the proof of (5.2) for this purpose. In fact the recent literature on the problem is quite vast. We cite only the paper by Mazet [190], dealing with constant mean curvature surfaces, and the result of Rosenberg et al. [247], contained in the next

Theorem 5.3 *Let $\mathbb{P}$ be a complete, parabolic manifold and let $N = \mathbb{R}^+ \times \mathbb{P}$ with the product metric. Assume that the sectional curvatures of $\mathbb{P}$ are bounded between two given constants. Let $f : \Sigma \to \mathbb{R}^+ \times \mathbb{P}$ be a properly immersed minimal hypersurface. Then $f(\Sigma) \subseteq \{c\} \times \mathbb{P}$ for some $c \in \mathbb{R}^+$.*

We will consider some related results in Chap. 7.

The next application of Theorem 5.2 has a topological flavor. This result, which is interesting when φ is not proper, ensures that some kind of "patological" gathering around points of $\overline{\varphi(M)} \setminus \varphi(M)$ does not occur when the map is sufficiently well behaved. To make the corollary more transparent, we state it using the sufficient conditions (5.20) and (5.21). First we introduce the following

Definition 5.1 Let $\mathscr{S}$ be a convex subset of $\mathbb{R}^n$. A point $p \in \overline{\mathscr{S}}$ is called an *n-corner* of $\mathscr{S}$ if it is the vertex of a nondegenerate cone containing $\mathscr{S}$.

Corollary 5.5 *Let $(M, (\,,\,))$ be a complete Riemannian manifold and let $\varphi : M \to \mathbb{R}^n$ be a map of class C^2. Suppose that* (5.20) *holds, and that*

$$|\tau(\varphi)(x)| \leq \frac{\tilde{C}}{r(x)^{\mu}} \qquad \text{for } r(x) \gg 1, \tag{5.40}$$

for some $\tilde{C} > 0$ and $\mu \in \mathbb{R}$ as in (5.20)*. Assume also that* (5.21) *holds. Then, the convex envelope* $\mathrm{Conv}(\varphi(M))$ *contains no n-corners.*

Proof We reason by contradiction and let $p \in \mathrm{Conv}(\varphi(M))$ be an n-corner. If $p \in \mathrm{Conv}(\varphi(M))\backslash\overline{\varphi(M)}$ fix a small ball around p contained in $\mathbb{R}^n\backslash\overline{\varphi(M)}$, and cut the corner transversally with an hyperplane sufficiently near to p; it is immediate to see that in this way we produce a convex set containing $\varphi(M)$ and strictly smaller than $\mathrm{Conv}(\varphi(M))$, contradiction.

Suppose now $p \in \varphi(M)$, and let $x \in M$ such that $\varphi(x) = p$. Consider the map $d\varphi_{|x}$; by assumption, there exists a direction $v \in T_xM$ such that $|d\varphi_{|x}v| \neq 0$, thus by continuity we can take a curve

$$\gamma : (-\varepsilon, \varepsilon) \to M \quad , \quad \gamma(0) = x \quad , \quad \dot{\gamma}(0) = v$$

with ε small such that $|d\varphi_{|\gamma(t)}(\dot{\gamma}(t))| \neq 0$ on $(-\varepsilon, \varepsilon)$. Therefore, $\varphi \circ \gamma$ is an immersed curve in $\mathbb{R}^n$, and this fact contradicts the assumption that p is an n-corner.

If $p \in \overline{\varphi(M)}\backslash\varphi(M)$, choose π_ξ as the hyperplane orthogonal to the direction of the cone and passing through p. It follows that $d(\varphi(M), \pi_\xi) = 0$. By (5.20) and (5.40), we argue that $|\tau(\varphi)|/|d\varphi|^2$ is bounded above on M. By Remark 5.2, (5.20) and (5.21) ensure that M is L-stochastically complete, where $L = |d\varphi|^{-2}\Delta$. By Theorem 5.2 we conclude the validity of (5.19) which gives $\theta = \pi/2$, contradiction. □

5.3 Bounded Submanifolds and Jorge-Koutroufiotis Type Results

Given complete Riemannian manifolds M and N of dimensions respectively m and n, of $m < n$, the *isometric immersion problem* asks whether there exists an isometric immersion $f : M \to N$. When $N = \mathbb{R}^n$, the Euclidean space, the isometric problem is answered by the Nash embedding theorem [207], which asserts that there is an isometric embedding $f : M \to \mathbb{R}^n$ provided the codimension $n - m$ is sufficiently large. However, for sufficiently low codimension, meaning here that $n - m \leq m - 1$, the existence of isometric immersions imposes strong restrictions on the curvatures and the answer in general depends on the geometries of M and N. For instance, it is a classical fact that no compact Riemannian surface with nonpositive Gaussian curvature everywhere can be isometrically immersed in $\mathbb{R}^3$ (Corollary 5.1), while the famous Hilbert-Efimov theorem [108, 143] says that no complete Riemannian surface having negative Gaussian curvature $K \leq -\delta^2 < 0$ can be isometrically immersed in $\mathbb{R}^3$. For higher dimensions, a theorem of Tompkins [264] states that a compact, flat, m-dimensional Riemannian manifold cannot be isometrically immersed in $\mathbb{R}^{2m-1}$. Tompkins theorem was later extended by Chern and Kuiper [86] (for dimensions $m = 2, 3$) and by Otsuki [215] (for any dimension m) in the following way (see also [92, Sect. 3.1]).

Theorem 5.4 *Let $f : M \to \mathbb{R}^n$ be an isometric immersion of a compact Riemannian m-manifold M into the Euclidean space $\mathbb{R}^n$, with $n \leq 2m - 1$. Then the sectional* *curvatures of M satisfy*

$$\sup_M {}^M K > 0.$$

In particular, if M is a compact Riemannian manifold of dimension m with nonpositive sectional curvatures, then M cannot be isometrically immersed into any Euclidean space $\mathbb{R}^n$ with $n \leq 2m - 1$.

One of the basic tools for the proof of Theorem 5.4, as well as for the proof of other results in this and the next section, is the following algebraic result, known as Otsuki lemma [215].

Lemma 5.1 *Let $\beta : \mathbb{R}^k \times \mathbb{R}^k \to \mathbb{R}^q$, with $q \leq k - 1$, be a symmetric bilinear form satisfying $\beta(v, v) \neq 0$ for every $v \neq 0$. Then there exist linearly independent vectors v, w such that*

$$\beta(v, v) = \beta(w, w) \quad \textit{and} \quad \beta(v, w) = 0. \tag{5.41}$$

Proof First, we extend β to a complex bilinear symmetric form $\beta^{\mathbb{C}} : \mathbb{C}^k \times \mathbb{C}^k \to \mathbb{C}^q$ and we consider the equation $\beta^{\mathbb{C}}(z, z) = 0$, which is equivalent to the following system of q quadratic equations in $\mathbb{C}$

$$\beta_1^{\mathbb{C}}(z, z) = 0, \ldots \beta_q^{\mathbb{C}}(z, z) = 0. \tag{5.42}$$

Since $q < k$, (5.42) has a nonzero solution z. Note that $z \notin \mathbb{R}^k$ because β never vanishes. Thus $z = v + iw$ with $w \neq 0$. On the other hand,

$$0 = \beta^{\mathbb{C}}(z, z) = \beta(v, v) - \beta(w, w) + 2i\beta(v, w)$$

and therefore the validity of (5.41).

Next we observe that if there exists two vectors v, w satisfying (5.41) and at least one of the two, say v, is nonzero, then they are linearly independent. Indeed,

$$\beta(av + bw, av + bw) = (a^2 + b^2)\beta(v, v) \text{ and } \beta(v, v) \neq 0.$$

□

Proof (of Theorem 5.4) Recall that if $f : M \to \mathbb{R}^n$ is an isometric immersion of a compact Riemannian manifold into $\mathbb{R}^n$, by Proposition 5.1 there exists a point $x_0 \in M$ and a normal vector $\xi \in T_{x_0}M^{\perp}$ such that the second fundamental form at x_0 with respect to ξ is positive definite. In particular,

$$\langle \mathrm{II}_{x_0}(v, v), \xi \rangle \neq 0$$

for every $v \in T_{x_0}M$, $v \neq 0$, where $\mathrm{II}_{x_0} : T_{x_0}M \times T_{x_0}M \to T_{x_0}M^{\perp}$ denotes the second fundamental tensor at x_0; hence $\mathrm{II}_{x_0}(\nu, \nu) \neq 0$ for each $\nu \in T_{x_0}M$, $\nu \neq 0$. Observe that $T_{x_0}M$ and $T_{x_0}M^{\perp}$ are real vector spaces of dimensions respectively m and $n-m$ with $n-m \leq m-1$. Therefore, by applying Otsuki lemma to II_{x_0} we know that there exist linearly independent vectors $v, w \in T_{x_0}M$ such that

$$\mathrm{II}_{x_0}(v,v) = \mathrm{II}_{x_0}(w,w) \text{ and } \mathrm{II}_{x_0}(v,w) = 0.$$

Then by Gauss equations we conclude that

$$\begin{aligned}\sup_M {}^M K \geq {}^M K(v \wedge w) &= \frac{\langle \mathrm{II}_{x_0}(v,v), \mathrm{II}_{x_0}(w,w)\rangle - |\mathrm{II}_{x_0}(v,w)|^2}{|v|^2|w|^2 - \langle v,w\rangle^2} \\ &= \frac{|\mathrm{II}_{x_0}(v,v)|^2}{|v|^2|w|^2 - \langle v,w\rangle^2} > 0.\end{aligned}$$

□

Theorem 5.4 was subsequently extended in a series of papers by O'Neill [211], Stiel [259] and Moore [196]. Their results can be summarized in the following theorem.

Theorem 5.5 *Let $f : M \to N$ be an isometric immersion of a compact Riemannian manifold M into a Cartan-Hadamard manifold N, respectively of dimensions m and n, with $n \leq 2m-1$. Then the sectional curvatures of M and N satisfy*

$$\sup_M {}^M K > \inf_N {}^N K.$$

We recall here that a Cartan-Hadamard manifold is a simply connected, complete, Riemannian manifold with nonpositive sectional curvatures.

Theorem 5.5 was improved by Jorge and Koutroufiotis in [154] to bounded, complete submanifolds with scalar curvature bounded from below, and in the version presented by Pigola, Rigoli and Setti in [227, Theorem 1.15] to complete submanifolds with scalar curvature satisfying

$${}^M S(x) \geq -B^2 \varrho_M^2(x) \left(\prod_{j=1}^{k} \log^{(j)}(\varrho_M(x)) \right)^2, \quad \varrho_M(x) \gg 1 \tag{5.43}$$

for some constant $B > 0$, some integer $k \geq 1$, where ϱ_M is the distance function on M to a fixed point and $\log^{(j)}$ is the j-th iterate of the logarithm (see also [92, Sect. 3.2]). Specifically, their result states as follows (see [227, Theorem 1.15]).

Theorem 5.6 *Let M and N be complete Riemannian manifolds of dimensions m and n, respectively, with $n \leq 2m-1$ and let $f : M \to N$ be an isometric immersion with $f(M) \subset {}^N B_R(p)$, where ${}^N B_R(p)$ denotes a geodesic ball of N centered at a point $p \in N$ and of radius R. Assume that the radial sectional curvature ${}^N K_{rad}$ along the* *radial geodesics issuing from p satisfies*

$$^{N}K_{rad} \le b \ \ in \ ^{N}B_R(p)$$

and $0 < R < \min\{inj_N(p), \pi/2\sqrt{b}\}$, *where we replace* $\pi/2\sqrt{b}$ *by* $+\infty$ *if* $b \le 0$. *If the scalar curvature of M satisfies (5.43), then*

$$\sup_M {}^{M}K \ge C_b^2(r) + \inf_{^{N}B_R(p)} {}^{N}K, \tag{5.44}$$

where

$$C_b(t) = \begin{cases} \sqrt{b}\cot(\sqrt{b}\,t) & \text{if } b > 0 \text{ and } 0 < t < \pi/2\sqrt{b}, \\ 1/t & \text{if } b = 0 \text{ and } t > 0, \\ \sqrt{-b}\coth(\sqrt{-b}\,t) & \text{if } b < 0 \text{ and } t > 0. \end{cases} \tag{5.45}$$

Remark 5.5 It is worth pointing out that the estimates in Theorem 5.6 are sharp. Indeed, if N is one of the standard model manifolds of a simply connected space form of constant sectional curvature b and M is a geodesic sphere of radius r in N, then equality in (5.44) is achieved.

For a proof of Theorem 5.6, which is a somewhat simplified version of the original arguments by Jorge and Koutroufiotis in [154], see [227]. We shall however prove the next result related to Theorem 5.6, see [181]. This will provide an application of the weak maximum principle for the Hessian.

Theorem 5.7 *Let* $f : M \to \mathbb{R}^n$ *be an isometric immersion into a nondegenerate cone* $\mathscr{C} = \mathscr{C}_{o,\xi,\theta}$ *of an m-dimensional manifold satisfying the weak maximum principle for the Hessian. Assume the codimension restriction*

$$0 < n - m < m \tag{5.46}$$

and suppose that the sectional curvature of M satisfy

$$^{M}K \le \chi^2 \qquad on\ M \tag{5.47}$$

for some constant $\chi \ge 0$. *Then*

$$\cos\theta \le \sqrt{d(\pi_\xi, f(M))\frac{\chi}{A_1}}, \tag{5.48}$$

where A_1 *is as in* (5.18).

Proof We follow the proof of Theorem 5.2 *verbatim* replacing (5.23) with

$$\langle f(x_o), \zeta\rangle \chi < \left(\xi\alpha^2\sqrt{1-\alpha^2}\right)b^2 \tag{5.49}$$

for some $\xi \in (0, 1), \alpha \in (0, \sqrt{1-\xi})$; then we replace (5.24) with

$$T\chi < \frac{\xi a^2\sqrt{b^2-a^2}}{b} \tag{5.50}$$

and arrive up to inequality (5.31) included. Next, we fix $x \in \Omega_o$ and we let $X, Y \in T_xM$ be orthonormal vectors spanning the 2-plane π. From Gauss equations and (5.47) we have

$$\langle \mathrm{II}_x(X,X), \mathrm{II}_x(Y,Y)\rangle - |\mathrm{II}_x(X,Y)|^2 = {}^M K(\pi) \le \chi^2, \tag{5.51}$$

where II_x is the second fundamental tensor at x. Since $0 < n - m < m$, by Otsuki lemma Lemma 5.1, it follows that there exists a unit vector $W \in T_xM$ such that

$$|\mathrm{II}_x(W, W)| \le \chi$$

hence, from (5.50) and (5.31) we deduce

$$|\mathrm{II}_x(W,W)| < \frac{\xi a^2\sqrt{b^2-a^2}}{Tb} = \frac{\xi a^2}{\sqrt{T^2+a^2f_{max}^2}} \le \frac{\xi a^2}{\sqrt{T^2+a^2|f(x)|^2}}. \tag{5.52}$$

Next, we let $\gamma : [0, \varepsilon) \to M$, $\varepsilon > 0$, be the geodesic characterized by the initial data

$$\gamma(0) = x, \qquad \dot\gamma(0) = W.$$

Call $s \in [0, \varepsilon)$ the arc-length parameter and define the function

$$g : [0,\varepsilon) \to \mathbb{R} \qquad g(s) = u(\gamma(s)),$$

with u defined in (5.26), that is,

$$u(x) = \sqrt{T^2 + a^2|f(x)|^2} - \langle f(x), \zeta\rangle.$$

A simple computation, using the fact that f is an isometric immersion, gives:

$$g''(s) = \langle \frac{S}{|f(\gamma)|}f(\gamma) - \zeta, \mathrm{II}_x(\dot\gamma,\dot\gamma)\rangle + \frac{S}{|f(\gamma)|} - \frac{S^3}{a^2|f(\gamma)|^3}\langle df(\dot\gamma), f(\gamma)\rangle^2, \tag{5.53}$$

where S, not to be confused with the scalar curvature ${}^{M}S$, has the expression in (5.34), evaluated, with the notation there, at $x = \gamma(s)$. Since

$$\left|\frac{S}{|f|}f - \zeta\right|^2 \leq 1 + S^2 - 2bS \quad , \quad \langle df(\dot{\gamma}), f\rangle^2 \leq |df(\dot{\gamma})|^2|f|^2 = |f|^2$$

Setting $S_o = S(\gamma(0))$, evaluating at $s = 0$ we deduce

$$g''(0) \geq -|\mathrm{II}_x(W, W)|\sqrt{S_o^2 - 2bS_o + 1} + \frac{a^2 - S_o^2}{\sqrt{T^2 + a^2|f(\gamma)|^2}}. \tag{5.54}$$

Inserting (5.52) into (5.54) we get

$$g''(0) \geq \frac{a^2}{\sqrt{T^2 + a^2|f(\gamma)|^2}}\left[1 - \xi\sqrt{1 + S_o^2 - 2bS_o} - \frac{S_o^2}{a^2}\right]. \tag{5.55}$$

Proceeding as in the proof of Theorem 5.2, since $a \in (0, b\sqrt{1-\xi}) \subset (0, b)$

$$g''(0) \geq \frac{a^2}{\sqrt{T^2 + a^2|f(\gamma)|^2}}\left[1 - \xi - \frac{a^2}{b^2}\right] \geq \frac{a^2\sqrt{b^2 - a^2}}{bT}[1 - \xi - \frac{a^2}{b^2}] = \delta > 0,$$

where δ is independent of $x \in \Omega_o$ and W.

On the other hand, a standard computation using the fact that γ is a geodesic and the definition of the Hessian of a function, gives $g''(0) = \mathrm{Hess}_x(u)(W, W)$. Putting together the last two inequalities we obtain

$$\mathrm{Hess}(u)(x)(W, W) \geq \delta > 0. \tag{5.56}$$

If x_o is an absolute maximum of u, then from (5.56) we immediately contradict the finite maximum principle, otherwise

$$\mathrm{Int}(\Omega_o) = \{x \in M : u(x) > u(x_o)\} \neq \emptyset \tag{5.57}$$

and (5.56) gives

$$\inf_{x \in \mathrm{Int}(\Omega_o)} \sup_{\substack{Y \in T_xM \\ |Y| = 1}} \mathrm{Hess}(u)(x)(Y, Y) \geq \delta > 0, \tag{5.58}$$

contradicting the validity of the weak maximum principle for the Hessian operator since the function u in bounded above on M. This completes the proof of Theorem 5.7. □

As a consequence of Theorem 5.7, we get the following corollaries: the former generalizes results of Tompkins [264], Chern-Kuiper [86] and Jorge-Koutroufiotis [154], whereas the latter improves on Theorem 8.3 of [92].

Corollary 5.6 *Let $(M, \langle , \rangle)$ be a complete m-dimensional Riemannian manifold with sectional curvature satisfying*

$$- B^2(1 + r(x)^2) \left(\prod_{j=1}^{k} \log^{(j)} r(x) \right)^2 \leq {}^M K_x \leq 0, \tag{5.59}$$

for some $B > 0$, some integer $k \geq 1$ and where $\log^{(j)}$ stands for the j-iterated logarithm. Then, M cannot be isometrically immersed into a nondegenerate cone of $\mathbb{R}^{2m-1}$.

Proof By (5.59), using Theorem 2.5 we have the validity of the weak maximum principle for the Hessian. The result follows immediately setting $\chi = 0$ and $n = 2m - 1$ in Theorem 5.7. □

In the next result we use the theory of flat bilinear forms introduced by Moore [197, 198] as an outgrowth of E. Cartan's theory of exteriorly orthogonal quadratic forms [62, 63]. See the book of Dajczer [92], for a sound presentation.

Corollary 5.7 *Let $(M, \langle \, , \, \rangle, J)$ be a Kähler manifold of real dimension 2m such that the weak maximum principle holds for the Hessian. Then M cannot be isometrically immersed into a nondegenerate cone of $\mathbb{R}^{3m-1}$.*

Proof The proof follows the same lines as in [92], so we only sketch it. By contradiction, assume the existence of an isometric immersion f of M into a nondegenerate cone $\mathscr{C}_{o,\xi,\theta} \subset \mathbb{R}^{3m-1}$. From the assumptions, since the codimension is $m - 1 < m$, for every $x \in M$ the theory of flat bilinear forms ensure the existence of a vector $Z \in T_xM$, with $|Z| = 1$ and such that $\mathrm{II}(JZ, JZ) = -\mathrm{II}(Z, Z)$. We define u, Ω_o as in Theorem 5.7. Expression (5.53) gives at every point x, and for every $X \in T_xM$, $|X| = 1$

$$\mathrm{Hess}(u)(x)(X, X) \geq \langle \frac{S}{|f(x)|} f(x) - \xi, \mathrm{II}_x(X, X) \rangle + \frac{S}{|f(x)|}\left(1 - \frac{S^2}{a^2}\right).$$

This calculation is independent from the value of $a \in (0, b)$. If a is chosen to be sufficiently small that $S^2/a^2 < \delta < 1$ (note that, by definition, on Ω_o it holds $S = O(a^2)$ and $S/|f| \geq a^2/T$), evaluating along a sequence $\{x_k\}$ satisfying the weak maximum principle for the Hessian we deduce, for k sufficiently large,

$$\begin{aligned}\langle \frac{S}{|f(x_k)|} f(x_k) - \xi, \mathrm{II}_x(X_k, X_k) \rangle &\leq \mathrm{Hess}(u)(x_k)(X_k, X_k) - \frac{S}{|f(x_k)|}(1 - \delta) \\ &\leq \frac{1}{k} - \frac{a^2}{T}(1 - \delta) < 0\end{aligned}$$

for every $X_k \in T_{x_k}M$, $|X_k| = 1$. This fact contradicts the existence of Z. □

Remark 5.6 As we know from Theorem 2.5, inequality (5.59) provides a sharp sufficient condition for the Omori-Yau maximum principle to hold for the Hessian. As far as we know, it is an open problem to obtain other general sufficient conditions ensuring the validity of the weak maximum principle for the Hessian.

5.4 Cylindrically Bounded Submanifolds

5.4.1 Sectional Curvature Estimates

In this section we will introduce an extension of Theorem 5.6, recently given by Alías et al. [20], to the case of complete cylindrically bounded submanifolds of a Riemannian product $\mathbb{R}^\ell \times \mathbb{P}$, where $\mathbb{P}$ is a complete Riemannian manifold of dimension $n - \ell$. In this context, an isometric immersion $f : M \to \mathbb{R}^\ell \times \mathbb{P}$ of a Riemannian manifold M is said to be *cylindrically bounded* if there exists ${}^{\mathbb{P}}B_R(p)$, a geodesic ball of $\mathbb{P}$ centered at a point $p \in \mathbb{P}$ with radius $R > 0$, such that $f(M) \subset \mathbb{R}^\ell \times {}^{\mathbb{P}}B_R(p)$.

The main result in [20] deals with the sectional curvature of cylindrically bounded submanifolds and it can be stated as follows (see [20, Theorem 4]).

Theorem 5.8 *Let M and $\mathbb{P}$ be complete Riemannian manifolds respectively of dimensions m and $n - \ell$, with $n + \ell \leq 2m - 1$. Let $f : M \to \mathbb{R}^\ell \times \mathbb{P}$ be a cylindrically bounded isometric immersion with $f(M) \subset \mathbb{R}^\ell \times {}^{\mathbb{P}}B_R(p)$. Assume that the radial sectional curvature ${}^{\mathbb{P}}K_{rad}$ along the radial geodesics issuing from p satisfies ${}^{\mathbb{P}}K_{rad} \leq b$ in ${}^{\mathbb{P}}B_R(p)$ and $0 < R < \min\{inj_N(p), \pi/2\sqrt{b}\}$, where we replace $\pi/2\sqrt{b}$ by $+\infty$ if $b \leq 0$. Assume that either*

(i) the scalar curvature of M satisfies (5.43), or
(ii) the immersion $f : M \to \mathbb{R}^\ell \times \mathbb{P}$ is proper and

$$\sup_{f^{-1}(\partial^{\mathbb{R}^\ell} B_t(o) \times {}^{\mathbb{P}}B_R(p))} |\mathrm{II}| \leq \sigma(t), \tag{5.60}$$

where II is the second fundamental tensor of the immersion and $\sigma : \mathbb{R}^+_0 \to \mathbb{R}^+$ is a positive function satisfying $\frac{1}{\sigma} \notin L^1(+\infty)$. Then

$$\sup_M {}^M K \geq C_b^2(R) + \inf_{{}^{\mathbb{P}}B_r(p)} {}^N K, \tag{5.61}$$

where $C_b(R)$ is defined in (5.45).

Remark 5.7 It is worth pointing out that the codimension restriction $n+\ell \leq 2m-1$ cannot be relaxed. Actually, together with the bound $m \leq n - 1$, it implies that $n - \ell \geq 3$ and $m \geq \ell + 2$. In particular, for $n = 3$ we have that $\ell = 0$, and therefore

$f(M) \subset {}^{\mathbb{P}}B_R(p)$. In fact, the flat cylinder $\mathbb{R} \times \mathbb{S}^1(\tilde{R}) \subset \mathbb{R} \times {}^{\mathbb{R}^2}B_R(o)$, with $0 < \tilde{R} < R$, shows that the restriction $2m - 1 \geq n + \ell$ is necessary.

On the other hand, estimate (5.61) is sharp. Indeed, for every $n \geq 3$ and $\ell \leq n-3$ we can choose $\mathbb{P}$ to be one of the standard model manifolds of a simply connected space form of constant sectional curvature b and consider $M = \mathbb{R}^\ell \times \partial^{\mathbb{P}}B_{\tilde{R}}(p)$, where $\partial^{\mathbb{P}}B_{\tilde{R}}(p)$ is a geodesic sphere of radius $\tilde{R}$ in $\mathbb{P}$. Take $f: M \hookrightarrow \mathbb{R}^\ell \times {}^{\mathbb{P}}B_{\tilde{R}}(p)$ to be the canonical isometric immersion, with $0 < \tilde{R} < R$. Then, $\sup_M {}^MK$ is the constant sectional curvature of the geodesic sphere $\partial^{\mathbb{P}}B_{\tilde{R}}(p) \subset \mathbb{P}$, which is given by

$$\begin{cases} b/\sin^2(\sqrt{b}\,\tilde{R}) & \text{if } b > 0 \text{ and } 0 < \tilde{R} < \pi/2\sqrt{b}, \\ 1/\tilde{R}^2 & \text{if } b = 0 \text{ and } \tilde{R} > 0, \\ -b/\sinh^2(\sqrt{-b}\,\tilde{R}) & \text{if } b < 0 \text{ and } \tilde{R} > 0. \end{cases}$$

In particular, observe that

$$\sup_M {}^MK = C_b^2(\tilde{R}) + b.$$

Since in this case ${}^{\mathbb{P}}K = b$, then, for every $0 < \tilde{R} < R$, we have

$$\sup_M {}^MK = C_b^2(\tilde{R}) + b \geq C_b^2(R) + \inf_{\mathbb{P}} {}^{\mathbb{P}}K,$$

which shows that estimate (5.61) is sharp.

We also refer the reader to [20] for several applications of Theorem 5.8 as well as for an interesting improvement of the condition on the growth of the rate of the norm of the second fundamental tensor in (5.60) for the case of hypersurfaces (see Theorem 7 in [20]).

Remark 5.8 It should be observed that Hasanis and Koutroufiotis [138] established similar sectional curvature estimates for cylindrically bounded submanifolds of the Euclidean space $\mathbb{R}^n$, with scalar curvature bounded below. In a slightly more general situation, Giménez [127] established sectional curvature estimates for submanifolds with scalar curvature bounded below immersed in a tubular neighborhood of certain, (P-submanifolds), embedded submanifolds of Hadamard manifolds. Theorem 5.8, besides extending Hasanis and Koutroufiotis results to a larger class of submanifolds, can be easily adapted to reproduce Giménez's result.

For the proof of the main result in this section, Theorem 5.8, we will need the Hessian comparison result in Theorem 1.4 for the Riemannian manifold $\mathbb{P}$, in the particular case where ${}^{\mathbb{P}}K_{rad} \leq b$. Thus, following the notation in Theorem 1.4, one has

$$\frac{h'(t)}{h(t)} = C_b(t),$$

so that

$$\overline{\text{Hess}}\,\rho_{\mathbb{P}} \geq C_b(\rho_{\mathbb{P}})\left(\langle , \rangle - d\rho_{\mathbb{P}} \otimes d\rho_{\mathbb{P}}\right), \tag{5.62}$$

in the sense of symmetric bilinear forms, where $\rho_{\mathbb{P}} = \text{dist}_{\mathbb{P}}(\,,p)$ and $\overline{\text{Hess}}$ stands for the Hessian operator on $\mathbb{P}$.

Now we are ready for the

Proof (of Theorem 5.8) (i) Let $\phi_b(t)$ be the function given by

$$\phi_b(t) = \begin{cases} 1 - \cos(\sqrt{b}\,t) & \text{if } b > 0 \text{ and } 0 < t < \pi/2\sqrt{b}, \\ t^2 & \text{if } b = 0 \text{ and } t > 0, \\ \cosh(\sqrt{-b}\,t) & \text{if } b < 0 \text{ and } t > 0. \end{cases}$$

For later use note that $\phi_b'(t) \geq 0$ on the domain of definition. Set $\pi_{\mathbb{P}} : \mathbb{R}^\ell \times \mathbb{P} \to \mathbb{P}$ for the projection on the second factor. We define the function $u : M \to \mathbb{R}$ by setting

$$u = \phi_b(\rho_{\mathbb{P}}) \circ (\pi_{\mathbb{P}} \circ f).$$

Note that, since $\pi_{\mathbb{P}}(f(M)) \subset {}^{\mathbb{P}}B_R(p)$, $u^* = \sup_M u \leq \phi_b(R) < +\infty$. Now the idea of the proof is similar to the idea of Jorge and Koutroufiotis in [154]. We will apply the Omori-Yau maximum principle for the Hessian to the function u in order to control the second fundamental form of the immersion restricted to certain subspaces of the tangent space and apply Otsuki lemma in the estimate of the sectional curvature.

To show the validity of the Omori-Yau maximum principle we may suppose that $\sup{}^M K < +\infty$. Otherwise the estimate in (5.61) is trivially satisfied. In this case, since the scalar curvature is an average of sectional curvatures from (5.43) it follows that

$${}^M K_{rad}(x) \geq -\widehat{B}^2 \varrho_M^2(x) \left(\prod_{j=1}^{k} \log^{(j)}(\varrho_M(x)) \right)^2, \quad \varrho_M(x) \gg 1, \tag{5.63}$$

for some positive constant $\widehat{B} > 0$, where ${}^M K_{rad}$ denotes the radial sectional curvature of M. According to Theorem 2.5, this curvature decay and completeness of M suffice to conclude that the Omori-Yau maximum principle for the Hessian holds on M. Therefore, there exists a sequence of points $\{x_k\}$ in M with the properties

$$\text{(i) } u(x_k) > u^* - \frac{1}{k}, \text{ (ii) } |\nabla u(x_k)| < \frac{1}{k}, \text{ and (iii)}' \text{ Hess}(u)(x_k) < \frac{1}{k}\langle , \rangle. \tag{5.64}$$

To compute $\mathrm{Hess}(u)$ we use formula (1.177). For $x \in M$, setting $z = \pi_{\mathbb{P}}(f(x))$ we have

$$\mathrm{Hess}(u)(x) = \overline{\mathrm{Hess}}\,(\phi_b(\rho_{\mathbb{P}}))(z)(d(\pi_{\mathbb{P}} \circ f), d(\pi_{\mathbb{P}} \circ f)) + \big\langle \overline{\nabla}\phi_b(\rho_{\mathbb{P}}), \nabla d(\pi_{\mathbb{P}} \circ f)(x)\big\rangle_{\mathbb{P}}, \tag{5.65}$$

where, to clarify the writing, $\overline{\mathrm{Hess}}$ and $\overline{\nabla}$ denote respectively the Hessian and the gradient operator on $\mathbb{P}$. Now, since f is an isometric immersion, indicating with II its second fundamental tensor,

$$\nabla d(\pi_{\mathbb{P}} \circ f) = \nabla d\pi_{\mathbb{P}}(df, df) + d\pi_{\mathbb{P}}(\mathrm{II}(\,,\,)) = d\pi_{\mathbb{P}}(\mathrm{II}(\,,\,)), \tag{5.66}$$

where the last equality is due to the fact that $\pi_{\mathbb{P}}$, as immediately verified, is totally geodesic. We now estimate the term $\overline{\mathrm{Hess}}\,(\phi_b(\rho_{\mathbb{P}}))(z)(d(\pi_{\mathbb{P}} \circ f), d(\pi_{\mathbb{P}} \circ f))$. By Eqs. (5.62), (1.182) and $\phi_b' \geq 0$ we have

$$\begin{aligned}
&\overline{\mathrm{Hess}}\,(\phi_b(\rho_{\mathbb{P}}))(z)(d(\pi_{\mathbb{P}} \circ f), d(\pi_{\mathbb{P}} \circ f)) \\
&= \phi_b'(\rho_{\mathbb{P}})(z)\,\overline{\mathrm{Hess}}\,(\rho_{\mathbb{P}})(d(\pi_{\mathbb{P}} \circ f), d(\pi_{\mathbb{P}} \circ f)) \\
&+ \phi_b''(\rho_{\mathbb{P}})(z)(d\rho_{\mathbb{P}} \otimes d\rho_{\mathbb{P}})(d(\pi_{\mathbb{P}} \circ f), d(\pi_{\mathbb{P}} \circ f)) \\
&\geq \phi_b'(\rho_{\mathbb{P}})(z)C_b(\rho_{\mathbb{P}}(z))\,\{\langle d(\pi_{\mathbb{P}} \circ f), d(\pi_{\mathbb{P}} \circ f)\rangle_{\mathbb{P}} \\
&- d\rho_{\mathbb{P}} \otimes d\rho_{\mathbb{P}}(d(\pi_{\mathbb{P}} \circ f), d(\pi_{\mathbb{P}} \circ f))\} \\
&+ \phi_b''(\rho_{\mathbb{P}})(z)(d\rho_{\mathbb{P}} \otimes d\rho_{\mathbb{P}})(d(\pi_{\mathbb{P}} \circ f), d(\pi_{\mathbb{P}} \circ f)).
\end{aligned}$$

Taking into account that ϕ_b satisfies the differential equation

$$\phi_b'' - C_b(t)\phi_b' = 0,$$

the above inequality simplifies to

$$\begin{aligned}
&\overline{\mathrm{Hess}}\,(\phi_b(\rho_{\mathbb{P}}))(z)(d(\pi_{\mathbb{P}} \circ f), d(\pi_{\mathbb{P}} \circ f)) \\
&\quad \geq \phi_b'(\rho_{\mathbb{P}})(z)C_b(\rho_{\mathbb{P}}(z))\langle d(\pi_{\mathbb{P}} \circ f), d(\pi_{\mathbb{P}} \circ f)\rangle_{\mathbb{P}}.
\end{aligned} \tag{5.67}$$

Inserting (5.66) and (5.67) into (5.65) yields

$$\begin{aligned}
\mathrm{Hess}(u)(x) \geq\ &\phi_b'(\rho_{\mathbb{P}})(z)C_b(\rho_{\mathbb{P}}(z))\langle d(\pi_{\mathbb{P}} \circ f), d(\pi_{\mathbb{P}} \circ f)\rangle_{\mathbb{P}} \\
&+ \big\langle \overline{\nabla}\phi_b(\rho_{\mathbb{P}})(z), d\pi_{\mathbb{P}}(\mathrm{II}(\,,\,))\big\rangle_{\mathbb{P}}
\end{aligned} \tag{5.68}$$

in the sense of symmetric bilinear forms. Now, since $m \geq \ell + 2$, for each $x \in M$ we can choose a subspace $V_x \subset T_xM$ of a fixed dimension $\dim V_x \geq m - \ell$ (≥ 2) such that $df(V_x) \perp T_{f(x)}\mathbb{R}^\ell$, having canonically decomposed $T_{f(x)}(\mathbb{R}^\ell \times \mathbb{P})$ as $T_{f(x)}\mathbb{R}^\ell \oplus T_{f(x)}\mathbb{P}$.

Thus, for each (say) $v \in T_{f(x)}\mathbb{R}^{\ell}$, $\langle df(V_x), v\rangle_{\mathbb{R}^{\ell}\times\mathbb{P}} = 0$. In particular, for $X_x \in V_x$,

$$|(\pi_{\mathbb{P}} \circ df)(X_x)|_{\mathbb{P}} = |df(X_x)|_{\mathbb{R}^{\ell}\times\mathbb{P}}. \tag{5.69}$$

Evaluating (5.68) in such an $X = X_x$ and using (5.69) and Gauss lemma, we get

$$\begin{aligned}\mathrm{Hess}(u)(x)(X,X) &\geq \phi_b'(\rho_{\mathbb{P}}(z))C_b(\rho_{\mathbb{P}}(z))|df(X_x)|_{\mathbb{R}^{\ell}\times\mathbb{P}}\\ &\quad + \phi_b'(\rho_{\mathbb{P}}(z))\big\langle\overline{\nabla}\rho_{\mathbb{P}}, d\pi_{\mathbb{P}}(\mathrm{II}(X,X))\big\rangle_{\mathbb{P}}\\ &\geq \phi_b'(\rho_{\mathbb{P}}(z))\Big\{C_b(\rho_{\mathbb{P}}(z))|X|^2_M - \big|d\pi_{\mathbb{P}(\mathrm{II}(X,X))}\big|_{\mathbb{P}}\Big\}.\end{aligned}$$

From this inequality and (5.64) (iii)$'$ we obtain, for each x_k, $X \in V_{x_k}$, $z_k = \pi_{\mathbb{P}(f(x_k))}$, and $s_k = s(x_k)$, where $s(x) = \rho_{\mathbb{P}}\,(\pi_{\mathbb{P}}(f(x_k)))$,

$$\frac{|X|^2}{k} > \mathrm{Hess}(u)(x_k)(X,X) \geq \phi_b'(s_k)\Big\{C_b(s_k)|X|^2_M - \Big|d\pi_{\mathbb{P}(\mathrm{II}_{x_k}(X,X))}\Big|_{\mathbb{P}}\Big\},$$

hence

$$|d\pi_{\mathbb{P}}(\mathrm{II}_{x_k}(X,X))|_{\mathbb{P}} \geq \left\{C_b(s_k) - \frac{1}{k\phi_b'(s_k)}\right\}|X|^2_M,$$

and finally

$$|\mathrm{II}_{x_k}(X,X)|_{\mathbb{R}^{\ell}\times\mathbb{P}} \geq \left\{C_b(s_k) - \frac{1}{k\phi_b'(s_k)}\right\}|X|^2_M.$$

Now consider $\beta_{x_k}\colon V_{x_k}\times V_{x_k} \to T_{x_k}M^{\perp} \subseteq T(\mathbb{R}^{\ell}\times\mathbb{P})$, the restriction of the second fundamental tensor II_{x_k} to V_{x_k}. We have that

$$\dim T_{x_k}M^{\perp} = n - m \leq m - \ell - 1 \leq \dim V_{x_k} - 1$$

since $2m - 1 \geq n + \ell$, and therefore we can apply Lemma 5.1 to β_{x_k} to conclude that there exist linearly independent vectors $X_k, Y_k \in V_{x_k}$ such that

$$\mathrm{II}(X_k,X_k) = \mathrm{II}(Y_k,Y_k) \quad \text{and} \quad \mathrm{II}(X_k,Y_k) = 0;$$

furthermore, without loss of generality we can suppose $|X_k| \geq |Y_k| \geq 1$. We will now compare the sectional curvature ${}^{M}K(X_k \wedge Y_k)$ in M of the plane spanned by X_k and Y_k, with the sectional curvature ${}^{\mathbb{R}^{\ell}\times\mathbb{P}}K(X_k \wedge Y_k)$ in $\mathbb{R}^{\ell}\times\mathbb{P}$ of the same plane. Observe that, since $X_k, Y_k \in V_{x_k} \perp T\mathbb{R}^{\ell}$,

$${}^{\mathbb{R}^{\ell}\times\mathbb{P}}K(X_k \wedge Y_k) = {}^{\mathbb{P}}K(X_k \wedge Y_k).$$

Then, by Gauss equations we have

$$\begin{aligned} {}^{M}K(X_k\wedge Y_k) - {}^{\mathbb{P}}K(X_k\wedge Y_k) &= {}^{M}K(X_k\wedge Y_k) - {}^{\mathbb{R}^\ell\times\mathbb{P}}K(X_k\wedge Y_k) \\ &= \frac{\langle \mathrm{II}(X_k,X_k),\mathrm{II}(Y_k,Y_k)\rangle - |\mathrm{II}(X_k,Y_k)|^2}{|X_k|^2|Y_k|^2 - \langle X_k,Y_k\rangle^2} \\ &\geq \frac{|\mathrm{II}(X_k,X_k)|^2}{|X_k|^2|Y_k|^2} \geq \left(\frac{|\mathrm{II}(X_k,X_k)|}{|X_k|^2}\right)^2 \\ &\geq \left(C_b(s_k) - \frac{1}{k\phi_b'(s_k)}\right)^2. \end{aligned}$$

Thus

$$\sup_M {}^{M}K - \inf_{{}^{\mathbb{P}}B_R(p)} {}^{\mathbb{P}}K \geq \left(C_b(s_k) - \frac{1}{k\,\phi_b'(s_k)}\right)^2. \tag{5.70}$$

Observe that $u^* = \phi_b(s^*)$, where $s^* = \sup_M s$ and $s_k \to s^* \leq R$. Therefore, letting $k\to\infty$ we have that

$$\sup_M {}^{M}K - \inf_{{}^{\mathbb{P}}B_r(p)} {}^{\mathbb{P}}K \geq C_b^2(s^*) \geq C_b^2(R).$$

This finishes the proof of item (i) in Theorem 5.8.

Remark 5.9 It is interesting to realize that in the proof of item (i) of Theorem 5.8 we have used only the weak maximum principle for the Hessian, since condition (ii) in (5.64) is unnecessary.

(ii) In this case, we cannot apply directly the Omori-Yau maximum principle for the Hessian, but we may apply parts of the proof of its version given in [227, Theorem 1.9] by Pigola, Rigoli and Setti. It is worth pointing out that their approach in [227] is different from the one presented in Chap. 3.

Consider again the function $u : M \to \mathbb{R}$ given by $u = \phi_b(\rho_{\mathbb{P}}) \circ (\pi_{\mathbb{P}} \circ f)$; as we already know, $u^* = \sup_M u \leq \phi_b(R)$. Let $\psi : M \to \mathbb{R}_0^+$ be defined by

$$\psi(x) = \exp\left(\int_0^{|y(x)|} \frac{ds}{\sigma(s)}\right),$$

where $y(x) = \pi_{\mathbb{R}^\ell}(f(x))$. Since f is proper and $\pi_{\mathbb{P}}(f(M)) \subset {}^{\mathbb{P}}B_R(p)$, we have $|y(x)| \to +\infty$ as $x\to\infty$, and since $\frac{1}{\sigma} \notin L^1(+\infty)$, $\psi(x) \to +\infty$ as $x\to\infty$.

We let $x_0 \in M$ with $\pi_{\mathbb{P}}(f(x_0)) \neq p$ and set

$$u_k(x) = \frac{u(x) - u(x_0) + 1}{\psi(x)^{1/k}}.$$

Thus $u_k(x_0) > 0$ and since $u^* \le \phi_b(R) < +\infty$ and $\psi(x) \to +\infty$ as $x \to \infty$, we have that $\limsup_{x\to\infty} u_k(x) \le 0$. Hence u_k attains a positive absolute maximum at a point $x_k \in M$. In this way, we produce a sequence $\{x_k\} \subset M$. We begin by showing that

$$\limsup_{k\to+\infty} u(x_k) = u^*. \tag{5.71}$$

To prove this, assume by contradiction that there exists a point $\hat{x} \in M$ such that

$$u(\hat{x}) > u(x_k) + \delta$$

for some $\delta > 0$ and for each $k \ge k_0$ sufficiently large. If $\{x_k\}$ lies in a compact subset of M, then up to passing to a subsequence, $\{x_k\} \to \bar{x}$ so that

$$u(\hat{x}) \ge u(\bar{x}) + \delta > u(\bar{x}).$$

On the other hand, since for every k

$$u_k(x_k) = \frac{u(x_k) - u(x_0) + 1}{\psi(x_k)^{1/k}} \ge u_k(\hat{x}) = \frac{u(\hat{x}) - u(x_0) + 1}{\psi(\hat{x})^{1/k}},$$

letting $k \to +\infty$ we deduce that

$$u(\bar{x}) - u(x_0) + 1 = \lim_{k\to+\infty} u_k(x_k) \ge \lim_{k\to+\infty} u_k(\hat{x}) = u(\hat{x}) - u(x_0) + 1,$$

showing that

$$u(\bar{x}) \ge u(\hat{x}),$$

which is a contradiction. In the case where $\{x_k\}$ does not lie in any compact subset of M then, using $\psi(x_k) \to +\infty$ as $k \to +\infty$ on a subsequence, and for each k such that $\psi(x_k) > \psi(\hat{x})$, we have

$$u_k(\hat{x}) = \frac{u(\hat{x}) - u(x_0) + 1}{\psi(\hat{x})^{1/k}} > \frac{u(x_k) - u(x_0) + 1 + \delta}{\psi(x_k)^{1/k}} > u_k(x_k),$$

contradicting the definition of x_k. This proves (5.71) and, by passing to a subsequence if necessary, we may assume that

$$\lim_{k\to+\infty} u(x_k) = u^*.$$

Now consider first the case where $x_k \to \infty$ as $k \to +\infty$. Since u_k attains a positive maximum at x_k we have $\nabla u_k(x_k) = 0$ and $\mathrm{Hess}(u_k)(x_k)(X, X) \le 0$ for every

$X \in T_{x_k}M$. This yields

$$\nabla u(x_k) = \frac{u(x_k) - u(x_0) + 1}{k\psi(x_k)} \nabla\psi(x_k) \tag{5.72}$$

and

$$\begin{aligned} \operatorname{Hess}(u)(x_k) &\le \frac{u(x_k) - u(x_0) + 1}{k\psi(x_k)} \left(\operatorname{Hess}(\psi)(x_k) + \left(\frac{1}{k} - 1\right) \frac{1}{\psi(x_k)} d\psi \otimes d\psi \right) \\ &\le \frac{u(x_k) - u(x_0) + 1}{k\psi(x_k)} \operatorname{Hess}(\psi)(x_k). \end{aligned} \tag{5.73}$$

We now estimate the right-hand side of the above inequality from above. Since $\psi(x) = \zeta(y)$ where $y = y(x) = \pi_{\mathbb{R}^\ell}(f(x))$ and $\zeta(y) = \exp(\int_0^{|y|} ds/\sigma(s))$, from (1.182) we have

$$\begin{aligned} \operatorname{Hess}(\psi)(x)(X,X) = {}& {}^{\mathbb{R}^\ell}\operatorname{Hess}(\zeta)(y)(d(\pi_{\mathbb{R}^\ell} \circ f)(X), d(\pi_{\mathbb{R}^\ell} \circ f)(X)) \\ & + \langle {}^{\mathbb{R}^\ell}\nabla\zeta(y), \nabla d(\pi_{\mathbb{R}^\ell} \circ f)(X,X) \rangle \end{aligned} \tag{5.74}$$

for every vector $X \in T_xM$, where ${}^{\mathbb{R}^\ell}\nabla$ and ${}^{\mathbb{R}^\ell}\operatorname{Hess}$ denote, respectively, the gradient and the Hessian operators on $\mathbb{R}^\ell$. Observe also that

$${}^{\mathbb{R}^\ell}\nabla\zeta(y) = \frac{\zeta(y)}{\sigma(|y|)} {}^{\mathbb{R}^\ell}\nabla|y|,$$

and therefore

$$\nabla\psi(x) = \frac{\psi(x)}{\sigma(|y|)} {}^{\mathbb{R}^\ell}\nabla|y|. \tag{5.75}$$

Thus, for every $X \in T_xM$ such that $d(\pi_{\mathbb{R}^\ell} \circ f)(X) = 0$ from (5.74) it follows that

$$\operatorname{Hess}(\psi)(x)(X,X) = \frac{\psi(x)}{\sigma(|y(x)|)} \langle {}^{\mathbb{R}^\ell}\nabla|y|, \nabla d(\pi_{\mathbb{R}^\ell} \circ f)(X,X) \rangle \le \frac{\psi(x)}{\sigma(|y(x)|)} |\mathrm{II}(X,X)|.$$

Therefore, by (5.60) we obtain

$$\frac{1}{\psi(x)} \operatorname{Hess}\psi(x)(X,X) \le \frac{|\mathrm{II}_x(X,X)|}{\sigma(|y(x)|)} \le |X|^2 \tag{5.76}$$

for every $X \in T_xM$ with $d(\pi_{\mathbb{R}^\ell} \circ f)(X) = 0$.

As in the proof of item (i), since $m \ge \ell + 2$, we may choose for each $x_k \in M$ a subspace $V_{x_k} \subset T_{x_k}M$ with $\dim V_{x_k} \ge m - \ell \ge 2$ and such that $df(V_{x_k}) \perp T\mathbb{R}^\ell$. Then,

$d(\pi_{\mathbb{R}^\ell} \circ f)(X) = 0$ for every $X \in V_{x_k}$, and from (5.73) and (5.76) we get

$$\mathrm{Hess}(u)(x_k)(X,X) \leq \frac{u(x_k) - u(x_0) + 1}{k\psi(x_k)} \mathrm{Hess}(\psi)(x_k)(X,X) \leq \frac{\phi_b(r) + 1}{k}|X|^2,$$

for every $X \in V_{x_k}$. Moreover, using the Hessian comparison theorem, we also have

$$\mathrm{Hess}(u)(x)(X,X) \geq \phi_b'(s)\left(C_b(s)|X|^2 - |\mathrm{II}(X,X)|\right) \tag{5.77}$$

for every $X \in V_{x_k}$, since $d(\pi_{\mathbb{R}^\ell} \circ f)(X) = X$. Therefore, we obtain

$$\frac{\phi_b(r) + 1}{k}|X|^2 \geq \mathrm{Hess}(u)(x_k)(X,X) \geq \phi_b'(s_k)\left(C_b(s_k)|X|^2 - |\mathrm{II}_{x_k}(X,X)|\right)$$

for every x_k and every $X \in V_{x_k}$, where $z_k = \pi_{\mathbb{P}}(f(x_k))$ and $s_k = s(x_k) = \rho_{\mathbb{P}}(z_k)$. Hence

$$|\mathrm{II}_{x_k}(X,X)| \geq \left(C_b(s_k) - \frac{\phi_b(r) + 1}{k\phi_b'(s_k)}\right)|X|^2.$$

Reasoning now as in the last part of the proof of item (i), there exist linearly independent vectors $X_k, Y_k \in V_{x_k}$ such that, by Gauss equations,

$${}^M K(X_k \wedge Y_k) - {}^{\mathbb{P}}K(X_k \wedge Y_k) = \left(\frac{|\alpha(X_k, X_k)|}{|X_k|^2}\right)^2 \geq \left(C_b(s_k) - \frac{\phi_b(R) + 1}{k\phi_b'(s_k)}\right)^2.$$

From here we obtain

$$\sup_M {}^M K - \inf_{{}^{\mathbb{P}}B_R(p)} {}^{\mathbb{P}}K \geq \left(C_b(s_k) - \frac{\phi_b(R) + 1}{k\,\phi_b'(s_k)}\right)^2, \tag{5.78}$$

and letting $k \to \infty$ we conclude that

$$\sup_M {}^M K - \inf_{{}^{\mathbb{P}}B_r(p)} {}^{\mathbb{P}}K \geq C_b^2(s^*) \geq C_b^2(R),$$

where $s^* = \sup_M s$, $u^* = \phi_b(s^*)$ and $s_k \to s^* \leq R$.

To finish the proof of (ii), we need to consider the case where the sequence $\{x_k\} \subset M$ remains in a compact set. In that case, passing to a subsequence if necessary, we may assume that $x_k \to x_\infty \in M$ and u attains its absolute maximum at x_∞. Thus $\mathrm{Hess}(u)(x_\infty)(X,X) \leq 0$ for all $X \in T_{x_\infty}M$. In particular, it follows from (5.77) that for every $X \in V_{x_\infty}$

$$0 \geq \mathrm{Hess}(u)(x_\infty)(X,X) \geq \phi_b'(s_\infty)\left(C_b(s_\infty)|X|^2 - |\mathrm{II}_{x_\infty}(X,X)|\right),$$

where $s_\infty = \rho_{\mathbb{P}}(\pi_{\mathbb{P}}(f(x_\infty)))$. Therefore

$$|\mathrm{II}_{x_\infty}(X,X)| \geq C_b(s_\infty)|X|^2.$$

By applying Lemma 5.1 to $\beta_{x_\infty}\colon V_{x_\infty} \times V_{x_\infty} \to T_{x_\infty}M^\perp \subset T(\mathbb{R}^\ell \times \mathbb{P})$, the restriction of the second fundamental form II_{x_∞} to V_{x_∞}, and reasoning as in the last part of the proof of (i), we have that there exist linearly independent vectors $X_\infty, Y_\infty \in V_{x_\infty}$ such that, by Gauss equations,

$${}^M\!K(X_\infty, Y_\infty) - {}^{\mathbb{P}}\!K(X_\infty, Y_\infty) = \left(\frac{|\mathrm{II}(X_\infty, X_\infty)|}{|X_\infty|^2}\right)^2 \geq C_b^2(s_\infty).$$

Thus, we conclude that

$$\sup_M {}^M\!K - \inf_{{}^{\mathbb{P}}\!B_r(p)} {}^{\mathbb{P}}\!K \geq C_b^2(s_\infty) \geq C_b^2(R). \tag{5.79}$$

□

5.4.2 *Mean Curvature Estimates and Stochastic Completeness*

The Calabi problem in its original form, presented by Calabi [56] and promoted by Chern [84] about the same time, consisted of two conjectures on Euclidean minimal hypersurfaces. The first conjecture is that any complete minimal hypersurface of $\mathbb{R}^n$ must be unbounded. The second and more ambitious conjecture asserted that any complete, nonflat, minimal hypersurface in $\mathbb{R}^n$ has unbounded projections in every $(n-2)$-dimensional subspace.

Both conjectures turned out to be false for immersed surfaces in $\mathbb{R}^3$. First Jorge and Xavier [155] exhibit a nonflat complete minimal surface lying between two parallel planes. Later on Nadirashvili [205] constructed a complete minimal surface inside a round ball in $\mathbb{R}^3$. On the other hand, it was recently shown by Colding and Minicozzi [91] that both conjectures hold for minimal surfaces which are embedded in $\mathbb{R}^3$ with finite topology. Their work involves the close relation between the Calabi conjectures and properness of the immersion.

It is worth pointing out that the immersed counterexamples to Calabi's conjectures discussed above are not proper. Hence, as observed in [18], it is natural to ask if any possible higher dimensional counterexample to Calabi's second conjecture must be nonproper.

In the special case of minimal immersion, it follows from the main results of Alías, Bessa and Dajczer in [18] that a complete hypersurface of $\mathbb{R}^n$, $n \geq 3$, with bounded projection in a two dimensional subspace cannot be proper [18, Corollary 1] (see Corollary 5.8 below). On the other hand, as another application of the method in [18], one can also generalize the results by Markvorsen [184] and Bessa and Montenegro [42] about stochastic incompleteness of minimal submanifolds to submanifolds of bounded mean curvature.

In this section we will introduce the main result in [18], which deals with the mean curvature of cylindrically bounded submanifolds and can be stated as follows (see [18, Theorem 1]).

Theorem 5.9 *Let M and $\mathbb{P}$ be complete Riemannian manifolds of dimensions m and $n-\ell$ respectively, with $m \geq \ell+1$. Let $f: M \to \mathbb{R}^\ell \times \mathbb{P}$ be a cylindrically bounded isometric immersion with $f(M) \subset \mathbb{R}^\ell \times {}^{\mathbb{P}}B_R(p)$. Assume that the radial sectional curvature ${}^{\mathbb{P}}K_{rad}$ along the radial geodesics issuing from p satisfies ${}^{\mathbb{P}}K_{rad} \leq b$ in ${}^{\mathbb{P}}B_R(p)$ and $0 < R < \min\{inj_N(p), \pi/2\sqrt{b}\}$, where we replace $\pi/2\sqrt{b}$ by $+\infty$ if $b \leq 0$.*

(i) If

$$\sup_M |\mathbf{H}| < \frac{(m-\ell)}{m} C_b(R),$$

then M is stochastically incomplete.
(ii) If the immersion $f: M \to \mathbb{R}^\ell \times \mathbb{P}$ is proper, then

$$\sup_M |\mathbf{H}| \geq \frac{(m-\ell)}{m} C_b(R).$$

In particular, for Euclidean hypersurfaces one has the following consequence.

Corollary 5.8 *Let $f: M \to \mathbb{R}^n$ be a complete hypersurface with mean curvature* $\mathbf{H}$. *If $f(M) \subset \mathbb{R}^{n-2} \times {}^{\mathbb{R}^2}B_R(o)$ and $\sup_M |\mathbf{H}| < 1/(n-1)R$, then f cannot be proper.*

Therefore, a complete minimal hypersurface of $\mathbb{R}^n$, $n \geq 3$, with bounded projection in a two-dimensional subspace cannot be proper. Related to this, Sa Earp and Rosenberg proved earlier [106] the weaker fact that a complete minimal hypersurface of $\mathbb{R}^n$ with bounded projection in an $(n-1)$-dimensional subspace cannot be proper. As an application of Corollary 5.8, it follows that any possible counterexample M to the higher dimensional second Calabi conjecture (with $n \geq 4$) must be nonproper, since it should satisfy

$$f(M) \subset \mathbb{R}^2 \times {}^{\mathbb{R}^{n-2}}B_R(o) \subset \mathbb{R}^{n-2} \times {}^{\mathbb{R}^2}B_R(o).$$

In other words, the second Calabi conjecture is true for proper immersions when $n \geq 4$.

Remark 5.10 Observe that the assumption on the bound of the mean curvature in Corollary 5.8 cannot be weakened since $1/(n-1)R$ is the mean curvature of the cylinder $\mathbb{R}^{n-2} \times \mathbb{S}^1(R)$. On the other hand, Martín and Morales [186] constructed examples of complete minimal surfaces properly immersed in the interior of a cylinder $\mathbb{R} \times {}^{\mathbb{R}^2}B_R(o)$. By the above result these surfaces cannot be proper in $\mathbb{R}^3$.

Proof (of Theorem 5.9)
(i) Suppose that M is stochastically complete or, equivalently, that the weak maximum principle holds on M (see Theorem 2.8). As in the proof of (i) in

Theorem 5.8, let $u = \phi_b(\rho_{\mathbb{P}}) \circ (\pi_{\mathbb{P}} \circ f)$, where ϕ_b, $\rho_{\mathbb{P}}$ and $\pi_{\mathbb{P}}$ are defined at the beginning of that proof. Since $\pi_{\mathbb{P}}(f(M)) \subset {}^{\mathbb{P}}B_R(p)$, we have that $u^* = \sup_M u \leq \phi_b(R) < +\infty$. The idea of the proof is to apply the weak maximum principle to the function u.

Let $x \in M$ and let $\{e_1, \dots, e_m\}$ be an orthonormal basis for T_xM. Proceeding as in the proof of Theorem 5.8 (i), and using the same notation, we have

$$\sum_{i=1}^{m} \mathrm{Hess}(u)(x)(e_i, e_i) = \phi_b'(s) \sum_{i=1}^{m} \overline{\mathrm{Hess}}\,(\rho_{\mathbb{P}})(z)(d(\pi_{\mathbb{P}} \circ f)(e_i), d(\pi_{\mathbb{P}} \circ f)(e_i)) \tag{5.80}$$

$$+ C_b(s)\phi_b'(s) \sum_i \left\langle \overline{\nabla}\rho_{\mathbb{P}}, d(\pi_{\mathbb{P}} \circ f)(e_i)\right\rangle_{\mathbb{P}}^2$$

$$+ m\phi_b'(s)\left\langle \overline{\nabla}\rho_{\mathbb{P}}, d\pi_{\mathbb{P}}(\mathbf{H})\right\rangle_{\mathbb{P}},$$

where $s = \rho_{\mathbb{P}}(z)$ and $z = \pi_{\mathbb{P}}(f(x))$. Therefore, by the Hessian comparison theorem, that is, using (5.62) we have

$$\sum_{i=1}^{m} \overline{\mathrm{Hess}}(\rho_{\mathbb{P}})(z)(d(\pi_{\mathbb{P}} \circ f)(e_i), d(\pi_{\mathbb{P}} \circ f)(e_i))$$

$$\geq C_b(s)\left\{ \sum_{i=1}^{m} \left[|d(\pi_{\mathbb{P}} \circ f)(e_i)|^2 - \left\langle \overline{\nabla}\rho_{\mathbb{P}}, d(\pi_{\mathbb{P}} \circ f)(e_i)\right\rangle^2 \right] \right\}.$$

From here and (5.80) we obtain

$$\Delta u(x) = \sum_{i=1}^{m} \mathrm{Hess}(u)(x)(e_i, e_i) \geq \phi_b'(s)C_b(s) \sum_{i=1}^{m} |d(\pi_{\mathbb{P}} \circ f)(e_i)|^2$$

$$+ m\phi_b'(s)\left\langle \overline{\nabla}\rho_{\mathbb{P}}(z), d\pi_{\mathbb{P}}(\mathbf{H})\right\rangle_{\mathbb{P}}$$

$$\geq \phi_b'(s)C_b(s) \sum_{i=1}^{m} |d(\pi_{\mathbb{P}} \circ f)(e_i)|^2 - m\phi_b'(s) \sup_M |\mathbf{H}|.$$

Moreover, since $m = \sum_{i=1}^{m} |d(\pi_{\mathbb{P}} \circ f)(e_i)|^2 + |d(\pi_{\mathbb{R}^\ell} \circ f)(e_i)|^2$, we have

$$\Delta u(x) \geq \phi_b'(s)\left\{ (m - \ell)C_b(s) - m \sup_M |\mathbf{H}| \right\}.$$

Since the weak maximum principle holds on M and $u^* \leq \phi_b(R) < +\infty$, there exists a sequence of points $\{x_k\} \subset M$ satisfying

$$\text{(i) } u(x_k) > u^* - \frac{1}{k}, \text{ and (ii) } \Delta u(x_k) < \frac{1}{k}.$$

Then, setting $s_k = s(x_k)$ we deduce

$$\frac{1}{k} > \Delta u(x) \geq \phi_b'(s_k)\left((m-\ell)C_b(s_k) - m\sup_M |\mathbf{H}|\right).$$

Finally, since $\lim_{k\to\infty}\phi_b'(s_k) > 0$, letting $k\to\infty$ we have

$$\sup_M |\mathbf{H}| \geq \frac{(m-\ell)}{m}\, C_b(R).$$

This finishes the proof of item (i) in Theorem 5.9.

(ii) After item (i), it suffices to prove that the weak maximum principle holds on M. Indeed, we will show something stronger, namely, the validity of the Omori-Yau maximum principle. According to Theorem 2.4, it is enough to find a function $\gamma : M \to \mathbb{R}_0^+$ satisfying all the three requirements (i)–(iii) of the Theorem itself.

Since f is proper and $f(M) \subset \mathbb{R}^\ell \times {}^{\mathbb{P}}B_R(p)$, the function $\gamma(x) = |y(x)|$, where $y(x) = \pi_{\mathbb{R}^\ell}(f(x))$, satisfies $\gamma(x) \to +\infty$ as $x\to\infty$. Moreover, outside of a compact set, we now have

$$|\nabla\gamma(x)| \leq |{}^{\mathbb{R}^\ell\times\mathbb{P}}\nabla|y(x)|| = |{}^{\mathbb{R}^\ell}\nabla|y|| = 1.$$

Therefore, γ satisfies conditions (i) and (ii). Let us now check that the validity of (iii).

To compute $\Delta\gamma$, that is, the Laplacian of $\gamma(x) = \rho_{\mathbb{R}^\ell}(\pi_{\mathbb{R}^\ell}(f(x)))$, we proceed similarly to what we did before: fixed an orthonormal frame $\{e_i\}$, $i = 1,\ldots,m$, on M we have

$$\begin{aligned}
\Delta\gamma &= \sum_{i=1}^m \overline{\mathrm{Hess}}\,\rho_{\mathbb{R}^\ell}(d(\pi_{\mathbb{R}^\ell}\circ f)(e_i), d(\pi_{\mathbb{R}^\ell}\circ f)(e_i)) + m\left\langle\overline{\nabla}\rho_{\mathbb{R}^\ell}, d\pi_{\mathbb{R}^\ell}(\mathbf{H})\right\rangle_{\mathbb{R}^\ell}\\
&= \frac{1}{\rho_{\mathbb{R}^\ell}}\left\{\sum_{i=1}^m |d(\pi_{\mathbb{R}^\ell}\circ f)(e_i)|^2 - \sum_{i=1}^m \left\langle\overline{\nabla}\rho_{\mathbb{R}^\ell}, d(\pi_{\mathbb{R}^\ell}\circ f)(e_i)\right\rangle^2\right\}\\
&\quad + m\left\langle\overline{\nabla}\rho_{\mathbb{R}^\ell}, d\pi_{\mathbb{R}^\ell}(\mathbf{H})\right\rangle_{\mathbb{R}^\ell}.
\end{aligned}$$

We may assume that

$$\sup_M m|\mathbf{H}|(x) < +\infty,$$

otherwise, $\sup_M |\mathbf{H}| = +\infty$ and there is nothing to prove. Hence, from the above we have

$$\Delta\gamma(x) \leq \frac{m}{\gamma(x)} + m|\mathbf{H}|(x) \leq \Lambda$$

outside a compact set for some constant Λ sufficiently large. Summing up, the function $\gamma : M \to \mathbb{R}_0^+$ satisfies all the three requirements in Theorem 2.4 and therefore the Omori-Yau maximum principle for the Laplacian holds on M. The result then follows from (i). □

5.5 Consequences on the Gauss Map of Submanifolds of Euclidean Space

The same kind of problems can be considered also for smooth maps $\varphi : M \to N$ with (M, g) and (N, h) generic Riemannian manifolds. For a general N not splitting into a product of the type $N_1 \times N_2$, it only makes sense to consider the case where $\varphi(M)$ is bounded in N. We recall the following notion already mentioned in Sect. 1.9.

Definition 5.2 A geodesic ball $B_R(q) \subset (N, h)$ is called a *regular ball* if it is contained in the complement of the cut locus of q and, denoting with ${}^N K_p$ the supremum of the sectional curvatures of N at p, we have

$$\max\{0, \sup_{p \in B_R(q)} {}^N K_p\}^{1/2} < \frac{\pi}{2R}.$$

In particular, if $B_R(q)$ is a regular ball the distance function $\rho(y) = \mathrm{dist}_{(N,h)}(y, q)$ is smooth on $B_R(q) \setminus \{q\}$.

Remark 5.11 Since the injectivity radius of any point p is positive, regular balls do always exist. Strictly speaking in what follows we are only interested in the smoothness of $\rho(y)$. However, the nontrivial result of [118] contained in Lemma 5.2 below, and that will be essential in Theorem 5.11, is stated in terms of regular geodesic balls. This justifies our choice.

Next result, in case of an isometric immersion into a Euclidean ball, yields the validity of the first Calabi conjecture if (M, g) is stochastically complete. We need the following extended form, proved in [225], in the proof of Theorem 5.11 below. The argument is quite elementary.

Proposition 5.2 *Let (M, g) be a stochastically complete manifold and $\varphi : M \to (N, h)$ a smooth map with image $\varphi(M)$ contained in a regular geodesic ball $B_R(q)$ of N. Suppose that*

$$ {}^N K \le b \text{ on } B_R(q) \tag{5.81}$$

for some $b \in \mathbb{R}$. Furthermore, assume that

$$|\tau(\varphi)| \le \tau_0 \tag{5.82}$$

for some $\tau_0 \in \mathbb{R}^+$, where $\tau(\varphi)$ is the tension field of the map φ. Then, having set

$$e = \inf_M e(\varphi) \tag{5.83}$$

we have

$$\begin{cases} (i) \quad R \geq \frac{1}{\sqrt{b}} \arctan(2\sqrt{b}\frac{e}{\tau_0}), & \text{if } b > 0; \\ (ii) \quad R \geq 2\frac{e}{\tau_0}, & \text{if } b = 0; \\ (iii) \quad R \geq \frac{1}{\sqrt{-b}} \operatorname{arctanh}(2\sqrt{-b}\frac{e}{\tau_0}), & \text{if } b < 0. \end{cases} \tag{5.84}$$

Remark 5.12 For the definition of $\tau(\varphi)$ see Sect. 1.7 of Chap. 1.

Proof We limit ourselves to the case $b < 0$, the other cases being similar. Let $\rho(y) = \mathrm{dist}_{(N,h)}(y, q)$. By the Hessian comparison theorem and (5.81) we have

$$\mathrm{Hess}(\rho) \geq \sqrt{-b}\coth(\sqrt{-b}\rho)(h - d\rho \otimes d\rho). \tag{5.85}$$

To simplify the writing let $b = -1$ and set

$$u = \frac{1}{2}\cosh \rho \circ \varphi.$$

Then, if $m = \dim M$, by the composition law (1.181) of the Laplacian,

$$\Delta u = \sum_{i=1}^{m} \mathrm{Hess}(\frac{1}{2}\cosh \rho)(d\varphi(e_i), d\varphi(e_i)) + d(\frac{1}{2}\cosh \rho)(\tau(\varphi)),$$

where $\{e_i\}$, $i = 1, \ldots, m$, is a local orthonormal frame on M. From (5.85) we deduce

$$\Delta u \geq (2e(\varphi) + \tanh(\rho \circ \varphi)h(\nabla \rho, \tau(\varphi)))\, u. \tag{5.86}$$

Since $u \geq 1/2$ and

$$-\tanh(R)\tau_0 \leq \tanh(\rho \circ \varphi)h(\nabla \rho, \tau(\varphi)),$$

using Cauchy-Schwarz inequality we obtain

$$\Delta u \geq 2(e - \frac{1}{2}\tanh(R)\tau_0)u. \tag{5.87}$$

Since $u \leq \frac{1}{2}\cosh(R)$, applying the weak maximum principle to (5.87) we obtain

$$e - \frac{1}{2}\tanh(R)\tau_0 \leq 0,$$

which is equivalent to the validity of (5.84) (iii). □

Now we will consider an isometric immersion $f : M \to \mathbb{R}^n$ together with its Gauss map $\gamma_f : M \to G_m(\mathbb{R}^n))$, from M into the *Grassmann manifold* of m-planes through the origin of $\mathbb{R}^n$, where $m = \dim M$. Since we shall need to perform some computations, we next describe the Riemannian geometry of $G_m(\mathbb{R}^n)$ with its canonical metric. Towards this aim we consider the usual action of the (proper) rotation group $SO(n)$ on $\mathbb{R}^n$. This action induces an action on $G_m(\mathbb{R}^n)$ which is clearly transitive. We let $\{\varepsilon_1, \ldots, \varepsilon_n\}$ be the canonical basis of $\mathbb{R}^n$ and we fix as an origin of $G_m(\mathbb{R}^n)$ the point $o = \text{span}\{\varepsilon_1, \ldots, \varepsilon_m\}$, that is, the m-dimensional vector space generated by $\varepsilon_1, \ldots, \varepsilon_m$. The isotropy subgroup of the action of $SO(n)$ on $G_m(\mathbb{R}^n)$ fixing the origin o is the subgroup $SO_o(n))$ given by

$$SO_o(n) = \text{diag}(A, B) \tag{5.88}$$

with $A \in O(m)$, $B \in O(n-m)$ and $\det A \cdot \det B = 1$. We can thus realize $G_m(\mathbb{R}^n)$ as the homogeneous space

$$SO(n)/SO_o(n).$$

We fix the indices convention

$$1 \leq a, b, \ldots \leq n, \quad 1 \leq i, j, \ldots \leq m, \quad m+1 \leq \alpha, \beta, \ldots \leq n.$$

To describe the Riemannian structure of $G_m(\mathbb{R}^n)$ we let $\{\theta^a_b\}$ be the Maurer-Cartan forms of $SO(n)$ and let σ be a local section of the principal fiber bundle

$$\pi : SO(n) \overset{SO_o(n)}{\longrightarrow} G_m(\mathbb{R}^n),$$

where $\pi : A \to \text{span}\{A_1, \ldots, A_m\}$, with A_i, $i = 1, \ldots, n$ the columns of the matrix A. In what follows, throughout this paragraph we shall omit the pullback notation. Since a change of local section is of the type

$$\tilde{\sigma} = \sigma \cdot K, \quad K : U \subset G_m(\mathbb{R}^n) \to SO_o(n),$$

$U \subset G_m(\mathbb{R}^n)$ open, from the transformation law of the Maurer-Cartan form of $SO(n)$ (see [152]) given by

$$\tilde{\sigma}^{-1}d\tilde{\sigma} = {}^T\tilde{\sigma}d\tilde{\sigma} = {}^TK({}^T\sigma d\sigma)K + {}^TKdK = K^{-1}(\sigma^{-1}d\sigma)K + K^{-1}dK,$$

we immediately see that the quadratic form

$$ds^2 = \sum_{i,\alpha} (\theta_i^\alpha)^2 \tag{5.89}$$

is independent of the choice of σ and therefore defines a metric on $G_m(\mathbb{R}^n)$ with local orthonormal coframe given by the forms

$$\varphi^{\alpha,i} = \theta_i^\alpha, \quad \text{with } 1 \leq i \leq m, \quad m+1 \leq \alpha \leq n. \tag{5.90}$$

The corresponding Levi-Civita connection forms are

$$\varphi_{\beta,j}^{\alpha,i} = \delta_\beta^\alpha \theta_j^i + \delta_j^i \theta_\beta^\alpha, \tag{5.91}$$

as they can be easily found by following the procedure in Proposition 1.1 or simply guessed by the Maurer-Cartan equations $d\theta_b^a = -\theta_c^a \wedge \theta_b^c$ on $SO(n)$, verifying that they satisfy Eqs. (1.4) and (1.5).

In a way analogous to what we have done for the Grassmannian, we describe $\mathbb{R}^n$ as the homogeneous space

$$\mathbb{R}^n = \mathbb{R}^n \ltimes SO(n)/SO_o(n), \tag{5.92}$$

where the semidirect product $\mathbb{R}^n \ltimes SO(n)$ is the group of (proper) rigid motions of $\mathbb{R}^n$ and $SO(n)$ is the isotropy group at the origin $0 \in \mathbb{R}^n$. We denote by

$$\pi_1 : \mathbb{R}^n \ltimes SO(n) \overset{SO(n)}{\longrightarrow} \mathbb{R}^n \tag{5.93}$$

the projection of the principal bundle, that is, $\pi_1 : (x, A) \to x$. Let ζ be a local section of (5.93) and let $\{\theta^a, \theta_b^a\}$ be the Maurer-Cartan forms of $\mathbb{R}^n \ltimes SO(n)$. The quadratic form

$$\langle\, ,\, \rangle_{\mathbb{R}^n} = \sum_a (\theta^a)^2 \qquad \text{on } \mathbb{R}^n \tag{5.94}$$

is independent of the choice of ζ and defines the Euclidean metric of $\mathbb{R}^n$. The corresponding Levi-Civita connection forms are

$$\{\theta_b^a\}. \tag{5.95}$$

We let $F : \mathbb{R}^n \ltimes SO(n) \to SO(n)$ be the map

$$F : (x, A) \to A. \tag{5.96}$$

Given the isometric immersion $f : M \to \mathbb{R}^n$ and the Darboux frame (f, e) along f, where $e = \{e_1, \ldots, e_m, e_{m+1}, \ldots, e_n\}$, the Gauss map $\gamma_f : M \to G_m(\mathbb{R}^n)$ is defined by

$$\gamma_f(p) = \mathrm{span}_p\{e_1, \ldots, e_m\} \tag{5.97}$$

in such a way that the following diagram is commutative:

$$\begin{array}{ccccc} & & \mathbb{R}^n \ltimes SO(n) & \xrightarrow{F} & SO(n) \\ & \swarrow^{\pi_1} & \uparrow^{(f,e)} & & \downarrow^{\pi} \\ \mathbb{R}^n & \xleftarrow[f]{} & U \subset M & \xrightarrow{\gamma_f} & G_m(\mathbb{R}^n) \end{array} \tag{5.98}$$

For later use we shall reprove here the well known result of Ruh and Vilms [249].

Theorem 5.10 *Let $f : M \to \mathbb{R}^n$ be an isometric immersion. Then f has parallel mean curvature vector if and only if $\gamma_f : M \to G_m(\mathbb{R}^n)$ is a harmonic map.*

Proof Recall that, with respect to a Darboux frame, (f, e), along f, the coefficients of the second fundamental tensor II are given by h^α_{ij} where

$$\theta^\alpha_i = h^\alpha_{ij}\theta^j, \quad h^\alpha_{ij} = h^\alpha_{ji}.$$

Using the commutativity of the diagram (5.98) we obtain

$$\gamma_f^*\varphi^{\alpha,i} = \theta^\alpha_i = h^\alpha_{ij}\theta^j, \tag{5.99}$$

so that, with our notation (see Chap. 1),

$$(\gamma_f)^{\alpha,i}_j = h^\alpha_{ij}. \tag{5.100}$$

We then compute the generalized second fundamental tensor of γ_f according to (5.100). We have

$$\begin{aligned} (\gamma_f)^{\alpha,i}_{jk}\theta^k &= d(\gamma_f)^{\alpha,i}_j - (\gamma_f)^{\alpha,i}_t\theta^t_j + (\gamma_f)^{\beta,t}_j\varphi^{\alpha,i}_{\beta,t} \\ &= dh^\alpha_{ij} - h^\alpha_{it}\theta^t_j + h^\beta_{tj}(\delta^\alpha_\beta\theta^i_t + \delta^i_t\theta^\alpha_\beta) \\ &= h^\alpha_{ijk}\theta^k. \end{aligned}$$

From Codazzi equations

$$h^\alpha_{ijk} = h^\alpha_{jik} = h^\alpha_{jki},$$

so that

$$(\gamma_f)^{\alpha,i}_{kk} = h^{\alpha}_{ikk} = h^{\alpha}_{kki}.$$

In other words, $\tau(\gamma_f) = 0$ if and only if $\nabla \mathbf{H} = 0$. □

Let $G^+_m(\mathbb{R}^n)$ be the Grassmann manifold of oriented m-planes in $\mathbb{R}^n$, so that $G^+_m(\mathbb{R}^n) = SO(n)/SO(n) \times SO(n-m)$. Having fixed $q \in G^+_m(\mathbb{R}^n)$, a geodesic ball $B_R(q)$ is regular if

$$R < \begin{cases} \frac{\pi}{2}, & \text{if } n-m=1; \\ \frac{\pi}{2\sqrt{2}}, & \text{if } n-m>1. \end{cases} \tag{5.101}$$

To describe a regular geodesic ball in algebraic terms is far from being trivial. The next result is due to Fischer-Colbrie [118], and we refer to her paper for the proof.

Lemma 5.2 *Let $\Lambda_m(\mathbb{R}^n)$ be the algebra of m-multivectors of $\mathbb{R}^n$ with inner product $\langle , \rangle$ induced by the usual inner product of $\mathbb{R}^n$. Consider the Plücker embedding*

$$\mathscr{P} : G^+_m(\mathbb{R}^n) \hookrightarrow \Lambda_m(\mathbb{R}^n)$$

associating to the oriented plane $\Pi \in G^+_m(\mathbb{R}^n)$ spanned by the oriented orthonormal vectors $e_1, \dots, e_m$ the unit m-vector $e_1 \wedge \cdots \wedge e_m$. Let $R > 0$ be as in (5.101) *and define*

$$\mathscr{B}_R(q) = \{\Pi \in G^+_m(\mathbb{R}^n) : \langle \Pi, q\rangle \geq \cos^m(R/\sqrt{m})\}.$$

Then $\mathscr{B}_R(q)$ is contained in the regular geodesic ball $B_R(q)$.

We are now ready to prove

Theorem 5.11 *Let $f : M \to \mathbb{R}^n$ be an isometric immersion of a stochastically complete oriented manifold of dimension m into $\mathbb{R}^n$ with parallel mean curvature vector. Suppose there exists a decomposable m-vector q such that for each $x \in M$*

$$\langle \gamma_f(x), q\rangle \geq \cos^m(R/\sqrt{m})$$

with R as in (5.101). *Then f is minimal.*

Remark 5.13 For a previous version of the result see [153].

Proof From the Ruh-Vilms theorem, Theorem 5.10, γ_f is harmonic. Thus, for any $\tau_0 > 0$

$$|\tau(\gamma_f)| \leq \tau_0.$$

Because of Lemma 5.2,

$$\gamma_f(M) \subset B_R(q)$$

with $B_R(q)$ a geodesic regular ball in $G_m^+(\mathbb{R}^n)$. Because of Proposition 5.2

$$\inf_M e(\gamma_f) \leq \begin{cases} \frac{\tau_0}{2} \tan R, & \text{if } n-m=1; \\ \frac{\tau_0}{2\sqrt{2}} \tan(\sqrt{2}R), & \text{if } n-m>1. \end{cases}$$

In both cases, due to the arbitrariness of τ_0 we deduce

$$\inf_M e(\gamma_f) = 0. \tag{5.102}$$

On the other hand,

$$\gamma_f^*(ds^2) = \sum_{\alpha,i} (\varphi^{\alpha,i})^2 = \sum_{\alpha,i} (h_{ij}^\alpha \theta^j)^2$$

with respect to an oriented Darboux frame (f, e) along f. Thus making explicit the computations, with the help of Gauss equations, we obtain

$$\gamma_f^*(ds^2) = -\mathrm{Ric}_M + m\langle \mathrm{II}, \mathbf{H} \rangle.$$

Tracing the above with respect to the metric g of M induced by f, we obtain

$$2e(\gamma_f) = -S + m^2|\mathbf{H}|^2.$$

A further application of Gauss equation yields

$$|\mathrm{II}|^2 = m^2|\mathbf{H}|^2 - S.$$

Hence, from (5.102) we deduce

$$\inf_M |\mathrm{II}|^2 = 0.$$

But

$$|\mathbf{H}|^2 \leq |\mathrm{II}|^2$$

and since $|\mathbf{H}|$ is constant it follows that $\mathbf{H} = 0$. □

Let us consider again the Plücker embedding $\mathscr{P} : G_m^+(\mathbb{R}^n) \hookrightarrow \Lambda_m(\mathbb{R}^n)$. We recall that

$$\dim \Lambda_m(\mathbb{R}^n) = \binom{n}{m}$$

so that the unit sphere in $\Lambda_m(\mathbb{R}^n)$ has dimension $d = \binom{n}{m} - 1$ and it will be indicated by $\mathbb{S}^d$. Identifying an oriented m-plane of $\mathbb{R}^n$ with a unitary m-multivector we may think of γ_f as a map with values in $\mathbb{S}^d$. Having fixed an oriented m-plane Π with oriented orthonormal basis $\{v_1, \dots, v_m\}$ we define the angle Θ between Π and $\gamma_f(x)$ via

$$\cos\Theta = \langle v_1 \wedge \cdots \wedge v_m, e_1 \wedge \cdots \wedge e_m \rangle \tag{5.103}$$

where $\{e_i\}$, $i = 1, \dots, m$ is the part of the oriented Darboux frame (f, e) along f tangent to M at x.

Definition 5.3 We shall say that $\gamma_f(x)$ is contained in the open spherical cap $C_{\Theta_0}(V)$ centered at $V = v_1 \wedge \cdots \wedge v_m$ with radius Θ_0, $0 \le \Theta_0 \le \pi$ if and only if

$$\cos\Theta > \cos\Theta_0. \tag{5.104}$$

$\gamma_f(x)$ will be contained in the closure $\overline{C_{\Theta_0}(V)}$ if and only if

$$\cos\Theta \ge \cos\Theta_0. \tag{5.105}$$

In the proof of Theorem 5.12 below we shall need the following computational result.

Lemma 5.3 *Let $f : M \to \mathbb{R}^n$ be an isometric immersion of an oriented m-dimensional manifold with associated Gauss map $\gamma_f : M \to \mathbb{S}^d \subset \Lambda_m(\mathbb{R}^n)$. Fix $V = v_1 \wedge \cdots \wedge v_m \in \mathbb{S}^d$ a unit m-multivector. Set $\gamma_f(x) = e_1 \wedge \cdots \wedge e_m$ for an oriented Darboux frame (f, e) along f, and define*

$$u(x) = \langle \gamma_f(x), V \rangle, \tag{5.106}$$

so that $-1 \le u \le 1$. If f has parallel mean curvature vector then

$$\Delta u \le -|\mathrm{II}|^2 \left(u - \sqrt{\frac{2(n-m-1)}{n-m}} \sqrt{1-u^2} \right), \tag{5.107}$$

where II is the second fundamental tensor of f.

Proof Let $\{\theta^a\}$ be a local, oriented Darboux coframe along f. We use the index convention fixed above. Then

$$u = \langle e_1 \wedge \cdots \wedge e_m, V \rangle \tag{5.108}$$

and therefore if h^α_{ij} are the coefficients of the second fundamental tensor of f we have

$$\theta^\alpha_i = h^\alpha_{ij} \theta^j. \tag{5.109}$$

Differentiating (5.108) we obtain

$$
\begin{aligned}
du &= \sum_{i=1}^{m} \langle e_1 \wedge \cdots \wedge de_i \wedge \cdots \wedge e_m, V\rangle \\
&= \sum_{i=1}^{m} (-1)^{i-1} \langle e_\alpha \wedge e_1 \wedge \cdots \wedge \hat{e}_i \wedge \cdots \wedge e_m, V\rangle h^\alpha_{ij}\theta^j = u_j\theta^j,
\end{aligned}
$$

where, as usual, the symbol ^ means omitted. Indicating with u_{jk} the coefficients of Hess(u), that is,

$$
u_{jk}\theta^k = du_j - u_k\theta^k_j,
$$

after an elementary but tedious computation using Codazzi equations we have

$$
\begin{aligned}
u_{jk}\theta^k &= d\left(\sum_{i=1}^{m} (-1)^{i-1} \langle e_\alpha \wedge e_1 \wedge \cdots \wedge \hat{e}_i \wedge \cdots \wedge e_m, V\rangle h^\alpha_{ij}\right) \\
&\quad - \left(\sum_{t=1}^{m} (-1)^{t-1} \langle e_\alpha \wedge e_1 \wedge \cdots \wedge \hat{e}_t \wedge \cdots \wedge e_m, V\rangle h^\alpha_{tk}\right)\theta^k_j \\
&= \Big[(-1)^{i-1} h^\alpha_{ijk} \langle e_\alpha \wedge e_1 \wedge \cdots \wedge \hat{e}_i \wedge \cdots \wedge e_m, V\rangle \\
&\quad - (-1)^{i-1} h^\alpha_{ij} h^\alpha_{tk} \langle e_t \wedge e_1 \wedge \cdots \wedge \hat{e}_i \wedge \cdots \wedge e_m, V\rangle \\
&\quad + \sum_{t<i} (-1)^i (-1)^t h^\alpha_{ij} h^\beta_{tk} \langle e_\alpha \wedge e_\beta \wedge e_1 \wedge \cdots \wedge \hat{e}_t \wedge \cdots \wedge \hat{e}_i \wedge \cdots \wedge e_m, V\rangle \\
&\quad + \sum_{t>i} (-1)^{i-1} (-1)^t h^\alpha_{ij} h^\beta_{tk} \langle e_\alpha \wedge e_\beta \wedge e_1 \wedge \cdots \wedge \hat{e}_i \wedge \cdots \wedge \hat{e}_t \wedge \cdots \wedge e_m, V\rangle \Big]\theta^k
\end{aligned}
$$

Now the fact that $\mathbf{H}$ is parallel is equivalent to

$$
h^\alpha_{iik} = 0
$$

which, by Codazzi equations, turns out to be equivalent to $h^\alpha_{kii} = 0$. Tracing u_{jk} with respect to j and k and using the above we obtain

$$
\Delta u + |\mathrm{II}|^2 u = 2\sum_{\alpha<\beta}\sum_{t<i} (-1)^{i+t} [h^\alpha_{ik}h^\beta_{tk} - h^\alpha_{tk}h^\beta_{ik}] \langle e_\alpha \wedge e_\beta \wedge e_1 \wedge \cdots \wedge \hat{e}_t \wedge \cdots \wedge \hat{e}_i \wedge \cdots \wedge e_m, V\rangle. \tag{5.110}
$$

Next we set

$$
Q = \sum_{\alpha<\beta}\sum_{t<i} (-1)^{i+t} [h^\alpha_{ik}h^\beta_{tk} - h^\alpha_{tk}h^\beta_{ik}] \ell_{\alpha\beta ti} \tag{5.111}
$$

with

$$\ell_{\alpha\beta ti} = \langle e_\alpha \wedge e_\beta \wedge e_1 \wedge \cdots \wedge \hat{e}_t \wedge \cdots \wedge \hat{e}_i \wedge \cdots \wedge e_m, V\rangle. \tag{5.112}$$

To estimate Q from above we modify an idea of Reilly [241]. First of all we apply Cauchy-Schwarz inequality to obtain

$$Q^2 \leq \sum_{\alpha<\beta} \sum_{t<i} [h^\alpha_{ik} h^\beta_{tk} - h^\alpha_{tk} h^\beta_{ik}]^2 \sum_{\alpha<\beta} \sum_{t<i} \ell^2_{\alpha\beta ti}. \tag{5.113}$$

To bound from above the term

$$\sum_{\alpha<\beta} \sum_{t<i} \ell^2_{\alpha\beta ti}$$

we simply observe that, since V is a unit m-multivector

$$\sum_{\alpha<\beta} \sum_{t<i} \ell^2_{\alpha\beta ti} \leq 1 - u^2. \tag{5.114}$$

To estimate from above the remaining term in (5.113) we recall the following algebraic result due to Chern et al. [87].

Lemma 5.4 *Let $C = (c_{ij})$ and $D = (d_{ij})$ be symmetric $n \times n$ matrices. Then*

$$\sum_{i,k} \left(\sum_j (c_{ij} d_{jk} - c_{kj} d_{ji}) \right)^2 \leq 2(\sum_{i,j} c^2_{ij})(\sum_{k,\ell} d^2_{k\ell}). \tag{5.115}$$

In other words, if we set $|C|$ to denote the Hilbert-Schmidt norm of the $n \times n$ matrix C, we have

$$|CD - DC|^2 \leq 2|C|^2|D|^2.$$

Proof Observing that $|C|^2 = \left|T^{-1}CT\right|^2$ for every orthogonal $n \times n$ matrix T, without loss of generality we may assume that D, for instance, is a diagonal matrix, with entries $d_1, \ldots, d_n$. An easy computation now shows that

$$|CD - DC|^2 = \sum_{i \neq j} (c_{ij})^2 (d_i - d_j)^2.$$

Since $\left(d_i - d_j\right)^2 \le 2\left(d_i + d_j\right)^2$ we deduce

$$|CD - DC|^2 = \sum_{i \neq j} \left(c_{ij}\right)^2 \left(d_i - d_j\right)^2 \le \sum_{i \neq j} 2\left(c_{ij}\right)^2 \left(d_i + d_j\right)^2$$

$$\le 2\left[\sum_{i \neq j} \left(c_{ij}\right)^2\right]\left(\sum_i d_i^2\right) = 2|C|^2|D|^2$$

and the lemma follows. □

Next, we fix the indices α and β. Using the previous lemma we obtain

$$\sum_{t<i} [h_{ik}^{\alpha} h_{tk}^{\beta} - h_{tk}^{\alpha} h_{ik}^{\beta}]^2 = \frac{1}{2} \sum_{t,i} [h_{ik}^{\alpha} h_{tk}^{\beta} - h_{tk}^{\alpha} h_{ik}^{\beta}]^2 \le \sum_{i,k} (h_{ik}^{\alpha})^2 \sum_{i,k} (h_{ik}^{\beta})^2 = |\mathrm{II}^{\alpha}|^2 |\mathrm{II}^{\beta}|^2,$$

where we have set

$$\mathrm{II}^{\alpha} = \langle \mathrm{II}, e_{\alpha} \rangle.$$

From the previous inequality we deduce

$$\sum_{\alpha<\beta} \sum_{t<i} [h_{ik}^{\alpha} h_{tk}^{\beta} - h_{tk}^{\alpha} h_{ik}^{\beta}]^2 \le \sum_{\alpha<\beta} |\mathrm{II}^{\alpha}|^2 |\mathrm{II}^{\beta}|^2. \tag{5.116}$$

On the other hand, Newton's inequalities give

$$\sum_{\alpha<\beta} |\mathrm{II}^{\alpha}|^2 |\mathrm{II}^{\beta}|^2 \le \frac{(n-m)(n-m-1)}{2} \left(\sum_{\alpha} \frac{|\mathrm{II}^{\alpha}|^2}{n-m}\right)^2 = \frac{n-m-1}{2(n-m)} |\mathrm{II}|^4.$$

From (5.115) we finally obtain

$$\sum_{\alpha<\beta} \sum_{t<i} [h_{ik}^{\alpha} h_{tk}^{\beta} - h_{tk}^{\alpha} h_{ik}^{\beta}]^2 \le \frac{n-m-1}{2(n-m)} |\mathrm{II}|^4. \tag{5.117}$$

Putting together (5.117), (5.114), (5.113) yields

$$Q \le \sqrt{\frac{n-m-1}{2(n-m)}} |\mathrm{II}|^2 \sqrt{1-u^2}. \tag{5.118}$$

Using (5.118) together with (5.110) we obtain (5.107). □

We are now ready to state our result (see [239] for another version).

Theorem 5.12 *Let $f : M \to \mathbb{R}^n$ be a complete, oriented, m-dimensional isometric immersion with parallel mean curvature vector* $\mathbf{H}$. *Assume that the scalar curvature $S(x)$ of M satisfies*

$$S(x) \le m^2|\mathbf{H}|^2 - \frac{B}{(1+r(x))^{\mu}} \tag{5.119}$$

for some $B > 0$ and $0 \le \mu < 2$. Suppose furthermore that

$$\liminf_{r\to+\infty} \frac{\log \mathrm{vol} B_r}{r^{2-\mu}} < +\infty. \tag{5.120}$$

Then $\gamma_f(M)$ is not contained in any closed spherical cap in $\mathbb{S}^{\binom{n}{m}-1} \subset \Lambda_m(\mathbb{R}^n)$ of radius

$$\Theta < \arccos\sqrt{\frac{2(n-m-1)}{3n-3m-2}}. \tag{5.121}$$

Proof Given a unit multivector $V = v_1 \wedge \cdots \wedge v_m$, we set

$$u = \langle \gamma_f, V\rangle$$

By Lemma 5.3

$$\Delta u \le -b(x)f(u)$$

where we have set

$$b(x) = |\mathrm{II}|^2(x)$$

and

$$f(u) = u - \sqrt{\frac{2(n-m-1)}{n-m}}\sqrt{1-u^2}.$$

Since $b(x) = |\mathrm{II}|^2(x) = m^2|\mathbf{H}|^2 - S(x)$, from (5.119) we see that

$$b(x) \ge \frac{B}{(1+r(x))^{\mu}} > 0 \text{ on } M.$$

Since u is clearly bounded from below, it follows from Theorem 4.1 that $f(u_*) \le 0$. Solving the inequality one obtains

$$u_* \le \sqrt{\frac{2(n-m-1)}{3n-3m-2}}$$

as required. □

5.6 An Application of the Open Weak Maximum Principle

We now perform a computation similar to what we did to obtain Eq. (1.180). Suppose that M and N are manifolds with poles, respectively o_M and o_N. Let $\rho_M(y) = \mathrm{dist}_{(M,\langle\,,\rangle_M)}(y, o_M)$ and $\rho_N(z) = \mathrm{dist}_{(N,\langle\,,\rangle_N)}(z, o_N)$. We consider an isometric immersion

$$f : \Sigma \to M \times N.$$

Letting π_M and π_N be the two canonical projections of the product manifold onto its factors, we want to compute Δu and Δv, where u and v are respectively given by

$$u = \varphi_M(\rho_M) \circ \pi_M \circ f, \qquad v = \varphi_N(\rho_N) \circ \pi_N \circ f,$$

with $\varphi_M, \varphi_N : \mathbb{R}_0^+ \to \mathbb{R}$ smooth functions to be chosen later. Of course it will be enough to compute, for instance, the first. Towards this aim we recall that from (1.181) we have

$$\Delta u = \sum_{i=1}^{s=\dim\Sigma} \mathrm{Hess}\,(\varphi_M(\rho_M))(d(\pi_M \circ f)(e_i), d(\pi_M \circ f)(e_i)) + \langle {}^M\nabla \varphi_M(\rho_M), \tau(\pi_M \circ f)\rangle. \tag{5.122}$$

where $\{e_i\}$ is a local orthonormal frame on Σ. We therefore need to compute $d(\pi_M \circ f)$ and $\tau(\pi_M \circ f)$. In order to do this we fix the index convention $1 \leq i, j, \ldots \leq s = \dim \Sigma$, $1 \leq \alpha, \beta, \ldots \leq m = \dim M$, $1 \leq a, b, \ldots \leq n = \dim N$. We can apply formula (1.180) again, but for our purposes we need to make explicit the terms appearing in it. Thus it is in fact worth to redo the computation. Towards this aim we let

$$\{\theta^i\}, \{\theta^i_j\}, \{e_i\}, \quad \{\psi^\alpha\}, \left\{\psi^\alpha_\beta\right\}, \{\varepsilon_\alpha\}, \quad \{\omega^a\}, \{\omega^a_b\}, \{E_a\}$$

be local orthonormal coframes, with corresponding Levi-Civita connection forms and dual frames, respectively on Σ, M and N; computations follow the same formalism of Sect. 1.7 in Chap. 1. Recall that, for the product manifold structure on $M \times N$, $\{\psi^\alpha, \omega^a\}$ form a local orthonormal coframe, with corresponding connection forms $\varphi^\alpha_\beta = \psi^\alpha_\beta$, $\varphi^\alpha_a = \varphi^a_\alpha = 0$, $\varphi^a_b = \omega^a_b$; now

$$\pi_M^* \psi^\alpha = \psi^\alpha,$$

so that

$$(\pi_M \circ f)^\alpha_i \theta^i = (\pi_M \circ f)^* \psi^\alpha = f^*\left(\pi_M^* \psi^\alpha\right) = f^* \psi^\alpha = f^\alpha_i \theta^i,$$

while

$$\pi_M^* \omega^a = 0,$$

implying

$$(\pi_M \circ f)_i^a = 0.$$

It follows that

$$d(\pi_M \circ f) = f_i^\alpha \theta^i \otimes \varepsilon_\alpha. \tag{5.123}$$

We now compute $\tau(\pi_M \circ f)$, for $\pi_M \circ f : \Sigma \to M$. We have

$$(\pi_M \circ f)_{ij}^\alpha \theta^j = df_i^\alpha - f_k^\alpha \theta_i^k + f_i^\beta \psi_\beta^\alpha = f_{ij}^\alpha \theta^j,$$

while

$$(\pi_M \circ f)_{ij}^a \theta^j = 0.$$

Hence,

$$\tau(\pi_M \circ f) = f_{ii}^\alpha \varepsilon_\alpha. \tag{5.124}$$

For later use (see Theorem 5.13) we note that, setting $f^* \omega^a = f_i^a \theta^i$, since f is an isometric immersion we have

$$|d(\pi_M \circ f)|^2 + |d(\pi_N \circ f)|^2 = \Sigma_{i,a,\alpha} \left(f_i^\alpha\right)^2 + \left(f_i^a\right)^2 = s = \dim \Sigma. \tag{5.125}$$

We now insert (5.124) and (5.123) into (5.122) to obtain

$$\Delta u = \sum_{i,\alpha,\beta} \text{Hess}\,(\varphi_M(\rho_M)) \left(f_i^\alpha \varepsilon_\alpha, f_i^\beta \varepsilon_\beta\right) + \sum_{i,\alpha} \langle {}^M\nabla \varphi_M(\rho_M), f_{ii}^\alpha \varepsilon_\alpha \rangle. \tag{5.126}$$

Similarly

$$\Delta v = \sum_{i,\alpha,\beta} \text{Hess}\,(\varphi_N(\rho_N)) (f_i^a E_a, f_i^b E_b) + \sum_{i,\alpha} \langle {}^N\nabla \varphi_N(\rho_N), f_{ii}^a E_a \rangle. \tag{5.127}$$

Again, since f is an isometric immersion,

$$s\mathbf{H} = f_{ii}^\alpha \varepsilon_\alpha + f_{ii}^a E_a$$

where $\mathbf{H}$ is the mean curvature of the immersion. Then the last term in the right-hand side of (5.126) and (5.127) can be written respectively in the form

$$\langle {}^M\nabla \varphi_M(\rho_M), s\mathbf{H} \rangle \quad \text{and} \quad \langle {}^N\nabla \varphi_N(\rho_N), s\mathbf{H} \rangle. \tag{5.128}$$

We are now ready to prove the following lemma, extending some previous result of Dierkes [101]. This calculation has been performed also by de Lira and Medeiros [98].

Lemma 5.5 *Let M, N be complete manifolds with poles respectively o_M and o_N and distance functions ρ_M, ρ_N as above. Assume*

$$ {}^{M}K_{\mathrm{rad}}(y) \geq -G_M(\rho_M(y)) \quad \textit{and} \quad {}^{N}K_{\mathrm{rad}}(z) \leq -G_N(\rho_N(z)). \tag{5.129} $$

Let g_M and g_N be respectively the positive (if any) solutions on $\mathbb{R}^+$ of the Cauchy problems

$$ \begin{cases} g_M''(t) - G_M(t)g_M(t) = 0 \ \ on \ \mathbb{R}_0^+ \\ g_M(0) = 0, \ g_M'(0) = 1 \end{cases} \qquad \begin{cases} g_N''(t) - G_N(t)g_N(t) = 0 \ \ on \ \mathbb{R}_0^+ \\ g_N(0) = 0, \ g_N'(0) = 1 \end{cases} $$

and define the functions

$$ \varphi_M(\rho) = \int_0^\rho g_M(s)\, ds, \quad \varphi_N(\rho) = \int_0^\rho g_N(s)\, ds. \tag{5.130} $$

Let $f : \Sigma \to M \times N$ be an isometric immersion of an s-dimensional manifold Σ into the product manifold $M \times N$ and, setting π_M, π_N for the canonical projections of $M \times N$ onto its factors, define

$$ u = \varphi_M(\rho_M) \circ \pi_M \circ f, \qquad v = \varphi_N(\rho_N) \circ \pi_N \circ f. \tag{5.131} $$

Then

$$ \begin{cases} \Delta u \leq g_M'(\rho_M(\pi_M \circ f)) \sum_{i=1}^s |d(\pi_M \circ f)(e_i)|^2 + s\langle {}^{M}\nabla \varphi_M(\rho_M), \mathbf{H} \rangle_{M\times N} \\ \Delta v \geq g_N'(\rho_N(\pi_N \circ f)) \sum_{i=1}^s |d(\pi_N \circ f)(e_i)|^2 + s\langle {}^{N}\nabla \varphi_N(\rho_N), \mathbf{H} \rangle_{M\times N}, \end{cases} \tag{5.132} $$

where $\{e_i\}$ is a local orthonormal frame on Σ.

Proof By the Hessian comparison theorem, Theorem 1.4, we have

$$ \operatorname{Hess}(\rho_M) \leq \frac{g_M'(\rho_M)}{g_M(\rho_M)} \{\langle\, , \,\rangle_M - d\rho_M \otimes d\rho_M\} $$

in the sense of symmetric bilinear forms. Then by (1.168) and the definition (5.130) of φ_M we deduce

$$ \operatorname{Hess}(\varphi_M(\rho_M)) \leq g_M'(\rho_M)\langle\, , \,\rangle_M. $$

Substituting this expression in (5.126) and using (5.128) we obtain

$$\Delta u \leq g_M'(\rho_M) \sum_{i=1}^{s} |d(\pi_M \circ f)(e_i)|^2 + s\langle {}^M\nabla \varphi_M(\rho_M), \mathbf{H} \rangle_{M \times N}.$$

Similarly one obtains the second of (5.132). □

Remark 5.14 One can give bounds on G_M and G_N to guarantee that g_M and g_N be positive on $\mathbb{R}^+$. A detailed discussion on this can be found for instance in [44].

Since here we are interested only in showing an application of the open form of the weak maximum principle, we limit ourselves to the simplest case where $G_M \equiv G_N \equiv 0$. We prove the following

Theorem 5.13 *Let M and N be manifolds with poles o_M and o_N and dimensions m and n respectively. Assume that their radial sectional curvatures satisfy ${}^MK_{\mathrm{rad}} \geq 0$ and ${}^NK_{\mathrm{rad}} \leq 0$. Consider $M \times N$ with the product metric and the real function*

$$\sigma = \mu(\rho_M^2 \circ \pi_M) + (\rho_N^2 \circ \pi_N)$$

defined on $M \times N$, with $\mu \in \left(\frac{m-s}{m}, 0\right)$ and where the integers s and m satisfy $s > m$. Let $f : \Sigma \to M \times N$ be an s-dimensional, stochastically complete, minimal submanifold and $\Omega \subset \Sigma$ an open set with $\partial\Omega \neq \emptyset$ such that, for some $\Lambda \in \mathbb{R}_0^+$,

$$f(\partial\Omega) \subset \{(y, z) \in M \times N : \sigma(y, z) = \Lambda\} \tag{5.133}$$

and $\sigma \circ f$ is bounded above on Ω. Then

$$f(\Omega) \subset \{(y, z) \in M \times N : \sigma(y, z) \leq \Lambda\}. \tag{5.134}$$

Proof On Σ we consider the function $w = \sigma \circ f$. Specializing (5.132) to this case we have

$$\Delta w \geq \mu \sum_{i=1}^{s} |d(\pi_M \circ f)(e_i)|^2 + \sum_{i=1}^{s} |d(\pi_N \circ f)(e_i)|^2$$

and therefore, by (5.125),

$$\Delta w \geq s + (\mu - 1)|d(\pi_M \circ f)|^2;$$

but

$$|d(\pi_M \circ f)|^2 \leq m$$

so that we finally obtain

$$\Delta w \geq s + m(\mu - 1) > 0, \tag{5.135}$$

where the last strict inequality is due to our choice of the parameter μ. We now reason by contradiction and we suppose that (5.134) is not satisfied. This means that there exists at least a point $x_0 \in \Omega$ such that $w(x_0) = \sigma(f(x_0)) > \Lambda$. Note that, by (5.133), $w(x) = (\sigma \circ f)(x) \equiv \Lambda$ for each $x \in \partial\Omega$. Therefore, $\sup_{\partial\Omega} w < w(x_0) \leq \sup_{\Omega} w$. By Theorem 4.6, that is, the open form of the weak maximum principle, we necessarily have

$$s + m(\mu - 1) \leq 0,$$

a contradiction. □

Chapter 6
Applications to Hypersurfaces

The chapter begins with some introductory considerations on surfaces with constant mean curvature into 3-dimensional space forms based on, by now classical, works of Klotz, Osserman, Hoffman, Tribuzy... with the purpose to motivate their appropriate extensions to the higher dimensional case. In particular, we analyze their classification, lower and upper estimates on the Gaussian curvature and their relative sharpness. The proofs of these classical results do strongly depend on the conformal structure of the surfaces motivating the need of an alternative approach in higher dimensions. Following Alías and García-Martínez [14, 15] we provide new arguments based on the maximum principle. See for instance the proofs of Theorems 6.4 and 6.5 below. We also provide a further approach based on the principal curvature theorem (Theorem 6.7) of Smyth and Xavier [258].

We then focus our attention on the constant scalar curvature case with the aid of the well-known Cheng and Yau operator $\square$ (that is, the differential operator associated to the first Newton operator of a 2-sided hypersurface). The main result is given in Theorem 6.10. Proceeding we introduce, in some detail, the general Newton operators and briefly discuss the ellipticity of the associated differential operators. These material will be used also, for instance, in Chap. 7. In order to achieve a proof of Theorem 6.10 we Taylor an appropriate form of the Omori-Yau maximum principle for trace operators under curvature assumptions, see Theorem 6.13. Its proof follows the lines of that of Theorem 2.5, but we decided to report here some details because of the existence of the cut locus.

In Sect. 6.3 we consider hypersurfaces Σ whose image is contained into a non-degenerate Euclidean cone. Motivated by the results of Chap. 5 we give a lower bound estimate for $\sup_\Sigma |H_{k+1}|/H_k$, H_k the k-th mean curvature, in terms of the width of the cone.

In the final section of the chapter we give the same type of estimates but in case the image of Σ is contained in a regular geodesic ball of a generic complete Riemannian manifold N.

L.J. Alías et al., *Maximum Principles and Geometric Applications*,
Springer Monographs in Mathematics, DOI 10.1007/978-3-319-24337-5_6

6.1 Constant Mean Curvature Hypersurfaces in Space Forms

In a classical paper, Klotz and Osserman [160] characterized totally umbilical spheres and circular cylinders as the only complete surfaces immersed into the Euclidean 3-space $\mathbb{R}^3$ with constant mean curvature $H \neq 0$ and whose Gaussian curvature does not change sign. Later on, Hoffman [145] and Tribuzy [265] gave an extension of that result to the case of surfaces with constant mean curvature in the Euclidean 3-sphere $\mathbb{S}^3$ and in the hyperbolic space $\mathbb{H}^3$, respectively. Specifically, putting together the results of these authors in a single statement, one gets the following result (see also [76, Proposition 3.3]).

Theorem 6.1 *Let Σ be a complete surface immersed into a 3-dimensional space form with constant mean curvature H. If its Gaussian curvature K does not change sign, then Σ is either a totally umbilical surface or $K = 0$ and*

(a) Σ is a circular cylinder $\mathbb{R} \times \mathbb{S}^1(r) \subset \mathbb{R}^3$, $r > 0$,
(b) Σ is a flat torus $\mathbb{S}^1(\sqrt{1-r^2}) \times \mathbb{S}^1(r) \subset \mathbb{S}^3$, $0 < r < 1$,
(c) Σ is a hyperbolic cylinder $\mathbb{H}^1(-\sqrt{1+r^2}) \times \mathbb{S}^1(r) \subset \mathbb{H}^3$, $r > 0$.

As a nice application of Theorem 6.1, one gets the following consequence for the infimum of the Gaussian curvature of Σ.

Theorem 6.2 *Let Σ be a complete surface immersed into a 3-dimensional space form with constant mean curvature H such that $H^2 + c > 0$, where c denotes the constant sectional curvature of the ambient space ($c = 0, 1, -1$). Then either*

(i) $\inf_\Sigma K = H^2 + c$, and Σ is a totally umbilical surface, or
(ii) $\inf_\Sigma K \leq 0$, with equality if and only if

  (a) Σ is a circular cylinder $\mathbb{R} \times \mathbb{S}^1(r) \subset \mathbb{R}^3$, $r > 0$,
  (b) Σ is a flat torus $\mathbb{S}^1(\sqrt{1-r^2}) \times \mathbb{S}^1(r) \subset \mathbb{S}^3$, $0 < r < 1$,
  (c) Σ is a hyperbolic cylinder $\mathbb{H}^1(-\sqrt{1+r^2}) \times \mathbb{S}^1(r) \subset \mathbb{H}^3$, $r > 0$.

Actually, it follows from the Gauss equation of the surface that $K \leq H^2 + c$ on Σ, with equality at the umbilical points of Σ. Therefore, $\inf_\Sigma K \leq H^2 + c$ with equality if and only if Σ is totally umbilical. This proves part (i). Moreover, if $\inf_\Sigma K < H^2 + c$ then it must be $\inf_\Sigma K \leq 0$ necessarily. Otherwise, one would have $K \geq \inf_\Sigma K > 0$ which is not possible by Theorem 6.1, since the nontotally umbilical surfaces in (a), (b) and (c) are all flat. This shows that $\inf_\Sigma K \leq 0$. Finally, if equality holds, $\inf_\Sigma K = 0$, then $K \geq 0$ and the result follows from Theorem 6.1.

As another nice application of Theorem 6.1, one also gets the following consequence for the supremum of the Gaussian curvature of Σ.

Theorem 6.3 *Let Σ be a complete surface immersed into a 3-dimensional space form with constant mean curvature H. Then either*

(i) $\sup_\Sigma K = H^2 + c$, *or*
(ii) $0 \le \sup_\Sigma K < H^2 + c$, *with equality* $\sup_\Sigma K = 0$ *if and only if*

(a) Σ *is a circular cylinder* $\mathbb{R} \times \mathbb{S}^1(r) \subset \mathbb{R}^3$, $r > 0$,
(b) Σ *is a flat torus* $\mathbb{S}^1(\sqrt{1-r^2}) \times \mathbb{S}^1(r) \subset \mathbb{S}^3$, $0 < r < 1$,
(c) Σ *is a hyperbolic cylinder* $\mathbb{H}^1(-\sqrt{1+r^2}) \times \mathbb{S}^1(r) \subset \mathbb{H}^3$, $r > 0$.

In fact, one knows from the Gauss equation of the surface that $\sup_\Sigma K \le H^2 + c$. Moreover, if $\sup_\Sigma K < H^2 + c$ then it must be $\sup_\Sigma K \ge 0$ necessarily. Otherwise, if one assumes that $\sup_\Sigma K < 0$ then it would follow that $K \le \sup_\Sigma K < 0$ which is not possible by Theorem 6.1, since the nontotally umbilical surfaces in (a), (b) and (c) are all flat. This shows that either $\sup_\Sigma K = H^2 + c$ or $0 \le \sup_\Sigma K < H^2 + c$. Finally, if equality $\sup_\Sigma K = 0$ holds, then $K \le 0$ and the result follows again from Theorem 6.1.

Rotational surfaces show that the estimates in Theorems 6.2 and 6.3 are sharp. For instance, let us consider the Delaunay rotational surfaces in the Euclidean space. For a given constant $H \ne 0$, we may consider the family of unduloids in $\mathbb{R}^3$ with constant mean curvature H, which are given by the following parametrization

$$(s,\theta) \mapsto (x_B(s), y_B(s)\cos\theta, y_B(s)\sin\theta), \quad (s,\theta) \in \mathbb{R} \times [0, 2\pi],$$

where $0 < B < 1$ and

$$x_B(s) = \int_0^s \frac{1 + B\sin(2Ht)}{\sqrt{1 + B^2 + 2B\sin(2Ht)}} dt$$

$$y_B(s) = \frac{\sqrt{1 + B^2 + 2B\sin(2Hs)}}{2|H|}$$

(see [158] for the details). The first fundamental form of these surfaces is $ds^2 + y_B(s)^2 d\theta^2$ and the Gaussian curvature is then

$$K_B(s,\theta) = K_B(s) = -\frac{y_B''(s)}{y_B(s)} = \frac{4H^2 B(B + \sin(2Hs))(1 + B\sin(2Hs))}{(1 + B^2 + 2B\sin(2Hs))^2}.$$

Therefore, for these examples we have

$$\inf_\Sigma K_B = -\frac{4H^2 B}{(1-B)^2} < 0$$

and

$$\sup_\Sigma K_B = \frac{4H^2 B}{(1+B)^2} > 0.$$

Then, for a given $\varepsilon > 0$ there exists $0 < B < 1$ such that $\inf_\Sigma K_B = -\varepsilon < 0$, showing that the estimate $\inf_\Sigma K \leq 0$ in Theorem 6.2 is sharp. On the other hand, for a given $\varepsilon > 0$ one may also find $B_1, B_2 \in (0, 1)$ such that $\sup_\Sigma K_{B_1} = \varepsilon$ and $\sup_\Sigma K_{B_2} = H^2 - \varepsilon$, respectively, showing that the estimate $0 \leq \sup_\Sigma K < H^2$ in Theorem 6.3 is also sharp (with $c = 0$).

It is worth pointing out that the proof of Theorem 6.1 (and hence Theorems 6.2 and 6.3) strongly depends on the conformal structure of the 2-dimensional surface Σ, and cannot be extended to higher dimensions. Our objective in this section is to introduce extensions of Theorems 6.2 and 6.3 to the case of m-dimensional hypersurfaces, $m \geq 3$, using an alternative approach by Alías and García-Martínez which is based on the maximum principles.

Specifically, we will prove the following extension of Theorem 6.2 (see [14, Theorem 3]).

Theorem 6.4 *Let Σ be a stochastically complete hypersurface immersed into an $(m+1)$-dimensional space form, $m \geq 3$, with constant mean curvature H such that $H^2 + c > 0$, where c denotes the constant sectional curvature of the ambient space ($c = 0, 1, -1$). If S stands for the scalar curvature of Σ, then*

(i) either

$$\inf_\Sigma S = m(m-1)(c + H^2)$$

and Σ is a totally umbilical hypersurface,

(ii) or

$$\inf_\Sigma S \leq \widehat{B}_{|H|,c} \tag{6.1}$$

where

$$\widehat{B}_{|H|,c} = \frac{m(m-2)}{2(m-1)}\left(2(m-1)c + mH^2 + |H|\sqrt{m^2H^2 + 4(m-1)c}\right). \tag{6.2}$$

Moreover, the equality $\inf_\Sigma S = \widehat{B}_{|H|,c}$ holds and this infimum is attained at some point of Σ if and only if Σ is a (stochastically complete) open piece of

(a) a circular cylinder $\mathbb{R} \times \mathbb{S}^{m-1}(r) \subset \mathbb{R}^{m+1}$, $r > 0$,

(b) a minimal Clifford torus $\mathbb{S}^k(\sqrt{k/m}) \times \mathbb{S}^{m-k}(\sqrt{(m-k)/m}) \subset \mathbb{S}^{m+1}$, with $k = 1, \ldots, m-1$, or a constant mean curvature torus $\mathbb{S}^1(\sqrt{1-r^2}) \times \mathbb{S}^{m-1}(r) \subset \mathbb{S}^{m+1}$, with $0 < r < \sqrt{(m-1)/m}$,

(c) a hyperbolic cylinder $\mathbb{H}^1(-\sqrt{1+r^2}) \times \mathbb{S}^{m-1}(r) \subset \mathbb{H}^{m+1}$, $r > 0$.

In the particular case that Σ is complete, we obtain the following consequence (see [14, Corollary 4]).

Corollary 6.1 *Let Σ be a complete hypersurface immersed into an $(m+1)$-dimensional space form, $m \geq 3$, with constant mean curvature H such that $H^2 + c > 0$, where c denotes the constant sectional curvature of the ambient space $(c = 0, 1, -1)$. Then*

(i) either

$$\inf_{\Sigma} S = m(m-1)(c+H^2)$$

and Σ is a totally umbilical hypersurface,

(ii) or

$$\inf_{\Sigma} S \leq \widehat{B}_{|H|,c}.$$

Moreover, the equality $\inf_{\Sigma} S = \widehat{B}_{|H|,c}$ holds and this infimum is attained at some point of Σ if and only if Σ is

(a) a circular cylinder $\mathbb{R} \times \mathbb{S}^{m-1}(r) \subset \mathbb{R}^{m+1}$, $r > 0$,
(b) a minimal Clifford torus $\mathbb{S}^k(\sqrt{k/m}) \times \mathbb{S}^{m-k}(\sqrt{(m-k)/m}) \subset \mathbb{S}^{m+1}$, with $k = 1, \ldots, m-1$, or a constant mean curvature torus $\mathbb{S}^1(\sqrt{1-r^2}) \times \mathbb{S}^{m-1}(r) \subset \mathbb{S}^{m+1}$, with $0 < r < \sqrt{(m-1)/m}$,
(c) a hyperbolic cylinder $\mathbb{H}^1(-\sqrt{1+r^2}) \times \mathbb{S}^{m-1}(r) \subset \mathbb{H}^{m+1}$, $r > 0$.

On the other hand, Theorem 6.3 admits the following extension to the m-dimensional case (see [15, Theorem 6]).

Theorem 6.5 *Let Σ be a hypersurface immersed into an $(m+1)$-dimensional space form, $m \geq 3$, with constant mean curvature H and with two distinct principal curvatures, one of them being simple. Assume that the Omori-Yau maximum principle holds on Σ, and let c denote the constant sectional curvature of the ambient space $(c = 0, 1, -1)$.*

(i) If $H^2 + c \geq 0$ then

$$B_{|H|,c} \leq \sup_{\Sigma} S \leq m(m-1)(c+H^2),$$

where

$$B_{|H|,c} = \frac{m(m-2)}{2(m-1)}\left(2(m-1)c + mH^2 - |H|\sqrt{m^2H^2 + 4(m-1)c}\right). \tag{6.3}$$

(ii) If $H^2 + c < 0$ (necessarily with $c = -1$) then either

$$\sup_{\Sigma} S = m(m-1)(-1+H^2)$$

or $4(m-1)/m^2 \leq H^2 < 1$ *and*

$$B_{|H|,-1} \leq \sup_{\Sigma} S \leq \widehat{B}_{|H|,-1} < m(m-1)(-1+H^2),$$

where $\widehat{B}_{|H|,c}$ *is given by* (6.2).

Moreover, the equality $\sup_{\Sigma} S = B_{|H|,c}$ *holds and this supremum is attained at some point of* Σ *if and only if* Σ *is an open piece of*

(a) *a circular cylinder* $\mathbb{R}^{m-1} \times \mathbb{S}^1(r) \subset \mathbb{R}^{m+1}$, $r > 0$,
(b) *a constant mean curvature torus* $\mathbb{S}^1(\sqrt{1-r^2}) \times \mathbb{S}^{m-1}(r) \subset \mathbb{S}^{m+1}$, *with* $r \geq \sqrt{(m-1)/m}$,
(c) *a hyperbolic cylinder* $\mathbb{H}^{m-1}(-\sqrt{1+r^2}) \times \mathbb{S}^1(r) \subset \mathbb{H}^{m+1}$, *with either* $r = 1/\sqrt{m(m-2)}$ *if* $H^2 = 1$, *or* $0 < r < 1/\sqrt{m(m-2)}$ *in the case* $H^2 > 1$, *or* $1/\sqrt{m(m-2)} < r \leq 1/\sqrt{m-2}$ *in the case* $H^2 < 1$.

In particular, when Σ is properly immersed, we have the following result (see [14, Theorem 4]).

Corollary 6.2 *Let* Σ *be a hypersurface which is properly immersed into an* $(m+1)$-*dimensional space form,* $m \geq 3$, *with constant mean curvature* H *and with two distinct principal curvatures, one of them being simple.*

(i) *If* $H^2 + c \geq 0$ *then*

$$B_{|H|,c} \leq \sup_{\Sigma} S \leq m(m-1)(c+H^2).$$

(ii) *If* $H^2 + c < 0$ *(necessarily with* $c = -1$*) then either*

$$\sup_{\Sigma} S = m(m-1)(-1+H^2)$$

or $4(m-1)/m^2 \leq H^2 < 1$ *and*

$$B_{|H|,-1} \leq \sup_{\Sigma} S \leq \widehat{B}_{|H|,-1} < m(m-1)(-1+H^2).$$

Moreover, the equality $\sup_{\Sigma} S = B_{|H|,c}$ *holds and this supremum is attained at some point of* Σ *if and only if* Σ *is*

(a) *a circular cylinder* $\mathbb{R}^{m-1} \times \mathbb{S}^1(r) \subset \mathbb{R}^{m+1}$, $r > 0$,
(b) *a constant mean curvature torus* $\mathbb{S}^1(\sqrt{1-r^2}) \times \mathbb{S}^{m-1}(r) \subset \mathbb{S}^{m+1}$, *with* $r \geq \sqrt{(m-1)/m}$,
(c) *a hyperbolic cylinder* $\mathbb{H}^{m-1}(-\sqrt{1+r^2}) \times \mathbb{S}^1(r) \subset \mathbb{H}^{m+1}$, *with either* $r = 1/\sqrt{m(m-2)}$ *if* $H^2 = 1$, *or* $0 < r < 1/\sqrt{m(m-2)}$ *in the case* $H^2 > 1$, *or* $1/\sqrt{m(m-2)} < r \leq 1/\sqrt{m-2}$ *in the case* $H^2 < 1$.

Remark 6.1 Regarding the condition of having two distinct principal curvatures, it is well known, since the pioneering work of Otsuki [215], that if both principal curvatures have multiplicity greater than 1, then the distributions of the space of principal vectors corresponding to each principal curvature are completely integrable and each principal curvature is constant on each of the integral leaves of the corresponding distribution. In particular, if the mean curvature is constant, then the two principal curvatures are also constant and the hypersurface is an isoparametric hypersurface with exactly two constant principal curvatures, with multiplicities k and $m-k$, and $1<k<m-1$. Then, by the classical results on isoparametric hypersurfaces in Riemannian space forms [64, 172, 253] the hypersurface must be an open piece of one of the three following standard product embeddings: $\mathbb{R}^k \times \mathbb{S}^k(r) \subset \mathbb{R}^{m+1}$ with $r>0$, if $c=0$; $\mathbb{S}^k(\sqrt{1-r^2}) \times \mathbb{S}^{m-k}(r) \subset \mathbb{S}^{m+1}$ with $0<r<1$, if $c=1$; and $\mathbb{H}^k(-\sqrt{1+r^2}) \times \mathbb{S}^{m-k}(r) \subset \mathbb{H}^{m+1}$ with $r>0$, if $c=-1$. Therefore, under the condition of having two distinct principal curvatures, the interesting case for studying constant mean curvature hypersurfaces is the case where one of the principal curvatures is simple, that is, with multiplicity 1.

In [273], Wei studied complete hypersurfaces in the Euclidean sphere with constant mean curvature and with two distinct principal curvartures, one of them being simple, deriving a characterization of the tori $\mathbb{S}^1(\sqrt{1-r^2}) \times \mathbb{S}^{m-1}(r)$ in terms of the behavior of the squared norm of the second fundamental form (see also [140] for a previous corresponding result for the case of minimal hypersurfaces in $\mathbb{S}^{m+1}$ given by Hasanis et al.). It is worth pointing out that the estimates in Corollary 6.1 for the infimum of the scalar curvature [equivalently, for the supremum of the squared norm of the second fundamental form, see (6.10) below] and in Corollary 6.2 for the supremum of the scalar curvature [equivalently, for the infimum of the squared norm of the second fundamental form, see (6.11) below], when written in terms of the second fundamental form, are equivalent to Wei's estimates, with the advantage that the new approach here works for hypersurfaces in every Riemannian space form and that the estimate in Corollary 6.1 does not need the condition of having two distinct principal curvatures. We also refer the readers to [79, 139, 217] or [270] for other previous results about minimal compact hypersurfaces with two distinct principal curvatures in the Euclidean sphere $\mathbb{S}^{m+1}$.

6.1.1 Proof of the Main Results

Let Σ be an oriented hypersurface isometrically immersed into an $(m+1)$-dimensional Riemannian space form of constant sectional curvature $c=0,1,-1$, and denote by $A:\mathfrak{X}(\Sigma) \to \mathfrak{X}(\Sigma)$ its second fundamental form (with respect to a globally defined normal unit vector field ν) and by H its mean curvature, $H=(1/m)\,\mathrm{Tr}(A)$. In the general m-dimensional case, instead of the scalar curvature, it will be more appropriate to deal with the so called traceless second fundamental form of the hypersurface, which is given by $\Phi=A-HI$, where I denotes the identity operator on $\mathfrak{X}(\Sigma)$. Observe that $\mathrm{Tr}(\Phi)=0$ and $|\Phi|^2=\mathrm{Tr}(\Phi^2)=|A|^2-mH^2 \geq 0$, with equality if and only if Σ is totally umbilical. For that reason, Φ is also called the total umbilicity tensor of Σ.

As is well known, the curvature tensor R of the hypersurface is given by Gauss equations, which can be written both in term of A as

$$R(X,Y)Z = c(-\langle X,Z\rangle Y + \langle Y,Z\rangle X) - \langle AX,Z\rangle AY + \langle AY,Z\rangle AX \tag{6.4}$$

and in terms of Φ as

$$\begin{aligned} R(X,Y)Z = &(c+H^2)(-\langle X,Z\rangle Y + \langle Y,Z\rangle X) - \langle \Phi X,Z\rangle \Phi Y + \langle \Phi Y,Z\rangle \Phi X \\ &+H(-\langle \Phi X,Z\rangle Y + \langle Y,Z\rangle \Phi X - \langle X,Z\rangle \Phi Y + \langle \Phi Y,Z\rangle X) \end{aligned} \tag{6.5}$$

for $X, Y, Z \in \mathfrak{X}(\Sigma)$. In particular, the Ricci and the scalar curvatures of Σ are given, respectively, by

$$\begin{aligned} \mathrm{Ric}(X,Y) &= (m-1)c\langle X,Y\rangle + mH\langle AX,Y\rangle - \langle AX,AY\rangle \\ &= (m-1)(c+H^2)\langle X,Y\rangle + (m-2)H\langle \Phi X,Y\rangle - \langle \Phi X,\Phi Y\rangle, \end{aligned} \tag{6.6}$$

for $X, Y \in \mathfrak{X}(\Sigma)$, and

$$S = m(m-1)R = m(m-1)c + m^2H^2 - |A|^2 = m(m-1)(c+H^2) - |\Phi|^2. \tag{6.7}$$

Here, and in what follows, with R we indicate the *normalized* scalar curvature. From (6.7) we obtain the identities

$$m^2H^2 = |A|^2 + m(m-1)(R-c), \tag{6.8}$$

and

$$|\Phi|^2 = \frac{m-1}{m}|A|^2 - (m-1)(R-c) = m(m-1)H^2 - m(m-1)(R-c). \tag{6.9}$$

In particular, if H is constant it follows from here that

$$\inf_\Sigma S = m(m-1)(c+H^2) - \sup_\Sigma |\Phi|^2 \tag{6.10}$$

and

$$\sup_\Sigma S = m(m-1)(c+H^2) - \inf_\Sigma |\Phi|^2. \tag{6.11}$$

For the proof of the main results we will use the following Simons type formula for the Laplace-Beltrami operator of $|A|^2$.

Lemma 6.1 *Let Σ be a hypersurface immersed into an $(m+1)$-dimensional Riemannian space form (with constant sectional curvature c) and let A stand for its second fundamental form. Then*

$$\frac{1}{2}\Delta|A|^2 = |\nabla A|^2 + m\,\mathrm{Tr}(A\circ \mathrm{hess}\,H) - cm^2H^2 + (cm - |A|^2)|A|^2 + mH\,\mathrm{Tr}(A^3) \tag{6.12}$$

where $\nabla A : \mathfrak{X}(\Sigma)\times\mathfrak{X}(\Sigma)\to\mathfrak{X}(\Sigma)$ denotes the covariant differential of A,

$$\nabla A(X,Y) = (\nabla_Y A)X = \nabla_Y(AX) - A(\nabla_Y X), \quad X,Y\in\mathfrak{X}(\Sigma).$$

Formula (6.12) follows from the more general formula (1.149) in the particular case where the ambient space has constant sectional curvature c. For the sake of completeness, we include here another derivation of it, following Nomizu and Smyth [208].

Proof A standard tensor computation implies that

$$\frac{1}{2}\Delta|A|^2 = \frac{1}{2}\Delta\langle A,A\rangle = |\nabla A|^2 + \langle A,\Delta A\rangle. \tag{6.13}$$

Here $\Delta A : \mathfrak{X}(\Sigma)\to\mathfrak{X}(\Sigma)$ is the rough Laplacian,

$$\Delta A(X) = \mathrm{Tr}(\nabla^2 A(X,\cdot,\cdot)) = \sum_{i=1}^{m}\nabla^2 A(X,e_i,e_i),$$

where $\{e_1,\dots,e_m\}$ is a local orthonormal frame on Σ. Recall that ∇A is symmetric by the Codazzi equation of the hypersurface and, hence, $\nabla^2 A$ is also symmetric in its two first variables,

$$\nabla^2 A(X,Y,Z) = \nabla^2 A(Y,X,Z), \quad X,Y,Z\in\mathfrak{X}(\Sigma).$$

With respect to the symmetries of $\nabla^2 A$ in the other variables, it is not difficult to see that

$$\nabla^2 A(X,Y,Z) = \nabla^2 A(X,Z,Y) + R(Z,Y)AX - A(R(Z,Y)X).$$

Thus, using Gauss Eq. (6.4) it follows from here that

$$\begin{aligned}\Delta A(X) &= \sum_{i=1}^{m}\left(\nabla^2 A(e_i,e_i,X) + R(e_i,X)Ae_i - A(R(e_i,X)e_i)\right) \qquad (6.14)\\ &= \mathrm{Tr}(\nabla_X(\nabla A)) - cmHX + (cm-|A|^2)AX + mHA^2X\\ &= m\nabla_X\nabla H - cmHX + (cm-|A|^2)AX + mHA^2X,\end{aligned}$$

where we have used the facts that trace commutes with ∇_X and that $\mathrm{Tr}(\nabla A) = m\nabla H$ because of Codazzi equations (1.145) (see Remark 6.2 below). Therefore, by (6.13) we conclude that

$$\begin{aligned}\frac{1}{2}\Delta|A|^2 &= |\nabla A|^2 + m\sum_{i=1}^{m}\langle \nabla_{e_i}\nabla H, Ae_i\rangle - cm^2H^2 + (cm - |A|^2)|A|^2 + mH\,\mathrm{Tr}(A^3)\\ &= |\nabla A|^2 + m\,\mathrm{Tr}(A\circ \mathrm{hess}\, H) - cm^2H^2 + (cm - |A|^2)|A|^2 + mH\,\mathrm{Tr}(A^3).\end{aligned}$$

□

Remark 6.2 For a hypersurface Σ isometrically immersed into a general $(m+1)$-dimensional Riemannian manifold N, Codazzi equation (1.145) is equivalent, in Koszul notation, to

$$(\nabla_Y A)X - (\nabla_X A)Y = \left({}^N R(X,Y)\nu\right)^{\top} \tag{6.15}$$

for every $X, Y \in \mathfrak{X}(\Sigma)$, where A denotes the Weingarten operator with respect to ν. Therefore in general, and using the fact that $\nabla_{e_i}A$ is self-adjoint, we have for every $X \in \mathfrak{X}(\Sigma)$

$$\langle (\nabla_{e_i}A)e_i, X\rangle = \langle (\nabla_{e_i}A)X, e_i\rangle = \langle (\nabla_X A)e_i, e_i\rangle + \langle {}^N R(X, e_i)\nu, e_i\rangle,$$

so that

$$\begin{aligned}\langle \mathrm{Tr}(\nabla A), X\rangle &= \sum_{i=1}^{m}\langle (\nabla_{e_i}A)e_i, X\rangle\\ &= \sum_{i=1}^{m}\langle (\nabla_X A)e_i, e_i\rangle + \sum_{i=1}^{m}\langle {}^N R(X, e_i)\nu, e_i\rangle\\ &= \mathrm{Tr}(\nabla_X A) - {}^N\mathrm{Ric}(X,\nu)\\ &= \nabla_X(\mathrm{Tr}\, A) - {}^N\mathrm{Ric}(X,\nu)\\ &= m\langle \nabla H, X\rangle - {}^N\mathrm{Ric}(X,\nu).\end{aligned}$$

In particular, it is enough for N to be Einstein to have $\mathrm{Tr}(\nabla A) = m\nabla H$.

When the mean curvature H is constant, then $\nabla\Phi = \nabla A$ and $\Delta|\Phi|^2 = \Delta|A|^2$, and one can rewrite (6.12) in terms of Φ as follows.

Corollary 6.3 *Let Σ a be hypersurface immersed into an $(m+1)$-dimensional Riemannian space form (with constant sectional curvature c) and let Φ stand for its total umbilicity tensor. If the mean curvature H is constant, then*

$$\frac{1}{2}\Delta|\Phi|^2 = |\nabla\Phi|^2 + mH\,\mathrm{Tr}(\Phi^3) - |\Phi|^2(|\Phi|^2 - m(c + H^2)), \tag{6.16}$$

where $\nabla\Phi : \mathfrak{X}(\Sigma) \times \mathfrak{X}(\Sigma) \to \mathfrak{X}(\Sigma)$ *denotes the covariant differential of* Φ*,*

$$\nabla\Phi(X, Y) = (\nabla_Y \Phi)X = \nabla_Y(\Phi X) - \Phi(\nabla_Y X), \quad X, Y \in \mathfrak{X}(\Sigma).$$

We will also need the following auxiliary result, known as Okumura's lemma, which can be found in [209] and [8, Lemma 2.6].

Lemma 6.2 *Let* $a_1, \dots, a_m$ *be real numbers such that* $\sum_{i=1}^m a_i = 0$*. Then*

$$-\frac{(m-2)}{\sqrt{m(m-1)}} \left(\sum_{i=1}^m a_i^2\right)^{3/2} \le \sum_{i=1}^m a_i^3 \le \frac{(m-2)}{\sqrt{m(m-1)}} \left(\sum_{i=1}^m a_i^2\right)^{3/2}.$$

Moreover, equality holds in the right-hand (respectively, left-hand) side if and only if $(m-1)$ *of the* a_i*'s are nonpositive (respectively, nonnegative) and equal.*

Proof To simplify the notation, let us define $\sum_{i=1}^m a_i^2 = b^2 \ge 0$; we have thus to prove that

$$-\frac{(m-2)}{\sqrt{m(m-1)}} b^3 \le \sum_{i=1}^m a_i^3 \le \frac{(m-2)}{\sqrt{m(m-1)}} b^3.$$

We follow the proof of [8, Lemma 2.6]. If $b = 0$ we have nothing to prove, so we can assume $b > 0$. Now we exploit the method of Lagrange's multipliers to find the critical point of the function $F = \sum_{i=1}^m a_i^3$ with the constraints $\sum_{i=1}^m a_i = 0$ and $\sum_{i=1}^m a_i^2 = b^2 > 0$. A simple computation shows that the critical points are solutions of a quadratic equation of the form

$$x^2 - \mu x - \frac{b^2}{m} = 0,$$

where μ is a real constant. Since the solutions of previous equation are given by

$$x_{+,-} = \frac{\mu \pm \sqrt{\mu^2 + \frac{4}{m} b^2}}{2},$$

it follows that (after reordering if necessary) the critical points are given by

$$a_1 = a_2 = \dots = a_p = x_+ > 0, \quad a_{p+1} = a_{p+2} = \dots = a_m = x_- < 0.$$

Evaluating F and the constraints at critical points gives

$$\begin{aligned} p x_+ + (m-p) x_- &= 0, \\ p(x_+)^2 + (m-p)(x_-)^2 &= b^2, \\ p(x_+)^3 + (m-p)(x_-)^3 &= F; \end{aligned}$$

this implies that

$$
\begin{aligned}
(x_+)^2 &= \frac{m-p}{mp}b^2, \\
(x_-)^2 &= \frac{p}{m(m-p)}b^2, \\
F &= \Big[\Big(\frac{m-p}{m}\Big)x_+ + \frac{p}{m}x_-\Big]b^2 = \Big[x_+ - \frac{p}{m}x_+ - \frac{p}{m}|x_-|\Big]b^2.
\end{aligned}
$$

Since F decreases as p increases, F reaches its maximum F_{max} for $p = 1$, and we have, using previous equations,

$$F_{max} = (x_+)^3 + (m-1)(x_-)^3 = \frac{(m-2)}{\sqrt{m(m-1)}}b^3,$$

while the symmetry of F implies that the minimum F_{min} is equal to $-\frac{(m-2)}{\sqrt{m(m-1)}}$. □

We are now ready to give the proof of the first main result of this chapter.

Proof (of Theorem 6.4) Since $\mathrm{Tr}(\Phi) = 0$, we may use Lemma 6.2 to estimate $\mathrm{Tr}(\Phi^3)$ as follows

$$|\,\mathrm{Tr}(\Phi^3)| \leq \frac{(m-2)}{\sqrt{m(m-1)}}|\Phi|^3,$$

and then

$$mH\,\mathrm{Tr}(\Phi^3) \geq -m|H||\,\mathrm{Tr}(\Phi^3)| \geq -\frac{m(m-2)}{\sqrt{m(m-1)}}|H||\Phi|^3.$$

Using this in (6.16), we find

$$
\begin{aligned}
\frac{1}{2}\Delta|\Phi|^2 &\geq |\nabla\Phi|^2 - \frac{m(m-2)}{\sqrt{m(m-1)}}|H||\Phi|^3 - |\Phi|^2(|\Phi|^2 - m(c+H^2)) \\
&\geq -|\Phi|^2 P_{|H|,c}(|\Phi|),
\end{aligned} \tag{6.17}
$$

where

$$P_{|H|,c}(x) = x^2 + \frac{m(m-2)}{\sqrt{m(m-1)}}|H|x - m(c+H^2).$$

Observe that, since $H^2 + c > 0$, the polynomial $P_{|H|,c}(x)$ has a unique positive root given by

$$\alpha_{|H|,c} = \frac{\sqrt{m}}{2\sqrt{m-1}}\left(\sqrt{m^2H^2 + 4(m-1)c} - (m-2)|H|\right).$$

If $\sup_\Sigma |\Phi| = +\infty$, then by (6.10) we have $\inf_\Sigma S = -\infty$, so that (6.1) holds trivially and there is nothing to prove. If $\sup_\Sigma |\Phi| < +\infty$, then by applying the weak maximum principle to the function $|\Phi|^2$ we know that there exists $\{x_k\}_{k\in\mathbb{N}}$ in Σ such that

$$\lim_{k\to\infty} |\Phi|(x_k) = \sup_\Sigma |\Phi|, \quad \text{and} \quad \Delta|\Phi|^2(x_k) < 1/k,$$

which jointly with (6.17) implies

$$1/k > \Delta|\Phi|^2(x_k) \geq -2|\Phi|^2(x_k)P_{|H|,c}(|\Phi|(x_k)).$$

Taking limits here, we get $0 \geq -2(\sup_\Sigma |\Phi|)^2 P_{|H|,c}(\sup_\Sigma |\Phi|)$, that is

$$(\sup_\Sigma |\Phi|)^2 P_{|H|,c}(\sup_\Sigma |\Phi|) \geq 0.$$

It follows from here that either $\sup_\Sigma |\Phi| = 0$ or $\sup_\Sigma |\Phi| > 0$ and then $P_{|H|,c}(\sup_\Sigma |\Phi|) \geq 0$. In the former case, which by (6.10) is equivalent to

$$\inf_\Sigma S = m(m-1)(c + H^2),$$

it means that $|\Phi| \equiv 0$ and the hypersurface is totally umbilical. In the latter, it must be $\sup_\Sigma |\Phi| \geq \alpha_{|H|,c}$ which by (6.10) is equivalent to inequality (6.1) since

$$\inf_\Sigma S = m(m-1)(c + H^2) - \sup_\Sigma |\Phi|^2 \leq m(m-1)(c + H^2) - \alpha^2_{|H|,c} = \widehat{B}_{|H|,c}.$$

Moreover, assume that equality $\inf_\Sigma S = \widehat{B}_{|H|,c}$ holds; equivalently, $\sup_\Sigma |\Phi| = \alpha_{|H|,c}$. In that case, $P_{|H|,c}(|\Phi|) \leq 0$ on Σ, which jointly with (6.17) implies that $|\Phi|^2$ is a subharmonic function on Σ. Therefore, if there exists a point $x_0 \in \Sigma$ at which this supremum is attained, then $|\Phi|^2$ is a subharmonic function on Σ which attains its supremum at some point of Σ and, by the strong maximum principle for the Laplace-Beltrami operator, it must be constant, $|\Phi| = \text{constant} = \alpha_{|H|,c}$. Thus, (6.17) becomes trivially an equality,

$$\frac{1}{2}\Delta|\Phi|^2 = 0 = -|\Phi|^2 P_{|H|,c}(|\Phi|).$$

From here we obtain that $\nabla\Phi = \nabla A = 0$, that is, the second fundamental form of the hypersurface is parallel. If $H = 0$ (which can occur only when $c = 1$) then by a classical local rigidity result by Lawson [168, Proposition 1] we know that Σ is an open piece of a minimal Clifford torus of the form $\mathbb{S}^k(\sqrt{k/m}) \times \mathbb{S}^{m-k}(\sqrt{(m-k)/m}) \subset \mathbb{S}^{m+1}$, with $k = 1, \dots, m-1$, which trivially satisfies $|\Phi| = \text{constant} = \alpha_{0,1} = \sqrt{m}$. If $H \neq 0$ then from the equality in (6.17) we also obtain the equality in Okumura's lemma (Lemma 6.2), which implies that the hypersurface has exactly two constant principal curvatures, with multiplicities $(m-1)$ and 1. Then, by the classical results on isoparametric hypersurfaces of Riemannian space forms [64, 172, 253] we know that Σ must be an open piece of one of the three following standard product embeddings:

(a) $\mathbb{R}^{m-1} \times \mathbb{S}^1(r) \subset \mathbb{R}^{m+1}$ or $\mathbb{R} \times \mathbb{S}^{m-1}(r) \subset \mathbb{R}^{m+1}$ with $r > 0$, if $c = 0$;
(b) $\mathbb{S}^1(\sqrt{1-r^2}) \times \mathbb{S}^{m-1}(r) \subset \mathbb{S}^{m+1}$, with $0 < r < 1$, if $c = 1$; and
(c) $\mathbb{H}^{m-1}(-\sqrt{1+r^2}) \times \mathbb{S}^1(r) \subset \mathbb{H}^{m+1}$, with $0 < r < 1/\sqrt{m(m-2)}$ (recall that $H^2 > -c = 1$), or $\mathbb{H}^1(-\sqrt{1+r^2}) \times \mathbb{S}^{m-1}(r) \subset \mathbb{H}^{m+1}$, with $r > 0$, if $c = -1$.

Obviously, in all the examples above $|\Phi| = \text{constant}$ and

$$S = m(m-1)(c + H^2) - |\Phi|^2 = \text{constant}.$$

A detailed analysis of the value of the constant S for these examples shows that when $c = 0$ then $S = 0 < \widehat{B}_{|H|,0}$ for the standard products $\mathbb{R}^{m-1} \times \mathbb{S}^1(r)$, whereas $S = m^2(m-2)H^2/(m-1) = \widehat{B}_{|H|,0}$ for the standard products $\mathbb{R} \times \mathbb{S}^{m-1}(r)$, with $r > 0$. On the other hand, when $c = 1$ we can see that

$$S = \frac{m(m-2)}{2(m-1)} \left(2(m-1) + mH^2 - |H|\sqrt{m^2H^2 + 4(m-1)}\right) < \widehat{B}_{|H|,1}$$

for the standard products $\mathbb{S}^1(\sqrt{1-r^2}) \times \mathbb{S}^{m-1}(r)$ if $r > \sqrt{(m-1)/m}$, whereas

$$S = \widehat{B}_{|H|,1}$$

if $0 < r < \sqrt{(m-1)/m}$. Finally, when $c = -1$ we have that

$$S = \frac{m(m-2)}{2(m-1)} \left(-2(m-1) + mH^2 - |H|\sqrt{m^2H^2 - 4(m-1)}\right) < \widehat{B}_{|H|,-1}$$

for the standard products $\mathbb{H}^{m-1}(-\sqrt{1+r^2}) \times \mathbb{S}^1(r)$, in the case where $0 < r < 1/\sqrt{m(m-2)}$, whereas

$$S = \widehat{B}_{|H|,-1}$$

for the standard products $\mathbb{H}^1(-\sqrt{1+r^2}) \times \mathbb{S}^{m-1}(r)$, with $r > 0$. For the details, see Appendix A in [14]. This finishes the proof of Theorem 6.4. □

Proof (of Corollary 6.1) Obviously, if $\sup_\Sigma |\Phi| = +\infty$, then by (6.10) we have $\inf_\Sigma S = -\infty$, so that (6.1) holds trivially and there is nothing to prove. If $\sup_\Sigma |\Phi| < +\infty$, then we can estimate

$$H\langle \Phi X, X\rangle \geq -|H||\langle \Phi X, X\rangle| \geq -|H||\Phi||X|^2 \geq -|H| \sup_\Sigma |\Phi||X|^2,$$

and

$$\langle \Phi X, \Phi X\rangle \leq |\Phi|^2|X|^2 \leq (\sup_\Sigma |\Phi|)^2|X|^2,$$

for $X \in \mathfrak{X}(\Sigma)$. Then, by (6.6) we obtain for every $X \in \mathfrak{X}(\Sigma)$,

$$\begin{aligned}\mathrm{Ric}(X,X) &= (m-1)(c+H^2)|X|^2 + (m-2)H\langle \Phi X, X\rangle - \langle \Phi X, \Phi X\rangle \\ &\geq \left((m-1)(c+H^2) - (m-2)|H| \sup_\Sigma |\Phi| - (\sup_\Sigma |\Phi|)^2\right)|X|^2.\end{aligned}$$

Therefore, if $\sup_\Sigma |\Phi| < +\infty$ then the Ricci curvature of Σ is bounded from below by the constant

$$C = (m-1)(c+H^2) - (m-2)|H| \sup_\Sigma |\Phi| - (\sup_\Sigma |\Phi|)^2.$$

Since Σ is complete, by Theorem 2.3 the classical Omori-Yau maximum principle holds on Σ and the result follows directly from Theorem 6.4. □

Remark 6.3 Let us recall from Theorem 2.7 that every parabolic Riemannian manifold is stochastically complete. Therefore, Theorem 6.4 remains valid for parabolic hypersurfaces, with the advantage that if Σ is assumed to be parabolic, then it is not necessary to assume that the infimum of S is attained at some point of Σ in order to conclude the characterization of the equality $\inf_\Sigma S = \widehat{B}_{|H|,c}$ [14, Corollary 6]. Indeed, if Σ is parabolic and the equality $\inf_\Sigma S = \widehat{B}_{|H|,c}$ holds, then we have $\sup_\Sigma |\Phi| = \alpha_{|H|,c}$ which implies that $P_{|H|,c}(|\Phi|) \leq 0$ on Σ. Then, from (6.17) we have that $|\Phi|^2$ is a subharmonic function on Σ which is bounded from above. Since Σ is parabolic, it must be constant, $|\Phi| = \text{constant} = \alpha_{|H|,c}$. The proof then finishes as in Theorem 6.4.

For the proof of Theorem 6.5, we will also need the following auxiliary result, which can be found in [15] (see also [22, Lemma 8]).

Lemma 6.3 *Let Σ be an m-dimensional Riemannian manifold and consider $T : \mathfrak{X}(\Sigma) \to \mathfrak{X}(\Sigma)$ a symmetric tensor on Σ with two distinct eigenvalues, one of them being simple, such that* $\mathrm{Tr}(T) = 0$ *and its covariant differential ∇T is symmetric. Then*

$$|\nabla T|^2 = \frac{m+2}{m}|\nabla|T||^2. \tag{6.18}$$

Proof Let us denote by λ and μ the two eigenvalues of T, with multiplicities $(m-1)$ and 1, respectively. Observe that λ and μ are smooth functions on Σ with $\mu = -(m-1)\lambda$, and

$$|T|^2 = m(m-1)\lambda^2. \tag{6.19}$$

Let D_λ and D_μ denote, respectively, the smooth distributions of the eigenspace corresponding to each eigenvalue. It then follows from the fact that T and ∇T are symmetric that $D_\lambda = D_\mu^\perp$ and D_λ is an involutive distribution, that is, $[X, Y] \in D_\lambda$ for every $X, Y \in D_\lambda$. This implies that $X(\lambda) = 0$ for every $X \in D_\lambda$. Actually, if $X, Y \in D_\lambda$, then $\nabla T(X, Y) = \nabla T(Y, X)$ implies that

$$Y(\lambda)X - X(\lambda)Y = \lambda[X, Y] - T([X, Y]) = 0.$$

Since $\dim(D_\lambda) = m - 1 \geq 2$, this yields $X(\lambda) = 0$ for every $X \in D_\lambda$, and hence $X(\mu) = 0$ for every $X \in D_\lambda$.

Let $\{e_1, \ldots, e_m\}$ be a local orthonormal frame on Σ diagonalizing the tensor T, so that $T(e_i) = \lambda e_i$ for every $1 \leq i \leq m-1$ and $T(e_m) = \mu e_m$. In particular,

$$\nabla\lambda = e_m(\lambda)e_m, \quad \text{and} \quad |\nabla\lambda|^2 = e_m(\lambda)^2. \tag{6.20}$$

Then, denoting by $T_{\alpha,\beta,\gamma} = \langle \nabla T(e_\alpha, e_\beta), e_\gamma \rangle$, we have that

$$\begin{aligned}
|\nabla T|^2 &= \sum_{\alpha,\beta=1}^{m} |\nabla T(e_\alpha, e_\beta)|^2 = \sum_{\alpha,\beta,\gamma=1}^{m} T^2_{\alpha,\beta,\gamma} \\
&= \sum_{i,j,k=1}^{m-1} T^2_{i,j,k} + \sum_{i,j=1}^{m-1} \left(T^2_{i,j,m} + T^2_{i,m,j} + T^2_{m,i,j}\right) \\
&\quad + \sum_{i=1}^{m-1} \left(T^2_{i,m,m} + T^2_{m,i,m} + T^2_{m,m,i}\right) + T^2_{m,m,m}.
\end{aligned}$$

From the symmetries of T and ∇T we know that $T_{\alpha,\beta,\gamma} = T_{\gamma,\beta,\alpha}$ and $T_{\alpha,\beta,\gamma} = T_{\beta,\alpha,\gamma}$, respectively, for every $1 \leq \alpha, \beta, \gamma \leq m$, which in turns yields $T_{\alpha,\beta,\gamma} = T_{\alpha,\gamma,\beta}$. Using this, we may write

$$|\nabla T|^2 = \sum_{i,j,k=1}^{m-1} T^2_{i,j,k} + 3\sum_{i,j=1}^{m-1} T^2_{i,j,m} + 3\sum_{i=1}^{m-1} T^2_{i,m,m} + T^2_{m,m,m}.$$

We claim that

$$\begin{cases} T_{i,j,k} = 0 & \text{for every } 1 \leq i,j,k \leq m-1, \\ T_{i,j,m} = 0 & \text{for every } 1 \leq i,j \leq m-1, i \neq j, \\ T_{i,i,m} = e_m(\lambda) & \text{for every } 1 \leq i \leq m-1, \\ T_{i,m,m} = 0 & \text{for every } 1 \leq i \leq m-1, \text{ and} \\ T_{m,m,m} = -(m-1)e_m(\lambda). \end{cases} \tag{6.21}$$

The proof of (6.21) is a straightforward computation using the symmetries of $T_{\alpha,\beta,\gamma}$ and (6.20). Therefore, by (6.19) and (6.20) we conclude that

$$|\nabla T|^2 = 3\sum_i^{m-1} e_m(\lambda)^2 + (m-1)^2 e_m(\lambda)^2 = (m-1)(m+2)|\nabla\lambda|^2 = \frac{m+2}{m}|\nabla|T||^2.$$

□

Proof (of Theorem 6.5) Since $\mathrm{Tr}(\Phi) = 0$ and Σ has two distinct principal curvatures with multiplicities $m-1$ and 1, it then follows that $|\Phi|$ is a positive smooth function on Σ, $|\Phi| > 0$, and

$$\mathrm{Tr}(\Phi^3) = \pm\frac{(m-2)}{\sqrt{m(m-1)}}|\Phi|^3.$$

Besides, $\nabla\Phi = \nabla A$ is symmetric by Codazzi equation, and by Lemma 6.3 we also have that

$$|\nabla\Phi|^2 = \frac{m+2}{m}|\nabla|\Phi||^2. \tag{6.22}$$

Therefore, using (6.16) we obtain that

$$\begin{aligned} |\Phi|\Delta|\Phi| &= \frac{1}{2}\Delta|\Phi|^2 - |\nabla|\Phi||^2 \\ &= \frac{2}{m}|\nabla|\Phi||^2 \pm \frac{m(m-2)}{\sqrt{m(m-1)}}H|\Phi|^3 - |\Phi|^2(|\Phi|^2 - m(c+H^2)) \\ &\leq \frac{2}{m}|\nabla|\Phi||^2 + \frac{m(m-2)}{\sqrt{m(m-1)}}|H||\Phi|^3 - |\Phi|^2(|\Phi|^2 - m(c+H^2)) \\ &= \frac{2}{m}|\nabla|\Phi||^2 - |\Phi|^2 Q_{|H|,c}(|\Phi|), \end{aligned}$$

where

$$Q_{|H|,c}(x) = x^2 - \frac{m(m-2)}{\sqrt{m(m-1)}}|H|x - m(c+H^2).$$

That is,

$$|\Phi|\Delta|\Phi| \leq \frac{2}{m}|\nabla|\Phi||^2 - |\Phi|^2 Q_{|H|,c}(|\Phi|). \tag{6.23}$$

Applying the Omori-Yau maximum principle to the function $|\Phi|$ we know that there exists $\{x_k\}_{k\in\mathbb{N}}$ in Σ such that

$$\lim_{k\to\infty} |\Phi|(x_k) = \inf_{\Sigma}|\Phi|, \quad |\nabla|\Phi|(x_k)| < 1/k \quad \text{and} \quad \Delta|\Phi|(x_k) > -1/k,$$

which jointly with (6.23) implies

$$\begin{aligned} -\frac{1}{k}|\Phi|(x_k) < |\Phi|(x_k)\Delta|\Phi|(x_k) &\leq \frac{2}{m}|\nabla|\Phi|(x_k)|^2 - |\Phi|^2(x_k)Q_{|H|,c}(|\Phi|(x_k)) \\ &< \frac{2}{mk^2} - |\Phi|^2(x_k)Q_{|H|,c}(|\Phi|(x_k)). \end{aligned}$$

Letting $k \to \infty$ here, we get

$$(\inf_{\Sigma}|\Phi|)^2 Q_{|H|,c}(\inf_{\Sigma}|\Phi|) \leq 0.$$

It follows from here that either $\inf_{\Sigma}|\Phi| = 0$, which by (6.11) is equivalent to $\sup_{\Sigma} S = m(m-1)(c+H^2)$, or $\inf_{\Sigma}|\Phi| > 0$ and then $Q_{|H|,c}(\inf_{\Sigma}|\Phi|) \leq 0$.

Observe that when $H^2 + c > 0$ the polynomial $Q_{|H|,c}(x)$ has a unique positive root given by

$$\beta_{|H|,c} = \frac{\sqrt{m}}{2\sqrt{m-1}}((m-2)|H| + \sqrt{m^2H^2 + 4(m-1)c}).$$

Therefore in this case $Q_{|H|,c}(\inf_{\Sigma}|\Phi|) \leq 0$ means that $\inf_{\Sigma}|\Phi| \leq \beta_{|H|,c}$, which by (6.11) is equivalent to

$$\sup_{\Sigma} S \geq m(m-1)(c+H^2) - \beta^2_{|H|,c} = B_{|H|,c}.$$

On the other hand, when $H^2 + c = 0$ and $c = 0$, then $H = 0$ and $Q_{0,0}(x) = x^2$ so that $Q_{0,0}(\inf_{\Sigma}|\Phi|) > 0$ for every $\inf_{\Sigma}|\Phi| > 0$. Therefore, in this case it must be $\inf_{\Sigma}|\Phi| = 0 = \beta_{0,0}$ and $\sup_{\Sigma} S = 0 = B_{0,0}$. In the case $H^2 + c = 0$ and

$c = -1$, then $|H| = 1$ and $Q_{1,-1}(x)$ has a unique positive root given by $\beta_{1,-1} = m(m-2)/\sqrt{m(m-1)}$. Therefore in this case $Q_{1,-1}(\inf_\Sigma |\Phi|) \leq 0$ means also that $\inf_\Sigma |\Phi| \leq \beta_{1,-1}$, which by (6.11) is equivalent to

$$\sup_\Sigma S \geq -\beta_{1,-1}^2 = B_{1,-1}.$$

In the case $H^2 + c < 0$ (with $c = -1$ necessarily) the polynomial $Q_{|H|,-1}(x) > 0$ for every $x \in \mathbb{R}$ if $H^2 < 4(m-1)/m^2$. Therefore, if $\inf_\Sigma |\Phi| > 0$ (or, equivalently, $\sup_\Sigma S < m(m-1)(-1+H^2)$) it must be necessarily $4(m-1)/m^2 \leq H^2 < 1$. In this case, the polynomial $Q_{|H|,-1}(x)$ has two positive roots (which in fact becomes a double root when $H^2 = 4(m-1)/m^2$) given by

$$\hat{\beta}_{|H|,-1} = \frac{\sqrt{m}}{2\sqrt{m-1}}((m-2)|H| - \sqrt{m^2H^2 - 4(m-1)})$$

and

$$\beta_{|H|,-1} = \frac{\sqrt{m}}{2\sqrt{m-1}}((m-2)|H| + \sqrt{m^2H^2 - 4(m-1)}).$$

Therefore, in this case $Q_{|H|,-1}(\inf_\Sigma |\Phi|) \leq 0$ means that

$$\hat{\beta}_{|H|,-1} \leq \inf_\Sigma |\Phi| \leq \beta_{|H|,-1},$$

which by (6.11) is equivalent to

$$\begin{aligned} B_{|H|,-1} = m(m-1)(-1+H^2) - \beta_{|H|,-1}^2 &\leq \sup_\Sigma S \\ &\leq m(m-1)(-1+H^2) - \hat{\beta}_{|H|,-1}^2 = \widehat{B}_{|H|,-1}. \end{aligned}$$

This finishes the proof of the first part of Theorem 6.5.

Let us now see what happens when the equality $\sup_\Sigma S = B_{|H|,c}$ holds and this supremum is attained at a point $x_0 \in \Sigma$. Equivalently, by (6.11), the equality $\inf_\Sigma |\Phi| = \beta_{|H|,c}$ holds and this infimum is attained at a point $x_0 \in \Sigma$. In that case, $|\Phi| \geq \beta_{|H|,c}$ and, therefore, $Q_{|H|,c}(|\Phi|) \geq 0$ on Σ. Observe that

$$\Delta \log |\Phi| = \frac{1}{|\Phi|}\Delta|\Phi| - \frac{1}{|\Phi|^2}|\nabla|\Phi||^2. \tag{6.24}$$

From (6.23) we have

$$\frac{1}{|\Phi|}\Delta|\Phi| \leq \frac{2}{m|\Phi|^2}|\nabla|\Phi||^2 - Q_{|H|,c}(|\Phi|),$$

which jointly with (6.24) gives

$$\begin{aligned}\Delta \log|\Phi| &\leq -\frac{(m-2)}{m|\Phi|^2}|\nabla|\Phi||^2 - Q_{|H|,c}(|\Phi|)\\ &= -\frac{(m-2)}{m}|\nabla \log|\Phi||^2 - Q_{|H|,c}(|\Phi|).\end{aligned}$$

That is,

$$\Delta \log|\Phi| + \frac{(m-2)}{m}|\nabla \log|\Phi||^2 \leq -Q_{|H|,c}(|\Phi|).$$

Thus, since $Q_{|H|,c}(|\Phi|) \geq 0$ on Σ, we obtain that

$$\Delta \log|\Phi| + \frac{(m-2)}{m}|\nabla \log|\Phi||^2 \leq 0 \quad \text{on } \Sigma.$$

Therefore, since there exists a point $x_0 \in \Sigma$ at which the infimum of $\log|\Phi|$ is attained then, by applying a strong maximum principle for the operator

$$L(u) = \Delta u + \frac{(m-2)}{m}|\nabla u|^2$$

we conclude that $\log|\Phi|$ is constant on Σ, and hence $|\Phi| = \beta_{|H|,c}$ is also constant. Since the mean curvature H is constant and Σ has two distinct principal curvatures, then they are necessarily constant and Σ is an isoparametric hypersurface with exactly two constant principal curvatures, with multiplicities $(m-1)$ and 1. Then, by the classical results on isoparametric hypersurfaces of Riemannian space forms [64, 172, 253] we conclude that Σ must be an open piece of one of the three following standard product embeddings:

(a) $\mathbb{R}^{m-1} \times \mathbb{S}^1(r) \subset \mathbb{R}^{m+1}$ or $\mathbb{R} \times \mathbb{S}^{m-1}(r) \subset \mathbb{R}^{m+1}$ with $r > 0$, if $c = 0$;
(b) $\mathbb{S}^1(\sqrt{1-r^2}) \times \mathbb{S}^{m-1}(r) \subset \mathbb{S}^{m+1}$, with $0 < r < 1$, if $c = 1$; and
(c) $\mathbb{H}^{m-1}(-\sqrt{1+r^2}) \times \mathbb{S}^1(r) \subset \mathbb{H}^{m+1}$ or $\mathbb{H}^1(-\sqrt{1+r^2}) \times \mathbb{S}^{m-1}(r) \subset \mathbb{H}^{m+1}$, with $r > 0$, if $c = -1$.

As in the proof of Theorem 6.4, the proof then finishes by doing a detailed analysis of the value of the constant S for these examples. For further details, see [15]. □

For the proof of Corollary 6.2 simply recall that the Omori-Yau maximum principle holds for every constant mean curvature hypersurface which is properly immersed into a Riemannian space form (Theorem 2.6).

6.1.2 Alternative Approaches to Corollary 6.2

In this section, we introduce alternative approaches to a version of Corollary 6.2 for the more general case of complete hypersurfaces in Euclidean space and in the Euclidean sphere. Observe that our more general version in Theorem 6.5 holds true for hypersurfaces satisfying the Omori-Yau maximum principle, which, in principle, does not imply completeness of the hypersurface.

First of all, for the case of hypersurfaces in Euclidean space ($c = 0$), Corollary 6.2 states that if Σ is a properly immersed hypersurface in $\mathbb{R}^{m+1}$ ($m \geq 3$) with constant mean curvature H and with two distinct principal curvatures, one of them being simple, then

$$\sup_{\Sigma} S \geq 0.$$

Moreover, the equality $\sup_{\Sigma} S = 0$ holds and this supremum is attained at some point of Σ if and only if Σ is a circular cylinder $\mathbb{R}^{m-1} \times \mathbb{S}^1(r) \subset \mathbb{R}^{m+1}$, with $r = 1/m|H| > 0$. Using an argument based on the so called *principal curvature theorem*, by Smyth and Xavier [258] and which we recall below, one can prove the following result, under the more general notion of completeness.

Theorem 6.6 *Let Σ be a complete hypersurface in $\mathbb{R}^{m+1}$ ($m \geq 3$) with constant mean curvature H and with two distinct principal curvatures, one of them being simple. Then*

$$\sup_{\Sigma} S \geq 0.$$

Moreover, the equality $\sup_{\Sigma} S = 0$ holds if and only Σ is either a circular cylinder $\mathbb{R}^{m-1} \times \mathbb{S}^1(r) \subset \mathbb{R}^{m+1}$, with $r = 1/m|H| > 0$, if $H \neq 0$, or a higher dimensional catenoid, if $H = 0$.

Theorem 6.7 (Principal Curvature Theorem) *Let Σ be a complete immersed orientable hypersurface in $\mathbb{R}^{m+1}$, which is not a hyperplane, with second fundamental form A. Let $\Lambda \subset \mathbb{R}$ be the set of nonzero values assumed by the eigenvalues of A, and set $\Lambda^{\pm} = \Lambda \cap \mathbb{R}^{\pm}$. Then*

(i) If Λ^{+} and Λ^{-} are both nonempty, then $\inf \Lambda^{+} = \sup \Lambda^{-} = 0$.
(ii) If Λ^{+} or Λ^{-} is empty, then the closure of Λ is connected.

Proof (of Theorem 6.6) Let λ and μ be the two distinct principal curvatures of Σ with multiplicities $(m-1)$ and 1, respectively. Observe that λ and μ are smooth functions on Σ with $mH = (m-1)\lambda + \mu$ and $|A|^2 = (m-1)\lambda^2 + \mu^2$. From the Gauss equation (6.7) (with $c = 0$) we find

$$\begin{aligned} S = m^2H^2 - |A|^2 &= -m(m-1)\lambda^2 + 2m(m-1)\lambda H \\ &= m(m-1)\lambda(2H - \lambda). \end{aligned} \tag{6.25}$$

Let $\Lambda \subset \mathbb{R}$ be the set of nonzero values assumed by λ and μ, and set $\Lambda^{\pm} = \Lambda \cap \mathbb{R}^{\pm}$. If $\sup_{\Sigma} S = -\tau^2 < 0$, then $S \leq -\tau^2 < 0$ and thus, by (6.25),

$$\lambda^2 - 2H\lambda - \frac{\tau^2}{m(m-1)} \geq 0.$$

Observe that, independently of the value of H, the polynomial $x^2 - 2Hx - \tau^2/m(m-1)$ has a positive root, given by

$$H + \sqrt{H^2 + \frac{\tau^2}{m(m-1)}} > 0,$$

and a negative root, given by

$$H - \sqrt{H^2 + \frac{\tau^2}{m(m-1)}} < 0.$$

Therefore, either

$$\lambda \geq H + \sqrt{H^2 + \frac{\tau^2}{m(m-1)}} > 0 \tag{6.26}$$

or

$$\lambda \leq H - \sqrt{H^2 + \frac{\tau^2}{m(m-1)}} < 0. \tag{6.27}$$

In the first case, by (6.26) we also have

$$\mu = mH-(m-1)\lambda \leq H-(m-1)\sqrt{H^2 + \frac{\tau^2}{m(m-1)}} < H-\sqrt{H^2 + \frac{\tau^2}{m(m-1)}} < 0.$$

In the second case, by (6.27) we also have

$$\mu = mH-(m-1)\lambda \geq H+(m-1)\sqrt{H^2 + \frac{\tau^2}{m(m-1)}} > H+\sqrt{H^2 + \frac{\tau^2}{m(m-1)}} > 0.$$

Therefore, in any case we have that Λ^+ and Λ^- are both nonempty, with

$$\inf \Lambda^+ \geq H + \sqrt{H^2 + \frac{\tau^2}{m(m-1)}} > 0$$

and

$$\sup \Lambda^- \leq H - \sqrt{H^2 + \frac{\tau^2}{m(m-1)}} < 0.$$

which contradicts the principal curvature theorem. As a consequence, it must be $\sup_\Sigma S \geq 0$.

Suppose now that $\sup_\Sigma S = 0$. If $H = 0$, since Σ is a minimal hypersurface in $\mathbb{R}^{m+1}$ with two distinct principal curvatures, one of them being simple, we know by a result due to do Carmo and Dajczer [103, Corollary 4.4] that Σ is part of a higher dimensional catenoid. But Σ being complete and the higher dimensional catenoid being simply connected (because $m \geq 3$), Σ is the catenoid (for further details, see the last part of the proof of Theorem 3.1 by Tam and Zhou in [262]). Observe also that the scalar curvature of a higher dimensional catenoid in $\mathbb{R}^{m+1}$ is given by $S = -m(m-1)\lambda^2 < 0$ and it does satisfy $\sup_\Sigma S = 0$, since $\sup_\Sigma S < 0$ cannot happen.

On the other hand, if $\sup_\Sigma S = 0$ and $H \neq 0$ (say $H > 0$) by (6.25) we have

$$\lambda(2H - \lambda) \leq 0.$$

This implies that either

$$\lambda \leq 0 \tag{6.28}$$

or

$$\lambda \geq 2H > 0. \tag{6.29}$$

Observe that the second case cannot happen. Actually, if (6.29) holds, then we would also have

$$\mu = mH - (m-1)\lambda \leq -(m-2)H < 0,$$

which contradicts again the principal curvature theorem, since $\inf \Lambda^+ \geq 2H > 0$ and $\sup \Lambda^- \leq -(m-2)H < 0$. Therefore, it must hold necessarily (6.28) and hence

$$\mu = mH - (m-1)\lambda \geq mH > 0.$$

This implies that $\inf \Lambda^+ \geq mH > 0$ and hence, again by the principal curvature theorem, Λ^- must be empty, which means that $\lambda = \text{constant} = 0$. Hence, $\mu = \text{constant} = mH > 0$ is also constant and, by the classical results on isoparametric hypersurfaces in Euclidean space [172, 253], we conclude that Σ is a circular cylinder $\mathbb{R}^{m-1} \times \mathbb{S}^1(r) \subset \mathbb{R}^{m+1}$, with $r = 1/mH > 0$. □

On the other hand, our estimate in Corollary 6.2 for the supremum of the scalar curvature, when written in terms of the squared norm of the second fundamental form, is equivalent to Wei's estimate in [273, Theorem 1.2]. Therefore, using Wei's results one can also derive the following result, under the more general notion of completeness.

Theorem 6.8 *Let Σ be a complete hypersurface in $\mathbb{S}^{m+1}$ $(m \geq 3)$ with constant mean curvature H and with two distinct principal curvatures, one of them being simple. Then*

$$\sup_{\Sigma} S \geq B_{|H|,1} = \frac{m(m-2)}{2(m-1)}\left(2(m-1) + mH^2 - |H|\sqrt{m^2H^2 + 4(m-1)}\right).$$

Moreover, the equality $\sup_{\Sigma} S = B_{|H|,1}$ holds if and only Σ is a constant mean curvature torus $\mathbb{S}^1(\sqrt{1-r^2}) \times \mathbb{S}^{m-1}(r) \subset \mathbb{S}^{m+1}$, with radius $r \geq \sqrt{m-1/m}$.

6.2 Constant Scalar Curvature Hypersurfaces

In this section we consider the geometry of complete constant scalar curvature hypersurfaces into space forms. The first results in this direction were obtained in the seminal paper by Cheng and Yau [83], where they introduced an appropriate differential operator, denoted by $\square$, for studying such hypersurfaces. When the ambient space is the Euclidean sphere $\mathbb{S}^{m+1}$, they showed that the only compact hypersurfaces in $\mathbb{S}^{m+1}$ with constant normalized scalar curvature $R \geq 1$ and nonnegative sectional curvature are either totally umbilical or isometric to a Riemannian product $\mathbb{S}^k(\sqrt{1-r^2}) \times \mathbb{S}^{m-k}(r) \subset \mathbb{S}^{m+1}$, $1 \leq k \leq m-1$. On the other hand, for the Euclidean space they also proved that the only complete noncompact hypersurfaces in $\mathbb{R}^{m+1}$ with constant normalized scalar curvature $R \geq 0$ and nonnegative sectional curvature are generalized cylinders of the form $\mathbb{R}^{m-k} \times \mathbb{S}^k(r) \subset \mathbb{R}^{m+1}$, $1 \leq k \leq m-1$. Since then, a number of papers appeared on the subject establishing rigidity results for such hypersurfaces under various assumptions (for instance, to quote a few, see [173, 271, 274] and the references therein). Here we state the following result (see [25, Theorem 1]) with the aid of a form of the Omori-Yau maximum principle for the Cheng and Yau operator $\square$ given in Theorem 6.12.

Theorem 6.9 *Let Σ be a complete oriented hypersurface isometrically immersed into the Euclidean sphere $\mathbb{S}^{m+1}$, $m \geq 3$, with constant (normalized) scalar curvature R satisfying $R \geq 1$. In the case where $R = 1$, assume further that the mean curvature function H does not change sign. Let Φ stand for the total umbilicity tensor of the* *immersion. Then*

(i) either $\sup_\Sigma |\Phi|^2 = 0$ *and* Σ *is a totally umbilical hypersurface,*
(ii) or

$$\sup_\Sigma |\Phi|^2 \geq \alpha_{m,1}(R) = \frac{m(m-1)R^2}{(m-2)(mR-(m-2))} > 0.$$

Moreover, if $R > 1$ *the equality* $\sup_\Sigma |\Phi|^2 = \alpha_{m,1}(R)$ *holds and this supremum is attained at some point of* Σ *if and only if* Σ *is a torus* $\mathbb{S}^1(\sqrt{1-r^2}) \times \mathbb{S}^{m-1}(r) \subset \mathbb{S}^{m+1}$, *with* $0 < r = \sqrt{(m-2)/mR} < \sqrt{(m-2)/m}$.

Equivalently, using (6.9) one can also state Theorem 6.9 either in terms of the squared norm of the second fundamental form $|A|^2$ or in terms of H^2. In terms of $|A|^2$, (i) and (ii) become

(i) either $\sup_\Sigma |A|^2 = m(R-1)$ and Σ is a totally umbilical hypersurface,
(ii) or

$$\sup_\Sigma |A|^2 \geq C_m(R) = (m-1)\frac{mR-(m-2)}{m-2} + \frac{m-2}{mR-(m-2)}.$$

On the other hand, in terms of H^2, (i) and (ii) become

(i) either $\sup_\Sigma H^2 = R-1$ and Σ is a totally umbilical hypersurface,
(ii) or

$$\sup_\Sigma H^2 \geq \frac{1}{m^2}\left((m-1)^2\frac{mR-(m-2)}{m-2} - 2(m-1) + \frac{m-2}{mR-(m-2)}\right).$$

Our approach here allows us to consider in general the case of hypersurfaces with constant scalar curvature in Riemannian space forms and to state the following result for the Euclidean and hyperbolic cases (see [25, Theorem 2]).

Theorem 6.10 *Let* Σ *be a complete oriented hypersurface isometrically immersed into an* $(m+1)$*-dimensional form* ($c = 0, -1$, *and* $m \geq 3$) *with constant (normalized) scalar curvature* R *satisfying* $R > 0$. *Let* Φ *stand for the total umbilicity tensor of the immersion. Then*

(i) either $\sup_\Sigma |\Phi|^2 = 0$ *and* Σ *is a totally umbilical hypersurface,*
(ii) or

$$\sup_\Sigma |\Phi|^2 \geq \alpha_{m,c}(R) = \frac{m(m-1)R^2}{(m-2)(mR-(m-2)c)} > 0.$$

Moreover, the equality $\sup_\Sigma |\Phi|^2 = \alpha_{m,c}(R)$ *holds and this supremum is attained at some point of* Σ *if and only if*

(a) $c = 0$ *and* Σ *is a circular cylinder* $\mathbb{R} \times \mathbb{S}^{m-1}(r) \subset \mathbb{R}^{m+1}$,
(b) $c = -1$ *and* Σ *is a hyperbolic cylinder* $\mathbb{H}^1(-\sqrt{1+r^2}) \times \mathbb{S}^{m-1}(r) \subset \mathbb{H}^{m+1}$,

where $r = \sqrt{(m-2)/mR} > 0$.

As in Theorem 6.9, we may also state Theorem 6.10 either in terms of $|A|^2$ or in terms of H^2. In the former, (i) and (ii) become

(i) either $\sup_\Sigma |A|^2 = m(R-c)$ and Σ is a totally umbilical hypersurface,
(ii) or

$$\sup_\Sigma |A|^2 \geq C_{m,c}(R) = (m-1)\frac{mR-(m-2)c}{m-2} + \frac{(m-2)c^2}{mR-(m-2)c},$$

while in the latter they become

(i) either $\sup_\Sigma H^2 = R - c$ and Σ is a totally umbilical hypersurface,
(ii) or

$$\sup_\Sigma H^2 \geq \frac{1}{m^2}\left((m-1)^2\frac{mR-(m-2)c}{m-2} - 2(m-1)c + \frac{(m-2)c^2}{mR-(m-2)}\right).$$

Finally, our approach allows also to state the following result where, under the assumption of $\Box$-parabolicity, we are able to improve the characterization of the equality $\sup_\Sigma |\Phi|^2 = \alpha_{m,c}(R)$, since there is no need to assume that the supremum is attained at any point (see [25, Theorem 3]).

Theorem 6.11 *Let* Σ *be a complete oriented hypersurface isometrically immersed into an* $(m+1)$*-dimensional form* ($c = 0, 1, -1$, *and* $m \geq 3$) *with constant (normalized) scalar curvature* R *satisfying* $R \geq c$ *and* $R > 0$. *In the case where* $c = 1$ *and* $R = 1$, *assume further that the mean curvature function* H *does not change sign. Let* Φ *stand for the total umbilicity tensor of the immersion and assume that the hypersurface is not totally umbilical. If* Σ *is* $\Box$*-parabolic, then*

$$\sup_\Sigma |\Phi|^2 \geq \alpha_{m,c}(R) = \frac{m(m-1)R^2}{(m-2)(mR-(m-2)c)} > 0,$$

with equality if and only if

(a) $c = 0$ *and* Σ *is a circular cylinder* $\mathbb{R} \times \mathbb{S}^{m-1}(r) \subset \mathbb{R}^{m+1}$,
(b) $c = 1$ *and* Σ *is a torus* $\mathbb{S}^1(\sqrt{1-r^2}) \times \mathbb{S}^{m-1}(r) \subset \mathbb{S}^{m+1}$,
(c) $c = -1$ *and* Σ *is a hyperbolic cylinder* $\mathbb{H}^1(-\sqrt{1+r^2}) \times \mathbb{S}^{m-1}(r) \subset \mathbb{H}^{m+1}$,

where $r = \sqrt{(m-2)/mR} > 0$.

6.2.1 *Hypersurfaces and Newton Operators*

The proof of our results for constant scalar curvature hypersurfaces in space forms is based on an Omori-Yau maximum principle for the Cheng and Yau operator $\square$. This operator is, in fact, the first (or, better, the second) of a series of second order linear differential operators which can be defined for hypersurfaces in general Riemannian ambient spaces. Since we will make use of these operators in the remaining sections of this chapter, as well as in Chap. 7, we describe them in detail here.

Consider, in general, a two-sided hypersurface Σ isometrically immersed into an $(m+1)$-dimensional Riemannian manifold N and let A denote the second fundamental form of the hypersurface with respect to a globally defined unit normal field ν. Recall that the k-mean curvatures of the hypersurface are given by

$$H_k = \binom{m}{k}^{-1} S_k,$$

where $S_0 = 1$ and, for $k = 1, \ldots, m$, S_k is the k-th elementary symmetric function of the principal curvatures $\kappa_1, \ldots, \kappa_m$ of the hypersurface, that is,

$$S_k = \sigma_k(\kappa_1, \ldots, \kappa_m) = \sum_{i_1 < \cdots < i_k} \kappa_{i_1} \cdots \kappa_{i_k}, \quad 1 \le k \le m.$$

In particular, when $k = 1$, H_1 is the usual mean curvature H of Σ. We also observe that

$$|A|^2 = m^2 H^2 - m(m-1) H_2. \tag{6.30}$$

Observe also that the characteristic polynomial of A can be written in terms of the H_k as

$$\det(tI - A) = \sum_{k=0}^{m} (-1)^k \binom{m}{k} H_k t^{m-k}. \tag{6.31}$$

The Newton operators $P_k : \mathfrak{X}(\Sigma) \to \mathfrak{X}(\Sigma)$ associated to the hypersurface are defined inductively by $P_0 = I$ and

$$P_k = S_k I - A \circ P_{k-1}, \quad 1 \le k \le m.$$

Equivalently,

$$P_k = \binom{m}{k} H_k I - \binom{m}{k-1} H_{k-1} A + \cdots + (-1)^{k-1} m H_1 A^{k-1} + (-1)^k A^k.$$

In particular, by the Cayley-Hamilton theorem and (6.31) we have $P_m = 0$. Observe that the Newton operators P_k are all self-adjoint operators which commute with the shape operator A. Even more, if $\{e_1, \dots, e_n\}$ is an orthonormal frame on $T_p\Sigma$ which diagonalizes A_p, with $A_p(e_i) = \kappa_i(p)e_i$, then

$$(P_k)_p(e_i) = \mu_{i,k}(p)e_i \tag{6.32}$$

where

$$\mu_{i,k} = \sum_{i_1<\dots<i_k, i_j\neq i} \kappa_{i_1}\cdots\kappa_{i_k}.$$

It follows from here that for each k, $0 \leq k \leq m-1$,

$$\mathrm{Tr}(P_k) = (m-k)S_k = c_k H_k \tag{6.33}$$

and

$$\mathrm{Tr}(A \circ P_k) = (k+1)S_{k+1} = c_k H_{k+1}, \tag{6.34}$$

where

$$c_k = (m-k)\binom{m}{k} = (k+1)\binom{m}{k+1}.$$

Associated to each Newton operator P_k one has the second order linear differential operator $L_k : C^2(\Sigma) \to C(\Sigma)$ for $k = 0, 1, \dots, m-1$, given by

$$L_k(u) = \mathrm{Tr}(P_k \circ \mathrm{hess}\ u).$$

In particular $L_0 = \Delta$ is the Laplace-Beltrami operator, while $L_1 = \Box$ is nothing but the Cheng and Yau operator.

Observe that

$$\begin{aligned} L_k(u) = \mathrm{Tr}(P_k \circ \mathrm{hess}\ u) &= \sum_{i=1}^m \langle P_k(\nabla_{e_i}\nabla u), e_i\rangle \\ &= \sum_{i=1}^m \langle \nabla_{e_i}\nabla u, P_k(e_i)\rangle = \sum_{i=1}^m \langle \nabla_{P_k(e_i)}\nabla u, e_i\rangle = \mathrm{Tr}(\mathrm{hess}\ u \circ P_k), \end{aligned}$$

where $\{e_1, \ldots, e_m\}$ is a (local) orthonormal frame on Σ. Moreover, we have

$$\begin{aligned}\operatorname{div}(P_k(\nabla u)) &= \sum_{i=1}^{m} \langle (\nabla_{e_i} P_k)(\nabla u), e_i \rangle + \sum_{i=1}^{m} \langle P_k(\nabla_{e_i} \nabla u), e_i \rangle \\ &= \langle \operatorname{div} P_k, \nabla u \rangle + L_k(u),\end{aligned}$$

where the divergence of P_k on Σ is given by

$$\operatorname{div} P_k = \operatorname{Tr}(\nabla P_k) = \sum_{i=1}^{m} (\nabla_{e_i} P_k)(e_i).$$

That is,

$$L_k(u) = \operatorname{Tr}(P_k \circ \text{hess } u) = \operatorname{div}(P_k(\nabla u)) - \langle \operatorname{div} P_k, \nabla u \rangle. \tag{6.35}$$

Remark 6.4 From Eq. (6.35), we conclude that the operator L_k is elliptic (respectively, semi-elliptic) if, and only if, P_k is positive definite (respectively, positive semi-definite). We observe that $L_0 = \Delta$ is always elliptic. In this respect, it is worth pointing out that the ellipticity of the operator $L_1 = \square$ is guaranteed by the assumption $H_2 > 0$. Indeed, if this happens the mean curvature does not vanish on Σ, because of the basic inequality $H_1^2 \geq H_2$. Therefore, we can choose the normal unit vector ν on Σ so that $H_1 > 0$. Furthermore

$$m^2 H_1^2 = \sum_{j=1}^{m} \kappa_j^2 + m(m-1)H_2 > \kappa_i^2$$

for every $i = 1, \ldots, m$, and then the eigenvalues of P_1 satisfy $\mu_{i,1} = mH_1 - \kappa_i > 0$ for every i (see, for instance, Lemma 3.10 in [111]). This shows ellipticity of $\square$. Regarding the operator L_j when $j \geq 2$, a natural hypothesis to guarantee ellipticity is the existence of an elliptic point in Σ, that is, a point $x \in \Sigma$ at which the second fundamental form A is positive definite (with respect to the appropriate orientation). In fact, it follows from the proof of [37, Proposition 3.2] that if Σ has an elliptic point and $H_{k+1} \neq 0$ on Σ, then each L_j, $1 \leq j \leq k$ is elliptic.

On the other hand, the divergence of the Newton operators are given in the following result (see also Lemma 3.1 in [23], paying attention to the different convention for the sign of ${}^N R$).

Lemma 6.4 *Let $f : \Sigma \to N$ be an isometrically immersed hypersurface into an $(m+1)$-dimensional Riemannian manifold N. Let $e_1, \ldots, e_m$ be a local orthonormal frame on Σ and ν be a local unit normal. Then*

$$\sum_{i=1}^{m} \langle (\nabla_{e_i} P_k)X, e_i \rangle = \sum_{j=0}^{k-1} \sum_{i=1}^{m} (-1)^{k-1-j} \left\langle {}^N R(e_i, A^{k-1-j} X)\nu, P_j e_i \right\rangle \tag{6.36}$$

for every vector field $X \in \mathfrak{X}(\Sigma)$, where A is the Weingarten operator in the direction of ν.

Proof We will prove Eq. (6.36) by induction on k, $1 \leq k \leq m-1$. Using Codazzi equations (1.145) [see also (6.15)] and the definition of P_1 it is not difficult to prove that this is true for $k=1$. Actually, since $P_1 = S_1 I - A$ we have

$$(\nabla_{e_i} P_1)X = e_i(S_1)X - (\nabla_{e_i} A)X = e_i(S_1)X - (\nabla_X A)e_i + \left({}^N R(e_i, X)\nu\right)^{\top},$$

with $^{\top}$ denoting the part tangential to Σ. Then,

$$\begin{aligned}\sum_{i=1}^{m} \langle(\nabla_{e_i} P_1)X, e_i\rangle &= \langle\nabla S_1, X\rangle - \mathrm{Tr}(\nabla_X A) + \sum_{i=1}^{m} \langle\langle^N R(e_i, X)\nu, e_i\rangle \\ &= \sum_{i=1}^{m} \langle\langle^N R(e_i, X)\nu, e_i\rangle,\end{aligned}$$

since $\mathrm{Tr}(\nabla_X A) = \nabla_X \mathrm{Tr}(A) = \langle\nabla S_1, X\rangle$. Thus assume that the equation holds for $k-1$. Then, from the very definition of the Newton operator $P_k = S_k I - P_{k-1} \circ A$ it follows that

$$(\nabla_{e_i} P_k)X = e_i(S_k)X - (\nabla_{e_i} P_{k-1})AX - P_{k-1}((\nabla_{e_i} A)X),$$

from which we deduce

$$\langle(\nabla_{e_i} P_k)X, e_i\rangle = e_i(S_k)\langle X, e_i\rangle - \langle(\nabla_{e_i} P_{k-1})AX, e_i\rangle - \langle(\nabla_{e_i} A)X, P_{k-1}e_i\rangle.$$

Using again Codazzi equations in the last term of the above equation we have

$$\langle(\nabla_{e_i} A)X, P_{k-1}e_i\rangle = \langle(\nabla_X A)e_i, P_{k-1}e_i\rangle - \langle^N R(e_i, X)\nu, P_{k-1}e_i\rangle$$

and then

$$\begin{aligned}\sum_{i=1}^{m} \langle(\nabla_{e_i} P_k)X, e_i\rangle &= \langle\nabla S_k, X\rangle - \sum_{i=1}^{m} \langle(\nabla_{e_i} P_{k-1})AX, e_i\rangle \\ &\quad - \sum_{i=1}^{m} \langle(\nabla_X A)e_i, P_{k-1}e_i\rangle + \sum_{i=1}^{m} \langle^N R(e_i, X)\nu, P_{k-1}e_i\rangle \\ &= \langle\nabla S_k, X\rangle - \sum_{i=1}^{m} \langle(\nabla_{e_i} P_{k-1})AX, e_i\rangle - \\ &\quad \mathrm{Tr}((\nabla_X A) \circ P_{k-1}) + \sum_{i=1}^{m} \langle^N R(e_i, X)\nu, P_{k-1}e_i\rangle.\end{aligned}$$

We claim that

$$\operatorname{Tr}((\nabla_X A) \circ P_{k-1}) = \langle \nabla S_k, X \rangle. \tag{6.37}$$

Using (6.37) and the induction hypothesis we conclude from here that

$$\begin{aligned}
\sum_{i=1}^{m} \langle (\nabla_{e_i} P_k) X, e_i \rangle &= -\sum_{i=1}^{m} \langle (\nabla_{e_i} P_{k-1}) AX, e_i \rangle + \sum_{i=1}^{m} \langle {}^N R(e_i, X)\nu, P_{k-1} e_i \rangle \\
&= -\sum_{j=0}^{k-2} \sum_{i=1}^{m} (-1)^{k-2-j} \langle {}^N R(e_i, A^{k-1-j} X)\nu, P_j e_i \rangle \\
&\quad + \sum_{i=1}^{m} \langle {}^N R(e_i, X)\nu, P_{k-1} e_i \rangle \\
&= \sum_{j=0}^{k-1} \sum_{i=1}^{m} (-1)^{k-1-j} \langle {}^N R(e_i, A^{k-1-j} X)\nu, P_j e_i \rangle,
\end{aligned}$$

that is, (6.36). It remains to prove (6.37). We will prove it by performing the computations in a local orthonormal frame on Σ that diagonalizes A. It is worth pointing out that such a frame does not always exist in the smooth category; problems occur when the multiplicity of the principal curvatures changes (also the principal curvatures are not necessarily everywhere differentiable). For this reason, we will work on the subset Σ_0 of Σ consisting of points at which the number of distinct principal curvatures is locally constant. Let us recall that Σ_0 is an open dense subset of Σ, and in every connected component of Σ_0, the principal curvatures form mutually distinct smooth principal curvature functions and, for such a principal curvature κ, the assignment $p \mapsto V_{\kappa(p)}(p)$ defines a smooth distribution, where $V_{\kappa(p)}(p) \subset T_p\Sigma$ denotes the eigenspace associated to $\kappa(p)$ (see for instance Paragraph 16.10 in [41]). Therefore, for every $p \in \Sigma_0$ there exists a local smooth orthonormal frame defined on a neighbourhood of p that diagonalizes A, that is , $\{e_1, \ldots, e_m\}$ are such that $Ae_i = \kappa_i e_i$, with each κ_i smooth. In this case,

$$(\nabla_X A) e_i = X(\kappa_i) e_i + \sum_{j \neq i} (\kappa_i - \kappa_j) \omega_i^j(X) e_j,$$

where, as usual, $\omega_i^j(X) = \langle \nabla_X e_i, e_j \rangle$. Observe also that for every i

$$P_{k-1}(e_i) = \mu_{i,k-1} e_i \tag{6.38}$$

with

$$\mu_{i,k-1} = \sum_{i_1 < \cdots < i_{k-1}, i_j \neq i} \kappa_{i_1} \cdots \kappa_{i_{k-1}}.$$

Then, by (6.38) we have

$$\begin{aligned}\mathrm{Tr}(\nabla_X A \circ P_{k-1}) &= \sum_{i=1}^{m} \mu_{i,k-1} X(\kappa_i) \\ &= \sum_{i=1}^{m} X(\kappa_i) \sum_{i_1<\cdots<i_{k-1}, i_j\neq i} \kappa_{i_1}\cdots\kappa_{i_{k-1}} \\ &= X\left(\sum_{i_1<\cdots<i_k} \kappa_{i_1}\cdots\kappa_{i_k}\right) = \langle \nabla S_k, X\rangle.\end{aligned}$$

This proves (6.37) on Σ_0, and by continuity, on Σ. □

In particular, when the ambient space has constant sectional curvature one has $\mathrm{div}P_k = 0$ for every $0 \leq k \leq m-1$ and (6.35) reduces to

$$L_k(u) = \mathrm{Tr}(P_k \circ \mathrm{hess}\, u) = \mathrm{div}(P_k(\nabla u)). \tag{6.39}$$

6.2.2 Some Preliminary Results

Let Σ be an oriented hypersurface isometrically immersed into an $(m+1)$-dimensional space form with curvature c, and let $P = P_1$ denote its first Newton operator. That is, $P : \mathfrak{X}(\Sigma) \to \mathfrak{X}(\Sigma)$ is the operator given by $P = mHI - A$. Recall that P is also a self-adjoint linear operator which commutes with A, and $\mathrm{Tr}(P) = m(m-1)H$. For $u \in C^2(\Sigma)$ set

$$\Box u = L_1 u = \mathrm{Tr}(P \circ \mathrm{hess}\, u) = \mathrm{div}(P(\nabla u)). \tag{6.40}$$

As we already know, $\Box$ defines a second order differential operator which, in general, is not elliptic. It is clear from the definition that $\Box$ is elliptic if and only if P is positive definite. Note that

$$\Box(uv) = u\Box v + v\Box u + 2\langle P(\nabla u), \nabla v\rangle$$

for every $u, v \in C^2(\Sigma)$. The operator $\Box$ arises naturally as the linearized operator of the scalar curvature for normal variations of the hypersurface (see for instance [242]). The following lemma will be essential for our computations.

Lemma 6.5 *Let Σ be an oriented isometrically immersed hypersurface into an $(m+1)$-dimensional space form with curvature c. Then*

$$\Box(mH) = \frac{m(m-1)}{2}\Delta R + |\nabla A|^2 - m^2|\nabla H|^2 + mH\,\mathrm{Tr}(A^3) - |A|^4 + mc(|A|^2 - mH^2).$$

In particular, if Σ has constant scalar curvature

$$\Box(mH) = |\nabla A|^2 - m^2|\nabla H|^2 + mH\,\mathrm{Tr}(A^3) - |A|^4 + mc(|A|^2 - mH^2). \tag{6.41}$$

Proof It follows from (6.40) that

$$\Box u = mH\,\mathrm{Tr}(\text{hess } u) - \mathrm{Tr}(A \circ \text{hess } u) = mH\Delta u - \mathrm{Tr}(A \circ \text{hess } u).$$

Setting $u = mH$ here we have

$$\begin{aligned}\Box(mH) &= mH\Delta(mH) - m\,\mathrm{Tr}(A \circ \text{hess } H)\\ &= \frac{1}{2}\Delta(m^2H^2) - m^2|\nabla H|^2 - m\,\mathrm{Tr}(A \circ \text{hess } H).\end{aligned}$$

From the identity (6.8) and Simons formula (6.12) we have

$$\begin{aligned}\frac{1}{2}\Delta(m^2H^2) - m\,\mathrm{Tr}(A \circ \text{hess } H) &= \frac{m(m-1)}{2}\Delta R + \frac{1}{2}\Delta|A|^2 - m\,\mathrm{Tr}(A \circ \text{hess } H)\\ &= \frac{m(m-1)}{2}\Delta R + |\nabla A|^2 + mH\,\mathrm{Tr}(A^3)\\ &\quad -|A|^4 + mc(|A|^2 - mH^2).\end{aligned}$$

Therefore we conclude from here that

$$\Box(mH) = \frac{m(m-1)}{2}\Delta R + |\nabla A|^2 - m^2|\nabla H|^2 + mH\,\mathrm{Tr}(A^3) - |A|^4 + mc(|A|^2 - mH^2).$$

Finally, (6.41) follows at once since R is constant. □

Lemma 6.6 *Let Σ be an oriented isometrically immersed hypersurface into an $(m+1)$-dimensional space form with curvature c. Assume that the mean curvature function H does not change sign, so that, without loss of generality, we may assume $H \geq 0$ on Σ. Let μ_- and μ_+ be, respectively, the minimum and the maximum of the eigenvalues of P at every point $p \in \Sigma$. If $R > c$ on Σ (resp., $R \geq c$ on Σ), then*

$$\mu_- > 0 \;(\textit{resp.}, \mu_- \geq 0)$$

and

$$\mu_+ < 2mH \;(\textit{resp.}, \mu_+ \leq 2mH).$$

Proof We follow the same argument as in the proof of Lemma 4.2 in [58]. From (6.8), if $R > c$ we have

$$m^2H^2 = |A|^2 + m(m-1)(R-c) > |A|^2.$$

Thus, indicating with $\kappa_1, \ldots, \kappa_m$ the principal curvatures of the hypersurface, we get

$$-mH < \kappa_i < mH, \quad i = 1, \ldots, m.$$

Therefore, for every i

$$0 < mH - \kappa_i < 2mH.$$

But $\mu_i = mH - \kappa_i$ are precisely the eigenvalues of the operator $P = mHI - A$. In particular, $\mu_- > 0$ and $\mu_+ < 2mH$. Similarly if $R \geq c$. $\square$

Remark 6.5 Observe that if $R > c$ on Σ, it follows from (6.8) that H does not vanish. Thus, connectedness of Σ implies that H does not change sign. Moreover, Lemma 6.6 implies that when $R > c$ the operator $\square$ is elliptic.

For the proof of our main computational result (see Lemma 6.8 below) we will need the following auxiliary result, which can be found in [9, Lemma 4.1] (see also [58, Lemma 2.5]).

Lemma 6.7 *Let Σ be an isometrically immersed hypersurface into an $(m+1)$-dimensional space form with curvature c, and assume that Σ has constant (normalized) scalar curvature $R \geq c$. Then*

$$|\nabla A|^2 \geq m^2 |\nabla H|^2. \tag{6.42}$$

Proof Since we are assuming that R is constant, from (6.8) we get $\nabla |A|^2 = \nabla(m^2 H^2) = 2m^2 H \nabla H$ and

$$|\nabla |A|^2|^2 = 4m^4 H^2 |\nabla H|^2. \tag{6.43}$$

Following the notation and the formalism of Sects. 1.5 and 1.6, we have

$$|A|^2 = \sum_{i,j=1}^m h_{ij}^2, \quad |\nabla A|^2 = \sum_{i,j,k=1}^m h_{ijk}^2, \quad \text{and} \quad |\nabla |A|^2|^2 = 4 \sum_{k=1}^m \left(\sum_{i,j=1}^m h_{ij} h_{ijk} \right)^2.$$

Therefore, using Cauchy-Schwarz inequality, we obtain

$$\begin{aligned} 4|A|^2 |\nabla A|^2 &= 4 \left(\sum_{i,j=1}^m h_{ij}^2 \right) \left(\sum_{i,j,k=1}^m h_{ijk}^2 \right) \\ &\geq 4 \sum_{k=1}^m \left(\sum_{i,j=1}^m h_{ij} h_{ijk} \right)^2 \\ &= |\nabla |A|^2|^2 = 4m^4 H^2 |\nabla H|^2, \end{aligned}$$

which jointly with (6.8) gives

$$|A|^2|\nabla A|^2 \geq m^2|A|^2|\nabla H|^2 + m^3(m-1)(R-c)|\nabla H|^2. \tag{6.44}$$

In particular, if $R \geq c$ we have

$$|A|^2(|\nabla A|^2 - m^2|\nabla H|^2) \geq 0. \tag{6.45}$$

Let $\Sigma_0 = \{x \in \Sigma : |A|^2(x) = 0\}$. It is clear from (6.45) that

$$|\nabla A|^2(x) \geq m^2|\nabla H|^2(x) \tag{6.46}$$

for every $x \in \Sigma \setminus \Sigma_0$ and, by continuity, for every $x \in \Sigma \setminus \mathrm{int}(\Sigma_0)$. Therefore, if $\mathrm{int}(\Sigma_0) = \emptyset$ inequality (6.46) holds true for every $x \in \Sigma$. On the other hand, if $\mathrm{int}(\Sigma_0) \neq \emptyset$ then $\nabla H \equiv 0$ and $\nabla A \equiv 0$ on $\mathrm{int}(\Sigma_0)$, and inequality (6.46) holds trivially also on $\mathrm{int}(\Sigma_0)$. □

Lemma 6.8 *Let Σ be an oriented isometrically immersed hypersurface into an $(m+1)$-dimensional space form with curvature c, and assume that Σ has constant (normalized) scalar curvature $R \geq c$. In the case where $R > c$, choose the orientation such that $H > 0$ on Σ. In the case where $R = c$, assume further that the mean curvature function H does not change sign, and choose the orientation such that $H \geq 0$ on Σ. Then*

$$\frac{1}{2}\square(|\Phi|^2) \geq \frac{1}{\sqrt{m(m-1)}}|\Phi|^2 Q_R(|\Phi|)\sqrt{|\Phi|^2 + m(m-1)(R-c)} \tag{6.47}$$

where

$$Q_R(x) = -(m-2)x^2 - (m-2)x\sqrt{x^2 + m(m-1)(R-c)} + m(m-1)R. \tag{6.48}$$

Proof Since R is constant, it follows from (6.9) that

$$\frac{m}{2(m-1)}\square(|\Phi|^2) = \frac{1}{2}\square(m^2H^2) = mH\square(mH) + m^2\langle P(\nabla H), \nabla H\rangle,$$

since $\square(u^2) = 2u\square(u) + 2\langle P(\nabla u), \nabla u\rangle$ for every $u \in C^2(\Sigma)$. From Lemma 6.6 we know that P is positive semi-definite. Therefore, using Lemma 6.5 we get

$$\begin{aligned}\frac{m}{2(m-1)}\square(|\Phi|^2) \geq mH\square(mH) &= mH(|\nabla A|^2 - m^2|\nabla H|^2) + m^2H^2\,\mathrm{Tr}(A^3)\\ &\quad -mH|A|^4 + m^2cH(|A|^2 - mH^2).\end{aligned} \tag{6.49}$$

From Lemma 6.7 we know that

$$|\nabla A|^2 - m^2|\nabla H|^2 \geq 0,$$

and since $H \geq 0$ we conclude from here and (6.49) that

$$\frac{1}{2(m-1)}\Box(|\Phi|^2) \geq mH^2\,\mathrm{Tr}(A^3) - H|A|^4 + mcH(|A|^2 - mH^2). \tag{6.50}$$

Recall that $|\Phi|^2 = |A|^2 - mH^2$, so that (6.50) becomes

$$\frac{1}{2(m-1)}\Box(|\Phi|^2) \geq mH^2\,\mathrm{Tr}(A^3) - H(|\Phi|^2 + mH^2)^2 + mcH|\Phi|^2. \tag{6.51}$$

On the other hand, a direct computation yields

$$\mathrm{Tr}(A^3) = \mathrm{Tr}(\Phi^3) + 3H|\Phi|^2 + mH^3,$$

and substituting this into (6.51) gives

$$\frac{1}{2(m-1)}\Box(|\Phi|^2) \geq mH^2\,\mathrm{Tr}(\Phi^3) - H|\Phi|^4 + mH(H^2 + c)|\Phi|^2. \tag{6.52}$$

Since $\mathrm{Tr}(\Phi) = 0$, when $m = 2$ one has $\Phi^2 = (1/2)|\Phi|^2 I$. Thus, $\Phi^3 = (1/2)|\Phi|^2\Phi$ and $\mathrm{Tr}(\Phi^3) = 0$ also. When $m \geq 3$, we may use Lemma 6.2 to estimate $\mathrm{Tr}(\Phi^3)$ as follows

$$|\,\mathrm{Tr}(\Phi^3)| \leq \frac{m-2}{\sqrt{m(m-1)}}|\Phi|^3, \tag{6.53}$$

and then

$$mH^2\,\mathrm{Tr}(\Phi^3) \geq -mH^2|\,\mathrm{Tr}(\Phi^3)| \geq -\frac{m(m-2)}{\sqrt{m(m-1)}}H^2|\Phi|^3.$$

Inserting this into (6.52) gives

$$\begin{aligned}\frac{1}{2(m-1)}\Box(|\Phi|^2) &\geq -\frac{m(m-2)}{\sqrt{m(m-1)}}H^2|\Phi|^3 - H|\Phi|^4 + mH(H^2+c)|\Phi|^2 \\ &= -H|\Phi|^2\left(|\Phi|^2 + \frac{m(m-2)}{\sqrt{m(m-1)}}H|\Phi| - m(H^2+c)\right).\end{aligned} \tag{6.54}$$

Besides, from (6.9) we get

$$H^2 = \frac{1}{m(m-1)}|\Phi|^2 + (R-c),$$

and therefore, taking into account that $H \geq 0$, we may write

$$H = \frac{1}{\sqrt{m(m-1)}}\sqrt{|\Phi|^2 + m(m-1)(R-c)}.$$

Finally, replacing H by this expression into (6.54), we get (6.47). $\square$

6.2.3 *An Omori-Yau Maximum Principle for the Cheng and Yau Operator*

For the proof of our main results we shall need the following version of an Omori-Yau maximum principle for the operator $\square$.

Theorem 6.12 *Let Σ be a complete oriented isometrically immersed hypersurface into an $(m+1)$-dimensional space form with curvature c, and assume that Σ has constant (normalized) scalar curvature $R \geq c$. In the case where $R = c$, assume further that the mean curvature function H does not change sign. If $\sup_\Sigma |\Phi|^2 < +\infty$, then the Omori-Yau maximum principle holds on Σ for the Cheng and Yau operator $\square$.*

Theorem 6.12 is a consequence of a much more general intrinsic result (see Theorem 6.13 below) which will be used also in the rest of this chapter and in Chap. 7. To derive Theorem 6.12 from Theorem 6.13 we first observe that the sectional curvature of Σ is bounded from below by a constant. In fact, from (6.9) and $\sup_\Sigma |\Phi|^2 < +\infty$ we have also $\sup_\Sigma |A|^2 < +\infty$. Therefore, by Gauss equation (6.4) the sectional curvature of the plane $X \wedge Y$ is given by

$$K(X \wedge Y) = c + \langle AX, X\rangle\langle AY, Y\rangle - \langle AX, Y\rangle^2 \geq c - 2\sup_\Sigma |A|^2 > -\infty.$$

On the other hand, it follows also from (6.9) and $\sup_\Sigma |\Phi|^2 < +\infty$ that $\sup_\Sigma H^2 < +\infty$ and hence $\sup_\Sigma \mathrm{Tr}(P) = m(m-1)\sup_\Sigma H < +\infty$. Then, we can apply item (ii) of Theorem 6.13 with $t = P$ to obtain the desired conclusion.

Theorem 6.13 *Let $(M, \langle\ ,\ \rangle)$ be a complete, noncompact, Riemannian manifold; let $o \in M$ be a reference point and denote by $r(x)$ the Riemannian distance function from o. Assume that the sectional curvature of M satisfies*

$$^M K(x) \geq -G^2(r(x)), \tag{6.55}$$

with $G \in C^1(\mathbb{R}_0^+)$ satisfying

$$i)\; G(0) > 0, \quad ii)\; G'(t) \geq 0, \quad iii)\; \frac{1}{G(t)} \notin L^1(+\infty). \tag{6.56}$$

Let T be a symmetric, positive semi-definite $(0, 2)$-tensor field on M.

- *(i) If $\operatorname{Tr}(T) > 0$ on M, then the $(1/\operatorname{Tr}(T))$-Omori Yau maximum principle holds on M for the associated semi-elliptic operator $L = \operatorname{Tr}(t \circ \operatorname{hess})$.*
- *(ii) If $\sup_M \operatorname{Tr}(T) < +\infty$, then Omori Yau maximum principle holds on M for the associated semi-elliptic operator $L = \operatorname{Tr}(t \circ \operatorname{hess})$.*

Proof The proof of Theorem 6.13 follows the ideas of that of Theorem 2.5. Keeping the same notation and doing a similar reasoning as there, we obtain here

$$Lr(x) \leq \operatorname{Tr}(T)(x)G(r(x) + 1)\frac{e^{\int_0^{r(x)} G(s)ds}}{e^{\int_1^{r(x)} G(s)ds} - 1} \tag{6.57}$$

for $r(x) \geq 2$. Define, as in the proof of Theorem 2.5,

$$\varphi(t) = \int_0^t \frac{ds}{G(s+1)} \tag{6.58}$$

so that

$$\varphi'(t) = \frac{1}{G(t+1)} \quad \text{and} \quad \varphi''(t) \leq 0.$$

Set $\gamma(x) = \varphi(r(x))$ on $M \setminus \bar{B}_2$ and note that

$$\gamma(x) \to +\infty \text{ as } x \to \infty \tag{6.59}$$

because $\varphi(t) \to +\infty$ as $t \to +\infty$ since $1/G \notin L^1(+\infty)$.

Using the formula $L\varphi(u) = \varphi'(u)Lu + \varphi''(u)\langle T\nabla u, \nabla u\rangle$ and that T is positive semi-definite, from (6.57) we obtain

$$L\gamma(x) \leq \varphi'(r(x))Lr(x) = \frac{1}{G(r(x)+1)}Lr(x) \leq \operatorname{Tr}(T)(x)\frac{e^{\int_0^{r(x)} G(s)ds}}{e^{\int_1^{r(x)} G(s)ds} - 1}.$$

Since $G \notin L^1(+\infty)$ we have

$$\sup_{t \geq 2} \frac{e^{\int_0^t G(s)ds}}{e^{\int_1^t G(s)ds} - 1} = \Lambda < +\infty. \tag{6.60}$$

In case (i), we deduce that $L\gamma(x) \leq \mathrm{Tr}(T)(x)\Lambda$, that is,

$$q(x)L\gamma(x) \leq \Lambda \quad \text{on} \quad D_o \cap (M \setminus \bar{B}_2), \tag{6.61}$$

with $q(x) = 1/\mathrm{Tr}(T)(x)$. In case (ii) we also have (6.61) with $q(x) \equiv 1$ and replacing Λ by the new constant $\sup_M \mathrm{Tr}(T)\Lambda$. Inequality (6.61) replaces (2.40) in the proof of Theorem 2.5. Furthermore, we also have

$$|\nabla\gamma| = \frac{1}{G(r+1)} \leq \frac{1}{G(0)} \leq \Lambda, \tag{6.62}$$

up to choosing Λ in (6.61) sufficiently large.

Let now $u \in C^2(M)$ with

$$u^* = \sup_M u < +\infty. \tag{6.63}$$

For a fixed $\eta > 0$ consider the sets

$$A_\eta = \{x \in M : u(x) > u^* - \eta\} \tag{6.64}$$

and

$$B_\eta = \{x \in A_\eta : |\nabla u(x)| < \eta\}. \tag{6.65}$$

Since $(M, \langle\, , \rangle)$ is complete, from Ekeland quasi-minimum principle, we deduce $B_\eta \neq \emptyset$. We have to show that

$$\inf_{x \in B_\eta} q(x)Lu(x) \leq 0, \tag{6.66}$$

which is equivalent to the claim of the theorem. To prove (6.66) we proceed as in the proof of (2.45) in Theorem 2.5. Precisely, we reason by contradiction and we suppose that there exists $\sigma_0 > 0$ such that for each $x \in B_\eta$

$$q(x)Lu(x) \geq \sigma_0. \tag{6.67}$$

Proceeding as in the proof of Theorem 2.5 from (2.46) we arrive to (2.49) and (2.50) that have to be substituted with

$$q(x)L\gamma_\sigma(x) = \sigma q(x)L\gamma(x) \leq \sigma\Lambda < \sigma_0 \qquad \text{on } D_o \cap \Omega_{T_1}. \tag{6.68}$$

$$|\nabla\gamma_\sigma| = \sigma|\nabla\gamma| \leq \sigma\Lambda < \eta \qquad \text{on } D_o \cap \Omega_{T_1}. \tag{6.69}$$

We proceed again as in the proof of Theorem 2.5 until we arrive to the existence of $z_0 \in A_\eta \cap \Omega_{T_1}$. Again we have to distinguish two cases according to $z_0 \in D_o$ or not.

If $z_0 \in D_o$, since z_0 is a maximum for $u - \gamma_\sigma$, we get $\nabla(u - \gamma_\sigma)(z_0) = 0$. Using this fact we infer that $z_0 \in B_\eta$ since, by (6.69),

$$|\nabla u(z_0)| = |\nabla \gamma_\sigma(z_0)| < \sigma \Lambda < \eta.$$

Thus $z_0 \in B_\eta \cap \Omega_{T_1}$. Again since z_0 is a maximum for $u - \gamma_\sigma$ and using the fact that T is positive semi-definite, we have

$$Lu(z_0) \leq L\gamma_\sigma(z_0)$$

and this, jointly with (6.68), yields

$$0 < \sigma_0 \leq q(z_0)Lu(z_0) \leq q(z_0)L\gamma_\sigma(z_0) < \sigma_0,$$

contradicting (6.67). This concludes the proof when $z_0 \in D_0$.

In case $z_0 \notin D_o$ we proceed as in the proof of Theorem 2.5 until (2.58), that remains the same, that is

$$\nabla u(z_0) = \nabla \gamma_\sigma^\varepsilon(z_0) \tag{6.70}$$

and (2.59) that has to be replaced by

$$Lu(z_0) \leq L\gamma_\sigma^\varepsilon(z_0). \tag{6.71}$$

From (6.70) we deduce

$$\begin{aligned} |\nabla u(z_0)| &= \sigma|\nabla \gamma^\varepsilon(z_0)| = \sigma\varphi'(r_\varepsilon(z_0) + \varepsilon)|\nabla r_\varepsilon(z_0)| \\ &= \frac{\sigma}{G(r(z_0) + 1)} \leq \frac{\sigma}{G(1)} < \eta. \end{aligned}$$

Since we already know that $z_0 \in A_\eta$ we conclude that $z_0 \in B_\eta$. Now we analyze (6.71). Recall that, by the triangle inequality, we have

$$r(x) \leq r_\varepsilon(x) + \varepsilon, \tag{6.72}$$

equality holding at z_0. Because of the Hessian comparison theorem, (6.72) and $G' \geq 0$, we have

$${}^{M}K(x) \geq -G^2(r(x)) \geq -G^2(r_\varepsilon(x) + \varepsilon).$$

Set $G_\varepsilon(t) = G(t + \varepsilon)$ and consider the following Cauchy problem

$$\begin{cases} g''(t) - G_\varepsilon^2(t)g(t) = 0 & \text{on } \mathbb{R}_0^+, \\ g(0) = 0, \quad g'(0) = 1. \end{cases} \tag{6.73}$$

Again by the Hessian comparison theorem, on D_{O_ε} we have

$$Lr_\varepsilon(x) \le \mathrm{Tr}(T)(x)\frac{\psi_\varepsilon'(r_\varepsilon(x))}{\psi_\varepsilon(r_\varepsilon(x))}$$

where

$$\psi_\varepsilon(t) = \frac{1}{G_\varepsilon(0)}\left(e^{\int_0^t G_\varepsilon(s)ds} - 1\right).$$

Observing that $z_0 \in D_{O_\varepsilon}$, we obtain using (6.72) and (6.60) that

$$\begin{aligned}
L\gamma^\varepsilon(z_0) &\le \varphi'(r_\varepsilon(z_0)+\varepsilon)Lr_\varepsilon(z_0)\\
&= \frac{1}{G(r_\varepsilon(z_0)+\varepsilon+1)}Lr_\varepsilon(z_0)\\
&= \frac{1}{G(r(z_0)+1)}Lr_\varepsilon(z_0)\\
&\le \frac{\mathrm{Tr}(T)(z_0)}{G(r(z_0)+1)}\frac{\psi_\varepsilon'(r_\varepsilon(z_0))}{\psi_\varepsilon(r_\varepsilon(z_0))}\\
&= \frac{\mathrm{Tr}(T)(z_0)}{G(r(z_0)+1)}G_\varepsilon(r_\varepsilon(z_0))\frac{e^{\int_0^{r_\varepsilon(z_0)} G(s+\varepsilon)ds}}{e^{\int_0^{r_\varepsilon(z_0)} G(s+\varepsilon)ds}-1}\\
&= \mathrm{Tr}(T)(z_0)\frac{G(r_\varepsilon(z_0)+\varepsilon)}{G(r(z_0)+1)}\frac{e^{\int_\varepsilon^{r_\varepsilon(z_0)+\varepsilon} G(s)ds}}{e^{\int_\varepsilon^{r_\varepsilon(z_0)+\varepsilon} G(s)ds}-1}\\
&= \mathrm{Tr}(T)(z_0)\frac{G(r(z_0))}{G(r(z_0)+1)}\frac{e^{\int_\varepsilon^{r(z_0)} G(s)ds}}{e^{\int_\varepsilon^{r(z_0)} G(s)ds}-1}\\
&\le \mathrm{Tr}(T)(z_0)\frac{e^{\int_0^{r(z_0)} G(s)ds}}{e^{\int_1^{r(z_0)} G(s)ds}-1}\\
&\le \mathrm{Tr}(T)(z_0)\Lambda.
\end{aligned}$$

Thus,

$$L\gamma_\sigma^\varepsilon(z_0) = \sigma L\gamma^\varepsilon(z_0) \le \mathrm{Tr}(T)(z_0)\sigma\Lambda < \mathrm{Tr}(T)(z_0)\sigma_0.$$

From (6.61) and (6.71) we deduce that

$$0 < \sigma_0 \le q(z_0)Lu(z_0) \le q(z_0)L\gamma_\sigma^\varepsilon(z_0) \le \sigma\Lambda < \sigma_0,$$

and this is a contradiction. □

6.2.4 *Proof of the Main Results*

6.2.4.1 Proof of Theorems 6.9 and 6.10

Since the arguments of the proof are common for Theorems 6.9 and 6.10, we will prove both of them jointly.

If $\sup_\Sigma |\Phi|^2 = +\infty$, then (ii) of Theorems 6.9 and 6.10 is trivially satisfied and there is nothing to prove. If $\sup_\Sigma |\Phi|^2 = 0$ (that is, Σ is totally umbilical) then (i) holds and there is also nothing to prove. Then, let us assume that $0 < \sup_\Sigma |\Phi|^2 < +\infty$.

Let $u = |\Phi|^2$. Because of Lemma 6.8 we have

$$\square(u) \geq \frac{2}{\sqrt{m(m-1)}} u\sqrt{u + m(m-1)(R-c)} Q_R(\sqrt{u}) = f(u) \tag{6.74}$$

where $Q_R(x)$ is given by (6.48).

If Σ is compact, there exists a point $x_0 \in \Sigma$ such that $u(x_0) = u^*$. Consequently, $\nabla u(x_0) = 0$ and $\square u(x_0) \leq 0$. Therefore, from (6.74) we get

$$f(u^*) \leq 0.$$

Now assume that Σ is complete and noncompact. Since $\sup_\Sigma |\Phi|^2 = u^* < +\infty$, Theorem 6.12 guarantees the validity of the Omori-Yau maximum principle on Σ for the Cheng and Yau operator; hence, there exists a sequence of points $\{x_k\}_{k\in\mathbb{N}}$ in Σ satisfying

$$u(x_k) > u^* - \frac{1}{k} \quad \text{and} \quad f(u(x_k)) \leq \square u(x_k) < \frac{1}{k} \tag{6.75}$$

for every $k \in \mathbb{N}$. Letting $k \to +\infty$ here we also have

$$f(u^*) = \frac{2}{\sqrt{m(m-1)}} u^* \sqrt{u^* + m(m-1)(R-c)} Q_R(\sqrt{u^*}) \leq 0.$$

In any case we deduce that $f(u^*) \leq 0$. Taking into account that $u^* > 0$ and $R \geq c$, we obtain that

$$Q_R(\sqrt{u^*}) \leq 0. \tag{6.76}$$

Since $R > 0$, we have $Q_R(0) = m(m-1)R > 0$ and the function $Q_R(x)$ is strictly decreasing for $x \geq 0$, with $Q_R(a_0) = 0$ at

$$a_0 = R\sqrt{\frac{m(m-1)}{(m-2)(mR-(m-2)c)}} > 0.$$

Therefore (6.76) implies

$$u^* \geq a_0^2 = \frac{m(m-1)R^2}{(m-2)(mR-(m-2)c)}.$$

In other words,

$$\sup_{\Sigma} |\Phi|^2 \geq \alpha_{m,c}(R) = \frac{m(m-1)R^2}{(m-2)(mR-(m-2)c)}.$$

This proves the inequality in (ii) in both theorems.

Moreover, equality $\sup_{\Sigma} |\Phi|^2 = \alpha_{m,c}(R)$ holds if and only if $\sqrt{u^*} = a_0$, and then $Q_R(\sqrt{u}) \geq 0$ on Σ, which jointly with (6.74) implies that

$$\square(u) \geq 0 \quad \text{on } \Sigma.$$

By Remark 6.5, when $R > c$ the operator $\square$ is elliptic. Therefore, if there exists a point $x_0 \in \Sigma$ at which this supremum is attained, then by the maximum principle the function $u = |\Phi|^2$ must be constant, $|\Phi| \equiv a_0$. Thus, (6.47) becomes trivially an equality

$$\frac{1}{2}\square(|\Phi|^2) = 0 = \frac{1}{\sqrt{m(m-1)}}|\Phi|^2 Q_R(|\Phi|)\sqrt{|\Phi|^2 + m(m-1)(R-c)}.$$

Therefore, all the inequalities in the proof of Lemma 6.8 must be equalities. In particular, (6.49) must be an equality, which means that H is constant. Besides, (6.50) must be also an equality or, equivalently, $|\nabla A|^2 - m^2|\nabla H|^2 = 0$. Since we already know that H is constant, this means that $\nabla A = 0$. That is, the second fundamental form is parallel. Finally, (6.53) must be also an equality, so that we obtain the equality in Lemma 6.2. This implies that the hypersurface has exactly two constant principal curvatures with multiplicities $(m-1)$ and 1. Then, by the classical results on isoparametric hypersurfaces of Riemannian space forms [64, 172, 253] we know that Σ must be one of the three following standard product embeddings:

(a) $\mathbb{R}^{m-1} \times \mathbb{S}^1(r) \subset \mathbb{R}^{m+1}$ or $\mathbb{R} \times \mathbb{S}^{m-1}(r) \subset \mathbb{R}^{m+1}$ with $r > 0$, if $c = 0$;
(b) $\mathbb{S}^1(\sqrt{1-r^2}) \times \mathbb{S}^{m-1}(r) \subset \mathbb{S}^{m+1}$, with $0 < r < 1$, if $c = 1$; and
(c) $\mathbb{H}^{m-1}(-\sqrt{1+r^2}) \times \mathbb{S}^1(r) \subset \mathbb{H}^{m+1}$ or $\mathbb{H}^1(-\sqrt{1+r^2}) \times \mathbb{S}^{m-1}(r) \subset \mathbb{H}^{m+1}$, with $r > 0$, if $c = -1$.

In the spherical case ($c = 1$) and for a given radius $0 < r < 1$ we know that the standard product embedding

$$\mathbb{S}^1(\sqrt{1-r^2}) \times \mathbb{S}^{m-1}(r) \hookrightarrow \mathbb{S}^{m+1} \subset \mathbb{R}^2 \times \mathbb{R}^m = \mathbb{R}^{m+2}$$

has constant principal curvatures given by

$$\kappa_1 = \frac{r}{\sqrt{1-r^2}}, \quad \kappa_2 = \cdots = \kappa_m = -\frac{\sqrt{1-r^2}}{r}.$$

Therefore,

$$H = \frac{mr^2 - (m-1)}{mr\sqrt{1-r^2}} \quad \text{and} \quad |\Phi|^2 = \frac{m-1}{mr^2(1-r^2)},$$

and its constant scalar curvature, which is given by (6.7), is $R = (m-2)/mr^2 > 0$. In particular, $R > 1$ if and only if $r < \sqrt{(m-2)/m}$. Thus,

$$|\Phi|^2 = \text{constant} = \sup_{\Sigma} |\Phi|^2 = \frac{m(m-1)R^2}{(m-2)(mR-(m-2))} = \alpha_{m,1}(R),$$

and equality holds, giving the characterization of the equality $\sup_{\Sigma} |\Phi|^2 = \alpha_{m,1}(R)$ in Theorem 6.9. This finishes the proof of Theorem 6.9.

On the other hand, in the Euclidean case ($c = 0$) and for a given radius $r > 0$, $\mathbb{R}^{m-1} \times \mathbb{S}^1(r) \hookrightarrow \mathbb{R}^{m+1}$ has constant principal curvatures given by

$$\kappa_1 = \cdots = \kappa_{m-1} = 0, \quad \kappa_m = \frac{1}{r}.$$

In this case,

$$H = \frac{1}{mr} \quad \text{and} \quad |\Phi|^2 = \frac{m-1}{mr^2},$$

so that, by (6.7), its constant scalar curvature is $R = 0$. Therefore this example does not satisfy the hypothesis of our result ($R > 0$). On the other hand, for a given radius $r > 0$, $\mathbb{R}^1 \times \mathbb{S}^{m-1}(r) \hookrightarrow \mathbb{R}^{m+1}$ has constant principal curvatures

$$\kappa_1 = 0, \quad \kappa_2 = \cdots = \kappa_m = \frac{1}{r}.$$

In this case,

$$H = \frac{m-1}{mr} \quad \text{and} \quad |\Phi|^2 = \frac{m-1}{mr^2},$$

and its constant scalar curvature is given by (6.7), $R = (m-2)/mr^2 > 0$. Thus,

$$|\Phi|^2 = \text{constant} = \sup_{\Sigma} |\Phi|^2 = \frac{(m-1)R}{m-2} = \alpha_{m,0}(R),$$

and equality holds, giving the characterization of the equality $\sup_\Sigma |\Phi|^2 = \alpha_{m,0}(R)$ in Theorem 6.10 when $c = 0$.

In the hyperbolic case ($c = -1$) and for a given $r > 0$ we have that the standard product embedding

$$\mathbb{H}^{m-1}(-\sqrt{1+r^2}) \times \mathbb{S}^1(r) \hookrightarrow \mathbb{H}^{m+1} \subset \mathbb{R}^m_1 \times \mathbb{R}^2 = \mathbb{R}^{m+2}_1$$

has constant principal curvatures

$$\kappa_1 = \cdots = \kappa_{m-1} = \frac{r}{\sqrt{1+r^2}}, \qquad \kappa_m = \frac{\sqrt{1+r^2}}{r}.$$

In this case,

$$H = \frac{mr^2+1}{mr\sqrt{1+r^2}} \quad \text{and} \quad |\Phi|^2 = \frac{m-1}{mr^2(1+r^2)},$$

so that, by (6.7), its constant scalar curvature is $R = -(m-2)/m(1+r^2) < 0$. Therefore this example does not satisfy the hypothesis of our result ($R > 0$). On the other hand, the standard product embedding

$$\mathbb{H}^1(-\sqrt{1+r^2}) \times \mathbb{S}^{m-1}(r) \hookrightarrow \mathbb{H}^{m+1} \subset \mathbb{R}^m_1 \times \mathbb{R}^2 = \mathbb{R}^{m+2}_1$$

has constant principal curvatures

$$\kappa_1 = \frac{r}{\sqrt{1+r^2}}, \qquad \kappa_2 = \cdots = \kappa_m = \frac{\sqrt{1+r^2}}{r}.$$

In this case,

$$H = \frac{mr^2+m-1}{mr\sqrt{1+r^2}} \quad \text{and} \quad |\Phi|^2 = \frac{m-1}{mr^2(1+r^2)},$$

and its constant scalar curvature, as given by (6.7), is $R = (m-2)/mr^2 > 0$. Thus,

$$|\Phi|^2 = \text{constant} = \sup_\Sigma |\Phi|^2 = \frac{m(m-1)R^2}{(m-2)(mR+(m-2))} = \alpha_{m,-1}(R),$$

and equality holds, giving the characterization of the equality $\sup_\Sigma |\Phi|^2 = \alpha_{m,c}(R)$ in Theorem 6.10 when $c = -1$. This finishes the proof of Theorem 6.10.

6.2.4.2 Proof of Theorem 6.11

Recall that a Riemannian manifold Σ is said to be parabolic if the only subharmonic functions on Σ which are bounded from above are constant; that is, for a function $u \in C^2(\Sigma)$

$$\Delta u \geq 0 \quad \text{and} \quad u \leq u^* < +\infty \quad \text{implies} \quad u = \text{constant}.$$

Let Σ be a Riemannian manifold and let T be a symmetric, positive semi-definite $(0, 2)$ tensor field on Σ. As in Theorem 6.13, consider the differential operator $L(u) = \mathrm{Tr}(t \circ \mathrm{hess}\, u)$. Following the terminology above, we will say that Σ is L-parabolic if the only solutions of the inequality $L(u) \geq 0$ which are bounded from above are constant.

To prove Theorem 6.11, observe first that if $\sup_\Sigma |\Phi|^2 = +\infty$ then there is nothing to prove. On the other hand, if $\sup_\Sigma |\Phi|^2 < +\infty$ then we may apply Lemma 6.8 and Theorem 6.12 as in the first part of the proof of Theorems 6.9 and 6.10 to conclude that $\sup_\Sigma |\Phi|^2 \geq \alpha_{m,c}(R)$. Moreover, if equality holds then we have that $\Box(|\Phi|^2) \geq 0$ on Σ. Therefore, by the $\Box$-parabolicity of Σ we conclude that the function $u = |\Phi|^2$ must be constant and equal to $\alpha_{m,c}(R)$. The rest of the proof follows as in the previous proof of Theorems 6.9 and 6.10.

After Theorem 6.11, it would be interesting to find some criteria for the $\Box$-parabolicity of Σ. In this respect and as an application of Theorem 4.15 we may state the following result.

Lemma 6.9 *Let $(M, \langle\, ,\, \rangle)$ be a complete Riemannian manifold, $o \in M$ a fixed origin and $r(x) = \mathrm{dist}_M(x, o)$. Let $L = \mathrm{div}\left(T(\nabla u, \,)^\sharp\right) = \mathrm{div}\,(t(\nabla u))$ and suppose that, for some $T_+ \in C^0\left(\mathbb{R}_0^+\right)$,*

$$0 \leq T(Y, Y) \leq T_+(r)$$

for every $Y \in T_xM$, $|Y| = 1$, and every $x \in \partial B_r$, where B_r denotes the geodesic ball of radius r centered at o. If

$$\frac{1}{T_+(r)\,\mathrm{vol}\,\partial B_r} \notin L^1(+\infty) \tag{6.77}$$

then L is parabolic on M.

In particular, for the case of the Cheng and Yau operator $\Box$ we can give the following parabolicity criterium.

Corollary 6.4 *Let Σ be an oriented isometrically immersed hypersurface into an $(m+1)$-dimensional space form with curvature c, and assume that Σ has constant (normalized) scalar curvature $R \geq c$. In the case where $R = c$, assume further* *that the mean curvature function H does not change sign. Assume that* $\sup_\Sigma |\Phi|^2 < +\infty$. *If for some point $o \in \Sigma$*

$$\frac{1}{\mathrm{vol}(\partial B_r)} \notin L^1(+\infty) \tag{6.78}$$

then Σ is $\Box$-parabolic.

Corollary 6.4 follows from Lemma 6.9 by observing that in this case $T_+(r) \leq 2m \sup_\Sigma H < +\infty$. Actually, since $T = P = mHI - A$, it follows from Lemma 6.6 that

$$T_+(r) = 2m \sup_{\partial B_r} H \leq 2m \sup_\Sigma H < +\infty.$$

6.3 Hypersurfaces into Nondegenerate Euclidean Cones

In this section we will apply Theorem 6.13 when t is the k-th Newton tensor of an isometrically immersed hypersurface $f : \Sigma \to \mathbb{R}^{m+1}$ oriented by a globally defined normal unit vector field ν.

As we did in Sect. 5.1, fixed an origin $o \in \mathbb{R}^{m+1}$ and a unit vector $\zeta \in \mathbb{S}^m$, for $\theta \in (0, \pi/2)$, we denote by $\mathscr{C} = \mathscr{C}_{o,\zeta,\theta}$ the nondegenerate cone with vertex o, direction ζ and width θ, that is,

$$\mathscr{C} = \mathscr{C}_{o,\zeta,\theta} = \{p \in \mathbb{R}^{m+1} \backslash \{o\} : \langle \frac{p-o}{|p-o|}, \zeta \rangle \geq \cos\theta\}.$$

By nondegenerate we mean that it is strictly smaller than a half-space. Assuming that the image $f(\Sigma)$ is inside a nondegenerate cone of $\mathbb{R}^{m+1}$, as an application of Theorem 6.13 and motivated by the results in Sect. 5.2, we provide a lower bound for the width of the cone in terms of higher order mean curvatures of the hypersurface. Specifically, we obtain the following result.

Theorem 6.14 *Let $f : \Sigma \to \mathbb{R}^{m+1}$ be an oriented isometric immersion of a complete noncompact hypersurface with sectional curvatures satisfying*

$$K \geq -G(r)^2$$

with $G \in C^1(\mathbb{R}_0^+)$ such that

$$(i)\ G(0) > 0; \quad (ii)\ G'(t) \geq 0; \quad (iii)\ \frac{1}{G(t)} \notin L^1(+\infty). \tag{6.79}$$

Assume that P_k *is positive semi-definite and* H_k *does not vanish on* Σ. *If* $f(\Sigma)$ *is contained into a nondegenerate cone* $\mathscr{C} = \mathscr{C}_{o,\zeta,\theta}$ *as above with vertex at* $o \in \mathbb{R}^{m+1} \setminus f(\Sigma)$, *then*

$$\sup_{\Sigma}\left(\frac{|H_{k+1}|}{H_k}\right) \geq A_0 \frac{\cos^2\theta}{d(\Pi_\zeta, f(\Sigma))}, \tag{6.80}$$

where $A_0 = \frac{6\sqrt{3}}{25\sqrt{5}} \approx 0.186$, Π_ζ *denote the hyperplane orthogonal to* ζ *passing through* o *and* $d(\Pi_\zeta, f(\Sigma))$ *is the Euclidean distance between this hyperplane and* $f(\Sigma)$.

Proof To prove the theorem we shall follow the ideas and use some of the computations performed in the proof of Theorem 5.1. We may assume without loss of generality that the vertex of the cone is the origin $o \in \mathbb{R}^{m+1}$, so that there exists $\zeta \in \mathbb{S}^m$ and $0 < \theta < \pi/2$ such that

$$\langle \frac{f(x)}{|f(x)|}, \zeta \rangle \geq \cos\theta \tag{6.81}$$

for every $x \in \Sigma$. Observe that

$$d(\Pi_\zeta, f(\Sigma)) = \inf_{x\in\Sigma} \langle f(x), \zeta \rangle.$$

We reason by contradiction and we suppose that (6.80) is false. Hence, there exists $x_0 \in \Sigma$ such that

$$\langle f(x_0), \zeta \rangle \sup_{\Sigma}\left(\frac{|H_{k+1}|}{H_k}\right) < A\cos^2\theta$$

for a positive constant $A < A_0$. For the ease of notation we set $\alpha = \langle f(x_0), \zeta \rangle > 0$, let $\beta \in (0, 1)$ and on Σ define the function

$$u(x) = \sqrt{\alpha^2 + \beta^2\cos^2\theta |f(x)|^2} - \langle f(x), \zeta \rangle.$$

Note that, by construction, $u(x_0) > 0$. We claim that

$$u(x) \leq \alpha$$

for every $x \in \Sigma$. An algebraic manipulation shows that the above inequality is equivalent to

$$\langle f(x), \zeta \rangle^2 + 2\alpha \langle f(x), \zeta \rangle - \beta^2 \cos^2\theta |f(x)|^2 \geq 0,$$

and using (6.81) we have

$$\langle f(x), \zeta\rangle^2 + 2\alpha\langle f(x), \zeta\rangle - \beta^2 \cos^2\theta |f(x)|^2 \geq \langle f(x), \zeta\rangle^2 - \cos^2\theta |f(x)|^2 \geq 0.$$

Next, we consider the closed nonempty set

$$\overline{\Omega_0} = \{x \in \Sigma : u(x) \geq u(x_0)\}.$$

Using again (6.81), for every $x \in \overline{\Omega_0}$ one has

$$\sqrt{\alpha^2 + \beta^2 \cos^2\theta |f(x)|^2} \geq u(x_0) + \langle f(x), \zeta\rangle \geq u(x_0) + \cos\theta |f(x)| > 0.$$

Squaring this inequality yields

$$(1-\beta^2)\cos^2\theta |f(x)|^2 + 2u(x_0)\cos\theta |f(x)| + u(x_0)^2 - \alpha^2 \leq 0$$

for every $x \in \overline{\Omega_0}$. The left-hand side of the above inequality is a quadratic polynomial in $|f(x)|$ with two distinct roots $\alpha_- < 0 < \alpha_+$ given by

$$\alpha_\pm = \frac{\pm\sqrt{\beta^2 u(x_0)^2 + (1-\beta^2)\alpha^2} - u(x_0)}{(1-\beta^2)\cos\theta}.$$

Therefore, for every $x \in \overline{\Omega_0}$ we have

$$0 < |f(x)| \leq \alpha_+ = \frac{\sqrt{\beta^2 u(x_0)^2 + (1-\beta^2)\alpha^2} - u(x_0)}{(1-\beta^2)\cos\theta}.$$

Now using the elementary inequality $\sqrt{1+t^2} \leq 1+t$ for $t \geq 0$, we obtain

$$\begin{aligned}
\alpha_+ &= \frac{1}{(1-\beta^2)\cos\theta}\left(\sqrt{\beta^2 u(x_0)^2\left(1 + \frac{(1-\beta^2)\alpha^2}{\beta^2 u(x_0)^2}\right)} - u(x_0)\right) \\
&= \frac{\beta u(x_0)}{(1-\beta^2)\cos\theta}\sqrt{1 + \frac{(1-\beta^2)\alpha^2}{\beta^2 u(x_0)^2}} - \frac{u(x_0)}{(1-\beta^2)\cos\theta} \\
&\leq \frac{\beta u(x_0)}{(1-\beta^2)\cos\theta}\left(1 + \frac{\sqrt{1-\beta^2}\alpha}{\beta u(x_0)}\right) - \frac{u(x_0)}{(1-\beta^2)\cos\theta} \\
&= \frac{\alpha}{\sqrt{1-\beta^2}\cos\theta} - \frac{u(x_0)}{(1+\beta)\cos\theta} \\
&\leq \frac{\alpha}{\sqrt{1-\beta^2}\cos\theta}.
\end{aligned}$$

Therefore,

$$|f(x)| \leq \frac{\alpha}{\sqrt{1-\beta^2\cos\theta}} \quad \text{on } \overline{\Omega}_0. \tag{6.82}$$

Next step is to compute $L_k u = \mathrm{Tr}(P_k \circ \mathrm{hess}(u))$ when P_k is the k-th Newton tensor, we first observe that

$$\nabla u = -\zeta^\top + \frac{\beta^2\cos^2\theta}{\sqrt{\alpha^2+\beta^2\cos^2\theta|f|^2}} f^\top, \tag{6.83}$$

where, as usual, $^\top$ denotes the tangential component along the immersion f. That is,

$$\zeta = \zeta^\top + \langle \zeta, \nu\rangle \nu \;\text{ and }\; f = f^\top + \langle f, \nu\rangle \nu.$$

Then a computation similar to that performed in the proof of Theorem 5.1 gives

$$\begin{aligned}
\mathrm{Hess}(u)(X,Y) = {} & \frac{\beta^2\cos^2\theta}{\sqrt{\alpha^2+\beta^2\cos^2\theta|f|^2}}\langle X, Y\rangle \\
& + \langle \frac{\beta^2\cos^2\theta}{\sqrt{\alpha^2+\beta^2\cos^2\theta|f|^2}} f - \zeta, \nu\rangle \langle AX, Y\rangle \\
& + \frac{-\beta^4\cos^4\theta}{(\alpha^2+\beta^2\cos^2\theta|f|^2)^{3/2}}\langle X, f^\top\rangle\langle Y, f^\top\rangle,
\end{aligned} \tag{6.84}$$

for every $X, Y \in \mathfrak{X}(\Sigma)$. Hence, having fixed a local orthonormal frame $\{e_i\}$ on Σ,

$$\begin{aligned}
L_k u = \mathrm{Tr}(P_k \circ \mathrm{hess}(u)) &= \langle P_k \circ \mathrm{hess}(u)(e_i), e_i\rangle \\
&= \mathrm{Hess}(u)(e_i, P_k e_i) \\
&= \langle \frac{\xi}{|f|} f - \zeta, \nu\rangle \,\mathrm{Tr}(A \circ P_k) + \frac{\xi}{|f|}\,\mathrm{Tr}(P_k) \\
&\quad - \frac{\xi^2}{|f|^2}\frac{1}{\sqrt{\alpha^2+\beta^2\cos^2\theta|f|^2}}\langle P_k f^\top, f^\top\rangle,
\end{aligned} \tag{6.85}$$

where

$$\xi(x) = \frac{\beta^2\cos^2\theta|f(x)|}{\sqrt{\alpha^2+\beta^2\cos^2\theta|f(x)|^2}}.$$

Therefore, using (6.33) and (6.34),

$$L_k u = c_k \langle \frac{\xi}{|f|} f - \zeta, \nu \rangle H_{k+1} + c_k \frac{\xi}{|f|} H_k - \frac{\xi^2}{|f|^2} \frac{1}{\sqrt{\alpha^2 + \beta^2 \cos^2\theta |f|^2}} \langle P_k f^\top, f^\top \rangle. \tag{6.86}$$

Observe that, by (6.81),

$$\left| \frac{\xi}{|f|} f - \zeta \right|^2 = \xi^2 - 2\xi \frac{\langle f, \zeta \rangle}{|f|} + 1 \leq \xi^2 - 2\cos\theta \xi + 1 \leq 1, \tag{6.87}$$

since $0 < \xi(x) < \beta \cos\theta$ for every $x \in \Sigma$. On the other hand, since P_k is positive semi-definite we have

$$0 \leq \langle P_k f^\top, f^\top \rangle \leq \mathrm{Tr}(P_k) |f^\top|^2 \leq c_k H_k |f|^2. \tag{6.88}$$

Since, by our hypothesis, $H_k > 0$ on Σ, from here we obtain

$$\begin{aligned} \frac{1}{c_k H_k} L_k u &\geq -\frac{|H_{k+1}|}{H_k} + \frac{\xi}{|f|} - \frac{\xi^2}{\sqrt{\alpha^2 + \beta^2 \cos^2\theta |f|^2}} \\ &\geq -\sup_\Sigma \frac{|H_{k+1}|}{H_k} + \frac{\alpha^2 \beta^2 \cos^2\theta}{(\alpha^2 + \beta^2 \cos^2\theta |f|^2)^{3/2}} \end{aligned} \tag{6.89}$$

on Σ. Recall that, because of our choice of x_0, we have

$$\sup_\Sigma \frac{|H_{k+1}|}{H_k} < A \frac{\cos^2\theta}{\alpha}$$

for a positive constant $A < A_0 = \frac{6\sqrt{3}}{25\sqrt{5}}$. On the other hand, by (6.82) we also have

$$|f|^2 < \frac{\alpha^2}{(1-\beta^2)\cos^2\theta} \tag{6.90}$$

on $\overline{\Omega}_0$ and therefore

$$\frac{\alpha^2 \beta^2 \cos^2\theta}{(\alpha^2 + \beta^2 \cos^2\theta |f|^2)^{3/2}} \geq \frac{\cos^2\theta}{\alpha} \beta^2 (1-\beta^2)^{3/2}$$

on $\overline{\Omega}_0$. Choose $\beta = \sqrt{2/5}$. Then, $\beta^2(1-\beta^2)^{3/2} = A_0$ and

$$\frac{1}{c_k H_k} L_k u \geq \frac{\cos^2\theta}{\alpha} (A_0 - A) > 0 \quad \text{on } \overline{\Omega}_0. \tag{6.91}$$

There are now two possibilities:

(i) x_0 is an absolute maximum for u on Σ. Then, $L_k u(x_0) \leq 0$, contradicting (6.91).
(ii) $\Omega_0 = \{x \in \Sigma : u(x) > u(x_0)\} \neq \emptyset$. In this case, since $u(x)$ is bounded above on Σ it is enough to evaluate inequality (6.91) along a sequence $\{x_k\}$ realizing the $1/c_k H_k$-weak maximum principle for the operator L_k on Σ. The latter holds because of item (i) of Theorem 6.13 and the present assumptions. We thus have $u(x_k) > u^* - 1/k$ and therefore $x_k \in \Omega_0$ for k sufficiently large and

$$0 < \frac{\cos^2\theta}{\alpha}(A_0 - A) \leq \frac{1}{c_k H_k} L_k u(x_k) < \frac{1}{k}.$$

By letting $k \to \infty$ we obtain the desired contradiction.

This completes the proof of the theorem. □

Corollary 6.5 *Let $f : \Sigma \to \mathbb{R}^{m+1}$ be an oriented isometric immersion of a complete noncompact hypersurface with sectional curvatures satisfying*

$$K \geq -G(r)^2$$

with $G \in C^1(\mathbb{R}_0^+)$ such that (6.79) holds. Assume that P_k is positive semi-definite. If $f(\Sigma)$ is contained into a nondegenerate cone $\mathscr{C} = \mathscr{C}_{o,\zeta,\theta}$ as above with vertex at $o \in \mathbb{R}^{m+1} \setminus f(\Sigma)$, then

$$\sup_\Sigma |H_{k+1}| \geq A_0 \frac{\cos^2\theta}{d(\Pi_\zeta, f(\Sigma))} \inf_\Sigma H_k, \tag{6.92}$$

where $A_0 = \frac{6\sqrt{3}}{25\sqrt{5}} \approx 0.186$, Π_ζ denote the hyperplane orthogonal to ζ passing through o and $d(\Pi_\zeta, f(\Sigma))$ is the Euclidean distance between this hyperplane and $f(\Sigma)$.

For the proof of Corollary 6.5 observe that (6.92) holds trivially if $\inf_\Sigma H_k = 0$. If $\inf_\Sigma H_k > 0$, then $H_k > 0$ everywhere and the result follows directly from Theorem 6.14 since the estimate (6.92) is weaker than (6.80).

On the other hand, in the case $k = 1$ we can slightly improve our Theorem 6.14, both regarding the condition on the ellipticity of P_1 and the value of the constant A_0 in (6.80). Specifically we prove the following.

Corollary 6.6 *Let $f : \Sigma \to \mathbb{R}^{m+1}$ be an oriented isometric immersion of a complete noncompact hypersurface with sectional curvatures satisfying*

$$K \geq -G(r)^2$$

with $G \in C^1(\mathbb{R}_0^+)$ such that (6.79) holds. If $H_2 > 0$ or, equivalently, the scalar curvature of Σ is positive, and $f(\Sigma)$ is contained into a nondegenerate cone

$\mathscr{C} = \mathscr{C}_{o,\zeta,\theta}$ as above with vertex at $o \in \mathbb{R}^{m+1} \setminus f(\Sigma)$, then

$$\sup_{\Sigma} \sqrt{H_2} \geq \sup_{\Sigma} \left(\frac{H_2}{H_1} \right) \geq B_m \frac{\cos^2 \theta}{d(\Pi_\zeta, f(\Sigma))}, \tag{6.93}$$

where $B_2 = B_3 = A_0 = \frac{6\sqrt{3}}{25\sqrt{5}} \approx 0.186$, and

$$B_m = \max_{0<\varrho<1} \left(\varrho^2 \sqrt{1-\varrho^2}(1 - \frac{3}{m}\varrho^2) \right)$$

for $m \geq 4$.

Remark 6.6 We emphasize that $B_m > A_0$ and $B_m \sim 2/(3\sqrt{3}) \approx 0.385$ when m goes to infinity.

Proof According to Remark 6.4, the assumption $H_2 > 0$ and $m^2 H_1^2 - |A|^2 = m(m-1)H_2 > 0$ guarantee that P_1 is positive definite for an appropriate choice of the unit normal ν, so that $H_1 > 0$ and $mH_1 - |A| > 0$ on Σ.

By Cauchy-Schwarz inequality,

$$H_1^2 - H_2 = \frac{1}{m(m-1)} \left(\sum_{i=1}^{m} \kappa_i^2 - \frac{1}{m} \left(\sum_{i=1}^{m} \kappa_i \right)^2 \right) \geq 0.$$

This immediately yields $H_2/H_1 \leq \sqrt{H_2}$ and gives the first inequality in (6.93).

As for the second inequality in (6.93), arguing as in the proof of Theorem 6.14, we reason by contradiction and assume that there exists a point $x_0 \in \Sigma$ such that

$$\alpha \sup_{\Sigma} \left(\frac{H_2}{H_1} \right) < A \cos^2 \theta \tag{6.94}$$

for a positive constant $A < B_m$, where $\alpha = \langle f(x_0), \zeta \rangle$. We then follow the proof of Theorem 6.14 until we reach (6.86), which jointly with (6.87) yields

$$L_1 u \geq -c_1 H_2 + c_1 \frac{\xi}{|f|} H_1 - \frac{\xi^2}{|f|^2} \frac{1}{\sqrt{\alpha^2 + \beta^2 \cos^2 \theta |f|^2}} \langle P_1 f^\top, f^\top \rangle.$$

The idea to improve the value of the constant A_0 in (6.80) is to improve the estimate (6.88) in the following way. Using $P_1 = mH_1 I - A$ we have

$$\langle P_1 f^\top, f^\top \rangle = mH_1 |f^\top|^2 - \langle A f^\top, f^\top \rangle \leq 2mH_1 |f|^2, \tag{6.95}$$

where the last inequality is due to the fact that

$$|\langle A f^\top, f^\top \rangle| \leq |A| |f^\top|^2 \leq mH_1 |f|^2.$$

Note that, for $k = 1$ and $m \geq 4$, (6.95) gives a better estimate than (6.88). In this case, using (6.95) we obtain

$$\begin{aligned}\frac{1}{c_1 H_1} L_1 u &\geq -\frac{H_2}{H_1} + \frac{\xi}{|f|} - \frac{2}{m-1} \frac{\xi^2}{\sqrt{\alpha^2 + \beta^2 \cos^2\theta |f|^2}} \\ &\geq -\sup_{\Sigma} \frac{H_2}{H_1} + \frac{\alpha^2\beta^2\cos^2\theta + \frac{m-3}{m}\beta^4\cos^4\theta |f|^2}{(\alpha^2 + \beta^2\cos^2\theta |f|^2)^{3/2}}\end{aligned}$$

on Σ, instead of (6.89). From (6.90) it follows that

$$\frac{\alpha^2\beta^2\cos^2\theta + \frac{m-3}{m}\beta^4\cos^4\theta |f|^2}{(\alpha^2 + \beta^2\cos^2\theta |f|^2)^{3/2}} \geq \frac{\cos^2\theta}{\alpha}\beta^2\sqrt{1-\beta^2}(1 - \frac{3}{m}\beta^2)$$

on $\overline{\Omega}_0$. Choose $\beta \in (0, 1)$ to maximize $\varrho^2\sqrt{1-\varrho^2}\big(1 - \frac{3}{m}\varrho^2\big)$, that is,

$$\beta^2 = \frac{4 + m - \sqrt{(4+m)^2 - 40m/3}}{10}$$

and

$$B_m = \beta^2\sqrt{1-\beta^2}\left(1 - \frac{3}{m}\beta^2\right).$$

Then,

$$\frac{1}{c_1 H_1} L_1 u \geq \frac{\cos^2\theta}{\alpha}(B_m - A) > 0 \quad \text{on } \overline{\Omega}_0. \tag{6.96}$$

The proof then finishes as in Theorem 6.14. □

For the case $k \geq 2$ there is an inequality corresponding to the first in (6.93), given by

$$\sup_{\Sigma} \sqrt[k+1]{H_{k+1}} \geq \sup_{\Sigma}\left(\frac{H_{k+1}}{H_k}\right).$$

However, to guarantee its validity ones needs to assume the existence of an elliptic point (see the next section for details).

6.4 Higher Order Mean Curvature Estimates

Several estimates for the k-mean curvatures H_k of a compact hypersurface in a complete Riemannian manifold have been subsequently obtained by Vlachos [269], Veeravalli [268], Fontenele and Silva [119], Roth [248] and Ranjbar-Motlagh [238].

In this section, we generalize a result given in the above last reference and that we now describe.

Let $f : \Sigma \to N$ denote a complete isometrically immersed hypersurface into a complete Riemannian manifold of dimension $m + 1$ whose image lies inside a closed geodesic ball ${}^N B_r(o)$ of radius r and center $o \in N$. Assume that $0 < r < \min\{\mathrm{inj}_N(o), \pi/2\sqrt{b}\}$ where $\mathrm{inj}_N(o)$ is the injectivity radius at o and $\pi/2\sqrt{b}$ is replaced by $+\infty$ in case $b \leq 0$. Assume also that there exists $x_0 \in \Sigma$ such that $f(x_0) \in \partial {}^N B_r(o)$. Of course, this is a slightly weaker assumption than asking Σ to be compact.

Let ${}^N K_{rad}$ denote the radial sectional curvatures in ${}^N B_r(o)$ along geodesics issuing from the center and assume that ${}^N K_{rad} \leq b$ for some constant $b \in \mathbb{R}$. Assume also that $H_{k+1} \neq 0$ everywhere for some $2 \leq k \leq m - 1$. In this situation, it turns that the x_0 is an elliptic point. More precisely, the second fundamental form of f at x_0 with respect to the inner pointing orientation is positive definite. From Gårding inequalities, [124], it follows that $H_j > 0$ for $1 \leq j \leq k + 1$.

In the above situation, it was shown in Theorem 4.2 in [238] that

$$\sup_{\Sigma} \left(\frac{H_{j+1}}{H_j} \right) \geq C_b(r)$$

for any $1 \leq j \leq k$, where the constant $C_b(r)$ given by (6.98) below is the mean curvature of a geodesic sphere of radius r in a simply connected space form of sectional curvature b. Moreover, if equality holds for some j then it follows that $f(\Sigma) = \partial {}^N B_r(o)$.

Our main goal in this section, following the results presented in [21], is to replace the assumption of compactness of the submanifold by the weaker completeness with the tools we have introduced in Theorem 6.13. The following is a corollary of the more general result given in Theorem 6.16 below. Here we express the more general, but technical, assumptions of Theorem 6.16 in a simpler geometric way.

Theorem 6.15 *Let $f : \Sigma \to N$ be a complete isometrically immersed hypersurface into a complete Riemannian manifold of dimension $m + 1$ such that $f(\Sigma) \subset {}^N B_r(o)$. Assume that, for some $2 \leq k \leq m - 1$, $H_{k+1} \neq 0$ everywhere and that the sectional curvatures satisfy ${}^\Sigma K \geq K$ for some constant $K \in \mathbb{R}$ and ${}^N K_{rad} \leq b$ for some constant $b \in \mathbb{R}$. If f has an elliptic point, then, for each $1 \leq j \leq k$,*

$$\sup_{\Sigma} \sqrt[j+1]{H_{j+1}} \geq \sup_{\Sigma} \left(\frac{H_{j+1}}{H_j} \right) \geq C_b(r). \tag{6.97}$$

Moreover, if there exists a point $x_0 \in \Sigma$ such that $f(x_0) \in \partial {}^N B_r(o)$ and $\sup_{\Sigma} \left(H_{j+1}/H_j\right) = C_b(r)$ for some j then $f(\Sigma) = \partial {}^N B_r(o)$.

It is a standard fact that if N has constant sectional curvature b, then the mean curvature of the geodesic sphere $\partial^N B_r(o)$ is

$$C_b(r) = \begin{cases} \sqrt{b}\cot(\sqrt{b}\,r) & \text{if } b > 0, \\ 1/r & \text{if } b = 0, \\ \sqrt{-b}\coth(\sqrt{-b}\,r) & \text{if } b < 0. \end{cases} \tag{6.98}$$

In the following result, it is convenient to think of $\partial^N B_r(o)$ as the smallest possible geodesic sphere centered at o enclosing the hypersurface.

Theorem 6.16 *Let $f : \Sigma \to N$ be a two-sided hypersurface isometrically immersed into a complete Riemannian manifold N of dimension $m + 1$, where Σ is complete and satisfies ${}^{\Sigma}K \geq K$ for some constant $K \in \mathbb{R}$. Assume that P_k is positive semi-definite for some $0 \leq k \leq m - 1$ and that $P_k \not\equiv 0$ everywhere outside a compact set. If $f(\Sigma) \subset {}^N B_r(o)$, with ${}^N B_r(o)$ a geodesic ball as above, then*

$$\sup_{\Sigma} \left(\frac{|H_{k+1}|}{H_k} \right) \geq C_b(r). \tag{6.99}$$

Moreover, if P_k is positive definite and there exists a point $x_0 \in \Sigma$ such that $f(x_0) \in \partial^N B_r(o)$, then equality in (6.99) implies $f(\Sigma) = \partial^N B_r(o)$.

In particular, we have the following consequence.

Corollary 6.7 *Let $f : \Sigma \to N$ be as above. Assume that P_k is positive semi-definite for some $0 \leq k \leq m - 1$. If $f(\Sigma) \subset {}^N B_r(o)$ for a geodesic ball ${}^N B_r(o)$ as in Theorem 6.16, then*

$$\sup_{\Sigma} |H_{k+1}| \geq C_b(r) \inf_{\Sigma} H_k. \tag{6.100}$$

For the proof of Corollary 6.7 observe that (6.100) holds trivially if $\inf_\Sigma H_k = 0$. If $\inf_\Sigma H_k > 0$, then $P_k \not\equiv 0$ everywhere and the result follows directly from Theorem 6.16 since the estimate (6.100) is weaker than (6.99).

Proof (of Theorem 6.16) We denote by $\rho : N \to \mathbb{R}$ the distance function to the reference point o and set $u = \rho \circ f$. Along Σ, indicating with ν the unit normal vector field of Σ,

$$\nabla\rho = \nabla u + \langle \nabla\rho, \nu \rangle \nu.$$

Furthermore, for each $X, Y \in \mathfrak{X}(\Sigma)$,

$$\text{Hess}\,(u)(X, Y) = \text{Hess}\,(\rho)(X, Y) + \langle \nabla\rho, \nu \rangle \langle AX, Y \rangle.$$

Let $e_1, \ldots, e_m$ be an orthonormal basis of principal directions at a point of Σ. We obtain

$$L_k u = \sum_{i=1}^{m} \text{Hess}\,(u)(e_i, P_k e_i) = \sum_{i=1}^{m} \text{Hess}\,(\rho)(e_i, P_k e_i) + \langle \nabla\rho, \nu\rangle \,\text{Tr}(A \circ P_k)$$
$$= \sum_{i=1}^{m} \text{Hess}\,(\rho)(e_i, P_k e_i) + c_k H_{k+1} \langle \nabla\rho, \nu\rangle.$$

By assumption, we have

$$P_k e_i = \mu_{i,k} e_i \quad \text{with} \quad \mu_{i,k} \geq 0.$$

Using the Hessian comparison theorem, for any fixed index i we obtain

$$\begin{aligned} \text{Hess}\,(\rho)(e_i, P_k e_i) &= \mu_{i,k} \,\text{Hess}\,(\rho)(e_i, e_i) \\ &\geq \mu_{i,k} C_b(u)(1 - \langle \nabla u, e_i\rangle^2) \\ &= C_b(u)(\mu_{i,k} - \langle \nabla u, e_i\rangle \langle P_k \nabla u, e_i\rangle). \end{aligned}$$

Summing over i we get

$$\begin{aligned} \sum_{i=1}^{m} \text{Hess}\,(\rho)(e_i, P_k e_i) &\geq C_b(u)(\text{Tr}(P_k) - \langle \nabla u, P_k \nabla u)\rangle \\ &= C_b(u)(c_k H_k - \langle \nabla u, P_k \nabla u\rangle) \end{aligned}$$

and therefore

$$L_k u \geq C_b(u)(c_k H_k - \langle \nabla u, P_k \nabla u\rangle) + c_k H_{k+1} \langle \nabla\rho, \nu\rangle. \tag{6.101}$$

Consider the function

$$\phi_b(t) = \begin{cases} 1 - \cos(\sqrt{b}\,t) & \text{if } b > 0, \\ t^2 & \text{if } b = 0, \\ \coth(\sqrt{-b}\,t) & \text{if } b < 0. \end{cases}$$

solution of

$$\phi_b''(t) - C_b(t)\phi_b'(t) = 0. \tag{6.102}$$

Then $\phi_b''(t) > 0$ since $\phi_b'(t) > 0$. We have

$$L_k \phi_b(u) = \phi_b''(u) \langle \nabla u, P_k \nabla u\rangle + \phi_b'(u) L_k u.$$

It follows from (6.101) and (6.102) that

$$L_k\phi_b(u) \geq c_k\phi_b'(u)(C_b(u)H_k + \langle\nabla\rho, \nu\rangle H_{k+1}).$$

Hence,

$$L_k\phi_b(u) \geq c_k\phi_b'(u)\left(C_b(u)H_k - |H_{k+1}|\right).$$

Since $\sup_\Sigma \phi_b(u) \leq \phi_b(r) < +\infty$, by Theorem 6.13 there exists a sequence $\{x_j\} \subset \Sigma$ such that

$$\phi_b(u(x_j)) > \sup_\Sigma \phi_b(u) - \frac{1}{j} \qquad \text{and} \qquad \frac{1}{c_kH_k(x_j)}L_k\phi_b(u)(x_j) < \frac{1}{j}.$$

Since $\sup_\Sigma \phi_b(u) = \phi_b(\sup_\Sigma u)$, then $\lim_{j\to\infty} u(x_j) = u^* = \sup_\Sigma u$. Thus,

$$\begin{aligned}\frac{1}{j} > \frac{1}{c_kH_k(x_j)}L_k\phi_b(u)(x_j) &\geq \phi_b'(u(x_j))\left(C_b(u(x_j)) - \frac{|H_{k+1}|}{H_k}(x_j)\right)\\ &\geq \phi_b'(u(x_j))\left(C_b(r) - \sup_\Sigma\left(\frac{|H_{k+1}|}{H_k}\right)\right)\end{aligned}$$

since $C_b(u(x_j) \geq C_b(r)$. Taking $j \to +\infty$ we conclude that

$$C_b(r) - \sup_\Sigma\left(\frac{|H_{k+1}|}{H_k}\right) \leq 0.$$

For the proof of the second part of the statement, first observe that equality in (6.99) yields $L_k\phi_b(u) \geq 0$. Since $\phi_b(u) \leq \phi_b(r) < +\infty$, it follows from the maximum principle for the elliptic operator L_k that $\phi(u)$, and hence u is constant. □

In the sequel, we want to replace some assumptions in Theorem 6.16 by simpler ones of geometrical nature. This, of course, is the case of Theorem 6.15 above. But first we consider the special case of H_2.

Corollary 6.8 *Let $f : \Sigma \to N$ be a hypersurface isometrically immersed into into a complete Riemannian manifold of dimension $m + 1$. Assume that Σ is complete with sectional curvature ${}^\Sigma K \geq K$ for some $K \in \mathbb{R}$. If $H_2 > 0$ and $f(\Sigma) \subset {}^N B_r(o)$ for a geodesic ball ${}^N B_r(o)$ as above, then*

$$\sup_\Sigma \sqrt{H_2} \geq \sup_\Sigma\left(\frac{H_2}{H}\right) \geq C_b(r). \tag{6.103}$$

Moreover, if there exists $x_0 \in \Sigma$ such that $f(x_0) \in \partial^N B_r(o)$ and $\sup_\Sigma(H_2/H) = C_b(r)$ then $f(\Sigma) = \partial^N B_r(o)$.

Proof In term of the principal curvatures $\kappa_1, \ldots, \kappa_m$ of f we have that

$$m^2H^2 = \sum_{j=1}^{m} \kappa_j^2 + m(m-1)H_2 > \kappa_i^2.$$

In particular, the immersion is two-sided since $H^2 > 0$. Moreover, the eigenvalues $\mu_{j,1}$ of P_1 satisfy $\mu_{j,1} = mH - \kappa_j > 0$ for any j and therefore L_1 is elliptic. Then, the second inequality and the characterization of equality follows from Theorem 6.16. For the first inequality, just observe that $H^2 - H_2 \geq 0$ yields $H_2/H \leq \sqrt{H_2}$. □

Remark 6.7 If the ambient space has constant curvature b, then by (1.142) the scalar curvature ${}^{\Sigma}S$ of Σ is related to H_2 by ${}^{\Sigma}S = m(m-1)(b + H_2)$. In this case inequality (6.103) reads as

$$\sup_{\Sigma} {}^{\Sigma}S \geq b + C_b(r) \inf_{\Sigma} H.$$

Proof (of Theorem 6.15) The existence of an elliptic point implies that H_{k+1} is positive at that point, and hence on Σ. The well-known Gårding inequalities yield, for the appropriate orientation,

$$H_1 \geq H_2^{1/2} \geq \cdots \geq H_k^{1/k} \geq H_{k+1}^{1/(k+1)} > 0. \tag{6.104}$$

Thus, the immersion is two-sided and $H_1 > 0$. Moreover, since Σ has an elliptic point and $H_{k+1} \neq 0$ on Σ, it follows from Remark 6.4 that, for any $1 \leq j \leq k$, the operators L_j are elliptic. Then, the second inequality and the characterization of the equality case follows from Theorem 6.16. For the first inequality observe that $H_{j+1}/H_j \leq \sqrt[j+1]{H_{j+1}}$ follows from (6.104). □

Chapter 7
Hypersurfaces in Warped Products

A classical result of Alexandrov [10] states that a compact hypersurface with constant mean curvature embedded in Euclidean space must be a round sphere. The original proof is based on a clever use of the maximum principle for elliptic partial differential equations. This method, now called the *Alexandrov's reflexion method*, also works for hypersurfaces in ambient spaces having a sufficiently large number of isometric reflexions, for instance in the hyperbolic space. It is worth to observe that, in an analytical context, this is the root of what has been called the "*moving plane*" technique, initiated by the pioneering work of Serrin, [254], and over and over successfully applied to prove special symmetries of solutions to certain PDE's.

To extend results of the type of Alexandrov to a larger class of Riemannian manifolds it appears convenient to consider manifolds with a sufficiently large family of complete embedded constant mean curvature hypersurfaces. Such a family plays the role of the umbilical hypersurfaces in spaces of constant sectional curvature, like the spheres do in Euclidean space. In this setting, given an immersed hypersurface, the main step is to look for geometric assumptions that force the hypersurface to be one of the selected family. In the compact case, this was first done by Montiel [194] that considers as a natural class of ambient manifolds that of warped products $\mathbb{R}\times_\rho\mathbb{P}$, where $\mathbb{P}$ is a complete m-dimensional Riemannian manifold and $\rho : \mathbb{R} \to \mathbb{R}^+$ is a smooth warping function. Then each leaf $\mathbb{P}_t = \{t\} \times \mathbb{P}$ (called here a *slice*) of the foliation $t \in \mathbb{R} \mapsto \mathbb{P}_t$ of $\mathbb{R} \times_\rho \mathbb{P}$ is a complete hypersurface with constant mean curvature. This approach was later considered in [12] where Alías and Dajczer generalized Montiel's results. Some of these generalizations hold even for complete, not necessarily compact, hypersurfaces.

The aim of this chapter is to extend the investigation to hypersurfaces with constant higher order mean curvatures, both in the compact and in the complete case. Typically, in this general setting, the differential operators we shall be dealing with are of trace type.

L.J. Alías et al., *Maximum Principles and Geometric Applications*,
Springer Monographs in Mathematics, DOI 10.1007/978-3-319-24337-5_7

7.1 Preliminaries

In this chapter, we consider the case where the ambient space N is a warped product $I \times_\rho \mathbb{P}$, where $I \subseteq \mathbb{R}$ is an open interval, $\mathbb{P}$ is a complete m-dimensional Riemannian manifold and $\rho : I \to \mathbb{R}^+$ is a smooth function. The product manifold $I \times \mathbb{P}$ is endowed with the Riemannian metric

$$\langle\, ,\, \rangle = \pi_I^*(dt^2) + \rho^2(\pi_I)\pi_{\mathbb{P}}^*(\langle\, ,\, \rangle_{\mathbb{P}}).$$

Here π_I and $\pi_{\mathbb{P}}$ denote the projections onto the corresponding factor and $\langle\, ,\, \rangle_{\mathbb{P}}$ is the Riemannian metric on $\mathbb{P}$. In particular, $I \times_\rho \mathbb{P}$ is complete if and only if $I = \mathbb{R}$. We also recall (see Sect. 1.8) that each leaf $\mathbb{P}_t = \{t\} \times \mathbb{P}$ of the foliation $t \to \mathbb{P}_t$ of $I \times_\rho \mathbb{P}$ is a complete totally umbilical hypersurface with constant k-mean curvature

$$\mathscr{H}_k(t) = \left(\frac{\rho'(t)}{\rho(t)}\right)^k, \qquad 0 \leq k \leq n,$$

with respect to the unit normal $-\mathscr{T} = -\frac{\partial}{\partial t}$.

Let $f : \Sigma \to I \times_\rho \mathbb{P}$ be an isometrically immersed hypersurface. We define the *height function* $h \in C^\infty(\Sigma)$ by setting $h = \pi_I \circ f$. Following the terminology introduced in [11], we will say that the hypersurface is *contained in a slab* if $f(\Sigma)$ lies between two leaves $\mathbb{P}_{t_1}, \mathbb{P}_{t_2}$ of the foliation, with $t_1 < t_2$ (in other words, $h(p) \in [t_1, t_2]$ for each $p \in \Sigma$).

The function theoretic approach to the generalized Omori-Yau maximum principle given in Theorem 3.2 allows us to apply it in different situations, where the choices of the functions γ and G of the quoted theorem are suggested by the geometric setting. The following example, of extrinsic nature, will be useful in the sequel for the case of *properly immersed* hypersurfaces into warped products. Assume the existence of a pole o in $\mathbb{P}$, and denote by $\hat{r} = \rho_{\mathbb{P}}$ the distance function on $\mathbb{P}$ from o. We will assume that the radial sectional curvature of $\mathbb{P}$ satisfies the condition

$$^{\mathbb{P}}K_{rad}(x) \geq -G^2(\hat{r}(x)), \tag{7.1}$$

where, without loss of generality, $G \in C^1(\mathbb{R}_0^+)$ satisfies

$$\text{(i) } G(0) > 0; \quad \text{(ii) } G'(t) \geq 0; \quad \text{(iii) } \frac{1}{G(t)} \notin L^1(+\infty). \tag{7.2}$$

Observe that if Σ is compact then every immersion $f : \Sigma \to I \times_\rho \mathbb{P}$ is proper and contained in a slab, and the Omori-Yau maximum principle trivially holds on Σ for any semi-elliptic operator. Assume then that Σ is noncompact and let $f : \Sigma \to I \times_\rho \mathbb{P}$ be a properly immersed hypersurface with image contained in the slab $[t_1, t_2] \times \mathbb{P}$.

Let $\hat{\gamma} : \mathbb{P} \to \mathbb{R}$ be the function given by $\hat{\gamma}(x) = \hat{r}^2(x)$ for every $x \in \mathbb{P}$, and set $\gamma : \Sigma \to \mathbb{R}$ for the associated function on Σ, defined as

$$\gamma = \tilde{\gamma} \circ f = \hat{r}^2(\pi_{\mathbb{P}} \circ f),$$

where $\tilde{\gamma} : I \times_\rho \mathbb{P} \to \mathbb{R}$ is given by $\tilde{\gamma}(t, x) = \hat{\gamma}(x)$. Since f is proper, if $p \to \infty$ in Σ, then $f(p) \to \infty$ in $N = I \times_\rho \mathbb{P}$, but being f contained in a slab, this means that $(\pi_{\mathbb{P}} \circ f)(p) = x(p) \to \infty$ in $\mathbb{P}$. Hence $\gamma(p) = \hat{r}^2(x(p)) \to +\infty$ as $p \to \infty$ in Σ, and γ satisfies condition (Γ_B) (i) in Theorem 3.2.

On the other hand, regarding the gradient of γ we have the following. Denote by $\tilde{\nabla}$, $\hat{\nabla}$ and ∇ the Levi-Civita connection (and the gradient operators) in $N = I \times_\rho \mathbb{P}$, $\mathbb{P}$ and Σ, respectively. Since $\gamma = \tilde{\gamma} \circ f$, along the immersion f we have

$$\tilde{\nabla}\tilde{\gamma} = \nabla\gamma + \left\langle \tilde{\nabla}\tilde{\gamma}, \nu \right\rangle \nu \tag{7.3}$$

where ν is a (local) smooth unit normal field along f. On the other hand, from $\tilde{\gamma}(t, x) = \hat{\gamma}(x)$ we have

$$\left\langle \tilde{\nabla}\tilde{\gamma}, \mathscr{T} \right\rangle = 0,$$

where, as usual, $\mathscr{T}$ stands for (the lift of) $\frac{\partial}{\partial t}$ to the product $I \times \mathbb{P}$, and

$$\langle \tilde{\nabla}\tilde{\gamma}, V \rangle = \left\langle \hat{\nabla}\hat{\gamma}, V \right\rangle_{\mathbb{P}}$$

for every V, where V denotes the lift of a vector field $V \in \mathfrak{X}(\mathbb{P})$ to $I \times \mathbb{P}$. Since

$$\langle \tilde{\nabla}\tilde{\gamma}, V \rangle = \rho^2 \left\langle \tilde{\nabla}\tilde{\gamma}, V \right\rangle_{\mathbb{P}},$$

we conclude from here that

$$\tilde{\nabla}\tilde{\gamma} = \frac{1}{\rho^2}\hat{\nabla}\hat{\gamma} = \frac{2\hat{r}}{\rho^2}\hat{\nabla}\hat{r}. \tag{7.4}$$

Therefore, since $|\hat{\nabla}\hat{r}| = \rho|\hat{\nabla}\hat{r}|_{\mathbb{P}} = \rho$ and $\rho(h) \geq \min_{[t_1, t_2]} \rho(t) > 0$, along the immersion we have

$$|\nabla\gamma| \leq |\tilde{\nabla}\tilde{\gamma}| = \frac{2\sqrt{\gamma}}{\rho(h)} \leq c\sqrt{\gamma} \tag{7.5}$$

for a constant $c > 0$. In particular Σ is complete (see the beginning of the proof of Theorem 3.2).

Next we will see that, under appropriate extrinsic restrictions, also condition (Γ_B) (ii) in the same theorem is satisfied. From (7.3) it follows that

$$\mathrm{Hess}(\gamma)(X,X) = \mathrm{Hess}\,(\tilde{\gamma})(X,X) + \left\langle \tilde{\nabla}\tilde{\gamma}, \nu \right\rangle \langle AX, X\rangle$$

for every tangent vector field $X \in \mathfrak{X}(\Sigma)$. From (7.4)

$$\tilde{\nabla}_{\mathcal{T}}\tilde{\nabla}\tilde{\gamma} = -\frac{\rho'}{\rho^3}\hat{\nabla}\hat{\gamma} = -\mathcal{H}\tilde{\nabla}\tilde{\gamma}, \tag{7.6}$$

where we recall that $\mathcal{H}(t) = \rho'(t)/\rho(t)$. In particular, $\mathrm{Hess}\,(\tilde{\gamma})(\mathcal{T},\mathcal{T}) = 0$. Then, writing $X = \hat{X} + \langle X, \mathcal{T}\rangle\mathcal{T}$, where $\hat{X} = (\pi_{\mathbb{P}})_*X$, we have

$$\mathrm{Hess}\,(\tilde{\gamma})(X,X) = \mathrm{Hess}\,(\tilde{\gamma})(\hat{X},\hat{X}) + 2\langle X, \mathcal{T}\rangle \mathrm{Hess}\,(\tilde{\gamma})(\hat{X},\mathcal{T}).$$

From (7.6) we have that

$$\mathrm{Hess}\,(\tilde{\gamma})(\hat{X},\mathcal{T}) = -\mathcal{H}(h)\left\langle \tilde{\nabla}\tilde{\gamma}, X\right\rangle = -\mathcal{H}(h)\langle \nabla\gamma, X\rangle.$$

On the other hand, using

$$\tilde{\nabla}_{\hat{X}}\tilde{\nabla}\tilde{\gamma} = \frac{1}{\rho^2}\hat{\nabla}_{\hat{X}}\hat{\nabla}\hat{\gamma} - \frac{\rho'}{\rho^3}\left\langle \hat{\nabla}\hat{\gamma}, \hat{X}\right\rangle \mathcal{T}$$

we also have

$$\mathrm{Hess}\,(\tilde{\gamma})(\hat{X},\hat{X}) = \frac{1}{\rho^2}\left\langle \hat{\nabla}_{\hat{X}}\hat{\nabla}\hat{\gamma}, \hat{X}\right\rangle = \left\langle \hat{\nabla}_{\hat{X}}\hat{\nabla}\hat{\gamma}, \hat{X}\right\rangle_{\mathbb{P}} = \mathrm{Hess}\,(\hat{\gamma})(\hat{X},\hat{X}).$$

Summing up,

$$\mathrm{Hess}(\gamma)(X,X) = \mathrm{Hess}\,(\hat{\gamma})(\hat{X},\hat{X}) - 2\mathcal{H}(h)\langle \nabla\gamma, X\rangle\langle \mathcal{T}, X\rangle + \left\langle \tilde{\nabla}\tilde{\gamma}, \nu\right\rangle\langle AX, X\rangle \tag{7.7}$$

for every tangent vector field $X \in \mathfrak{X}(\Sigma)$.

Observe that, using (7.5),

$$|\mathcal{H}(h)\langle \nabla\gamma, X\rangle\langle \mathcal{T}, X\rangle| \le |\mathcal{H}(h)|\,|\nabla\gamma||X|^2 \le c\sqrt{\gamma}|X|^2.$$

for a constant $c > 0$, since $|\mathcal{H}(h)| \le \max_{[t_1,t_2]}|\mathcal{H}(t)|$. On the other hand, using the general Hessian comparison theorem (Theorem 1.4) one has

$$\mathrm{Hess}\,(\hat{r}) \le \frac{g'(\hat{r})}{g(\hat{r})}\left(\langle,\rangle_{\mathbb{P}} - d\hat{r}\otimes d\hat{r}\right), \tag{7.8}$$

where $g(t)$ is the (positive on $\mathbb{R}^+$) solution of the Cauchy problem

$$\begin{cases} g''(t) - G(t)^2 g(t) = 0, \\ g(0) = 0, \quad g'(0) = 1. \end{cases} \tag{7.9}$$

As we did in Theorem 2.5, let

$$\psi(t) = \frac{1}{G(0)}\left(e^{\int_0^t G(s)ds} - 1\right).$$

Then $\psi(0) = 0$, $\psi'(0) = 1$ and

$$\psi''(t) - G(t)^2 h(t) = \frac{1}{G(0)}\left(G(t)^2 + G'(t)\, e^{\int_0^t G(s)ds}\right) \geq 0, \tag{7.10}$$

that is, ψ is a subsolution of (7.9). Hence, by Sturm comparison theorem

$$\frac{g'(t)}{g(t)} \leq \frac{\psi'(t)}{\psi(t)} \leq CG(t), \tag{7.11}$$

where the last inequality holds for a constant $C > 0$ and t sufficiently large. From (7.8) and for $\hat{r}$ sufficiently large it follows that

$$\mathrm{Hess}\,(\hat{r}) \leq CG(\hat{r})\,(\langle , \rangle_{\mathbb{P}} - d\hat{r} \otimes d\hat{r})\,. \tag{7.12}$$

Since $\mathrm{Hess}\,(\hat{\gamma}) = 2\hat{r}\,\mathrm{Hess}\,(\hat{r}) + 2d\hat{r} \otimes d\hat{r}$, we obtain from here that

$$\begin{aligned} \mathrm{Hess}\,(\hat{\gamma}) &\leq 2C\sqrt{\hat{\gamma}}G(\sqrt{\hat{\gamma}})\langle , \rangle_{\mathbb{P}} + 2\left(1 - C\sqrt{\hat{\gamma}}G((\sqrt{\hat{\gamma}})\right) d\hat{r} \otimes d\hat{r} \\ &\leq c\sqrt{\hat{\gamma}}G(\sqrt{\hat{\gamma}})\langle , \rangle_{\mathbb{P}} \end{aligned}$$

for a positive constant c and $\hat{\gamma}$ sufficiently large. Hence, if γ is sufficiently large we have

$$\mathrm{Hess}\,(\hat{\gamma})(\hat{X}, \hat{X}) \leq c\sqrt{\gamma}G(\sqrt{\gamma})|X|^2 \tag{7.13}$$

for every $X \in \mathfrak{X}(\Sigma)$ and for a certain positive constant c, where we are using the fact that

$$|\hat{X}|_{\mathbb{P}} \leq \frac{1}{\inf_M \rho(h)}|X| \leq \frac{1}{\min_{[t_1,t_2]} \rho(t)}|X|.$$

Therefore, since $\lim_{t\to+\infty} G(t) = +\infty$, from (7.7) we conclude

$$\text{Hess}\,(\gamma)(X,X) \le c\sqrt{\gamma}G(\sqrt{\gamma})|X|^2 + \left\langle \tilde{\nabla}\tilde{\gamma}, \nu\right\rangle\langle AX, X\rangle \tag{7.14}$$

for every tangent vector field $X \in \mathfrak{X}(\Sigma)$, outside a compact subset of Σ.

First assume that $\sup_\Sigma |H| < +\infty$. Tracing (7.14) we obtain

$$\Delta\gamma \le mc\sqrt{\gamma}G(\sqrt{\gamma}) + mH\left\langle \tilde{\nabla}\tilde{\gamma}, \nu\right\rangle$$

outside a compact set. Furthermore, by (7.5)

$$|H\left\langle \tilde{\nabla}\tilde{\gamma}, \nu\right\rangle| \le \sup_\Sigma |H||\tilde{\nabla}\tilde{\gamma}| \le c\sqrt{\gamma} \le c\sqrt{\gamma}G(\sqrt{\gamma})$$

for some constant $c > 0$. Thus, we conclude that, outside a compact subset of Σ,

$$\Delta\gamma \le c\sqrt{\gamma}G(\sqrt{\gamma})$$

for some constant $c > 0$. From the latter and (7.5), since

$$c\sqrt{\gamma} \le \tilde{c}\sqrt{\gamma}G(\sqrt{\gamma})$$

for some constant $\tilde{c} > 0$ and γ sufficiently large, to guarantee the validity of Theorem 3.2 (via Remark 3.3) for the Laplace-Beltrami operator Δ, we only need to verify that the function $\sqrt{t}G(\sqrt{t})$, positive on $\mathbb{R}^+$, satisfies (3.6), that is,

$$\frac{1}{\sqrt{t}G(\sqrt{t})} \notin L^1(+\infty)$$

and

$$(\sqrt{t}G(\sqrt{t}))' \ge -A(\log t + 1), t \gg 1,$$

for some $A \ge 0$. But it is a simple matter to check that both requirements follow from (7.2), the second with $A = 0$.

We now assume instead that $\sup_\Sigma |A|^2 < +\infty$. Using (7.5) again, we have

$$|\left\langle \tilde{\nabla}\tilde{\gamma}, \nu\right\rangle\langle AX, X\rangle| \le |\tilde{\nabla}\tilde{\gamma}||A||X|^2 \le c\sqrt{\gamma}G(\sqrt{\gamma})|X|^2$$

for a positive constant c and for γ sufficiently large. From (7.14) we therefore obtain

$$\text{Hess}\,(\gamma)(X,X) \le c\sqrt{\gamma}G(\sqrt{\gamma})|X|^2, \tag{7.15}$$

for every tangent vector field $X \in \mathfrak{X}(\Sigma)$, outside a compact subset of Σ. Thus, if T is a positive semi-definite $(0, 2)$-tensor field on Σ, we have from (7.15) that

$$L\gamma \leq mc \ \mathrm{Tr}(T) \sqrt{\gamma} G(\sqrt{\gamma}),$$

where $L = \mathrm{Tr}(t \circ \mathrm{hess})$ (recall that $t : \mathfrak{X}(\Sigma) \to \mathfrak{X}(\Sigma)$ denotes the corresponding $(1, 1)$-tensor field metrically equivalent to T). Therefore, in the case where $\sup_\Sigma \mathrm{Tr}(T) < +\infty$, we conclude from here that

$$L\gamma \leq mc \sup_\Sigma \mathrm{Tr}(T) \sqrt{\gamma} G(\sqrt{\gamma})$$

with γ sufficiently large. Again condition (3.6) is fulfilled and by Theorem 3.2 (via Remark 3.3) we know that the Omori-Yau maximum principle holds on Σ for the operator L. Finally, in the case where $\mathrm{Tr}(T) > 0$ on M, we have instead that

$$\frac{1}{\mathrm{Tr}(T)} L\gamma \leq mc \sqrt{\gamma} G(\sqrt{\gamma})$$

with γ sufficiently large. Similarly, we conclude then from Theorem 3.2 (via Remark 3.3) that the q-Omori-Yau maximum principle holds on Σ for the operator L with $q(x) = 1/\mathrm{Tr}(T)(x)$.

We summarize the above discussion in the following:

Theorem 7.1 *Let $\mathbb{P}$ be a complete, noncompact, Riemannian manifold with a pole o and radial sectional curvature satisfying condition ${}^{\mathbb{P}}K_{rad}(x) \geq -G^2(\hat{r}(x))$, where $\hat{r} = \rho_{\mathbb{P}}$ denotes the distance function on $\mathbb{P}$ from o, and $G \in C^1(\mathbb{R}_0^+)$ satisfies*

$$(i)\, G(0) > 0; \quad (ii)\, G'(t) \geq 0; \quad (iii)\, \frac{1}{G(t)} \notin L^1(+\infty).$$

Let $f : \Sigma \to N = I \times_\rho \mathbb{P}$ be a properly immersed hypersurface contained in a slab.

(i) If $\sup_\Sigma |H| < +\infty$, then Σ is complete and the Omori-Yau maximum principle holds on Σ for the Laplace-Beltrami operator.

Let T be a positive semi-definite operator $(0, 2)$-tensor field on Σ.

(ii) If $\sup_\Sigma |A| < +\infty$, then Σ is complete and the Omori-Yau maximum principle holds on Σ for any semi-elliptic operator $L = \mathrm{Tr}(t \circ \mathrm{hess})$ with $\sup_\Sigma \mathrm{Tr}(T) < +\infty$.

(iii) If $\sup_\Sigma |A| < +\infty$, then Σ is complete and the $\frac{1}{\mathrm{Tr}(T)}$-Omori-Yau maximum principle holds on Σ for any semi-elliptic operator $L = \mathrm{Tr}(t \circ \mathrm{hess})$ with $\mathrm{Tr}(T) > 0$ on Σ.

Remark 7.1 From the equality

$$|A|^2 = m^2H_1^2 - m(m-1)H_2$$

it follows that under the assumption $\inf_\Sigma H_2 > -\infty$ the condition $\sup_\Sigma |A|^2 < +\infty$ is equivalent to $\sup_\Sigma |H_1| < +\infty$.

7.2 Curvature Estimates for Hypersurfaces in Warped Products

The aim of this section is to obtain some estimates for the k-mean curvatures of hypersurfaces with image contained in a slab of a warped product space. Towards this aim we shall need the next computational

Proposition 7.1 *Let $f : \Sigma \to I \times_\rho \mathbb{P}$ be an isometrically immersed, oriented hypersurface with unit normal ν into a warped product space with $m = \dim \mathbb{P}$. Let $h = \pi_I \circ f$ be the height function and define*

$$\sigma(t) = \int_{t_0}^{t} \rho(s)\, ds. \tag{7.16}$$

Then, for each $k = 0, \ldots, m$,

$$L_k h = \mathscr{H}(h)(c_k H_k - \langle P_k \nabla h, \nabla h\rangle) + c_k \Theta H_{k+1}, \tag{7.17}$$

and

$$L_k \sigma(h) = c_k \rho(h)(\mathscr{H}(h) H_k + \Theta H_{k+1}), \tag{7.18}$$

where $c_k = (m-k)\binom{m}{k} = (k+1)\binom{m}{k+1}$, $\mathscr{H}(t) = \rho'(t)/\rho(t)$ and $\Theta = \langle \nu, \mathscr{T}\rangle$ is the angle function.

Proof From the calculations in Sect. 1.8 [see (1.206)] we have

$$\mathrm{Hess}(h) = \mathscr{H}(h)(\langle\, ,\, \rangle_\Sigma - dh \otimes dh) + \Theta A, \tag{7.19}$$

where A is the second fundamental tensor of f in the direction of the unit normal ν. In particular, for any $X \in \mathfrak{X}(\Sigma)$,

$$\mathrm{hess}(h)(X) = \mathscr{H}(h)(X - \langle \nabla h, X\rangle \nabla h) + \Theta AX. \tag{7.20}$$

Fix a local orthonormal frame $\{e_1, \ldots, e_m\}$ on Σ. Then, using (7.20) and the expressions of the traces of P_k and $P_k \circ A$, that is, (6.33) and (6.34), we have

$$\begin{aligned} L_k h =& \mathrm{Tr}(P_k \circ \mathrm{hess}(h)) = \sum_i \langle P_k \circ \mathrm{hess}(h)(e_i), e_i \rangle \\ =& \mathcal{H}(h)\big(\mathrm{Tr} P_k - \langle P_k \nabla h, \nabla h \rangle\big) + \Theta \mathrm{Tr}(P_k A) \\ =& \mathcal{H}(h)(c_k H_k - \langle P_k \nabla h, \nabla h \rangle) + c_k \Theta H_{k+1}. \end{aligned}$$

On the other hand, from (1.182)

$$\mathrm{Hess}(\sigma(h)) = \rho'(h) dh \otimes dh + \rho(h)\, \mathrm{Hess}(h),$$

so that, for each vector field X on Σ,

$$\mathrm{hess}(\sigma(h))(X) = \rho'(h) \langle \nabla h, X \rangle \nabla h + \rho(h)\, \mathrm{hess}(h)(X).$$

Therefore,

$$\begin{aligned} L_k \sigma(h) =& \mathrm{Tr}(P_k \circ \mathrm{hess}(\sigma(h))) = \sum_i \langle P_k \circ \mathrm{hess}(\sigma(h))(e_i), e_i \rangle \\ =& \rho'(h) \langle P_k \nabla h, \nabla h \rangle + \rho(h) \mathcal{H}(h)\big(\mathrm{Tr} P_k - \langle P_k \nabla h, \nabla h \rangle\big) + \rho(h) \Theta \mathrm{Tr}(P_k A) \\ =& c_k \rho(h)(\mathcal{H}(h) H_k + \Theta H_{k+1}). \end{aligned}$$

□

As a first application of the above computations, we deduce the following:

Theorem 7.2 *Let $f : \Sigma \to I \times_\rho \mathbb{P}$ be an immersed hypersurface. If the Omori-Yau maximum principle holds for the Laplacian on Σ and $h^* = \sup_\Sigma h < +\infty$, then*

$$\sup_\Sigma |H| \geq \inf_\Sigma \mathcal{H}(h).$$

In particular, and as an application of Theorem 7.1, we deduce the following result, that generalizes Theorem 2 in [11].

Corollary 7.1 *Let $\mathbb{P}$ be a complete, noncompact, Riemannian manifold with a pole $o_\mathbb{P}$ whose radial sectional curvature satisfies condition*

$$^{\mathbb{P}}K_{rad} \geq -G^2(\rho_\mathbb{P}), \tag{7.21}$$

with $G \in C^1(\mathbb{R}_0^+)$ satisfying (7.2) and $\rho_{\mathbb{P}} = \operatorname{dist}(\,,o_{\mathbb{P}})$. If $f : \Sigma \to I \times_\rho \mathbb{P}$ is a properly immersed hypersurface contained in a slab, then

$$\sup_\Sigma |H| \geq \inf_\Sigma \mathscr{H}(h). \tag{7.22}$$

As a consequence, there is no properly immersed hypersurface contained in a slab $[t_1, t_2] \times \mathbb{P}$ with

$$\sup_\Sigma |H| < \inf_{[t_1,t_2]} \mathscr{H}(t).$$

To prove Corollary 7.1, observe that if $\sup_\Sigma |H| = +\infty$ then inequality (7.22) trivially holds. Thus let us assume that $\sup_\Sigma |H| < +\infty$; then by Theorem 7.1, item (i), we know that the OYMP holds for the Laplacian on Σ and the result follows from Theorem 7.2.

Proof (of Theorem 7.2) Since h is bounded from above, we apply the OYMP using Eq. (7.17) for $k = 0$. Thus we find a sequence $\{x_k\} \subset \Sigma$ such that

$$\lim_{j \to +\infty} h(x_k) = h^* = \sup_\Sigma h,$$

$$|\nabla h(x_k)|^2 = 1 - \Theta^2(x_k) < \left(\frac{1}{k}\right)^2,$$

$$\Delta h(x_k) = \mathscr{H}(h(x_k))(m - |\nabla h(x_k)|^2) + mH(x_k)\Theta(x_k) < \frac{1}{k},$$

with $m = \dim \mathbb{P}$, and where in the second of the above we have used (1.208). Thus

$$\frac{1}{k} > \Delta h(x_k) \geq \mathscr{H}(h(x_k))(m - |\nabla h(x_k)|^2) - m \sup_\Sigma |H|.$$

Letting $k \to +\infty$ we get

$$0 \geq \mathscr{H}(h^*) - \sup_\Sigma |H|,$$

so that

$$\sup_\Sigma |H| \geq \mathscr{H}(h^*) \geq \inf_\Sigma \mathscr{H}(h),$$

completing the proof. □

Remark 7.2 If instead of the validity of the OYMP we assume that of the WMP, we easily see that we obtain the conclusion

$$\sup_{\Sigma} |H| \geq \frac{m-1}{m} \mathscr{H}\left(h^*\right) \geq \frac{m-1}{m} \inf_{\Sigma} \mathscr{H}(h),$$

which is weaker than (7.22). Is it possible to obtain the latter in the weaker assumption of the validity of the WMP? See also the next Theorem 7.3.

Corollary 7.2 *Let $\mathbb{P}$ be a complete, noncompact, Riemannian manifold with a pole whose radial sectional curvature satisfies condition* (7.21). *If $f : \Sigma \to I \times_{e^t} \mathbb{P}$ is a parabolic, properly immersed hypersurface with constant mean curvature $|H| \leq 1$ contained in a slab, then $f(\Sigma)$ is slice.*

Remark 7.3 Recall from Sect. 1.8 that $I \times_{e^t} \mathbb{P}$, with $\mathbb{P} = \mathbb{R}^m$, is the standard hyperbolic space foliated by horospheres.

Proof Observe that from (7.22) of Corollary 7.1, since $\mathscr{H}(h) \equiv 1$ we have $|H| = 1$; in particular Σ is orientable. Choose the orientation so that $H = 1$. In this case $\sigma(h) = e^h$ and by (7.18) with $k = 0$ we have

$$\Delta e^h = me^h(1 + \Theta) \geq 0.$$

Therefore, since $e^h \leq e^{h^*}$ it follows that e^h is a subharmonic function on Σ which is bounded from above. The conclusion now follows from parabolicity. □

For the next results, we will require $H_2 > 0$. Because of the basic inequality $H_1^2 \geq H_2$ we can suppose to have chosen a unit normal ν to the hypersurface such that $H_1 > 0$. With this in mind the requirement on the validity of the $\frac{1}{H_1}$-WMP for L_1 in what follows is clear.

Theorem 7.3 *Let $f : \Sigma \to I \times_\rho \mathbb{P}$ be an immersed hypersurface with $H_2 > 0$. If the $\frac{1}{H_1}$-WMP holds for L_1 on Σ and $h^* = \sup_\Sigma h < +\infty$, then*

$$\sup_{\Sigma} H_2^{1/2} \geq \inf_{\Sigma} \mathscr{H}(h).$$

Proof We may assume without loss of generality that $\sup_\Sigma H_2 < +\infty$ and $\inf_\Sigma \mathscr{H}(h) \geq 0$, otherwise the desired conclusion trivially holds. Since h is bounded from above and $\sup_\Sigma \sigma(h) = \sigma(h^*)$, we can find a sequence $\{x_k\} \subset \Sigma$ such that

$$\lim_{k \to +\infty} \sigma(h)(x_k) = \sigma(h^*) = \sup \sigma(h),$$

$$L_1(\sigma(h))(x_k) = < \frac{1}{k}.$$

Observe that the first implies $\lim_{k\to+\infty} h(x_k) = h^*$, because $\sigma(t)$ is strictly increasing, while from the second, using (7.18), we deduce

$$\begin{aligned}\frac{1}{k} >& \frac{1}{H_1}L_1(\sigma(h))(x_k) = m(m-1)\rho(h(x_k))\left(\mathscr{H}(h(x_k)) + \Theta(x_k)\frac{H_2}{H_1}(x_k)\right)\\ \geq& m(m-1)\rho(h(x_k))\left(\mathscr{H}(h(x_k)) - \frac{H_2}{H_1}(x_k)\right)\\ \geq& m(m-1)\rho(h(x_k))\left(\mathscr{H}(h(x_k)) - \sqrt{H_2}(x_k)\right)\end{aligned}$$

where $m = \dim \mathbb{P}$. Letting $k \to +\infty$, and, if necessary, up to passing to a subsequence we get

$$0 \geq \mathscr{H}(h^*) - \sup_{\Sigma} \sqrt{H_2},$$

that is,

$$\sup_{\Sigma} \sqrt{H_2} \geq \mathscr{H}(h^*) \geq \inf_{\Sigma} \mathscr{H}(h).$$

□

As a consequence of the previous theorem and of Theorem 7.1, item (iii), together with Remark 7.1, we have

Corollary 7.3 *Let $\mathbb{P}$ be a complete, noncompact, Riemannian manifold with a pole whose radial sectional curvature satisfies condition* (7.21). *If $f : \Sigma \to I \times_\rho \mathbb{P}$ is a properly immersed hypersurface with $H_2 > 0$, $\sup_\Sigma |H_1| < +\infty$ and contained in a slab, then*

$$\sup_{\Sigma} \sqrt{H_2} \geq \inf_{\Sigma} \mathscr{H}(h).$$

As a consequence, there is no properly immersed hypersurface with $H_2 > 0$ and $\sup_\Sigma |H_1| < +\infty$ contained in a slab $[t_1, t_2] \times \mathbb{P}$ with

$$\sup_{\Sigma} \sqrt{H_2} < \inf_{[t_1,t_2]} \mathscr{H}(t).$$

In the next theorem, the existence of an elliptic point enables us to guarantee both that H_{k-1} is strictly positive and the ellipticity of the operator L_{k-1} (see Remark 6.4). With this observation and reasoning as in the previous results we obtain the analogue of Theorem 7.3 for higher order mean curvatures.

Theorem 7.4 *Let $f : \Sigma \to I \times_\rho \mathbb{P}$ be an immersed hypersurface having an elliptic point, and for which $H_k > 0$ on Σ, with $3 \le k \le m$. If the $\frac{1}{H_{k-1}}$-WMP holds for L_{k-1} on Σ and $h^* = \sup_\Sigma h < +\infty$ then*

$$\sup_\Sigma H_k^{1/k} \ge \inf_\Sigma \mathscr{H}(h).$$

Companion to Corollary 7.3 above we have

Corollary 7.4 *Let $\mathbb{P}$ be a complete, noncompact, Riemannian manifold with a pole whose radial sectional curvature satisfies condition (7.21). Assume that $f : \Sigma \to I \times_\rho \mathbb{P}$ is a properly immersed hypersurface having an elliptic point and for which $H_k > 0$ on Σ, with $3 \le k \le m$, and $\sup_\Sigma H_1 < +\infty$. If $f(\Sigma)$ is contained in a slab, then*

$$\sup_\Sigma H_k^{1/k} \ge \inf_\Sigma \mathscr{H}(h).$$

As a consequence, there is no properly immersed hypersurface having an elliptic point, with $H_k > 0$ on Σ, $\sup_\Sigma H_1 < +\infty$ and contained in a slab $[t_1, t_2] \times \mathbb{P}$ with

$$\sup_\Sigma H_k^{1/k} < \inf_{[t_1,t_2]} \mathscr{H}(t).$$

Remark 7.4 Note that, in the assumption on H_1, differently from what required in Remark 7.1 we get rid of the modulus of H_1 since, using Gårding's inequalities one can prove the validity of

$$H_1 \ge H_2^{1/2} \ge \ldots \ge H_k^{1/k} > 0 \quad \text{on } \Sigma.$$

For more details see [195] or Chap. 6.

7.3 Hypersurfaces with Constant 2-Mean Curvature

In this section we consider some applications to hypersurfaces with positive constant 2-mean curvature H_2. Before stating the main results, let us introduce an auxiliary lemma that will be useful in the sequel.

Lemma 7.1 *Let $f : \Sigma \to I \times_\rho \mathbb{P}$ be a hypersurface with nonvanishing mean curvature and image contained in a slab. Assume that $\mathscr{H}' \ge 0$ and that the angle function Θ does not change sign. Choose on Σ the orientation so that $H_1 > 0$ and suppose that the OYMP for the Laplacian holds on Σ. We have*

$$\text{(i) } \textit{if } \Theta \le 0, \qquad \textit{then } \mathscr{H}(h) \ge 0,$$

$$\text{(ii) } \textit{if } \Theta \ge 0, \qquad \textit{then } \mathscr{H}(h) \le 0.$$

Proof Since h is bounded from below and the OYMP for the Laplacian holds on Σ, using (7.17) with $k = 0$ we can find a sequence $\{x_k\} \subset \Sigma$ such that

$$\lim_{k\to+\infty} h(x_k) = h_* = \inf_\Sigma h,$$
$$|\nabla h(x_k)|^2 = 1 - \Theta^2(x_k) < \left(\frac{1}{k}\right)^2,$$
$$\Delta h(x_k) = \mathcal{H}(h(x_k))(m - |\nabla h(x_k)|^2) + mH_1(x_k)\Theta(x_k) > -\frac{1}{k}.$$

Then

$$-mH_1(x_k)\Theta(x_k) < \frac{1}{k} + \mathcal{H}(h(x_k))(m - |\nabla h(x_k)|^2). \quad (7.23)$$

Similarly, since h is bounded from above, we can also find a second sequence $\{y_k\} \subset \Sigma$ such that

$$\lim_{k\to+\infty} h(y_k) = h^* = \sup h,$$
$$|\nabla h(y_k)|^2 = 1 - \Theta^2(y_k) < \left(\frac{1}{k}\right)^2,$$
$$\Delta h(y_k) = \mathcal{H}(h(y_k))(m - |\nabla h(y_k)|^2) + mH_1(y_k)\Theta(y_k) < \frac{1}{k}.$$

Hence

$$-mH_1(y_k)\Theta(y_k) > -\frac{1}{k} + \mathcal{H}(h(y_k))(m - |\nabla h(y_k)|^2). \quad (7.24)$$

Assume first that $\Theta \leq 0$. Since $\lim_{k\to+\infty} -\Theta(x_k) = -\operatorname{sgn}\Theta = 1 > 0$, we have $-\Theta(x_k) > 0$ for k sufficiently large. Furthermore, since $H_1(x_k) > 0$, using (7.23) it follows that

$$0 \leq \liminf_{k\to+\infty}\Big(-H_1(x_k)\Theta(x_k)\Big) \leq \mathcal{H}(h_*).$$

Therefore $\mathcal{H}(h_*) \geq 0$ and,since $\mathcal{H}$ is nondecreasing, we conclude that

$$\mathcal{H}(h) \geq \mathcal{H}(h_*) \geq 0.$$

Assume now that $\Theta \geq 0$ then $\lim_{k\to+\infty} \Theta(y_k) = \operatorname{sgn}\Theta = 1 > 0$, so that $\Theta(y_k) > 0$ for k sufficiently large. Therefore, since $H_1(y_k) > 0$, from (7.24) we deduce

$$0 \leq \liminf_{k\to+\infty}\Big(H_1(y_k)\Theta(y_k)\Big) \leq -\mathcal{H}(h^*).$$

Therefore $\mathcal{H}(h^*) \leq 0$ and again by $\mathcal{H}' \geq 0$, we deduce that

$$\mathcal{H}(h) \leq \mathcal{H}(h^*) \leq 0.$$

This completes the proof. □

In the rest of this section we will work basically with the operator L_1. We will assume that H_2 is a positive constant; recall that this implies that the immersion is two-sided. We can choose the normal unit vector ν on Σ such that $H_1 > 0$ and the operator L_1 is elliptic (see Remark 6.4).

Let $\sigma(t)$ be as in (7.16), that is, $\sigma(t) = \int_{t_0}^{t} \rho(s)\,ds$. By Proposition 7.1, and using the notation there, we know that

$$\begin{cases} \Delta\sigma(h) & = m\rho(h)(\mathcal{H}(h) + \Theta H_1), \\ L_1\sigma(h) & = m(m-1)\rho(h)(\mathcal{H}(h)H_1 + \Theta H_2). \end{cases} \tag{7.25}$$

Therefore,

$$\mathcal{L}_1\sigma(h) = m(m-1)\rho(h)(\mathcal{H}(h)^2 - \Theta^2 H_2), \tag{7.26}$$

where $\mathcal{L}_1$ is the operator given by

$$\mathcal{L}_1 = (m-1)\mathcal{H}(h)\Delta - \Theta L_1 = \mathrm{Tr}(\mathcal{P}_1 \circ \mathrm{hess}), \tag{7.27}$$

with

$$\mathcal{P}_1 = (m-1)\mathcal{H}(h)I - \Theta P_1.$$

Let us now state the first main result of this section, which extends Theorem 2.4 in [12] to the case of constant 2-mean curvature.

Theorem 7.5 *Let $f : \Sigma \to I \times_\rho \mathbb{P}$ be a compact hypersurface of constant positive 2-mean curvature H_2. If $\mathcal{H}'(t) \geq 0$ and the angle function Θ has constant sign, then $\mathbb{P}$ is necessarily compact and $f(\Sigma)$ is a slice.*

Proof As indicated above, Σ is orientable and we choose the orientation of Σ so that $H_1 > 0$. Since Σ is compact, we may apply Lemma 7.1. Let us consider first the case where $\Theta \leq 0$, for which $\mathcal{H}(h) \geq 0$. Thus, the operator $\mathcal{P}_1$ is positive semi-definite or, equivalently, $\mathcal{L}_1$ is semi-elliptic.

Since Σ is compact, there exist points $p^*, p_* \in \Sigma$ such that

$$h(p^*) = h^* = \max_{\Sigma} h \quad \text{and} \quad h(p_*) = h_* = \min_{\Sigma} h.$$

Therefore, $|\nabla h(p^*)| = |\nabla h(p_*)| = 0$, from which we deduce

$$\Theta(p^*) = \Theta(p_*) = -1$$

because of (1.208). Observe that

$$(\sigma(h))^* = \max_{\Sigma}(\sigma(h)) = \sigma(h^*) = \sigma(h(p^*))$$

and

$$(\sigma(h))_* = \min_{\Sigma}(\sigma(h)) = \sigma(h_*) = \sigma(h(p_*)),$$

because $\sigma(t)$ is strictly increasing. Taking into account that $\mathscr{P}_1$ is positive semi-definite, (7.26) gives

$$\mathscr{L}_1\sigma(h)(p^*) = m(m-1)\rho(h^*)(\mathscr{H}(h^*)^2 - H_2) \leq 0$$

and

$$\mathscr{L}_1\sigma(h)(p_*) = m(m-1)\rho(h_*)(\mathscr{H}(h_*)^2 - H_2) \geq 0.$$

Then, using $\mathscr{H}(h) \geq 0$ on Σ, we obtain

$$\mathscr{H}(h_*) \geq H_2^{1/2} \geq \mathscr{H}(h^*).$$

On the other hand, since $\mathscr{H}$ is nondecreasing, we also have $\mathscr{H}(h_*) \leq \mathscr{H}(h^*)$. Thus the validity of the equality $\mathscr{H}(h_*) = \mathscr{H}(h^*)$ and $\mathscr{H}(h) = H_2^{1/2}$ is constant on Σ. Note that, in particular, $\mathscr{L}_1$ is elliptic. By (7.25), using the basic inequality $H_1 \geq H_2^{1/2}$ and the fact that $\Theta \geq -1$, we obtain

$$\begin{aligned} L_1\sigma(h) &= m(m-1)\rho(h)H_2^{1/2}(H_1 + \Theta H_2^{1/2}) \\ &\geq m(m-1)\rho(h)H_2^{1/2}(H_1 - H_2^{1/2}) \geq 0. \end{aligned}$$

Hence, $L_1\sigma(h) \geq 0$ on the compact manifold Σ. Thus, since in this case L_1 is elliptic, by the maximum principle applied to L_1 we conclude that $\sigma(h)$, and hence h, is constant.

Finally, in the case where $\Theta \geq 0$ we know from Lemma 7.1 that $\mathscr{H}(h) \leq 0$ on Σ, so that the operator $-\mathscr{L}_1$ is semi-elliptic. The proof then follows as in the case $\Theta \leq 0$, working with $-\mathscr{L}_1$ instead of $\mathscr{L}_1$. □

In our next result, we consider complete (and noncompact) hypersurfaces, extending Theorem 2.9 in [12] to the case of constant 2-mean curvature hypersurfaces.

Theorem 7.6 *Let $f : \Sigma \to I \times_\rho \mathbb{P}$ be a complete hypersurface of constant positive 2-mean curvature H_2 such that condition* (2.26) *is satisfied, that is,*

$$^{\Sigma}K \geq -G^2(r), \tag{7.28}$$

where without loss of generality we can suppose that G is a smooth function on $\mathbb{R}_0^+$ even at the origin and satisfying conditions (7.2). Assume that $\sup_\Sigma |H_1| < +\infty$ and that $f(\Sigma)$ is contained in a slab. If $\mathscr{H}'(t) > 0$ and the angle function Θ has constant sign, then $f(\Sigma)$ is a slice.

Proof Choose the orientation of Σ so that $H_1 > 0$. By Theorem 2.5 we know that the OYMP holds for the Laplacian on Σ, so that we may apply Lemma 7.1. As a consequence, in the case where $\Theta \leq 0$ we have $\mathscr{H}(h) \geq 0$, and the operator $\mathscr{P}_1$ is positive semi-definite. In other words, the differential operator $\mathscr{L}_1$ is semi-elliptic. Furthermore, $\mathrm{Tr}(\mathscr{P}_1)$ is bounded above, indeed

$$\mathrm{Tr}(\mathscr{P}_1) = m(m-1)\mathscr{H}(h) - m(m-1)H_1\Theta \leq m(m-1)(\mathscr{H}(h^*) + H_1^*),$$

where $h^* = \sup_\Sigma h < +\infty$ and $H_1^* = \sup_\Sigma H_1 < +\infty$. Hence by Theorem 6.13, item (ii), we know that the OYMP holds for the operator $\mathscr{L}_1$ on Σ.

Since $\sup_\Sigma \sigma(h) = \sigma(h^*) < +\infty$, there exists a sequence $\{x_k\} \subset \Sigma$ such that

$$(i) \quad \lim_{k\to+\infty} \sigma(h(x_k)) = \sup_\Sigma \sigma(h) = \sigma(h^*),$$

$$(ii) \quad |\nabla(\sigma(h))(x_k)| = \rho(h(x_k))|\nabla h(x_k)| < \frac{1}{k},$$

$$(iii) \quad \mathscr{L}_1(\sigma(h))(x_k) < \frac{1}{k}.$$

Observe that condition (i) implies that $\lim_{k\to+\infty} h(x_k) = h^*$, because $\sigma(t)$ is strictly increasing. Thus by condition (ii) we also have $\lim_{k\to+\infty} |\nabla h(x_k)| = 0$. Therefore from (7.26)

$$\mathscr{L}_1\sigma(h)(x_k) = m(m-1)\rho(h(x_k))(\mathscr{H}(h(x_k))^2 - \Theta^2(x_k)H_2) < \frac{1}{k}.$$

Taking the limit for $k \to +\infty$ and observing that $\Theta^2(x_k) = 1 - |\nabla h(x_k)|^2 \to 1$ as $k \to +\infty$, we find

$$\mathscr{H}(h^*)^2 - H_2 \leq 0.$$

On the other hand, since h is also bounded from below, $\inf_\Sigma \sigma(h) = \sigma(h_*) > -\infty$, where $h_* = \inf_\Sigma h > -\infty$. Thus, we can find a sequence $\{y_k\} \subset \Sigma$ such that

$$(i) \quad \lim_{j\to+\infty} \sigma(h(y_k)) = \inf_M \sigma(h) = \sigma(h_*),$$

$$(ii) \quad |\nabla(\sigma(h))(y_k)| = \rho(h(y_k))|\nabla h(y_k)| < \frac{1}{k},$$

$$(iii) \quad \mathscr{L}_1(\sigma(h))(y_k) > -\frac{1}{k}.$$

Hence, proceeding as above and using

$$\mathscr{L}_1\sigma(h)(y_k) = m(m-1)\rho(h(y_k))(\mathscr{H}(h(y_k))^2 - \Theta^2(y_k)H_2) > -\frac{1}{k},$$

we find

$$\mathscr{H}(h_*)^2 - H_2 \geq 0.$$

Thus $\mathscr{H}(h_*)^2 \geq \mathscr{H}(h^*)^2$ and, since $\mathscr{H}(h_*), \mathscr{H}(h^*) \geq 0$, we deduce $\mathscr{H}(h_*) \geq \mathscr{H}(h^*)$. But $\mathscr{H}(t)$ is an increasing function, and it follows $h^* = h_*$.

Finally, let us consider the case where $\Theta \geq 0$. By Lemma 7.1 we then have $\mathscr{H}(h) \leq 0$ and the operator $-\mathscr{L}_1$ is semi-elliptic. Moreover

$$\mathrm{Tr}(-\mathscr{P}_1) = -m(m-1)\mathscr{H}(h) + m(m-1)H_1\Theta \leq m(m-1)(-\mathscr{H}(h_*) + H_1^*).$$

Hence the trace of $-\mathscr{P}_1$ is bounded from above and by Theorem 6.13 the OYMP holds for the operator $-\mathscr{L}_1$. Proceeding as above we arrive at the two inequalities

$$H_2 - \mathscr{H}(h_*)^2 \geq 0 \quad \text{and} \quad H_2 - \mathscr{H}(h^*)^2 \leq 0,$$

giving $\mathscr{H}(h_*)^2 \leq \mathscr{H}(h^*)^2$. Since $\mathscr{H}(h_*)$ and $\mathscr{H}(h^*)$ are nonpositive, this implies $\mathscr{H}(h_*) \geq \mathscr{H}(h^*)$. But $\mathscr{H}(t)$ is increasing, so that $h_* = h^*$ concluding the proof. □

In particular, Theorem 7.6 remains true if we replace condition (7.28) by the stronger condition of Σ having sectional curvature bounded from below by a constant. This happens, for instance, when the sectional curvature of $\mathbb{P}$ is itself bounded from below. This observation yields the next

Corollary 7.5 *Let $\mathbb{P}$ be a complete Riemannian manifold with sectional curvature bounded from below and let $f : \Sigma \to I \times_\rho \mathbb{P}$ be a complete hypersurface of constant positive 2-mean curvature H_2. Assume that $\sup_M |H_1| < +\infty$ and that $f(\Sigma)$ is contained in a slab. If $\mathscr{H}'(t) > 0$ and the angle function Θ does not change sign, then $f(\Sigma)$ is a slice.*

As already observed, for the proof of Corollary 7.5, it suffices to show that ${}^\Sigma K$ is bounded from below by a constant. Towards this aim we prove the following slightly stronger result, that will be useful in the sequel.

Lemma 7.2 *Let $\mathbb{P}$ be a Riemannian manifold with sectional curvature bounded from below and let $f : \Sigma \to I \times_\rho \mathbb{P}$ be an immersed hypersurface. Assume that $\sup_\Sigma |A|^2 < +\infty$, where A is the Weingarten operator of the immersion, and that $f(\Sigma)$ is contained in a slab. Then the sectional curvature of Σ is bounded from below by a constant.*

Proof (of Corollary 7.5) From Remark 7.1, under the assumptions of Corollary 7.5 we immediately obtain

$$\sup_\Sigma |A|^2 < +\infty.$$

Thus, given the validity of Lemma 7.2, ${}^\Sigma K$ is bounded from below by a constant. □

Proof (of Lemma 7.2) Recall that the Gauss equation for a hypersurface $f : \Sigma \to I \times_\rho \mathbb{P}$, according to (1.140), is given by

$$\langle \mathrm{R}(X,Y)Z,V\rangle = \left\langle \overline{\mathrm{R}}(X,Y)Z,V\right\rangle - \langle AX,Z\rangle\langle AY,V\rangle + \langle AY,Z\rangle\langle AX,V\rangle,$$

for $X,Y,Z,V \in \mathfrak{X}(M)$, where R and $\overline{\mathrm{R}}$ are the curvature tensors of Σ and $I \times_\rho \mathbb{P}$, respectively. Then, if $\{X,Y\}$ is an orthonormal basis for an arbitrary 2-plane tangent to Σ, we have

$$\begin{aligned} {}^\Sigma K(X\wedge Y) &= \overline{K}(X\wedge Y) + \langle AX,X\rangle\langle AY,Y\rangle - \langle AX,Y\rangle^2 \\ &\geq \overline{K}(X\wedge Y) - |AX||AY| - |AX|^2 \\ &\geq \overline{K}(X\wedge Y) - 2|A|^2, \end{aligned} \tag{7.29}$$

where the last inequality follows from the fact that

$$|AX|^2 \leq \mathrm{Tr}(A^2)|X|^2 = |A|^2$$

for every unit vector X tangent to Σ. Since we are assuming that $\sup_\Sigma |A|^2 < +\infty$, to obtain the desired conclusion it suffices to show that $\overline{K}(X\wedge Y)$ is bounded from below. The curvature tensor of $I \times_\rho \mathbb{P}$ expressed in terms of the curvature tensor of $\mathbb{P}$ is

$$\begin{aligned} \overline{\mathrm{R}}(U,V)W =& {}^{\mathbb{P}}\mathrm{R}(\hat{U},\hat{V})\hat{W} - \mathscr{H}^2(\pi_I)(\langle V,W\rangle U - \langle U,W\rangle V) \\ &+ \mathscr{H}'(\pi_I)\langle W,\mathscr{T}\rangle(\langle U,\mathscr{T}\rangle V - \langle V,\mathscr{T}\rangle U) \\ &- \mathscr{H}'(\pi_I)(\langle V,W\rangle\langle U,\mathscr{T}\rangle - \langle U,W\rangle\langle V,\mathscr{T}\rangle)\mathscr{T}, \end{aligned}$$

for every $U,V,W \in \mathfrak{X}\big(I \times_\rho \mathbb{P}\big)$, where $\mathscr{T} = \frac{\partial}{\partial_t}$ and we are using the notation $\hat{U}$ to denote $\pi_{\mathbb{P}*}U$ for an arbitrary $U \in \mathfrak{X}\big(I \times_\rho \mathbb{P}\big)$. Then, for the orthonormal basis

$\{X, Y\}$, since $|\mathcal{T}| = 1$ we find that

$$\overline{K}(X \wedge Y) = \frac{1}{\rho^2(h)} {}^{\mathbb{P}}K(\hat{X} \wedge \hat{Y}) \left|\hat{X} \wedge \hat{Y}\right|^2 - \mathcal{H}^2(h) - \mathcal{H}'(h)(\langle X, \mathcal{T}\rangle^2 + \langle Y, \mathcal{T}\rangle^2).$$

But, by (1.207), $\mathcal{T} = \nabla h + \Theta\nu$, where ν is any local unit normal to the hypersurface and $\Theta = \langle \mathcal{T}, \nu\rangle$. Therefore orthogonality of X and ν yields $\langle X, \mathcal{T}\rangle = \langle X, \nabla h\rangle + \langle X, \nu\rangle\Theta = \langle X, \nabla h\rangle$. Similarly $\langle Y, \mathcal{T}\rangle = \langle Y, \nabla h\rangle$, and since $\langle X, \nabla h\rangle^2 + \langle Y, \nabla h\rangle^2 \le |\nabla h|^2 = 1 - \Theta^2 \le 1$, from the above we deduce

$$\overline{K}(X \wedge Y) \ge \frac{1}{\rho^2(h)} {}^{\mathbb{P}}K(\hat{X} \wedge \hat{Y}) \left|\hat{X} \wedge \hat{Y}\right|^2 - \mathcal{H}^2(h) - |\mathcal{H}'(h)|. \tag{7.30}$$

On the other hand,

$$\left|\hat{X} \wedge \hat{Y}\right|^2 = \left|\hat{X}\right|^2 \left|\hat{Y}\right|^2 - \left\langle \hat{X}, \hat{Y}\right\rangle^2 = 1 - \langle X, \mathcal{T}\rangle^2 - \langle Y, \mathcal{T}\rangle^2 \le 1.$$

Therefore, if ${}^{\mathbb{P}}K \ge c$ for some constant c, we deduce

$$\frac{1}{\rho^2(h)} {}^{\mathbb{P}}K(\hat{X} \wedge \hat{Y}) \left|\hat{X} \wedge \hat{Y}\right|^2 \ge -\frac{|c|}{\rho^2(h)}. \tag{7.31}$$

Finally, since $f(\Sigma)$ is contained in a slab, h a bounded function, and we conclude from (7.29)–(7.31) that the sectional curvature ${}^{\Sigma}K(X \wedge Y)$ is bounded from below by an absolute constant. □

We observe that condition (7.28) has been used in the proof of Theorem 7.6 only to guarantee that the Omori-Yau maximum principle holds on Σ for the Laplacian and for the semi-elliptic operator $\mathcal{L}_1$ (or $-\mathcal{L}_1$). Therefore, the theorem remains true under any other hypothesis guaranteeing this latter fact. As a consequence we also state the following:

Theorem 7.7 *Let $\mathbb{P}$ be a complete, noncompact, Riemannian manifold with a pole whose radial sectional curvature satisfies condition* (7.21). *Let $f : \Sigma \to I \times_\rho \mathbb{P}$ be a properly immersed hypersurface of constant positive* 2*-mean curvature H_2. Assume that $\sup_\Sigma |H_1| < +\infty$ and that $f(\Sigma)$ is contained in a slab. If $\mathcal{H}'(t) > 0$ and the angle function Θ has constant sign, then $f(\Sigma)$ is a slice.*

As pointed out before, for the validity of Theorem 7.7 it suffices to show that the OYMP holds for the Laplacian and for the semi-elliptic operator $\mathcal{L}_1$ (or $-\mathcal{L}_1$) on Σ. But this follows directly from Theorem 7.1 and Remark 7.1 in the present assumptions.

7.4 Hypersurfaces with Constant Higher Order Mean Curvature

In this section we will extend our previous results to the case of m-dimensional hypersurfaces with nonzero constant k-mean curvature H_k, where $3 \leq k \leq m$. To this end, we will work with the operator L_{k-1}, and we will assume that there exists an elliptic point in Σ. Note that the existence of an elliptic point is always guaranteed when Σ is compact and $\rho' \neq 0$ on Σ (see the proof of Theorem 7.11 below). Recall from the discussion in Sect. 6.4 (see also Remark 6.4) that the existence of an elliptic point implies that H_k is positive, the immersion is two-sided and $H_1 > 0$ for the chosen orientation. Moreover, it also implies that, for every $1 \leq j \leq k-1$, the operators L_j are elliptic or, equivalently, the operators P_j are positive definite.

To extend our results to higher order mean curvatures, we introduce a family of operators, whose definition is suggested by that of $\mathscr{L}_1$ in (7.27). For $2 \leq k \leq m$, we set

$$\mathscr{L}_{k-1} = \mathrm{Tr}\Big(\Big[\sum_{j=0}^{k-1}(-1)^j \frac{c_{k-1}}{c_j}\mathscr{H}(h)^{k-1-j}\Theta^j P_j\Big] \circ \mathrm{hess}\Big) = \mathrm{Tr}(\mathscr{P}_{k-1} \circ \mathrm{hess}),$$

where

$$\mathscr{P}_{k-1} = \sum_{j=0}^{k-1}(-1)^j \frac{c_{k-1}}{c_j}\mathscr{H}(h)^{k-1-j}\Theta^j P_j. \tag{7.32}$$

We claim that

$$\mathscr{L}_{k-1}\sigma(h) = c_{k-1}\rho(h)(\mathscr{H}(h)^k + (-1)^{k-1}\Theta^k H_k). \tag{7.33}$$

We have already seen in (7.26) that the claim is true for $k = 2$; we now proceed by induction on k. For $k \geq 3$, we observe that

$$\mathscr{P}_{k-1} = \frac{c_{k-1}}{c_{k-2}}\mathscr{H}(h)\mathscr{P}_{k-2} + (-1)^{k-1}\Theta^{k-1}P_{k-1}$$

and consequently

$$\mathscr{L}_{k-1} = \frac{c_{k-1}}{c_{k-2}}\mathscr{H}(h)\mathscr{L}_{k-2} + (-1)^{k-1}\Theta^{k-1}L_{k-1}.$$

Hence, if $k \geq 3$ and we assume the claim true for $\mathscr{L}_{k-2}$, using (7.18) we infer

$$\begin{aligned}\mathscr{L}_{k-1}\sigma(h) &= \frac{c_{k-1}}{c_{k-2}}\mathscr{H}(h)\mathscr{L}_{k-2}\sigma(h) + (-1)^{k-1}\Theta^{k-1}L_{k-1}\sigma(h)\\ &= c_{k-1}\rho(h)(\mathscr{H}(h)^k + (-1)^{k-2}\mathscr{H}(h)\Theta^{k-1}H_{k-1}\\ &\quad +(-1)^{k-1}\mathscr{H}(h)\Theta^{k-1}H_{k-1} + (-1)^{k-1}\Theta^k H_k)\\ &= c_{k-1}\rho(h)(\mathscr{H}(h)^k + (-1)^{k-1}\Theta^k H_k),\end{aligned}$$

proving the claim.

We are now ready to present the following extension of Theorem 7.5.

Theorem 7.8 *Let $f : \Sigma \to I \times_\rho \mathbb{P}$ be a compact m-dimensional hypersurface with constant k-mean curvature H_k, for some $3 \leq k \leq m$ and with an elliptic point. If $\mathscr{H}'(t) \geq 0$ and the angle function Θ has constant sign, then $\mathbb{P}$ is necessarily compact and $f(\Sigma)$ is a slice.*

Proof Choose the orientation of Σ so that $H_1 > 0$. Since Σ is compact, we may apply Lemma 7.1. First let us consider the case where $\Theta \leq 0$, so that $\mathscr{H}(h) \geq 0$ and therefore $(-1)^j\mathscr{H}(h)^{k-1-1}\Theta^j \geq 0$ for each $j = 0, \ldots, k-1$. Since the operators $P_0 = I, P_1, \ldots, P_{k-1}$ are all positive definite, from (7.32) it follows that the operator $\mathscr{P}_{k-1}$ is positive semi-definite or, equivalently, $\mathscr{L}_{k-1}$ is semi-elliptic. Reasoning as in the proof of Theorem 7.5, yields

$$\mathscr{L}_{k-1}\sigma(h)(p^*) = c_{k-1}\rho(h^*)(\mathscr{H}(h^*)^k - H_k) \leq 0$$

and

$$\mathscr{L}_{k-1}\sigma(h)(p_*) = c_{k-1}\rho(h_*)(\mathscr{H}(h_*)^k - H_k) \geq 0,$$

with $p^*, p_* \in \Sigma$ satisfying $h(p^*) = h^* = \max_\Sigma h$ and $h(p_*) = h_* = \min_\Sigma h$.

Then, since $\mathscr{H}(h) \geq 0$ on Σ, we obtain

$$\mathscr{H}(h_*) \geq H_k^{1/k} \geq \mathscr{H}(h^*).$$

On the other hand, $\mathscr{H}' \geq 0$ implies $\mathscr{H}(h_*) \leq \mathscr{H}(h^*)$. Thus, $\mathscr{H}(h_*) = \mathscr{H}(h^*)$ and $\mathscr{H}(h) \equiv H_k^{1/k}$ is constant on Σ. Therefore, by (7.18), using Gårding inequality $H_{k-1} \geq H_k^{(k-1)/k}$ and $\Theta \geq -1$, we obtain

$$\begin{aligned}L_{k-1}\sigma(h) &= c_{k-1}\rho(h)(H_k^{1/k}H_{k-1} + \Theta H_k^{(k-1)/k})\\ &\geq c_{k-1}\rho(h)H_k^{1/k}(H_{k-1} - H_k^{(k-1)/k}) \geq 0.\end{aligned}$$

In other words, $L_{k-1}\sigma(h) \geq 0$ on the compact manifold Σ. By the maximum principle applied to the elliptic operator L_{k-1} we conclude that $\sigma(h)$, and hence h, is constant.

Finally, in case $\Theta \geq 0$ we know from Lemma 7.1 that $\mathscr{H}(h) \leq 0$ on Σ, so that the operator $(-1)^{k-1}\mathscr{L}_{k-1}$ is semi-elliptic. The proof then follows as in the case $\Theta \leq 0$, working with $(-1)^{k-1}\mathscr{L}_{k-1}$ instead of $\mathscr{L}_{k-1}$. □

For the case of complete (noncompact) hypersurfaces, we prove the following extension of Theorem 7.6.

Theorem 7.9 *Let $f : \Sigma \to I \times_\rho \mathbb{P}$ be a complete m-dimensional hypersurface with constant k-mean curvature H_k, for some $3 \leq k \leq m$, and with sectional curvature satisfying condition* (7.28)*. Assume the existence of an elliptic point in Σ, $\sup_\Sigma |H_1| < +\infty$ and that $f(\Sigma)$ is contained in a slab. If $\mathscr{H}'(t) > 0$ and the angle function Θ has constant sign, then $f(\Sigma)$ is a slice.*

Proof By Theorem 6.13, we know that the Omori-Yau maximum principle holds for the Laplacian on Σ, so that we may apply Lemma 7.1. Thus, in the case where $\Theta \leq 0$ we have $\mathscr{H}(h) \geq 0$ and therefore, proceeding as in Theorem 7.8, the differential operator $\mathscr{L}_{k-1}$ is semi-elliptic. Furthermore, since $0 \leq -\Theta \leq 1$, $\mathscr{H}(h) \geq 0$, $H_j \geq 0$ for $j = 0, \ldots, k-1$, we deduce

$$\mathrm{Tr}(\mathscr{P}_{k-1}) = c_{k-1}\sum_{j=0}^{k-1}(-1)^j\mathscr{H}(h)^{k-1-j}\Theta^j H_j \leq c_{k-1}\sum_{j=0}^{k-1}\mathscr{H}(h^*)^{k-1-j}H_j^*,$$

where $h^* = \sup_\Sigma h < +\infty$ and $H_j^* = \sup_\Sigma H_j \leq (\sup_\Sigma H_1)^j < +\infty$, and where the latter inequality is due to Gårding's inequalities. Hence by Theorem 6.13, the Omori-Yau maximum principle holds for the operator $\mathscr{L}_{k-1}$ and, proceeding as in the proof of Theorem 7.6, we find two sequences $\{x_j\} \subset \Sigma$ and $\{y_j\} \subset \Sigma$ satisfying

$$\lim_{j\to+\infty} h(x_j) = h^*, \quad \text{and} \quad \lim_{j\to+\infty} h(y_j) = h_*,$$

$$\lim_{j\to+\infty}\Theta(x_j) = \lim_{j\to+\infty}\Theta(y_j) = -1,$$

$$\mathscr{L}_{k-1}\sigma(h)(x_j) = c_{k-1}\rho(h(x_j))(\mathscr{H}(h(x_j))^k + (-1)^{k-1}\Theta^k(x_j)H_k) < \frac{1}{j},$$

and

$$\mathscr{L}_{k-1}\sigma(h)(y_j) = c_{k-1}\rho(h(y_j))(\mathscr{H}(h(y_j))^k + (-1)^{k-1}\Theta^k(y_j)H_k) > -\frac{1}{j}.$$

Letting $j \to +\infty$ in the above inequalities, we deduce

$$\mathcal{H}(h^*)^k \leq H_k \leq \mathcal{H}(h_*)^k,$$

and therefore $h_* = h^*$, as in the proof of Theorem 7.6.

Finally, in the case where $\Theta \geq 0$ we proceed again as in the proof of Theorem 7.6, working now with the operator $(-1)^{k-1}\mathcal{L}_{k-1}$, that, in this case, is semi-elliptic and with $\mathrm{Tr}((-1)^{k-1}\mathcal{P}_{k-1})$ bounded from above. □

As it happened for Theorem 7.7 in the previous section, Theorem 7.9 remains true if we replace condition (7.28) by the stronger condition of Σ having sectional curvature bounded from below by a constant. This leads to the next

Corollary 7.6 *Let $\mathbb{P}$ be a complete Riemannian manifold with sectional curvature bounded from below and let $f : \Sigma \to I \times_\rho \mathbb{P}$ be a complete hypersurface with constant k-mean curvature H_k, for some $3 \leq k \leq m$. Assume the existence of an elliptic point in Σ, $\sup_\Sigma |H_1| < +\infty$ and that $f(\Sigma)$ is contained in a slab. If $\mathcal{H}'(t) > 0$ and the angle function Θ has constant sign, then $f(\Sigma)$ is a slice.*

Indeed, by Gårding inequalities we know that $H_2 > 0$, so that (see Remark 7.1) $\sup_\Sigma |A|^2 \leq m^2(\sup_\Sigma H_1)^2 < +\infty$ and we can apply Lemma 7.2 to conclude that the sectional curvature of Σ is bounded from below. The result then follows from Theorem 7.9.

Finally, similarly to what happened in the previous section, condition (7.28) has been used in the proof of Theorem 7.9 only to guarantee that the Omori-Yau maximum principle holds on Σ for the Laplacian and for the semi-elliptic operator $\mathcal{L}_{k-1}$ (or $-\mathcal{L}_{k-1}$). Therefore, the theorem remains true under any other hypothesis guaranteeing that property. Then, and as a consequence of Theorem 7.1, we state the following:

Theorem 7.10 *Let $\mathbb{P}$ be a complete, noncompact, Riemannian manifold with a pole whose radial sectional curvature satisfies condition* (7.21)*. Let $f : \Sigma \to I \times_\rho \mathbb{P}$ be a properly immersed m-dimensional hypersurface of constant k-mean curvature, for some $3 \leq k \leq m$. Assume that there exists an elliptic point in Σ, that $\sup_\Sigma |H_1| < +\infty$ and that $f(\Sigma)$ is contained in a slab. If $\mathcal{H}'(t) > 0$ and the angle function Θ has constant sign, then $f(\Sigma)$ is a slice.*

7.4.1 Further Results for Hypersurfaces with Constant Higher Order Mean Curvatures

In this section we go on with the study of hypersurfaces with constant higher order mean curvatures and we describe some further results. We begin by proving next

Theorem 7.11 *Let $f : \Sigma \to I \times_\rho \mathbb{P}$ be a compact m-dimensional hypersurface of constant k-mean curvature, for some $2 \leq k \leq m$, and suppose that $\mathcal{H}$ does not*

vanish. Assume that the sectional curvature of $\mathbb{P}$ *satisfies*

$$^{\mathbb{P}}K \geq \sup_I\{(\rho')^2 - \rho''\rho\}, \tag{7.34}$$

and that the angle function Θ *has constant sign. Then either* $f(\Sigma)$ *is a slice and* $\mathbb{P}$ *is compact or* $I \times_\rho \mathbb{P}$ *has constant sectional curvature and* $f(\Sigma)$ *is a geodesic hypersphere. The latter case cannot occur if inequality* (7.34) *is strict.*

First we proceed with the proof of two important auxiliary results that will be essential in the proof of Theorem 7.11.

Lemma 7.3 *Let* $f : \Sigma \to I \times_\rho \mathbb{P}$ *be a m-dimensional immersed hypersurface and assume that* $\mathbb{P}$ *has constant sectional curvature* κ. *Then*

$$\mathrm{div} P_k = -(m-k)\Theta\Big(\frac{\kappa}{\rho^2(h)} + \mathscr{H}'(h)\Big)P_{k-1}\nabla h, \tag{7.35}$$

where $h = \pi_I \circ f$ *is the height function.*

Proof Let $e_1, \ldots, e_m$ be a local orthonormal frame on Σ and observe that

$$\begin{aligned}\langle \mathrm{div} P_k, X\rangle &= \sum_{i=1}^m \langle (\nabla_{e_i} P_k)X, e_i\rangle \\ &= \sum_{j=0}^{k-1}\sum_{i=1}^m (-1)^{k-1-j}\big\langle {}^N R(e_i, A^{k-1-j}X)\nu, P_j e_i\big\rangle\end{aligned} \tag{7.36}$$

for every vector field $X \in \mathfrak{X}(\Sigma)$. From (1.191), using the notation $\hat{U}$ for $\pi_{\mathbb{P}*}U$ and setting $N = I \times_\rho \mathbb{P}$, we know that

$$^N R(X,Y)\nu = {}^{\mathbb{P}}R(\hat{X},\hat{Y})\hat{\nu} + \mathscr{H}'(h)\Theta(\langle X, \nabla h\rangle Y - \langle Y, \nabla h\rangle X) \tag{7.37}$$

for every tangent vector fields $X, Y \in \mathfrak{X}(\Sigma)$, where $^{\mathbb{P}}R$ stands for the curvature tensor of the fiber $\mathbb{P}$. Then,

$$\begin{aligned}\big\langle {}^N R(e_i, A^{k-1-j}X)\nu, P_j e_i\big\rangle &= \Big\langle {}^{\mathbb{P}}R\big(\hat{e}_i, \big(\widehat{A^{k-1-j}X}\big)\big)\hat{\nu}, P_j e_i\Big\rangle \\ &\quad + \mathscr{H}'(h)\Theta\langle \nabla h, e_i\rangle\langle P_j \circ A^{k-1-j}X, e_i\rangle \\ &\quad - \mathscr{H}'(h)\Theta\langle A^{k-1-j}(\nabla h), X\rangle\langle P_j e_i, e_i\rangle.\end{aligned}$$

Therefore, for every fixed $j \in \{0, \ldots, k-1\}$ we get

$$\sum_{i=1}^{m} \langle {}^{N}R(e_i, A^{k-1-j}X)\nu, P_j e_i \rangle = \sum_{i=1}^{m} \left\langle {}^{\mathbb{P}}R(\hat{e}_i, \left(\widehat{A^{k-1-j}X}\right))\hat{\nu}, P_j e_i \right\rangle$$
$$+\mathscr{H}'(h)\Theta\langle \mathscr{S}_{k,j}(\nabla h), X \rangle, \tag{7.38}$$

where

$$\mathscr{S}_{k,j} = P_j \circ A^{k-1-j} - c_j H_j A^{k-1-j}.$$

Thus, from (7.36) we obtain

$$\langle \operatorname{div} P_k, X \rangle = \sum_{j=0}^{k-1} (-1)^{k-1-j} \sum_{i=1}^{m} \left\langle {}^{\mathbb{P}}R(\hat{e}_i, (\widehat{A^{k-1-j}X}))\hat{\nu}, P_j e_i \right\rangle$$
$$+\mathscr{H}'(h)\Theta \sum_{j=0}^{k-1} (-1)^{k-1-j} \langle \mathscr{S}_{k,j}(\nabla h), X \rangle. \tag{7.39}$$

We claim that, for every $k = 1, \ldots, m-1$,

$$\mathscr{A}_k = \sum_{j=0}^{k-1} (-1)^{k-1-j} \mathscr{S}_{k,j} = -(m-k)P_{k-1}. \tag{7.40}$$

In fact, we will prove the claim by induction on k, $k = 1, \ldots, m-1$. The case $k = 1$ is trivial, since $P_0 = I$ and

$$\mathscr{A}_1 = \mathscr{S}_{1,0} = I - mI = -(m-1)I.$$

Assume we have proved (7.40) for $k-1$. Then, since

$$P_j = S_j I - A \circ P_{j-1}$$

we get

$$\begin{aligned} \mathscr{A}_k &= P_{k-1} - c_{k-1}H_{k-1}I - \mathscr{A}_{k-1} \circ A \\ &= P_{k-1} - c_{k-1}H_{k-1}I + (m-k+1)P_{k-2} \circ A \\ &= -(m-k)P_{k-1}. \end{aligned}$$

Therefore, using (7.40), the expression in (7.39) reduces to

$$\langle \mathrm{div} P_k, X\rangle = \sum_{j=0}^{k-1}(-1)^{k-1-j}\sum_{i=1}^{m}\left\langle {}^{\mathbb{P}}R(\hat{e}_i, \widehat{(A^{k-1-j}X)})\hat{\nu}, P_j e_i\right\rangle$$
$$-(m-k)\mathscr{H}'(h)\Theta\langle P_{k-1}(\nabla h), X\rangle. \tag{7.41}$$

On the other hand, since we are assuming that the fiber $\mathbb{P}$ has constant sectional curvature κ, we have that

$${}^{\mathbb{P}}R(\hat{e}_i, \widehat{\left(A^{k-1-j}X\right)})\hat{\nu} = \kappa\langle \widehat{\left(A^{k-1-j}X\right)}, \hat{\nu}\rangle_{\mathbb{P}}\hat{e}_i$$
$$-\kappa\langle \hat{e}_i, \hat{\nu}\rangle_{\mathbb{P}}\widehat{(A^{k-1-j}X)}.$$

Hence a direct computation shows that

$$\sum_{i=1}^{m}\left\langle {}^{\mathbb{P}}R(\hat{e}_i, \widehat{(A^{k-1-j}X)})\hat{\nu}, P_j e_i\right\rangle = \frac{\kappa}{\rho^2(h)}\Theta\langle \mathscr{S}_{k,j}(\nabla h), X\rangle.$$

Thus, using again (7.40), we obtain

$$\sum_{j=0}^{k-1}(-1)^{k-1-j}\sum_{i=1}^{m}\left\langle {}^{\mathbb{P}}R(\hat{e}_i, \widehat{(A^{k-1-j}X)})\hat{\nu}, P_j e_i\right\rangle = -(m-k)\frac{\kappa}{\rho^2(h)}\Theta\langle P_{k-1}\nabla h, X\rangle.$$

Finally, using the latter in Eq. (7.41) we conclude that

$$\langle \mathrm{div} P_k, X\rangle = -(m-k)\left(\frac{\kappa}{\rho^2(h)} + \mathscr{H}'(h)\right)\Theta\langle P_{k-1}\nabla h, X\rangle \tag{7.42}$$

for every $X \in \mathfrak{X}(\Sigma)$, proving (7.35). □

Lemma 7.4 *Let $f : \Sigma \to I \times_\rho \mathbb{P}$ be an immersed hypersurface of dimension m with angle function Θ, height function h and local unit normal ν. Let $\hat{\Theta} = \rho(h)\Theta$. Then, for every $0 \le k \le m-1$ we have*

$$L_k\hat{\Theta} = -\binom{m}{k+1}\rho(h)\langle \nabla h, \nabla H_{k+1}\rangle - \rho'(h)c_k H_{k+1}$$
$$-\hat{\Theta}\mathscr{H}'(h)(|\nabla h|^2 c_k H_k - \langle P_k\nabla h, \nabla h\rangle) - \frac{\hat{\Theta}}{\rho(h)^2}\beta_k$$
$$-\hat{\Theta}\binom{m}{k+1}(mH_1H_{k+1} - (m-k-1)H_{k+2}),$$

where

$$\beta_k = \sum_{i=1}^{m} \mu_{i,k}{}^{\mathbb{P}}K(\hat{e}_i, \hat{\nu})|\hat{e}_i \wedge \hat{\nu}|^2.$$

Here the $\mu_{i,k}$'s are the eigenvalues of P_k and $\{e_1, \dots, e_m\}$ is a local orthonormal frame on Σ diagonalizing A, the Weingarten operator in the direction of ν.

Proof Since $\rho(t)\mathcal{T}$ is a conformal vector field,

$$\nabla\hat{\Theta} = -\rho(h)A\nabla h.$$

Therefore, using Eq. (7.20) we find

$$\nabla_X\nabla\hat{\Theta} = -\rho(h)(\nabla_X A)\nabla h - \rho'(h)AX - \hat{\Theta}A^2X.$$

Hence,

$$\begin{aligned} L_k\hat{\Theta} =& - \rho(h)\sum_{i=1}^{m} \langle P_k(\nabla_{e_i}A)\nabla h, e_i\rangle \\ & - \rho'(h)c_kH_{k+1} - \binom{m}{k+1}\hat{\Theta}(H_1H_{k+1} - (m-k-1)H_{k+2}). \end{aligned}$$

Using the expression of the covariant derivative of a tensor field we get

$$\begin{aligned} -P_k(\nabla_{e_i}A)\nabla h =& (\nabla_{e_i}P_k)A\nabla h - (\nabla_{e_i}P_k \circ A)\nabla h \\ =& (\nabla_{e_i}P_k)A\nabla h + (\nabla_{e_i}P_{k+1})\nabla h - e_i(S_{k+1})\nabla h. \end{aligned}$$

By Eq. (6.36) it follows that, setting $N = I \times_\rho \mathbb{P}$,

$$\begin{aligned} -\sum_{i=1}^{m} \langle P_k(\nabla_{e_i}A)\nabla h, e_i\rangle =& \sum_{i=1}^{m} \langle (\nabla_{e_i}P_k)A\nabla h, e_i\rangle \\ & + \sum_{i=1}^{m} \langle (\nabla_{e_i}P_{k+1})\nabla h, e_i\rangle - \nabla h(S_{k+1}) \\ =& \sum_{i=1}^{m} \left\langle {}^N R(e_i, \nabla h)\nu, P_k e_i\right\rangle - \nabla h(S_{k+1}). \end{aligned}$$

Since $\nabla h = \mathcal{T} - \Theta\nu$, we can write

$${}^N R(e_i, \nabla h)\nu = {}^N R(e_i, \mathcal{T})\nu - \Theta {}^N R(e_i, \nu)\nu.$$

Using Gauss equations and observing that $\hat{\mathscr{T}} = 0$ we get

$$^{N}R(e_i, \mathscr{T})\nu = -(\mathscr{H}(h)^2 + \mathscr{H}'(h))\Theta e_i = -\frac{\rho''(h)}{\rho(h)}\Theta e_i.$$

Therefore,

$$\sum_{i=1}^{m} \langle {}^{N}R(e_i, \mathscr{T})\nu, P_k e_i\rangle = -\frac{\rho''(h)}{\rho(h)}\Theta c_k H_k.$$

Again by Gauss equations

$$\begin{aligned} {}^{N}R(e_i, \nu)\nu =& {}^{\mathbb{P}}\mathrm{R}(\hat{e}_i, \hat{\nu})\hat{\nu} - \mathscr{H}(h)^2 e_i \\ &+ \mathscr{H}'(h)\Theta(\langle e_i, \nabla h\rangle \nu - \Theta e_i) - \mathscr{H}'(h)\langle e_i, \nabla h\rangle \mathscr{T}. \end{aligned}$$

Assume that the orthonormal basis $\{e_i\}_1^m$ diagonalizes A and hence P_k, that is, $P_k e_i = \mu_{i,k} e_i$ (no sum over i). Then

$$\begin{aligned} \sum_{i=1}^{m} \langle {}^{N}R(e_i, \nu)\nu, P_k e_i\rangle =& \frac{1}{\rho(h)^2} \sum_{i=1}^{m} \mu_{i,k} {}^{\mathbb{P}}K(\hat{e}_i, \hat{\nu})|\hat{e}_i \wedge \hat{\nu}|^2 \\ &- \frac{\rho''(h)}{\rho(h)} c_k H_k + \mathscr{H}'(h)(|\nabla h|^2 c_k H_k - \langle P_k \nabla h, \nabla h\rangle). \end{aligned}$$

Thus,

$$\begin{aligned} \sum_{i=1}^{m} \langle {}^{N}R(e_i, \nabla h)\nu, P_k e_i\rangle =& \sum_{i=1}^{m} \langle {}^{N}R(e_i, T)\nu, P_k e_i\rangle - \Theta \sum_{i=1}^{m} \langle {}^{N}R(e_i, \nu)\nu, P_k e_i\rangle \\ =& -\frac{\Theta}{\rho(h)^2} \sum_{i=1}^{m} \mu_{i,k} {}^{\mathbb{P}}K(\hat{e}_i, \hat{\nu})|\hat{e}_i \wedge \hat{\nu}|^2 \\ & -\Theta \mathscr{H}'(h)(|\nabla h|^2 c_k H_k - \langle P_k \nabla h, \nabla h\rangle) \end{aligned}$$

concluding the proof of the lemma. □

As an immediate consequence we deduce

Corollary 7.7 *Let $f : \Sigma \to I \times_\rho \mathbb{P}$ be an immersed hypersurface of dimension m, with angle function Θ and height function h. Assume that $\mathbb{P}$ has constant sectional* *curvature* κ *and let* $\hat{\Theta} = \rho(h)\Theta$. *Then, for every* $0 \leq k \leq m-1$ *we have*

$$L_k\hat{\Theta} = -\binom{m}{k+1}\rho(h)\langle\nabla h, \nabla H_{k+1}\rangle - \rho'(h)c_kH_{k+1}$$

$$-\hat{\Theta}\Big(\frac{\kappa}{\rho^2(h)} + \mathscr{H}'(h)\Big)(|\nabla h|^2 c_k H_k - \langle P_k\nabla h, \nabla h\rangle)$$

$$-\hat{\Theta}\binom{m}{k+1}(mH_1H_{k+1} - (m-k-1)H_{k+2}).$$

We are now ready to give the

Proof (of Theorem 7.11) We may assume without loss of generality that $\mathscr{H}(h) > 0$ on Σ. Since Σ is compact, there exists a point $p_0 \in \Sigma$ where the height function attains its maximum. Then $\nabla h(p_0) = 0$, $\Theta(p_0) = \pm 1$ and by (7.20)

$$\mathrm{Hess}(h)(p_0)(v,v) = \mathscr{H}(h^*)\langle v, v\rangle + \Theta(p_0)\langle Av, v\rangle(p_0) \leq 0.$$

If $\Theta(p_0) = -1$, then

$$\langle Av, v\rangle(p_0) \geq \mathscr{H}(h^*)\langle v, v\rangle > 0,$$

for any $v \neq 0$. Thus p_0 is an elliptic point, H_k is a positive constant and by Gårding inequalities

$$H_1 \geq H_2^{\frac{1}{2}} \geq \cdots \geq H_k^{\frac{1}{k}} > 0,$$

equality holding only at umbilical points. In particular, Σ is two-sided and then $\Theta \leq 0$. If $\Theta(p_0) = 1$, changing the orientation we reach the same conclusion.

Consider the function

$$\phi = \sigma(h)H_k^{\frac{1}{k}} + \rho(h)\Theta.$$

Let us prove that $L_{k-1}\phi \geq 0$. Since H_k is constant, using Lemma 7.4 and Proposition 7.1 and setting $\hat{U}$ for $\pi_{\mathbb{P}*}U$, we have

$$\begin{aligned}L_{k-1}\phi =& H_k^{\frac{1}{k}}L_{k-1}\sigma(h) + L_{k-1}\hat{\Theta}\\ =& c_{k-1}H_k^{\frac{1}{k}}(\rho'(h)H_{k-1} + \hat{\Theta}H_k) - c_{k-1}H_{k-1}\hat{\Theta}\mathscr{H}'(h)|\nabla h|^2\\ &+ \hat{\Theta}\mathscr{H}'(h)\langle P_{k-1}\nabla h, \nabla h\rangle - \hat{\Theta}\binom{m}{k}(mH_1H_k - (m-k)H_{k+1})\end{aligned}$$

$$
\begin{aligned}
&- \rho'(h)c_{k-1}H_k - \frac{\hat{\Theta}}{\rho(h)^2}\sum_{i=1}^{m}\mu_{i,k-1}{}^{\mathbb{P}}K(\hat{e}_i \wedge \hat{\nu})|\hat{e}_i \wedge \hat{\nu}|^2 \\
=& U + V + Z,
\end{aligned}
$$

where $\{e_i\}$ is a local orthonormal frame on Σ diagonalizing the Weingarten operator A in the direction of the unit normal ν, the $\mu_{i,k}$'s are the corresponding eigenvalues of P_{k-1},

$$
U = -\hat{\Theta}\binom{n}{k}(mH_1H_k - (m-k)H_{k+1} - kH_k^{\frac{k+1}{k}}),
$$

$$
V = c_{k-1}\rho'(h)(H_{k-1}H_k^{\frac{1}{k}} - H_k)
$$

and

$$
\begin{aligned}
Z =& -\hat{\Theta}\mathcal{H}'(h)(|\nabla h|^2 c_{k-1}H_{k-1} - \langle P_{k-1}\nabla h, \nabla h\rangle) \\
& - \frac{\hat{\Theta}}{\rho(h)^2}\sum_{i=1}^{m}\mu_{i,k-1}{}^{\mathbb{P}}K(\hat{e}_i \wedge \hat{\nu})|\hat{e}_i \wedge \hat{\nu}|^2.
\end{aligned}
$$

Then by Gårding inequalities,

$$
H_{k-1}H_k^{\frac{1}{k}} - H_k = H_k^{\frac{1}{k}}(H_{k-1} - H_k^{\frac{k-1}{k}}) \geq 0.
$$

Moreover,

$$
mH_1H_k - kH_k^{\frac{k+1}{k}} \geq mH_k^{\frac{k+1}{k}} - kH_k^{\frac{k+1}{k}} = (m-k)H_k^{\frac{k+1}{k}},
$$

and thus

$$
mH_1H_k - kH_k^{\frac{k+1}{k}} - (m-k)H_{k+1} \geq (m-k)(H_k^{\frac{k+1}{k}} - H_{k+1}) \geq 0. \tag{7.43}
$$

Finally, let $\alpha = \sup_I\{(\rho')^2 - \rho''\rho\}$. Since

$$
|\hat{e}_i \wedge \hat{\nu}|^2 = |\nabla h|^2 - \langle e_i, \nabla h\rangle^2,
$$

taking into account that the $\mu_{i,k-1}$'s are positive, we have

$$
\sum_{i=1}^{m}\mu_{i,k-1}{}^{\mathbb{P}}K(\hat{e}_i, \hat{\nu})|\hat{e}_i \wedge \hat{\nu}|^2
$$

$$\geq \alpha \sum_{i=1}^{m} \mu_{i,k-1} |\hat{e}_i \wedge \hat{v}|^2$$
$$= \alpha (c_{k-1} H_{k-1} |\nabla h|^2 - \langle P_{k-1} \nabla h, \nabla h \rangle).$$

Hence,

$$\frac{1}{\rho(h)^2} \sum_{i=1}^{m} \mu_{i,k-1} {}^{\mathbb{P}}K(\hat{e}_i, \hat{v}) |\hat{e}_i \wedge \hat{v}|^2$$
$$+ \mathscr{H}'(h)(|\nabla h|^2 c_{k-1} H_{k-1} - \langle P_{k-1} \nabla h, \nabla h \rangle)$$
$$\geq \Big(\frac{\alpha}{\rho(h)^2} + \mathscr{H}'(h) \Big) (|\nabla h|^2 c_{k-1} H_{k-1} - \langle P_{k-1} \nabla h, \nabla h \rangle) \geq 0,$$

where the last inequality follows from $\alpha = \sup_I \{-\rho^2 \mathscr{H}'\}$ and the positivity of the operator P_{k-1}. Thus, $L_{k-1}\phi \geq 0$ and since L_{k-1} is an elliptic operator and Σ is compact, we conclude, by the maximum principle that ϕ must be constant. Hence $L_{k-1}\phi = 0$ and the three terms U, V and Z in $L_{k-1}\phi$ vanish on Σ.

In particular $V = 0$ implies that Σ is a totally umbilical hypersurface. Moreover, since H_k is a positive constant and Σ is totally umbilical, all the higher order mean curvatures are constant. Thus H_1 is constant and the conclusion follows by Theorem 3.4 in [12] (see also Proposition 7.2 below). □

Note that, for the complete case, we can give the following version of Theorem 3.4 in [12].

Proposition 7.2 *Let $f : \Sigma \to I \times_\rho \mathbb{P}$ be a complete, parabolic, two-sided hypersurface with constant mean curvature. Assume that the Ricci curvature of $\mathbb{P}$ satisfies*

$$\mathrm{Ric}_{\mathbb{P}} > \sup_I \Big\{ (\rho')^2 - \rho'' \rho \Big\}. \tag{7.44}$$

Suppose that $f(\Sigma)$ is contained in a slab. If the angle function Θ has constant sign, then $f(\Sigma)$ is a slice.

Proof We consider the function ϕ defined in the proof of Theorem 7.11 for $k = 1$, that is, $\phi = H\sigma(h) + \hat{\Theta}$ and proceeding as we did there we compute

$$\Delta\phi = -\rho(h)\Theta \Big\{ |A|^2 - mH_1^2 + (m-1)\Big(\mathrm{Ric}_{\mathbb{P}}(\hat{v}, \hat{v}) + \mathscr{H}'(h)|\nabla h|^2 \Big) \Big\}.$$

Therefore, since $|A|^2 \geq mH_1^2$ and

$$|\nabla h|^2 = \rho(h)^2 |\hat{v}|^2,$$

from assumption (7.44) and the fact that Θ has constant sign the right-hand side of the above equation has constant sign too. But $f(\Sigma)$ is contained in a slab, so that ϕ is bounded. Parabolicity of Σ implies that ϕ is constant; it follows that

$$\Theta\left\{|A|^2 - mH_1^2 + (m-1)\Big(\mathrm{Ric}_{\mathbb{P}}(\hat{\nu}, \hat{\nu}) + \mathscr{H}'(h)|\nabla h|^2\Big)\right\} = 0 \quad \text{on } \Sigma.$$

We claim that

$$\mathscr{K} = \{x \in \Sigma : \Theta(x) = 0\}$$

has empty interior. Assume the contrary and let $V \subset \mathscr{K}$ be open. Then, on V, $\phi = H_1\sigma(h)$ is constant, thus if $H_1 \neq 0$, h is constant. But this is not possible since $|\nabla h|^2 = 1 - \Theta^2 = 1$ on V. Hence $H_1 = 0$ on Σ and $\phi = \rho(h)\Theta$ is constant on Σ. Since ϕ vanishes on V, $\phi \equiv 0$ on Σ. It follows that $\mathscr{K} = \Sigma$, and then $|\nabla h| \equiv 1$ on Σ, but this is not possible. Indeed, since Σ is complete and h is bounded, by Ekeland principle (Proposition 2.2) there exists $x_k \in \Sigma$ with $k \in \mathbb{N}$ such that $|\nabla h|(x_k) < \frac{1}{k}$. Hence $\Theta^2(x_k) = 1 - |\nabla h|^2(x_k) \geq 1 - \frac{1}{k^2} > 0$ contradicting $\mathscr{K} = \Sigma$. Summing up, $\mathscr{K}$ has empty interior. In particular, it follows that

$$\mathrm{Ric}_{\mathbb{P}}(\hat{\nu}, \hat{\nu}) + \mathscr{H}'(h)|\nabla h|^2 \equiv 0 \quad \text{on } \Sigma.$$

Hence, since the inequality in (7.44) is strict this is possible only if $\hat{\nu} \equiv 0$, which implies $|\nabla h|^2 \equiv 0$ on Σ and $f(\Sigma)$ is a slice. □

In what follows, we extend this result to higher order mean curvatures. Towards this aim, we let $\mathfrak{L}_k$ be the operator

$$\mathfrak{L}_k u = \mathrm{div}(P_k \nabla u), \tag{7.45}$$

for $u \in C^1(\Sigma)$. Note that

$$\mathfrak{L}_k u = \langle \mathrm{div} P_k, \nabla u\rangle + L_k u.$$

Following Definition 4.3 of Sect. 4.4, we introduce

Definition 7.1 We will say that the manifold Σ is $\mathfrak{L}_k$-*parabolic* if the only bounded above C^1 solutions of the inequality

$$\mathfrak{L}_k u \geq 0$$

are constant.

The following theorem is a special case of Theorem 2.6 in [226] and it can be recovered from Theorem 4.14 with the choice $\varphi(x,t) = t$ and $T = P_{k-1}$, noting that $\mathrm{Tr}(T) = c_{k-1}H_{k-1}$, with $c_{k-1} = k\binom{m}{k}$ as in Sect. 6.2.1.

Theorem 7.12 *Let $f : \Sigma \to I \times_\rho \mathbb{P}$ be a complete hypersurface. Fix an origin $o \in \Sigma$. If*

$$\left(\sup_{\partial B_t} H_{k-1} \mathrm{vol}(\partial B_t)\right)^{-1} \notin L^1(+\infty), \tag{7.46}$$

where ∂B_t is the geodesic sphere of radius t centered at o, then Σ is $\mathfrak{L}_{k-1}$-parabolic.

We are ready to state the last result of this section.

Theorem 7.13 *Let $f : \Sigma \to I \times_\rho \mathbb{P}$ be a complete hypersurface with $\sup_\Sigma |H_1| < +\infty$ satisfying condition (7.46). Suppose that f has constant k-mean curvature, for some $2 \leq k \leq m$, and that $f(\Sigma)$ is contained in a slab. Suppose that $\mathbb{P}$ has constant sectional curvature κ satisfying*

$$\kappa > \sup_I \{(\rho')^2 - \rho''\rho\}. \tag{7.47}$$

Assume that either $k = 2$ and H_2 is positive or $k \geq 3$ and there exists an elliptic point $p \in \Sigma$. If $\mathscr{H}(h)$ and the angle function Θ have constant sign, then $f(\Sigma)$ is a slice.

Remark 7.5 Comparing with Theorem 7.11 we have relaxed the condition on $\mathscr{H}$ but we are requiring, on the other hand, the existence of an elliptic point. This, on a compact manifold is guaranteed by the assumption $\mathscr{H} > 0$. Moreover, we observe that the angle function is indeed well defined because Σ is two-sided. For $k = 2$, this follows from the positivity of H_2 since $H_1^2 \geq H_2 > 0$. For the remaining values of k this property follows from Gårding inequalities, as in the compact setting. In any case we choose the orientation so that $H_1 > 0$.

Proof It follows from the assumptions that $\sup_\Sigma |A| < +\infty$ and therefore by Lemma 7.2 the sectional curvature of Σ is bounded from below. Thus the validity of the Omori-Yau maximum principle for the Laplacian. Assume $\mathscr{H}(h) \geq 0$; applying the latter to Eq. (7.17) with $k = 0$ we find, for an appropriate sequence $\{x_j\}$,

$$-\operatorname{sgn}\Theta \liminf_{j\to+\infty} H_1(x_j) \geq \mathscr{H}(h^*) \geq 0.$$

Therefore for the chosen orientation, $\operatorname{sgn}\Theta = -1$ and $\Theta \leq 0$ on Σ. Consider the operator $\mathfrak{L}_{k-1}$ introduced in (7.45) and the function

$$\phi = H_k^{\frac{1}{k}} \sigma(h) + \hat{\Theta},$$

where $\hat{\Theta} = \rho(h)\Theta$. Since $\mathbb{P}$ has constant sectional curvature κ, by (7.35) of Lemma 7.3 it follows that

$$\begin{aligned}
\mathfrak{L}_{k-1}\phi = &- (m-k+1)\Theta\Big(\frac{\kappa}{\rho^2(h)} + \mathscr{H}'(h)\Big)\langle P_{k-2}\nabla h, \nabla\phi\rangle + L_{k-1}\phi \\
= &- (m-k+1)\hat{\Theta}\Big(\frac{\kappa}{\rho^2(h)} + \mathscr{H}'(h)\Big)\langle P_{k-2}\nabla h, \nabla h\rangle H_k^{\frac{1}{k}} \\
&+ (m-k+1)\hat{\Theta}\Big(\frac{\kappa}{\rho^2(h)} + \mathscr{H}'(h)\Big)\langle P_{k-2}A\nabla h, \nabla h\rangle \\
&+ H_k^{\frac{1}{k}} L_{k-1}\sigma(h) + L_{k-1}\hat{\Theta}.
\end{aligned}$$

Using (7.18) and Corollary 7.7 we finally compute

$$\begin{aligned}
\mathfrak{L}_{k-1}\phi = &c_{k-1}\rho'(h)H_k^{\frac{1}{k}}(H_{k-1} - H_k^{\frac{k-1}{k}}) \\
&- \binom{n}{k}\hat{\Theta}\left(mH_1H_k - (n-k)H_{k+1} - kH_k^{\frac{k+1}{k}}\right) \\
&- (n-k)\hat{\Theta}\Big(\frac{\kappa}{\rho^2(h)} + \mathscr{H}'(h)\Big)\langle P_{k-1}\nabla h, \nabla h\rangle \\
&- (n-k+1)\hat{\Theta}H_k^{\frac{1}{k}}\Big(\frac{\kappa}{\rho^2(h)} + \mathscr{H}'(h)\Big)\langle P_{k-2}\nabla h, \nabla h\rangle.
\end{aligned} \tag{7.48}$$

Using Gårding inequalities as in Theorem 7.11, it is easy to prove that the first and the second terms on the right-hand side of (7.48) are nonnegative. By the fact that each P_j is positive definite for $j = 0, \ldots, k-1$, and by assumption (7.47), it also follows that all the remaining terms on the right-hand side of the previous equation are nonnegative. Thus $\mathfrak{L}_{k-1}\phi \geq 0$. Since, by assumption (7.46), Σ is $\mathfrak{L}_{k-1}$-parabolic, we conclude that ϕ has to be constant. In particular, $\mathfrak{L}_{k-1}\phi = 0$ and the four terms on the right-hand side of (7.48) vanish. Let us prove that $\mathscr{U} = \{p \in \Sigma : \Theta(p) = 0\}$ has empty interior. Indeed, assume the contrary and let $\mathscr{V} \neq \emptyset$ be an open set contained in $\mathscr{U}$. On $\mathscr{V}$ the function $\phi = \sigma(h)H_k^{1/k}$ is constant. Hence, since $H_k \neq 0$, $\sigma(h)$ and, equivalently h, is constant on $\mathscr{V}$. But this is impossible because $|\nabla h|^2 = 1 - \Theta^2 = 1$ on $\mathscr{V}$. Therefore, the third term on the right-hand of (7.48) vanishes identically and, due to the strict inequality in (7.47), we have

$$\langle P_{k-1}\nabla h, \nabla h\rangle = 0.$$

Since P_{k-1} is positive definite, this means that h has to be constant and $f(\Sigma)$ is a slice. $\square$

7.5 Height Estimates

In this section we present some geometric consequences of the open form of the WMP, that is, Theorem 4.6, and of Theorem 4.11 (in other words, Ahlfors parabolicity), related to the mean curvature and to the height function introduced in Proposition 7.1 at the beginning of the chapter. Firstly we consider the case of warped product spaces.

7.5.1 *Warped Product Spaces*

Let $I \times_\rho \mathbb{P}$ be a warped product manifold. Given a smooth function $u : \mathbb{P} \to I \subseteq \mathbb{R}$, we consider the immersion $\Gamma_u : \mathbb{P} \to I \times_\rho \mathbb{P}$ given by the graph of u, that is, $\Gamma_u(x) = (u(x), x)$, and we denote by $\Gamma_u(\mathbb{P})$ its image in $I \times_\rho \mathbb{P}$. The metric induced on $\mathbb{P}$ from the warped metric in the ambient space is, with the usual simplified notation, given by

$$\langle\, , \rangle = du^2 + \rho(u)^2 \langle\, , \rangle_{\mathbb{P}}$$

and the vector field

$$\nu = \frac{\rho(u)}{\sqrt{\rho(u)^2 + |Du|^2_{\mathbb{P}}}} \left(\frac{1}{\rho(u)^2} Du - \mathscr{T} \right) \tag{7.49}$$

defines a unit normal to the graph Γ_u satisfying $-1 \leq \langle \nu, \mathscr{T} \rangle < 0$. The mean curvature function H of $\Gamma_u : \mathbb{P} \to I \times_\rho \mathbb{P}$ with respect to this orientation is given by

$$\mathrm{div}_{\mathbb{P}} \left(\frac{Du}{\sqrt{\rho(u)^2 + |Du|^2_{\mathbb{P}}}} \right) = m\rho(u) \left(\frac{\rho'(u)}{\sqrt{\rho(u)^2 + |Du|^2_{\mathbb{P}}}} - H \right). \tag{7.50}$$

Here $\mathrm{div}_{\mathbb{P}}$, D and $|\cdot|_{\mathbb{P}}$ are respectively the divergence, the covariant derivative and the norm with respect to the original metric $\langle\, , \rangle_{\mathbb{P}}$ of $\mathbb{P}$. We fix an origin $o \in \mathbb{P}$ and for $u_0 = u(o)$ we set

$$\phi(t) = u_0 + \int_{u_0}^{t} \frac{ds}{\rho(s)}. \tag{7.51}$$

Defining

$$w(x) = \phi(u(x)), \tag{7.52}$$

a computation shows that w satisfies

$$\operatorname{div}_{\mathbb{P}}\left(\frac{Dw}{\sqrt{1+|Dw|^2_{\mathbb{P}}}}\right) = m\rho(\phi^{-1}(w))\left(\frac{\mathcal{H}(\phi^{-1}(w))}{\sqrt{1+|Dw|^2_{\mathbb{P}}}} - H\right). \tag{7.53}$$

Note that the above change of variable has a geometric interpretation in viewing the warped product metric as a metric conformal to the standard product metric on $J\times\mathbb{P}$, where $J \subseteq \mathbb{R}$ is the open interval $J = \phi(I)$, possibly coinciding with $\mathbb{R}$. Indeed, let $\mathcal{I} : I \times_\rho \mathbb{P} \to J \times \mathbb{P}$ be the map defined by

$$\mathcal{I}(t,x) = (\phi(t), x).$$

On $J \times \mathbb{P}$ we consider the metric $(\,,\,)$ conformal to the product metric of $J \times \mathbb{P}$ given by (in loose notation)

$$(\,,\,) = \lambda^2(s)\big(ds^2 + \langle\,,\,\rangle_{\mathbb{P}}\big)$$

with s the canonical coordinate on J. Then $\mathcal{I}$ is an isometry if and only if

$$\lambda(s) = \rho\big(\phi^{-1}(s)\big).$$

In this case the inverse map $\mathcal{I}^{-1} : J \times \mathbb{P} \to I \times_\rho \mathbb{P}$ is given by $\mathcal{I}^{-1}(s,x) = \big(\phi^{-1}(s), x\big)$, where

$$\phi^{-1}(s) = u_0 + \int_{u_0}^{s} \lambda(\tau)\, d\tau.$$

See [13] for complete details. We will refer to the operator in the left-hand side of (7.53) as to the mean curvature operator on $(\mathbb{P}, \langle\,,\,\rangle_{\mathbb{P}})$.

We are ready to prove the next

Theorem 7.14 *Let $\mathbb{P}$ be an m-dimensional Riemannian manifold and assume that the WMP holds on $\mathbb{P}$ for the mean curvature operator. For a smooth function $u : \mathbb{P} \to I \subseteq \mathbb{R}$, let $\Gamma_u : \mathbb{P} \to I \times_\rho \mathbb{P}$ be the graph of u and suppose that its mean curvature satisfies $H \le 0$ on $\mathbb{P}$. Assume that u and $|Du|_{\mathbb{P}}$ are bounded above. Then either $\Gamma_u(\mathbb{P})$ is a slice $\mathbb{P}_{u_0}$ (with $\mathcal{H}(u_0) = H^* \equiv H$) or $\mathcal{H}(u^*) \le H^*$, with $u^* = \sup_{\mathbb{P}} u$ and $H^* = \sup_{\mathbb{P}} H$.*

Proof Assume that $\Gamma_u(\mathbb{P})$ is not a slice, that is, u is nonconstant. We reason by contradiction and we suppose that $\mathcal{H}(u^*) > H^*$. By continuity of $\mathcal{H}(t)$ we can choose a regular value $\gamma < u^*$ of u, sufficiently near to u^* such that $\mathcal{H}(t) > H^*$ for

$t \in [\gamma, u^*]$. Next we define w as in (7.52) and we observe that $w^* = \sup_{\mathbb{P}} w = \phi(u^*)$, being ϕ obviously increasing. Furthermore,

$$\Omega_\gamma = \{x \in \mathbb{P} : u(x) > \gamma\} = \Omega = \{x \in \mathbb{P} : w(x) > \phi(\gamma)\}.$$

Since γ is a regular value of u, $\partial\Omega \neq \emptyset$. According to (7.53) w satisfies the equation

$$\operatorname{div}_{\mathbb{P}}\left(\frac{Dw}{\sqrt{1+|Dw|_{\mathbb{P}}^2}}\right) = m\rho(u)\left(\frac{\mathscr{H}(u)}{\sqrt{1+|Dw|_{\mathbb{P}}^2}} - H(x)\right) \quad \text{on } \Omega, \tag{7.54}$$

with

$$u = \phi^{-1}(w).$$

Since, by assumption, $H(x) \leq 0$, we have

$$\frac{H(x)}{\sqrt{1+|Dw|_{\mathbb{P}}^2}} \geq H(x),$$

so that

$$m\rho(u)\left(\frac{\mathscr{H}(u)}{\sqrt{1+|Dw|_{\mathbb{P}}^2}} - H(x)\right) \geq m\rho(u)\frac{\mathscr{H}(u) - H(x)}{\sqrt{1+|Dw|_{\mathbb{P}}^2}}.$$

On the other hand, observe that on Ω

$$\mathscr{H}(u) > H^* \geq H(x).$$

Since $|Du|_{\mathbb{P}}^2$ is bounded above and $\rho(u)$ is bounded away from 0 on $\overline{\Omega}$,

$$|Dw|_{\mathbb{P}}^2 = \frac{|Du|_{\mathbb{P}}^2}{\rho(u)}$$

is bounded above. Therefore there exists a constant $C > 0$ such that

$$\frac{m\rho(u)}{\sqrt{1+|Dw|_{\mathbb{P}}^2}} \geq C \quad \text{on } \Omega,$$

and

$$\begin{cases} \operatorname{div}_{\mathbb{P}}\left(\frac{Dw}{\sqrt{1+|Dw|^2_{\mathbb{P}}}}\right) \geq C(\mathscr{H}(u) - H(x)) \geq C(\mathscr{H}(u) - H^*) > 0 \text{ on } \Omega, \\ \\ \sup_{\Omega} w = w^* < +\infty. \end{cases}$$

We now apply the open form of the WMP, Theorem 4.6. Since $\mathscr{H}(u^*) - H^* > 0$, alternative (4.97) cannot occur. On the other hand,

$$\sup_{\Omega} w = w^* > \phi(\gamma) = \sup_{\partial\Omega} w,$$

and the second alternative cannot occur too. This gives the desired contradiction. □

The following corollary is a direct consequence of Theorem 7.14.

Corollary 7.8 *Let $\mathbb{P}$ be an m-dimensional Riemannian manifold and assume that the WMP holds on $\mathbb{P}$ for the mean curvature operator. For a smooth function $u : \mathbb{P} \to I \subseteq \mathbb{R}$, let $\Gamma_u : \mathbb{P} \to I \times_\rho \mathbb{P}$ be a minimal graph. Assume that u and $|Du|_{\mathbb{P}}$ are bounded above. Then either $\Gamma_u(\mathbb{P})$ is a slice $\mathbb{P}_{u_0}$ (with $\rho'(u_0) = 0$) or $\rho'(u^*) \leq 0$, with $u^* = \sup_{\mathbb{P}} u$.*

In the next result we estimate H from below.

Theorem 7.15 *Let $\mathbb{P}$ be an m-dimensional Riemannian manifold; assume that the WMP holds on $\mathbb{P}$ for the mean curvature operator; let $U \subset \mathbb{P}$ be an open subset with $\partial U \neq \emptyset$. For $u \in C^0(\overline{U}) \cap C^\infty(U)$ with $u(\overline{U}) \subset I$, let $\Gamma_u : U \to I \times_\rho \mathbb{P}$ be a graph with $\sup_U H \leq 0$. Assume that u and $|Du|_{\mathbb{P}}$ are bounded above. If $\sup_U u > \sup_{\partial U} u$ then $\mathscr{H}(\sup_U u) \leq \sup_U H$.*

Proof Since $\sup_U u > \sup_{\partial U} u$, we know that u is nonconstant. We reason by contradiction and we suppose that $\mathscr{H}(\sup_U u) > \sup_U H$. Now we proceed as in the proof of Theorem 7.14 by choosing $\gamma < \sup_U u$, sufficiently near to $\sup_U u$ such that $\mathscr{H}(t) \leq \sup_U H$ for $t \in [\gamma, \sup_U u]$ and $\overline{\Omega_\gamma} \subset U$, where

$$\Omega_\gamma = \{x \in U : u(x) > \gamma\}.$$

□

Similarly we have

Corollary 7.9 *Let $\mathbb{P}$ be an m-dimensional Riemannian manifold and assume that the WMP holds on $\mathbb{P}$ for the mean curvature operator; let $U \subset \mathbb{P}$ be an open subset with $\partial U \neq \emptyset$. For $u \in C^0(\overline{U}) \cap C^\infty(U)$ with $u(\overline{U}) \subset I$, let $\Gamma_u : U \to I \times_\rho \mathbb{P}$ be a minimal graph. Assume $\rho' > 0$ on $u(\overline{U})$ and suppose that u and $|Du|_{\mathbb{P}}$ are bounded above. Then $\sup_U u = \sup_{\partial U} u$.*

In the above results we have assumed the validity of the WMP for the mean curvature operator. A sufficient condition for this to happen is given by the completeness of the Riemannian manifold $(\mathbb{P}, \langle\, ,\rangle_{\mathbb{P}})$ together with the following volume growth condition

$$\liminf_{R\to+\infty} \frac{\log \mathrm{vol} B_R}{R^2} < +\infty, \tag{7.55}$$

where B_R is the geodesic ball in $\mathbb{P}$ centered at o and with radius R. To see this apply Theorem 4.4 with $\sigma = 0 = \mu$, $T = \langle\, ,\rangle$, $\delta = 1$, $f \equiv 0$ and $\varphi(x,t) = \frac{t}{\sqrt{1+t^2}}$ (see also Theorem 4.1 in [225]). Therefore, as another application of Theorem 7.14 we have the next

Corollary 7.10 *Let $\mathbb{P}$ be a complete m-dimensional Riemannian manifold. Fix an origin $o \in \mathbb{P}$ and suppose that condition (7.55) holds. For $u \in C^\infty(\mathbb{P})$ let $\Gamma_u : \mathbb{P} \to I \times_\rho \mathbb{P}$ be a graph with $H \leq 0$ on $\mathbb{P}$. Assume that u and $|Du|_{\mathbb{P}}$ are bounded above. Then either $\Gamma_u(\mathbb{P})$ is a slice $\mathbb{P}_{u_0}$ (with $\mathcal{H}(u_0) = H^* \equiv H$) or $\mathcal{H}(u^*) \leq H^*$, with $u^* = \sup_{\mathbb{P}} u$ and $H^* = \sup_{\mathbb{P}} H$.*

In our next result we only require the validity of the WMP for the Laplace-Beltrami operator, or equivalently the stochastic completeness of the manifold.

Theorem 7.16 *Let $f : \Sigma \to I \times_\rho \mathbb{P}$ be a stochastically complete, constant mean curvature hypersurface such that, for a correct orientation of the normal ν, $H \geq 0$. Suppose that the height function h is bounded above on Σ. If $\tau \in \mathbb{R}$ is such that $\mathcal{H}(\tau) > H$ and $\mathcal{H}'(t) \geq 0$ for $t > \tau$, then $h(x) \leq \tau$ on Σ.*

Proof First of all observe that because of our assumptions $\mathcal{H}(t) \geq 0$ for $t > \tau$. Now we reason by contradiction and suppose that $h^* > \tau$. Observe that h cannot be constant on Σ, because otherwise $h \equiv h^*$ and $f(\Sigma)$ is the slice $\{h^*\} \times \mathbb{P}$ with constant mean curvature $H = \mathcal{H}(h^*)$. But $\mathcal{H}$ is nondecreasing for $t > \tau$, and $h^* > \tau$ implies $H = \mathcal{H}(h^*) \geq \mathcal{H}(\tau)$, contradicting the hypothesis $H < \mathcal{H}(\tau)$. Therefore, h is nonconstant and we can choose a regular value γ, with $\tau < \gamma < h^*$, so that $\partial\Omega_\gamma \neq \emptyset$, where

$$\Omega_\gamma = \{x \in \Sigma : h(x) > \gamma\}.$$

Letting, as in (7.16),

$$\sigma(t) = \int_{t_0}^{t} \rho(s)\, ds \tag{7.56}$$

from (7.18) of Proposition 7.1 with $k = 0$ we have

$$\Delta\sigma(h) = m\rho(h)(\mathcal{H}(h) + \Theta H),$$

where Θ is the angle function. Because of the assumptions on $\mathscr{H}$ and $\mathscr{H}'$, we infer $\rho(h) \geq \rho(\gamma) > 0$ on Ω_γ and $\mathscr{H}(h) \geq \mathscr{H}(\gamma) \geq \mathscr{H}(\tau) > H$. Since $H \geq 0$, we deduce $\Theta H \geq -H$ and

$$\mathscr{H}(h) + \Theta H \geq \mathscr{H}(h) - H \geq \mathscr{H}(\gamma) - H > 0,$$

so that

$$\Delta\sigma(h) \geq m\rho(\gamma)(\mathscr{H}(\gamma) - H) \quad \text{on } \Omega_\gamma.$$

From (7.56), $\sigma(t)$ is an increasing function and therefore

$$\Lambda_{\sigma(\gamma)} = \{x \in \Sigma : \sigma(h(x)) > \sigma(\gamma)\} = \Omega_\gamma \quad \text{and} \quad \partial\Lambda_{\sigma(\gamma)} = \partial\Omega_\gamma.$$

We set $\Omega = \Lambda_{\sigma(\tau_0)}$ and $v = \sigma(h)|_{\overline{\Omega}}$, to deduce from the above

$$\Delta v \geq m\rho(\gamma)(\mathscr{H}(\gamma) - H) > 0 \quad \text{on } \Omega$$

and

$$\sup_{\Omega} v = \sigma(h^*) < +\infty.$$

Applying Theorem 4.6, either $\mathscr{H}(\gamma) - H \leq 0$ or $\sup_\Omega v = \sup_{\partial\Omega} v$. But $\mathscr{H}(\gamma) \geq \mathscr{H} > H$ and $\sup_\Omega v = \sigma(h^*) > \sigma(\gamma) = \sup_{\partial\Omega} v$, obtaining the desired contradiction. □

We now focus our attention on higher order mean curvatures. In order to guarantee the validity of the WMP for the type of operators that we will use in the next result, that is, trace type operators that cannot be put in divergence form, we will go through the validity of the strong maximum principle *via* a lower bound assumption on the sectional curvature ${}^\Sigma K$. This could have been done also in Theorem 7.16 relaxing the assumption on ${}^\Sigma K$ to a corresponding inequality for the Ricci curvature.

Theorem 7.17 *Let $f : \Sigma \to I \times_\rho \mathbb{P}$ be a complete, oriented, constant nonzero k-mean curvature hypersurface for some $2 \leq k \leq m$ with $\sup_\Sigma |H_1| < +\infty$. Assume the existence of an elliptic point on Σ so that, for a correct orientation of the normal ν, $H_k > 0$. Suppose*

$${}^\Sigma K(x) \geq -G^2(r(x))$$

for some $G \in C^1(\mathbb{R}^+)$ satisfying

$$(i)\ G(0) > 0; \qquad (ii)\ G'(t) \geq 0; \qquad (iii)\ \frac{1}{G(t)} \notin L^1(+\infty).$$

Assume that $\mathcal{H}(t) > 0$ *and that there exists* $\tau \in \mathbb{R}$ *such that* $\mathcal{H}(\tau) > H_k^{1/k}$, *with* $\mathcal{H}'(t) \geq 0$ *for* $t > \tau$, *and let* $\Omega_\tau = \{x \in \Sigma : h(x) > \tau\}$. *If the height function* h *is bounded above on* Σ, *then either* $\sup_{\Omega_\tau} \Theta > 0$ *or* $\Omega_\tau = \emptyset$, *that is,* $h(x) \leq \tau$ *on* Σ.

Proof Assume that $\Omega_\tau \neq \emptyset$ and by contradiction suppose that $\Theta \leq 0$ on Ω_τ. For the time being, assume the validity of the WMP on Σ for the operator $\tilde{\mathcal{L}}_{k-1}$ defined on functions of class $C^2(\Sigma)$ by

$$\tilde{\mathcal{L}}_{k-1}u = \mathrm{Tr}(\tilde{\mathcal{P}}_{k-1} \circ \mathrm{hess}(u)), \tag{7.57}$$

where

$$\tilde{\mathcal{P}}_{k-1} = \sum_{j=0}^{k-1} \frac{c_{k-1}}{c_j} \mathcal{H}(h)^{k-1-j} |\Theta|^j P_j,$$

with $c_k = (m-k)\binom{m}{k} = (k+1)\binom{m}{k+1}$.

Since $\Omega_\tau \neq \emptyset$, we have $h^* > \tau$. If h is constant on Σ, then $h \equiv h^*$ and $f(\Sigma)$ is the slice $\{h^*\} \times \mathbb{P}$ with k-mean curvature $H_k = \mathcal{H}(h^*)^k$. But $\mathcal{H}(t)$ is nondecreasing for $t > \tau$, and $h^* > \tau$ implies

$$H_k^{\frac{1}{k}} = \mathcal{H}(h^*) \geq \mathcal{H}(\tau),$$

which contradicts the hypothesis $\mathcal{H}(\tau) > H_k^{\frac{1}{k}}$. Hence, h is nonconstant and we can fix a regular value $\tau < \tau_0 < h^*$ for which $\partial\Omega_\gamma \neq \emptyset$. Note that, since $\Omega_\gamma \subset \Omega_\tau$, we have $\Theta \leq 0$ on Ω_γ. Thus, on Ω_γ the operator $\tilde{\mathcal{P}}_{k-1}$ becomes

$$\tilde{\mathcal{P}}_{k-1} = \mathcal{P}_{k-1} = \sum_{j=0}^{k-1} (-1)^j \frac{c_{k-1}}{c_j} \mathcal{H}(h)^{k-1-j} \Theta^j P_j.$$

With this observation, from Eq. (7.33) we have

$$\tilde{\mathcal{L}}_{k-1}\sigma(h) = c_{k-1}\rho(h)\left(\mathcal{H}(h)^k + (-1)^{k-1}\Theta^k H_k\right) \quad \text{on } \Omega_\gamma. \tag{7.58}$$

Because of the assumptions on $\mathcal{H}$ and $\mathcal{H}'$, we deduce $\rho(h) \geq \rho(\gamma) > 0$ on Ω_γ and $\mathcal{H}(h)^k \geq \mathcal{H}(\gamma)^k \geq \mathcal{H}(\tau)^k > H_k$. Since $H_k > 0$, it follows that $(-1)^{k-1}\Theta^k H_k \geq -H_k$ and

$$\mathcal{H}(h)^k + (-1)^{k-1}\Theta^k H_k \geq \mathcal{H}(h)^k - H_k \geq \mathcal{H}(\gamma)^k - H_k > 0.$$

From this it follows

$$\tilde{\mathcal{L}}_{k-1}\sigma(h) \geq c_{k-1}\rho(\tau_0)(\mathcal{H}(\tau_0)^k - H_k) \quad \text{on } \Omega_\gamma.$$

The definition of σ, given in (7.56), implies that $\sigma(t)$ is an increasing function and therefore

$$\Lambda_{\sigma(\gamma)} = \{x \in \Sigma : \sigma(h(x)) > \sigma(\gamma)\} = \Omega_\gamma \quad \text{and} \quad \partial\Lambda_{\sigma(\gamma)} = \partial\Omega_\gamma.$$

We set $\Omega = \Lambda_{\sigma(\gamma)}$ and $v = \sigma(h)|_{\overline{\Omega}}$, so that

$$\tilde{\mathscr{L}}_{k-1} v \geq c_{k-1}\rho(\gamma)(\mathscr{H}(\gamma)^k - H_k) > 0 \quad \text{on } \Omega$$

and

$$\sup_\Omega v = \sigma(h^*) < +\infty.$$

Applying Theorem 4.6, either $\mathscr{H}(\gamma)^k - H_k \leq 0$, which is impossible, or $\sup_\Omega v = \sup_{\partial\Omega} v$, which is also impossible because $\sup_\Omega v = \sigma(h^*) > \sigma(\gamma) = \sup_{\partial\Omega} v$. This gives the desired contradiction.

It remains to prove the validity of the WMP on Σ for $\tilde{\mathscr{L}}_{k-1}$. In the assumptions of the theorem we know that

$$H_1 \geq H_j^{1/j} \geq H_k^{1/k} > 0, j = 1, \ldots, k-1. \tag{7.59}$$

Since $\mathscr{H}(h) > 0$ on Σ, $\tilde{\mathscr{P}}_{k-1}$ is positive definite. Furthermore we have

$$\begin{aligned} \mathrm{Tr}(\tilde{\mathscr{P}}_{k-1}) &= c_{k-1}\sum_{j=0}^{k-1} |\Theta|^j \mathscr{H}(h)^{k-1-j} H_j \\ &\geq c_{k-1}\mathscr{H}(h)^{k-1} > 0 \quad \text{on } \Sigma, \end{aligned}$$

and

$$\mathrm{Tr}(\tilde{\mathscr{P}}_{k-1}) \leq c_{k-1}\sum_{j=0}^{k-1} \mathscr{H}(h)^{k-1-j} H_j^*.$$

Hence from the requirement

$$\sup_\Sigma |H_1| < +\infty$$

using (7.59) we have

$$0 < \mathrm{Tr}(\tilde{\mathscr{P}}_{k-1})(x) \leq \Lambda \tag{7.60}$$

on Σ, for some positive constant Λ. By the assumption on the sectional curvature of Σ, from Theorem 6.13 we deduce that the $\frac{1}{\mathrm{Tr}(\tilde{\mathscr{P}}_{k-1}(x))}$-WMP holds on Σ for the operator $\tilde{\mathscr{L}}_{k-1}$. However, because of (7.60), $\frac{1}{\mathrm{Tr}(\tilde{\mathscr{P}}_{k-1}(x))}$ is bounded from below by a positive constant and therefore the WMP holds for $\tilde{\mathscr{L}}_{k-1}$. □

Remark 7.6 Theorem 7.17 complements Theorem 6.2 of [27] and it extends the first part of Proposition 4 of [123] to the noncompact case.

7.5.2 *Products*

In what follows we shall consider the case of a Riemannian product $\mathbb{R} \times \mathbb{P}$. From now on, if the angle function Θ of a two-sided hypersurface has constant sign, the orientation ν will be chosen so that $\Theta \leq 0$. Observe that if the hypersurface is a local graph over $\mathbb{P}$, then either $\Theta > 0$ or $\Theta < 0$. Thus, requiring Θ not to change sign is an assumption weaker than that of being a local graph (compare, for instance, with [194]).

We begin by considering the case of constant mean curvature. Thus let $f : \Sigma \to \mathbb{R} \times \mathbb{P}$ be a two-sided hypersurface with constant mean curvature H and define

$$\phi = hH + \Theta. \tag{7.61}$$

By Lemma 7.4 and (7.17) in Proposition 7.1 with $k = 0$ and $\rho \equiv 1$ we obtain

$$\Delta\phi = -\Theta\left(|A|^2 - mH^2 + {}^{\mathbb{P}}\mathrm{Ric}(\hat{\nu}, \hat{\nu})\right), \tag{7.62}$$

where A is the Weingarten operator of the hypersurface in the direction of ν and $\hat{\nu}$ denotes the projection of ν onto the fiber $\mathbb{P}$, that is,

$$\nu = \langle \nu, \mathscr{T} \rangle \mathscr{T} + \hat{\nu}.$$

In particular,

$$|\hat{\nu}|^2_{\mathbb{P}} = |\nabla h|^2 \leq 1. \tag{7.63}$$

We also record, for later use, that from (7.17)

$$\Delta h = mH\Theta. \tag{7.64}$$

Since the WMP for Δ on Σ is equivalent to the stochastic completeness of $(\Sigma, \langle\ ,\ \rangle)$, we have

Theorem 7.18 *Let $f : \Sigma \to \mathbb{R} \times \mathbb{P}$ be a stochastically complete hypersurface of dimension m with constant mean curvature $H > 0$. Suppose that for some $\alpha > 0$*

$$\mathrm{Ric}_{\mathbb{P}} \geq -m\alpha \tag{7.65}$$

and

$$H^2 > \alpha. \tag{7.66}$$

Let $\Omega \subset \Sigma$ be an open set with $\partial\Omega \neq \emptyset$ for which $f(\Omega)$ is contained in a slab and $f(\partial\Omega) \subset \mathbb{P}_0 = \{0\} \times \mathbb{P}$. If

$$\beta = \sup_{\Omega} \Theta < 0 \tag{7.67}$$

then

$$f(\Omega) \subset \left[0, \frac{(1+\beta)H}{H^2 - \alpha}\right] \times \mathbb{P}. \tag{7.68}$$

Remark 7.7 It is clear that the choice $f(\partial\Omega) \subset \mathbb{P}_0$ is only a matter of "normalization". One can possibly choose $f(\partial\Omega) \subset \mathbb{P}_{t_0}$ for some $t_0 \in I$ changing (7.68) accordingly.

Proof For any fixed $\delta > 0$ such that

$$\alpha < \alpha + \frac{\delta}{m} \leq H^2$$

we consider the function

$$\psi = \phi - \frac{\alpha + \delta/m}{H} h = \Theta + \frac{H^2 - \alpha - \delta/m}{H} h. \tag{7.69}$$

Using (7.62) and (7.64) we obtain

$$\Delta\psi = -\Theta(|A|^2 - mH^2 + \mathrm{Ric}_{\mathbb{P}}(\hat{\nu}, \hat{\nu}) + m\alpha + \delta).$$

From (7.65), using also (7.63) and the fact that $\alpha > 0$, we have

$$\mathrm{Ric}_{\mathbb{P}}(\hat{\nu}, \hat{\nu}) \geq -m\alpha|\hat{\nu}|^2_{\mathbb{P}} = -m\alpha|\nabla h|^2 \geq -m\alpha \quad \text{on } \Sigma.$$

Since $|A|^2 \geq mH^2$, this yields $\Delta\psi \geq -\Theta\delta$ on Σ, and by (7.67)

$$\Delta\psi \geq -\Theta\delta \geq -\beta\delta > 0 \quad \text{on } \Omega.$$

We define $v = \psi|_{\overline{\Omega}}$. Then, since $f(\Omega)$ is contained in a slab we deduce

$$\begin{cases} \Delta v \geq -\beta\delta > 0 \text{ on } \Omega; \\ \sup_{\Omega} v < +\infty. \end{cases} \tag{7.70}$$

Since Σ is stochastically complete and alternative (4.97) of Theorem 4.6 cannot occur we obtain

$$\sup_{\Omega} v = \sup_{\partial\Omega} v.$$

But $f(\partial\Omega) \subset \{0\} \times \mathbb{P}$ so that $h \equiv 0$ on $\partial\Omega$ and then $v \equiv \psi \equiv \Theta \leq \beta$ on $\partial\Omega$, so that

$$\beta \geq \sup_{\partial\Omega} v = \sup_{\Omega} v.$$

We thus have

$$\beta \geq v = \psi = \Theta + \frac{H^2 - \alpha - \delta/m}{H} h \geq -1 + \frac{H^2 - \alpha - \delta/m}{H} h \quad \text{on } \Omega.$$

That is,

$$h(x) \leq \frac{(1+\beta)H}{H^2 - \alpha - \delta/m} \quad \text{on } \Omega$$

for each $\delta > 0$ such that $\alpha < \alpha + \delta/m \leq H^2$. Letting $\delta \to 0^+$ we conclude

$$h(x) \leq \frac{(1+\beta)H}{H^2 - \alpha} \quad \text{on } \Omega. \tag{7.71}$$

On the other hand, from (7.64) and (7.67)

$$\Delta h \leq mH\beta < 0 \quad \text{on } \Omega$$

and since $f(\Omega)$ is contained in a slab, the function $w = h|_{\overline{\Omega}}$ is bounded below. Reasoning as above, using now the dual statement of Theorem 4.6 in Remark 4.6, we deduce

$$\inf_{\Omega} w = \inf_{\partial\Omega} w = 0,$$

that is,

$$h(x) \geq 0 \quad \text{on } \Omega. \tag{7.72}$$

Putting (7.71) and (7.72) together we obtain (7.68). □

In Theorem 7.18 if we assume that Σ is parabolic for the Laplace-Beltrami operator Δ, then (7.67) can be relaxed to

$$\Theta \leq 0 \quad \text{on } \Omega$$

conclusion (7.68) holding with no changes. To see this, simply observe that since β could be 0, instead of (7.70) we have

$$\begin{cases} \Delta v \geq 0 \text{ on } \Omega; \\ \sup_\Omega v < +\infty. \end{cases} \tag{7.73}$$

By Ahlfors parabolicity either

$$\sup_\Omega v = \sup_{\partial\Omega} v \tag{7.74}$$

or v is constant on Ω, and in this latter case (7.74) still holds. The rest of the proof is as in Theorem 7.18. The same applies to the reasoning for the lower bound $h(x) \geq 0$.

Next we observe that also the "limit" case $\alpha = 0$, in other words $\mathrm{Ric}_{\mathbb{P}} \geq 0$, can be easily treated. Indeed, fix $\hat{\alpha} > 0$ sufficiently small that (7.66) holds. Then (7.65) is obviously true with $\hat{\alpha}$ instead of α. Applying Theorem 7.18 and letting $\hat{\alpha} \downarrow 0^+$ instead of (7.68) we deduce the improved height estimate

$$f(\Omega) \subset \left[0, (1+\beta)\frac{1}{H}\right] \times \mathbb{P}.$$

Thus putting together the above observations, we have that if we strengthen the assumption of stochastic completeness to parabolicity for the operator Δ and we require $\mathrm{Ric}_{\mathbb{P}} \geq 0$ we can get rid of (7.66) and relax assumption (7.67) to $\Theta \leq 0$ on Ω to obtain the height estimate

$$f(\Omega) \subset \left[0, \frac{1}{H}\right] \times \mathbb{P}. \tag{7.75}$$

In other words we have

Corollary 7.11 *Let $f : \Sigma \to \mathbb{R} \times \mathbb{P}$ be a parabolic hypersurface with constant mean curvature $H > 0$ and assume $\mathrm{Ric}_{\mathbb{P}} \geq 0$. Let $\Omega \subset \Sigma$ be an open set with $\partial\Omega \neq \emptyset$ for which $f(\Omega)$ is contained in a slab and $f(\partial\Omega) \subset \mathbb{P}_0 = \{0\} \times \mathbb{P}$. If*

$\Theta \leq 0$ on Ω then

$$f(\Omega) \subset \left[0, \frac{1}{H}\right] \times \mathbb{P}.$$

This result directly compares with the height estimates obtained by Cheng and Rosenberg, for Σ compact, in [80] (see also [147]).

Next theorem extends Theorem 7.18 to higher order mean curvatures.

Theorem 7.19 *Let $f : \Sigma \to \mathbb{R} \times \mathbb{P}$ be an immersed hypersurface with constant, nonzero, k-mean curvature H_k, for some $k = 2, \ldots, m$ and with an elliptic point. Choose the normal ν so that $H_k > 0$, suppose that for some $\alpha > 0$ the sectional curvature ${}^{\mathbb{P}}K$ of $\mathbb{P}$ satisfies*

$$ {}^{\mathbb{P}}K \geq -\alpha \tag{7.76}$$

and, having set $H^*_{k-1} = \sup_\Sigma H_{k-1}(x)$, assume

$$H_k^{\frac{k+1}{k}} > \alpha H^*_{k-1}. \tag{7.77}$$

Suppose that the WMP holds on Σ for the operator L_{k-1}. Let $\Omega \subset \Sigma$ be an open set with $\partial\Omega \neq \emptyset$ for which $f(\Omega)$ is contained in a slab and $f(\partial\Omega) \subset \mathbb{P}_0 = \{0\} \times \mathbb{P}$. If

$$\beta = \sup_\Omega \Theta < 0 \tag{7.78}$$

then

$$f(\Omega) \subset \left[0, \frac{(1+\beta)H_k}{H_k^{\frac{k+1}{k}} - \alpha H^*_{k-1}}\right] \times \mathbb{P}. \tag{7.79}$$

Proof Let us consider the function

$$\phi = H_k^{1/k} h + \Theta.$$

We know from Eq. (7.17) (see also Proposition 4.1 in [27]) that

$$L_{k-1} h = c_{k-1} H_k \Theta, \tag{7.80}$$

where $c_{k-1} = k\binom{m}{k}$. On the other hand, since H_k is constant from Lemma 7.4 we also have

$$L_{k-1}\Theta = -\Theta\binom{m}{k}(mH_1H_k - (m-k)H_{k+1}) - \Theta\sum_{i=1}^{m}\mu_{i,k-1}{}^{\mathbb{P}}K(\hat{e}_i \wedge \hat{v})|\hat{e}_i \wedge \hat{v}|^2, \tag{7.81}$$

where the $\mu_{i,k-1}$'s are the eigenvalues of P_{k-1} and $\{e_1, \ldots, e_m\}$ is a local orthonormal frame on Σ diagonalizing A (and with the above meaning for the notation $\hat{e}_i$). Recall that L_{k-1} is elliptic or, equivalently, the eigenvalues $\mu_{i,k-1}$ are all positive. It follows from (7.80) and (7.81) that

$$\begin{aligned} L_{k-1}\phi = &-\Theta\binom{m}{k}(mH_1H_k - (m-k)H_{k+1} - kH_k^{\frac{k+1}{k}}) \\ &-\Theta\sum_{i=1}^{m}\mu_{i,k-1}{}^{\mathbb{P}}K(\hat{e}_i \wedge \hat{v})|\hat{e}_i \wedge \hat{v}|^2. \end{aligned}$$

Using Gårding inequalities as in the proof of Theorem 7.11 we obtain (7.43), that is

$$mH_1H_k - kH_k^{\frac{k+1}{k}} - (m-k)H_{k+1} \geq (m-k)(H_k^{\frac{k+1}{k}} - H_{k+1}) \geq 0,$$

and therefore,

$$L_{k-1}\phi \geq -\Theta\sum_{i=1}^{m}\mu_{i,k-1}{}^{\mathbb{P}}K(\hat{e}_i, \hat{v})|\hat{e}_i \wedge \hat{v}|^2. \tag{7.82}$$

For any fixed $\delta > 0$ such that

$$\alpha H_{k-1}^* < \alpha H_{k-1}^* + \delta \leq H_k^{\frac{k+1}{k}}$$

we consider the function

$$\psi = \phi - \frac{\alpha H_{k-1}^* + \delta}{H_k}h = \Theta + \frac{H_k^{\frac{k+1}{k}} - \alpha H_{k-1}^* - \delta}{H_k}h. \tag{7.83}$$

Using (7.80) and (7.82) we obtain

$$L_{k-1}\psi \geq -\Theta\sum_{i=1}^{m}\mu_{i,k-1}{}^{\mathbb{P}}K(\hat{e}_i \wedge \hat{v})|\hat{e}_i \wedge \hat{v}|^2 - \Theta c_{k-1}(\alpha H_{k-1}^* + \delta). \tag{7.84}$$

Observe that

$$|\hat{e}_i \wedge \hat{v}|^2 = |\nabla h|^2 - \langle e_i, \nabla h\rangle^2 \le |\nabla h|^2 \le 1. \tag{7.85}$$

From (7.76), using also (7.85), the fact that $\alpha > 0$ and each $\mu_{i,k-1} > 0$, we have

$$\begin{aligned}\sum_{i=1}^{m} \mu_{i,k-1}{}^{\mathbb{P}}K(\hat{e}_i \wedge \hat{v})|\hat{e}_i \wedge \hat{v}|^2 &\ge -\alpha \sum_{i=1}^{m} \mu_{i,k-1}|\nabla h|^2 \ge -\alpha \mathrm{Tr}(P_{k-1}) \\ &= -\alpha c_{k-1} H_{k-1} \ge -\alpha c_{k-1} H^*_{k-1},\end{aligned}$$

that is,

$$\sum_{i=1}^{m} \mu_{i,k-1}{}^{\mathbb{P}}K(\hat{e}_i \wedge \hat{v})|\hat{e}_i \wedge \hat{v}|^2 \ge -\alpha c_{k-1} H^*_{k-1}. \tag{7.86}$$

Putting together (7.84) and (7.86), and using (7.78), we finally obtain

$$L_{k-1}\psi \ge -c_{k-1}\Theta\delta \ge -c_{k-1}\beta\delta \quad \text{on } \Omega. \tag{7.87}$$

We define $v = \psi|_{\overline{\Omega}}$. Then, $f(\Omega)$ contained in a slab implies that v satisfies

$$\begin{cases} L_{k-1}v \ge -c_{k-1}\beta\delta > 0 \text{ on } \Omega; \\ \sup_{\Omega} v < +\infty. \end{cases}$$

Since the WMP holds on Σ for L_{k-1} and alternative (4.97) of Theorem 4.6 cannot occur, we have

$$\sup_{\Omega} v = \sup_{\partial\Omega} v.$$

But $f(\partial\Omega) \subset \{0\} \times \mathbb{P}$, hence $h \equiv 0$ on $\partial\Omega$ and then $v \equiv \psi \equiv \Theta \le \beta$ on $\partial\Omega$, so that

$$\beta \ge \sup_{\partial\Omega} v = \sup_{\Omega} v.$$

We thus have

$$\beta \ge v = \psi = \Theta + \frac{H_k^{\frac{k+1}{k}} - \alpha H^*_{k-1} - \delta}{H_k} h \ge -1 + \frac{H_k^{\frac{k+1}{k}} - \alpha H^*_{k-1} - \delta}{H_k} h \quad \text{on } \Omega,$$

in other words,

$$h(x) \leq \frac{(1+\beta)H_k}{H_k^{\frac{k+1}{k}} - \alpha H_{k-1}^* - \delta} \quad \text{on } \Omega$$

for each $\delta > 0$ such that $\alpha H_{k-1}^* < \alpha H_{k-1}^* + \delta \leq H_k^{\frac{k+1}{k}}$. Letting $\delta \to 0^+$ we conclude

$$h(x) \leq \frac{(1+\beta)H_k}{H_k^{\frac{k+1}{k}} - \alpha H_{k-1}^*} \quad \text{on } \Omega. \tag{7.88}$$

On the other hand, from (7.78) and (7.80)

$$L_{k-1}h \leq c_{k-1}H_k\beta < 0 \quad \text{on } \Omega,$$

and we conclude as in Theorem 7.18 that

$$h(x) \geq 0 \quad \text{on } \Omega. \tag{7.89}$$

Putting (7.88) and (7.89) together we obtain (7.79), completing the proof. □

As we know, there are geometric conditions that imply the validity of the WMP for the Laplace operator. For instance, completeness of Σ and the volume growth condition (7.55) imply the validity of the WMP for Δ on Σ. Therefore, Theorem 7.18 remains true if one changes stochastic completeness by completeness and condition (7.55).

On the other hand, if we assume instead of (7.65) in Theorem 7.18 that

$$^{\mathbb{P}}K \geq -\alpha$$

for some $\alpha > 0$, then obviously $\mathrm{Ric}_{\mathbb{P}} \geq -n\alpha$. Moreover, from the Gauss equations for the hypersurface Σ we have that

$$\begin{aligned} {}^{\Sigma}K(X \wedge Y) &= \overline{K}(X \wedge Y) + \langle AX, X\rangle\langle AY, Y\rangle - \langle AX, Y\rangle^2 \\ &\geq \overline{K}(X \wedge Y) - 2|A|^2, \end{aligned}$$

where $\{X, Y\}$ is an orthonormal basis for an arbitrary 2-plane tangent to Σ. Here $\overline{K}(X \wedge Y)$ denotes the sectional curvature in the ambient space $\mathbb{R} \times \mathbb{P}$ of the 2-plane spanned by $\{X, Y\}$. Taking into account that

$$\overline{K}(X \wedge Y) = {}^{\mathbb{P}}K(\hat{X} \wedge \hat{Y})|\hat{X} \wedge \hat{Y}|^2$$

and

$$|\hat{X} \wedge \hat{Y}|^2 \leq |X \wedge Y|^2 = 1$$

we obtain $\overline{K}(X \wedge Y) \geq -\alpha$ and therefore

$$^{\Sigma}K(X \wedge Y) \geq -\alpha - 2|A|^2. \tag{7.90}$$

Therefore, fixing an origin $o \in \Sigma$ and denoting with $r(x)$ the distance from o in Σ, if

$$|A(x)| \leq G(r(x))$$

we conclude from (7.90) that the sectional curvatures satisfy

$$^{\Sigma}K \geq -\alpha - 2G(r(x))^2. \tag{7.91}$$

Therefore, if Σ is complete and we assume $G \in C^1([0, +\infty))$ satisfying

$$\text{(i) } G(0) > 0, \quad \text{(ii) } G'(t) \geq 0 \quad \text{and} \quad \text{(iii) } 1/G(t) \notin L^1(+\infty)$$

by Theorem 6.13 the Omori-Yau maximum principle (hence the WMP) holds on Σ for Δ. This provides a proof of the following version of Theorem 7.18.

Theorem 7.20 *Let $f : \Sigma \to \mathbb{R} \times \mathbb{P}$ be a complete hypersurface with constant mean curvature $H > 0$. Assume that*

$$\beta = \sup_{\Sigma} \Theta < 0$$

and suppose that $^{\mathbb{P}}K \geq -\alpha$ and $H^2 > \alpha$, for some $\alpha > 0$. Furthermore, assume that

$$|A(x)| \leq G(r(x))$$

for some $G \in C^1([0, +\infty))$ satisfying

$$\text{(i) } G(0) > 0, \quad \text{(ii) } G'(t) \geq 0 \quad \textit{and} \quad \text{(iii) } 1/G(t) \notin L^1(+\infty),$$

where $r(x)$ denotes the distance in Σ from some fixed origin o.

If $\Omega \subset \Sigma$ is an open set with $\partial\Omega \neq \emptyset$ for which $f(\Omega)$ is contained in a slab and $f(\partial\Omega) \subset \mathbb{P}_0 = \{0\} \times \mathbb{P}$, then

$$f(\Omega) \subset \left[0, \frac{(1+\beta)H}{H^2 - \alpha}\right] \times \mathbb{P}.$$

As for Theorem 7.19, here is an alternative version.

Theorem 7.21 *Let $f : \Sigma \to \mathbb{R} \times \mathbb{P}$ be a complete immersed hypersurface of dimension m with constant, nonzero, k-mean curvature H_k, for some $k = 2, \ldots, m$ and with an elliptic point. Choose the normal ν so that $H_k > 0$, assume that*

$$\beta = \sup_{\Sigma} \Theta < 0 \tag{7.92}$$

and suppose that ${}^{\mathbb{P}}K \geq -\alpha$ and $H_k^{\frac{k+1}{k}} > \alpha H_{k-1}^*$, for some $\alpha > 0$. Furthermore, let

$$|A(x)| \leq G(r(x))$$

for some $G \in C^1([0, +\infty))$ satisfying

$$\text{(i) } G(0) > 0, \quad \text{(ii) } G'(t) \geq 0 \quad \text{and} \quad \text{(iii) } 1/G(t) \notin L^1(+\infty),$$

where $r(x)$ denotes the distance in Σ from some fixed origin o.

If $\Omega \subset \Sigma$ is an open set with $\partial\Omega \neq \emptyset$ for which $f(\Omega)$ is contained in a slab and $f(\partial\Omega) \subset \mathbb{P}_0 = \{0\} \times \mathbb{P}$, then

$$f(\Omega) \subset \left[0, \frac{(1+\beta)H_k}{H_k^{\frac{k+1}{k}} - \alpha H_{k-1}^*}\right] \times \mathbb{P}. \tag{7.93}$$

Proof It is enough to prove the validity of the WMP for the operator L_{k-1} on Σ. Now observe that (7.91), completeness and the fact that $H_{k-1}(x) > 0$ on Σ imply, by Theorem 6.13, the validity of the q-Omori-Yau maximum principle on Σ for L_{k-1} with $q(x) = \dfrac{1}{c_{k-1}H_{k-1}(x)}$. In particular, that of the q-WMP. However, since $H_{k-1}(x)$ is bounded from above on Σ by (7.77), then $q(x)$ is bounded from below by a positive constant, and by the Remark 3.1 this implies the validity of the WMP for L_{k-1} on Σ. □

The following version of Theorem 7.21 can be obtained with a reasoning similar to that in the proof Corollary 7.11. We leave the details to the interested reader.

Theorem 7.22 *Let $f : \Sigma \to \mathbb{R} \times \mathbb{P}$ be an immersed hypersurface of dimension m with constant k-mean curvature H_k, for some $k = 2, \ldots, m$, with an elliptic point (in particular, $H_k \neq 0$), and let the sectional curvature of $\mathbb{P}$ satisfy ${}^{\mathbb{P}}K \geq 0$. Chosen the normal ν so that $H_k > 0$, assume that Σ is L_{k-1}-parabolic. Let $\Omega \subset \Sigma$ be an open set with $\partial\Omega \neq \emptyset$ for which $f(\Omega)$ is contained in a slab and $f(\partial\Omega) \subset \mathbb{P}_0 = \{0\} \times \mathbb{P}$. If $\Theta \leq 0$ on Ω then*

$$f(\Omega) \subset \left[0, \frac{1}{H_k^{\frac{1}{k}}}\right] \times \mathbb{P}.$$

Here is a further result related to Theorem 7.18.

Theorem 7.23 *Let $f : \Sigma \to \mathbb{R} \times \mathbb{P}$ be a hypersurface of dimension m with constant mean curvature $H > 0$, and assume that $h : \Sigma \to \mathbb{R}$ goes to $+\infty$ as $x \to \infty$. Suppose that for some $\alpha > 0$*

$$\mathrm{Ric}_{\mathbb{P}} \geq -m\alpha \tag{7.94}$$

and

$$H^2 > \alpha. \tag{7.95}$$

Let $\Omega \subset \Sigma$ be a relatively compact open set with $\partial\Omega \neq \emptyset$ such that $f(\partial\Omega) \subset \mathbb{P}_0 = \{0\} \times \mathbb{P}$. If

$$\beta = \sup_{\Omega} \Theta < 0 \tag{7.96}$$

then

$$f(\Omega) \subset \left[0, \frac{(1+\beta)H}{H^2 - \alpha}\right] \times \mathbb{P}. \tag{7.97}$$

Proof Since $H > 0$ is constant, it follows from (7.64) that

$$\Delta h \leq mH < +\infty.$$

By Theorem 3.1, with $q(x) \equiv 1$, $\gamma = h$ and $L = \Delta$, we derive the validity of the WMP on Σ for Δ. Equivalently, Σ is stochastically complete. Since Ω is relatively compact, $f(\Omega)$ is contained in a slab and we can apply Theorem 7.18. □

The key of the above proof is to apply Kash'minskii test *via* the function h to obtain the stochastic completeness of Σ. In Theorem 3.1 we proved that a similar test yields the validity of the WMP for a wide class of operators including the L_{k-1}'s operators considered above. Hence the validity of the next

Theorem 7.24 *Let $f : \Sigma \to \mathbb{R} \times \mathbb{P}$ be an immersed hypersurface of dimension m with constant, nonzero, k-mean curvature H_k, for some $k = 2, \ldots, m$ and with an elliptic point. Chosen the normal ν so that $H_k > 0$, suppose that for some $\alpha > 0$*

$${}^{\mathbb{P}}K \geq -\alpha \tag{7.98}$$

and, having set $H_{k-1}^* = \sup_\Sigma H_{k-1}(x)$, assume that

$$H_k^{\frac{k+1}{k}} > \alpha H_{k-1}^*. \tag{7.99}$$

Suppose that $h : \Sigma \to \mathbb{R}$ *goes to* $+\infty$ *as* $x \to \infty$. *Let* $\Omega \subset \Sigma$ *be a relatively compact open set with* $\partial\Omega \neq \emptyset$ *such that* $f(\partial\Omega) \subset \mathbb{P}_0 = \{0\} \times \mathbb{P}$. *If*

$$\beta = \sup_{\Omega} \Theta < 0 \tag{7.100}$$

then

$$f(\Omega) \subset \left[0, \frac{(1+\beta)H_k}{H_k^{\frac{k+1}{k}} - \alpha H_{k-1}^*} \right] \times \mathbb{P}. \tag{7.101}$$

Proof Since $H_k > 0$ is constant, it follows from (7.80) that

$$L_{k-1}h \leq c_{k-1}H_k < +\infty.$$

Theorem 3.1 with $q(x) \equiv 1$, $\gamma = h$ and $L = L_{k-1}$, gives the validity of the WMP on Σ for L_{k-1}. Since Ω is relatively compact, $f(\Omega)$ is contained in a slab and we can apply Theorem 7.19. □

7.6 Killing Graphs

We now consider the case where the $(m+1)$-dimensional ambient manifold $(N, \langle\, ,\, \rangle)$ is endowed with a nonsingular Killing vector field Y with complete flow lines and integrable orthogonal distribution. Let $\mathbb{P}$ be a fixed integral leaf. Note that the leaves of the foliation are totally geodesic hypersurfaces of N. The flow $\Phi : \mathbb{R} \times \mathbb{P} \to N$ generated by Y takes isometrically $\mathbb{P} = \mathbb{P}_0$ to the leaf $\mathbb{P}_s = \Phi_s(\mathbb{P})$ for any $s \in \mathbb{R}$, where $\Phi_s = \Phi(s, \cdot)$. We now consider an immersion $\Gamma : \mathbb{P} \to N$ of the form

$$\Gamma(x) = \Gamma_u(x) = \Phi(u(x), x) \tag{7.102}$$

for some smooth function $u : \mathbb{P} \to \mathbb{R}$. In this case the hypersurface $\Gamma(\mathbb{P}) = \Gamma_u(\mathbb{P})$ is called the *Killing graph of* u, [95]. Since Y is nonsingular we can define $\gamma = |Y|^{-2} > 0$. The unit normal to the graph is then given by

$$\nu(x) = \frac{1}{\sqrt{\gamma(x) + |Du|^2(x)}} \left(\gamma(x)Y(x) - \Phi_{u(x)*}(Du(x)) \right), \tag{7.103}$$

where D denotes the covariant derivative on $\mathbb{P}$, and where, for simplicity of notation, we are denoting again by γ and Y the restrictions of γ and Y on $\mathbb{P}$ along Γ. The Killing graph Γ has constant mean curvature H, in the direction of the normal ν, if and only if (see[24])

$$Lu = \mathrm{div}_{\log\sqrt{\gamma}}\left(\frac{Du}{W}\right) = mH \text{ on } \mathbb{P}, \tag{7.104}$$

where

$$W = \sqrt{\gamma + |\nabla u|^2} \tag{7.105}$$

and L is the operator

$$Lu = \mathrm{div}\left(\frac{Du}{W}\right) - \left\langle \frac{D\gamma}{2\gamma}, \frac{Du}{W}\right\rangle. \tag{7.106}$$

Here div is the divergence on $\mathbb{P}$. We have the following:

Theorem 7.25 *Let N be a complete Riemannian manifold endowed with a complete nonsingular Killing field Y and let $\mathbb{P}$ be an integral leaf of the Killing foliation. Let $\varGamma = \varGamma_u : \mathbb{P} \to N$ be a Killing graph with constant mean curvature $H \geq 0$. Assume that*

$$\sup_{\mathbb{P}} |Y| < +\infty \tag{7.107}$$

and

$$\liminf_{R\to+\infty} \frac{\log \int_{B_R} |Y|}{R^2} < +\infty, \tag{7.108}$$

where $B_R = B_R(o)$ stands for the geodesic ball in $\mathbb{P}$ centered at a fixed origin o with radius R.

If there exists a regular value τ of u such that u is bounded above on some connected component of the upper level

$$\varOmega_\tau = \{x \in \mathbb{P} : u(x) > \tau\}$$

then the Killing graph is minimal.

Proof First of all we observe that, according to Theorem 4.4 (see also Theorem 3.2 of [24]), conditions (7.107), (7.108) and completeness of $\mathbb{P}$, imply the validity of the WMP for the operator L of (7.106) on $\mathbb{P}$. Let $\varOmega$ be the connected component of $\varOmega_\tau$ on which u is bounded above. Note that $\emptyset \neq \partial\varOmega \subseteq \{x \in \mathbb{P} : u(x) = \tau\}$. Set $v = u|_{\overline{\varOmega}}$. By contradiction, suppose $H > 0$. From (7.104)

$$\begin{cases} Lv = mH > 0 \text{ on } \varOmega; \\ \\ \sup_{\varOmega} v < +\infty. \end{cases}$$

Applying Theorem 4.6 and noting that, since $H > 0$, alternative (4.97) cannot occur, we deduce that

$$\sup_{\Omega} v = \sup_{\partial\Omega} v = \tau$$

so that $u \equiv \tau$ on Ω. Hence $\Gamma_u(x) = \Phi(\tau, x) \subseteq \mathbb{P}_\tau$ on Ω. Thus Ω with the induced metric is isometric to an open set of $\mathbb{P}_\tau$ which is totally geodesic in N. Therefore $H = 0$, contradiction. □

From the above theorem we deduce the following corollary related to the results given in [24].

Corollary 7.12 *In the assumptions of Theorem 7.25 if u is bounded above then the Killing graph $\Gamma = \Gamma_u : \mathbb{P} \to N$ is minimal.*

Chapter 8
Applications to Ricci Solitons

In this chapter we apply maximum principles techniques to the study of the geometry of *Ricci solitons*. Ricci solitons have become the subject of a rapidly increasing investigation since the appearance of the seminal works of Hamilton [133] and Perelman [217]; this investigation has been mainly directed towards two goals, *classification* and *triviality* in the sense we shall explain below; among the enormous literature on the subject we only quote, as a few examples, the papers [59, 60, 67, 68, 70, 112, 188, 220–222, 224, 231] and references therein; see also [69] for classification results on a wide class of structures generalizing the concept of Ricci soliton.

First of all we describe, in some detail, the setting. Given a Riemannian manifold $(M, \langle\, ,\, \rangle)$, a *(generic) Ricci soliton structure* $(M, \langle\, ,\, \rangle, X)$, or a *soliton structure* for short,is the choice, if any, of a smooth vector field $X \in \mathfrak{X}(M)$ on M and of a real constant λ such that

$$\mathrm{Ric} + \frac{1}{2}\mathscr{L}_X\langle\, ,\, \rangle = \lambda\langle\, ,\, \rangle, \tag{8.1}$$

where, as in Chap. 1, $\mathscr{L}_X\langle\, ,\, \rangle$ is the Lie derivative of the metric in the direction of X. In what follows we shall refer to λ as to the *soliton constant*. The soliton is called *expanding*, *steady* or *shrinking* if, respectively, $\lambda < 0$, $\lambda = 0$ or $\lambda > 0$ (this terminology comes from the dynamical context, where solitons have been first considered: see for instance [61]). If X is the gradient of a *potential* $f \in C^\infty(M)$, then (8.1) takes the form

$$\mathrm{Ric} + \mathrm{Hess}(f) = \lambda\langle\, ,\, \rangle; \tag{8.2}$$

indeed, as it is well known (see (1.112)),

$$\frac{1}{2}\mathscr{L}_{\nabla f}\langle\, ,\, \rangle = \mathrm{Hess}(f). \tag{8.3}$$

L.J. Alías et al., *Maximum Principles and Geometric Applications*,
Springer Monographs in Mathematics, DOI 10.1007/978-3-319-24337-5_8

Note that the left-hand side of (8.2) is the *Bakry-Emery Ricci tensor* Ric_f, that is, (8.2) can be re-written as

$$\mathrm{Ric}_f = \lambda\langle\, , \,\rangle. \tag{8.4}$$

Note also that when X or ∇f are Killing vector fields (8.1) and (8.2) reduce to the *Einstein equation*

$$\mathrm{Ric} = \lambda\langle\, , \,\rangle, \tag{8.5}$$

thus both (8.1) and (8.2) could be interpreted as a perturbation of the latter. When X is Killing or $X = 0$, or f is constant, we call the underlying Einstein manifold a *trivial* Ricci soliton.

In the rest of the chapter we shall adopt the *elliptic* point of view, focusing only on the defining Eqs. (8.1) and (8.2) and on consequent properties. The interested reader may consult for instance [61] for the original parabolic setting related to the dynamical context of the Ricci flow. In the case of (8.2), the soliton is called a *gradient Ricci soliton*; as we shall see below, in this latter case more sophisticated technical tools are available to reach geometric conclusions. However, at first, we study the *generic* case (i.e. when the vector field is not necessarily a gradient of some potential f) and then we specialize to gradient solitons improving, in this other setting, the general conclusions. We observe that Naber [204, Theorem 1.3] has shown that a generic Ricci soliton $(M, \langle\, , \,\rangle, X)$ which is complete, shrinking and with bounded curvature is in fact a gradient shrinking Ricci soliton, that is, for some function $f \in C^\infty(M)$ and some Killing field Y, $X = \nabla f + Y$. However, in the complete noncompact case, there exist Ricci solitons that are not gradient. For instance, they are explicitly constructed by Baird and Danielo [35], Baird [34], Lott [179] and Lauret [166, 167]. This justifies the study of the general case.

8.1 Basic Formulas for Generic Ricci Solitons

The proofs of our results on generic Ricci solitons are based on three interesting formulas. The first we are going to present is due to Bochner [53] (see also [221, 232]); to perform computations we shall use the method of the moving frames referring to a local orthonormal coframe $\{\theta^i\}$ for the metric with corresponding Levi-Civita connection and curvature forms indicated, respectively, with $\{\theta^i_j\}$ and $\{\Theta^i_j\}$, $1 \le i, j \le m$, $m = \dim M$; as usual, the Einstein summation convention is in force.

Lemma 8.1 (Generalized Bochner Formula) *Let $Y \in \mathfrak{X}(M)$. Then*

$$\mathrm{div}\,(\mathcal{L}_Y\langle\, , \,\rangle)\,(Y) = \frac{1}{2}\Delta|Y|^2 - |\nabla Y|^2 + \mathrm{Ric}(Y, Y) + \nabla_Y(\mathrm{div}\, Y). \tag{8.6}$$

Remark 8.1 Note that $\operatorname{div}(\mathscr{L}_Y\langle\, ,\,\rangle)$ is a 1-form (see Remark 1.22 in Chap. 1).

Remark 8.2 Note that (8.6) is valid also when Y is only a C^2 vector field on M.

Remark 8.3 Equation (8.6) indeed generalizes the usual Bochner formula (see (1.176)). To see this, for $u \in C^3(M)$ let $Y = \nabla u$. Then, because of (8.3), (8.6) becomes

$$2\operatorname{div}(\operatorname{Hess}(u))(\nabla u) = \frac{1}{2}\Delta|\nabla u|^2 - |\operatorname{Hess}(u)|^2 + \operatorname{Ric}(\nabla u, \nabla u) + \langle \nabla \Delta u, \nabla u\rangle.$$

Since

$$\operatorname{div}(\operatorname{Hess}(u))(\nabla u) = \operatorname{Ric}(\nabla u, \nabla u) + \langle \nabla \Delta u, \nabla u\rangle,$$

from the above we immediately deduce that (8.6) is, in this case, equivalent to

$$\frac{1}{2}\Delta|\nabla u|^2 = |\operatorname{Hess}(u)|^2 + \operatorname{Ric}(\nabla u, \nabla u) + \langle \nabla \Delta u, \nabla u\rangle. \tag{8.7}$$

Proof (of Lemma 8.1) Let $\{e_i\}$ be the orthonormal frame dual to $\{\theta^i\}$. Then $Y = Y^i e_i = Y_i e_i$, and setting Y_{ij} for the coefficients of the covariant derivative ∇Y of Y we have, according to Sect. 1.2,

$$Y_{ij}\theta^j = dY_i - Y_k\theta^k_i. \tag{8.8}$$

Differentiating (8.8), using the definition of covariant derivative, the structure equations (1.4) and (1.32) and the components R^i_{jkt} of the Riemann curvature tensor (1.34) we obtain

$$Y_{ikj}\theta^k \wedge \theta^j = \frac{1}{2} Y_t R^t_{ikj}\theta^k \wedge \theta^j.$$

Thus, skew-symmetrizing we deduce

$$Y_{ijk} - Y_{ikj} = Y_t R^t_{ijk} = Y_t R_{tijk}. \tag{8.9}$$

Since (see Eq. (1.31))

$$\mathscr{L}_Y\langle\, ,\,\rangle = (Y_{ik} + Y_{ki})\theta^i \otimes \theta^k,$$

we have

$$\operatorname{div}(\mathscr{L}_Y\langle\, ,\,\rangle)(Y) = Y_i Y_{ikk} + Y_i Y_{kik}. \tag{8.10}$$

From the commutation relations (8.9), tracing with respect to i and k, we obtain

$$Y_{kik} = Y_{kki} + Y_t R_{tkik} = Y_{kki} + Y_t R_{ti}, \tag{8.11}$$

where, as usual, with R_{ti} we have indicated the components of the Ricci tensor. It follows that

$$Y_i Y_{kik} = \nabla_Y(\operatorname{div} Y) + \operatorname{Ric}(Y, Y). \tag{8.12}$$

On the other hand, from $|Y|^2 = Y_i Y_i$, we deduce

$$d|Y|^2 = 2Y_i Y_{ik} \theta^k$$

and

$$\Delta|Y|^2 = 2Y_{ik}Y_{ik} + 2Y_i Y_{ikk},$$

or, in other words,

$$\frac{1}{2}\Delta|Y|^2 = |\nabla Y|^2 + Y_i Y_{ikk}. \tag{8.13}$$

Substituting (8.12) and (8.13) into (8.10) we immediately obtain (8.6). □

Recalling the definition of Δ_X, that is the *X-Laplacian* $\Delta_X = \Delta - \langle X, \nabla \rangle$, (see Chap. 3), as a direct consequence of Lemma 8.1 we obtain the following (see also [221])

Proposition 8.1 *Let* $(M, \langle\, ,\, \rangle, X)$ *be a generic Ricci soliton structure on* $(M, \langle\, ,\, \rangle)$. *Then*

$$\frac{1}{2}\Delta|X|^2 = |\nabla X|^2 - \operatorname{Ric}(X, X) \tag{8.14}$$

or, equivalently,

$$\frac{1}{2}\Delta_X|X|^2 = |\nabla X|^2 - \lambda|X|^2. \tag{8.15}$$

Proof We trace the soliton equation (8.1) to obtain

$$S + \operatorname{div} X = m\lambda$$

where S is the scalar curvature of $(M, \langle\, ,\, \rangle)$. From here we deduce

$$\nabla S = -\nabla \operatorname{div} X. \tag{8.16}$$

Next, by (1.68)

$$2R_{ik,i} = S_k \tag{8.17}$$

or, in other words,

$$\nabla S = 2 \operatorname{div} \operatorname{Ric} \tag{8.18}$$

(note the little abuse of notation: more precisely, we should write $\nabla S = 2(\operatorname{div} \operatorname{Ric})^\sharp$, or equivalently $dS = 2 \operatorname{div} \operatorname{Ric}$). Comparing (8.16) and (8.18) we obtain

$$\nabla \operatorname{div} X = -2 \operatorname{div} \operatorname{Ric} . \tag{8.19}$$

Now, taking the divergence of (8.1) and using the fact that $\operatorname{div}(\lambda \langle\, , \rangle) = 0$, we get

$$\operatorname{div}(\mathscr{L}_X \langle\, , \rangle) = -2 \operatorname{div} \operatorname{Ric},$$

and (8.19) yields

$$\nabla \operatorname{div} X = \operatorname{div}(\mathscr{L}_X \langle\, , \rangle).$$

In particular,

$$\nabla_X \operatorname{div} X = \operatorname{div}(\mathscr{L}_X \langle\, , \rangle)(X). \tag{8.20}$$

Thus applying (8.6) of Lemma 8.1 we immediately deduce (8.14). As for (8.15) observe that

$$\mathscr{L}_X \langle\, , \rangle (X, X) = \langle X, \nabla |X|^2 \rangle$$

and use the soliton equation (8.1) into (8.14). □

Note that in the compact case formula (8.14) immediately yields results via integration. For instance, generalizing on Petersen and Wylie [221], Barros and Ribeiro [38] have proved the following

Proposition 8.2 *Let $(M, \langle\, , \rangle, X)$ be a compact generic Ricci soliton of dimension $m \geq 3$. If $\int_M \operatorname{Ric}(X, X) \leq 0$ then X is a Killing vector field and the soliton is trivial.*

Proof Integrating (8.14) we have

$$\int_M |\nabla X|^2 = \int_M \operatorname{Ric}(X, X) \leq 0,$$

so that $\nabla X \equiv 0$ and X, in particular, is Killing (see e.g. [232, Proposition 5.12]). □

For later purposes it is worth to give another form of (8.15) in case $X = \nabla f$. Let us recall the notation Δ_f for the *f-Laplacian*, that is, the diffusion operator acting, say, on $u \in C^2(M)$ as

$$\Delta_f u = \Delta_{\nabla f} u = \Delta u - \langle \nabla f, \nabla u \rangle = e^f \operatorname{div}(e^{-f} \nabla u)$$

(see also Sect. 3.1 in Chap. 3).

Corollary 8.1 *Let* $(M, \langle\, ,\,\rangle, \nabla f)$ *be a gradient soliton on* $(M, \langle\, ,\,\rangle)$. *Then*

$$\frac{1}{2}\Delta_f |\nabla f|^2 = |\operatorname{Hess}(f)|^2 - \lambda |\nabla f|^2. \tag{8.21}$$

Proof Formula (8.21) follows directly from (8.15) by setting $X = \nabla f$. □

Before proceeding to the next proposition we need to determine some more "commutation rules".

Lemma 8.2 *Let* $Y \in \mathfrak{X}(M)$. *Then*

$$Y_{tkkt} - Y_{kktt} = \frac{1}{2}\langle \nabla S, Y \rangle + \frac{1}{2}\operatorname{Tr}\left(\ell_Y\langle\, ,\,\rangle \circ \operatorname{ric}\right), \tag{8.22}$$

where $\ell_Y\langle\, ,\,\rangle$ *and* ric *are the* $(1,1)$*-versions, respectively, of* $\mathscr{L}_Y\langle\, ,\,\rangle$ *and* Ric.

Proof We start from the commutation rule (8.9). By taking covariant derivative we deduce

$$Y_{ijkt} - Y_{ikjt} = Y_{st} R_{sijk} + Y_s R_{sijk,t}. \tag{8.23}$$

Next, we recall that, by definition of covariant derivative,

$$Y_{ijk}\theta^k = dY_{ij} - Y_{tj}\theta^t_i - Y_{it}\theta^t_j. \tag{8.24}$$

Thus, differentiating both members of (8.24), using the structure equations and (8.24) itself, we arrive at

$$Y_{ijkl}\theta^l \wedge \theta^k = -\frac{1}{2}(Y_{tj}R_{tilk} + Y_{it}R_{tjlk})\theta^l \wedge \theta^k$$

from which, skew-symmetrizing with respect to l and k, we obtain

$$Y_{ijkl} - Y_{ijlk} = Y_{tj}R_{tikl} + Y_{it}R_{tjkl}. \tag{8.25}$$

Now (8.22) follows immediately from (8.25), (8.23), (8.18) and tracing. □

For later use we put together the next commutation rules (see Chap. 1 for details):

Lemma 8.3 *For the Ricci tensor we have the following*

$$R_{ij,k} = R_{ji,k} \tag{8.26}$$

$$R_{ij,k} - R_{jk,j} = -R_{tijk,t} \tag{8.27}$$

$$R_{ij,kl} - R_{ij,lk} = R_{is}R_{sjkl} + R_{js}R_{sikl}. \tag{8.28}$$

Proof Equation (8.26) is obvious by the symmetry of the Ricci tensor. Equation (8.27) is (1.67) and comes from the second Bianchi identity, while (8.28) is (1.123). □

We are now ready to prove

Proposition 8.3 *Let $(M, \langle\, ,\, \rangle, X)$ be a Ricci soliton structure on $(M, \langle\, ,\, \rangle)$ with soliton constant λ, and let S be the scalar curvature. Then*

$$\frac{1}{2}\Delta_X R_{ij} = \lambda R_{ij} - R_{ikjt}R_{kt} + \frac{1}{4}R_{it}(X_{tj} - X_{jt}) + \frac{1}{4}R_{tj}(X_{ti} - X_{it}) \tag{8.29}$$

and

$$\frac{1}{2}\Delta_X S = \lambda S - |\operatorname{Ric}|^2 = \lambda S - \frac{S^2}{m} - \left|\operatorname{Ric} - \frac{S}{m}\langle\, ,\, \rangle\right|^2. \tag{8.30}$$

Proof We start from the soliton equation (8.1) which, in components, reads

$$R_{ij} + \frac{1}{2}(X_{ij} + X_{ji}) = \lambda\delta_{ij}. \tag{8.31}$$

Differentiating (8.31) we get

$$R_{ji,k} = -\frac{1}{2}(X_{ijk} + X_{jik}); \tag{8.32}$$

skew-symmetrizing and using the commutation relation (8.9) we deduce

$$\begin{aligned} R_{ji,k} - R_{jk,i} &= -\frac{1}{2}(X_{ijk} + X_{jik} - X_{jki} - X_{kji}) \\ &= -\frac{1}{2}X_sR_{sjik} + \frac{1}{2}\left(X_{kji} - X_{ijk}\right). \end{aligned} \tag{8.33}$$

Now we observe that, using the first Bianchi identity (1.44),

$$X_{kji} - X_{ijk} = X_{kij} + X_sR_{skji} - \left(X_{ikj} + X_sR_{sijk}\right) = X_{kij} - X_{ikj} - X_sR_{sjik},$$

thus (8.33) becomes

$$R_{ji,k} = R_{jk,i} - X_s R_{sjik} + \frac{1}{2}\left(X_{kij} - X_{ikj}\right). \tag{8.34}$$

From the definition of ΔR_{ij}, using (8.34), (8.27), (8.28) and (8.17) (that is (1.68)) we have

$$\begin{aligned} \Delta R_{ij} &= R_{ij,kk} = R_{ji,kk} = \left(R_{ji,k}\right)_k \qquad (8.35)\\ &= R_{jk,ik} + \left[-X_s R_{sjik} + \frac{1}{2}\left(X_{kij} - X_{ikj}\right)\right]_k \qquad (8.36)\\ &= R_{kj,ik} - X_{sk} R_{sjik} - X_s R_{sjik,k} + \frac{1}{2}\left(X_{kijk} - X_{ikjk}\right). \end{aligned}$$

From (8.28) we have that $R_{kj,ik} = R_{kj,ki} + R_{sj}R_{si} + R_{ks}R_{sjik}$, while the second Bianchi identity (see (1.51)) implies $-R_{sjik,k} = -R_{iksj,k} = -R_{is,j} + R_{ij,s}$, thus, inserting the previous relations in (8.35) and using again (8.17) and the soliton equation (8.31), we have

$$\begin{aligned} \Delta R_{ij} &= R_{kj,ki} + R_{sj}R_{si} + R_{ks}R_{sjik} + X_s R_{ij,s} - X_s R_{is,j} - X_{sk}R_{sjik} + \frac{1}{2}\left(X_{kijk} - X_{ikjk}\right)\\ &= \frac{1}{2}S_{ji} + R_{it}R_{tj} + R_{sk}R_{iksj} + \langle X, \nabla R_{ij}\rangle - X_s R_{is,j} - (2\lambda\delta_{sk} - 2R_{sk} - X_{ks})R_{sjik}\\ &+ \frac{1}{2}\left(X_{kijk} - X_{ikjk}\right), \end{aligned}$$

which implies

$$\Delta_X R_{ij} = 2\lambda R_{ij} - 3R_{sk}R_{isjk} + R_{it}R_{tj} + \frac{1}{2}S_{ji} - X_s R_{is,j} + X_{ks}R_{sjik} + \frac{1}{2}\left(X_{kijk} - X_{ikjk}\right). \tag{8.37}$$

Now we have, tracing (8.31), taking the covariant derivative and using the commutation relations (8.9) and (8.25),

$$\begin{aligned} \frac{1}{2}S_{ji} &= -\frac{1}{2}X_{kkji} = -\frac{1}{2}\left(X_{kjki} + X_t R_{tkkj,i} + X_{ti}R_{tkkj}\right) = -\frac{1}{2}X_{kjki} + \frac{1}{2}X_t R_{tj,i} + \frac{1}{2}X_{ti}R_{tj}\\ &= -\frac{1}{2}X_{kjik} + \frac{1}{2}R_{is}X_{sj} - \frac{1}{2}X_{ks}R_{kisj} + \frac{1}{2}X_{ti}R_{tj} + \frac{1}{2}X_t R_{tj,i}. \end{aligned}$$

Using the latter in (8.37) and (8.23) we deduce

$$\begin{aligned} \Delta_X R_{ij} &= 2\lambda R_{ij} - 3R_{sk}R_{isjk} + R_{it}R_{tj} - \frac{1}{2}X_{kjik} + \frac{1}{2}R_{it}X_{tj} - \frac{1}{2}X_{ks}R_{kisj} \qquad (8.38)\\ &+ \frac{1}{2}X_{ti}R_{tj} + \frac{1}{2}X_t R_{tj,i} - X_t R_{ti,j} + X_{ks}R_{sjik} + \frac{1}{2}\left(X_{kijk} - X_{ikjk}\right) \end{aligned}$$

$$
\begin{aligned}
&= 2\lambda R_{ij} - 3R_{sk}R_{isjk} + R_{it}R_{tj} + \frac{1}{2}\big(X_{kijk} - X_{kjik}\big) + \frac{1}{2}R_{it}X_{tj} + \frac{3}{2}X_{ks}R_{sjik} \\
&+ \frac{1}{2}X_{ti}R_{tj} + \frac{1}{2}X_tR_{tj,i} - X_tR_{ti,j} - \frac{1}{2}X_{ikjk} \\
&= 2\lambda R_{ij} - 3R_{sk}R_{isjk} + R_{it}R_{tj} + \frac{1}{2}X_{tk}R_{tkij} + \frac{1}{2}X_tR_{tkij,k} + \frac{1}{2}R_{it}X_{tj} \\
&+ \frac{3}{2}X_{kt}R_{tjik} + \frac{1}{2}X_{ti}R_{tj} + \frac{1}{2}X_tR_{tj,i} - X_tR_{ti,j} \\
&+ \frac{1}{2}X_{tj}R_{ti} + \frac{1}{2}X_tR_{ti,j} + \frac{1}{2}X_{tk}R_{tikj} - \frac{1}{2}X_{it}R_{tj} \\
&= 2\lambda R_{ij} - 3R_{sk}R_{isjk} + R_{it}R_{tj} + \frac{1}{2}X_{tk}R_{tkij} + R_{it}X_{tj} + X_{kt}R_{tjik} + \frac{1}{2}R_{tj}(X_{ti} - X_{it}).
\end{aligned}
$$

Now we observe that, using the first Bianchi identity and the soliton equation,

$$
\begin{aligned}
\frac{1}{2}X_{tk}R_{tkij} + X_{kt}R_{tjik} &= -\frac{1}{2}X_{tk}R_{tijk} - \frac{1}{2}X_{tk}R_{tjki} + X_{kt}R_{tjik} \\
&= -\frac{1}{2}X_{kt}R_{kijt} - \frac{1}{2}X_{tk}R_{tjki} + X_{kt}R_{tjik} \\
&= -\frac{1}{2}X_{kt}R_{tjik} - \frac{1}{2}X_{tk}R_{tjki} + X_{kt}R_{tjik} \\
&= \frac{1}{2}X_{kt}R_{tjik} - \frac{1}{2}X_{tk}R_{tjki} \\
&= \frac{1}{2}R_{tjik}(X_{kt} + X_{tk}) \\
&= -\lambda R_{ij} + R_{kt}R_{ikjt}.
\end{aligned}
$$

Using the latter in (8.38) together with the soliton equation we get

$$
\begin{aligned}
\Delta_X R_{ij} &= 2\lambda R_{ij} - 3R_{sk}R_{isjk} + R_{it}R_{tj} - \lambda R_{ij} + R_{kt}R_{ikjt}R_{it}X_{tj} \\
&\quad + \frac{1}{2}R_{tj}(X_{ti} - X_{it}) + \frac{1}{2}R_{tj}(X_{ti} - X_{it}) \\
&= \lambda R_{ij} - 2R_{kt}R_{ikjt} + R_{it}\left[\lambda\delta_{tj} - \frac{1}{2}(X_{tj} + X_{jt})\right] + R_{it}X_{tj} + \frac{1}{2}R_{tj}(X_{ti} - X_{it}) \\
&= 2\lambda R_{ij} - 2R_{kt}R_{ikjt} + \frac{1}{2}R_{it}\big(X_{tj} - X_{jt}\big) + \frac{1}{2}R_{tj}(X_{ti} - X_{it}),
\end{aligned}
$$

that is, Eq. (8.29). Tracing with respect to i and j we immediately deduce Eq. (8.30). □

As an immediate application we have (see [39])

Proposition 8.4 *Let $(M, \langle\,,\,\rangle, X)$ be a compact generic Ricci soliton of dimension $m \geq 2$. If the scalar curvature is constant the metric is Einstein.*

Proof Tracing equation (8.1) we have

$$m\lambda - S = \operatorname{div} X. \tag{8.39}$$

Thus from (8.30) we get

$$\left|\operatorname{Ric} - \frac{S}{m}\langle\,,\,\rangle\right|^2 = \frac{S}{m}\operatorname{div} X + \frac{1}{2}\langle\nabla S, X\rangle - \frac{1}{2}\Delta S.$$

Integrating over M and using integration by parts we obtain

$$\begin{aligned}\int_M \left|\operatorname{Ric} - \frac{S}{m}\langle\,,\,\rangle\right|^2 &= \frac{1}{m}\int_M S\operatorname{div} X + \frac{1}{2}\int_M \langle\nabla S, X\rangle \\ &= \frac{m-2}{2m}\int_M \langle\nabla S, X\rangle.\end{aligned}$$

Since S is constant it follows that

$$\operatorname{Ric} = \frac{S}{m}\langle\,,\,\rangle$$

and the metric is Einstein. Moreover, from the soliton equation we also have

$$\frac{1}{2}\mathcal{L}_X\langle\,,\,\rangle = \left(\lambda - \frac{S}{m}\right)\langle\,,\,\rangle.$$

Thus if X is not Killing $\lambda - \frac{S}{m}$ is a nonzero constant and by a result of Yano and Nagano [278] we conclude that M is isometric to a Euclidean sphere of dimension m. □

Corollary 8.2 *Let $(M, \langle\,,\,\rangle, \nabla f)$ be a gradient Ricci soliton on $(M, \langle\,,\,\rangle)$ with soliton constant λ, and let S be the scalar curvature. Then*

$$\frac{1}{2}\Delta_f S = \lambda S - |\operatorname{Ric}|^2 = \lambda S - \frac{S^2}{m} - \left|\operatorname{Ric} - \frac{S}{m}\langle\,,\,\rangle\right|^2. \tag{8.40}$$

Our aim is now to compute the Laplacian of $|T|^2$, where T is the traceless Ricci tensor defined in (1.70); we recall that, in components,

$$T_{ij} = R_{ij} - \frac{S}{m}\delta_{ij} \tag{8.41}$$

and a simple computation shows that

$$|T|^2 = |\operatorname{Ric}|^2 - \frac{S^2}{m}. \tag{8.42}$$

First we need the following simple lemma.

Lemma 8.4 *Let Z be a $(0,2)$-tensor, with (local) components Z_{ij}. Then*

$$\frac{1}{2}\Delta|Z|^2 = |\nabla Z|^2 + Z_{ij}Z_{ij,kk}. \tag{8.43}$$

Remark 8.4 Note that the quantity $Z_{ij}Z_{ij,kk}$ is globally defined, since it is the difference between the two globally defined quantities $\frac{1}{2}\Delta|Z|^2$ and $|\nabla Z|^2$. Some authors write $Z_{ij}Z_{ij,kk}$ as $\langle Z, \Delta Z\rangle$: this is a slight abuse of notation, since the quantity $Z_{ij,kk} = \Delta Z$ is *not* globally defined.

Proof Since $|Z|^2 = Z_{ij}Z_{ij}$ we have

$$\left(|Z|^2\right)_k = 2Z_{ij,k}Z_{ij}$$

and

$$\Delta|Z|^2 = \left(|Z|^2\right)_{kk} = \left(2Z_{ij,k}Z_{ij}\right)_k = 2Z_{ij,kk}Z_{ij} + 2Z_{ij,k}Z_{ij,k},$$

which easily implies Eq. (8.43) since $Z_{ij,k}Z_{ij,k} = |\nabla Z|^2$. □

Remark 8.5 The same formula can be used, with obvious modifications, to compute the Laplacian of the squared norm of tensors of any type.

Using (8.42), Lemma 8.4 and the fact that $\Delta(uv) = u\Delta v + 2\langle\nabla u, \nabla v\rangle + v\Delta u$ for u and v at least $C^2(M)$, it follows that

$$\frac{1}{2}\Delta|T|^2 = \frac{1}{2}\Delta|\operatorname{Ric}|^2 - \frac{1}{2m}\Delta S^2 = |\nabla \operatorname{Ric}|^2 + R_{ij}R_{ij,kk} - \frac{1}{m}S\Delta S - \frac{1}{m}|\nabla S|^2. \tag{8.44}$$

We have

Proposition 8.5 *Let $(M, \langle\, ,\, \rangle, X)$ be a generic Ricci soliton structure on $(M, \langle\, ,\, \rangle)$, $m \geq 3$, with soliton constant λ. Let S be the scalar curvature, T the traceless Ricci tensor and W the Weyl tensor. Then*

$$\frac{1}{2}\Delta_X|T|^2 = |\nabla T|^2 + 2\left(\lambda - \frac{m-2}{m(m-1)}S\right)|T|^2 + \frac{4}{m-2}\operatorname{Tr}(t^3) - 2T_{ik}T_{sj}W_{ksij}, \tag{8.45}$$

where t is the $(1,1)$-version of T.

Proof Using (8.29) we have

$$\begin{aligned}2R_{ij}R_{ij,kk} &= 2R_{ik}\Delta R_{ik}\\ &= 2\lambda|\operatorname{Ric}|^2 + 2\operatorname{Tr}(\operatorname{ric}^3) + \frac{1}{2}\langle\nabla|\operatorname{Ric}|^2, X\rangle\\ &\quad + R_{ik}S_{ik} - X_{js}R_{skji}R_{ik} - X_sR_{ik}R_{is,k}\\ &\quad -4R_{ik}R_{sj}R_{skji} + X_{jkij}R_{ik} - X_{ikjj}R_{ik},\end{aligned} \tag{8.46}$$

where ric is the $(1,1)$-version of Ric. First we analyze the term $X_{jkij}R_{ik}$. Towards this aim we consider the soliton equation (8.31); tracing with respect to i and j we obtain

$$S + X_{tt} = m\lambda,$$

so that, taking covariant derivatives,

$$S_i = -X_{tti} \tag{8.47}$$

and similarly, from (8.47)

$$S_{ik} = -X_{ttik}. \tag{8.48}$$

It follows that

$$R_{ik}S_{ik} = -X_{ttik}R_{ik}. \tag{8.49}$$

From the commutation rules (8.25) and (8.23) we get

$$X_{jkij} = X_{jkji} + X_{tk}R_{ti} + X_{jt}R_{tkij} = X_{jjki} + X_{si}R_{sk} + X_sR_{sk.i} + X_{tk}R_{ti} + X_{jt}R_{tkij}$$

and therefore, using (8.49) and Eq. (8.31),

$$\begin{aligned}R_{ik}X_{jkij} &= -S_{ik}R_{ik} + R_{ik}(X_{kt} + X_{tk})R_{ti} + X_sR_{ik}R_{sk,i} + X_{jt}R_{ik}R_{tkij}\\ &= -S_{ik}R_{ik} + 2\lambda|\operatorname{Ric}|^2 - 2\operatorname{Tr}(\operatorname{ric}^3) + X_sR_{ik}R_{sk,i} + X_{jt}R_{ik}R_{tkij}.\end{aligned}$$

Substituting the latter into (8.46) and simplifying we obtain

$$2R_{ik}\Delta R_{ik} = 4\lambda|\operatorname{Ric}|^2 + \frac{1}{2}\langle\nabla|\operatorname{Ric}|^2, X\rangle - 2X_{js}R_{ik}R_{skji} - 4R_{ik}R_{sj}R_{skji} - X_{iktt}R_{ik}. \tag{8.50}$$

Next, we analyze the term $X_{iktt}R_{ik}$. Towards this aim we take covariant derivative of the soliton equation (8.31) and get

$$R_{ij,k} = -\frac{1}{2}(X_{ijk} + X_{jik}).$$

Tracing with respect to j and k yields

$$R_{ik,k} = -\frac{1}{2}(X_{ikk} + X_{kik}),$$

so that, using (8.11), (8.17) and (8.47),

$$S_k = -X_{ktt} - X_{tkt} = -X_{ktt} - X_{ttk} - X_s R_{sk} = S_k - X_{ktt} - X_s R_{sk},$$

that is,

$$X_{itt} = -X_s R_{si}.$$

Taking covariant derivative of the latter

$$X_{ittk} = -X_{sk} R_{si} - X_s R_{si,k}. \tag{8.51}$$

Now, from (8.23) and (8.25) we obtain

$$X_{ittk} = X_{itkt} + X_{st} R_{sitk} + X_{is} R_{sttk} = X_{iktt} + 2X_{st} R_{sitk} - X_{is} R_{sk} + X_s R_{sitk,t}.$$

Hence, using (8.51) and (8.27) we deduce

$$\begin{aligned} R_{ik} X_{iktt} &= -X_s R_{ik} R_{si,k} - X_s R_{ik} R_{sitk,t} - 2X_{st} R_{sitk} R_{ik} \\ &= -X_s R_{ik} R_{si,k} - 2X_{st} R_{sitk} R_{ik} + X_s R_{ik}(R_{ks,i} - R_{ki,s}) \\ &= -\frac{1}{2}\langle \nabla |\operatorname{Ric}|^2, X\rangle - 2X_{st} R_{sitk} R_{ik}. \end{aligned} \tag{8.52}$$

We substitute (8.52) in (8.50) to get

$$2R_{ik}\Delta R_{ik} = 4\lambda |\operatorname{Ric}|^2 + \langle \nabla |\operatorname{Ric}|^2, X\rangle + 4R_{ik} R_{sj} R_{skij}. \tag{8.53}$$

Thus, from (8.44), (8.53) and (8.30) we finally have

$$\begin{aligned} \Delta |T|^2 &= 2|\nabla \operatorname{Ric}|^2 - \frac{2}{m}|\nabla S|^2 + 4\lambda |\operatorname{Ric}|^2 + \langle \nabla |\operatorname{Ric}|^2, X\rangle \\ &\quad + 4R_{ik} R_{sj} R_{skij} - 4\frac{\lambda}{m} S^2 - \frac{2}{m} S\langle \nabla S, X\rangle \\ &\quad + \frac{4}{m^2} S^3 + \frac{4}{m} S |T|^2. \end{aligned} \tag{8.54}$$

An immediate computation shows that

$$|\nabla T|^2 = |\nabla \operatorname{Ric}|^2 - \frac{1}{m}|\nabla S|^2.$$

Using this fact and (8.42) after some algebraic manipulations from (8.54) we obtain

$$\frac{1}{2}\Delta_X|T|^2 = |\nabla T|^2 + 2\lambda|T|^2 + \frac{2}{m}\left(\frac{S^2}{m} + |T|^2\right)S + 2R_{ik}R_{sj}R_{skij}. \tag{8.55}$$

To analyze the last term on the right-hand side of (8.55) we use the decomposition of the Riemann curvature tensor into its reducible components given in (1.84). A simple computation yields

$$\begin{aligned} R_{ik}R_{js}R_{ijks} &= W_{ijks}R_{ik}R_{js} + \frac{2m-1}{(m-1)(m-2)}S|\operatorname{Ric}|^2 \\ &\quad - \frac{2}{m-2}\operatorname{Tr}(\operatorname{ric}^3) - \frac{S^3}{(m-1)(m-2)}. \end{aligned}$$

Next, we observe that

$$\operatorname{Tr}(\operatorname{ric}^3) = \operatorname{Tr}(t^3) + \frac{3}{m}S|\operatorname{Ric}|^2 - \frac{2}{m^2}S^3,$$

and that

$$W_{ijks}R_{ik}R_{js} = W_{ijks}T_{ik}T_{js}$$

since all the traces of the Weyl tensor vanish. Inserting these relations into (8.55) and using (8.42) we obtain (8.45), completing the proof. □

8.2 The Validity of the Maximum Principle on Solitons

In analogy with the Bakry-Emery Ricci tensor we define the tensor Ric_X as

$$\operatorname{Ric}_X = \operatorname{Ric} + \frac{1}{2}\mathscr{L}_X\langle\,,\,\rangle; \tag{8.56}$$

indeed, note that, if $X = \nabla f$ for some $f \in C^\infty(M)$, then $\operatorname{Ric}_X = \operatorname{Ric}_{\nabla f} = \operatorname{Ric}_f$ (see Eq. (8.4)). The next is a key result.

Theorem 8.1 *Let $(M, \langle\,,\,\rangle)$ be a complete manifold of dimension m and let $X \in \mathfrak{X}(M)$ be a vector field satisfying the growth condition*

$$|X| \leq \sqrt{F(r)} \tag{8.57}$$

for some positive, nondecreasing function $F \in C^1(\mathbb{R}_0^+)$, where $r(x)$ is the distance from a fixed origin o. Suppose that

$$\mathrm{Ric}_X(\nabla r, \nabla r) \geq -(m-1)G(r) \tag{8.58}$$

for a positive $G \in C^1(\mathbb{R}_0^+)$ such that

$$\inf_{\mathbb{R}_0^+} \frac{G'}{G^{3/2}} > -\infty. \tag{8.59}$$

Then there exists $A = A(m) > 0$ sufficiently large such that

$$\Delta_X r \leq A\sqrt{G(r)} + \sqrt{F(r)} \tag{8.60}$$

pointwise in $M \setminus (\{o\} \cup \mathrm{cut}(o))$ and weakly on all of M.

Proof Let h be the solution on $\mathbb{R}_0^+$ of the Cauchy problem

$$\begin{cases} h'' - G(s)h = 0 \\ h(0) = 0, h'(0) = 1; \end{cases} \tag{8.61}$$

note that, since $G \geq 0$ then $h > 0$ on $\mathbb{R}^+$. Fix $x \in M \setminus (\{o\} \cup \mathrm{cut}(o))$ and let $\gamma : [0, \ell] \to M$, $\ell = \mathrm{length}(\gamma)$, be a minimizing geodesic with $\gamma(0) = o$ and $\gamma(\ell) = x$. Note that $G(r(\gamma(t))) = G(t)$. From Bochner formula (8.7) applied to the distance function r (outside $\{o\} \cup \mathrm{cut}(o)$), we have

$$0 = |\,\mathrm{Hess}(r)|^2 + \mathrm{Ric}(\nabla r, \nabla r) + \langle \nabla \Delta r, \nabla r \rangle \tag{8.62}$$

so that, using the inequality

$$|\,\mathrm{Hess}(r)|^2 \geq \frac{(\Delta r)^2}{m-1}$$

it follows that the function $\psi(t) = (\Delta r) \circ \gamma(t)$, $t \in (0, \ell]$, satisfies the Riccati differential inequality

$$\psi' + \frac{1}{m-1}\psi^2 \leq -\mathrm{Ric}(\dot\gamma, \dot\gamma) \tag{8.63}$$

on $(0, \ell]$. With h as in (8.61), using the definition (8.56) of Ric_X, (8.58) and (8.63) we compute

$$\begin{aligned}(h^2\psi)' &= 2hh'\psi + h^2\psi' \\ &\leq 2hh'\psi - \frac{h^2}{m-1}\psi^2 + (m-1)G(t)h^2 + \frac{1}{2}\mathscr{L}_X\langle\,,\,\rangle(\dot\gamma, \dot\gamma)\end{aligned}$$

$$= -\left(\frac{h\psi}{\sqrt{m-1}} - \sqrt{m-1}h'\right)^2 + (m-1)(h')^2$$
$$+(m-1)G(t)h^2 + (\langle X, \dot{\gamma}\rangle)'h^2.$$

Note that in passing from the second to the third line of the above inequality we have used (1.31) of Chap. 1. We define

$$\psi_G(t) = (m-1)\frac{h'}{h}(t)$$

so that, using (8.61), we have

$$(h^2\psi_G)' = (m-1)(h')^2 + (m-1)G(t)h^2.$$

Inserting the latter into the above inequality we obtain

$$(h^2\psi)' \leq (h^2\psi_G')^2 + h^2(\langle X, \dot{\gamma}\rangle)'.$$

Integrating on $[0, r]$ and using (8.61) yields

$$h^2(r)\psi(r) \leq h^2(r)\psi_G(r) + \int_0^r h^2(\langle X, \dot{\gamma}\rangle)'. \tag{8.64}$$

Next, we define

$$\psi_X = (\Delta_X r) \circ \gamma = \Delta_X r \circ \gamma - \langle X, \nabla r\rangle \circ \gamma = \psi - \langle X, \nabla r\rangle. \tag{8.65}$$

Thus, using (8.64), (8.61) and integrating by parts, we compute

$$\begin{aligned} h^2\psi_X(r) &\leq h^2\psi_G(r) - h^2\langle X, \nabla r\rangle \circ \gamma(r) + \int_0^r h^2(\langle X, \dot{\gamma}\rangle)' \\ &= h^2\psi_G(r) - h^2\langle X, \nabla r\rangle \circ \gamma(r) + \left(h^2 \left.\langle X, \dot{\gamma}\rangle\right|_o^r\right) - \int_0^r (h^2)'\langle X, \dot{\gamma}\rangle \\ &= h^2\psi_G(r) - \int_0^r (h^2)'\langle X, \dot{\gamma}\rangle, \end{aligned}$$

that is,

$$h^2\psi_X(r) \leq h^2\psi_G(r) - \int_0^r (h^2)'\langle X, \dot{\gamma}\rangle \tag{8.66}$$

on $(0, \ell]$. Observe now that by Cauchy-Schwarz inequality and (8.57)

$$-\langle X, \dot{\gamma}\rangle \leq |X| \leq \sqrt{F(r)},$$

while from (8.61) and $G \geq 0$ we deduce

$$(h^2)' = 2hh' \geq 0.$$

Thus, inserting the above into (8.66), using $(h^2)' \geq 0$, integrating by parts and recalling that $F' \geq 0$, we obtain

$$h^2\psi_X(r) \leq h^2\psi_G(r) + h^2\sqrt{F(r)} - \int_0^r \frac{h^2F'}{2\sqrt{F}} \leq h^2\psi_G(r) + h^2\sqrt{F(r)}$$

on $(0, \ell]$. It follows that

$$\psi_X(r) \leq \psi_G(r) + \sqrt{F(r)}$$

on $(0, \ell]$. In particular

$$\Delta_X r(x) \leq (m-1)\frac{h'(r(x))}{h(r(x))} + \sqrt{F(r(x))} \tag{8.67}$$

pointwise on $M \setminus (\{o\} \cup \mathrm{cut}(o))$. Proceeding as in the proof of Lemma 1.6 (see also Theorem 2.4 of [230]) one shows that (8.67) holds weakly on all of M. Next, we fix $D > 0$ and we define

$$g(t) = \frac{1}{D\sqrt{G(0)}}\left(e^{D\int_0^t \sqrt{G(s)}} - 1\right)$$

so that $g(0) = 0$ and $g'(0) = 1$. Furthermore

$$g'' - G(t)g \geq \frac{G}{\sqrt{G(0)}}\left(e^{D\int_0^t \sqrt{G(s)}}\left(\inf_{t\in\mathbb{R}_0^+} \frac{(G')^{3/2}}{2G}(t) + D - \frac{1}{D}\right)\right) \geq 0$$

for D sufficiently large, because of (8.59). Therefore, by Sturm comparison (see Lemma 1.4 in Chap. 1)

$$\frac{h'}{h}(t) \leq \frac{g'}{g}(t) = D\sqrt{G(t)}\frac{e^{D\int_0^t \sqrt{G(s)}}}{e^{D\int_0^t \sqrt{G(s)}} - 1} \leq D\sqrt{G(t)}. \tag{8.68}$$

Thus, from (8.67) and (8.68) we obtain (8.60). □

Corollary 8.3 *Let $(M, \langle\,,\,\rangle)$ be a complete manifold of dimension m and $X \in \mathfrak{X}(M)$ a vector field satisfying the growth condition*

$$|X| \leq \sqrt{F(r)} \tag{8.69}$$

for some positive, nondecreasing function $F \in C^1(\mathbb{R}_0^+)$. Suppose that

$$\mathrm{Ric}_X(\nabla r, \nabla r) \geq -(m-1)G(r) \tag{8.70}$$

for a positive $G \in C^1(\mathbb{R}_0^+)$ such that

$$\inf_{\mathbb{R}_0^+} \frac{G'}{G^{3/2}} > -\infty, \tag{8.71}$$

and assume that

$$\frac{1}{\sqrt{F}+\sqrt{G}} \notin L^1(+\infty), \quad (\sqrt{F}+\sqrt{G})'(t) \geq -B(\log t + 1) \tag{8.72}$$

for $t \gg 1$ and some constant $B \geq 0$. Then the Omori-Yau maximum principle holds on M for the operators Δ_X and Δ.

Proof Let $\gamma(x) = r(x)$; by (8.60)

$$\Delta_X r \leq C(\sqrt{F(r)} + \sqrt{G(r)})$$

for $r \gg 1$ and a positive constant C. Moreover

$$|\nabla r| = 1 \leq C(\sqrt{F(r)} + \sqrt{G(r)})$$

for some $C > 0$ and $r(x) \to +\infty$ as $x \to \infty$. Hence the validity of the Omori-Yau maximum principle for the operator Δ_X follows directly from Theorem 3.2 and Remark 3.3.

As for Δ, simply observe that, with the notations of the proof of Theorem 8.1,

$$\psi = \psi_X + \langle \nabla r, X \rangle \circ \gamma,$$

so that (8.57) and (8.60) give

$$\Delta r \leq B(\sqrt{F(r)} + \sqrt{G(r)})$$

for $r \gg 1$ and some constant $B > 0$ sufficiently large. Then the proof proceeds as before. □

We note that in both estimates for $\Delta_X r$ and Δr we have the dependence on the lower bound for Ric_X and the upper bound for $|X|$. It is worth, in view of our applications to solitons, to derive an upper estimate for $\Delta_X r$ which depends only on the lower bound for Ric_X. Towards this aim we go back to the proof of Theorem 8.1 to show the validity of the next

Proposition 8.6 *Let $(M, \langle\, , \,\rangle)$ be a complete manifold of dimension m and $X \in \mathfrak{X}(M)$. Suppose that* Ric_X *satisfies (8.58) for some $G \in C^0(\mathbb{R}_0^+)$. Then, there exist a sufficiently small geodesic ball B_R and a constant $C = C(B_R) > 0$ such that*

$$\Delta_X r(x) \le C + (m-1)\int_0^{r(x)} G(t)dt \tag{8.73}$$

weakly on $M \setminus B_R$.

Proof We reason as in the proof of Theorem 8.1 to arrive at

$$\psi' \le -\frac{1}{m-1}\psi^2 - \mathrm{Ric}(\dot\gamma, \dot\gamma). \tag{8.74}$$

We then define ψ_X as in (8.65) so that

$$\psi_X' = \psi' - (\langle X, \nabla r\rangle \circ \gamma)' = \psi' - \frac{1}{2}\mathscr{L}_X\langle\, , \,\rangle(\dot\gamma, \dot\gamma).$$

Thus, using (8.74), we obtain

$$\psi_X' \le -\frac{\psi^2}{m-1} - \mathrm{Ric}(\dot\gamma, \dot\gamma) - \frac{1}{2}\mathscr{L}_X(\dot\gamma, \dot\gamma),$$

and using (8.56)

$$\psi_X' \le -\mathrm{Ric}_X(\dot\gamma, \dot\gamma).$$

From (8.58) it follows that

$$\psi_X'(r) \le (m-1)G(r),$$

in $M \setminus (\{o\} \cup \mathrm{cut}(o))$. Choosing $\varepsilon > 0$ so small that B_ε is inside the domain of the normal coordinates at o and setting $C_\varepsilon = \max_{\partial B_\varepsilon} \Delta_X r$, integration over $[\varepsilon, r(x)]$ gives

$$\Delta_X r(x) \le C_\varepsilon + (m-1)\int_\varepsilon^{r(x)} G(t)dt \tag{8.75}$$

pointwise in $M \setminus (B_\varepsilon \cup \mathrm{cut}(o))$ and weakly on $M \setminus B_\varepsilon$. One can verify this second claim proceeding as in the proof of Lemma 1.6. Here is a second way suggested by the argument of Theorem 2.1 in [236].

Consider an exhaustion $\{\Omega_n\}$ of $M \setminus \mathrm{cut}(o)$ by bounded domains with smooth boundaries starshaped with respect to o with $\overline{B}_\varepsilon \subset \Omega_1$. Fix n and let ν be the outward unit normal to $\partial\Omega_n$; denote by $\rho(x) = \mathrm{dist}\,(x, \partial\Omega_n)$, with the convention that $\rho(x) > 0$ if $x \in \Omega_n$ and $\rho(x) < 0$ if $x \notin \Omega_n$. Thus ρ is the radial coordinate for the Fermi coordinates (see also Chap. 2, Sect. 2.3 and [72]) relative to $\partial\Omega_n$. By Gauss lemma $|\nabla\rho| = 1$ and $\nabla\rho = -\nu$ on $\partial\Omega_n$. Let

$$\Omega_{n,\delta} = \{x \in \Omega_n : \rho(x) > \delta\}$$

for some $\delta > 0$ sufficiently small and define the Lipschitz function

$$\psi_\delta(x) = \begin{cases} 1 & \text{if} \quad x \in \Omega_{n,\delta} \\ \rho(x)/\delta & \text{if} \quad x \in \Omega_n \setminus \Omega_{n,\delta} \\ 0 & \text{if} \quad x \in M \setminus \Omega_n. \end{cases}$$

Let $\varphi \in C_c^\infty(M \setminus B_\varepsilon)$, $\varphi \geq 0$; then $\varphi\psi_\delta \in W_0^{1,2}(\Omega_n \setminus \overline{B}_\varepsilon)$ and $\varphi\psi_\delta \geq 0$. Because of the pointwise validity of (8.75) in $\Omega_n \setminus (B_\varepsilon \cup \text{cut}(o))$, and therefore of its validity in the weak sense there, having set $G(x)$ for the right-hand side of (8.75) and using Gauss lemma we have

$$\begin{aligned} \int_{\Omega_n\setminus\overline{B}_\varepsilon} G(x)\varphi\psi_\delta &\geq \int_{\Omega_n\setminus\overline{B}_\varepsilon} -\langle\nabla r, \nabla(\varphi\psi_\delta)\rangle - \langle X, \nabla r\rangle(\varphi\psi_\delta) \\ &= -\left(\int_{\Omega_n\setminus\overline{B}_\varepsilon} (\langle\nabla r, \nabla\varphi\rangle + \langle X, \nabla r\rangle\varphi)\psi_\delta\right) - \frac{1}{\delta}\int_{\Omega_n\setminus\Omega_{n,\delta}} \langle\nabla r, \nabla\rho\rangle\varphi, \end{aligned}$$

where in the last equality we have used the fact that $\overline{B}_\varepsilon \subset \Omega_{n,\delta}$. Therefore, by the coarea formula,

$$\int_{\Omega_n\setminus\overline{B}_\varepsilon} G(x)\varphi\psi_\delta \geq -\left(\int_{\Omega_n\setminus\overline{B}_\varepsilon} (\langle\nabla r, \nabla\varphi\rangle + \langle X, \nabla r\rangle)\varphi\psi_\delta\right) - \frac{1}{\delta}\int_0^\delta dt \int_{\partial\Omega_{n,t}} \langle\nabla r, \nabla\rho\rangle\varphi.$$

Letting $\delta \downarrow 0^+$ we get

$$\int_{\Omega_n\setminus\overline{B}_\varepsilon} G(x)\varphi \geq -\left(\int_{\Omega_n\setminus\overline{B}_\varepsilon} \langle\nabla r, \nabla\varphi\rangle + \langle X, \nabla r\rangle\varphi\right) + \int_{\partial\Omega_n} \langle\nabla r, \nabla\rho\rangle\varphi,$$

and since Ω_n is starshaped,

$$\int_{\Omega_n\setminus\overline{B}_\varepsilon} G(x)\varphi \geq -\left(\int_{\Omega_n\setminus\overline{B}_\varepsilon} \langle\nabla r, \nabla\varphi\rangle + \langle X, \nabla r\rangle\varphi\right).$$

By letting $n \to +\infty$, observing that cut(o) has measure 0 and supp φ is compact, using Fatou's Lemma the above yields

$$\int_{M\setminus\overline{B}_\varepsilon} G(x)\varphi \geq -\left(\int_{M\setminus\overline{B}_\varepsilon} \langle\nabla r, \nabla\varphi\rangle + \langle X, \nabla r\rangle\varphi\right),$$

showing the validity of (8.73) with $B = B_\varepsilon$ and $C = C_\varepsilon$. □

Let us suppose now that $(M, \langle\, ,\, \rangle, X)$ is a Ricci soliton structure on the complete manifold $(M, \langle\, ,\, \rangle)$. Thus

$$\mathrm{Ric}_X = \lambda \langle\, ,\, \rangle,$$

and independently of the sign of λ we can choose $G(t)$ to be an appropriate positive constant in such a way that Ric_X satisfies (8.58). Then, from (8.73)

$$\Delta_X r \leq Ar + B$$

for some constants $A, B > 0$ outside a compact set. Hence applying Theorem 3.2 and the subsequent discussion we have

Proposition 8.7 *Let $(M, \langle\, ,\, \rangle, X)$ be a soliton structure on the complete manifold $(M, \langle\, ,\, \rangle)$. Then the Omori-Yau maximum principle holds for the operator Δ_X. Furthermore, if*

$$|X| \leq \sqrt{F(r)}$$

for some positive, nondecreasing function $G \in C^1(\mathbb{R}_0^+)$ with the property that

$$\frac{1}{\sqrt{F(t)}} \notin L^1(+\infty),$$

then it also holds for the operator Δ.

Proof We only have to prove the second part of the proposition which however follows immediately since

$$\Delta r = \Delta_X r + \langle X, \nabla r \rangle \leq \Delta_X r + |X| \leq Ar + B + \sqrt{F(r)}.$$

□

8.3 Statements and Proofs of the Main Results

8.3.1 The Generic Case

In the next result we deal with a generic soliton structure. We will refine next theorem later on when dealing with gradient solitons.

Theorem 8.2 *Let $(M, \langle\, ,\, \rangle)$ be a complete manifold of dimension m and scalar curvature S, and let $S_* = \inf_{x \in M} S(x)$. Let $(M, \langle\, ,\, \rangle, X)$ be a soliton structure on M with soliton constant λ.*

(i) *If* $\lambda < 0$ *then* $m\lambda \le S_* \le 0$. *Furthermore, if* $S(x_0) = S_* = m\lambda$ *for some* $x_0 \in M$, *then* $(M, \langle\, ,\, \rangle)$ *is Einstein and* X *is a Killing field; while if* $S(x_0) = S_* = 0$, *for some* $x_0 \in M$, *then* $(M, \langle\, ,\, \rangle)$ *is Ricci flat and* X *is a homothetic vector field.*
(ii) *If* $\lambda = 0$ *then* $S_* = 0$. *Furthermore, if* $S(x_0) = S_* = 0$ *for some* $x_0 \in M$, *then* $(M, \langle\, ,\, \rangle)$ *is Ricci flat and* X *is a Killing field.*
(iii) *If* $\lambda > 0$ *then* $0 \le S_* \le m\lambda$. *Furthermore, if* $S(x_0) = S_* = 0$ *for some* $x_0 \in M$, *then* $(M, \langle\, ,\, \rangle)$ *is flat and* X *is a homothetic vector field; while if* $S(x_0) = S_* = m\lambda$, *for some* $x_0 \in M$, *then* $(M, \langle\, ,\, \rangle)$ *is compact, Einstein and* X *is a Killing field.*

Proof First of all we observe that since $(M, \langle\, ,\, \rangle, X)$ is a soliton structure, the Omori-Yau maximum principle holds for Δ_X. We want to apply Theorem 3.6 in Chap. 3 with $q(x) \equiv 1$ and $T = \langle\, ,\, \rangle$, so that $L_{T,X} = \Delta_X$. Note that in the present case (3.56) is automatically satisfied. We proceed observing that, by Proposition 8.3, we have the validity of

$$\frac{1}{2}\Delta_X S = \lambda S - \frac{S^2}{m} - \left|\mathrm{Ric} - \frac{S}{m}\langle\, ,\, \rangle\right|^2, \tag{8.76}$$

from which, setting $u = -S$ we immediately deduce the differential inequality

$$\frac{1}{2}\Delta_X u \ge \lambda u + \frac{u^2}{m}. \tag{8.77}$$

We apply Theorem 3.6 with the choices $F(t) = t^2$,

$$\varphi(u, |\nabla u|) = \lambda u + \frac{u^2}{m}.$$

Then $u^* < +\infty$ and

$$\lambda u^* + \frac{(u^*)^2}{m} \le 0. \tag{8.78}$$

But $u^* = -S_*$ so that the claimed bounds on S_* in the statement of Theorem 8.2 follow immediately from (8.78).

Case (i). Suppose now $\lambda < 0$ and that, for some $x_0 \in M$, $S(x_0) = S_* = m\lambda$. In particular $S(x) \ge m\lambda$ on M and the function $w(x) = S(x) - m\lambda \ge 0$ on M. From (8.76)

$$\frac{1}{2}\Delta_X S \le \lambda S - \frac{S^2}{m} \tag{8.79}$$

and thus, we immediately see that

$$\Delta w - \langle X, \nabla w\rangle + 2\lambda w \leq \Delta w - \langle X, \nabla w\rangle + 2\frac{S}{m}w \leq 0. \tag{8.80}$$

We let

$$\Omega_0 = \{x \in M : w(x) = 0\}.$$

Ω_0 is closed and nonempty since $x_0 \in \Omega_0$; let now $y \in \Omega_0$. By the maximum principle applied to (8.80) (see [125, p. 35]), $w \equiv 0$ in a neighborhood of y so that Ω_0 is open. Connectedness of M yields $\Omega_0 = M$ and $S(x) \equiv m\lambda$ on M. From Eq. (8.76) we then deduce

$$\left|\mathrm{Ric} - \frac{S}{m}\langle\, , \rangle\right| \equiv 0,$$

that is, $(M, \langle\, , \rangle)$ is Einstein and from the soliton equation (8.1), since $S = m\lambda$, X is a Killing field. Analogously, if $S(x_0) = S_* = 0$ for some $x_0 \in M$ we deduce that $S(x) \equiv 0$ and therefore that $(M, \langle\, , \rangle)$ is Ricci flat and X is a homothetic vector field.

Case (ii). Suppose $\lambda = 0$ and that, for some $x_0 \in M$, $S(x_0) = S_* = 0$. From (8.79)

$$\Delta S - \langle X, \nabla S\rangle \leq -\frac{S^2}{m} \leq 0.$$

Since $S(x) \geq S_* = 0$ on M, by the maximum principle we conclude $S(x) \equiv 0$ and, by (8.76), $(M, \langle\, , \rangle)$ is Ricci flat. Again, from (8.1), X is a Killing field.

Case (iii). Finally suppose $\lambda > 0$. Then $S(x) \geq S_* \geq 0$. From (8.79)

$$\Delta S - \langle X, \nabla S\rangle - 2\lambda S \leq 0.$$

If $S(x_0) = S_* = 0$ for some $x_0 \in M$, then again by the maximum principle $S(x) \equiv 0$. By (8.76), $(M, \langle\, , \rangle)$ is Ricci flat and from (8.1) we have $\mathcal{L}_X\langle\, , \rangle = 2\lambda\langle\, , \rangle$, so that X is a homothetic vector field and by Tashiro [263, Theorem 4.1] (see also [161, 277]), M is flat. Suppose now that $S(x_0) = S_* = m\lambda$ for some $x_0 \in M$. From (8.79)

$$\Delta S - \langle X, \nabla S\rangle \leq 2\frac{S}{m}(m\lambda - S)$$

and since $S(x) \geq S_* = m\lambda > 0$

$$\Delta S - \langle X, \nabla S\rangle \leq 0$$

on M. By the maximum principle $S(x) \equiv m\lambda$; from (8.79) $(M, \langle\, ,\, \rangle)$ is Einstein and (8.1) implies that X is a Killing field. Furthermore, since $\lambda > 0$ $(M, \langle\, ,\, \rangle)$ is compact by Myer's theorem. □

From Theorem 8.2 we have the following

Corollary 8.4 *Let $(M, \langle\, ,\, \rangle)$ be a complete manifold of dimension m and scalar curvature S such that*

$$S_* = \inf_{x \in M} S(x) < 0 \textit{ (respectively } > 0).$$

Then $(M, \langle\, ,\, \rangle)$ support no shrinking or steady (respectively, expanding or steady) Ricci soliton structure $(M, \langle\, ,\, \rangle, X)$.

The next result shows that the existence of a soliton structure implies some sort of rigidity.

Corollary 8.5 *Let $(M, \langle\, ,\, \rangle)$ be a complete manifold of dimension m admitting a shrinking or steady soliton structure $(M, \langle\, ,\, \rangle, X)$. Then any isometric minimal immersion of $(M, \langle\, ,\, \rangle)$ into $\mathbb{R}^n$, $n > m$, is totally geodesic.*

Proof Indicating with II, as usual, the second fundamental tensor of the immersion, by Gauss equation and minimality we have

$$S(x) = -|\mathrm{II}|^2(x).$$

Thus, if M is not totally geodesic $S_* < 0$, contradicting (ii) or (iii) of Theorem 8.2. □

The next is a gap result for the length of the traceless Ricci tensor T; to prove it we shall need the following estimate due to Huisken [150, Lemma 3.4].

Proposition 8.8 *Let $(M, \langle\, ,\, \rangle)$ be a Riemannian manifold of dimension $m \geq 2$ with traceless Ricci tensor T and Weyl tensor W. Then*

$$\left|T_{ik}T_{jt}W_{ijkt}\right| \leq \frac{\sqrt{2}}{2}\sqrt{\frac{m-2}{m-1}}|W||T|^2. \tag{8.81}$$

Proof We provide the proof of the estimate for the sake of completeness. First we observe that Eq. (8.81) can be rewritten as

$$\left|\frac{6}{m-2}T_{ik}T_{jt}W_{ijkt}\right| \leq \frac{3\sqrt{2}}{\sqrt{(m-1)(m-2)}}|W|^2|T|^2.$$

Using the definition of the tensor V in the orthogonal decomposition of the Riemann curvature tensor given in Eq. (1.95), a long but simple computation shows that

$$\left|\frac{6}{m-2}T_{ik}T_{jt}W_{ijkt}\right| = \frac{3}{4}(m-2)W_{ijkt}V_{ijrs}V_{rskt} = \frac{3}{4}(m-2)W_{ijkt}(V\circ V)_{ijkt}, \tag{8.82}$$

where $(V\circ V)_{ijkt} = V_{ijrs}V_{rskt}$ are the components of the $(0,4)$-tensor $V\circ V$; in terms of the components of T,

$$\begin{aligned}(V\circ V)_{ijkt} &= \frac{4}{(m-2)^2}\left(T_{ik}T_{jt} - T_{it}T_{jk}\right)\\ &\quad + \frac{2}{(m-2)^2}\left(T_{is}T_{sk}\delta_{jt} - T_{is}T_{st}\delta_{jk} + T_{js}T_{st}\delta_{ik} - T_{js}T_{sk}\delta_{it}\right).\end{aligned}$$

A simple inspection shows that $V\circ V$ has the same symmetries of Riem, therefore following Remark 1.12 it can be decomposed into three orthogonal parts; more explicitly we have

$$(V\circ V) = T_1 + T_2 + T_3,$$

where $T_1 \perp T_2 \perp T_3$ and T_1 is the "scalar" part, T_2 is the "traceless Ricci" part and T_3 is the "Weyl" part. Now we compute

$$(V\circ V)_{kjkt} = \frac{2}{(m-2)^2}\left[(m-4)T_{jk}T_{kt} + |T|^2\delta_{jt}\right] = (V\circ V)_{jktk} \tag{8.83}$$

and

$$(V\circ V)_{ktkt} = \frac{4}{m-2}|T|^2, \tag{8.84}$$

so that, using (1.104) and (1.105), we deduce

$$(T_1)_{ijkt} = \frac{4}{m(m-1)(m-2)}|T|^2\left(\delta_{ik}\delta_{jt} - \delta_{it}\delta_{jk}\right) \tag{8.85}$$

and

$$(T_2)_{ijkt} = \frac{1}{m-2}\left(\xi_{ik}\delta_{jt} - \xi_{jk}\delta_{it} + \xi_{jt}\delta_{ik} - \xi_{it}\delta_{jk}\right), \tag{8.86}$$

with

$$\xi_{ik} = (V\circ V)_{isks} - \frac{(V\circ V)_{lsls}}{m}\delta_{ik} = \frac{2(m-4)}{(m-2)^2}\left(T_{it}T_{tk} - \frac{|T|^2}{m}\delta_{ik}\right). \tag{8.87}$$

Using the Cauchy-Schwarz inequality, the conclusion of the proposition follows from (8.82) if we show that

$$|T_3|^2 \leq \frac{32}{(m-1)(m-2)^3}|T|^4,$$

since T_1 and T_2 are by construction orthogonal to W. We have

$$|T_1|^2 = (T_1)_{ijkt}(T_1)_{ijkt} = \frac{32}{m(m-1)(m-2)^2}|T|^4$$

and, from (8.87),

$$|T_2|^2 = \frac{4}{(m-2)^2}\xi_{ik}\xi_{ik} = \frac{16(m-4)^2}{(m-2)^5}\left(Z - \frac{|T|^4}{m}\right),$$

with $Z = T_{it}T_{tk}T_{ks}T_{si}$; note that $Z \geq \frac{|T|^4}{m}$, and thus $-mZ \leq -|T|^4$. Moreover, from the definition of $(V \circ V)$ we get

$$\begin{aligned} |(V \circ V)|^2 &= (V \circ V)_{ijkt}(V \circ V)_{ijkt} \qquad (8.88)\\ &= \frac{32}{(m-2)^4}\left(|T|^4 - Z\right) + \frac{16}{(m-2)^4}\left[(m-2)Z + |T|^4\right] - \frac{64}{(m-2)^4}Z \\ &= \frac{16}{(m-2)^4}\left[(m-8)Z + 3|T|^4\right]. \end{aligned}$$

Now we can conclude, since

$$\begin{aligned} |T_3|^2 &= |(V \circ V)|^2 - |T_1|^2 - |T_2|^2 \\ &= \frac{16}{(m-2)^4}\left[(m-8)Z + 3|T|^4\right] \\ &\quad - \frac{32}{m(m-1)(m-2)^2}|T|^4 - \frac{16(m-4)^2}{(m-2)^5}\left(Z - \frac{|T|^4}{m}\right) \\ &= \frac{32}{(m-1)(m-2)^5}\left(m^2 - 3m + 3\right)|T|^4 - \frac{32m}{(m-2)^5}Z \\ &= \frac{32}{(m-2)^5}\left[-mZ + \frac{(m^2 - 3m + 3)}{m-1}|T|^4\right] \\ &\leq \frac{32}{(m-1)(m-2)^3}|T|^4. \end{aligned}$$

□

We are now ready to prove

Theorem 8.3 *Let $(M, \langle\, ,\, \rangle)$ be a complete manifold of dimension $m \geq 3$, scalar curvature $S(x)$, traceless Ricci tensor T and Weyl tensor W. Suppose that*

$$(i)\ S^* = \sup_M S(x) < +\infty, \quad (ii)\ |W|^* = \sup_M |W|(x) < +\infty. \tag{8.89}$$

Let $(M, \langle\, ,\, \rangle, X)$ be a Ricci soliton structure on M with soliton constant λ. Then, either $(M, \langle\, ,\, \rangle)$ is Einstein or $|T|^* = \sup_M |T|^*(x)$ satisfies

$$|T|^* \geq \frac{1}{2}\left(\sqrt{m(m-1)}\lambda - S^* \frac{m-2}{\sqrt{m(m-1)}} - \sqrt{\frac{m(m-2)}{2}}|W|^*\right). \tag{8.90}$$

In particular if $(M, \langle\, ,\, \rangle)$ is conformally flat, then either $(M, \langle\, ,\, \rangle)$ has constant sectional curvature or

$$|T|^* \geq \frac{1}{2}\left(\sqrt{m(m-1)}\lambda - S^* \frac{m-2}{\sqrt{m(m-1)}}\right). \tag{8.91}$$

Remark 8.6 If $(M, \langle\, ,\, \rangle)$ is Einstein and in addition it is a shrinking soliton which is not Ricci flat, by Theorem 8.2, S is a positive constant and thus $(M, \langle\, ,\, \rangle)$ is compact by Myers' theorem. In this latter case, if $m \geq 4$ and $|W|$ is sufficiently small, precisely if

$$|W|^2 < \frac{1}{30} S^2 \text{ for } m = 4, \quad |W|^2 < \frac{1}{100} S^2 \text{ for } m = 5,$$

and

$$|W|^2 < \frac{4}{(m+1)m(m-1)(m-2)} S^2, \text{ for } m \geq 6,$$

then, by Corollary 2.5 in [150], or by Proposition 8.9 below, using the facts that M is compact and the above inequalities are strict, $(M, \langle\, ,\, \rangle)$ has positive curvature operator in the sense of (2.117). Hence, from a result of Tachibana [261], or Theorem 2.17 in the compact case, $(M, \langle\, ,\, \rangle)$ has positive constant sectional curvature and it is therefore a finite quotient of $\mathbb{S}^m$.

We now give a proof, with Proposition 8.9 below, of Corollary 2.5 in [150], since it will be used also later in Theorems 8.8 and 8.9.

Let $(V, \langle\, ,\, \rangle)$ be an m-dimensional inner product vector space and let $\{\theta^i\}$, $i = 1, \dots m$, be an orthonormal basis of V^*. Let $Z = Z_{ijkl}\theta^l \otimes \theta^k \otimes \theta^j \otimes \theta^i$ be a tensor having the same symmetries of the Riemann curvature tensor. Then we know that Z defines a symmetric operator $\mathfrak{Z} : \Lambda^2(V^*) \to \Lambda^2(V^*)$ by setting, for $\omega = \omega_{ij}\theta^i \wedge \theta^j$,

$$\mathfrak{Z}\omega = \frac{1}{2}(Z_{ijkl}\omega_{ij})\theta^k \wedge \theta^l$$

(see also Sect. 2.4). Hence λ is an eigenvalue of $\mathfrak{Z}$ if, for some $\omega \in \Lambda^2(V^*)$, $\omega \neq 0$,

$$\mathfrak{Z}\omega = \lambda\omega$$

or, in other words,

$$Z_{ijkl}\omega_{ij} = 2\lambda\omega_{kl}.$$

Furthermore, note that

$$|\mathfrak{Z}|^2 = \frac{1}{4}|Z|^2,$$

where $|\mathfrak{Z}|$ is the norm in $\Lambda^2(V^*)$ with the inner product induced by $\langle\, ,\, \rangle$. We have

Lemma 8.5 *Let λ be an eigenvalue of a symmetric traceless operator $\mathfrak{Z}$: $\Lambda^2(V^*) \to \Lambda^2(V^*)$ as defined above. Then, if $N = \binom{m}{2} = \dim \Lambda^2(V^*)$,*

$$\lambda^2 \leq \frac{N-1}{N}|\mathfrak{Z}|^2 = \frac{(m-2)(m+1)}{4m(m-1)}|Z|^2. \tag{8.92}$$

Proof More generally, if T is a symmetric, traceless, $(0,2)$-tensor on W, with $\dim W = N$ and λ_i, $i = 1, \ldots, N$ its eigenvalues, we have

$$\lambda_i^2 \leq \frac{N-1}{N}|T|^2 \qquad \text{for } i = 1, \ldots, N. \tag{8.93}$$

To see this, since $\sum_{i=1}^N \lambda_i = 0$ and $|T|^2 = \sum_{i=1}^N \lambda_i^2$, observe that

$$(N-1)|T|^2 - N\lambda_i^2 = (N-1)\sum_{j=1, j\neq i}^{N} \lambda_j^2 - \lambda_i^2 = (N-1)\sum_{j=1, j\neq i}^{N}\left(\lambda_j + \frac{\lambda_i}{m-1}\right)^2 \geq 0.$$

Now (8.92) follows immediately from (8.93). □

Now we let the tensor field Z to be $\overset{\circ}{\text{Riem}}$, defined, according to (1.94), by

$$\overset{\circ}{\text{Riem}} = W + V, \tag{8.94}$$

where W is the Weyl tensor and V, as defined in (1.96), has components

$$V_{ijkt} = \frac{1}{m-2}\left(T_{ik}\delta_{jt} - T_{jk}\delta_{it} + T_{jt}\delta_{ik} - T_{it}\delta_{jk}\right)$$

with respect to a local orthonormal coframe and where T is the traceless Ricci tensor. We set $\overset{\circ}{\mathfrak{R}} : \Lambda^2(M) \to \Lambda^2(M)$ and note that $\overset{\circ}{\mathfrak{R}}$ is a symmetric traceless linear operator so that, is λ is any of its eigenvalues, by Lemma 8.5

$$\lambda^2 \le \frac{(m-2)(m+1)}{4m(m-1)}\left|\overset{\circ}{\mathrm{Riem}}\right|^2,$$

and using the fact that (8.94) is an orthogonal decomposition we finally have

$$\lambda^2 \le \frac{(m-2)(m+1)}{4m(m-1)}\left[|W|^2 + |V|^2\right]. \tag{8.95}$$

We now let

$$\delta_4 = \frac{1}{5}, \quad \delta_5 = \frac{1}{10}, \quad \delta_m = \frac{2}{(m-2)(m+1)} \quad \text{for } m \ge 6.$$

From inequality (8.95) and with the above notation we obtain the following

Proposition 8.9 *Let $m \ge 4$ and suppose that for some $\varepsilon > 0$*

$$|W|^2 + |V|^2 = |W|^2 + \frac{4}{m-2}|T|^2 \le \delta_m(1-\varepsilon)^2\frac{2}{m(m-1)}S^2, \tag{8.96}$$

where S is the scalar curvature. Then, for each $\omega \in \Lambda^2(M)$,

$$\langle \mathfrak{R}\omega, \omega\rangle \ge \varepsilon|\omega|^2. \tag{8.97}$$

Proof From (8.95) and (8.96), if λ is an eigenvalue of $\overset{\circ}{\mathfrak{R}}$ then

$$|\lambda| \le \frac{1-\varepsilon}{m(m-1)}S. \tag{8.98}$$

Now according to (1.94)

$$\mathrm{Riem} = \overset{\circ}{\mathrm{Riem}} + U,$$

where, from Eq. (1.95), U has components

$$U_{ijkt} = \frac{S}{m(m-1)}(\delta_{ik}\delta_{jt} - \delta_{it}\delta_{jk})$$

so that the corresponding eigenvalues of $\mathfrak{U} : \Lambda^2(M) \to \Lambda^2(M)$ are all equal to $\frac{S}{m(m-1)}$. Hence, since $\mathfrak{R} = \overset{\circ}{\mathfrak{R}} + \mathfrak{U}$, from (8.98) and this last fact the validity of (8.97) follows immediately. □

Proof (of Theorem 8.3) From Eq. (8.45), Okumura's lemma (that is, Lemma 6.2) and Huisken estimate (8.81), we obtain

$$\frac{1}{2}\Delta_X|T|^2 \geq |\nabla T|^2 + 2\left(\lambda - \frac{m-2}{m(m-1)}S - \frac{1}{\sqrt{2}}\sqrt{\frac{m-2}{m-1}}|W|\right)|T|^2$$
$$-\frac{4}{\sqrt{m(m-1)}}|T|^3.$$

We set $u = |T|^2$ to deduce from the above

$$\frac{1}{2}\Delta_X u \geq 2\left(\lambda - \frac{m-2}{m(m-1)}S^* - \frac{1}{\sqrt{2}}\sqrt{\frac{m-2}{m-1}}|W|^*\right)u$$
$$-\frac{4}{\sqrt{m(m-1)}}u^{3/2}. \tag{8.99}$$

We observe that, if $|T|^* = +\infty$, then (8.90) is obviously satisfied, otherwise $u^* < +\infty$ and, in the assumptions of the theorem, by Proposition 8.7 we have the validity of the Omori-Yau maximum principle for Δ_X. It follows that

$$\left[\frac{1}{2}\left(\lambda\sqrt{m(m-1)} - \frac{m-2}{\sqrt{m(m-1)}}S^* - \sqrt{\frac{m(m-2)}{2}}|W|^*\right) - \sqrt{u^*}\right]u^* \leq 0,$$

from which we deduce that either $u^* = 0$, that is, $T \equiv 0$ on M, or $|T|^*$ satisfies (8.90). In the first case, $(M, \langle\, ,\,\rangle)$ is Einstein. Note that if $(M, \langle\, ,\,\rangle)$ is Einstein and conformally flat an immediate checking using decomposition of the Riemann curvature tensor (1.84) shows that $(M, \langle\, ,\,\rangle)$ has constant sectional curvature. □

8.3.2 Gradient Solitons

In this subsection we consider the case of *gradient* Ricci solitons. For the proof of our next result, Theorem 8.4 below, we shall first need the following proposition, which is a consequence of estimate (8.60) of Theorem 8.1.

Proposition 8.10 *Let* $(M, \langle\, ,\, \rangle)$ *be a complete Riemannian manifold of dimension* m *and* $X \in \mathfrak{X}(M)$ *a vector field satisfying the growth condition*

$$|X| \le \sqrt{F(r)} \tag{8.100}$$

for some positive, nondecreasing function $F \in C^1(\mathbb{R}_0^+)$. *Assume*

$$\mathrm{Ric}_X(\nabla r, \nabla r) \ge -(m-1)G(r) \tag{8.101}$$

for a positive $G \in C^1(\mathbb{R}_0^+)$ *such that*

$$\inf_{\mathbb{R}_0^+} \frac{G'}{G^{3/2}} > -\infty. \tag{8.102}$$

Then

$$vol(\partial B_r) \le Ce^{B\int_o^r(\sqrt{G(s)}+\sqrt{F(s)})ds} \tag{8.103}$$

for almost every r, *and, as a consequence,*

$$vol(B_r) \le C\int_o^r \left(e^{B\int_o^t(\sqrt{G(s)}+\sqrt{F(s)})ds}\right)dt + D \tag{8.104}$$

for some sufficiently large positive constants B, C, D.

Proof Recalling (8.60) we set

$$h(r) = e^{\frac{B}{m-1}\int_o^r(\sqrt{G(s)}+\sqrt{F(s)})ds} - 1 \tag{8.105}$$

with $B \ge \max\{A, 2\}$ and A as in (8.60). Since $\Delta r = \Delta_X r + \langle X, \nabla r\rangle$, from (8.60), (8.57) and the choice of h we have

$$\Delta r \le (m-1)\frac{h'(r)}{h(r)}$$

pointwise on $M \setminus (\{o\} \cup \mathrm{cut}\{o\})$ and weakly on all of M. This means that for each $\varphi \in \mathrm{Lip}_c(M)$, $\varphi \geq 0$,

$$-\int_M \langle \nabla r, \nabla \varphi \rangle \leq (m-1) \int_M \frac{h'(r(x))}{h(r(x))} \varphi. \tag{8.106}$$

Next, we fix $0 < s < R$ and $\varepsilon > 0$, we let ρ_ε be the piecewise linear function

$$\rho_\varepsilon(t) = \begin{cases} 0, & \text{if } t \in [0, s) \\ \frac{t-s}{\varepsilon}, & \text{if } t \in [s, s+\varepsilon) \\ 1, & \text{if } t \in [s+\varepsilon, R-\varepsilon) \\ \frac{R-t}{\varepsilon}, & \text{if } t \in [R-\varepsilon, R) \\ 0, & \text{if } t \in [R, +\infty), \end{cases}$$

and we define the radial cut off function

$$\varphi_\varepsilon(x) = \rho_\varepsilon(r(x)) h(r(x))^{1-m}.$$

Indicating again with $\chi_{a,b}$, $a < b$, the characteristic function of the annulus $B_b \backslash B_a$ we have

$$\nabla \varphi_\varepsilon(x) = \left(\frac{1}{\varepsilon} \chi_{s,s+\varepsilon}(x) - \frac{1}{\varepsilon} \chi_{R-\varepsilon,R}(x) - (m-1) \frac{h'(r(x))}{h(r(x))} \rho_\varepsilon(r(x)) \right) h(r(x))^{1-m} \nabla r(x)$$

almost everywhere on M. Hence, using φ_ε into (8.106) and simplifying we get

$$\frac{1}{\varepsilon} \int_{B_R \backslash B_{R-\varepsilon}} h(r(x))^{1-m} \leq \frac{1}{\varepsilon} \int_{B_{s+\varepsilon} \backslash B_s} h(r(x))^{1-m}.$$

From the co-area formula (1.252)

$$\frac{1}{\varepsilon} \int_{R-\varepsilon}^{R} h(t)^{1-m} \mathrm{vol}(\partial B_t) dt \leq \frac{1}{\varepsilon} \int_s^{s+\varepsilon} h(t)^{1-m} \mathrm{vol}(\partial B_t) dt,$$

and letting $\varepsilon \to 0^+$

$$\frac{\mathrm{vol}(\partial B_R)}{h(R)^{m-1}} \leq \frac{\mathrm{vol}(\partial B_s)}{h(s)^{m-1}}$$

for almost every $0 < s < R$. Letting $s \to 0$, recalling that $\mathrm{vol}(\partial B_s) \sim c_m s^{m-1}$ and $h(s) \sim cs$, for c_m, c appropriate positive constants, we obtain

$$\mathrm{vol}(\partial B_R) \leq C e^{B \int_0^R (\sqrt{G(s)} + \sqrt{F(s)}) ds}$$

for some constant $C > 0$ and almost every R. Using again the co-area formula we also deduce the validity of (8.104). □

Remark 8.7 In the assumptions of the proposition suppose that $X = \nabla f$ for some $f \in C^\infty(M)$. From (8.60) we then have

$$\Delta_f r \leq A\sqrt{G(r)} + \sqrt{F(r)}.$$

Hence, for $B \geq \max\{A, 1\}$, the function $h(r)$ defined in (8.105) yields

$$\Delta_f r \leq (m-1)\frac{h'(r)}{h(r)}. \tag{8.107}$$

Considering the weighted manifold $(M, \langle\, ,\,\rangle, e^{-f})$, where the *weighted volume* $\mathrm{vol}_f(B_r)$ of the geodesic ball B_r is defined by

$$\mathrm{vol}_f(B_r) = \int_{B_r} e^{-f}, \tag{8.108}$$

from (8.107) we deduce: for each $\varphi \in \mathrm{Lip}_c(M)$, $\varphi \geq 0$,

$$-\int_M \langle\nabla r, \nabla\varphi\rangle e^{-f} \leq (m-1)\int_M \frac{h'(r(x))}{h(r(x))}\varphi e^{-f}. \tag{8.109}$$

Then, the above proof shows that

$$\mathrm{vol}_f(\partial B_r) \leq Ce^{B\int_0^r(\sqrt{G(s)}+\sqrt{F(s)})ds} \tag{8.110}$$

and, as a consequence,

$$\mathrm{vol}_f(B_r) \leq C\int_0^r \left(e^{B\int_0^r(\sqrt{G(s)}+\sqrt{F(s)})ds}\right)dt + D \tag{8.111}$$

for some sufficiently large constants $B, C, D > 0$.

Observe that, instead of estimate (8.60) of Theorem 8.1, we could have used estimate (8.75) appearing in the proof of Proposition 8.6 to obtain the following

Proposition 8.11 *Let $(M, \langle\, ,\,\rangle)$ be a complete manifold of dimension m, $f \in C^\infty(M)$ and suppose that*

$$\mathrm{Ric}_f(\nabla r, \nabla r) \geq -(m-1)G(r) \tag{8.112}$$

for some $G \in C^0(\mathbb{R}_0^+)$. Then

$$\mathrm{vol}_f(\partial B_r) \leq e^{C(r-\varepsilon)+\int_\varepsilon^r\left(\int_\varepsilon^t(m-1)G(s)ds\right)dt} \tag{8.113}$$

for $r \geq \varepsilon$, $\varepsilon > 0$ sufficiently small, and an appropriate $C \in \mathbb{R}$. As a consequence

$$vol_f(B_r) \leq \int_0^r e^{C(x-\varepsilon)+\int_\varepsilon^x \left(\int_\varepsilon^t (m-1)G(s)ds\right)dt} dx + D, \tag{8.114}$$

with C, ε as above and $D > 0$ a positive constant.

Observe that (8.111) and (8.114) are indeed different and each one of them has its own advantages. For instance, let us consider the case of a gradient soliton $(M, \langle\, ,\, \rangle, \nabla f)$ with soliton constant $\lambda \in \mathbb{R}$. For the validity of (8.111) we have to choose

$$(m-1)G(r) = -\lambda, \quad \text{if } \lambda < 0$$

and

$$G(r) = \alpha \in \mathbb{R}^+, \quad \text{if } \lambda \geq 0.$$

So (8.111) is certainly far from being sharp in case $\lambda \geq 0$. However, it is sensitive to the growth of $|\nabla f|$ which in some instances could be assigned or known for some special reasons. On the contrary (8.114) totally ignores the growth of $|\nabla f|$ and has some indetermination in the factor $C(t-\varepsilon)$ since C, as we know from the proof of Proposition 8.6, is given by

$$C = \max_{\partial B_\varepsilon} \Delta_f r, \quad \varepsilon > 0 \text{ sufficiently small.}$$

However in (8.114) we can choose

$$(m-1)G(r) = -\lambda$$

irrespectively of the sign of λ. This latter fact tells us that for any gradient soliton we have

$$\text{vol}_f(B_r) \leq D + \int_0^r e^{Ct - \frac{\lambda}{2}t^2} dt. \tag{8.115}$$

Therefore, $\text{vol}_f(B_r)$ grows at most as $r^{-1}e^{\frac{|\lambda| r^2}{2}}$, as $r \to +\infty$.

In the next result we shall use estimate (8.111).

Theorem 8.4 *Let $(M, \langle\, ,\, \rangle, \nabla f)$ be an expanding gradient soliton on the complete manifold $(M, \langle\, ,\, \rangle)$, and let $0 \leq \sigma \leq \frac{2}{3}$. Assume*

$$\limsup_{r(x)\to+\infty} \frac{|\nabla f|^2(x)}{r(x)^\sigma} \begin{cases} = 0, & \text{if } \sigma \in (0, \frac{2}{3}] \\ < +\infty & \text{if } \sigma = 0. \end{cases} \tag{8.116}$$

Then the soliton is trivial.

Proof We know, from (8.21), that $|\nabla f|^2$ satisfies the differential inequality

$$\Delta_f |\nabla f|^2 \geq -2\lambda |\nabla f|^2. \tag{8.117}$$

Next, from (8.116), (8.111) and the soliton equation (8.4)

$$\begin{aligned} \mathrm{vol}_f(B_r) &\leq C \int_0^r e^{B\left(\sqrt{\frac{|\lambda|}{m-1}}t + \frac{2}{\sigma+2}\Lambda t^{\frac{\sigma}{2}+1}\right)} dt \\ &\leq Cre^{B\left(\sqrt{\frac{|\lambda|}{m-1}}r + \frac{2}{\sigma+2}\Lambda r^{\frac{\sigma}{2}+1}\right)} \end{aligned}$$

for some constants $B, C, \Lambda > 0$. It follows that

$$\frac{\log \mathrm{vol}_f(B_r)}{r^{2-\sigma}} \leq C \frac{r^{\frac{\sigma}{2}+1}}{r^{2-\sigma}} = Cr^{\frac{3}{2}\sigma - 1}$$

for $r \gg 1$ and some constant $C > 0$. Thus under the assumptions of the theorem

$$\liminf_{\to+\infty} \frac{\log \mathrm{vol}_f(B_r)}{r^{2-\sigma}} < +\infty. \tag{8.118}$$

Assume, by contradiction, that $|\nabla f| \not\equiv 0$ and choose $\gamma > 0$ such that

$$\Omega_\gamma = \{x \in M : |\nabla f|^2(x) > \gamma\} \neq \emptyset.$$

Applying Theorem 4.4 to (8.117) we immediately obtain the desired contradiction in case $0 < \sigma \leq \frac{2}{3}$. When $\sigma = 0$ we use the Δ_f-stochastic completeness of $(M, \langle\, , \rangle)$ due to (8.115) instead of Theorem 4.4. □

It is interesting to observe that for gradient solitons there is a general upper bound for $|\nabla f|^2$. To see this first of all we recall that, from Theorem 8.2, the scalar curvature is always bounded from below. Precisely, in terms of the soliton constant λ, we have

$$S(x) \geq \begin{cases} m\lambda, & \text{if } \lambda < 0, \\ 0 & \text{if } \lambda \geq 0. \end{cases} \tag{8.119}$$

A second ingredient is given by *Hamilton's identity* that we are going to prove in the next

Lemma 8.6 *Let $(M, \langle\, , \rangle, \nabla f)$ be a gradient soliton with soliton constant λ. Then*

$$S + |\nabla f|^2 - 2\lambda f = \Lambda \tag{8.120}$$

for some constant $\Lambda \in \mathbb{R}$.

Proof We show that

$$\nabla\left(S+|\nabla f|^2-2\lambda f\right)\equiv 0.$$

Towards this end we know that

$$\left(S+|\nabla f|^2-2\lambda f\right)_i=S_i+(f_kf_k)_i-2\lambda f_i=S_i+2f_{ki}f_k-2\lambda f_i. \tag{8.121}$$

Next, we trace (8.2) to obtain

$$S+\Delta f=m\lambda$$

so that

$$S_i=-f_{kki}. \tag{8.122}$$

On the other hand, taking the covariant derivative of (8.2)

$$R_{ki,j}=-f_{kij}$$

so that, tracing with respect to k and j

$$R_{ki,k}=-f_{kik}.$$

From the commutation rules (8.9) that in this case read

$$f_{ikj}-f_{ijk}=f_tR_{tikj}$$

we obtain

$$-f_{kki}=R_{ki,k}+f_tR_{ti}.$$

Thus using Schur's identities $2R_{ki,k}=S_i$ we have

$$-f_{kki}=\frac{1}{2}S_i+f_tR_{ti}$$

that, together with (8.122), yields

$$S_i=2f_tR_{ti}.$$

Inserting (8.2) into the above we finally obtain

$$S_i=2\lambda f_i-2f_{ki}f_k$$

from which, substituting into (8.121), we immediately deduce the desired conclusion. □

The next result is due to Zhang [282]; the present proof is slightly different from the original argument.

Proposition 8.12 *Let* $(M, \langle\, ,\, \rangle, \nabla f)$ *be a gradient Ricci soliton on the m-dimensional complete manifold* $(M, \langle\, ,\, \rangle)$. *Then there exist constants* $a, b \geq 0$ *such that*

$$(i)\ |\nabla f|(x) \leq |\lambda| r(x) + a$$

and

$$(ii)\ |f|(x) \leq \frac{1}{2}|\lambda| r(x)^2 + ar(x) + b.$$

Proof We only prove (i) since (ii) follows immediately from it. For $x \in M$ let $\gamma(t)$, $t \in [0, r(x)]$ be a unit speed minimal geodesic connecting o to x. If $\lambda = 0$, from (8.119), $S(x) \geq 0$ and from (8.120) we deduce

$$S + |\nabla f|^2 = \Lambda \geq 0.$$

Therefore

$$|\nabla f| \leq \sqrt{\Lambda}. \tag{8.123}$$

If $\lambda > 0$, by adding a constant to f, the soliton equation (8.2) is still satisfied but we can normalized (8.120) in such a way that

$$S + |\nabla f|^2 - 2\lambda f \equiv 0. \tag{8.124}$$

Of course this process leaves $|\nabla f|$ unchanged. Since, from (8.119), $S(x) \geq 0$, (8.124) yields

$$2\lambda f \geq |\nabla f|^2 \geq 0. \tag{8.125}$$

We let $h(t) = f(\gamma(t))$. Then, using the above,

$$|h'(t)| = |\langle \nabla f, \dot{\gamma} \rangle(t)| \leq |\nabla f(\gamma(t))| \leq \sqrt{2\lambda}\sqrt{h(t)},$$

so that

$$(\sqrt{h(t)})' \leq \sqrt{\frac{\lambda}{2}}.$$

Integrating this last inequality on $[0, r(x)]$

$$|\sqrt{h(r(x))} - \sqrt{h(0)}| \le \sqrt{\frac{\lambda}{2}} r(x),$$

that is,

$$\sqrt{f(x)} \le \sqrt{\frac{\lambda}{2}} r(x) + \sqrt{f(o)},$$

and, from (8.125),

$$|\nabla f(x)| \le \lambda r(x) + \sqrt{2\lambda f(o)}. \tag{8.126}$$

Finally, we consider the case $\lambda < 0$. Again we use the normalized identity (8.124). From it, using (8.119), we obtain

$$0 \le |\nabla f|^2 = 2\lambda f - S \le 2\lambda f - m\lambda. \tag{8.127}$$

We let $h(t) = \frac{m}{2} - f(\gamma(t)) \ge 0$. Then, proceeding as above we get

$$\sqrt{\frac{m}{2} - f(x)} \le \sqrt{\frac{|\lambda|}{2}} r(x) + \sqrt{\frac{m}{2} - f(o)}.$$

From (8.127) we thus have

$$|\nabla f(x)| \le \sqrt{2|\lambda|}\sqrt{\frac{m}{2} - f(x)} \le |\lambda| r(x) + \sqrt{2\lambda f(o) - m\lambda}. \tag{8.128}$$

Inequalities (8.123), (8.126) and (8.128) prove (i). □

Our next aim is to improve the conclusions of Theorem 8.2 in case $(M, \langle\, ,\, \rangle, X)$ is a gradient soliton, that is, $X = \nabla f$ for some $f \in C^\infty(M)$. Towards this aim we prove the following classification result.

Proposition 8.13 *Let $(M, \langle\, ,\, \rangle)$ be a complete, connected, Einstein manifold of dimension $m \ge 3$ and (constant) scalar curvature S. Let $(M, \langle\, ,\, \rangle, \nabla f)$ be a gradient soliton on M with soliton constant λ. Then:*

(i) If $S = 0$ one of the following possibilities occurs:

(i1) $\lambda = 0$ and M is isometric to a cylinder $\mathbb{R} \times \Sigma$ over a totally geodesic, Ricci flat hypersurface $\Sigma \subset M$. Furthermore, on $\mathbb{R} \times \Sigma$ the potential function f can be expressed in the form $f(t, x) = at + b$ for some constants $a, b \in \mathbb{R}$.

(i2) $\lambda \neq 0$ *and* M *is isometric to* $\mathbb{R}^m$. *Furthermore, on* $\mathbb{R}^m$ *the potential function* f *can be expressed in the form* $f(x) = \frac{\lambda}{2}|x|^2 + \langle b, x\rangle + c$ *for some* $b \in \mathbb{R}^m$ *and* $c \in \mathbb{R}$.

(ii) If $S \neq 0$ *then the soliton is trivial.*

Proof We follow the argument in Theorem 2.3 of Pigola et al. [224], where the authors consider the more general case of *almost* Ricci solitons (that is, when $\lambda \in C^\infty(M)$ and not necessarily constant). By assumption, in a local orthonormal coframe we have

$$R_{ij} = \frac{S}{m}\delta_{ij} \tag{8.129}$$

and

$$R_{ij} + f_{ij} = \lambda\delta_{ij}, \tag{8.130}$$

thus we deduce

$$f_{ij} = \left(\lambda - \frac{S}{m}\right)\delta_{ij}. \tag{8.131}$$

Since the quantity $\lambda - \frac{S}{m}$ is constant, taking the covariant derivative of (8.131), skew-symmetrizing and using the commutation rule (1.116) we get

$$f_{ijk} - f_{ikj} = f_t R_{tijk} = 0.$$

Tracing with respect to i and k and using (8.129) we have

$$f_t R_{tk} = \frac{S}{m} f_k = 0. \tag{8.132}$$

Now we consider two cases, according to the statement of the Theorem.
Case (i). If $S = 0$ (and M is then Ricci flat), (8.131) implies

$$f_{ij} = \lambda\delta_{ij}. \tag{8.133}$$

Now, if $\lambda = 0$ (**case (i1)**), $f_{ij} = 0$ (i.e. f is affine); thus $|\nabla f|$ is constant, proving that either f is constant (and the soliton is trivial), or f has no critical point at all. In this latter case, we have $|\nabla f| = a$, for some $a \in \mathbb{R} \setminus \{0\}$. Then, a Cheeger-Gromoll type argument shows that the flow ϕ of the vector field ∇f establishes a Riemannian isometry $\phi : \mathbb{R} \times \Sigma \to M$, where Σ is any of the (totally geodesic) level sets of f and f is a linear function of the parameter $t \in \mathbb{R}$. Indeed, let $\Sigma = \{f(x) = b\}$, $b \in \mathbb{R}$, be a nonempty, smooth, level set hypersurface; then, the integral curves of the complete vector field $Y = \frac{\nabla f}{|\nabla f|}$ are unit speed geodesics orthogonal to Σ (see Step 1 in the proof of Theorem 8.5 below). Moreover, the flow of Y gives rise to a smooth map $\phi : \mathbb{R} \times \Sigma \to M$ which coincides with the normal exponential map $\exp^{\perp}$ of Σ; in particular, ϕ is surjective. Evaluating the equation $\mathrm{Hess}(f) = 0$ along the integral curve $\phi(t, x)$ issuing from $x \in \Sigma$ we deduce that $y(t) = f(\phi(t, x))$ satisfies

$$\begin{cases} y'' & = 0 \\ y(0) & = b \\ y'(0) & = |\nabla f|(x) = a, \end{cases}$$

which implies

$$f(\phi(t, x)) = at + b, \quad t \in \mathbb{R},$$

and since $a \neq 0, f$ is strictly monotone along the geodesic curves $\phi_x(t)$ issuing from $x \in \Sigma$. Then ϕ is also injective, hence a diffeomorphism. Note that $|\nabla f|$ is constant on Σ and that ϕ moves Σ onto every other level set of f. To conclude, we show that

$$\phi^* \langle \, , \, \rangle = dt^2 + a^2 \langle \, , \, \rangle_{\Sigma}, \tag{8.134}$$

where $\langle \, , \, \rangle_{\Sigma} = (\phi_0)^* \langle \, , \, \rangle$ denotes the metric induced by M on the smooth hypersurface Σ. Indeed, applying Gauss lemma we have

$$\phi^* \langle \, , \, \rangle = dt^2 + (\phi_t)^* \langle \, , \, \rangle;$$

furthermore, using $\mathrm{Hess}(f) = 0$, $\frac{df(\phi(t,x))}{dt} = a$ and the definition of the Lie derivative, we see that, on $T\Sigma_{\phi_t} = Y^{\perp}_{\phi_t}$,

$$\frac{d}{dt} (\phi_t)^* \langle \, , \, \rangle = 0.$$

Whence, integrating on $[0, t]$ we conclude the validity of (8.134). Summarizing, we have obtained that, if f has no critical point, then $(M, \langle \, , \, \rangle)$ is isometric to the warped product manifold

$$\left(\mathbb{R} \times \Sigma, dt^2 + a^2 \langle \, , \, \rangle_{\Sigma}\right),$$

with Σ a smooth hypersurface of M; moreover, since M is Ricci flat, then also Σ must be Ricci flat.

On the other hand, if $\lambda \neq 0$ (**case (i2)**), then it is known that M is isometric to $\mathbb{R}^m$ and the potential takes the form

$$f(x) = \frac{\lambda}{2} |x|^2 + \langle b, x \rangle + c$$

for some $b \in \mathbb{R}^m$ and $c \in \mathbb{R}$, see Theorem 8.5 and its proof below or [263].
Case (ii). If $S \neq 0$, from (8.132) we immediately deduce $f_k = 0$, which implies that f is constant and the soliton is trivial. □

For the sake of completeness we recall here the next

Theorem 8.5 *Let $(M, \langle \, , \, \rangle)$ be a complete m-dimensional Riemannian manifold. If there exists a smooth function $f : M \to \mathbb{R}$ such that* $\mathrm{Hess}(f) = \lambda \langle \, , \, \rangle$ *for some constant $\lambda \neq 0$, then M is isometric to $\mathbb{R}^m$.*

Proof We follow the proof in the Appendix of Pigola et al. [231] (see also [223]). Let $f \in C^\infty(M)$ be a solution of

$$\mathrm{Hess}(f) = \lambda \langle \, , \, \rangle \tag{8.135}$$

for some constant $\lambda \neq 0$; without loss of generality, we can assume λ to be positive. Now we divide the proof in four steps.

Step 1. We show first that f has a critical point. By contradiction, suppose that $|\nabla f| \neq 0$ on M and consider the vector field $Y = \frac{\nabla f}{|\nabla f|}$. Y is complete, since $|Y| \in L^\infty(M)$ and M is geodesically complete by assumption (see e.g. [171, Chap. 12]). Let $\gamma : \mathbb{R} \to M$ be an integral curve of Y, that is $Y_{\gamma(s)} = \dot{\gamma}(s)$ for every s in $\mathbb{R}$. A direct computation that uses (8.135) shows that, for every vector field X,

$$\left\langle \nabla_{\dot{\gamma}} \dot{\gamma}, X \right\rangle = \frac{1}{|\nabla f|} \mathrm{Hess}\,(f)\,(\dot{\gamma}, X) - \frac{1}{|\nabla f|} \mathrm{Hess}\,(f)\,(\dot{\gamma}, \dot{\gamma})\,\langle \dot{\gamma}, X \rangle = 0,$$

therefore γ is a unit speed geodesic of M. Evaluating (8.135) along γ we deduce that the smooth function $y\,(s) = (f \circ \gamma)\,(s)$ satisfies

$$y'' = \frac{d^2 y}{ds^2} = \lambda .$$

Integrating the previous equation on $[0, s]$ yields $y'(s) = \lambda s + y'(0)$, which implies $y'(s_0) = 0$, where $s_0 = -\lambda^{-1} y'(0)$. Then, recalling that γ is an integral curve of Y, we conclude

$$0 = y'\,(s_0) = \langle \nabla f\,(\gamma\,(s_0))\,, \dot{\gamma}\,(s_0) \rangle = |\nabla f\,(\gamma\,(s_0))| \neq 0,$$

a contradiction.
Step 2. Let $o \in M$ be a critical point of f and set $r\,(x) = \mathrm{dist}_{(M, \langle \, , \, \rangle)}\,(x, o)$. Now we fix $x \in M$ and we let $\gamma : [0, r(x)] \to M$ be a unit speed, minimizing geodesic emanating from $\gamma(0) = o$; then the function $y\,(s) = (f \circ \gamma)\,(s)$ satisfies the Cauchy

problem

$$\begin{cases} y'' = \lambda, \\ y'(0) = 0, \quad y(0) = f(o). \end{cases}$$

Integrating on $[0, r(x)]$ yields

$$f(x) = \alpha(r(x)), \tag{8.136}$$

where

$$\alpha(s) = \frac{\lambda}{2}s^2 + f(o);$$

in particular, f is a proper function with precisely one critical point.
Step 3. Since $f(x) = \alpha(r(x))$ is smooth and $\alpha(s)$ satisfies $\alpha'(s) \neq 0$ for every $s > 0$, we have that

$$r(x) = \alpha^{-1}(f(x))$$

is smooth on $M \setminus \{o\}$. By Bishop's density result (see the discussion in Sect. 1.9 and [48]) we deduce that cut $(o) = \emptyset$ and the exponential map $\exp_o : T_oM \approx \mathbb{R}^m \to M$ is a diffeomorphism. Let us introduce geodesic polar coordinates $(r, \theta) \in \mathbb{R}^+ \times \mathbb{S}^{m-1}$ on T_oM. Furthermore, let us consider a local orthonormal coframe $\{\theta^i\}$ on $\mathbb{S}^{m-1}$ with dual frame $\{E_i\}$; thus, the standard metric of $\mathbb{S}^{m-1}$ writes as $d\theta^2 = \sum \theta^i \otimes \theta^i$. We extend both $\{\theta^i\}$ and $\{E_i\}$ radially. Then, by Gauss lemma,

$$\langle\,,\,\rangle = dr \otimes dr + \sum_{i,j}^{m-1} \sigma_{i,j}(r, \theta)\, \theta^i \otimes \theta^j,$$

where, since the metric $\langle\,,\,\rangle$ is infinitesimally Euclidean and the standard metric of $\mathbb{R}^m \approx T_oM$ writes as

$$\langle\,,\,\rangle_{\mathbb{R}^m} = dr \otimes dr + r^2 \delta_{ij} \theta^i \otimes \theta^j,$$

we have the further condition

$$\sigma_{ij}(r, \theta) = \delta_{ij} r^2 + o\left(r^2\right) \quad \text{as } r \searrow 0. \tag{8.137}$$

Now we use the fact that, by the definition of Lie derivative,

$$\frac{\partial}{\partial r}\sigma_{ij} = (\mathcal{L}_{\nabla r}\langle\,,\,\rangle)\left(E_i, E_j\right) = 2\,\mathrm{Hess}\,(r)\left(E_i, E_j\right); \tag{8.138}$$

on the other hand, by (8.136), $\nabla r = \frac{\nabla f}{|\nabla f|}$, so that, using Eq. (8.135), we deduce that, for every $E_i, E_j \in \nabla r^{\perp}$,

$$\mathrm{Hess}\,(r)\left(E_i, E_j\right) = \left\langle \nabla_{E_i}\left(\frac{\nabla f}{|\nabla f|}\right), E_j \right\rangle = \frac{1}{r}\sigma_{ij}. \tag{8.139}$$

Combining (8.137)–(8.139) we conclude that the coefficients σ_{ij} are the (unique) solutions of the asymptotic Cauchy problems

$$\begin{cases} \frac{\partial}{\partial r}\sigma_{ij} = \frac{2}{r}\sigma_{ij} \\ \sigma_{ij}(r,\theta) = r^2\delta_{ij} + o\left(r^2\right), \quad \text{as } r \searrow 0. \end{cases}$$

Integrating finally gives

$$\sigma_{ij}(r,\theta) = r^2\delta_{ij}.$$

Since $\left(\mathbb{R}^+ \times \mathbb{S}^{m-1}, dr \otimes dr + r^2 \sum_i \theta^i \otimes \theta^i\right)$ is isometric to $\mathbb{R}^m \setminus \{0\}$, the proof is completed. □

Now we are ready to prove the next

Theorem 8.6 *Let $(M, \langle\,,\,\rangle)$ be a complete Riemannian manifold of dimension $m \geq 3$ and scalar curvature S, and let $S_* = \inf_{x\in M} S(x)$. Let $(M, \langle\,,\,\rangle, \nabla f)$ be a gradient soliton structure on M with soliton constant λ.*

(i) If $\lambda < 0$ then $m\lambda \leq S_* \leq 0$. Moreover, if $S(x_0) = S_* = m\lambda$ for some $x_0 \in M$ and $m \geq 3$, then $(M, \langle\,,\,\rangle)$ is Einstein and the soliton is trivial; while if $S(x_0) = S_* = 0$, for some $x_0 \in M$, then $(M, \langle\,,\,\rangle)$ is isometric to the standard Euclidean space $\mathbb{R}^m$. On the latter, the potential function f can be expressed in the form $f(x) = \frac{\lambda}{2}|x|^2 + \langle b, x\rangle + c$ for some $b \in \mathbb{R}^m$ and $c \in \mathbb{R}$.
(ii) If $\lambda = 0$ then $S_* = 0$. Furthermore, if $S(x_0) = S_* = 0$ for some $x_0 \in M$, then $(M, \langle\,,\,\rangle)$ is isometric to a cylinder $\mathbb{R} \times \Sigma$ over a totally geodesic, Ricci flat hypersurface $\Sigma \subset M$. On $\mathbb{R} \times \Sigma$ the potential function f can be expressed in the form $f(t, y) = at + b$ for some constants $a, b \in \mathbb{R}$ and $(t, y) \in \mathbb{R} \times \Sigma$.
(iii) If $\lambda > 0$ then $0 \leq S_* \leq m\lambda$. Furthermore, if $S(x_0) = S_* = 0$ for some $x_0 \in M$, then $(M, \langle\,,\,\rangle)$ is isometric to the standard Euclidean space $\mathbb{R}^m$. On the latter, the potential function f can be expressed in the form $f(x) = \frac{\lambda}{2}|x|^2 + \langle b, x\rangle + c$ for some $b \in \mathbb{R}^m$ and $c \in \mathbb{R}$, while if $S(x_0) = S_* = m\lambda$, for some $x_0 \in M$, then $(M, \langle\,,\,\rangle)$ is compact, Einstein and f is constant (i.e. the soliton is trivial).

Proof The proof of the result is based on Theorem 8.2 and Proposition 8.13 as follows.

Case (i). From Theorem 8.2 we know that $m\lambda \leq S_* \leq 0$. Suppose now that $S(x_0) = S_* = m\lambda$ for some $x_0 \in M$. Again by Theorem 8.2 we know that $S(x) \equiv m\lambda = S_*$ on M and that M is Einstein and the soliton is trivial; if $S(x_0) = S_* = 0$ for some $x_0 \in M$, then $(M, \langle\,,\,\rangle)$ is Ricci flat, hence Einstein. Since $\lambda \neq 0$, by Proposition 8.13

(i2) $(M, \langle\,,\,\rangle)$ is isometric to the standard Euclidean space $\mathbb{R}^m$. On the latter the potential function f can be expressed in the form $f(x) = \frac{\lambda}{2}|x|^2 + \langle b, x\rangle + c$ for some $b \in \mathbb{R}^m$ and $c \in \mathbb{R}$.

Case (ii). $S_* = 0$ follows from Theorem 8.2. If $S(x_0) = S_* = 0$ for some $x_0 \in M$, by the same theorem we know that $(M, \langle\,,\,\rangle)$ is Ricci flat and therefore Einstein. Using Proposition 8.13 (i1) we deduce that $(M, \langle\,,\,\rangle)$ is isometric to a cylinder $\mathbb{R} \times \Sigma$ over a totally geodesic Ricci flat hypersurface $\Sigma \subset M$. On $\mathbb{R} \times \Sigma$ the potential function f can be expressed in the form $f(t, y) = at + b$ for some constants $a, b \in \mathbb{R}$ and $(t, y) \in \mathbb{R} \times \Sigma$.

Case (iii). is dealt with in a similar manner. □

8.3.3 A Topological Result on Weighted Manifolds

We end this section with a topological result dealing with weighted manifolds. Gradient Ricci solitons are natural weighted manifolds with weight given *via* the potential function f in the form of the density

$$e^{-f}\, dvol,$$

where $dvol$ is the volume element of $(M\langle\,,\,\rangle)$. For more details on weighted manifolds see for instance [200, 201] and references therein. We leave to the interested reader the application of the next result to the special case of gradient Ricci solitons.

Theorem 8.7 *For $f \in C^\infty(M)$, let $\left(M, \langle\,,\,\rangle, e^{-f}\, dvol\right)$ be a geodesically complete weighted manifold and assume the existence of an origin $o \in M$ and of functions $\lambda \geq 0$ and g bounded such that, for each unit speed geodesic γ issuing from $\gamma(0) = o$ we have*

$$\mathrm{Ric}_f(\dot\gamma, \dot\gamma) = \mathrm{Ric}(\dot\gamma, \dot\gamma) + \mathrm{Hess}(f)(\dot\gamma, \dot\gamma) \geq \lambda(\gamma) + \langle(\nabla g)(\gamma), \dot\gamma\rangle \tag{8.140}$$

and

$$\lambda(\gamma(t)) \notin L^1(+\infty). \tag{8.141}$$

Then:

(i) $|\pi_1(M)| < +\infty$;
(ii) *if, in addition,* $\mathrm{Ric} \leq c\langle\,,\,\rangle$ *for some $c \in \mathbb{R}$ and $\lambda(x) = \lambda_0(r(x))$, with $r(x) =$* $\mathrm{dist}\,(x, o)$, *then M is diffeomorphic to the interior of a compact manifold N with $\partial N \neq \emptyset$.*

(iii) in the assumptions of (ii), if $\lambda(x) \geq \lambda_0 > 0$ *and* $\sup_M \{|\nabla f| + |g|\} \leq \Lambda < +\infty$*, then* M *is compact and*

$$\operatorname{diam}(M) \leq \frac{1}{\lambda_0}\left[2\Lambda + \sqrt{4\Lambda^2 + \pi^2(m-1)c}\right].$$

For the proof we shall need the following

Lemma 8.7 *Let* $(M, \langle\,,\,\rangle)$ *be a Riemannian manifold,* $a \in M$ *and* $r(x) = \operatorname{dist}(x, o)$*. Fix* $q \in M$ *and let* $\gamma : [0, r(q)] \to M$ *be a minimizing geodesic from* o *to* q *with* $|\dot{\gamma}| = 1$*. For* $p \in M$ *set*

$$\Gamma_p = \max\left\{0, \sup_{B_1(p)} \operatorname{Ric}\right\}.$$

Then, if $r(q) > 2$*, we have*

$$\int_0^{r(q)} \operatorname{Ric}(\dot{\gamma}, \dot{\gamma}) \leq 2(m-1) + \Gamma_o + \Gamma_q.$$

Proof First of all let us recall that, from (1.246) in Chap. 1, for each $h \in C^1([0, r(x)])$ with $h(o) = h(r(x)) = 0$ we have

$$\int_0^{r(x)} \operatorname{Ric}(\dot{\gamma}, \dot{\gamma}) \leq (m-1)\int_0^{r(x)} \left(h'\right)^2 + \int_0^{r(x)} \left(1 - h^2\right) \operatorname{Ric}(\dot{\gamma}, \dot{\gamma}).$$

Choosing

$$h(s) = \begin{cases} s & \text{on } [0, 1] \\ 1 & \text{on } [1, r(x) - 1] \\ r(x) - s & \text{on } [r(x) - 1, r(x)], \end{cases}$$

where $r(x) > 2$, we obtain

$$\int_0^{r(x)} \operatorname{Ric}(\dot{\gamma}, \dot{\gamma}) \leq 2(m-1) + \int_0^1 \left(1 - s^2\right) \operatorname{Ric}(\dot{\gamma}, \dot{\gamma})$$
$$+ \int_{r(x)-1}^{r(x)} \left(1 - (r(x) - s)^2\right) \operatorname{Ric}(\dot{\gamma}, \dot{\gamma}) \leq 2(m-1) + \Gamma_o + \Gamma_q.$$

□

Lemma 8.8 *Let* $\left(M, \langle\,,\,\rangle, e^{-f}\, dvol\right)$ *be a complete weighted Riemannian manifold,* $o \in M$ *and* $r(x) = \operatorname{dist}(x, o)$*. Suppose there exist functions* $\lambda \geq 0$ *and* g *bounded* *such that, for each unit speed geodesic γ issuing from o,*

$$\mathrm{Ric}_f\,(\dot\gamma,\dot\gamma) \geq \lambda(\gamma) + \langle \nabla g, \dot\gamma\rangle. \tag{8.142}$$

Then, for each such geodesic,

$$\begin{aligned}\int_0^t \mathrm{Ric}\,(\dot\gamma,\dot\gamma) &= \langle \nabla f, \dot\gamma(0)\rangle - \langle \nabla f, \dot\gamma(t)\rangle + \int_0^t \lambda(\gamma(s))\,ds + g(\gamma(t)) - g(0) \\ &\geq -|\nabla f|(\gamma(0)) - |\nabla f|(\gamma(t)) - 2\sup|g| + \int_0^t \lambda(\gamma(s))\,ds.\end{aligned} \tag{8.143}$$

Proof We rewrite (8.142) in the form

$$\mathrm{Ric}\,(\dot\gamma,\dot\gamma) + \mathrm{Hess}(f)(\dot\gamma,\dot\gamma) \geq \lambda(\gamma) + \langle \nabla g, \dot\gamma\rangle \tag{8.144}$$

and then in the form

$$\mathrm{Ric}\,(\dot\gamma,\dot\gamma) + \frac{d}{dt}\langle \nabla f(\gamma), \dot\gamma\rangle \geq \lambda(\gamma) + \frac{d}{dt} g(\gamma).$$

Now integrating on $[0, t]$

$$\int_0^t \mathrm{Ric}\,(\dot\gamma,\dot\gamma) + \langle \nabla f, \dot\gamma(t)\rangle - \langle \nabla f, \dot\gamma(0)\rangle + \int_0^t \lambda(\gamma(s))\,ds + g(\gamma(t)) - g(0)$$

from which the Lemma follows immediately. □

Proof (of Theorem 8.7) The idea of the proof is in [275]. Let $\pi : \tilde M \to M$ be the Riemannian universal covering of M. Defining $\tilde f = f \circ \pi$, and since π is a local isometry, $\tilde M$ becomes a complete weighted Riemannian manifold. Moreover, since every unit speed geodesic $\gamma = \pi \circ \tilde\gamma$ of $\tilde M$ projects to a unit speed geodesic $\gamma = \pi \circ \tilde\gamma$ we see that

$$\widetilde{\mathrm{Ric}}_{\tilde f}\left(\dot{\tilde\gamma},\dot{\tilde\gamma}\right) = \mathrm{Ric}_f\,(\dot\gamma,\dot\gamma) \geq \lambda(\gamma) + \frac{d}{dt} g(\gamma) = \tilde\lambda(\tilde\gamma) + \frac{d}{dt}\tilde g(\tilde\gamma),$$

where $\tilde g = g \circ \pi$ is bounded and $\tilde\lambda = \lambda \circ \pi \geq 0$ satisfies

$$\tilde\lambda(\tilde g)(t) \notin L^1(+\infty). \tag{8.145}$$

We identify

$$\pi_1(M, o) = \mathrm{Deck}\left(\tilde M\right),$$

the covering transformation group, and recall that there is a bijective correspondence between $\pi_1(M, o)$ and $\pi^{-1}(\{o\})$. Therefore, it suffices to show that $\pi^{-1}(\{o\}) \subset \tilde{B}_R(\tilde{o})$ for some $R \gg 1$, with $\tilde{o}$ a fixed preimage of o. Since $\pi_1(M, o) = \mathrm{Deck}\left(\tilde{M}\right)$ acts transitively on the fiber $\pi^{-1}(\{o\})$, we have

$$\pi^{-1}(\{o\}) = \left\{h(\tilde{o}) : h \in \mathrm{Deck}\left(\tilde{M}\right)\right\},$$

and we are reduced to showing that

$$\tilde{r}(h(\tilde{o})) \leq R < +\infty \qquad \text{for each } h \in \mathrm{Deck}\left(\tilde{M}\right),$$

where $\tilde{r}(\tilde{x}) = \mathrm{dist}\,(\tilde{x}, \tilde{o})$ for $\tilde{x} \in \tilde{M}$. Fix $h \in \mathrm{Deck}\left(\tilde{M}\right)$ and a unit speed minimizing geodesic $\tilde{\gamma}_{h(\tilde{o})} : [0, \tilde{r}(h(\tilde{o}))] \to \tilde{M}$, issuing from $\tilde{\gamma}_{h(\tilde{o})}(0) = \tilde{o}$. Recalling that

$$\widetilde{\mathrm{Ric}}\left(\dot{\tilde{\gamma}}, \dot{\tilde{\gamma}}\right) = \widetilde{\mathrm{Ric}}_{\tilde{f}}\left(\dot{\tilde{\gamma}}, \dot{\tilde{\gamma}}\right) - \frac{d}{dt}\left\langle\left(\tilde{\nabla}\tilde{f}\right)(\tilde{\gamma})\dot{\tilde{\gamma}}\right\rangle$$

and using Lemmas 8.7 and 8.8, we get

$$\int_0^{\tilde{r}(h(\tilde{o}))} \tilde{\lambda}\left(\tilde{\gamma}_{h(\tilde{o})}\right)(s)\, ds \leq 2(m-1) + \tilde{\gamma}_{\tilde{o}} + \tilde{\Gamma}_{h(\tilde{o})} + \left|\tilde{\nabla}\tilde{f}\right|(\tilde{o}) + \left|\tilde{\nabla}\tilde{f}\right|(h(\tilde{o})) + 2\sup_{\tilde{M}} |\tilde{g}|.$$

Since $\pi : \tilde{M} \to M$ is a local isometry and $\tilde{o}, h(\tilde{o}) \in \pi^{-1}(\{o\})$ we deduce

$$\left|\tilde{\nabla}\tilde{f}\right|(\tilde{o}) = |\nabla f|(o) = \left|\tilde{\nabla}\tilde{f}\right|(h(\tilde{o})).$$

On the other hand $\mathrm{Deck}\left(\tilde{M}\right) \subset \mathrm{Iso}\left(\tilde{M}\right)$, so $h\left(\tilde{B}_1(\tilde{o})\right)$ is isometric to $\tilde{B}_1(h(\tilde{o}))$ and we have

$$|\Gamma_{\tilde{o}}| = \left|\Gamma_{h(\tilde{o})}\right|.$$

Summarizing, we have obtained that, for each $h \in \mathrm{Deck}\left(\tilde{M}\right)$,

$$\int_0^{\tilde{r}(h(\gamma))} \lambda\left(\tilde{\gamma}_{h(\tilde{o})}\right)(s)\, ds \leq 2\{(m-1) + \Gamma_{\tilde{o}} + |\nabla f|(o)\} + 2\sup_M |g|. \tag{8.146}$$

We now argue by contradiction and we suppose the existence of a sequence of transformations $\{h_n\} \subset \mathrm{Deck}\left(\tilde{M}\right)$ such that

$$\tilde{r}(h_n(\tilde{o})) \to +\infty \quad \text{as } n \to +\infty. \tag{8.147}$$

Let $\tilde{\gamma}_{h_n(\tilde{o})}(s) = \exp_{\tilde{o}}\left(s\tilde{\xi}_n\right)$, where $\tilde{\xi}_n \in \mathbb{S}^{m-1}_{\tilde{o}} \subset T_{\tilde{o}}\tilde{M}$. Then, there exists a subsequence $\left\{\tilde{\xi}_{n_k}\right\} \to \tilde{\xi} \in \mathbb{S}^{m-1}_{\tilde{o}}$ as $k \to +\infty$, and by the Ascoli-Arzelà Theorem the sequence of minimizing geodesic $\left\{\tilde{\gamma}_{h_{n_k}(\tilde{o})}\right\}$ converges uniformly on compact subsets of $\mathbb{R}^+_0$ to the unit speed geodesic $\tilde{\gamma}(s) = \exp_{\tilde{o}}\left(s\tilde{\xi}\right)$. Since by (8.145)

$$\int_0^{+\infty} \tilde{\lambda}(\tilde{\gamma})(s)\,ds = +\infty$$

we can choose T sufficiently large such that

$$\int_0^{T} \tilde{\lambda}(\tilde{\gamma})(s)\,ds > 2\{(m-1) + \Gamma_{\tilde{o}} + |\nabla f|(o)\} + 2\sup_M |g|. \tag{8.148}$$

On the other hand, according to (8.147), we can find $k_0 > 0$ such that, for each $k \geq k_0$, $\tilde{r}(h_{n_k}(\tilde{o})) > T$. From this, from inequality (8.146) and the definition of $\tilde{\lambda}(\tilde{x}) = \lambda(\pi(\tilde{x})) \geq 0$ it follows that

$$\begin{aligned}\int_0^{T} \tilde{\lambda}\left(\tilde{\gamma}_{h_{n_k}(\tilde{o})}\right)(s)\,ds &\leq \int_0^{\tilde{r}\left(h_{n_k}(\tilde{o})\right)} \tilde{\lambda}\left(\tilde{\gamma}_{h_{n_k}(\tilde{o})}\right)(s)\,ds \\ &\leq 2\{(m-1) + \Gamma_{\tilde{o}} + |\nabla f|(o)\} + 2\sup_M |g|.\end{aligned}$$

Hence, letting $k \to +\infty$ we deduce

$$\int_0^{T} \tilde{\lambda}(\tilde{\gamma})(s)\,ds \leq 2\{(m-1) + \Gamma_{\tilde{o}} + |\nabla f|(o)\} + 2\sup_M |g|,$$

contradicting (8.148). This proves (i). To prove (ii) suppose $\mathrm{Ric} \leq c\langle\,,\,\rangle$. Fix $q \in M$ such that $r(q) = \mathrm{dist}\,(o,q) > 2$, and let γ_q be a minimizing geodesic joining o to q. Combining Lemmas 8.7 and 8.8 and recalling that $\lambda(x) = \lambda_0(r(x))$ is radial we obtain

$$\begin{aligned}&-|\nabla f|(o) - |\nabla f|(\gamma(r(q))) - 2\sup_M |g| \\ &\quad + \int_0^{r(q)} \lambda_0(s)\,ds \leq 2(m-1) + \Gamma_o + \Gamma_q \leq 2(m-1) + 2c,\end{aligned}$$

which implies

$$|\nabla f|(q) \geq \int_0^{r(q)} \lambda_0(s)\,ds - \{|\nabla f|(o) + 2(m-1) + 2c\}.$$

Since $0 < \lambda_0 \notin L^1(+\infty)$, if $r(q) \geq R_0$ sufficiently large we have $|\nabla f|(q) > 0$. Thus f has no critical points in $M \setminus B_{R_0}(o)$. Again from Lemmas 8.7 and 8.8, for each $0 \leq t \leq r(q)$,

$$\int_0^t \lambda_0(s)\, ds - \big\langle (\nabla f)(\gamma_q), \dot{\gamma}_q \big\rangle + \big\langle (\nabla f)(\gamma_q), \dot{\gamma}_q \big\rangle_{s=0} + g(q) - g(o) \leq 2(m-1) + 2c,$$

so that

$$\frac{d}{ds} f\big(\gamma_q\big)_{|s=t} \geq \int_0^t \lambda_0(s)\, ds - \left\{ |\nabla f|(o) + 2 \sup_M |g| + 2(m-1) + 2c \right\}.$$

Thus, integrating on $[2, r(q)]$,

$$\begin{aligned} f(q) &\geq \int_2^{r(q)} \int_0^t \lambda_0(s)\, ds - \big|f\big(\gamma_q(2)\big)\big| \\ &\quad - \left\{ |\nabla f|(o) + 2 \sup_M |g| + 2(m-1) + 2c \right\} (r(q) - 2) \\ &\geq \int_2^{r(q)} \int_0^{r(q)} \lambda_0(s)\, ds - \max_{\partial B_2(o)} |f| \\ &\quad - \left\{ |\nabla f|(o) + 2 \sup_M |g| + 2(m-1) + 2c \right\} (r(q) - 2) \to +\infty \end{aligned}$$

as $r(q) \to +\infty$. Therefore, f is a smooth exhaustion function whose critical points are confined in a compact set. By Morse theory (see e.g. the classical [191]) there exists a compact manifold N with boundary such that M is diffeomorphic to the interior of N.

Finally we prove (iii). Suppose that $\sup_M (|\nabla f| + |g|) \leq \Lambda < +\infty$; then by (8.142) in Lemma 8.8, for each unit speed geodesic γ issuing from o we have

$$\mathrm{Ric}\,(\dot{\gamma}, \dot{\gamma}) \geq \lambda_0 + \frac{d}{dt} G(\gamma),$$

where $G(\gamma) = -\langle (\nabla f)(\gamma), \dot{\gamma} \rangle + g(\gamma)$ satisfies

$$|G(\gamma)| \leq \sup_M (|\nabla f| + |g|).$$

From Proposition 8.14 below, or the original Theorem 1.2 in [122], we obtain the desired diameter estimate. □

We would like to observe that similar results are expected for generic Ricci solitons. However, it is not clear, given the soliton structure $(M, \langle\, , \,\rangle, X)$, what should be the "natural" weight associated to M when X is not a gradient vector field.

Proposition 8.14 *Let $(M, \langle\, , \,\rangle)$ be a complete Riemannian manifold, $p, q \in M$ and $\gamma : [0, \ell] \to M$, $\gamma(0) = p$, $\gamma(\ell) = q$ be a unit length minimizing geodesic. Assume that*

$$\mathrm{Ric}\,(\dot{\gamma}, \dot{\gamma}) \geq \lambda_0 + \frac{d}{dt} f(t), \tag{8.149}$$

with $|f(t)| \geq \Lambda$ on $[0, \ell]$ and $\lambda_0 \in \mathbb{R}^+$. Then

$$\ell(\gamma) \leq \frac{\pi}{\lambda_0}\Big(\sqrt{\Lambda^2 + (m-1)\lambda_0} + \Lambda\Big). \tag{8.150}$$

Proof Let γ be as in the statement of the proposition; then, given $h \in C^1([0, \ell])$ with $h(0) = 0 = h(\ell)$, from (1.246) of Chap. 1 we have

$$\int_0^\ell (h')^2 - \frac{\mathrm{Ric}\,(\dot{\gamma}, \dot{\gamma})}{m-1} h^2 \geq 0.$$

Choose $h(t) = \sin\left(\frac{\pi t}{\ell}\right)$, so that the above inequality yields

$$\frac{\pi^2}{2\ell} - \int_0^\ell \frac{\mathrm{Ric}\,(\dot{\gamma}, \dot{\gamma})}{m-1} \sin^2\left(\frac{\pi t}{\ell}\right) dt \geq 0. \tag{8.151}$$

We now estimate the integral from below with the aid of (8.149) and integration by parts. We have

$$\begin{aligned}
\int_0^\ell \frac{\mathrm{Ric}\,(\dot{\gamma}, \dot{\gamma})}{m-1} \sin^2\left(\frac{\pi t}{\ell}\right) dt &\geq \int_0^\ell \left[\frac{\lambda_0}{m-1} \sin^2\left(\frac{\pi t}{\ell}\right) + \frac{1}{m-1} \sin^2\left(\frac{\pi t}{\ell}\right) \frac{d}{dt} f\right] dt \\
&= \frac{\lambda_0}{m-1}\frac{\ell}{2} - \frac{\pi}{\ell}\frac{1}{m-1} \int_0^\ell f(t) \sin\left(\frac{2\pi t}{\ell}\right) dt \\
&\geq \frac{\lambda_0}{m-1}\frac{\ell}{2} - \frac{\pi}{\ell}\frac{1}{m-1} \int_0^\ell |f(t)| \left|\sin\left(\frac{2\pi t}{\ell}\right)\right| dt \\
&\geq \frac{\lambda_0}{m-1}\frac{\ell}{2} - \frac{\pi}{\ell}\frac{1}{m-1} \Lambda \ell = \frac{\lambda_0}{m-1}\frac{\ell}{2} - \frac{\pi}{2}\frac{\Lambda}{m-1}.
\end{aligned}$$

Thus from (8.151)

$$\frac{\pi^2}{2\ell} - \frac{\lambda_0}{m-1}\frac{\ell}{2} + \frac{\pi}{m-1}\Lambda \geq 0$$

from which (8.150) follows at once. □

8.4 A Further Result on Generic Ricci Solitons

We now go back to generic Ricci solitons for a further very recent classification result (see [68]). This will appeal to strong parabolicity as introduced in Sect. 4.4.

For the computations below we shall use the next formulas that can be verified by direct calculation. For any vector fields Y and Z and functions $u, v \in C^2(M)$, with $v(x) \neq 0$ for each $x \in M$, we have

$$\Delta_Y\left(\frac{u}{v}\right) = \frac{1}{v}\Delta_Y u - \frac{u}{v^2}\Delta_Y v - 2\left\langle \nabla\left(\frac{u}{v}\right), \frac{\nabla v}{v}\right\rangle, \tag{8.152}$$

and

$$\Delta_{Y+Z} u = \Delta_Y u - \langle Z, \nabla u\rangle. \tag{8.153}$$

In the next computational result we shall use Eq. (8.15), coming from the generalized Bochner formula, and (8.30).

Lemma 8.9 *Let $(M, \langle\, ,\, \rangle, X)$ be a generic Ricci soliton with scalar curvature $S > 0$ on M. Let $\alpha \in (0, 1]$; then*

$$\Delta_{X-2\nabla \log S}\left(\frac{|X|^2}{S^\alpha}\right) \le \frac{2}{S^\alpha}\left\{\left(\frac{2-\alpha}{\alpha}\right)|\nabla X|^2 - \left[(\alpha+1)\lambda - \alpha\frac{|\operatorname{Ric}|^2}{S}\right]|X|^2\right\}. \tag{8.154}$$

Proof Using Eqs. (8.15) and (8.30), with the aid of (8.152) and (8.153), we compute

$$\begin{aligned}\Delta_{X-2\nabla \log S}\left(\frac{|X|^2}{S^\alpha}\right) &= \frac{2}{S^\alpha}\left\{|\nabla X|^2 - \left[(\alpha+1)\lambda - \alpha\frac{|\operatorname{Ric}|^2}{S}\right]|X|^2\right\} \\ &+ \frac{2(1-\alpha)}{S^{\alpha+1}}\langle \nabla|X|^2, \nabla S\rangle + \frac{\alpha(\alpha-1)}{S^{\alpha+2}}|X|^2|\nabla S|^2.\end{aligned} \tag{8.155}$$

Next, from Kato's inequality we deduce

$$|\nabla|X|^2| \le 2|X||\nabla X|,$$

while from Cauchy-Schwarz and Young's inequalities with $\varepsilon > 0$ we get

$$\frac{1}{S^{\alpha+1}}\langle \nabla S, \nabla|X|^2\rangle \le \frac{2|X||\nabla X||\nabla S|}{S^{\alpha+1}} \le \frac{1}{\varepsilon}\frac{|\nabla X|^2}{S^\alpha} + \frac{\varepsilon}{S^{\alpha+2}}|X|^2|\nabla S|^2.$$

Hence, for any $\alpha \in (0, 1]$, inserting the previous inequalities into (8.155) and rearranging terms we have

$$\Delta_{X-2\nabla \log S}\left(\frac{|X|^2}{S^\alpha}\right) \leq \frac{2}{S^\alpha}\left\{\left[1+\frac{1-\alpha}{\varepsilon}\right]|\nabla X|^2 - \left[(\alpha+1)\lambda - \alpha\frac{|\operatorname{Ric}|^2}{S}\right]|X|^2\right\}$$
$$+\frac{1-\alpha}{S^{\alpha+2}}(2\varepsilon-\alpha)|X|^2|\nabla S|^2.$$

Choosing $\varepsilon = \frac{\alpha}{2}$ we finally obtain (8.154). □

Corollary 8.6 *Let $(M, \langle\, ,\,\rangle, X)$ be a complete, generic, shrinking Ricci soliton with scalar curvature $S > 0$, satisfying $S^* = \sup_M S < +\infty$ and $|\operatorname{Ric}| \leq \Lambda S$ for some constant $\Lambda > 0$. Assume that*

$$|\nabla X| = o(|X|) \quad \text{as } r(x) \to +\infty. \tag{8.156}$$

Then, there exists $\alpha \in (0, 1]$ and a compact $K = K_\alpha \subset M$ such that

$$\Delta_{X-2\nabla \log S}\left(\frac{|X|^2}{S^\alpha}\right) < 0 \quad \text{on } M \setminus K. \tag{8.157}$$

Proof An immediate consequence of the assumptions and of (8.154). □

The next inequality (8.158) will be crucial. It comes from Proposition 8.5 and Corollary 8.2.

Lemma 8.10 *Let $(M, \langle\, ,\,\rangle, X)$ be a generic Ricci soliton of dimension m and scalar curvature $S > 0$ on M. Then*

$$\frac{1}{2}\Delta_{X-2\nabla \log S}\left(\frac{|T|^2}{S^2}\right) \geq 2\frac{|T|^2}{S^3}\left(|T| - \frac{1}{\sqrt{m(m-1)}}S\right)^2 + \frac{1}{S^3}\left(\frac{|T|}{\sqrt{S}}|\nabla S| - \sqrt{S}|\nabla T|\right)^2$$
$$-\sqrt{\frac{2(m-2)}{m-1}}\frac{1}{S^2}|W||T|^2. \tag{8.158}$$

Proof We use Eqs. (8.45), (8.30), (8.152) and

$$|\operatorname{Ric}|^2 = |T|^2 + \frac{S^2}{m}$$

to compute

$$\Delta_{X-2\nabla \log S}\left(\frac{|T|^2}{S^2}\right) = A + B + C, \tag{8.159}$$

where

$$A = \frac{2}{S^3}\left(S|\nabla T|^2 + \frac{1}{S}|T|^2|\nabla S|^2 - \langle \nabla |T|^2, \nabla S\rangle\right),$$

$$B = \frac{8}{m-2}\frac{1}{S^2}\mathrm{Tr}(t^3) + \frac{4}{m(m-1)}\frac{|T|^2}{S} + 4\frac{|T|^4}{S^3},$$

$$C = -\frac{4}{S^2}T_{ik}T_{sj}W_{ksij}.$$

Next we use Cauchy-Schwarz inequality and

$$|\nabla |T|^2| \leq 2|\nabla T||T|$$

to obtain

$$A \geq \frac{2}{S^3}\left(\frac{|T|}{\sqrt{S}}|\nabla S| - \sqrt{S}|\nabla T|\right)^2. \tag{8.160}$$

Since T is trace free, by Okumura's lemma (Lemma 6.2),

$$\mathrm{Tr}(t^3) \geq -\frac{m-2}{\sqrt{m(m-1)}}|T|^3$$

with equality holding if and only if either $|T| = 0$ or $|T| = \frac{1}{\sqrt{m(m-1)}}S$. Therefore

$$B \geq 4\frac{|T|^2}{S^3}\left(|T| - \frac{1}{\sqrt{m(m-1)}}S\right)^2. \tag{8.161}$$

Finally, by Huisken's inequality of Proposition 8.8,,

$$C \geq -\frac{2\sqrt{2}}{S^2}\sqrt{\frac{m-2}{m-1}}|W||T|^2. \tag{8.162}$$

Inequality (8.158) now follows immediately by putting together (8.159)–(8.162). □

For the proof of Theorems 8.8 and 8.9 below we shall need the property that for a complete shrinking soliton $(M, \langle\,,\,\rangle, X)$ we have $|X| \to +\infty$ as $x \to \infty$, at least in case that the Ricci curvature is bounded above. This can be shown with a minor variation of the proof of Lemma 8.7 above. Indeed, we have

Lemma 8.11 *Let $(M, \langle\,,\,\rangle, X)$ be a complete Ricci soliton with* $\mathrm{Ric} \leq \Lambda\langle\,,\,\rangle$ *and let $o \in M$. Then there exists a constant a such that, for each $x \in M$,*

$$|X|(x) \geq \lambda r(x) + a,$$

where $r(x) = \mathrm{dist}\,(x, o)$.

Proof We fix a minimal unit speed geodesic γ connecting o to x; proceeding as in Lemma 8.7, with the same choice of function $h(t)$, we arrive at the inequality

$$\int_0^{r(x)} \mathrm{Ric}\,(\dot\gamma,\dot\gamma) \le 2(m-1) + \int_0^1 \left(1-s^2\right)\mathrm{Ric}\,(\dot\gamma,\dot\gamma)$$
$$+\int_{r(x)-1}^{r(x)} \left[1-(r(x)-s)^2\right]\mathrm{Ric}\,(\dot\gamma,\dot\gamma).$$

Using the assumption $\mathrm{Ric} \le \Lambda\langle\, ,\, \rangle$ we then deduce

$$\int_0^{r(x)} \mathrm{Ric}\,(\dot\gamma,\dot\gamma) \le 2(m-1) + 2\Lambda. \tag{8.163}$$

We now recall that, since γ is a geodesic,

$$\frac{1}{2}\mathcal{L}_X(\dot\gamma,\dot\gamma)(t) = \frac{d}{dt}\langle X,\dot\gamma\rangle(t);$$

hence, substituting $\mathrm{Ric}\,(\dot\gamma,\dot\gamma)$ in (8.163) *via* the soliton equation

$$\mathrm{Ric} + \frac{1}{2}\mathcal{L}_X\langle\, ,\, \rangle = \lambda\langle\, ,\, \rangle$$

we obtain

$$\lambda r(x) - \langle X,\dot\gamma\rangle(r(x)) + \langle X,\dot\gamma\rangle(0) \le 2[(m-1)+\Lambda].$$

Using the Cauchy-Schwarz inequality $|X| \ge |\langle X,\dot\gamma\rangle|$ we obtain the desired conclusion. □

Theorem 8.8 *Let $(M,\langle\, ,\, \rangle,X)$ be a 3-dimensional, complete generic shrinking Ricci soliton. Furthermore, if M is noncompact, assume that the scalar curvature S is bounded and $|\nabla X| = o(|X|)$ as $r\to\infty$. Then $(M,\langle\, ,\, \rangle)$ is isometric to a finite quotient of either $\mathbb{S}^3$, $\mathbb{R}\times\mathbb{S}^2$ or $\mathbb{R}^3$.*

Note that under the assumptions of Theorem 8.8 Ric is bounded above (see below in the proof of Theorems 8.8 and 8.9).

In higher dimensions Theorem 8.8 generalizes to

Theorem 8.9 *Let $(M,\langle\, ,\, \rangle,X)$ be a complete generic shrinking Ricci soliton of dimension $m \ge 4$. Furthermore, if M is noncompact, assume that the scalar curvature S is bounded and $|\nabla X| = o(|X|)$ as $r\to\infty$. If, for some $\Lambda > 0$,*

$|\operatorname{Ric}| \le \Lambda S$ and

$$|W|\, S \le \sqrt{\frac{2(m-1)}{m-2}}\left(|T| - \frac{1}{\sqrt{m(m-1)}} S\right)^2, \tag{8.164}$$

then $(M, \langle\,,\,\rangle)$ is isometric to a finite quotient of either $\mathbb{S}^m$, $\mathbb{R}\times\mathbb{S}^{m-1}$ or $\mathbb{R}^m$.

The above theorems extend to the nongradient case previous results of Perelman [218], Cao et al. [61], Catino [65] and Catino et al. [66].

Remark 8.8 Tracing the soliton equation (8.1) it follows that the previous theorems in particular hold simply assuming that $|\nabla X|$ is bounded and, if $m > 3$, assuming also inequality (8.164).

Proof (of Theorems 8.8 and 8.9) First of all, from Theorem 8.2 we know that the soliton is either flat or has scalar curvature $S > 0$. Moreover, we note that under the assumptions of Theorems 8.8 and 8.9 the metric has bounded Ricci curvature (and this fact will be crucial in the three dimensional case, see below). From the growth estimates on the vector field X proved in Lemma 8.11, we know that $|X| \to \infty$ as $r \to \infty$.

In dimension three, every complete shrinking soliton has nonnegative sectional curvature [75]. Moreover, by Hamilton's strong maximum principle (see [134]), either $\langle\,,\,\rangle$ has strictly positive sectional curvature or it splits a line. In this latter case, either the soliton is flat or it is isometric to a quotient of the round cylinder $\mathbb{R}\times\mathbb{S}^2$. So from now on, in dimension three, we can assume that the metric has strictly positive sectional curvature. In particular, from [133, Corollary 8.2], it holds $|\operatorname{Ric}|^2 < \frac{1}{2}S^2$. Moreover, the pinching condition (8.164) is automatically satisfied, since the Weyl tensor vanishes in three dimension. Thus, it is sufficient to prove Theorem 8.9, with $m \ge 3$, to conclude. Now, the proof follows the arguments in [65]. Under the assumptions of Theorem 8.9, Corollary 8.6 applies. Hence, from Lemma 8.10 and Theorem 4.12, we have that $\frac{|T|^2}{S^2}$ must be constant on M. Therefore, from the proof of Lemma 8.10, we get that $(M, \langle\,,\,\rangle)$ is either Einstein or satisfies the identity $|T| = \frac{1}{\sqrt{m(m-1)}}S$. Now, if $m = 3$, this latter case violates the fact that the metric has positive sectional curvature, since it would imply $|\operatorname{Ric}|^2 = \frac{1}{2}S^2$; so $(M, \langle\,,\,\rangle)$ is Einstein, hence it has constant positive sectional curvature and is a finite quotient of $\mathbb{S}^3$. On the other hand, if $m \ge 4$, the pinching assumption (8.164) on the Weyl curvature implies that $(M, \langle\,,\,\rangle)$ is either Einstein or has vanishing Weyl tensor in case $|T| = \frac{1}{\sqrt{m(m-1)}}S$. In the first case, since $m \ge 4$ the metric has constant positive scalar curvature; being M Einstein, by Myers theorem M is compact. Moreover, again from the pinching condition (8.164), we get that

$$|W|^2 \le \frac{2}{m^2(m-1)(m-2)}S^2 < \frac{4}{m(m-1)(m-2)(m+1)}S^2.$$

Thus, the pinching condition in Proposition 8.9 is satisfied for some $\varepsilon > 0$ since M is compact and the above inequality for $|W|^2$ is strict. Then $(M, \langle\, ,\, \rangle)$ has positive curvature operator. Hence, from a classical theorem of Tachibana [261], see Theorem 2.17 in the compact case, we conclude that $(M, \langle\, ,\, \rangle)$ has constant positive sectional curvature and is a finite quotient of $\mathbb{S}^m$. On the other hand, if the Weyl tensor vanishes, from the classification of locally conformally flat shrinking Ricci solitons given in [66] we obtain that if $(M, \langle\, ,\, \rangle)$ is nonflat and noncompact, then it must be a finite quotient of $\mathbb{R} \times \mathbb{S}^{m-1}$. This concludes the proof of Theorems 8.8 and 8.9. □

Chapter 9
Spacelike Hypersurfaces in Lorentzian Spacetimes

The aim of this chapter is to present a number of applications of the maximum principle to spacelike hypersurfaces in a Lorentzian ambient space. In doing so we first introduce a few basic notions and results of Lorentzian geometry that will be used later on and that constitute the geometric background of our analysis.

We recall that *maximal hypersurfaces*, in a general Lorentzian ambient space, are spacelike hypersurfaces with zero mean curvature. Their importance is well known, for instance, because of the role they play in different problems in General Relativity (see for instance [185] and references therein). From the mathematical point of view, the first important global result obtained has been the Lorentzian analogue of the Calabi-Bernstein theorem: it states that the only complete maximal hypersurfaces in the Lorentz-Minkowski space are spacelike hyperplanes; equivalently, the only maximal entire graphs in $\mathbb{L}^{m+1}$ are spacelike hyperplanes.

This theorem, inspired by the classical Bernstein theorem on minimal surfaces in $\mathbb{R}^3$, was first obtained by Calabi [55] under the dimensional restriction $m \leq 4$. Later on, Cheng and Yau [81] extended it to the general case providing the first application of a Simons-type formula in the context of spacelike hypersurfaces in a Lorentz ambient space. This result deeply contrast with the Euclidean case, since the Bernstein theorem for minimal hypersurfaces in $\mathbb{R}^{m+1}$ is false for $m > 7$ (see [50]). After the general proof given by Cheng and Yau, several authors have approached the two-dimensional version of the theorem from different perspectives, providing various extensions and new proofs of the result for the case of maximal surfaces in $\mathbb{L}^3$ [16, 17, 115, 116, 162, 246]. For a general dimension some other authors have developed different related Bernstein-type results on spacelike hypersurfaces in $\mathbb{L}^{m+1}$, looking, for instance, to characterize spacelike hyperplanes among the complete spacelike hypersurfaces with constant mean curvature [3, 7, 216, 276]. This topic is considered in the second half of the chapter, where we also provide some comparison results for the Lorentzian distance function from a fixed origin (see Lemmas 9.11 and 9.12).

L.J. Alías et al., *Maximum Principles and Geometric Applications*,
Springer Monographs in Mathematics, DOI 10.1007/978-3-319-24337-5_9

In the final part of the chapter we prove some properties for spacelike graphs in a generalized Robertson-Walker spacetime and we end with some discussion with the corresponding results in the Riemannian case of Chap. 7.

9.1 Foundations of Lorentzian Geometry

We begin by observing that the basic references for the material that follows are [40, 212] that can be consulted for a smooth introduction to the concepts and results below.

A *Lorentzian metric* $\langle\,,\,\rangle$ on an n-dimensional smooth manifold N ($n \geq 2$) is a symmetric nondegenerate $(0, 2)$-tensor field on N of constant index 1. In other words, $\langle\,,\,\rangle$ assigns smoothly to each point $p \in N$ a Lorentzian scalar product $\langle\,,\,\rangle_p$: $T_pN \times T_pN \to \mathbb{R}$, that is, a symmetric bilinear form on T_pN such that

(i) $\langle v, w\rangle_p = 0$ for all $w \in T_pN$ implies $v = 0$, in other words, $\langle\,,\,\rangle_p$ is nondegenerate;
(ii) $\max\{\dim V : V \leqslant T_pN, \quad \langle v, v\rangle_p < 0 \quad \text{for every } v \in V\} = 1.$

We are now ready for the next

Definition 9.1 A *Lorentzian manifold* is an n-dimensional smooth manifold N ($n \geq 2$) endowed with a Lorentzian metric $\langle\,,\,\rangle$.

Here are some linear algebra considerations. A tangent vector $v \in T_pN$ is said to be

(i) *spacelike* if $\langle v, v\rangle > 0$ or $v = 0$,
(ii) *timelike* if $\langle v, v\rangle < 0$,
(iii) *lightlike* (or *null*) if $\langle v, v\rangle = 0$ and $v \neq 0$.

The set of all lightlike vectors in T_pN is called the *lightcone* at $p \in N$. The category into which a given tangent vector falls is called its *causal character*.

More generally, a linear subspace $V \leq T_pN$ is said to be

(i) *spacelike* if the restriction of the Lorentzian metric $\langle\,,\,\rangle$ to V is positive definite; that is, $\langle\,,\,\rangle|_V$ is a Euclidean metric;
(ii) *timelike* if the restriction of the Lorentzian metric $\langle\,,\,\rangle$ to V is nondegenerate of index 1; that is, $\langle\,,\,\rangle|_V$ is a Lorentzian metric,
(iii) *lightlike* (or *null*) if the restriction of the Lorentzian metric $\langle\,,\,\rangle$ to V is degenerate.

In cases (i) and (ii) we simply say that V is *nondegenerate*. The category into which V falls is called its *causal character*. It can be easily seen that a subspace V is spacelike (resp. timelike) if and only if $V^\perp$ is timelike (resp. spacelike), where

$$V^\perp = \big\{w \in T_pN : \langle v, w\rangle = 0 \quad \text{for all } v \in V\big\}.$$

Lemma 9.1 *If v is a spacelike (resp. timelike) vector in T_pN, then the subspace $v^\perp(= span\{v\}^\perp)$ is timelike (resp. spacelike) and $T_pN = span\{v\} \oplus v^\perp$.*

Spacelike subspaces are the easiest to deal with. Classical Euclidean geometry holds on them and, in particular, the Cauchy-Schwarz inequality:

$$|\langle v, w\rangle| \leq |v||w| \text{ for all } v, w \in V,$$

equality holding if and only if v and w are linearly dependent.

Let $\mathcal{T}_p$ be the set of all timelike vectors of T_pN. For a given $u \in \mathcal{T}_p$ we set

$$C(u) = \{v \in \mathcal{T}_p : \langle u, v\rangle < 0\}$$

to denote the *timecone* of T_pN determined by u. Obviously, $u \in C(u) \neq \emptyset$ and

$$C(-u) = -C(u) = \{v \in \mathcal{T}_p : \langle u, v\rangle > 0\}.$$

Moreover, given another $v \in \mathcal{T}_p$, since $\langle u, v\rangle \neq 0$ then either $v \in C(u)$ or $v \in C(-u)$. In other words, $\mathcal{T}_p$ is the disjoint union these two timecones, $\mathcal{T}_p = C(u) \cup C(-u)$.

The following algebraic result is immediate.

Lemma 9.2 *Two timelike vectors v and w in T_pN belong to the same timecone if and only $\langle v, w\rangle < 0$.*

As a consequence, timecones are convex: If $v, w \in C(u)$ and $a, b \geq 0$ (not both zero) then $av + bw \in C(u)$. It also follows from Lemma 9.2 that for timelike vectors the three following statements are equivalent

(i) $u \in C(v)$,
(ii) $v \in C(u)$,
(iii) $C(u) = C(v)$.

For a timelike vector $u \in \mathcal{T}_p$ we set $|u| = \sqrt{-\langle u, u\rangle}$. We have

Proposition 9.1 *Let v and w be timelike vectors in T_pN. Then*

*(i) $|\langle v, w\rangle| \geq |v||w|$, equality holding if and only if v and w are linearly dependent (*backwards Cauchy-Schwarz inequality*).*
(ii) If v and w are in the same timecone, there is a unique number θ, called the hyperbolic angle between v and w, such that

$$\langle v, w\rangle = -|v||w| \cosh\theta.$$

As a consequence, we have the validity of the *backwards Minkowski inequality*: if v and w are in the same timecone, then

$$|v + w| \geq |v| + |w|,$$

with equality if and only if v and w are linearly dependent.

In each tangent space T_pN of a Lorentzian manifold N there are two timecones, and there is no intrinsic way to distinguish one from the other. A *time-orientation* of T_pN is a particular choice of one of them. Globalizing this notion we have the following: a *time-orientation of a Lorentzian manifold* N is a map τ on N that assigns to each point $p \in N$ a timecone τ_p in T_pN and that is smooth, in the sense that for each $p \in N$ there exists a neighborhood U of p and a (smooth) vector field X on U such that $X_q \in \tau_q$ for each $q \in U$.

Definition 9.2 A Lorentzian manifold N is said to be *time-orientable* if N admits a time-orientation. The choice of a specific time-orientation on N makes N time-oriented. A *spacetime* is a time-oriented Lorentzian manifold of dimension $n \geq 2$.

It is important to observe that for a Lorentzian manifold there is no relation between orientability and time-orientability.

We have the validity of the next

Lemma 9.3 *A Lorentzian manifold N is time-orientable if and only if there exists a timelike vector field X globally defined on N.*

If X is such a vector field, we can choose τ_p as the timecone of T_pN such that $X(p) \in \tau_p$.

A timelike vector field X defined on a spacetime N is said to be *future-directed* (or *future-pointing*) if $X(p) \in \tau_p$ for every $p \in N$.

Example 9.1 (Lorentz-Minkowski Spacetime) Let $\mathbb{L}^n$ denote the n-dimensional Lorentz-Minkowski space, that is, the real vector space $\mathbb{R}^n$ endowed with the Lorentzian metric

$$\langle\, , \rangle = dx_1^2 + \cdots + dx_{n-1}^2 - dx_n^2,$$

where $(x_1, \ldots, x_n)$ are the canonical coordinates in $\mathbb{R}^n$.

Observe that $(0, \ldots, 0, 1)$ is a unit timelike vector field globally defined on $\mathbb{L}^n$, which determines a time-orientation on $\mathbb{L}^n$.

Example 9.2 (de Sitter Spacetime) Let $\mathbb{S}_1^n$ denote the n-dimensional de Sitter space (or Lorentzian sphere), that is,

$$\mathbb{S}_1^n = \{x \in \mathbb{L}^{n+1} : \langle x, x \rangle = 1\} \subset \mathbb{L}^{n+1}.$$

where

$$\langle\, , \rangle = dx_1^2 + \cdots + dx_n^2 - dx_{n+1}^2$$

is the Lorentzian metric of $\mathbb{L}^{n+1}$.

It is not difficult to see that for each $x \in \mathbb{S}_1^n$

$$T_x\mathbb{S}_1^n = \{v \in \mathbb{L}^{n+1} : \langle v, x \rangle = 0\} = x^\perp.$$

Therefore, by Lemma 9.1, $T_x\mathbb{S}_1^n$ is a timelike hyperplane and, with the obvious identifications,

$$T_x\mathbb{L}^{n+1} = \mathbb{L}^{n+1} = T_x\mathbb{S}_1^n \oplus \mathrm{span}\{x\}.$$

In other words, $\mathbb{S}_1^n$ is a *Lorentzian hypersurface* of $\mathbb{L}^{n+1}$. For this reason, a vector field on $\mathbb{S}_1^n$ can be regarded as a map $X : \mathbb{S}_1^n \to \mathbb{L}^{n+1}$ such that at each point $x \in \mathbb{S}_1^n$, $X(x)$ is orthogonal to x.

Observe that the vector field

$$X(x) = (x_1 x_{n+1}, \dots, x_n x_{n+1}, 1 + x_{n+1}^2), \quad x \in \mathbb{S}_1^n,$$

is a timelike vector field globally defined on $\mathbb{S}_1^n$ which determines a time-orientation on $\mathbb{S}_1^n$. Actually, for each $x \in \mathbb{S}_1^n$ we have $\langle X(x), x\rangle = 0$, so that $X(x) \in T_x\mathbb{S}_1^n$ and

$$\langle X(x), X(x)\rangle = -(1 + x_{n+1}^2) \leq -1.$$

Example 9.3 (Anti-de Sitter Spacetime) Let $\mathbb{H}_1^n$ denote the n-dimensional anti-de Sitter space (or Lorentzian hyperbolic space), that is,

$$\mathbb{H}_1^n = \{x \in \mathbb{R}_2^{n+1} : \langle x, x\rangle = -1\} \subset \mathbb{R}_2^{n+1},$$

where

$$\langle\, ,\, \rangle = dx_1^2 + \cdots + dx_{n-1}^2 - dx_n^2 - dx_{n+1}^2$$

is the pseudo-Euclidean metric of $\mathbb{R}_2^{n+1}$ with index 2.

Similarly to de Sitter spacetime, for each $x \in \mathbb{H}_1^n$

$$T_x\mathbb{H}_1^n = \{v \in \mathbb{R}_2^{n+1} : \langle v, x\rangle = 0\} = x^\perp$$

and the restriction of $\langle\, ,\, \rangle$ to $T_x\mathbb{H}_1^n$ is a Lorentzian metric, because of the decomposition $\mathbb{R}_2^{n+1} = T_x\mathbb{S}_1^n \oplus \mathrm{span}\{x\}$ with $\langle x, x\rangle = -1$. Therefore, $\mathbb{H}_1^n$ is a Lorentzian hypersurface of $\mathbb{R}_2^{n+1}$.

A vector field on $\mathbb{H}_1^n$ can be regarded as a map $X : \mathbb{H}_1^n \to \mathbb{R}_2^{n+1}$ such that at each point $x \in \mathbb{H}_1^n$, $X(x)$ is orthogonal to x. In this case, the vector field $X(x) = (0, \dots, 0, x_{n+1}, -x_n)$ gives a timelike vector field globally defined on $\mathbb{H}_1^n$ which determines a time-orientation on $\mathbb{H}_1^n$. For each $x \in \mathbb{H}_1^n$ we have $\langle X(x), x\rangle = 0$, so that $X(x) \in T_x\mathbb{H}_1^n$ and

$$\langle X(x), X(x)\rangle = -x_n^2 - x_{n+1}^2 = -1 - (x_1^2 + \dots x_{n-1}^2) \leq -1.$$

Example 9.4 (Generalized Robertson-Walker Spacetimes) Let $(\mathbb{P}, \langle\, ,\, \rangle_{\mathbb{P}})$ be a Riemannian manifold of dimension $n - 1$, and let I be an open interval of $\mathbb{R}$. Denote with $-I \times_\rho \mathbb{P}$ the product manifold $I \times \mathbb{P}$ endowed with the Lorentzian metric

$$\langle\, , \rangle = -dt^2 + \rho^2(t)\langle\, , \rangle_{\mathbb{P}},$$

where $\rho > 0$ is a positive smooth function on I and where we are following the convention used in Sect. 1.8, omitting the projection maps.

That is, $-I \times_\rho \mathbb{P}$ is nothing but a Lorentzian warped product with Lorentzian base $(I, -dt^2)$, Riemannian fiber $(\mathbb{P}, \langle\, , \rangle_{\mathbb{P}})$, and warping function ρ. Following [30], the Lorentzian manifold $-I \times_\rho \mathbb{P}$ is called a *generalized Robertson-Walker spacetime*. In particular, when the Riemannian factor $\mathbb{P}$ has constant sectional curvature $-I \times_\rho \mathbb{P}$ is classically called a *Robertson-Walker spacetime*. The vector field

$$\partial_t = (\partial/\partial_t)_{(t,x)}, \quad (t, x) \in -I \times_\rho \mathbb{P},$$

is a unit timelike vector field globally defined on a generalized Robertson-Walker spacetime thus determining a time-orientation.

Let N be a Lorentzian manifold and consider $\widetilde{N}$ the set of all timecones in tangent spaces of N. If X is a timelike vector field locally defined on an open set $U \subset N$, we can consider the map $\tau_X : U \to \widetilde{N}$ such that for each $p \in U$, $\tau_X(p)$ is the timecone containing $X(p)$. If τ_Y is another such local time-orientation, then by Lemma 9.2 $\tau_X(p) = \tau_Y(p)$ if and only if $\langle X(p), Y(p)\rangle < 0$. Therefore, there is a unique smooth structure on $\widetilde{N}$ for which the natural two-to-one map $k : \widetilde{N} \to N$ is a double covering map. The pullbacked Lorentzian metric on $\widetilde{N}$ makes this a *Lorentzian covering*, called the *time-orientation covering* of N, which is time-orientable. Obviously N is time-orientable if and only if $k : \widetilde{N} \to N$ is trivial. We collect these observations in the following

Lemma 9.4 *Let N be a Lorentzian manifold. Then*

(i) $\widetilde{N}$ is time-orientable.
(ii) N is time-orientable if and only if $k : \widetilde{N} \to N$ is trivial

As a consequence

Corollary 9.1 *Every simply connected Lorentzian manifold is time-orientable.*

9.1.1 Levi-Civita Connection and Geodesics

As in the Riemannian case, on a Lorentzian manifold N there is a unique connection $\overline{\nabla}$ which is both compatible with the metric tensor and torsion free. That is, $\overline{\nabla}$ is the unique (linear) connection satisfying

$$X(\langle Y, Z\rangle) = \langle \overline{\nabla}_X Y, Z\rangle + \langle Y, \overline{\nabla}_X Z\rangle$$

and

$$\overline{\nabla}_X Y - \overline{\nabla}_Y X = [X, Y]$$

for all vector fields X, Y, Z on N. As in the Riemannian case, this connection is called the *Levi-Civita connection* of N and it is characterized by *Koszul formula*

$$\begin{aligned} 2\langle \overline{\nabla}_X Y, Z\rangle &= X(\langle Y, Z\rangle) + Y(\langle Z, X\rangle) - Z(\langle X, Y\rangle) \\ &\quad -\langle X, [Y, Z]\rangle + \langle Y, [Z, X]\rangle + \langle Z, [X, Y]\rangle. \end{aligned}$$

Therefore parallel transport and geodesics may be defined for Lorentzian manifolds as in the case of Riemannian manifolds. Specifically, a *geodesic* in a Lorentzian manifold N is a curve $\sigma : I \to N$ whose velocity vector field $\dot{\sigma}$ is parallel, that is,

$$\overline{\nabla}_{\dot{\sigma}} \dot{\sigma} = 0.$$

Since $\dot{\sigma}$ is parallel, every geodesic has constant causal character, in the sense that its velocity vectors $\dot{\sigma}(s)$ are spacelike, or timelike or lightlike.

Again as in the Riemannian setting, the *exponential map* of N at a point p collects the geodesics starting at p, and its defined by $\overline{\exp}_p : E_p \to N$,

$$\overline{\exp}_p(v) = \sigma_v(1),$$

where E_p is the subset of all vectors $v \in T_pN$ such the maximal geodesic starting at p with initial velocity v is defined at $t = 1$. Obviously, E_p is the largest subset of T_pN on which $\overline{\exp}_p$ can be defined, and if $v \in E_p$ then $\sigma_v(t) = \overline{\exp}_p(tv)$ whenever $tv \in E_p$. In this context, a Lorentzian manifold N is said to be *complete* if σ_v is defined for all t.

For each point $p \in N$ there exists a neighborhood $\tilde{U}$ of 0 in T_pN on which $\overline{\exp}_p$ is a diffeomorphism onto a neighborhood U of p in N. In that case, U is said to be a *normal neighbourhood* of $p \in N$ if $\tilde{U}$ is starshaped about 0. If U is a normal neighborhood of $p \in N$, then U is starshaped about p in the sense that for every point $q \in U$ there exists a unique geodesic $\sigma_q : [0, 1] \to U$ with $\sigma_q(0) = p$ and $\sigma_q(1) = q$. Besides, $\dot{\sigma}_q(0) = \overline{\exp}_p^{-1}(q) \in \tilde{U}$ and σ_q is the radial geodesic segment from p to q.

The *arc length* of a piecewise smooth curve $\alpha : [a, b] \to N$ is $L(\alpha) = \int_a^b |\alpha'(t)| dt \geq 0$ where $|\alpha'(t)| = \sqrt{|\langle \alpha'(t), \alpha'(t)\rangle|}$. For instance, $L(\alpha) = 0$ for every lightlike curve. Obviously, if U is a normal neighborhood of p, then for every $q \in U$

$$L(\sigma) = |\overline{\exp}_p^{-1}(q)|,$$

where $\sigma : [0, 1] \to U$ is the radial geodesic from p to q (with initial velocity $\overline{\exp}_p^{-1}(q)$).

The next result is Proposition 34 in Chap. 5 of [212].

Lemma 9.5 (Maximizing Property of Timelike Geodesics) *Let U be a normal neighborhood of p in a Lorentzian manifold N. If $q \in U$ and there exists a timelike curve in U from p to q, then the radial geodesic segment from p to q is the unique (up to reparameterizations) longest timelike curve in U from p to q.*

9.1.2 Curvature of a Lorentzian Manifold

For a Lorentzian manifold N with Levi-Civita connection $\overline{\nabla}$, the *curvature tensor* is the $(1,3)$-tensor field on N given by

$$\overline{R}(X,Y)Z = [\overline{\nabla}_X, \overline{\nabla}_Y]Z - \overline{\nabla}_{[X,Y]}Z = \overline{\nabla}_X\overline{\nabla}_Y Z - \overline{\nabla}_Y\overline{\nabla}_X Z - \overline{\nabla}_{[X,Y]}Z.$$

The *sectional curvature* is defined only for nondegenerate planes. Given a point $p \in N$ and a nondegenerate tangent plane $\Pi \leq T_pN$, we define the sectional curvature $\overline{K}(\Pi)$ of Π to be

$$\overline{K}(\Pi) = \frac{\langle \overline{R}_p(u,v)v, u\rangle}{\langle u,u\rangle\langle v,v\rangle - \langle u,v\rangle^2},$$

where $\{u, v\}$ is an (arbitrary) basis of Π and where, since Π is nondegenerate,

$$\langle u,u\rangle\langle v,v\rangle - \langle u,v\rangle^2 \neq 0.$$

For a Riemannian manifold N, the sectional curvature is defined on the Grassmann bundle $G_2(N)$ of all tangent 2-planes of N, while for a Lorentzian manifold N it is defined on the Grassmann subbundle $G_2^*(N)$ of nondegenerate tangent 2-planes of N, a proper subset of the ordinary Grassmann bundle $G_2(N)$.

A Lorentzian manifold is said to have *constant sectional curvature* if $\overline{K}$ is constant on $G_2^*(N)$. In particular, N is *flat* if $\overline{K} = 0$ or, equivalently, $\overline{R} = 0$. For example, the Lorentz-Minkowski spaces $\mathbb{L}^n$ are flat, the deSitter spaces $\mathbb{S}_1^n$ have constant sectional curvature $\overline{K} = 1$ and the anti-deSitter spaces $\mathbb{H}_1^n$ have constant sectional curvature $\overline{K} = -1$. If N has constant sectional curvature $\overline{K}$ then its curvature tensor is expressed as

$$\overline{R}(X,Y)Z = \overline{K}(\langle Y,Z\rangle X - \langle X,Z\rangle Y).$$

Similar to what happens in the Riemannian case, we also have Schur's theorem: if N is connected of dimension $n \geq 3$ and for each $p \in N$, $\overline{K}$ is constant on the nondegenerate 2-planes in T_pN, then $\overline{K}$ is constant.

In Riemannian geometry, curvature inequalities such as $\overline{K} \geq c$ or $a \leq \overline{K} \leq b$, usually called pinching conditions, have been intensively studied. On the contrary, in Lorentzian geometry inequalities of this type make no sense because of the following result due to [165] and [93, 94] (see Proposition 28 in Chap. 8 of [212]).

Theorem 9.1 *Let p be a point of a Lorentzian manifold N. The following conditions are equivalent*

(i) $\overline{K}(\Pi)$ *is constant for all nondegenerate tangent planes of* T_pN.
(ii) $\overline{K}(\Pi) \geq c$ *or* $\overline{K}(\Pi) \leq c$, $c \in \mathbb{R}$, *for all nondegenerate tangent planes of* T_pN.
(iii) $a \leq \overline{K}(\Pi) \leq b$, $a, b \in \mathbb{R}$, *for all spacelike tangent planes of* T_pN.
(iv) $a \leq \overline{K}(\Pi) \leq b$, $a, b \in \mathbb{R}$, *for all timelike tangent planes of* T_pN.

Obviously, if N is connected of dimension $n \geq 3$ and if at each point p of N one of the conditions of the above theorem holds, then by Schur's theorem, N has constant sectional curvature.

The *Ricci curvature* of a Lorentzian manifold N is the symmetric $(0, 2)$-tensor field defined by

$$\overline{\mathrm{Ric}}(X, Y) = \mathrm{Tr}(Z \to \overline{R}(Z, X)Y).$$

In other words, if $\{e_1, \ldots, e_n\}$ is a local orthonormal frame field on N, then

$$\overline{\mathrm{Ric}}(X, Y) = \sum_{i=1}^{n} \varepsilon_i \langle \overline{R}(e_i, X)Y, e_i \rangle,$$

where $\varepsilon_i = \langle e_i, e_i \rangle = \pm 1$. In particular, $\overline{\mathrm{Ric}}(X, X) = \langle X, X \rangle \sum_{i=1}^{n} \overline{K}(X \wedge e_i)$.

The *scalar curvature* of N is

$$\overline{S} = \mathrm{Tr}(\overline{\mathrm{Ric}}) = \sum_{i=1}^{n} \varepsilon_i \overline{\mathrm{Ric}}(e_i, e_i).$$

In this context, one defines the *Einstein gravitational tensor* of the Lorentzian manifold N to be

$$\overline{G} = \overline{\mathrm{Ric}} - \frac{1}{2}\overline{S}\langle \, , \, \rangle.$$

Note that the tensor $\overline{G}$ is divergence free due to the following Lorentzian version of Eq. (1.68),

$$\mathrm{div}(\overline{\mathrm{Ric}}) = \frac{1}{2} d\overline{S}. \tag{9.1}$$

In general relativity, an *observer* is a timelike future-directed unit vector field X. An observer measures gravity by the *tidal force operator* acting on spacelike vectors orthogonal to X, that is, the operator $F_X : X^{\perp} \to X^{\perp}$ given by

$$F_X(Y) = \overline{R}(X, Y)X.$$

Via the tidal force operator we express in a natural way the empirical fact that gravity attracts by

$$\langle F_X(Y), Y \rangle \leq 0, \tag{9.2}$$

that is, $\overline{K}(\Pi) \leq 0$ for all timelike planes.

We now introduce a condition weaker than (9.2), that is the so called *timelike convergence condition*,

$$\overline{\mathrm{Ric}}(X, X) \geq 0 \tag{9.3}$$

for all timelike vector fields X. Its mathematical interpretation is that, on average, gravity attracts.

9.2 Spacelike Hypersurfaces in Lorentzian Spacetimes

A smooth immersion $f : \Sigma \to N$ of an m-dimensional connected manifold Σ into a spacetime N of dimension $n = m + 1$ is said to be a *spacelike hypersurface* if $f_*(T_x\Sigma)$ is a spacelike hyperplane of $T_{f(x)}N$ for every $x \in \Sigma$. Equivalently, the pullback via f of the ambient Lorentzian metric is a Riemannian metric on Σ, which, as usual, will also be denoted by $\langle\, ,\, \rangle$.

As a first interesting property, let us remark that every spacelike hypersurface of a spacetime N admits a globally defined unit normal field. This follows from the fact that there exists a unit timelike vector field T globally defined on N, that determines a time-orientation on N. Since the tangent hyperplane is spacelike at every point $x \in \Sigma$, there exists a unique timelike unit normal field ν on Σ which is in the same time-orientation of T, and hence we may assume that Σ is oriented by ν. We will refer to ν as the *future-directed Gauss map* of Σ. In particular, every spacelike hypersurface of an orientable spacetime N is itself orientable.

Let ∇ denote the Levi-Civita connection of Σ. Then the Gauss and Weingarten formulas for the hypersurface in N are given, respectively, by

$$\overline{\nabla}_X Y = \nabla_X Y - \langle AX, Y \rangle \nu$$

and

$$AX = -\overline{\nabla}_X \nu$$

for all tangent vector fields $X, Y \in \mathfrak{X}(\Sigma)$. Here $A : \mathfrak{X}(\Sigma) \to \mathfrak{X}(\Sigma)$ is the *Weingarten (or shape) operator* of Σ with respect to ν. One defines the *(future) mean curvature* H of Σ by setting

$$H = -\frac{1}{m}\,\mathrm{Tr}(A). \tag{9.4}$$

The choice of the $-$ sign in our definition of H is motivated by the fact that the mean curvature vector is given by $\mathbf{H} = H\nu$. Thus, $H(x) > 0$ at a point $x \in \Sigma$ if and only if $\mathbf{H}(x)$ is future-directed.

The curvature tensor R of a spacelike hypersurface Σ is described in terms of $\overline{R}$, the curvature tensor of the ambient spacetime N, and the shape operator A of Σ by the Gauss equation, which can be written as

$$R(X,Y)Z = (\overline{R}(X,Y)Z)^{\top} + \langle AX, Z\rangle AY - \langle AY, Z\rangle AX \tag{9.5}$$

for all tangent vector fields $X, Y, Z \in \mathfrak{X}(\Sigma)$. Here $(\overline{R}(X,Y)Z)^{\top}$ denotes the tangential component of $\overline{R}(X,Y)Z$ along the immersion. In particular, if $\{v, w\}$ forms a basis of a (spacelike) tangent plane $\Pi \leq T_x\Sigma$, then

$$K(\Pi) = \overline{K}(\Pi) - \frac{\langle A_x v, v\rangle\langle A_x w, w\rangle - \langle A_x v, w\rangle^2}{\langle v, v\rangle\langle w, w\rangle - \langle v, w\rangle^2},$$

where $K(\Pi)$ denotes the sectional curvature of Π in Σ, $\overline{K}(\Pi)$ is the sectional curvature of Π in N. Note that

$$\langle v, v\rangle\langle w, w\rangle - \langle v, w\rangle^2 > 0$$

because Π is spacelike.

On the other hand, Codazzi equation of the hypersurface describes the normal component of $\overline{R}(X,Y)Z$ in terms of the derivative of the shape operator, and it is given by

$$\langle \overline{R}(X,Y)Z, \nu\rangle = \langle (\nabla_X A)Y, Z\rangle - \langle (\nabla_Y A)X, Z\rangle,$$

where $\nabla_X A$ denotes the covariant derivative of A. Equivalently

$$(\overline{R}(X,Y)\nu)^{\top} = (\nabla_Y A)X - (\nabla_X A)Y.$$

In particular, when the ambient spacetime N has constant sectional curvature, then $\overline{R}(X,Y)\nu = 0$ and Codazzi equation reduces to

$$(\nabla_X A)Y - (\nabla_Y A)X = 0. \tag{9.6}$$

Recall that the *divergence of the shape operator* A is given by

$$\operatorname{div} A = \operatorname{Tr}(\nabla A) = \sum_{i=1}^{m} \nabla A(e_i, e_i) = \sum_{i=1}^{m} (\nabla_{e_i} A)e_i,$$

where $\{e_1, \dots, e_m\}$ is a local orthonormal frame of tangent vector fields along the immersion. As an application of Codazzi equation, we may compute $\operatorname{div} A$ as follows

$$\begin{aligned}\langle \operatorname{div} A, X\rangle &= \sum_{i=1}^{m} \langle (\nabla_X A)e_i, e_i\rangle + \sum_{i=1}^{m} \langle \overline{R}(e_i, X)e_i, \nu\rangle \\ &= \operatorname{Tr}(\nabla_X A) - \sum_{i=1}^{m} \langle \overline{R}(e_i, X)\nu, e_i\rangle \\ &= -m\langle \nabla H, X\rangle - \overline{\operatorname{Ric}}(X, \nu),\end{aligned}$$

for every tangent vector field $X \in \mathfrak{X}(\Sigma)$. Here we are using the fact that Tr commutes with ∇_X. Therefore,

$$\langle \operatorname{div} A, X\rangle + m\langle \nabla H, X\rangle = -\overline{\operatorname{Ric}}(X, \nu). \tag{9.7}$$

In particular, if the ambient spacetime N is Einstein then $\overline{\operatorname{Ric}}(X, \nu) = 0$ and

$$\operatorname{div} A = -m\nabla H. \tag{9.8}$$

From (9.5) we can express the Ricci curvature of Σ as

$$\begin{aligned}\operatorname{Ric}(X, X) &= \sum_{i=1}^{m} \langle \overline{R}(e_i, X)X, e_i\rangle - \langle AX, X\rangle \sum_{i=1}^{m} \langle Ae_i, e_i\rangle + \sum_{i=1}^{m} \langle AX, e_i\rangle^2 \\ &= \sum_{i=1}^{m} \overline{K}(X \wedge e_i)(|X|^2 - \langle X, e_i\rangle^2) + mH\langle AX, X\rangle + |AX|^2,\end{aligned}$$

where $\overline{K}(X \wedge e_i)$ denotes the sectional curvature of the spacelike plane $X \wedge e_i$. Therefore,

$$\begin{aligned}\operatorname{Ric}(X, X) &= \sum_{i=1}^{m} \overline{K}(X \wedge e_i)(|X|^2 - \langle X, e_i\rangle^2) - \frac{m^2H^2}{4}|X|^2 + |AX + \frac{mH}{2}X|^2 \\ &\geq \sum_{i=1}^{m} \overline{K}(X \wedge e_i)(|X|^2 - \langle X, e_i\rangle^2) - \frac{m^2H^2}{4}|X|^2.\end{aligned}$$

In particular, if $\overline{K} \geq \bar{c}$ for all spacelike planes in N, then

$$\operatorname{Ric}(X, X) \geq \left((m-1)\bar{c} - \frac{m^2H^2}{4}\right)|X|^2.$$

These considerations yield the following

Proposition 9.2 *Let N be a spacetime such that $\overline{K} \geq \bar{c}$, $\bar{c} \in \mathbb{R}$, for all spacelike planes in N. Then, every spacelike hypersurface Σ with bounded mean curvature in N has Ricci curvature bounded from below.*

In particular, every spacelike hypersurface Σ with bounded mean curvature in a Lorentzian spacetime with constant sectional curvature has Ricci curvature bounded from below. As a consequence, by Theorem 2.3, one has the following.

Corollary 9.2 *The Omori-Yau maximum principle for the Laplace-Beltrami operator holds on every complete spacelike hypersurface with bounded mean curvature into a Lorentzian spacetime with constant sectional curvature.*

9.2.1 Maximal Hypersurfaces as Solutions of a Variational Problem

A maximal hypersurface is a spacelike hypersurface with $H = 0$. The terminology maximal comes from the fact that these hypersurfaces locally maximize area.

Actually, if $f : \Sigma \to N$ is a spacelike hypersurface, every smooth function with compact support $\varphi \in C_c^\infty(\Sigma)$ induces a normal variation of f of the original immersion f, given by $f_t(p) = \overline{\exp}_{f(p)}(t\varphi(p)\nu(p))$. Since φ has compact support and $f_0 = f$ is spacelike, there exists $\varepsilon > 0$ such that f_t is also spacelike, for every $|t| < \varepsilon$. Then we consider the *m-dimensional area function*, $\mathscr{A} : (-\varepsilon, \varepsilon) \to \mathbb{R}$, defined by

$$\mathscr{A}(t) = \text{Area}(\Sigma_t) = \text{Area}(\Sigma, f_t^*(\langle\, , \rangle)) = \int_\Sigma d\Sigma_t,$$

The first variation of the area is given by the following classical result.

Lemma 9.6 *Let $f : \Sigma \to N$ be a spacelike hypersurface immersed into a spacetime N, and let f_t be a normal variation as above, induced by a smooth function with compact support $\varphi \in C_c^\infty(\Sigma)$. Then*

$$\delta_\varphi \mathscr{A} = \frac{d\mathscr{A}}{dt}(0) = m \int_\Sigma \varphi H d\Sigma.$$

From the above formula, it is clear the following

Corollary 9.3 *Σ is a maximal hypersurface if and only if $\delta_\varphi \mathscr{A} = 0$ for every $\varphi \in C_c^\infty(\Sigma)$.*

The stability of this variational problem is given by the *second variation formula of the area*,

$$\delta_\varphi^2 \mathscr{A} = \frac{d^2\mathscr{A}}{dt^2}(0) = \int_\Sigma (\varphi\Delta\varphi - (|A|^2 + \overline{\text{Ric}}(\nu, \nu))\varphi^2) d\Sigma = \int_\Sigma \varphi J\varphi d\Sigma.$$

Here $J = \Delta - |A|^2 - \overline{\mathrm{Ric}}(\nu, \nu)$, where Δ stands, as usual, for the Laplace-Beltrami operator of Σ. We refer the reader to [54] for the first, second and higher order variational formulas of the area away from a maximal hypersurface (see also [120]).

A maximal hypersurface is said to be *stable* if $Q(\varphi) = \int_\Sigma \varphi J \varphi d\Sigma \leq 0$ for every $\varphi \in C_c^\infty(\Sigma)$.

Corollary 9.4 *Every maximal hypersurface in a spacetime obeying the timelike convergence condition (9.3) is stable.*

For a proof, simply observe that for every $\varphi \in C_c^\infty(\Sigma)$

$$Q(\varphi) = -\int_\Sigma (|\nabla\varphi|^2 + (|A|^2 + \overline{\mathrm{Ric}}(\nu, \nu))\varphi^2) d\Sigma \leq 0$$

because of $\overline{\mathrm{Ric}}(\nu, \nu) \geq 0$.

In contrast, if N does not obey the timelike convergence condition, then every totally geodesic hypersurface on which $\overline{\mathrm{Ric}}(\nu, \nu) < 0$ can be deformed through parallel hypersurfaces to spacelike hypersurfaces of greater area. This happens, for instance, for totally geodesic equators in de Sitter space.

9.2.2 *Spacelike Hypersurfaces and General Relativity*

Maximal hypersurfaces and, more generally, spacelike hypersurfaces with constant mean curvature are also interesting from the physical point of view because of their role in general relativity. For instance, they are convenient as initial data for solving the Cauchy problem of the Einstein equation as showed by Lichnerowicz in [176]. See also the excellent and recent book by Choquet-Bruhat [89] where, in Chaps. 6–8, the author discusses the local initial value problem on spacelike hypersurfaces and the corresponding constraint Einstein equations.

Let N^4 be a 4-dimensional spacetime. A *stress-energy tensor field* $\overline{T}$ on N is a symmetric $(0, 2)$ tensor field satisfying some reasonable conditions from a physical point of view such us, say, $\overline{T}(X, X) \geq 0$ for any timelike vector X.

N is said to obey the Einstein equation with source $\overline{T}$ (and with zero cosmological constant) if

$$\overline{\mathrm{Ric}} - \frac{1}{2}\overline{S}\langle\, , \rangle = 8\pi\overline{T},$$

where $\overline{S}$ stands for the scalar curvature of N. In this case,

$$\mathrm{Tr}(\overline{T}) = -\frac{1}{8\pi}\overline{S}$$

and

$$\operatorname{div} \overline{T} = 0,$$

because of Eq. (9.1).

In particular, when $\overline{T} = 0$ this equation is called the *Einstein vacuum equation*, and it is equivalent to $\overline{\mathrm{Ric}} = 0$.

Assume that N^4 obeys the Einstein equation. Then for every spacelike hypersurface Σ^3 one has

$$\overline{\mathrm{Ric}}(\nu, \nu) = 8\pi \overline{T}(\nu, \nu) - \frac{1}{2}\overline{S},$$

where ν denotes a chosen unit vector field normal to Σ. It follows from here that the Gauss equation for the scalar curvature of Σ can be written as

$$S - |A|^2 + (\operatorname{Tr} A)^2 = \psi, \tag{9.9}$$

where $\psi = 16\pi \overline{T}(\nu, \nu) \in C^\infty(\Sigma)$.

On the other hand, from the expression for $\operatorname{div} A$ derived in (9.7), we also get

$$\operatorname{div} A - 3\nabla \operatorname{Tr} A = Z \tag{9.10}$$

where Z is the tangent vector field on Σ determined by

$$\langle Z, X \rangle = -8\pi \overline{T}(X, \nu) \text{ for every } X \in \mathfrak{X}(\Sigma).$$

In particular, when $\overline{T} = 0$ then $\psi = 0$ and $Z = 0$.

Equations (9.9) and (9.10) are called the *Einstein constraint equations*. Solving the Cauchy problem for Einstein equation requires to previously solve the constraint equations (9.9) and (9.10) as a system of PDEs with unknowns g and A (the initial value problem).

An *initial data set* for the Einstein equation is a triple (Σ, g, A) where Σ is a 3-dimensional manifold, g is a Riemannian metric on Σ, and A is a self-adjoint $(1, 1)$-tensor field on Σ satisfying (9.9) and (9.10).

A *solution* of the Cauchy problem for the Einstein equation corresponding to the initial data set (Σ, g, A), is a spacetime N^4 obeying the Einstein equation for which there exists a spacelike isometric embedding $j : \Sigma \hookrightarrow N$ with A as its future directed Weingarten operator of Σ.

Using the conformal techniques introduced by Lichnerowicz [176], and developed by Choquet-Bruhat [88], on a maximal hypersurface the system of constraints (9.9) and (9.10) can be split into a linear system and a nonlinear equation. Then the solution of the initial value problem rests on the global solution of this nonlinear elliptic equation on the initial 3-manifold.

For instance, let us see how it works in the easiest case of the Einstein vacuum equation ($\overline{T} = 0$). Consider an arbitrary Riemannian metric g_0 on a 3-manifold Σ, and set

$$g = \phi^4 g_0, \quad \phi \in C^\infty(\Sigma), \quad \phi > 0 \quad \text{and} \quad B := \phi^6 \left(A - \frac{1}{3} (\operatorname{Tr} A) I \right).$$

Then Eqs. (9.9) and (9.10) become

$$\Delta_0 \phi - \frac{S_0}{8} \phi + \frac{|B|^2}{8} \frac{1}{\phi^7} - \frac{1}{12} \tau^2 \phi^5 = 0, \tag{9.11}$$

$$\operatorname{div}_0 B - \frac{2}{3} \phi^6 \nabla_0 \tau = 0, \tag{9.12}$$

where $\tau = \operatorname{Tr} A$, and Δ_0, S_0, div_0 and ∇_0 denote, respectively, the Laplace-Beltrami operator, the scalar curvature, the divergence and the gradient of g_0. Equation (9.11) is known as the *Lichnerowicz equation*.

Assume that B is a solution of the linear system

$$\operatorname{div}_0 B = 0, \quad \operatorname{Tr} B = 0,$$

and $\phi > 0$ is a solution of the elliptic equation (9.11). Then, setting $g = \phi^4 g_0$ and

$$A = \frac{1}{\phi^6} B + \frac{1}{3} \tau I, \quad \text{with } \tau \in \mathbb{R},$$

we obtain that (Σ, g, A) is an initial data set for the Einstein vacuum equation for the solution of the corresponding Cauchy problem, and Σ is a constant mean curvature spacelike hypersurface with $H = -(1/3)\tau$.

9.3 Spacelike Hypersurfaces in Lorentz-Minkowski Space

Let us consider the case of spacelike hypersurfaces in the flat Lorentz-Minkowski space $\mathbb{L}^{m+1}$, that is, the real vector space $\mathbb{R}^{m+1}$ endowed with the Lorentzian metric

$$\langle v, w \rangle = v_1 w_1 + \cdots + v_m w_m - v_{m+1} w_{m+1}.$$

In this case, the future-directed Gauss map can be regarded as a map $\nu : \Sigma \to \mathbb{H}^m_+$, where $\mathbb{H}^m_+$ denotes the future connected component of the m-dimensional hyperbolic space,

$$\mathbb{H}^m_+ = \{ p \in \mathbb{L}^{m+1} : \langle p, p \rangle = -1, \quad p_{m+1} \geq 1 \}.$$

The image $\nu(\Sigma) \subset \mathbb{H}^m_+$ will be called the *hyperbolic image* of Σ.

An interesting remark on the topology of spacelike hypersurfaces in Lorentz-Minkowski space is that every complete spacelike hypersurface is spatially entire, in the sense that the projection $\Pi : \Sigma \to \mathbb{R}^m$ of Σ onto any spacelike hyperplane is a diffeomorphism. To see this, assume that $\mathbb{R}^m = a^\perp$ for a future-directed unit vector a and let

$$\Pi(x) = f(x) + \langle f(x), a \rangle a, \quad x \in \Sigma,$$

be the orthogonal projection onto $a^\perp$. Then, for every $x \in \Sigma$ and every tangent vector $v \in T_x\Sigma$,

$$\begin{aligned}\langle d\Pi_x(v), d\Pi_x(v) \rangle &= \langle df_x(v), df_x(v) \rangle + \langle df_x(v), a \rangle^2 \\ &\geq \langle df_x(v), df_x(v) \rangle = \langle v, v \rangle.\end{aligned}$$

That is, $\Pi^*(\langle\, , \rangle_o) \geq \langle\, , \rangle$, where $\langle\, , \rangle_o$ stands for the Euclidean metric in $a^\perp$. This means that Π is a local diffeomorphism which increases the distance. The completeness of Σ implies then that $\Pi(\Sigma) = \mathbb{R}^m$ and that Π is a covering map [163, Lemma VIII.1]. Since $\mathbb{R}^m$ is simply connected, Π must be a global diffeomorphism and the hypersurface Σ can be seen as an entire graph over the spacelike hyperplane $a^\perp$.

Summarizing, we have the following result.

Proposition 9.3 *Let $f : \Sigma \to \mathbb{L}^{m+1}$ be a complete spacelike hypersurface in the Lorentz-Minkowski space. Then*

(i) Σ is diffeomorphic to $\mathbb{R}^m$.
(ii) The immersion $f : \Sigma \to \mathbb{L}^{m+1}$ is actually an embedding.
(iii) Its image $f(\Sigma)$ is a closed subset in $\mathbb{L}^{m+1}$.

In particular, we deduce that there exists no compact (without boundary) spacelike hypersurface in $\mathbb{L}^{m+1}$.

However, it is worth pointing out that, for instance, there exist examples of spacelike entire graphs in $\mathbb{L}^{m+1}$ which are not complete (see Example 9.5 below). This fact points out an interesting difference between the behavior of hypersurfaces in Euclidean space $\mathbb{R}^{m+1}$ and that of spacelike hypersurfaces in the Lorentz-Minkowski space. Actually, as it is well known every closed embedded hypersurface in Euclidean space $\mathbb{R}^{m+1}$ is necessarily complete, while there exist examples of complete embedded hypersurfaces in $\mathbb{R}^{m+1}$ which are not closed.

Example 9.5 (A Spacelike Entire Graph Which Is Not Complete) Let $u : \mathbb{R}^m \to \mathbb{R}$ be the real function defined by

$$u(x) = u(x_1, \dots, x_m) = \begin{cases} \int_0^{|x_1|} \sqrt{1 - e^{-s}}ds & \text{if } |x_1| \geq 1, \\ \\ \phi(x_1) & \text{if } |x_1| < 1. \end{cases}$$

where $\phi \in C^\infty(\mathbb{R})$ is a smooth extension of $\int_0^t \sqrt{1-e^{-s}}ds$ satisfying $\phi'(s)^2 < 1$ for all $s \in (-1, 1)$. The entire spacelike graph $\Gamma_u(\mathbb{R}^m)$ in $\mathbb{L}^{m+1}$ given by

$$\Gamma_u(\mathbb{R}^m) = \{(x, u(x)) : x \in \mathbb{R}^m\}$$

is not complete. To see this, observe that the curve $\alpha : \mathbb{R} \to \Gamma_u(\mathbb{R}^m)$ given by $\alpha(s) = (s, 0, \ldots, 0, u(s, 0))$ is a divergent curve with finite length,

$$\int_{-\infty}^{+\infty} |\alpha'(s)|ds = \int_{-1}^{1} \sqrt{1-\phi'(s)^2}ds + 2\int_1^{+\infty} e^{-s/2}ds < 2\left(1+\frac{2}{\sqrt{e}}\right).$$

It is worth pointing out that this situation cannot happen if the mean curvature is constant, due to the following result of Cheng and Yau [82] (see also [135–137] for more details on the subject).

Proposition 9.4 *Every closed embedded spacelike hypersurface with constant mean curvature in the Lorentz-Minkowski space is complete. In particular, every spacelike entire graph with constant mean curvature is complete.*

Recall that a maximal hypersurface in $\mathbb{L}^{m+1}$ is a spacelike hypersurface with zero mean curvature. The importance of maximal hypersurfaces (in general Lorentzian ambient spaces) is well known, not only from the mathematical point of view but also, as briefly remarked in Sect. 9.2.2, from a physics perspective, because of their role in different problems in general relativity (see for instance [185] and the references therein). The first application of the Omori-Yau maximum principle for the Laplace-Beltrami operator in the context of spacelike hypersurfaces was the proof, given by Cheng and Yau [82], of the version of Bernstein theorem for maximal hypersurfaces in $\mathbb{L}^{m+1}$, usually called the Calabi-Bersntein theorem, that is one of the most important global results about spacelike hypersurfaces. In its parametric version reads as follows [82].

Theorem 9.2 *The only complete maximal hypersurfaces in the Lorentz-Minkowski space are spacelike hyperplanes.*

Theorem 9.2 also admits a nonparametric version in terms of entire maximal graphs, first established by Calabi [57] in case $m \leq 4$. Later, Cheng and Yau [82] extended the result, both in the nonparametric and the parametric case, to a general dimension m. Let $\Omega \subseteq \mathbb{R}^m$ be a domain and $u : \Omega \to \mathbb{R}$ a smooth function on Ω. Then, the graph $\Gamma_u(\Omega)$ determined by u,

$$\Gamma_u(\Omega) = \{(x_1, \ldots, x_m, u(x_1, \ldots, x_m)) : (x_1, \ldots, x_m) \in \Omega\} \subset \mathbb{L}^{m+1},$$

defines a spacelike hypersurface in $\mathbb{L}^{m+1}$ if and only if the Euclidean gradient of u, Du, satisfies $|Du| < 1$ on Ω. In this case, the future-directed Gauss map of $\Gamma_u(\Omega)$

is given by

$$\nu = \frac{1}{\sqrt{1-|Du|^2}}\left(\frac{\partial u}{\partial x_1},\ldots,\frac{\partial u}{\partial x_m},1\right), \tag{9.13}$$

while the (future) mean curvature H of $\Gamma_u(\Omega)$ is given by

$$\mathrm{Div}\left(\frac{Du}{\sqrt{1-|Du|^2}}\right) = mH, \quad \text{with} \quad |Du| < 1, \tag{9.14}$$

where Div stands for the Euclidean divergence in $\mathbb{R}^m$.

A spacelike graph is said to be *entire* if $\Omega = \mathbb{R}^m$. Therefore, for every real number H, the solutions to (9.14) which are globally defined on $\mathbb{R}^m$ represent spacelike entire graphs in $\mathbb{L}^{m+1}$ with constant mean curvature H. The nonparametric version of the Calabi-Bernstein theorem can thus be stated as follows [82].

Theorem 9.3 *The only entire maximal graphs in the Lorentz-Minkowski space are spacelike hyperplanes. In other words, when $H = 0$ the only entire solutions of (9.14) are affine functions.*

Theorem 9.3 is a consequence of Proposition 9.4 and Theorem 9.2. This result deeply contrast with the Euclidean case, since the Bernstein theorem for entire minimal graphs in $\mathbb{R}^{m+1}$ is false for $m > 7$ (see [50]).

The proof of Theorem 9.2 is an application of the Omori-Yau maximum principle that makes use of the following Simons formula for constant mean curvature spacelike hypersurfaces in Lorentzian spacetimes with constant sectional curvature.

Lemma 9.7 *Let Σ be a constant mean curvature spacelike hypersurface immersed into a Lorentzian spacetime with constant sectional curvature $\bar{c}$. Then*

$$\frac{1}{2}\Delta|A|^2 = |\nabla A|^2 - m^2\bar{c}H^2 + (m\bar{c} + |A|^2)|A|^2 + mH\,\mathrm{Tr}(A^3), \tag{9.15}$$

where A is the Weingarten operator of the hypersurface.

The proof of (9.15) parallels that of (6.12) and it is left to the reader.

Proof (of Theorem 9.2) The idea of the proof is to apply the Omori-Yau maximum principle (for the infimum) to the positive function $u = 1/\sqrt{1+|A|^2}$. This holds because of Corollary 9.2

Since $u_* = \inf_\Sigma u \geq 0$, there exists a sequence of points $\{x_k\}$ in Σ such that

$$\text{(i) } u(x_k) < u_* + \frac{1}{k}, \quad \text{(ii) } |\nabla u(x_k)| < \frac{1}{k}, \quad \text{and (iii) } \Delta u(x_k) > -\frac{1}{k}.$$

Writing $|A|^2 = 1/u^2 - 1 = (1-u^2)/u^2$, we have that

$$\frac{1}{2}\Delta|A|^2 = -\frac{\Delta u}{u^3} + \frac{3|\nabla u|^2}{u^4}.$$

That is,

$$\frac{1}{2}u^4\Delta|A|^2 = -u\Delta u + 3|\nabla u|^2.$$

On the other hand, from Simons formula (9.15) we also know that

$$\frac{1}{2}\Delta|A|^2 = |\nabla A|^2 + |A|^4 \geq |A|^4 = \frac{(1-u^2)^2}{u^4}.$$

That is, $u^4\Delta|A|^2 \geq 2(1-u^2)^2$. This yields

$$-u\Delta u + 3|\nabla u|^2 = \frac{1}{2}u^4\Delta|A|^2 \geq (1-u^2)^2 \geq 0.$$

Evaluating this inequality at the points x_k we have

$$0 \leq (1-u(x_k)^2)^2 \leq -u(x_k)\Delta u(x_k) + 3|\nabla u(x_k)|^2 \leq \frac{u(x_k)}{k} + \frac{3}{k^2}$$

and letting $k \to \infty$ we obtain

$$u_* = \lim_{k\to\infty} u(x_k) = 1.$$

Since $0 < u = 1/\sqrt{1+|A|^2} \leq 1$ on Σ, this means that $u \equiv 1$ and then $|A|^2$ vanishes identically on Σ. Therefore Σ is a totally geodesic hypersurface in $\mathbb{L}^{m+1}$, but the only totally geodesic hypersurfaces in $\mathbb{L}^{m+1}$ are (open pieces of) spacelike hyperplanes. By completeness, Σ is a spacelike hyperplane, ending the proof of Theorem 9.2. □

In what follows, we will introduce other Bernstein-type results for constant mean curvature spacelike hypersurfaces in $\mathbb{L}^{m+1}$ which are also obtained via applications of the Omori-Yau maximum principle. The first of them was simultaneous and independently given by Aiyama [3] and Xin [276], extending a first weaker version of Palmer [216]. It reads as follows.

Theorem 9.4 *The only complete spacelike hypersurfaces with constant mean curvature in the Lorentz-Minkowski space having bounded hyperbolic image $\nu(\Sigma) \subset \mathbb{H}^m_+$ are spacelike hyperplanes.*

Proof Assume that the hyperbolic image of Σ is contained in a geodesic ball $\widetilde{B}_\varrho(a)$ in $\mathbb{H}^m_+$ of radius $\varrho > 0$ centered at a point $a \in \mathbb{H}^m_+$. Observe that

$$\tilde{B}_\varrho(a) = \{q \in \mathbb{H}^m_+ : 1 \leq -\langle q, a\rangle \leq \cosh \varrho\},$$

so that $1 \leq -\langle \nu, a\rangle \leq \cosh \varrho$ on Σ. Since H is constant, we know from Corollary 9.2 that the Omori-Yau maximum principle holds on Σ. Applying it to the bounded above function $u = -\langle \nu, a\rangle$, we deduce the existence of a sequence of points $\{x_k\}_{k\in\mathbb{N}}$ in Σ such that

$$\lim_{k\to\infty} u(x_k) = u^* = \sup_\Sigma u \leq \cosh \varrho \quad \text{and} \quad \Delta u(x_k) < \frac{1}{k}.$$

A standard computation shows that

$$\nabla\langle \nu, a\rangle = -Aa^\top,$$

where $a^\top \in \mathfrak{X}(\Sigma)$ denotes the tangential component of a along the immersion, that is,

$$a = a^\top - \langle \nu, a\rangle \nu.$$

Furthermore, using Codazzi equation (9.6), one also obtains

$$\text{Hess}\,\langle \nu, a\rangle(X, Y) = -\langle(\nabla_{a^\top} A)(X), Y\rangle + \langle \nu, a\rangle\langle AX, AY\rangle,$$

for every tangent vector fields $X, Y \in \mathfrak{X}(\Sigma)$, so that

$$\begin{aligned}\Delta\langle \nu, a\rangle &= \text{Tr}(\text{Hess}\,\langle \nu, a\rangle) = -\,\text{Tr}(\nabla_{a^\top} A) + |A|^2\langle \nu, a\rangle \\ &= m\langle a^\top, \nabla H\rangle + |A|^2\langle \nu, a\rangle.\end{aligned}$$

In particular, since the mean curvature is constant, we get

$$\Delta\langle \nu, a\rangle = |A|^2\langle \nu, a\rangle.$$

Therefore,

$$\Delta u(x_k) = |A|^2(x_k)u(x_k) < \frac{1}{k}$$

for each $k \in \mathbb{N}$. By Cauchy-Schwarz inequality, $|A|(x)^2 \geq mH^2$ at every $x \in \Sigma$, which jointly with the previous inequality yields

$$0 \leq mH^2 u(x_k) \leq |A|^2(x_k)u(x_k) < \frac{1}{k}.$$

Letting $k \to \infty$ we conclude that $H = 0$ and, by Theorem 9.2, the hypersurface must be a spacelike hyperplane. □

From the expression for the future Gauss map of a spacelike graph $\Gamma_u(\Omega)$ given in (9.13) it follows that the hyperbolic image of $\Gamma_u(\Omega)$ is bounded if and only if $\sqrt{1 - |Du|^2}$ is bounded away from zero on Ω. Thus we can also formulate Theorem 9.4 in nonparametric form as follows.

Corollary 9.5 *For any real number H, the only entire solutions to the constant mean curvature equation (9.16) with $|Du| \leq 1 - \varepsilon < 1$ on $\mathbb{R}^m$ are affine functions (and $H = 0$).*

On the other hand, let us recall that every complete spacelike hypersurface in $\mathbb{L}^{m+1}$ is spatially entire. In particular, they cannot be spatially bounded; thus, for instance, there is no complete spacelike hypersurface contained in the slab determined by two parallel timelike (or lightlike) hyperplanes. As for spacelike hyperplanes, we have the following result, due to Aledo and Alías [7].

Theorem 9.5 *The only complete spacelike hypersurfaces with constant mean curvature in the Lorentz-Minkowski space which are bounded between two parallel spacelike hyperplanes are (parallel) spacelike hyperplanes.*

It is worth pointing out that the corresponding result for minimal surfaces in Euclidean space $\mathbb{R}^3$ is false, since there are examples of complete nonflat minimal surfaces contained between two parallel planes [155].

Proof Let $f : \Sigma \to \mathbb{L}^{m+1}$ be a complete spacelike hypersurface and assume that for a future-directed unit vector $a \in \mathbb{L}^{m+1}$, $f(\Sigma)$ is bounded between the two parallel hyperplanes

$$\Pi_c = \{p \in \mathbb{L}^{m+1} : \langle p, a \rangle = c\}$$

and

$$\Pi_C = \{p \in \mathbb{L}^{m+1} : \langle p, a \rangle = C\}$$

with $c < C$. In other words the function $u(x) = \langle f(x), a \rangle$ satisfies

$$c \leq u(x) \leq C \quad \text{on } \Sigma.$$

A standard computation shows that

$$\nabla u = a^{\top}$$

and

$$\text{Hess}(u)(X, Y) = -\langle \nu, a \rangle \langle AX, Y \rangle,$$

for every tangent vector fields $X, Y \in \mathfrak{X}(\Sigma)$. Hence

$$\Delta u = mH\langle \nu, a\rangle.$$

Using the Omori-Yau maximum principle, there exist sequences $\{x_k\}_{k\in\mathbb{N}}$ and $\{y_k\}_{k\in\mathbb{N}}$ in Σ such that

$$u(x_k) > \sup_\Sigma u - \frac{1}{k}, \quad |\nabla u(x_k)| < \frac{1}{k}, \quad \text{and} \quad \Delta u(x_k) = mH\langle \nu, a\rangle(x_k) < \frac{1}{k},$$

$$u(y_k) < \inf_\Sigma u + \frac{1}{k}, \quad |\nabla u(y_k)| < \frac{1}{k}, \quad \text{and} \quad \Delta u(y_k) = mH\langle \nu, a\rangle(y_k) > -\frac{1}{k}.$$

Recall that $a = a^\top - \langle \nu, a\rangle\nu = \nabla u - \langle \nu, a\rangle\nu$, and

$$\langle \nu, a\rangle = -\sqrt{1 + |\nabla u|^2}.$$

Therefore, from the above inequalities we get

$$\begin{aligned}\frac{-1}{mk\sqrt{1+|\nabla u(x_k)|^2}} < \frac{-\Delta u(x_k)}{m\sqrt{1+|\nabla u(x_k)|^2}} = H = \frac{-\Delta u(y_k)}{m\sqrt{1+|\nabla u(y_k)|^2}} \\ < \frac{1}{mk\sqrt{1+|\nabla u(y_k)|^2}}\end{aligned}$$

and letting $k \to \infty$ we conclude that $H = 0$. Finally, Theorem 9.2 implies that the hypersurface must be a spacelike hyperplane. □

Using Proposition 9.4, Theorem 9.5 can be formulated in a nonparametric version as follows.

Corollary 9.6 *For any real number H, the only entire solutions to the constant mean curvature equation*

$$\mathrm{Div}\left(\frac{Du}{\sqrt{1-|Du|^2}}\right) = mH, \quad \textit{with} \quad |Du| < 1, \tag{9.16}$$

that are bounded on $\mathbb{R}^m$ are the constants (and $H = 0$).

9.3.1 Alternative Approaches in Dimension $m = 2$ Using Parabolicity

After the general proof of the Calabi-Bernstein theorem given by Cheng and Yau, several authors have approached the two-dimensional version of the theorem from different perspectives, providing various extensions and new proofs of the result for the case of maximal surfaces in $\mathbb{L}^3$ [16, 17, 115, 116, 162, 246].

We describe here two different approaches to the two-dimensional version of the theorem which are based on parabolicity. The first one is a simple approach, inspired by previous work of Chern [85], given by Romero in [246]. The second one is due to Alías and Palmer [17] and it provides a local upper bound for the total curvature of geodesic discs in a maximal surface in $\mathbb{L}^3$. This involves the local geometry of the surface and its hyperbolic image.

Remark 9.1 We note that the Calabi-Bernstein theorem is no longer true for entire timelike minimal graphs in $\mathbb{L}^{m+1}$, even in the simplest two-dimensional case. Actually, if x_3 stands for the timelike coordinate in $\mathbb{L}^3$, then the graph given by $x_2 = x_3 \tanh x_1$, with $(x_1, x_3) \in \mathbb{R}^2$, is an entire nonplanar timelike graph in $\mathbb{L}^3$, having zero mean curvature and positive Gaussian curvature [162, 192], as shown by a simple computation.

Nevertheless, in [192] Weinstein (formerly Milnor) obtained a very interesting conformal analogue of the Calabi-Bernstein theorem for timelike surfaces. Specifically, she proved that every timelike entire graph in $\mathbb{L}^3$ with zero mean curvature is conformally equivalent to the Lorentzian plane (see also [177, 178] for some extensions of this conformal analogue due to Lin and Weinstein). Moreover, Magid [180] and Weinstein [193] developed independently different approaches to the study of the Calabi-Bernstein problem for timelike surfaces in $\mathbb{L}^3$. In particular, in [180] Magid showed that every timelike entire graph with zero mean curvature over either a timelike or a spacelike plane in $\mathbb{L}^3$ is a global translation surface.

Let $f : \Sigma \to \mathbb{L}^3$ be a maximal surface oriented by its future-directed Gauss map ν. For each fixed vector $a \in \mathbb{L}^3$, we consider the smooth function $\langle \nu, a \rangle$ on Σ. Following the computations in the proof of Theorem 9.4 we have

$$\nabla \langle \nu, a \rangle = -Aa^{\top} \quad \text{and} \quad \Delta \langle \nu, a \rangle = |A|^2 \langle \nu, a \rangle, \tag{9.17}$$

where $a^{\top} = a + \langle \nu, a \rangle \nu$. Thus $|a^{\top}|^2 = \langle \nu, a \rangle^2 + \langle a, a \rangle$ and

$$|\nabla \langle \nu, a \rangle|^2 = \langle A^2(a^{\top}), a^{\top} \rangle = K|a^{\top}|^2 = K(\langle \nu, a \rangle^2 + \langle a, a \rangle), \tag{9.18}$$

where we use the fact that, because of $H = 0$, $A^2 = KI$. Furthermore

$$|A|^2 = 2K \tag{9.19}$$

and

$$\Delta \langle \nu, a \rangle = 2K \langle \nu, a \rangle. \tag{9.20}$$

In particular, if $a \in \mathbb{L}^3$ is chosen to be lightlike and past-directed, that is $\langle a, a\rangle = 0$, $a \neq 0$, with $\langle \nu, a\rangle > 0$, having set $u = \langle \nu, a\rangle$ from (9.18) and (9.20) we get

$$\Delta\left(\frac{1}{u}\right) = -\frac{\Delta u}{u^2} + \frac{2|\nabla u|^2}{u^3} = 0,$$

that is, $1/u$ is a positive harmonic function globally defined on Σ.

On the other hand, since $K = (1/2)|A|^2 \geq 0$ and Σ is complete, using the Bishop-Gromov comparison theorem (Theorem 1.3) we deduce that $\mathrm{vol}\,\partial B_r(o) \leq Cr$ where o is any fixed origin in Σ and C is a positive constant. From Theorem 2.23 we thus deduce that Σ is parabolic. Therefore, $1/u$ is a positive constant and, from (9.20), $K \equiv 0$ which means that the surface is a totally geodesic spacelike plane.

Romero's approach also allows us to obtain a simple direct proof of the following result.

Theorem 9.6 *The only maximal surfaces in $\mathbb{L}^3$ which are complete with respect to the metric induced from the Euclidean metric in $\mathbb{R}^3$ are spacelike planes.*

As a direct application of Theorem 9.6 we infer, without using Proposition 9.4 of Cheng and Yau on the completeness of constant mean curvature spacelike hypersurfaces, the following consequences.

Corollary 9.7 *The only maximal surfaces in $\mathbb{L}^3$ whose image is closed in $\mathbb{L}^3$ are spacelike planes.*

Corollary 9.8 *The only entire maximal graphs in $\mathbb{L}^3$ are spacelike planes.*

Proof of Theorem 9.6 For simplicity, denote by $g = \langle\, ,\, \rangle$ the Riemannian metric induced on Σ from the Lorentzian metric of $\mathbb{L}^3$. Choose $b \in \mathbb{L}^3$ a future-directed unit timelike vector, so that $\langle b, b\rangle = -1$ and $\langle \nu, b\rangle \leq -1 < 0$. Therefore $1 - \langle \nu, b\rangle \geq 2 > 0$ and we may introduce on Σ the conformal metric

$$\tilde{g} = (1 - \langle \nu, b\rangle)^2 g. \tag{9.21}$$

From Eq. (1.80), the Gaussian curvature $\tilde{K}$ of $\tilde{g}$ is given by

$$(1 - \langle \nu, b\rangle)^2 \tilde{K} = K - \Delta \log(1 - \langle \nu, b\rangle), \tag{9.22}$$

where K is the Gaussian curvature of g, which is given by (9.19). Using (9.17) we compute

$$\Delta\left(\log(1 - \langle \nu, b\rangle)\right) = \frac{\Delta\langle \nu, b\rangle}{\langle \nu, b\rangle - 1} - \frac{|\nabla\langle \nu, b\rangle|^2}{(\langle \nu, b\rangle - 1)^2} = K,$$

which means that the conformal metric (9.21) is flat. If we prove that it is also complete then, reasoning as before, we obtain that $(\Sigma, \tilde{g})$ is parabolic. Since parabolicity is a conformal invariant property in dimension $m = 2$ (see Remark 9.2 below), we deduce that (Σ, g) is parabolic and the proof follows at once as above.

It remains to prove that $(\Sigma, \tilde{g})$ is complete. Assume without loss of generality that $b = (0, 0, 1)$ and write $\nu = (\nu_1, \nu_2, \nu_3)$, so that

$$\langle \nu, \nu \rangle = \nu_1^2 + \nu_2^2 - \nu_3^2 = -1 \quad \text{and} \quad \langle \nu, b \rangle = -\nu_3 \leq -1. \tag{9.23}$$

For every $x \in \Sigma$ and $w \in T_x\Sigma$, set

$$df_x(w) = (w_1, w_2, w_3). \tag{9.24}$$

Therefore

$$g_x(w, w) = \langle df_x(w), df_x(w) \rangle = w_1^2 + w_2^2 - w_3^2, \tag{9.25}$$

and

$$\langle df_x(w), \nu \rangle = \nu_1 w_1 + \nu_2 w_2 - \nu_3 w_3 = 0. \tag{9.26}$$

We also have that

$$\tilde{g}_x(w, w) = (1 + \nu_3)^2 g_x(w, w) \geq \nu_3^2 g_x(w, w) = \nu_3^2(w_1^2 + w_2^2) - \nu_3^2 w_3^2. \tag{9.27}$$

From (9.23) and (9.26) we have

$$\nu_3^2 w_3^2 = (\nu_1 w_1 + \nu_2 w_2)^2 \leq (\nu_1^2 + \nu_2^2)(w_1^2 + w_2^2) = (\nu_3^2 - 1)(w_1^2 + w_2^2), \tag{9.28}$$

which jointly with (9.27) yields

$$\tilde{g}_x(w, w) \geq w_1^2 + w_2^2. \tag{9.29}$$

Let $\langle , \rangle_0$ denote here the Euclidean metric in $\mathbb{R}^3$ and let g' denote the Riemannian metric induced on Σ from $\langle , \rangle_0$. From (9.24) we have

$$g'_x(w, w) = \langle df_x(w), df_x(w) \rangle_0 = w_1^2 + w_2^2 + w_3^2, \tag{9.30}$$

which jointly with (9.25) gives

$$w_1^2 + w_2^2 = \frac{1}{2}(g_x(w, w) + g'_x(w, w)) \geq \frac{1}{2} g'_x(w, w). \tag{9.31}$$

Therefore, by (9.29) we get

$$\tilde{g}_x(w, w) \geq \frac{1}{2} g'_x(w, w). \tag{9.32}$$

This implies that $\widetilde{L} \geq (1/\sqrt{2})L'$, where $\widetilde{L}$ and L' denote the length of a curve on Σ with respect to the Riemannian metrics $\tilde{g}$ and g', respectively. As a consequence, since we are assuming that the metric g' is complete on Σ, it follows that $\tilde{g}$ is also complete. □

Remark 9.2 Let $(M, \langle\, , \rangle)$ be an m-dimensional Riemannian manifold. Using formula (1.74) we immediately obtain that, under a conformal change of the metric of the type

$$\widetilde{\langle\, , \rangle} = \varphi^2 \langle\, , \rangle,$$

for some strictly positive smooth function φ on M, the Laplace-Beltrami operator changes according to the formula

$$\varphi^2 \tilde{\Delta} u = \Delta u + \frac{(m-2)}{\varphi} \langle \nabla \varphi, \nabla u \rangle \tag{9.33}$$

where $u \in C^2(M)$. In particular, if $m = 2$ parabolicity of $(M, \langle\, , \rangle)$ is equivalent to that of $(M, \widetilde{\langle\, , \rangle})$.

The second approach to the Calabi-Bernstein theorem of this section is based on an upper bound for the total curvature of geodesic discs in a maximal surface in $\mathbb{L}^3$, involving the local geometry of the surface and its hyperbolic image. Specifically, we prove the following integral inequality for the Gaussian curvature.

Theorem 9.7 *Let $f : \Sigma \to \mathbb{L}^3$ be a maximal surface in the Lorentz-Minkowski space. Let p be a point of Σ and $R > 0$ be a positive real number such that the geodesic disc $B_R(p)$ of radius R about p is relatively compact in Σ. Then for all $0 < r < R$ we have*

$$0 \leq \int_{B_r(p)} K \leq \frac{c_r}{\log (R/r)}, \tag{9.34}$$

where

$$c_r = \frac{\pi^3}{4} \frac{(1 + \cosh^2 \varrho_r)^2}{\cosh \varrho_r \arctan (\cosh \varrho_r)} > 0.$$

Here ϱ_r denotes the radius of a geodesic disc in $\mathbb{H}^2_+$ containing the hyperbolic image of $B_r(p)$.

The integral inequality (9.34) clearly implies the parametric version of Calabi-Bernstein theorem. Indeed, if Σ is complete, then R can approach infinity in (9.34) for a fixed arbitrary $p \in \Sigma$ and a fixed r, implying that

$$\int_{B_r(p)} K = 0.$$

Taking into account that the Gaussian curvature of a maximal surface in $\mathbb{L}^3$ is always nonnegative, this yields $K \equiv 0$ on Σ.

The proof of Theorem 9.7 is an application of the following (intrinsic) local integral inequality, which is a consequence of Theorem 2.24 for the particular case where $m = 2$ and $G \equiv 0$.

Lemma 9.8 *Let M be a Riemannian surface with nonnegative Gaussian curvature. Let $u \in C^2(M)$ satisfy $u\Delta u \geq 0$. Let $B_R(p)$ be relatively compact in M. Then, for $0 < r < R$,*

$$\int_{B_r(p)} u\Delta u \leq \frac{4\pi}{\log(R/r)} \sup_{B_R(p)} u^2.$$

Proof In the assumptions of the lemma, $\mathrm{Ric} = K\langle\ ,\ \rangle \geq 0$, and therefore we can choose $h(r) = r$ in (2.168). Then the above inequality follows from (2.167) of Theorem 2.24. □

Proof (of Theorem 9.7) Let us assume that the hyperbolic image of $B_r(p)$ is contained in a geodesic disc $\widetilde{B}_{\varrho_r}(a)$ in $\mathbb{H}^2_+$ of radius ϱ_r centered at the point $a \in \mathbb{H}^2_+$. Recall that

$$\widetilde{B}_{\varrho_r}(a) = \{q \in \mathbb{H}^2_+ : 1 \leq -\langle q, a\rangle \leq \cosh\varrho_r\},$$

so that $1 \leq -\langle \nu(x), a\rangle \leq \cosh\varrho_r$ for all $x \in B_r(p)$.

Since Σ is a maximal surface in $\mathbb{L}^3$, it is a Riemannian surface with nonnegative Gaussian curvature, so that we can apply Lemma 9.8 to an appropriate smooth function u. Choosing $u = \arctan(-\langle\nu, a\rangle)$, by (9.18) and (9.20) u satisfies

$$\Delta u = -\frac{1}{1+\langle\nu, a\rangle^2}\Delta\langle\nu, a\rangle + \frac{2\langle\nu, a\rangle}{(1+\langle\nu, a\rangle^2)^2}|\nabla\langle\nu, a\rangle|^2 = \frac{-4K\langle\nu, a\rangle}{(1+\langle\nu, a\rangle^2)^2}.$$

It follows that

$$u\Delta u = \phi(-\langle\nu, a\rangle)K \geq 0, \tag{9.35}$$

where $\phi : \mathbb{R}\to\mathbb{R}$ is given by

$$\phi(t) = \frac{4t\arctan(t)}{(1+t^2)^2}.$$

Since $\phi(t)$ is strictly decreasing for $t \geq 1$, we have

$$\phi(t) \geq \phi(\cosh\varrho_r) = \frac{4\cosh\varrho_r \arctan(\cosh\varrho_r)}{(1+\cosh^2\varrho_r)^2}$$

on $[1, \cosh \varrho_r]$. Hence, from (9.35) we get

$$u\Delta u \geq \frac{4\cosh \varrho_r \arctan(\cosh \varrho_r)}{(1+\cosh^2 \varrho_r)^2} K \geq 0$$

on $B_r(p)$. Integrating this inequality over $B_r(p)$ and using Lemma 9.8 we conclude that

$$0 \leq \frac{4\cosh \varrho_r \arctan(\cosh \varrho_r)}{(1+\cosh^2 \varrho_r)^2} \int_{B_r(p)} K \leq \int_{B_r(p)} u\Delta u \leq \frac{\pi^3}{\log(R/r)},$$

that is,

$$0 \leq \int_{B_r(p)} K \leq \frac{c_r}{\log(R/r)}.$$

□

9.4 Comparison Theory for the Lorentzian Distance Function from a Point

In this section we will establish some comparison results for the Hessian and the Laplacian of the Lorentzian distance function from a point. We do this, differently to what we did in Chap. 1, by using the more canonical approach of Jacobi fields.

We start by introducing some basic concepts about the Lorentzian distance function in arbitrary spacetimes.

Consider an n-dimensional spacetime N, that is, a time-oriented Lorentzian manifold of dimension $n \geq 2$. Let p, q be points in N. Using the standard terminology and notation of Lorentzian geometry, one says that q is in the *chronological future* of p, written $p \ll q$, if there exists a future-directed timelike curve from p to q. Similarly, q is in the *causal future* of p, written $p < q$, if there exists a future-directed causal (i.e. nonspacelike) curve from p to q. Obviously, $p \ll q$ implies $p < q$. As usual, $p \leq q$ means that either $p < q$ or $p = q$.

For a subset $S \subset N$, one defines the *chronological future of S* as

$$I^+(S) = \{q \in N : p \ll q \text{ for some } p \in S\},$$

and the *causal future of S* as

$$J^+(S) = \{q \in N : p \leq q \text{ for some } p \in S\}.$$

Thus $S \cup I^+(S) \subset J^+(S)$. In a dual way,

$$I^-(S) = \{q \in N : q \ll p \text{ for some } p \in S\}$$

and

$$J^-(S) = \{q \in N : q \leq p \text{ for some } p \in S\}$$

are the *chronological past* and *causal past* of S, respectively.

In particular, the chronological future $I^+(p)$ and the causal future $J^+(p)$ of a point $p \in N$ are

$$I^+(p) = \{q \in N : p \ll q\}, \quad \text{and} \quad J^+(p) = \{q \in N : p \leq q\}.$$

As it is well-known, $I^+(p)$ is always open, but $J^+(p)$ is neither open nor closed, in general. For instance, for a point $p \in \mathbb{L}^n$ in the Lorentz-Minkowski space, $I^+(p)$ is just the future timecone of p,

$$I^+(p) = \{q \in \mathbb{L}^n : \langle q-p, q-p\rangle < 0 \text{ and } \langle q-p, e_n\rangle < 0\},$$

and

$$J^+(p) = \overline{I^+(p)} = \{p\} \cup \{q \in \mathbb{L}^n : \langle q-p, q-p\rangle \leq 0 \text{ and } \langle q-p, e_n\rangle < 0\}.$$

At the other extreme, the *Lorentzian cylinder*

$$\mathbb{R} \times \mathbb{S}^1_1 = \{p = (p_1, p_2, p_3) \in \mathbb{L}^3 : p_2^2 - p_3^2 = 1\}$$

has trivial causality: even for a single point, $I^+(p) = J^+(p)$ is the entire spacetime.

Now we introduce the Lorentzian distance function. Observe that it cannot be defined in a way similar to that of the Riemannian case. Indeed, consider $p, q \in N$ such that $p \ll q$. For any given timelike curve α from p to q, $L(\alpha) > 0$ and there exists a sequence $\{\alpha_k\}$ of piecewise smooth *almost lightlike* curves from p to q such that $\alpha_k \to \alpha$ but $L(\alpha_k) \to 0$. In particular, the infimum of the Lorentzian lengths of all piecewise smooth future-directed causal curves from p to q is always zero.

On the other hand, from the maximizing property of timelike geodesics in Lorentzian manifolds given in Lemma 9.5, we know that if $p \ll q$ and $q \in U$ is in a normal neighborhood of p, then the radial geodesic segment from p to q is the *longest* timelike curve in U from p to q. Therefore it is natural to introduce the following definition.

Definition 9.3 Let N be a spacetime. If $q \in J^+(p)$, then the *Lorentzian distance (or time separation)* $d(p, q)$ is the *supremum* of the Lorentzian lengths of all piecewise smooth future-directed causal curves from p to q (possibly, $d(p, q) = +\infty$). If $q \notin J^+(p)$, then the Lorentzian distance $d(p, q) = 0$ by definition.

In particular, $d(p,q) > 0$ if and only if $q \in I^+(p)$. Moreover, if $p \ll p$ (in other words, $p \in I^+(p)$) then there exists a timelike loop at p and, giving more and more rounds to it, one gets $d(p,p) = +\infty$. Otherwise, $d(p,p) = 0$.

The comparison between Riemannian distance and Lorentzian distance is more dual than direct: the former minimizes while the latter maximizes. Since it involves time orientation, the Lorentzian distance is symmetric only in trivial cases. The Lorentzian distance function $d : N \times N \to [0, +\infty]$ is always lower semicontinuous. However, for an arbitrary spacetime it may fail to be continuous in general, and may also fail to be finite-valued. We refer the reader to [212] and [40] for further details. As a matter of fact, globally hyperbolic spacetimes turn out to be the natural class of spacetimes for which the Lorentzian distance function is finite-valued and continuous. Recall that a spacetime N is said to be *globally hyperbolic* if

(i) it is causal, that is, there exists no causal loop in N, and
(ii) the intersections $J^+(p) \cap J^-(q)$ are compact for every $p, q \in N$.

Given a point $p \in N$, one can define the Lorentzian distance function from p, $d_p : N \to [0, +\infty]$, by $d_p(q) = d(p,q)$. In order to guarantee the smoothness of d_p, one needs to restrict this function on certain special subsets of N. Let $T_{-1}N|_p$ be the fiber of the unit future observer bundle of N at p, that is,

$$T_{-1}N|_p = \{v \in T_pN : v \text{ is a future-directed timelike unit vector}\},$$

and consider the *Lorentzian cut locus* function $s_p : T_{-1}N|_p \to [0, +\infty]$, given by

$$s_p(v) = \sup\{t \geq 0 : d_p(\sigma_v(t)) = t = L(\sigma_v|_{[0,t]})\},$$

where $\sigma_v : [0, a) \to N$ is the future maximal geodesic starting at p with initial velocity v. Then, one can define the subset $\tilde{\mathscr{I}}^+(p) \subset T_pN$ given by

$$\tilde{\mathscr{I}}^+(p) = \{tv : \text{ for all } v \in T_{-1}N|_p \text{ and } 0 < t < s_p(v)\}$$

and consider the subset $\mathscr{I}^+(p) \subset N$ given by

$$\mathscr{I}^+(p) = \overline{\exp}_p(\mathrm{int}(\tilde{\mathscr{I}}^+(p))) \subset I^+(p).$$

Observe that $\overline{\exp}_p : \mathrm{int}(\tilde{\mathscr{I}}^+(p)) \to \mathscr{I}^+(p)$ is a diffeomorphism, where $\overline{\exp}_p$ denotes the exponential map of N at p, and $\mathscr{I}^+(p)$ is an open subset (possibly empty).

Recall that a spacetime N is said to be *strongly causal at a point* p if for any given neighborhood U of p there exists a neighborhood $V \subset U$ of p such that any future-directed causal curve in N with endpoints in V is entirely contained in U. In particular, N is called *strongly causal* if it is strongly causal at any of its points. For instance, every globally hyperbolic spacetime is strongly causal (see [212] for further details). The following result summarizes the main properties that we need about the smoothness of the Lorentzian distance function and it can be found in [113, Sect. 3.1].

Lemma 9.9 *Let N be a spacetime and $p \in N$.*

(i) If N is strongly causal at p, then $s_p(v) > 0$ for all $v \in T_{-1}N|_p$ and $\mathscr{I}^+(p) \neq \emptyset$.
(ii) If $\mathscr{I}^+(p) \neq \emptyset$, then the Lorentzian distance function d_p is smooth on $\mathscr{I}^+(p)$ and its gradient $\overline{\nabla} d_p$ is a past-directed timelike (geodesic) unit vector field on $\mathscr{I}^+(p)$.

Indeed, $d_p(q) = |\overline{\exp}_p^{-1}(q)|$ for every $q \in \mathscr{I}^+(p)$ and $d_p(\sigma_v(t)) = t$, so that

$$\frac{d}{dt} d_p(\sigma_v(t)) = \langle \overline{\nabla} d_p(\sigma_v(t)), \dot{\sigma}_v(t) \rangle = 1$$

which implies $\overline{\nabla} d_p(\sigma_v(t)) = -\dot{\sigma}_v(t)$.

9.4.1 Hessian and Laplacian Comparison Theorems

Let N be an n-dimensional spacetime with a reference point $p \in N$ such that $\mathscr{I}^+(p) \neq \emptyset$, and let d_p denote the Lorentzian distance function from p.

Given a smooth even function $G : \mathbb{R} \to \mathbb{R}$, let h be the solution of the Cauchy problem

$$\begin{cases} h'' - Gh = 0 \\ h(0) = 0, \quad h'(0) = 1 \end{cases}$$

and let $I_0 = [0, r_0) \subseteq [0, +\infty)$ be the maximal interval on which h is positive. Observe that, in particular, when $G \equiv c$ is constant, $c \in \mathbb{R}$, then $h = h_c$ is given by

$$h_c(t) = \begin{cases} \frac{1}{\sqrt{c}} \sinh(\sqrt{c}\, t) & \text{if } c > 0 \text{ and } t \in I_0 = [0, +\infty) \\ t & \text{if } c = 0 \text{ and } t \in I_0 = [0, +\infty) \\ \frac{1}{\sqrt{-c}} \sin(\sqrt{-c}\, t) & \text{if } c < 0 \text{ and } t \in I_0 = [0, \pi/\sqrt{-c}). \end{cases} \tag{9.36}$$

The next Hessian comparison result assumes that the sectional curvatures of the timelike planes of N containing the radial direction $\overline{\nabla} d_p$ are bounded from below by a function G, and it can be stated as follows (see [151, Theorem 5]).

Lemma 9.10 *Assume that*

$$\overline{K}(v \wedge \overline{\nabla} d_p(q)) \geq G(d_p(q))$$

for every $q \in \mathscr{I}^+(p)$ *with* $d_p(q) < r_0$, *and for every spacelike vector* $v \in T_qN$ *orthogonal to* $\overline{\nabla} d_p(q)$. *Then*

$$\overline{\operatorname{Hess}}\, d_p(v, v) \leq -\frac{h'}{h}(d_p(q))\langle v, v\rangle, \quad q \in \mathscr{I}^+(p), \tag{9.37}$$

where $\overline{\operatorname{Hess}}$ *stands for the Hessian operator on* N.

On the other hand, under the assumption that the sectional curvatures of the timelike planes of N containing the radial direction are bounded from above by a function G, we have the following result (see also [151, Theorem 5]).

Lemma 9.11 *Assume that*

$$\overline{K}(v \wedge \overline{\nabla} d_p(q)) \leq G(d_p(q))$$

for every $q \in \mathscr{I}^+(p)$ *with* $d_p(q) < r_0$, *and for every spacelike vector* $v \in T_qN$ *orthogonal to* $\overline{\nabla} d_p(q)$. *Then*

$$\overline{\operatorname{Hess}}\, d_p(v, v) \geq -\frac{h'}{h}(d_p(q))\langle v, v\rangle, \quad q \in \mathscr{I}^+(p), \tag{9.38}$$

where $\overline{\operatorname{Hess}}$ *stands for the Hessian operator on* N.

Observe that if $q \in \mathscr{I}^+(p)$, with $d_p(q) < r_0$, and

$$\overline{K}(v \wedge \overline{\nabla} d_p(q)) \leq G(d_p(q))$$

for every spacelike vector $v \in T_qN$ orthogonal to $\overline{\nabla} d_p(q)$ (curvature hypothesis in Lemma 9.11), then

$$\overline{\operatorname{Ric}}(\overline{\nabla} d_p(q), \overline{\nabla} d_p(q)) = -\sum_{i=1}^{n-1} \overline{K}(e_i \wedge \overline{\nabla} d_p(q)) \geq -(n-1)G(d_p(q)), \tag{9.39}$$

where $\{e_1, \ldots, e_{n-1}, e_n = \overline{\nabla} d_p(q)\}$ is a local orthonormal basis. The next Laplacian comparison result holds under this weaker hypothesis on the radial Ricci curvature of N (see [151, Theorem 6]). In particular, when N obeys the so called timelike convergence condition, then condition (9.39) trivially holds with $G \equiv 0$.

Lemma 9.12 *Assume that*

$$\overline{\operatorname{Ric}}(\overline{\nabla} d_p(q), \overline{\nabla} d_p(q)) \geq -(n-1)G(d_p(q))$$

for every $q \in \mathscr{I}^+(p)$ with $d_p(q) < r_0$. Then

$$\overline{\Delta} d_p(q) \geq -(n-1)\frac{h'}{h}(d_p(q)), \quad q \in \mathscr{I}^+(p), \tag{9.40}$$

where $\overline{\Delta}$ stands for the (Lorentzian) Laplacian operator on N.

Remark 9.3 In particular, when $G \equiv c$ is constant then $h = h_c$ is given by (9.36) and

$$f_c(t) = \frac{h_c'}{h_c}(t) = \begin{cases} \sqrt{c}\coth(\sqrt{c}\,t) & \text{if } c > 0 \text{ and } t > 0, \\ 1/t & \text{if } c = 0 \text{ and } t > 0, \\ \sqrt{-c}\cot(\sqrt{-c}\,t) & \text{if } c < 0 \text{ and } 0 < t < \pi/\sqrt{-c}. \end{cases}$$

Therefore, from Lemmas 9.10, 9.11 and 9.12 we recover Lemmas 3.2, 3.1 and 3.3, respectively, in [26].

It is worth pointing out that the function f_c arises naturally when computing the index form of a timelike unit geodesic $\sigma_c : [0, s] \to N_c$ in a Lorentzian space form of constant curvature c. Indeed, if J_c is a Jacobi field along σ_c such that $J_c(0) = 0$ and $J_c(s) = v \perp \dot{\sigma}_c(s)$, then a direct computation using the Jacobi equation gives

$$\begin{aligned} I_{\sigma_c}(J_c, J_c) &= -\int_0^s \left(\langle J_c'(t), J_c'(t)\rangle + c\langle J_c(t), J_c(t)\rangle\right) dt \\ &= -\int_0^s \left(\mathrm{s}_c'(t)^2 + c\,\mathrm{s}_c(t)^2\right) dt\, \langle v, v\rangle = -f_c(s)\langle v, v\rangle, \end{aligned}$$

where

$$\mathrm{s}_c(t) = \begin{cases} \frac{\sinh(\sqrt{c}\,t)}{\sinh(\sqrt{c}\,s)} & \text{if } c > 0 \text{ and } 0 \leq t \leq s, \\ t/s & \text{if } c = 0 \text{ and } 0 \leq t \leq s, \\ \frac{\sin(\sqrt{-c}\,t)}{\sin(\sqrt{-c}\,s)} & \text{if } c < 0 \text{ and } 0 \leq t \leq s < \pi/\sqrt{-c}. \end{cases} \tag{9.41}$$

On the other hand, when $\mathscr{I}^+(p) \neq \emptyset$, $f_c(t)$ is the future mean curvature of the level set $\Sigma_c(t) = \{q \in \mathscr{I}^+(p) : d_p(q) = t\} \subset N_c$.

Lemmas 9.10–9.12 were first proved by Impera in [151] using an analytic approach inspired by Petersen [219], and which was also used in [230] for establishing the corresponding Riemannian comparison results. Below and for the reader's convenience we include an alternative geometric proof extracted from [19] which follows essentially the classical ideas of Greene and Wu [129], as already developed in [113] (see also [26]). We will give only the detailed proof of Lemma 9.12, the other two being similar.

Proof (of Lemma 9.12) For a given $q \in \mathscr{I}^+(p) \subset N$, set $v = \overline{\exp}_p^{-1}(q) \in \mathrm{int}(\tilde{\mathscr{I}}^+(p))$ and let $\sigma(t) = \overline{\exp}_p(tv)$ be, with $0 \leq t < s_p(v)$, the radial future directed unit timelike geodesic with $\sigma(0) = p$ and $\sigma(s) = q$, where $s = d_p(q)$. Let $\{e_1, \ldots, e_n\}$ be orthonormal vectors in $T_q\overline{M}$ orthogonal to $\dot{\sigma}(s) = -\overline{\nabla} d_p(q)$, so that

$$\overline{\Delta} d_p(q) = \sum_{j=1}^{n-1} \overline{\mathrm{Hess}}\, d_p(e_j, e_j). \tag{9.42}$$

From [113, Proposition 3.3] we know that, for every $j = 1, \ldots, n-1$,

$$\overline{\mathrm{Hess}}\, d_p(e_j, e_j) = I_\sigma(J_j, J_j),$$

where J_j is the unique Jacobi field along σ such that $J_j(0) = 0$ and $J_j(s) = e_j$. Since $\sigma : [0, s] \to \mathscr{I}^+(p)$ and $\overline{\exp}_p : \mathrm{int}(\tilde{\mathscr{I}}^+(p)) \to \mathscr{I}^+(p)$ is a diffeomorphism, then there is no conjugate point of $\sigma(0)$ along the geodesic segment $\sigma|_{[0,s]}$. Therefore, by the maximality of the index of Jacobi fields [40, Theorem 10.23] we have

$$\overline{\mathrm{Hess}}\, d_p(e_j, e_j) \geq I_\sigma(X_j, X_j)$$

for every vector field X_j along σ such that $X_j(0) = 0$, $X_j(s) = e_j$ and $X_j(t) \perp \dot{\sigma}(t)$ for every t. In particular,

$$\overline{\Delta} d_p(q) \geq \sum_{j=1}^{n-1} I_\sigma(X_j, X_j). \tag{9.43}$$

Let $\{E_1(t), \ldots, E_n(t)\}$ be an orthonormal frame of parallel vector fields along σ such that $E_j(s) = e_j$ for every $j = 1, \ldots, n-1$, and $E_n = \dot{\sigma}$, and define

$$X_j(t) = \frac{h(t)}{h(s)} E_j(t), \quad j = 1, \ldots, n-1.$$

Since X_j is orthogonal to σ and $X_j(0) = 0$ and $X_j(s) = e_j$, we may use X_j in (9.43). Observe that $\{X_1, \ldots, X_{n-1}\}$ are orthogonal along σ, and

$$\langle X_j(t), X_j(t) \rangle = \frac{h(t)^2}{h(s)^2} \quad \text{and} \quad \langle X_j'(t), X_j'(t) \rangle = \frac{h'(t)^2}{h(s)^2},$$

for every $j = 1, \ldots, n-1$. Therefore, for every j we get

$$\begin{aligned} I_\sigma(X_j, X_j) &= -\int_0^s \left(\langle X_j'(t), X_j'(t)\rangle - \langle \overline{R}(X_j(t), \dot{\sigma}(t))\dot{\sigma}(t), X_j(t)\rangle\right) dt \\ &= -\int_0^s \left(\frac{h'(t)^2}{h(s)^2} - \frac{h(t)^2}{h(s)^2}\langle \overline{R}(E_j(t), \dot{\sigma}(t))\dot{\sigma}(t), E_j(t)\rangle\right) dt, \end{aligned}$$

and then

$$\sum_{j=1}^{n-1} I_\sigma(X_j, X_j) = -(n-1)\int_0^s \left(\frac{h'(t)^2}{h(s)^2} - \frac{h(t)^2}{(n-1)h(s)^2}\overline{\mathrm{Ric}}(\dot\sigma(t), \dot\sigma(t))\right) dt$$

$$\geq -(n-1)\int_0^s \frac{h'(t)^2 + h(t)^2 G(t)}{h(s)^2} dt$$

$$= -(n-1)\frac{1}{h(s)^2}\int_0^s \frac{d}{dt}(h(t)h'(t))dt = -(n-1)\frac{h'(s)}{h(s)}.$$

Thus, from (9.43) we get the result. □

9.5 Spacelike Hypersurfaces Contained in the Chronological Future of a Point

In this section we will derive some applications of the Omori-Yau maximum principle for spacelike hypersurfaces contained in the chronological future of a point by working with the Lorentzian distance function restricted on the hypersurface. We refer the reader also to [19] for other applications to the case of trapped submanifolds.

Consider $f : \Sigma \to N$ a spacelike hypersurface immersed into an $m+1$ dimensional spacetime N with future-directed Gauss map ν, and assume that there exists a point $p \in N$ such that $\mathscr{I}^+(p) \neq \emptyset$ and $f(\Sigma) \subset \mathscr{I}^+(p)$. Let $r = d_p$ denote the Lorentzian distance function from p, and let $u = r \circ f : \Sigma \to (0, \infty)$ be the function r along the hypersurface, which is a smooth function on Σ. As usual, set $u_* = \inf_\Sigma u \geq 0$ and $u^* = \sup_\Sigma u \leq +\infty$.

Our first objective is to compute the Hessian and the Laplacian of u. Towards this aim, observe that

$$\overline{\nabla} r = \nabla u - \langle \overline{\nabla} r, \nu\rangle \nu$$

along Σ, where ∇u stands for the gradient of u on Σ. In particular, since $\langle \overline{\nabla} r, \overline{\nabla} r\rangle = -1$ and $\langle \overline{\nabla} r, \nu\rangle > 0$, we have that

$$\partial r/\partial \nu = \langle \overline{\nabla} r, \nu\rangle = \sqrt{1 + |\nabla u|^2} \geq 1.$$

Moreover, from Gauss and Weingarten formulas, we have

$$\overline{\nabla}_X \overline{\nabla} r = \nabla_X \nabla u + \sqrt{1+|\nabla u|^2} AX + \langle AX, \nabla u\rangle \nu - X(\sqrt{1+|\nabla u|^2})\nu,$$

for every tangent vector field $X \in \mathfrak{X}(\Sigma)$. Thus

$$\mathrm{Hess}(u)(X,X) = \overline{\mathrm{Hess}}(r)(X,X) - \sqrt{1+|\nabla u|^2}\langle AX, X\rangle, \tag{9.44}$$

where $\overline{\mathrm{Hess}}(r)$ and $\mathrm{Hess}(u)$ stand for the Hessian of r and u in N and Σ, respectively. Tracing this expression, one gets that the Laplacian of u is given by

$$\Delta u = \overline{\Delta} r + \overline{\mathrm{Hess}}(r)(\nu,\nu) + mH\sqrt{1+|\nabla u|^2}. \tag{9.45}$$

On the other hand, we have the following decomposition for X:

$$X = X^* - \langle X, \overline{\nabla} r\rangle \overline{\nabla} r = X^* - \langle X, \nabla u\rangle \overline{\nabla} r,$$

where X^* is spacelike and orthogonal to $\overline{\nabla} r$. In particular

$$|X^*|^2 = |X|^2 + \langle X, \nabla u\rangle^2.$$

Taking into account that

$$\overline{\nabla}_{\overline{\nabla}_r} \overline{\nabla} r = 0$$

one has

$$\overline{\mathrm{Hess}}(r)(X,X) = \overline{\mathrm{Hess}}(r)(X^*,X^*) \tag{9.46}$$

for every $X \in \mathfrak{X}(\Sigma)$.

Assume now that $u < r_0 \leq +\infty$ and that

$$\overline{K}(v \wedge \overline{\nabla} r(q)) \geq G(r(q))$$

for every $q \in \mathscr{I}^+(p)$ with $r(q) < r_0$, and for every spacelike vector $v \in T_qN$ orthogonal to $\overline{\nabla} r(q)$. Then, by the Hessian comparison result for r given in Lemma 9.10 and using (9.46), one gets that

$$\overline{\mathrm{Hess}}(r)(X,X) = \overline{\mathrm{Hess}}(r)(X^*,X^*) \leq -\frac{h'}{h}(u)(1+\langle X, \nabla u\rangle^2)$$

for every unit tangent vector field $X \in \mathfrak{X}(\Sigma)$. Therefore, from (9.44).

$$\mathrm{Hess}(u)(X,X) \leq -\frac{h'}{h}(u)(1+\langle X, \nabla u\rangle^2) - \sqrt{1+|\nabla u|^2}\langle AX, X\rangle.$$

Tracing this inequality, one gets the following inequality for the Laplacian of u

$$\Delta u \leq -\frac{h'}{h}(u)(m + |\nabla u|^2) + mH\sqrt{1 + |\nabla u|^2}.$$

Proceeding in a similar way in case

$$\overline{K}(v \wedge \overline{\nabla} r(q)) \leq G(r(q))$$

and using now Lemma 9.11, we arrive to the corresponding inequalities

$$\mathrm{Hess}(u)(X, X) \geq -\frac{h'}{h}(u)(1 + \langle X, \nabla u\rangle^2) - \sqrt{1 + |\nabla u|^2}\langle AX, X\rangle.$$

and

$$\Delta u \geq -\frac{h'}{h}(u)(m + |\nabla u|^2) + mH\sqrt{1 + |\nabla u|^2}.$$

Similarly, under the assumption $\overline{\mathrm{Ric}}(\overline{\nabla} r, \overline{\nabla} r) \geq -mG(r)$ on $\mathscr{I}^+(p)$, and using the Laplacian comparison result given in Lemma 9.12, one has that

$$\overline{\Delta} r \geq -m\frac{h'}{h}(u)$$

along the hypersurface. Therefore, we conclude from (9.45)

$$\Delta u \geq -m\frac{h'}{h}(u) + \overline{\mathrm{Hess}}(r)(\nu, \nu) + mH\sqrt{1 + |\nabla u|^2}.$$

Summarizing we have the following

Proposition 9.5 *Let N be an $(m+1)$-dimensional spacetime with a reference point $p \in N$ such that $\mathscr{I}^+(p) \neq \emptyset$, and let $r = d_p$. Given a smooth even function $G : \mathbb{R} \to \mathbb{R}$, let h be the solution of the Cauchy problem*

$$\begin{cases} h'' - Gh = 0 \\ h(0) = 0, \quad h'(0) = 1 \end{cases}$$

and let $I_0 = [0, r_0) \subseteq [0, +\infty)$, with $r_0 \leq +\infty$, be the maximal interval on which h is positive.

Consider a spacelike hypersurface $f : \Sigma \to N$ such that $f(\Sigma) \subset \mathscr{I}^+(p) \cap B^+(p, r_0)$, where

$$B^+(p, r_0) = \{q \in N : r(q) < r_0\}$$

stands for the future inner ball of radius r_0 at p. Let $u = r \circ f : \Sigma \to (0, \infty)$ be the function r along the hypersurface. If

$$\overline{K}(v \wedge \overline{\nabla} r(q)) \geq G(r(q)) \quad (resp. \leq) \tag{9.47}$$

for every $q \in \mathscr{I}^+(p) \cap B^+(p, r_0)$, and for every spacelike vector $v \in T_q N$ orthogonal to $\overline{\nabla} r(q)$, then

$$\Delta u \leq -\frac{h'}{h}(u)(m + |\nabla u|^2) + mH\sqrt{1 + |\nabla u|^2} \quad (resp. \geq).$$

Replacing assumption (9.47) with

$$\overline{\mathrm{Ric}}(\overline{\nabla} r, \overline{\nabla} r) \geq -mG(r) \ on \ \mathscr{I}^+(p) \tag{9.48}$$

we obtain the following inequality

$$\Delta u \geq -m\frac{h'}{h}(u) + \overline{\mathrm{Hess}}(r)(v, v) + mH\sqrt{1 + |\nabla u|^2}.$$

Now we are ready to give the first main result in this section, which is given as Theorem 13 in [151] (see also Theorem 4.2 in [26] for the case where $G \equiv c$).

Theorem 9.8 *Let N be an $(m+1)$-dimensional spacetime with a reference point $p \in N$ such that $\mathscr{I}^+(p) \neq \emptyset$, and let $r = d_p$. Given a smooth even function $G : \mathbb{R} \to \mathbb{R}$, let h be the solution of the Cauchy problem*

$$\begin{cases} h'' - Gh = 0 \\ h(0) = 0, \quad h'(0) = 1 \end{cases}$$

and let $I_0 = [0, r_0) \subseteq [0, +\infty)$, with $r_0 \leq +\infty$, be the maximal interval on which h is positive. Assume that

$$\overline{K}(v \wedge \overline{\nabla} r(q)) \geq G(r(q))$$

for every $q \in \mathscr{I}^+(p) \cap B^+(p, r_0)$, and for every spacelike vector $v \in T_q N$ orthogonal to $\overline{\nabla} r(q)$, and let $f : \Sigma \to N$ be a spacelike hypersurface such that $f(\Sigma) \subset \mathscr{I}^+(p) \cap B^+(p, r_0)$. If the Omori-Yau maximum principle for Δ holds on Σ, then its future mean curvature H satisfies

$$\sup_\Sigma H \geq \frac{h'}{h}(u_*),$$

where u denotes the Lorentzian distance r along the hypersurface. In particular, if $u_* = 0$ then $\sup_\Sigma H = +\infty$.

Proof We know from Proposition 9.5 that

$$\Delta u \leq -\frac{h'}{h}(u)(m + |\nabla u|^2) + mH\sqrt{1 + |\nabla u|^2}.$$

Applying the Omori-Yau maximum principle to the positive function u, with $u_* \geq 0$, there exists a sequence $\{x_k\}_{k\in\mathbb{N}}$ in Σ such that

$$u(x_k) < u_* + \frac{1}{k}, \quad |\nabla u(x_k)| < \frac{1}{k}, \quad \text{and } \Delta u(x_k) > -\frac{1}{k}.$$

Therefore,

$$-\frac{1}{k} < \Delta u(x_k) \leq -\frac{h'}{h}(u(x_k))(m + |\nabla u(x_k)|^2) + mH(x_k)\sqrt{1 + |\nabla u(x_k)|^2}.$$

It follows from here that

$$\sup_\Sigma H \geq H(x_k) \geq \frac{-1/k + \frac{h'}{h}(u(x_k))(m + |\nabla u(x_k)|^2)}{m\sqrt{1 + |\nabla u(x_k)|^2}},$$

and letting $k \to \infty$ we get the result. The last assertion follows from the fact that $\lim_{t\to 0} \frac{h'}{h}(t) = +\infty$. □

As a direct application of Theorem 9.8 we have the following consequence, given in [26, Corollary 4.3] for the case $G \equiv c$.

Corollary 9.9 *Under the assumptions of Theorem 9.8, if the Omori-Yau maximum principle holds on Σ and H is bounded from above, then there exists some $\delta > 0$ such that $f(\Sigma) \subset O^+(p, \delta)$, where $O^+(p, \delta)$ denotes the future outer ball in N of radius δ, that is,*

$$O^+(p, \delta) = \{q \in I^+(p) : r(q) > \delta\}.$$

For a proof, simply observe that $\sup_\Sigma H < +\infty$ implies that $\inf_\Sigma u > 0$.

On the other hand, when $G \equiv c$ we also have the following (see [26, Corollary 4.4])

Corollary 9.10 *Under the assumptions of Theorem 9.8 with $G \equiv c$, when $c \geq 0$ there exists no spacelike hypersurface Σ contained in $\mathscr{I}^+(p)$ satisfying the Omori-Yau maximum principle and having $H \leq \sqrt{c}$. When $c < 0$, there exists no such hypersurface with* $\inf_\Sigma u < \pi/2\sqrt{-c}$ *and $H \leq 0$.*

For a proof, observe that when $G \equiv c \geq 0$ Theorem 9.8 implies that for every spacelike hypersurface Σ contained in $\mathscr{I}^+(p)$ on which the Omori-Yau maximum principle holds, one gets

$$\sup_{\Sigma} H \geq \frac{h'_c}{h_c}(u_*) > \lim_{t\to+\infty} \frac{h'_c}{h_c}(u_*)(t) = \sqrt{c}.$$

Therefore, it cannot happen $\sup_\Sigma H \leq \sqrt{c}$. On the other hand, when $c < 0$ Theorem 9.8 also implies that every spacelike hypersurface Σ contained in $\mathscr{I}^+(p)$, with $u_* < \pi/2\sqrt{-c}$, on which the Omori-Yau maximum principle holds satisfies

$$\sup_{\Sigma} H \geq \frac{h'_c}{h_c}(u_*) > \frac{h'_c}{h_c}\left(\frac{\pi}{2\sqrt{-c}}\right) = 0.$$

Therefore, it cannot happen $\sup_\Sigma H \leq 0$.

On the other, under the assumption that the radial Ricci curvature is bounded from below by a function, we derive the following.

Theorem 9.9 *Let N be an $(m+1)$-dimensional spacetime with a reference point $p \in N$ such that $\mathscr{I}^+(p) \neq \emptyset$, and let $r = d_p$. Given a smooth even function $G : \mathbb{R} \to \mathbb{R}$, let h be the solution of the Cauchy problem*

$$\begin{cases} h'' - Gh = 0 \\ h(0) = 0, \quad h'(0) = 1 \end{cases}$$

and let $I_0 = [0, r_0) \subseteq [0, +\infty)$, with $r_0 \leq +\infty$, be the maximal interval on which h is positive. Assume that

$$\overline{\mathrm{Ric}}(\overline{\nabla} r, \overline{\nabla} r) \geq -mG(r) \text{ on } \mathscr{I}^+(p),$$

and let $f : \Sigma \to N$ be a spacelike hypersurface such that $f(\Sigma) \subset \mathscr{I}^+(p) \cap B^+(p, \delta)$, with $\delta < r_0$. If the Omori-Yau maximum principle for Δ holds on Σ, then its future mean curvature H satisfies

$$\inf_{\Sigma} H \leq \frac{h'}{h}(u^*),$$

where u denotes the Lorentzian distance r along the hypersurface.

Proof From Proposition 9.5 we know

$$\Delta u \geq -m\frac{h'}{h}(u) + \overline{\mathrm{Hess}}(r)(\nu, \nu) + mH\sqrt{1 + |\nabla u|^2}.$$

Now, since $u^* = \sup_\Sigma u \leq \delta$, by applying the Omori-Yau maximum principle to the function u, there exists a sequence of points $\{x_k\}_{k\in\mathbb{N}}$ in Σ such that

$$u(x_k) > u^* - \frac{1}{k}, \quad |\nabla u(x_k)| < \frac{1}{k}, \quad \text{and } \Delta u(x_k) < \frac{1}{k}.$$

Therefore

$$\begin{aligned}\frac{1}{k} &> \Delta u(x_k) \\ &\geq -m\frac{h'}{h}(u(x_k)) + \overline{\mathrm{Hess}}(r)(x_k)(\nu(x_k), \nu(x_k)) + mH(x_k)\sqrt{1+|\nabla u(x_k)|^2},\end{aligned}$$

so that

$$\inf_{\Sigma} H \leq H(x_k) \leq \frac{1/k + m\frac{h'}{h}(u(x_k)) - \overline{\mathrm{Hess}}(r)(x_k)(\nu(x_k), \nu(x_k))}{m\sqrt{1+|\nabla u(x_k)|^2}}. \tag{9.49}$$

On the other hand, we have the following decomposition for $\nu(x_k)$,

$$\nu(x_k) = \nu^*(x_k) - \langle \nu(x_k), \overline{\nabla} r(x_k)\rangle \overline{\nabla} r(x_k),$$

with $\nu^*(x_k)$ orthogonal to $\overline{\nabla} r(x_k)$. Since

$$\langle \overline{\nabla} r(x_k), \overline{\nabla} r(x_k)\rangle = \langle \nu(x_k), \nu(x_k)\rangle = -1$$

and

$$\overline{\nabla} r(x_k) = \nabla u(x_k) - \langle \overline{\nabla} r(x_k), \nu(x_k)\rangle \nu(x_k),$$

we have $|\nu^*(x_k)|^2 = |\nabla u(x_k)|^2$ and $\lim_{k\to+\infty} |\nu^*(x_k)|^2 = 0$. That is,

$$\lim_{k\to+\infty} \nu^*(x_k) = 0.$$

Now, taking into account that

$$\overline{\mathrm{Hess}}(r)(x_k)(\nu(x_k), \nu(x_k)) = \overline{\mathrm{Hess}}(r)(x_k)(\nu^*(x_k), \nu^*(x_k))$$

and letting $k \to \infty$ in (9.49), we conclude

$$\inf_{\Sigma} H \leq \lim_{k\to+\infty} H(x_k) \leq \frac{h'}{h}(u^*).$$

□

In particular, when the ambient spacetime has constant sectional curvature c, by putting together Theorems 9.9 and 9.8, we derive the following consequences [26, Theorem 4.5 and Corollary 4.6].

Theorem 9.10 *Let N_c be an spacetime with constant sectional curvature c and let $p \in N_c$. Let us consider $f : \Sigma \to N_c$ a spacelike hypersurface such that $f(\Sigma) \subset$*

$\mathscr{I}^+(p) \cap B^+(p, \delta)$ for some $\delta > 0$ (with $\delta \leq \pi/\sqrt{-c}$ if $c < 0$). If the Omori-Yau maximum principle holds on Σ, then its future mean curvature H satisfies

$$\inf_{\Sigma} H \leq f_c(u^*) \leq f_c(u_*) \leq \sup_{\Sigma} H,$$

where u denotes the Lorentzian distance r along the hypersurface.

Corollary 9.11 *Let N_c be an spacetime with constant sectional curvature c and let $p \in N_c$. If Σ is a complete spacelike hypersurface in N_c with constant mean curvature H which is contained in $\mathscr{I}^+(p)$ and bounded from above by a level set of the Lorentzian distance function $r = d_p$ (with $r < \pi/\sqrt{-c}$ if $c < 0$), then Σ is necessarily a level set of r.*

For a proof simply recall from Corollary 9.2 that the Omori-Yau maximum principle holds on every complete spacelike hypersurface with constant mean curvature into a Lorentzian spacetime with constant sectional curvature.

The last results above have a specially illustrative consequence when the ambient is the Lorentz-Minkowski spacetime. In that case, it can be easily seen that for a given $p \in \mathbb{L}^{m+1}$

$$\mathscr{I}^+(p) = \{q \in \mathbb{L}^{m+1} : \langle q-p, q-p\rangle < 0, \quad \text{and} \quad \langle q-p, e_{m+1}\rangle < 0\},$$

and the Lorentzian distance function from p is given by

$$d_p(q) = \sqrt{-\langle q-p, q-p\rangle}$$

for every $q \in \mathscr{I}^+(p)$. In particular, the level sets of d_p are precisely the future components of the hyperbolic spaces centered at p. Also, observe that the boundary of $\mathscr{I}^+(p)$ is nothing but the future component of the lightcone with vertex at p. Then, Corollary 9.9 implies that every complete spacelike hypersurface $f : \Sigma \to \mathbb{L}^{m+1}$ contained in $\mathscr{I}^+(p)$ and having bounded mean curvature is bounded away from the lightcone, in the sense that there exists some $\delta > 0$ such that

$$\langle f(x) - p, f(x) - p\rangle \leq -\delta^2 < 0$$

for every $x \in \Sigma$. Also, Corollary 9.8 implies that there exists no complete spacelike hypersurface contained in $\mathscr{I}^+(p)$ and having nonpositive bounded future mean curvature. In particular, there exists no complete hypersurface with constant mean curvature $H \leq 0$ contained in $\mathscr{I}^+(p)$. Finally, Corollary 9.10 allows one to improve Theorem 2 in [7] as follows [26, Corollary 4.7].

Corollary 9.12 *The only complete spacelike hypersurfaces with constant mean curvature in the Lorentz-Minkowski space $\mathbb{L}^{m+1}$ which are contained in $\mathscr{I}^+(p)$ (for some fixed $p \in \mathbb{L}^{m+1}$) and bounded from above by a hyperbolic space centered at p are precisely the hyperbolic spaces centered at p.*

9.6 Generalized Robertson-Walker Spacetimes

Generalized Robertson-Walker spacetimes are warped products of the type $-I \times_\rho \mathbb{P}$, with metric $\langle\, , \rangle$ of the form

$$\langle\, , \rangle = -\pi_I^*\left(dt^2\right) + \rho^2(\pi_I)\left(\pi_{\mathbb{P}}^*\langle\, , \rangle_{\mathbb{P}}\right), \tag{9.50}$$

where π_I and $\pi_{\mathbb{P}}$ are the projections, respectively, on the I and $\mathbb{P}$ factors of the product and t is the global coordinate on the interval $I \subseteq \mathbb{R}$. The notation $-I \times_\rho \mathbb{P}$ reminds of the minus sign in (9.50). From now on we will also drop the pullback notation in (9.50), as we did in the Riemannian case in Chap. 1. The family of generalized Robertson-Walker spacetimes is very large: it includes, for instance, the Lorentz-Minkowski spacetime, the Einstein-De Sitter spacetime, static Einstein spacetimes and the Robertson-Walker spacetimes, for which $\mathbb{P}$ has constant sectional curvature. From the point of view of Physics, since a generalized Robertson-Walker spacetime is not necessarily spatially homogeneous, they seem to provide a more adequate model for spacetime on the small scale; for more detailed considerations see [237].

As for their geometry, it is worth to observe that in a very recent paper, [77], B.-Y. Chen has proved that a Lorentzian manifold is (an open portion of) a generalized Robertson-Walker spacetime if and only if it admits a timelike closed conformal vector field (for the notion of closed conformal vector field and a similar characterization of Riemannian warped products see (1.202) and the discussion that follows there).

Given a spacelike hypersurface $f : \Sigma \to -I \times_\rho \mathbb{P}$ and denoting with ν the unique timelike normal globally defined on Σ, that is, the future directed Gauss map with the same orientation of $\partial_t = \frac{\partial}{\partial t}$, we set

$$\Theta = \langle \nu, \partial_t \rangle = -\cosh\theta \le -1, \quad \theta > 0, \tag{9.51}$$

to denote the opposite of the hyperbolic cosine of the hyperbolic angle θ between Σ and ∂_t. In case f is of the form

$$\Gamma_u : \mathbb{P} \to -I \times_\rho \mathbb{P},$$

with $u : \mathbb{P} \to I$ and

$$\Gamma_u(x) = (u(x), x),$$

that is, the hypersurface is a graph, the metric $\langle\, , \rangle$ induced on $\mathbb{P}$ from the Lorentzian metric of the ambient space *via* Γ_u is given by

$$\langle\, , \rangle = -du^2 + \rho(u)^2\langle\, , \rangle_{\mathbb{P}}.$$

Therefore Γ_u is spacelike if and only if $|Du|^2 < \rho(u)^2$ on $\mathbb{P}$, where Du denotes the gradient of u in $\mathbb{P}$ and $|Du|$ its norm with respect to the original metric $\langle\, ,\, \rangle_{\mathbb{P}}$ of $\mathbb{P}$. Assuming that $\Gamma_u : \mathbb{P} \to -I \times_\rho \mathbb{P}$ is spacelike, it is not difficult to see [compare with (7.49)] that the vector field

$$\nu = \frac{\rho(u)}{\sqrt{\rho(u)^2 - |Du|^2}}\left(\frac{1}{\rho(u)^2}Du + \partial_t\right) \tag{9.52}$$

defines the future-pointing Gauss map of Γ_u, for which

$$\cosh\theta = -\Theta = \frac{\rho(u)}{\sqrt{\rho(u)^2 - |Du|^2}};$$

in particular

$$\sinh\theta = \frac{|Du|}{\sqrt{\rho(u)^2 - |Du|^2}}.$$

The corresponding mean curvature H in the direction of ν is given by the differential equation (compare with (7.50) of the Riemannian case)

$$\operatorname{div}_{\mathbb{P}}\left(\frac{Du}{\sqrt{\rho(u)^2 - |Du|^2}}\right) = m\rho(u)\left(H - \frac{\rho'(u)}{\sqrt{\rho(u)^2 - |Du|^2}}\right), \tag{9.53}$$

where $\operatorname{div}_{\mathbb{P}}$ denotes the divergence operator in $(\mathbb{P}, \langle\, ,\, \rangle_{\mathbb{P}})$.

In what follows we shall need the open form of the weak maximum principle for the Lorentzian mean curvature operator

$$Lv = \operatorname{div}_{\mathbb{P}}\left(\frac{Dv}{\sqrt{1 - |Dv|^2}}\right), \tag{9.54}$$

with v in the class

$$A_1(\mathbb{P}) = \left\{ v \in \mathrm{Lip}_{loc}(\mathbb{P}) : |Dv| < 1 \quad \text{and} \quad \frac{1}{\sqrt{1 - |Dv|^2}} \in L^1_{loc}(\mathbb{P}) \right\}.$$

Since proofs are very similar to those presented in Chap. 4, we limit ourselves to introduce the appropriate setting and to state the results we shall need without details.

Thus, let $\varphi \in C^0([0,\omega)) \cap C^1((0,\omega))$ for some $0 < \omega < +\infty$ and suppose

$$(i)\ \varphi(0) = 0; \quad (ii)\ \varphi(t) > 0 \quad \text{on } (0,\omega),\ \varphi(t) \le At^\delta \quad \text{on } (0,\omega) \tag{9.55}$$

for some constants $A, \delta > 0$. For a Riemannian manifold $(M, \langle\, , \,\rangle)$ and

$$u \in A_\omega(M) = \left\{u \in \mathrm{Lip}_{loc}(M) : |\nabla u| < \omega \quad \text{and } |\nabla u|^{-1}\varphi(|\nabla u|) \in L^1_{loc}(M)\right\}$$

define

$$L_\varphi u = \mathrm{div}\left(|\nabla u|^{-1}\varphi(|\nabla u|)\nabla u\right) \tag{9.56}$$

in the weak sense.

We say that the weak maximum principle holds on M for L_φ if, for each $u \in A_\omega(M)$ with $u^* = \sup_\Sigma u < +\infty$ and for each $\gamma < u^*$ we have

$$\inf_{\Omega_\gamma} \{L_\varphi u\} \le 0 \tag{9.57}$$

in the weak sense [see (3.170)], where $\Omega_\gamma = \{x \in M : u(x) > \gamma\}$.

Similarly, we say that the open WMP holds on M for L_φ if for each $f \in C^0(\mathbb{R})$, for each open set $\Omega \subset M$ with $\partial\Omega \neq \emptyset$ and for each $v \in C^0(\overline{\Omega}) \cap A_\omega(\Omega)$ satisfying

$$\begin{cases} (i)\ L_\varphi v & \ge f(v) \quad \text{on } \Omega \\ (ii)\ \sup_\Omega v & < +\infty \end{cases} \tag{9.58}$$

we have that either

$$\sup_\Omega v = \sup_{\partial\Omega} v \tag{9.59}$$

or

$$f\left(\sup_\Omega v\right) \le 0. \tag{9.60}$$

Of course in (9.58) the differential inequality in (i) has to be understood in the weak sense. We have

Theorem 9.11 *In the above assumptions the validity of the WMP for the operator L_φ is equivalent to that of the open WMP.*

Next we give a geometric condition on the volume growth of geodesic balls that guarantees the validity of the WMP.

Theorem 9.12 *Let $(M, \langle\, , \,\rangle)$ be a complete Riemannian manifold. Assume that*

$$\liminf_{r \to +\infty} \frac{\log \mathrm{vol}\, B_r}{r^{1+\delta}} = d_0 < +\infty. \tag{9.61}$$

Let $u \in A_\omega(M)$ be such that $u^* < +\infty$. Then, for all $\gamma < u^*$ we have

$$\inf_{\Omega_\gamma} L_\varphi u \le 0 \tag{9.62}$$

in the weak sense, where $\Omega_\gamma = \{x \in M : u(x) > \gamma\}$.

Remark 9.4 Suppose that $q(x) \in C^0(M)$ with $q(x) > 0$ on M; results similar to Theorems 9.11 and 9.12 can be given for the q-WMP and the open q-WMP for the operator L_φ on $(M, \langle\,,\,\rangle)$.

In the statements of the next results we will assume the validity on the Riemannian manifold $(\mathbb{P}, \langle\,,\,\rangle_{\mathbb{P}})$ of the weak maximum principle for the Lorentzian mean curvature operator introduced in (9.54) above. According to Theorem 9.12 its validity is guaranteed by the volume growth condition

$$\liminf_{r\to+\infty} \frac{\log \operatorname{vol} B_r}{r^2} < +\infty. \tag{9.63}$$

In terms of curvature, by Theorem 1.3 and Proposition 1.7, the above requirement is implied by

$$\operatorname{Ric}_{\mathbb{P}} \ge -(m-1)B^2(1+r^2)\langle\,,\,\rangle_{\mathbb{P}} \tag{9.64}$$

for some constant $B > 0$.

We are now ready to prove the next

Theorem 9.13 *Let $-I \times_\rho \mathbb{P}$ be a generalized Robertson-Walker spacetime and assume the validity of the WMP for the Lorentzian mean curvature operator on $(\mathbb{P}, \langle\,,\,\rangle_{\mathbb{P}})$. Let Γ_u be an entire spacelike maximal graph in $-I \times_\rho \mathbb{P}$ with $I = (a, b)$, $-\infty \le a < b \le +\infty$, which is not a slice. Then*

$$\text{either } u^* = b \quad \text{or } u^* \le \inf\big\{\lambda \in I : \rho'(t) < 0 \quad \text{on } [\lambda, b)\big\}. \tag{9.65}$$

Similarly, if $u_* = \inf_{\mathbb{P}} u$,

$$\text{either } u_* = a \quad \text{or } u_* \ge \sup\big\{\mu \in I : \rho'(t) > 0 \quad \text{on } (a, \mu)\big\}, \tag{9.66}$$

with the convention $\inf \emptyset = +\infty$ and $\sup \emptyset = -\infty$.

Proof Let $m = \dim \mathbb{P}$ and let us prove (9.65); the proof of (9.66) is analogous. Suppose that $u^* < b$; if $\{\lambda \in I : \rho'(t) < 0 \quad \text{on } [\lambda, b)\} = \emptyset$ then (9.65) is trivially satisfied, otherwise, by contradiction, assume that

$$u^* > \inf\big\{\lambda \in I : \rho'(t) < 0 \quad \text{on } [\lambda, b)\big\}.$$

Choose $\lambda < u^*$ such that $\rho'(t) < 0$ on $[\lambda, b)$ and sufficiently near to u^* so that, if

$$\Lambda_\lambda = \{x \in \mathbb{P} : u(x) > \lambda\},$$

then $\partial\Lambda_\lambda \neq \emptyset$. We fix an origin $o \in \mathbb{P}$ and for $u_0 = u(o)$ we consider the function

$$\psi(s) = \int_{u_0}^{s} \frac{dt}{\rho(t)}$$

for which $\psi' = \frac{1}{\rho} > 0$. Setting $v(x) = \psi(u(x))$, a calculation using (9.53) with $H = 0$ shows that

$$\operatorname{div}_{\mathbb{P}}\left(\frac{Dv}{\sqrt{1-|Dv|^2}}\right) = -m\frac{\rho'\big(\psi^{-1}(v)\big)}{\sqrt{1-|Dv|^2}}.$$

Let $\gamma = \psi(\lambda)$ and observe that

$$\Omega_\gamma = \{x \in \mathbb{P} : v(x) > \gamma\} = \Lambda_\lambda.$$

Since $\rho'\big(\psi^{-1}(v)\big) < 0$ on Ω_γ we have

$$\operatorname{div}_{\mathbb{P}}\left(\frac{Dv}{\sqrt{1-|Dv|^2}}\right) \geq -m\rho'\big(\psi^{-1}(v)\big) \quad \text{on } \Omega_\gamma.$$

Now observe that

$$\sup_{\Omega_\gamma} v = \psi(u^*) > \gamma = \sup_{\partial\Omega_\gamma} v.$$

Therefore, by the open WMP we conclude that

$$-m\rho'(u^*) \leq 0,$$

yielding the desired contradiction. □

As an application of the above theorem we have

Corollary 9.13 *Let $-I \times_\rho \mathbb{P}$ be a generalized Robertson-Walker spacetime and assume the validity of the WMP for the Lorentzian mean curvature operator on $(\mathbb{P}, \langle\,,\,\rangle_{\mathbb{P}})$. For $a, b \in I$, $-\infty < a < b < +\infty$, let $(a, b) \times \mathbb{P}$ be an open slab in $-I \times_\rho \mathbb{P}$ and assume the existence of $t_0 \in (a, b)$ with the property that $\rho'(t) > 0$ on $[a, t_0)$ and $\rho'(t) < 0$ on $(t_0, b]$. Then the only entire maximal graph contained in $(a, b) \times \mathbb{P}$ is the slice $u(x) \equiv t_0$.*

Proof Choose $\varepsilon > 0$ sufficiently small that $\rho'(t) > 0$ on $(a-\varepsilon, t_0)$ and $\rho'(t) < 0$ on $(t_0, b+\varepsilon)$. Set $\alpha = a-\varepsilon$ and $\beta = b+\varepsilon$, and let $J = (\alpha, \beta)$. By contradiction suppose that $\Gamma_u(\mathbb{P})$ is not a slice. By applying Theorem (9.13) to the generalized Robertson-Walker spacetime $-J \times_\rho \mathbb{P}$ and taking into account that $u^* < \beta$ and $u_* > \alpha$ we conclude that

$$u^* \leq t_0 \leq u_*,$$

that is, $u \equiv t_0$, contradiction. Therefore $\Gamma_u(\mathbb{P})$ must be a maximal slice in the slab $(a,b) \times \mathbb{P}$, and the only maximal slice contained in that slab is $u \equiv t_0$. □

Next result yields a height estimate for spacelike graphs.

Theorem 9.14 *Let $-I \times_\rho \mathbb{P}$ be a generalized Robertson-Walker spacetime and assume the validity of the WMP for the Lorentzian mean curvature operator on $(\mathbb{P}, \langle\, ,\, \rangle_{\mathbb{P}})$. Let $u \in C^\infty(\mathbb{P})$, $u^* = \sup_{\mathbb{P}} u < +\infty$. Assume that the graph*

$$\Gamma_u : \mathbb{P} \to -I \times_\rho \mathbb{P}$$

is a spacelike hypersurface such that $H_* = \inf_{\mathbb{P}} H \leq 0$ [H in the direction of ν given in (9.52)]. Then either $\Gamma_u(\mathbb{P})$ is a slice $\mathbb{P}_{u_0} = \{u_0\} \times \mathbb{P}$ with $\mathscr{H}(u_0) = \frac{\rho'(u_0)}{\rho(u_0)} = H_* \equiv H$ or $\mathscr{H}(u^*) \geq H_*$.

Proof If $\Gamma_u(\mathbb{P})$ is a slice $\mathbb{P}_{u_0}$, then, since $Du \equiv 0$, from (9.53) it follows directly $\mathscr{H}(u_0) = H_* \equiv H$. So assume that u is nonconstant. We reason by contradiction and we suppose $\mathscr{H}(u^*) < H_*$. Define

$$\Omega_\gamma = \{x \in \mathbb{P} : u(x) > \gamma\},$$

having chosen $\gamma < u^*$ such that $\partial\Omega_\gamma \neq \emptyset$ and $\mathscr{H}(u) < H_*$ on Ω_γ. Note that this is always possible since $\mathscr{H}$ is continuous. Reasoning as in Theorem 9.13 we define

$$v(x) = \psi(u(x)) = \int_{u_0}^{u(x)} \frac{ds}{\rho(s)}$$

so that, from (9.53), v satisfies

$$\operatorname{div}_{\mathbb{P}}\left(\frac{Dv}{\sqrt{1-|Dv|^2}}\right) = m\rho(u)\left(H - \frac{\mathscr{H}(u)}{\sqrt{1-|Dv|^2}}\right).$$

Since $H_* \leq 0$ we have

$$H \geq H_* \geq \frac{H_*}{\sqrt{1-|Dv|^2}}$$

and therefore

$$\begin{aligned}\operatorname{div}_{\mathbb{P}}\left(\frac{Dv}{\sqrt{1-|Dv|^2}}\right) &\geq m\rho(u)\left(H_* - \frac{\mathscr{H}(u)}{\sqrt{1-|Dv|^2}}\right)\\ &\geq m\rho(u)(H_* - \mathscr{H}(u)) \geq C\big(H_* - \mathscr{H}(\psi^{-1}(v))\big)\end{aligned}$$

on Ω_γ, where $C = m\min_{[\gamma,u^*]}\rho$ and where we have used the fact that $\mathscr{H}(u) < H_*$ on Ω_γ. Since ψ is strictly increasing,

$$\sup_{\Omega_\gamma} v = v^* = \psi(u^*) > \psi(\gamma) = \sup_{\partial\Omega_\gamma} v.$$

Then, because of the open form of the WMP we have

$$H_* - \mathscr{H}(\psi^{-1}(v^*)) = H_* - \mathscr{H}(u^*) \leq 0,$$

contradicting the fact that $H_* - \mathscr{H}(u^*) > 0$. □

Remark 9.5 We note that the conclusion $\mathscr{H}(u^*) \geq H_*$ gives interesting information only if $\mathscr{H}(u^*) < 0$, that is $\rho'(u^*) < 0$.

We conclude the section with a result that directly compares with the Riemannian case given in Sect. 7.5.2. We restrict ourselves to the case of constant mean curvature, that corresponds to Theorem 7.18, but results similar to those contained for instance in 7.19 can be given also in the Lorentzian case, see [29]. We do this to point out differences, in particular the role played by the hyperbolic angle θ introduced in (9.51).

Let us consider the case of a hypersurface in a Lorentzian product.

Theorem 9.15 *Let $f : \Sigma \to -\mathbb{R} \times \mathbb{P}$ be a stochastically complete spacelike hypersurface with constant mean curvature $H > 0$ (with respect to ν as in Sect. 9.2). Suppose that for some $\alpha > 0$*

$$\operatorname{Ric}_{\mathbb{P}} \geq -m\alpha. \tag{9.67}$$

Let $\Omega \subset \Sigma$ be an open set with $\partial\Omega \neq \emptyset$ for which $f(\Omega)$ is contained in a slab and $f(\partial\Omega) \subset \{0\} \times \mathbb{P}$. Assume

$$\sup_{\Omega} \Theta^2 = \beta^2 < \frac{\alpha + H^2}{\alpha}. \tag{9.68}$$

Then

$$f(\Omega) \subset \left[\frac{(1-\beta)H}{H^2 - \alpha(\beta^2-1)}, 0\right] \times \mathbb{P}. \tag{9.69}$$

Proof If $\beta = 1$ then there is nothing to prove because, in this case, $\Theta \equiv -1$ is constant on Ω or, equivalently, the height function $h = \pi_{\mathbb{R}} \circ f$ is constant on Ω. Thus, $f(\Omega)$ is contained in the slice $\{0\} \times \mathbb{P}$. Let $\beta > 1$. From (9.68), we can choose $\delta > 0$ sufficiently small that

$$(\alpha + \delta)(\beta^2 - 1) < H^2.$$

We consider the function

$$\psi_\delta = \Theta + \frac{H^2 - (\alpha + \delta)(\beta^2 - 1)}{H} h.$$

Then a computation similar to that performed in Proposition 7.1 for $k = 0$ yields

$$\Delta h = -m\Theta H.$$

Similarly proceeding as in Lemma 7.4 with $\rho \equiv 1$ and $k = 0$ we obtain

$$\Delta\Theta = \Theta|A|^2 + \Theta \operatorname{Ric}_{\mathbb{P}}(\hat{\nu}, \hat{\nu}),$$

where $\hat{\nu}$ denotes the projection of ν onto the fibre $\mathbb{P}$. Therefore, using

$$|A|^2 = m^2 H^2 - m(m-1)H_2$$

with

$$H_2 = \binom{m}{2}^{-1} \sum_{1 \le i < j \le m} k_i k_j,$$

κ_i the eigenvalues of A, we obtain

$$\Delta\psi_\delta = \Theta\big\{m(m-1)(H^2 - H_2) + \operatorname{Ric}_{\mathbb{P}}(\hat{\nu}, \hat{\nu}) + m(\alpha + \delta)(\beta^2 - 1)\big\}. \tag{9.70}$$

From (9.67)

$$\operatorname{Ric}_{\mathbb{P}}(\hat{\nu}, \hat{\nu}) \ge -m\alpha|\hat{\nu}|^2 = -m\alpha(\Theta^2 - 1) \ge -m\alpha(\beta^2 - 1)$$

on Ω. Thus, using the basic inequality $H^2 \ge H_2$, (9.70) gives

$$\Delta\psi_\delta \le m\Theta\delta(\beta^2 - 1) \le -m\delta(\beta^2 - 1) < 0$$

on Ω, where the last inequality is due to $\beta > 1$. We define $w = \psi_{\delta_{|\overline{\Omega}}}$. Since $f(\Omega)$ is contained in a slab we have

$$\begin{cases} \Delta w & \leq -m\delta(\beta^2 - 1) \quad \text{on } \Omega \\ \inf_{\Omega} w & > -\infty. \end{cases}$$

Stochastic completeness and the open form of the WMP imply

$$\inf_{\Omega} w = \inf_{\partial\Omega} w.$$

By assumption $f(\partial\Omega) \subset \{0\} \times \mathbb{P}$ and thus $h \equiv 0$ on $\partial\Omega$, so that $w = \psi_\delta = \Theta \geq -\beta$ on $\partial\Omega$. We then have

$$-\beta \leq \Theta + \frac{H^2 - (\alpha + \delta)(\beta^2 - 1)}{H} h \leq -1 + \frac{H^2 - (\alpha + \delta)(\beta^2 - 1)}{H} h.$$

Dividing by the positive quantity $H^2 - (\alpha + \delta)(\beta^2 - 1)$,

$$h \geq \frac{(1 - \beta)H}{H^2 - (\alpha + \delta)(\beta^2 - 1)}.$$

Taking the limit as $\delta \downarrow 0$ we deduce

$$h \geq \frac{(1 - \beta)H}{H^2 - \alpha(\beta^2 - 1)}.$$

On the other hand

$$\begin{cases} \Delta h & = -mH\Theta \geq mH > 0 \\ \sup_{\Omega} h & < +\infty, \end{cases}$$

and reasoning as above we deduce $h \leq 0$ on Ω. This completes the proof of the theorem. □

Remark 9.6 We recall that, by definition, H is the future mean curvature of the hypersurface. The corresponding statement for the case of negative constant mean curvature is the same, up to replacing (9.69) with

$$f(\Omega) \subset \left[0, \frac{(1 - \beta)H}{H^2 - \alpha(\beta^2 - 1)}\right] \times \mathbb{P}. \tag{9.71}$$

The proof is analogous, it is enough to substitute the function ψ_δ with $-\psi_\delta$.

Similarly to what happens in the Riemannian case, but under the assumption of stochastic completeness, weaker than parabolicity (see Corollary 7.11), as a consequence of the above theorem we have

Corollary 9.14 *Let $f : \Sigma \to -\mathbb{R} \times \mathbb{P}$ be a stochastically complete spacelike hypersurface with constant mean curvature $H > 0$. Suppose that*

$$\mathrm{Ric}_{\mathbb{P}} \geq 0. \tag{9.72}$$

Let $\Omega \subset \Sigma$ be an open set with $\partial\Omega \neq \emptyset$ for which $f(\Omega)$ is contained in a slab and $f(\partial\Omega) \subset \{0\} \times \mathbb{P}$, and assume

$$\beta = \sup_{\Omega} |\Theta| < +\infty. \tag{9.73}$$

Then

$$f(\Omega) \subset \left[\frac{(1-\beta)}{H}, 0\right] \times \mathbb{P}. \tag{9.74}$$

The observation in Remark 9.6 applies here too.

In the present Lorentzian setting we can give a sufficient condition for stochastic completeness different, for instance, from that of the Riemannian case reported in Theorem 7.20.

Theorem 9.16 *Let $f : \Sigma \to -\mathbb{R} \times \mathbb{P}$ be a complete spacelike hypersurface with constant mean curvature $H > 0$. Assume that the height function $h = \pi_{\mathbb{R}} \circ f : \Sigma \to \mathbb{R}$ satisfies*

$$\lim_{x \to \infty} h(x) = -\infty.$$

Suppose that for some $\alpha > 0$

$$\mathrm{Ric}_{\mathbb{P}} \geq -m\alpha.$$

Let $\Omega \subset \Sigma$ be a relatively compact open set with $\partial\Omega \neq \emptyset$ such that $f(\partial\Omega) \subset \{0\} \times \mathbb{P}$. Assume

$$\sup_{\Omega} \Theta^2 = \beta^2 < \frac{\alpha + H^2}{\alpha}.$$

Then

$$f(\Omega) \subset \left[\frac{(1-\beta)H}{H^2 - \alpha(\beta^2 - 1)}, 0\right] \times \mathbb{P}.$$

Proof The result follows from Theorem 9.15 once we show the validity of the WMP on Σ for the Laplace-Beltrami operator. Towards this end we let $\gamma = -h$, so that it satisfies

$$\Delta\gamma = m\Theta H \leq -mH < 0$$

and

$$\gamma(x) \to +\infty \quad \text{as } x \to \infty.$$

We now apply Theorem 3.1 to deduce the validity of the WMP for Δ on Σ. □

References

1. L.V. Ahlfors, An extension of Schwarz's lemma. Trans. Am. Math. Soc. **43**(3), 359–364 (1938)
2. L.V. Ahlfors, L. Sario, *Riemann Surfaces*. Princeton Mathematical Series, vol. 26 (Princeton University Press, Princeton, NJ, 1960)
3. R. Aiyama, On the Gauss map of complete space-like hypersurfaces of constant mean curvature in Minkowski space. Tsukuba J. Math. **16**(2), 353–361 (1992)
4. G. Albanese, M. Rigoli, Lichnerowicz-type equations on complete manifolds. Adv. Nonlin. Anal. doi: 10.1515/anona-2015-0106, Published online November (2015, to appear)
5. G. Albanese, L.J. Alías, M. Rigoli, A general form of the weak maximum principle and some applications. Rev. Mat. Iberoam. **29**(4), 1437–1476 (2013)
6. G. Albanese, L. Mari, M. Rigoli, On the role of gradient terms in coercive quasilinear differential inequalities on Carnot groups. Nonlinear Anal. **126**, 234–261 (2015)
7. J.A. Aledo, L.J. Alías, On the curvatures of bounded complete spacelike hypersurfaces in the Lorentz-Minkowski space. Manuscripta Math. **101**(3), 401–413 (2000)
8. H. Alencar, M. do Carmo, Hypersurfaces with constant mean curvature in spheres. Proc. Am. Math. Soc. **120**(4), 1223–1229 (1994)
9. H. Alencar, M. do Carmo, A.G. Colares, Stable hypersurfaces with constant scalar curvature. Math. Z. **213**(1), 117–131 (1993)
10. A.D. Alexandrov, A characteristic property of spheres. Ann. Mat. Pura Appl. (4) **58**, 303–315 (1962)
11. L.J. Alías, M. Dajczer, Uniqueness of constant mean curvature surfaces properly immersed in a slab. Comment. Math. Helv. **81**(3), 653–663 (2006)
12. L.J. Alías, M. Dajczer, Constant mean curvature hypersurfaces in warped product spaces. Proc. Edinb. Math. Soc. (2) **50**(3), 511–526 (2007)
13. L.J. Alías, M. Dajczer, Normal geodesic graphs of constant mean curvature. J. Differ. Geom. **75**(3), 387–401 (2007)
14. L.J. Alías, S.C. García-Martínez, On the scalar curvature of constant mean curvature hypersurfaces in space forms. J. Math. Anal. Appl. **363**(2), 579–587 (2010)
15. L.J. Alías, S.C. García-Martínez, An estimate for the scalar curvature of constant mean curvature hypersurfaces in space forms. Geom. Dedicata **156**, 31–47 (2012)
16. L.J. Alías, B. Palmer, Zero mean curvature surfaces with non-negative curvature in flat Lorentzian 4-spaces. R. Soc. Lond. Proc. Ser. A Math. Phys. Eng. Sci. **455**(1982), 631–636 (1999)
17. L.J. Alías, B. Palmer, On the Gaussian curvature of maximal surfaces and the Calabi-Bernstein theorem. Bull. Lond. Math. Soc. **33**(4), 454–458 (2001)

L.J. Alías et al., *Maximum Principles and Geometric Applications*,
Springer Monographs in Mathematics, DOI 10.1007/978-3-319-24337-5

18. L.J. Alías, G.P. Bessa, M. Dajczer, The mean curvature of cylindrically bounded submanifolds. Math. Ann. **345**(2), 367–376 (2009)
19. L.J. Alías, G.P. Bessa, J.H.S. de Lira, Geometric analysis of Lorentzian distance function on trapped submanifolds (2015, preprint)
20. L.J. Alías, G.P. Bessa, J.F. Montenegro, An estimate for the sectional curvature of cylindrically bounded submanifolds. Trans. Am. Math. Soc. **364**(7), 3513–3528 (2012)
21. L.J. Alías, M. Dajczer, M. Rigoli, Higher order mean curvature estimates for bounded complete hypersurfaces. Nonlinear Anal. **84**, 73–83 (2013)
22. L.J. Alías, S.C. de Almeida, A. Brasil Jr., Hypersurfaces with constant mean curvature and two principal curvatures in $\mathbb{S}^{n+1}$. An. Acad. Brasil. Ciênc. **76**(3), 489–497 (2004)
23. L.J. Alías, J.H.S. de Lira, J.M. Malacarne, Constant higher-order mean curvature hypersurfaces in Riemannian spaces. J. Inst. Math. Jussieu **5**(4), 527–562 (2006)
24. L.J. Alías, J.H.S. de Lira, M. Rigoli, Geometric elliptic functionals and mean curvature. Ann. Sc. Norm. Super. Pisa Cl. Sci. To appear in Vol. XV (2016). Submitted: 2013-07-30 Accepted: 2014-07-09 pp. 44 DOI Number: 10.2422/2036-2145.201308_003
25. L.J. Alías, S.C. García-Martínez, M. Rigoli, A maximum principle for hypersurfaces with constant scalar curvature and applications. Ann. Global Anal. Geom. **41**(3), 307–320 (2012)
26. L.J. Alías, A. Hurtado, V. Palmer, Geometric analysis of Lorentzian distance function on spacelike hypersurfaces. Trans. Am. Math. Soc. **362**(10), 5083–5106 (2010)
27. L.J. Alías, D. Impera, M. Rigoli, Hypersurfaces of constant higher order mean curvature in warped products. Trans. Am. Math. Soc. **365**(2), 591–621 (2013)
28. L.J. Alías, J. Miranda, M. Rigoli, A new open form of the weak maximum principle and geometric applications. Commun. Anal. Geom. **24**(1) (2016, to appear)
29. L.J. Alías, M. Rigoli, S. Scoleri, Weak maximum principles and geometric estimates for spacelike hypersurfaces in generalized Robertson-Walker spacetimes. Nonlinear Anal. **129**, 119–142 (2015)
30. L.J. Alías, A. Romero, M. Sánchez, Uniqueness of complete spacelike hypersurfaces of constant mean curvature in generalized Robertson-Walker spacetimes. Gen. Relativ. Gravit. **27**(1), 71–84 (1995)
31. H. Amann, Supersolutions, monotone iterations, and stability. J. Differ. Equ. **21**(2), 363–377 (1976)
32. D.G. Aronson, H.F. Weinberger, Multidimensional nonlinear diffusion arising in population genetics. Adv. Math. **30**(1), 33–76 (1978)
33. A. Atsuji, Remarks on harmonic maps into a cone from a stochastically complete manifold. Proc. Jpn. Acad. Ser. A Math. Sci. **75**(7), 105–108 (1999)
34. P. Baird, A class of three-dimensional Ricci solitons. Geom. Topol. **13**(2), 979–1015 (2009)
35. P. Baird, L. Danielo, Three-dimensional Ricci solitons which project to surfaces. J. Reine Angew. Math. **608**, 65–91 (2007)
36. C. Bär, G.P. Bessa, Stochastic completeness and volume growth. Proc. Am. Math. Soc. **138**(7), 2629–2640 (2010)
37. J.L.M. Barbosa, A.G. Colares, Stability of hypersurfaces with constant r-mean curvature. Ann. Global Anal. Geom. **15**(3), 277–297 (1997)
38. A. Barros, E. Ribeiro, Some characterizations for compact almost Ricci solitons. Proc. Am. Math. Soc. **140**(3), 1033–1040 (2012)
39. A. Barros, R. Batista, E. Ribeiro, Compact almost Ricci solitons with constant scalar curvature are gradient. Monatsh. Math. **174**(1), 29–39 (2014)
40. J.K. Beem, P.E. Ehrlich, K.L. Easley, *Global Lorentzian Geometry*. Monographs and Textbooks in Pure and Applied Mathematics, vol. 202, 2nd edn. (Marcel Dekker, New York, 1996)
41. A.L. Besse, *Einstein Manifolds*. Classics in Mathematics (Springer, Berlin, 2008). Reprint of the 1987 edition
42. G.P. Bessa, J.F. Montenegro, Mean time exit and isoperimetric inequalities for minimal submanifolds of $N \times \mathbb{R}$. Bull. Lond. Math. Soc. **41**(2), 242–252 (2009)
43. G.P. Bessa, L.F. Pessoa, Maximum principle for semi-elliptic trace operators and geometric applications (2012). arXiv:1208.1322v1 [math.DG]

44. B. Bianchini, L. Mari, M. Rigoli, On some aspects of oscillation theory and geometry. Mem. Am. Math. Soc. **225**(1056), vi + 195 (2013)
45. B. Bianchini, L. Mari, P. Pucci, M. Rigoli, On the interplay among weak and strong maximum principles, compact support principles and Keller-Osserman conditions on complete manifolds (2015, preprint)
46. B. Bianchini, L. Mari, M. Rigoli, Semilinear and quasilinear Yamabe type equations with sign-changing nonlinearities on non-compact manifolds (2015, preprint)
47. M.-F. Bidaut-Véron, S. Pohozaev, Nonexistence results and estimates for some nonlinear elliptic problems. J. Anal. Math. **84**, 1–49 (2001)
48. R.L. Bishop, Decomposition of cut loci. Proc. Am. Math. Soc. **65**(1), 133–136 (1977)
49. S. Bochner. Vector fields and Ricci curvature. Bull. Am. Math. Soc. **52**, 776–797 (1946)
50. E. Bombieri, E. De Giorgi, E. Giusti, Minimal cones and the Bernstein problem. Invent. Math. **7**, 243–268 (1969)
51. W.M. Boothby, *An Introduction to Differentiable Manifolds and Riemannian Geometry*. Pure and Applied Mathematics, vol. 120, 2nd edn. (Academic, Orlando, FL, 1986)
52. A. Borbély, On minimal surfaces satisfying the Omori-Yau principle. Bull. Aust. Math. Soc. **84**(1), 33–39 (2011)
53. J.-P. Bourguignon, Les variétés de dimension 4 à signature non nulle dont la courbure est harmonique sont d'Einstein. Invent. Math. **63**(2), 263–286 (1981)
54. D. Brill, F. Flaherty, Isolated maximal surfaces in spacetime. Commun. Math. Phys. **50**(2), 157–165 (1976)
55. E. Calabi, An extension of E. Hopf's maximum principle with an application to Riemannian geometry. Duke Math. J. **25**, 45–56 (1958)
56. E. Calabi, Quasi-surjective mappings and a generalization of Morse theory, in *Proceedings of the U.S.-Japan Seminar in Differential Geometry (Kyoto, 1965)* (Nippon Hyoronsha, Tokyo, 1966), pp. 13–16
57. E. Calabi, Examples of Bernstein problems for some nonlinear equations, in *Global Analysis (Proceedings of the Symposium Pure Mathematics, Berkeley, CA, 1968)*, vol. XV (American Mathematical Society, Providence, RI, 1970), pp. 223–230
58. A. Caminha, On hypersurfaces into Riemannian spaces of constant sectional curvature. Kodai Math. J. **29**(2), 185–210 (2006)
59. H.-D. Cao, Q. Chen, On Bach-flat gradient shrinking Ricci solitons. Duke Math. J. **162**(6), 1149–1169 (2013)
60. H.-D. Cao, G. Catino, Q. Chen, C. Mantegazza, L. Mazzieri, Bach-flat gradient steady Ricci solitons. Calc. Var. Partial Differ. Equ. **49**(1–2), 125–138 (2014)
61. H.-D. Cao, B.-L. Chen, X.-P. Zhu, Recent developments on Hamilton's Ricci flow, in *Surveys in Differential Geometry. Geometric Flows*, vol. 12 (International Press, Somerville, MA, 2008), pp. 47–112
62. E. Cartan, Sur les variétés de courbure constante d'un espace euclidien ou non-Euclidien. Bull. Soc. Math. Fr. **47**, 125–160 (1919)
63. E. Cartan, Sur les variétés de courbure constante d'un espace euclidien ou non-Euclidien. Bull. Soc. Math. Fr. **48**, 132–208 (1920)
64. É. Cartan, Familles de surfaces isoparamétriques dans les espaces à courbure constante. Ann. Mat. Pura Appl. **17**(1), 177–191 (1938)
65. G. Catino, Complete gradient shrinking Ricci solitons with pinched curvature. Math. Ann. **355**(2), 629–635 (2013)
66. G. Catino, C. Mantegazza, L. Mazzieri, Locally conformally flat ancient ricci flows. Anal. PDE **8**(2), 365–371 (2015)
67. G. Catino, P. Mastrolia, D.D. Monticelli, Classification of expanding and steady ricci solitons with integral curvature decay. Geom. Topol. (2016, to appear)
68. G. Catino, P. Mastrolia, D.D. Monticelli, M. Rigoli, Analytic and geometric properties of generic Ricci solitons. Trans. Am. Math. Soc. (2016, to appear)
69. G. Catino, P. Mastrolia, D.D. Monticelli, M. Rigoli. On the geometry of gradient Einstein-type manifolds (2014). arXiv:1402.3453v1 [math.DG]

70. G. Catino, P. Mastrolia, D.D. Monticelli, M. Rigoli. Conformal Ricci solitons and related integrability conditions. Adv. Geom. (2016, to appear)
71. I. Chavel, *Eigenvalues in Riemannian Geometry*. Pure and Applied Mathematics, vol. 115 (Academic, Orlando, FL, 1984). Including a chapter by Burton Randol, With an appendix by Jozef Dodziuk
72. I. Chavel, *Riemannian Geometry*. Cambridge Studies in Advanced Mathematics, vol. 98, 2nd edn. (Cambridge University Press, Cambridge, 2006). A modern introduction
73. J. Cheeger, D.G. Ebin. *Comparison Theorems in Riemannian Geometry* (AMS Chelsea Publishing, Providence, RI, 2008). Revised reprint of the 1975 original
74. J. Cheeger, M. Gromov, M. Taylor, Finite propagation speed, kernel estimates for functions of the Laplace operator, and the geometry of complete Riemannian manifolds. J. Differ. Geom. **17**(1), 15–53 (1982)
75. B.-L. Chen, Strong uniqueness of the Ricci flow. J. Differ. Geom. **82**(2), 363–382 (2009)
76. B.-Y. Chen. *Geometry of Submanifolds*. Pure and Applied Mathematics, vol. 22 (Marcel Dekker, New York, 1973)
77. B.-Y. Chen, A simple characterization of generalized Robertson-Walker spacetimes. Gen. Relativ. Gravit. **46**(12), 5 (2014). Art. 1833
78. Q. Chen, Y.L. Xin, A generalized maximum principle and its applications in geometry. Am. J. Math. **114**(2), 355–366 (1992)
79. Q.-M. Cheng. The rigidity of Clifford torus $S^1(\sqrt{1/n}) \times S^{n-1}(\sqrt{(n-1)/n})$. Comment. Math. Helv. **71**(1), 60–69 (1996)
80. X. Cheng, H. Rosenberg, Embedded positive constant r-mean curvature hypersurfaces in $M^m \times \mathbf{R}$. An. Acad. Brasil. Ciênc. **77**(2), 183–199 (2005)
81. S.Y. Cheng, S.T. Yau, Differential equations on Riemannian manifolds and their geometric applications. Commun. Pure Appl. Math. **28**(3), 333–354 (1975)
82. S.Y. Cheng, S.T. Yau, Maximal space-like hypersurfaces in the Lorentz-Minkowski spaces. Ann. Math. (2) **104**(3), 407–419 (1976)
83. S.Y. Cheng, S.T. Yau, Hypersurfaces with constant scalar curvature. Math. Ann. **225**(3), 195–204 (1977)
84. S.S. Chern, The geometry of G-structures. Bull. Am. Math. Soc. **72**, 167–219 (1966)
85. S.-S. Chern. Simple proofs of two theorems on minimal surfaces. Enseignement Math. (2) **15**, 53–61 (1969)
86. S.-S. Chern, N.H. Kuiper, Some theorems on the isometric imbedding of compact Riemann manifolds in Euclidean space. Ann. Math. (2) **56**, 422–430 (1952)
87. S.S. Chern, M. do Carmo, S. Kobayashi, Minimal submanifolds of a sphere with second fundamental form of constant length, in *Functional Analysis and Related Fields (Proceedings of Conference for M. Stone, University Chicago, Chicago, IL, 1968)* (Springer, New York, 1970), pp. 59–75
88. Y. Choquet-Bruhat, The problem of constraints in general relativity: solution of the Lichnerowicz equation, in *Differential Geometry and Relativity*. Mathematical Physics and Applied Mathematics, vol. 3 (Reidel, Dordrecht, 1976), pp. 225–235
89. Y. Choquet-Bruhat, *General Relativity and the Einstein Equations*. Oxford Mathematical Monographs (Oxford University Press, Oxford, 2009)
90. Y. Choquet-Bruhat, J. Isenberg, D. Pollack, The constraint equations for the Einstein-scalar field system on compact manifolds. Class. Quant. Grav. **24**(4), 809–828 (2007)
91. T.H. Colding, W.P. Minicozzi II, The Calabi-Yau conjectures for embedded surfaces. Ann. Math. (2) **167**(1), 211–243 (2008)
92. M. Dajczer, *Submanifolds and Isometric Immersions*. Mathematics Lecture Series, vol. 13 (Publish or Perish, Houston, TX, 1990). Based on the notes prepared by Mauricio Antonucci, Gilvan Oliveira, Paulo Lima-Filho and Rui Tojeiro
93. M. Dajczer, K. Nomizu, On the boundedness of Ricci curvature of an indefinite metric. Bol. Soc. Brasil. Mat. **11**(1), 25–30 (1980)
94. M. Dajczer, K. Nomizu, On sectional curvature of indefinite metrics. II. Math. Ann. **247**(3), 279–282 (1980)

95. M. Dajczer, P.A. Hinojosa, J.H.S. de Lira, Killing graphs with prescribed mean curvature. Calc. Var. Partial Differ. Equ. **33**(2), 231–248 (2008)
96. E.N. Dancer, On the number of positive solutions of weakly nonlinear elliptic equations when a parameter is large. Proc. Lond. Math. Soc. (3) **53**(3), 429–452 (1986)
97. E.N. Dancer, Y. Du, Some remarks on Liouville type results for quasilinear elliptic equations. Proc. Am. Math. Soc. **131**(6), 1891–1899 (2003)
98. J.H.S. de Lira, A.A. Medeiros, Maximum principles for minimal submanifolds in Riemannian products (2014, preprint)
99. Ph. Delanoë, Obstruction to prescribed positive Ricci curvature. Pac. J. Math. **148**(1), 11–15 (1991)
100. D.M. DeTurck, N. Koiso, Uniqueness and nonexistence of metrics with prescribed Ricci curvature. Ann. Inst. H. Poincaré Anal. Non Linéaire **1**(5), 351–359 (1984)
101. U. Dierkes, Maximum principles and nonexistence results for minimal submanifolds. Manuscripta Math. **69**(2), 203–218 (1990)
102. M.P. do Carmo, *Riemannian Geometry*. Mathematics: Theory & Applications (Birkhäuser Boston, Boston, MA, 1992). Translated from the second Portuguese edition by Francis Flaherty
103. M.P. do Carmo, M. Dajczer, Rotation hypersurfaces in spaces of constant curvature. Trans. Am. Math. Soc. **277**(2), 685–709 (1983)
104. J. Dodziuk, Maximum principle for parabolic inequalities and the heat flow on open manifolds. Indiana Univ. Math. J. **32**(5), 703–716 (1983)
105. Y. Du, Z. Guo, Liouville type results and eventual flatness of positive solutions for p-Laplacian equations. Adv. Differ. Equ. **7**(12), 1479–1512 (2002)
106. R.S. Earp, H. Rosenberg, Some remarks on surfaces of prescribed mean curvature, in *Differential Geometry*. Pitman Monographs & Surveys in Pure & Applied Mathematics, vol. 52 (Longman Sci. Tech., Harlow, 1991), pp. 123–148
107. J. Eells, L. Lemaire, *Selected Topics in Harmonic Maps, CBMS Regional Conference Series in Mathematics*, vol. 50 (Published for the Conference Board of the Mathematical Sciences, Washington, DC, 1983)
108. N.V. Efimov, Hyperbolic problems in the theory of surfaces, in *Proceedings of the International Congress of Mathematicians (Moscow, 1966)* (Izdat. "Mir", Moscow, 1968), pp. 177–188
109. L.P. Eisenhart, *Riemannian Geometry*. Princeton Landmarks in Mathematics (Princeton University Press, Princeton, NJ, 1997). Eighth Printing, Princeton Paperbacks
110. I. Ekeland, Nonconvex minimization problems. Bull. Am. Math. Soc. (N.S.) **1**(3), 443–474 (1979)
111. M.F. Elbert, Constant positive 2-mean curvature hypersurfaces. Illinois J. Math. **46**(1), 247–267 (2002)
112. M. Eminenti, G. La Nave, C. Mantegazza, Ricci solitons: the equation point of view. Manuscripta Math., **127**, 345–367 (2008)
113. F. Erkekoğlu, E. García-Río, D.N. Kupeli, On level sets of Lorentzian distance function. Gen. Relativ. Gravit. **35**(9), 1597–1615 (2003)
114. J.-H. Eschenburg, E. Heintze, Comparison theory for Riccati equations. Manuscripta Math. **68**(2), 209–214 (1990)
115. F.J.M. Estudillo, A. Romero, Generalized maximal surfaces in Lorentz-Minkowski space L^3. Math. Proc. Camb. Philos. Soc. **111**(3), 515–524 (1992)
116. F.J.M. Estudillo, A. Romero, On the Gauss curvature of maximal surfaces in the 3-dimensional Lorentz-Minkowski space. Comment. Math. Helv. **69**(1), 1–4 (1994)
117. H. Federer, Geometric measure theory, in *Die Grundlehren der mathematischen Wissenschaften, Band 153* (Springer, New York, 1969)
118. D. Fischer-Colbrie, Some rigidity theorems for minimal submanifolds of the sphere. Acta Math. **145**(1–2), 29–46 (1980)
119. F. Fontenele, S.L. Silva, On the m-th mean curvature of compact hypersurfaces. Houst. J. Math. **32**(1), 47–57 (2006) (electronic)

120. T. Frankel, Applications of Duschek's formula to cosmology and minimal surfaces. Bull. Am. Math. Soc. **81**, 579–582 (1975)
121. S. Gallot, D. Hulin, J. Lafontaine, *Riemannian Geometry*. Universitext, 3rd edn. (Springer, Berlin, 2004)
122. G.J. Galloway, A generalization of Myers' theorem and an application to relativistic cosmology. J. Differ. Geom. **14**(1), 105–116 (1980, 1979)
123. S.C. García-Martínez, D. Impera, M. Rigoli, A sharp height estimate for compact hypersurfaces with constant k-mean curvature in warped product spaces. Proc. Edinb. Math. Soc. (2) **58**(2), 403–419 (2015)
124. L. Gårding, An inequality for hyperbolic polynomials. J. Math. Mech. **8**, 957–965 (1959)
125. D. Gilbarg, N.S. Trudinger, *Elliptic Partial Differential Equations of Second Order*. Classics in Mathematics (Springer, Berlin, 2001). Reprint of the 1998 edition
126. J.R. Giles, *Convex Analysis with Application in the Differentiation of Convex Functions*. Research Notes in Mathematics, vol. 58 (Pitman (Advanced Publishing Program), Boston, MA, 1982)
127. F. Giménez, Estimates for the curvature of a submanifold which lies inside a tube. J. Geom. **58**(1–2), 95–105 (1997)
128. S.I. Goldberg, *Curvature and Homology* (Dover Publications, Mineola, NY, 1998). Revised reprint of the 1970 edition.
129. R.E. Greene, H. Wu, *Function Theory on Manifolds Which Possess a Pole*. Lecture Notes in Mathematics, vol. 699 (Springer, Berlin, 1979)
130. A. Grigor′yan, Stochastically complete manifolds. Dokl. Akad. Nauk SSSR **290**(3), 534–537 (1986)
131. A. Grigor′yan, Analytic and geometric background of recurrence and non-explosion of the Brownian motion on Riemannian manifolds. Bull. Am. Math. Soc. (N.S.) **36**(2), 135–249 (1999)
132. A. Grigor'yan, S.-T. Yau, Isoperimetric properties of higher eigenvalues of elliptic operators. Am. J. Math. **125**(4), 893–940 (2003)
133. R.S. Hamilton, Three-manifolds with positive Ricci curvature. J. Differ. Geom. **17**(2), 255–306 (1982)
134. R.S. Hamilton, Four-manifolds with positive curvature operator. J. Differ. Geom. **24**(2), 153–179 (1986)
135. S.G. Harris, Complete codimension-one spacelike immersions. Class. Quant. Grav. **4**(6), 1577–1585 (1987)
136. S.G. Harris, Closed and complete spacelike hypersurfaces in Minkowski space. Class. Quant. Grav. **5**(1), 111–119 (1988)
137. S.G. Harris, Complete spacelike immersions with topology. Classical Quantum Gravity **5**(6), 833–838 (1988)
138. T. Hasanis, D. Koutroufiotis, Immersions of Riemannian manifolds into cylinders. Arch. Math. (Basel) **40**(1), 82–85 (1983)
139. T. Hasanis, T. Vlachos, A pinching theorem for minimal hypersurfaces in a sphere. Arch. Math. (Basel) **75**(6), 469–471 (2000)
140. T. Hasanis, A. Savas-Halilaj, T. Vlachos, Complete minimal hypersurfaces in a sphere. Monatsh. Math. **145**(4), 301–305 (2005)
141. E. Hebey, *Nonlinear Analysis on Manifolds: Sobolev Spaces and Inequalities*. Courant Lecture Notes in Mathematics, vol. 5 (New York University Courant Institute of Mathematical Sciences, New York, 1999)
142. J. Heinonen, T. Kilpeläinen, O. Martio, *Nonlinear Potential Theory of Degenerate Elliptic Equations* (Dover Publications, Mineola, NY, 2006). Unabridged republication of the 1993 original
143. D. Hilbert, Ueber Flächen von constanter Gaussscher Krümmung. Trans. Am. Math. Soc. **2**(1), 87–99 (1901)

144. E. Hopf, *Elementare Bemerkungen über die Lösungen partieller Differentialgleichungen zweiter Ordnung vom elliptischen Typus* Sitzungsber. Berliner Akad. Wiss., Math. Phys. Kl. 19, 147–152 (1927)
145. D.A. Hoffman, Surfaces of constant mean curvature in manifolds of constant curvature. J. Differ. Geom. **8**, 161–176 (1973)
146. D. Hoffman, W.H. Meeks III, The strong halfspace theorem for minimal surfaces. Invent. Math. **101**(2), 373–377 (1990)
147. D. Hoffman, J.H.S. de Lira, H. Rosenberg, Constant mean curvature surfaces in $M^2 \times \mathbf{R}$. Trans. Am. Math. Soc. **358**(2), 491–507 (2006)
148. I. Holopainen, Volume growth, Green's functions, and parabolicity of ends. Duke Math. J. **97**(2), 319–346 (1999)
149. P. Hsu, Heat semigroup on a complete Riemannian manifold. Ann. Probab. **17**(3), 1248–1254 (1989)
150. G. Huisken, Ricci deformation of the metric on a Riemannian manifold. J. Differ. Geom. **21**(1), 47–62 (1985)
151. D. Impera, Comparison theorems in Lorentzian geometry and applications to spacelike hypersurfaces. J. Geom. Phys. **62**(2), 412–426 (2012)
152. G.R. Jensen, *Higher Order Contact of Submanifolds of Homogeneous Spaces*. Lecture Notes in Mathematics, vol. 610 (Springer, Berlin, New York, 1977)
153. G.R. Jensen, M. Rigoli, Harmonic Gauss maps. Pac. J. Math. **136**(2), 261–282 (1989)
154. L. Jorge, D. Koutroufiotis, An estimate for the curvature of bounded submanifolds. Am. J. Math., **103**(4), 711–725 (1981)
155. L. Jorge, F. Xavier, A complete minimal surface in $\mathbf{R}^3$ between two parallel planes. Ann. Math. (2) **112**(1), 203–206 (1980)
156. J. Jost, *Riemannian Geometry and Geometric Analysis*. Universitext, 6th edn. (Springer, Heidelberg, 2011)
157. J.B. Keller, On solutions of $\Delta u = f(u)$. Commun. Pure Appl. Math. **10**, 503–510 (1957)
158. K. Kenmotsu, *Surfaces with Constant Mean Curvature*. Translations of Mathematical Monographs, vol. 221(American Mathematical Society, Providence, RI, 2003). Translated from the 2000 Japanese original by Katsuhiro Moriya and revised by the author
159. R.Z. Khas′minskiĭ, Ergodic properties of recurrent diffusion processes and stabilization of the solution of the Cauchy problem for parabolic equations. Teor. Verojatnost. i Primenen. **5**, 196–214 (1960)
160. T. Klotz, R. Osserman, Complete surfaces in E^3 with constant mean curvature. Comment. Math. Helv. **41**, 313–318 (1966/1967)
161. S. Kobayashi, A theorem on the affine transformation group of a Riemannian manifold. Nagoya Math. J. **9**, 39–41 (1955)
162. O. Kobayashi, Maximal surfaces in the 3-dimensional Minkowski space L^3. Tokyo J. Math. **6**(2), 297–309 (1983)
163. S. Kobayashi, K. Nomizu, *Foundations of Differential Geometry*. Wiley Classics Library, vol. II (Wiley, New York, 1996). Reprint of the 1969 original, A Wiley-Interscience Publication
164. A. Korn, Zwei anwendungen der methode der sukzessiven annäherungen, in *Mathematische Abhandlungen Hermann Amandus Schwarz*, ed. by C. Carathéodory, G. Hessenberg, E. Landau, L. Lichtenstein (Springer, Berlin, Heidelberg, 1914), pp. 215–229
165. R.S. Kulkarni, The values of sectional curvature in indefinite metrics. Comment. Math. Helv. **54**(1), 173–176 (1979)
166. J. Lauret, Ricci soliton homogeneous nilmanifolds. Math. Ann. **319**(4), 715–733 (2001)
167. J. Lauret, Ricci soliton solvmanifolds. J. Reine Angew. Math. **650**, 1–21 (2011)
168. H.B. Lawson Jr., Local rigidity theorems for minimal hypersurfaces. Ann. Math. **89**(2), 187–197 (1969)
169. V.K. Le, On some equivalent properties of sub- and supersolutions in second order quasilinear elliptic equations. Hiroshima Math. J. **28**(2), 373–380 (1998)

170. J.M. Lee. *Riemannian Manifolds*. Graduate Texts in Mathematics, vol. 176 (Springer, New York, 1997). An introduction to curvature
171. J.M. Lee, *Introduction to Smooth Manifolds*. Graduate Texts in Mathematics, vol. 218, 2nd edn. (Springer, New York, 2013)
172. T. Levi-Civita, Famiglie di superficie isoparametriche nell'ordinario spazio euclideo. Atti Accad. Naz. Lincei Rend. Cl. Sci. Fis. Mat. Natur. **26**, 355–362 (1937)
173. H. Li, Hypersurfaces with constant scalar curvature in space forms. Math. Ann. **305**(4), 665–672 (1996)
174. A. Lichnerowicz, L'intégration des équations de la gravitation relativiste et le problème des n corps. J. Math. Pures Appl. (9) **23**, 37–63 (1944)
175. A. Lichnerowicz, *Géométrie des groupes de transformations* Travaux et Recherches Mathématiques, III (Dunod, Paris, 1958) ix+193 pp
176. L. Lichtenstein, Zur Theorie der konformen Abbildung: konforme Abbildung nicht analytischer, singularitäten freier Flächenstücke auf ebene Gebiete. Bull. Acad. Sci. Cracovie A 192–217 (1916)
177. S. Lin, T. Weinstein, E^3-complete timelike surfaces in E_1^3 are globally hyperbolic. Michigan Math. J. **44**(3), 529–541 (1997)
178. S. Lin, T. Weinstein, A better conformal Bernstein's theorem. Geom. Dedicata **81**(1–3), 53–59 (2000)
179. J. Lott, On the long-time behavior of type-III Ricci flow solutions. Math. Ann. **339**(3), 627–666 (2007)
180. M.A. Magid, The Bernstein problem for timelike surfaces. Yokohama Math. J. **37**(2), 125–137 (1990)
181. L. Mari, M. Rigoli, Maps from Riemannian manifolds into non-degenerate Euclidean cones. Rev. Mat. Iberoam. **26**(3), 1057–1074 (2010)
182. L. Mari, D. Valtorta, On the equivalence of stochastic completeness and Liouville and Khas'minskii conditions in linear and nonlinear settings. Trans. Am. Math. Soc. **365**(9), 4699–4727 (2013)
183. L. Mari, M. Rigoli, A.G. Setti, Keller-Osserman conditions for diffusion-type operators on Riemannian manifolds. J. Funct. Anal. **258**(2), 665–712 (2010)
184. S. Markvorsen, On the mean exit time from a minimal submanifold. J. Differ. Geom. **29**(1), 1–8 (1989)
185. J.E. Marsden, F.J. Tipler, Maximal hypersurfaces and foliations of constant mean curvature in general relativity. Phys. Rep. **66**(3), 109–139 (1980)
186. F. Martín, S. Morales, A complete bounded minimal cylinder in $\mathbb{R}^3$. Michigan Math. J. **47**(3), 499–514 (2000)
187. P. Mastrolia, D.D. Monticelli, M. Rigoli, A note on curvature of Riemannian manifolds. J. Math. Anal. Appl., **399**(2), 505–513 (2013)
188. P. Mastrolia, M. Rigoli, M. Rimoldi, Some geometric analysis on generic Ricci solitons. Commun. Contemp. Math. **15**(3), 1250058, 25 (2013)
189. P. Mastrolia, M. Rigoli, A.G. Setti, *Yamabe-Type Equations on Complete, Noncompact Manifolds*. Progress in Mathematics, vol. 302 (Birkhäuser/Springer Basel AG, Basel, 2012)
190. L. Mazet, A general halfspace theorem for constant mean curvature surfaces. Am. J. Math. **135**(3), 801–834 (2013)
191. J. Milnor, *Morse Theory*. Based on lecture notes by M. Spivak and R. Wells. Annals of Mathematics Studies, vol. 51 (Princeton University Press, Princeton, NJ, 1963)
192. T.K. Milnor, A conformal analog of Bernstein's theorem for timelike surfaces in Minkowski 3-space, in *The Legacy of Sonya Kovalevskaya (Cambridge, MA, and Amherst, MA, 1985)*. Contemporary Mathematics, vol. 64 (American Mathematical Society, Providence, RI, 1987), pp. 123–132
193. T.K. Milnor, Entire timelike minimal surfaces in $E^{3,1}$. Michigan Math. J. **37**(2), 163–177 (1990)
194. S. Montiel, Unicity of constant mean curvature hypersurfaces in some Riemannian manifolds. Indiana Univ. Math. J. **48**(2), 711–748 (1999)

195. S. Montiel, A. Ros, Compact hypersurfaces: the Alexandrov theorem for higher order mean curvatures, in *Differential Geometry*. Pitman Monographs & Surveys in Pure & Applied Mathematics, vol. 52 (Longman Scientific & Technical, Harlow, 1991), pp. 279–296
196. J.D. Moore, An application of second variation to submanifold theory. Duke Math. J. **42**, 191–193 (1975)
197. J.D. Moore, Conformally flat submanifolds of Euclidean space. Math. Ann. **225**(1), 89–97 (1977)
198. J.D. Moore, Submanifolds of constant positive curvature. I. Duke Math. J. **44**(2), 449–484 (1977)
199. K. Motomiya, On functions which satisfy some differential inequalities on Riemannian manifolds. Nagoya Math. J. **81**, 57–72 (1981)
200. O. Munteanu, J. Wang, Analysis of weighted Laplacian and applications to Ricci solitons. Commun. Anal. Geom. **20**(1), 55–94 (2012)
201. O. Munteanu, J. Wang, Geometry of manifolds with densities. Adv. Math. **259**, 269–305 (2014)
202. M. Murata, Uniqueness and nonuniqueness of the positive Cauchy problem for the heat equation on Riemannian manifolds. Proc. Am. Math. Soc. **123**(6), 1923–1932 (1995)
203. S.B. Myers, Riemannian manifolds with positive mean curvature. Duke Math. J. **8**, 401–404 (1941)
204. A. Naber, Noncompact shrinking four solitons with nonnegative curvature. J. Reine Angew. Math. **645**, 125–153 (2010)
205. N. Nadirashvili, Hadamard's and Calabi-Yau's conjectures on negatively curved and minimal surfaces. Invent. Math. **126**(3), 457–465 (1996)
206. N.S. Nadirashvili, A theorem of Liouville type on a Riemannian manifold. Usp. Mat. Nauk **40**(5(245)), 259–260 (1985)
207. J. Nash, The imbedding problem for Riemannian manifolds. Ann. Math. (2) **63**, 20–63 (1956)
208. K. Nomizu, B. Smyth, A formula of Simons' type and hypersurfaces with constant mean curvature. J. Differ. Geom. **3**, 367–377 (1969)
209. M. Okumura, Hypersurfaces and a pinching problem on the second fundamental tensor. Am. J. Math. **96**, 207–213 (1974)
210. H. Omori, Isometric immersions of Riemannian manifolds. J. Math. Soc. Jpn. **19**, 205–214 (1967)
211. B. O'Neill, Immersion of manifolds of nonpositive curvature. Proc. Am. Math. Soc. **11**, 132–134 (1960)
212. B. O'Neill, *Semi-Riemannian Geometry*. Pure and Applied Mathematics, vol. 103 (Academic [Harcourt Brace Jovanovich Publishers], New York, 1983). With applications to relativity
213. R. Osserman, On the inequality $\Delta u \geq f(u)$. Pac. J. Math. **7**, 1641–1647 (1957)
214. R. Osserman, *A Survey of Minimal Surfaces*, 2nd edn. (Dover Publications, New York, 1986)
215. T. Ôtsuki, Isometric imbedding of Riemann manifolds in a Riemann manifold. J. Math. Soc. Jpn. **6**, 221–234 (1954)
216. B. Palmer, The Gauss map of a spacelike constant mean curvature hypersurface of Minkowski space. Comment. Math. Helv. **65**(1), 52–57 (1990)
217. O. Perdomo, Rigidity of minimal hypersurfaces of spheres with constant Ricci curvature. Rev. Colombiana Mat. **38**(2), 73–85 (2004)
218. G. Perelman, The entropy formula for the Ricci flow and its geometric applications (2002). arXiv:math/0211159v1 [math.DG]
219. P. Petersen, *Riemannian Geometry*. Graduate Texts in Mathematics, vol. 171, 2nd edn. (Springer, New York, 2006)
220. P. Petersen, W. Wylie, On gradient Ricci solitons with symmetry. Proc. Am. Math. Soc. **137**, 2085–2092 (2009)
221. P. Petersen, W. Wylie, Rigidity of gradient Ricci solitons. Pac. J. Math. **241**(2), 329–345 (2009)
222. P. Petersen, W. Wylie, On the classification of gradient Ricci solitons. Geom. Topol. **14**(4), 2277–2300 (2010)

223. S. Pigola, M. Rimoldi, Characterizations of model manifolds by means of certain differential systems. Can. Math. Bull. **55**(3), 632–645 (2012)
224. S. Pigola, M. Rigoli, M. Rimoldi, A.G. Setti, Ricci almost solitons. Ann. Sc. Norm. Super. Pisa Cl. Sci. (5) **10**(4), 757–799 (2011)
225. S. Pigola, M. Rigoli, A.G. Setti, A remark on the maximum principle and stochastic completeness. Proc. Am. Math. Soc. **131**(4), 1283–1288 (2003) (electronic)
226. S. Pigola, M. Rigoli, A.G. Setti, A Liouville-type result for quasi-linear elliptic equations on complete Riemannian manifolds. J. Funct. Anal. **219**(2), 400–432 (2005)
227. S. Pigola, M. Rigoli, A.G. Setti, Maximum principles on Riemannian manifolds and applications. Mem. Am. Math. Soc. **174**(822), x+99 (2005)
228. S. Pigola, M. Rigoli, A.G. Setti, Maximum principles at infinity on Riemannian manifolds: an overview. Mat. Contemp. **31**, 81–128 (2006). Workshop on Differential Geometry (Portuguese)
229. S. Pigola, M. Rigoli, A.G. Setti, Some non-linear function theoretic properties of Riemannian manifolds. Rev. Mat. Iberoam. **22**(3), 801–831 (2006)
230. S. Pigola, M. Rigoli, A.G. Setti, *Vanishing and Finiteness Results in Geometric Analysis*. Progress in Mathematics, vol. 266 (Birkhäuser Verlag, Basel, 2008). A generalization of the Bochner technique
231. S. Pigola, M. Rimoldi, A.G. Setti, Remarks on non-compact gradient Ricci solitons. Math. Z. **268**(3–4), 777–790 (2011)
232. W.A. Poor, *Differential Geometric Structures* (McGraw-Hill Book, New York, 1981)
233. M.H. Protter, H.F. Weinberger, *Maximum Principles in Differential Equations* (Springer, New York, 1984). Corrected reprint of the 1967 original
234. P. Pucci, J. Serrin, The strong maximum principle revisited. J. Differ. Equ. **196**(1), 1–66 (2004)
235. P. Pucci, J. Serrin, *The Maximum Principle*. Progress in Nonlinear Differential Equations and Their Applications, vol. 73 (Birkhäuser Verlag, Basel, 2007)
236. P. Pucci, M. Rigoli, J. Serrin, Qualitative properties for solutions of singular elliptic inequalities on complete manifolds. J. Differ. Equ. **234**(2), 507–543 (2007)
237. M. Rainer, H.-J. Schmidt, Inhomogeneous cosmological models with homogeneous inner hypersurface geometry. Gen. Relat. Gravit. **27**(12), 1265–1293 (1995)
238. A. Ranjbar-Motlagh, Rigidity of spheres in Riemannian manifolds and a non-embedding theorem. Bol. Soc. Brasil. Mat. (N.S.) **32**(2), 159–171 (2001)
239. A. Ratto, M. Rigoli, A. G. Setti, On the Omori-Yau maximum principle and its applications to differential equations and geometry. J. Funct. Anal. **134**(2), 486–510 (1995)
240. A. Ratto, M. Rigoli, L. Véron, Scalar curvature and conformal deformation of hyperbolic space. J. Funct. Anal. **121**(1), 15–77 (1994)
241. R.C. Reilly, Extrinsic rigidity theorems for compact submanifolds of the sphere. J. Differ. Geom. **4**, 487–497 (1970)
242. R.C. Reilly, Variational properties of functions of the mean curvatures for hypersurfaces in space forms. J. Differ. Geom. **8**, 465–477 (1973)
243. M. Rigoli, A.G. Setti, Liouville type theorems for ϕ-subharmonic functions. Rev. Mat. Iberoam. **17**(3), 471–520 (2001)
244. M. Rigoli, S. Zamperlin, "A priori" estimates, uniqueness and existence of positive solutions of Yamabe type equations on complete manifolds. J. Funct. Anal. **245**(1), 144–176 (2007)
245. M. Rigoli, M. Salvatori, M. Vignati, Some remarks on the weak maximum principle. Rev. Mat. Iberoam. **21**(2), 459–481 (2005)
246. A. Romero, Simple proof of Calabi-Bernstein's theorem on maximal surfaces. Proc. Am. Math. Soc. **124**(4), 1315–1317 (1996)
247. H. Rosenberg, F. Schulze, J. Spruck, The half-space property and entire positive minimal graphs in $M \times \mathbb{R}$. J. Differ. Geom. **95**(2), 321–336 (2013)
248. J. Roth, Extrinsic radius pinching in space forms of nonnegative sectional curvature. Math. Z. **258**(1), 227–240 (2008)

249. E.A. Ruh, J. Vilms, The tension field of the Gauss map. Trans. Am. Math. Soc. **149**, 569–573 (1970)
250. D.H. Sattinger, *Topics in Stability and Bifurcation Theory*. Lecture Notes in Mathematics, vol. 309 (Springer, Berlin, 1973)
251. R. Schoen, S.-T. Yau, Complete three-dimensional manifolds with positive Ricci curvature and scalar curvature, in *Seminar on Differential Geometry*. Annals of Mathematics Studies, vol. 102 (Princeton University Press, Princeton, NJ, 1982), pp. 209–228
252. R. Schoen, S.-T. Yau, *Lectures on Differential Geometry*. Conference Proceedings and Lecture Notes in Geometry and Topology, I (International Press, Cambridge, MA, 1994). Lecture notes prepared by Wei Yue Ding, Kung Ching Chang [Gong Qing Zhang], Jia Qing Zhong and Yi Chao Xu, Translated from the Chinese by Ding and S. Y. Cheng, Preface translated from the Chinese by Kaising Tso
253. B. Segre, Famiglie di ipersuperficie isoparametriche negli spazi euclidei ad un qualunque numero di dimensioni. Atti Accad. Naz. Lincei Cl. Sci. Fis. Mat. Natur. Rend. **27**, 203–207 (1938)
254. J. Serrin, A symmetry problem in potential theory. Arch. Ration. Mech. Anal. **43**, 304–318 (1971)
255. J. Serrin, H. Zou, Cauchy-Liouville and universal boundedness theorems for quasilinear elliptic equations and inequalities. Acta Math. **189**(1), 79–142 (2002)
256. R.W. Sharpe, *Differential Geometry*. Graduate Texts in Mathematics, vol. 166 (Springer, New York, 1997). Cartan's generalization of Klein's Erlangen program, With a foreword by S.S. Chern
257. G.E. Shilov, B.L. Gurevich, *Integral, Measure and Derivative: A Unified Approach*, English edition (Dover Publications, New York, 1977). Translated from the Russian and edited by Richard A. Silverman, Dover Books on Advanced Mathematics
258. B. Smyth, F. Xavier, Efimov's theorem in dimension greater than two. Invent. Math. **90**(3), 443–450 (1987)
259. E.F. Stiel, Immersions into manifolds of constant negative curvature. Proc. Am. Math. Soc. **18**, 713–715 (1967)
260. K.-T. Sturm, Analysis on local Dirichlet spaces. I. Recurrence, conservativeness and L^p-Liouville properties. J. Reine Angew. Math. **456**, 173–196 (1994)
261. S. Tachibana, A theorem of Riemannian manifolds of positive curvature operator. Proc. Japan Acad. **50**, 301–302 (1974)
262. L.-F. Tam, D. Zhou, Stability properties for the higher dimensional catenoid in $\mathbb{R}^{n+1}$. Proc. Am. Math. Soc. **137**(10), 3451–3461 (2009)
263. Y. Tashiro, Complete Riemannian manifolds and some vector fields. Trans. Am. Math. Soc. **117**, 251–275 (1965)
264. C. Tompkins, Isometric embedding of flat manifolds in Euclidean space. Duke Math. J. **5**(1), 58–61 (1939)
265. R. Tribuzy, Hopf's method and deformations of surfaces preserving mean curvature. An. Acad. Brasil. Ciênc. **50**(4), 447–450 (1978)
266. N.T. Varopoulos, The Poisson kernel on positively curved manifolds. J. Funct. Anal. **44**(3), 359–380 (1981)
267. N.T. Varopoulos, *Potential Theory and Diffusion on Riemannian Manifolds*, in Conference on Harmonic Analysis in Honor of Antoni Zygmund, Vol. I, II (Chicago, IL, 1981). Wadsworth Mathematics Series (Wadsworth, Belmont, CA, 1983), pp. 821–837
268. A.R. Veeravalli, On the mean curvatures sharp estimates of hypersurfaces. Expo. Math. **20**(3), 255–261 (2002)
269. T. Vlachos, A characterization for geodesic spheres in space forms. Geom. Dedicata **68**(1), 73–78 (1997)
270. Q. Wang, Rigidity of Clifford minimal hypersurfaces. Monatsh. Math. **140**(2), 163–167 (2003)
271. Q. Wang, C. Xia, Rigidity theorems for closed hypersurfaces in space forms. Q. J. Math. **56**(1), 101–110 (2005)

272. F.W. Warner, *Foundations of Differentiable Manifolds and Lie Groups*. Graduate Texts in Mathematics, vol. 94 (Springer, New York, Berlin, 1983). Corrected reprint of the 1971 edition
273. G. Wei, Complete hypersurfaces with constant mean curvature in a unit sphere. Monatsh. Math. **149**(3), 251–258 (2006)
274. G. Wei, Y.J. Suh, Rigidity theorems for hypersurfaces with constant scalar curvature in a unit sphere. Glasg. Math. J. **49**(2), 235–241 (2007)
275. W. Wylie, Complete shrinking Ricci solitons have finite fundamental group. Proc. Am. Math. Soc. **136**(5), 1803–1806 (2008)
276. Y.L. Xin, On the Gauss image of a spacelike hypersurface with constant mean curvature in Minkowski space. Comment. Math. Helv. **66**(4), 590–598 (1991)
277. K. Yano, T. Nagano, The de Rham decomposition, isometries and affine transformations in Riemannian spaces. Jpn. J. Math. **29**, 173–184 (1959)
278. K. Yano, T. Nagano, Einstein spaces admitting a one-parameter group of conformal transformations. Ann. Math. (2) **69**, 451–461 (1959)
279. S.-T. Yau, Harmonic functions on complete Riemannian manifolds. Commun. Pure Appl. Math. **28**, 201–228 (1975)
280. S.-T. Yau, Some function-theoretic properties of complete Riemannian manifold and their applications to geometry. Indiana Univ. Math. J. **25**(7), 659–670 (1976)
281. S.-T. Yau, A general Schwarz lemma for Kähler manifolds. Am. J. Math. **100**(1), 197–203 (1978)
282. Z.-H. Zhang, On the completeness of gradient Ricci solitons. Proc. Am. Math. Soc. **137**(8), 2755–2759 (2009)

Index

L.J. Alías et al., *Maximum Principles and Geometric Applications*,
Springer Monographs in Mathematics, DOI 10.1007/978-3-319-24337-5

GPSR Compliance
The European Union's (EU) General Product Safety Regulation (GPSR) is a set of rules that requires consumer products to be safe and our obligations to ensure this.

If you have any concerns about our products, you can contact us on

ProductSafety@springernature.com

In case Publisher is established outside the EU, the EU authorized representative is:

Springer Nature Customer Service Center GmbH
Europaplatz 3
69115 Heidelberg, Germany

www.ingramcontent.com/pod-product-compliance
Ingram Content Group UK Ltd.
Pitfield, Milton Keynes, MK11 3LW, UK
UKHW021935200726
13853UKWH00011B/2192

* 9 7 8 3 3 1 9 7 9 6 0 5 5 *